THE

HISTORY

& PRACTICE

OF ANCIENT

ASTRONOMY

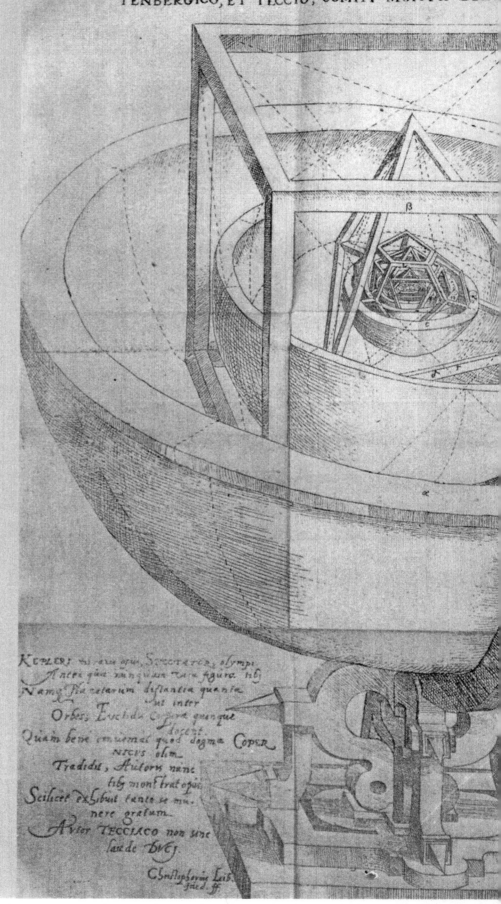

TABVLA III. ORBIVM PLANETARVM DIMENSIONE
REGVLARIA CORPORA GEOMETRI
ILLVSTRISS: PRINCIPI, AC DÑO. DÑO. FR
TENBERGICO, ET TECCIO, COMITI MONTIS BELG

Kepler hi orin opur, Spectator, olympi.
   Antea quæ nunquam vera figura tibi
Namq Planetarum distantia quanta
                 ut inter
   Orbess Euclidis Corpora quinque
                 docent.
Quàm bene convenies quod dogma Coper
         nicus olim
   Tradidit, Auctor nunc
            tibi monstrat opus.
Scilicet exhibui tanto se mu-
      nere gratum.
   Autor Tecciaco non sine
         laude Duci.

         Christophorus Leib
            fecit ff.

The
History
& Practice
of Ancient
Astronomy

✳ ✳ ✳

JAMES EVANS

New York   Oxford

Oxford University Press

1998

Oxford University Press

Oxford   New York
Athens   Auckland   Bangkok   Bogotá   Buenos Aires   Calcutta
Cape Town   Chennai   Dar es Salaam   Delhi   Florence   Hong Kong   Istanbul
Karachi   Kuala Lumpur   Madrid   Melbourne   Mexico City   Mumbai
Nairobi   Paris   São Paulo   Singapore   Taipei   Tokyo   Toronto   Warsaw

and associated companies in
Berlin   Ibadan

Copyright © 1998 by Oxford University Press, Inc.

Published by Oxford University Press, Inc.
198 Madison Avenue, New York, New York 10016

Oxford is a registered trademark of Oxford University Press

Library of Congress Cataloging-in-Publication Data
Evans, James, 1948–
The history and practice of ancient astronomy / James Evans.
p.   cm.
Includes bibliographical references and index.
ISBN 978-0-19-509539-5

1. Astronomy, Ancient.   I. Title.
QB16.E93   1998
520'.938—dc21   97-16539

For permission to reprint, I gratefully acknowledge the following:

Archiv für Orientforschung, for permission to quote from Hermann
Hunger and David Pingree, MUL.APIN: *An Astronomical
Compendium in Cuneiform.*

Gerald Duckworth & Co., for permission to quote from
G. J. Toomer, *Ptolemy's Algamest.*

Harvard University Press, for permission to quote from the
following volumes in the Loeb Classical Library *Aristotle: On the
Heavens*, W. K. C. Guthrie, trans. *Strabo: Geography*, Horace Leonard
Jones, trans. *Cicero: De re publica*, Clinton W. Keys, trans.
Pliny: Natural History, H. Rackham, trans.

The University of Chicago Press, for permission to quote from
Richmond Lattimore's translation, *The Iliad of Homer.*

The University of Wisconsin Press, for permission to reproduce
data from W. D. Stahlman and O. Gingerich, *Solar and Planetary
Longitudes for Years −2500 to +2000 by 10-day intervals.*

13 15 17 19 20 18 16 14 12

Printed in the United States of America
on acid-free paper

Being asked to what end he had been born, he replied,
"To study the Sun and Moon and the heavens."

Diogenes Laërtius, speaking of Anaxagoras.
Lives and Opinions of Eminent Philosophers II, 10.

I know that my day's life is marked for death.
But when I search into the close, revolving spirals of stars,
my feet no longer touch the Earth. Then,
by the side of Zeus himself, I take my share of immortality.

Epigram attributed to Ptolemy.
Palatine Anthology IX, 577.

The ancient Western astronomical tradition is one of great richness and impressive duration. It begins with records of planet observations made by the Babylonians in the second millennium B.C. It includes the development of an astronomy based on geometrical methods and philosophical principles by the Greeks between the time of Aristotle (fourth century B.C.) and the time of Ptolemy (second century A.D.). After a period of decline, or at least of quiescence, astronomy underwent a renaissance in the Islamic Middle East in the ninth century A.D. For the next several centuries the language of astronomical learning was Arabic, as Greek had been before, and as Akkadian had been before that. This astronomical tradition culminated with the astronomical revolution of the sixteenth century in central Europe, where Latin was the language of scientific discourse. This history of nearly 3,000 years therefore involves contributions by the Babylonian, Greek, Arabic, and medieval Latin cultures. But it was the Greek period that determined the fundamental character of this endeavor.

This book is called *The History and Practice of Ancient Astronomy*. In the largest sense, its subject is the ancient astronomical tradition of the West, which I take to encompass the period and the cultures named. But the focus of this book is the Greek period. One cannot really understand what medieval Arabic and Latin astronomy were about, nor can one understand what Copernicus and Kepler did in the Renaissance, without understanding Ptolemy.

Of course, Greek astronomy did not develop in a vacuum. Indeed, in our century scholars have come to appreciate how important an influence Babylonian astronomical practice exerted on the Greeks of the Hellenistic and Roman periods. Babylonian astronomy is a complex subject, intellectually and historically rich, and fully worthy of study in its own right. I have not been able to devote space to Babylonian astronomy that would be commensurate with its intrinsic significance. However, I have tried to include enough to give the reader an insight into the essential character of Babylonian astronomy, its historical development, and the nature of its influence on the Greeks.

In the same way, I have not attempted to write a history of medieval Arabic astronomy or of medieval or Renaissance European astronomy. Each of these subjects, if treated in adequate detail, would require a book of its own. However, I have often illustrated the continuity of the Western astronomical tradition by showing what becomes of some aspect of Greek astronomy (e.g., astronomical tables) in the Middle Ages. Some subjects, such as the astrolabe, that show a rich development in the Middle Ages are treated in considerable detail. And, of course, no treatment of Greek planetary theory could be considered adequate if it omitted a discussion of its radical transformation by Copernicus in the sixteenth century.

In calling this book *History and Practice* I pledged to stay as close and as true as possible to both. Staying close to history means bringing the reader into direct contact with the ancient sources. I have tried always to tell not only *what* but also *how* we know about the astronomy of the ancient past. Throughout the book, many extracts from ancient writers are reproduced, to allow the reader to form his or her own impression of the ancient astronomical discourse. While scholars can agree about the main outlines of the history of Western astronomy, opinion is often divided on details, and occasionally even on issues of major importance. Where the evidence is conflicting, I have not tried to hide our ignorance but have presented the case as I see it.

The material culture of ancient astronomy is an important part of its history. The instruments used by the ancient astronomers are a part of the story, no less than the texts they wrote and studied. Many illustrations are reproduced here to provide a visual impression of the nature of the evidence on which our reconstruction of the past must be based.

In our time, knowledge is fragmented into hundreds of specialities and subspecialities. No one science occupies a central place. But in ancient Greece and medieval Islam, as well as in medieval Europe, astronomy held a privileged place, with important connections to philosophy and religion, as well as to art and literature. For the ancient Pythagoreans, astronomy was one of the four chief branches of mathematics, along with arithmetic (i.e., number theory), geometry, and music theory. In the medieval universities these same four arts became the quadrivium—the upper-level sequence of courses in the arts curriculum. Thus, an introduction to astronomy remained a central part of the experience deemed essential for an adequate education. A complete history of the astronomical tradition certainly cannot leave out of account the relation of astronomy to the broader culture.

Staying close to the practice of astronomy means explaining a subject in enough detail for the reader to understand what the ancient astronomers actually *did*. Nearly every subject that is treated in this book is treated in enough detail to permit the reader to practice the art of astronomy as it was practiced in antiquity. After working through chapter 3, the reader should be able to make a sundial by methods approximating those used by Greek and Roman astronomers. After working through chapter 7, the reader should be able to predict the next retrogradation of Jupiter, either by the methods of the Babylonian scribes or by the methods of Ptolemy.

The decision to focus on astronomical practice entailed a number of compromises. For example, topics that seemed too complex to be treated in adequate detail without extravagant demands for space and on the reader's patience have been omitted. The best example of such an omission is the ancient lunar theory. Thus, while both the Babylonian and Greek planetary theories are discussed in detail, I have chosen to let the Moon go. But I am confident that the reader who has mastered Ptolemy's theories of the Sun and of Mars in this book will have no trouble with the lunar theory if he or she should pursue it elsewhere.

In focusing on practice, the question naturally arises of what astronomical knowledge the reader can be assumed already to possess. I have not assumed that the reader knows any astronomy. The basic astronomical facts required for understanding the ancient texts are developed as the book progresses.

But perhaps the most serious choice to be made in writing a book about astronomical practice is the selection of the appropriate level of mathematics. For, in both Greece and Babylonia, astronomy was already a thoroughly mathematical subject. My goal has been to treat the astronomical concepts rigorously and accurately, but to minimize the mathematical tedium as much as possible. This is done by several different methods.

First, I have followed the ancient and medieval practice of emphasizing *astronomical tables*. Already in Ptolemy's day handy tables were produced to make astronomy more user-friendly. These tables (for problems associated with the daily revolution of the celestial sphere and for the more complex motions of the planets) in fact served to define the practice of astronomy. Wherever in the medieval world there were tables, real astronomy was practiced; where tables were lacking there were only dilettantes and dabblers. So the reader of this book will learn to use tables. And thus the reader will be prepared for any further study of Greek, Arabic, or medieval or Renaissance Latin astronomy.

A second way I have found of minimizing the mathematical labor is to rely on graphical methods and on models (such as the astrolabe) whenever possible. So, for example, the reader can construct a sundial purely by graphical methods, without any computation at all. The reader can predict the position of Mars according to Ptolemy's theory by manipulating an instrument (the Ptolemaic slats) rather than by performing a tedious trigonometrical calculation. Some of the necessary models can be assembled from the patterns found in the appendix to this book.

When a more detailed mathematical treatment of some topic seems desirable, I usually place it in a special section or separate it off in a *Mathematical Postscript*, after a less mathematical treatment. This will allow readers who are on friendly terms with trigonometry to pursue a subject in more detail, without subjecting other readers to unnecessary abuse. Those who wish to skip the mathematical postscripts can do so without fear that they are missing concepts essential to later developments.

In the sciences, it is common to encounter monographs in which the author interrupts the development from time to time by posing problems and exercises for the reader. This is the author's way of saying, You can't be sure you understand this material unless you can use it. But the exercises and suggestions for observations that are interspersed throughout this book are unusual features for a historical work. These are meant to give the reader the chance to practice the art of the ancient astronomer. Any attempt at a grand historical synthesis or a philosophical analysis of the Greek view of nature that is not underpinned with a sound understanding of how Greek astronomy actually worked is headed for trouble. I hope that the attention to detail and the provision of exercises will also make the book useful for teaching. But every reader of the book—the general reader, the classicist who wants to know more about Greek planetary theory, the astronomer who wants to understand the early history of his or her field—is urged to work as many of the exercises as possible. There is all the difference in the world between *knowing about* and *knowing how to do*.

In translations from ancient writers, pointed brackets < > enclose conjectural restorations to the text. Square brackets [ ] enclose words added for the sake of clarity but that have no counterparts in the original text. When the translator is not identified, the translation is my own.

ACKNOWLEDGMENTS

Over the years, many people helped in many ways to make my work easier and more enjoyable. In the United States: Arnold Arons, Bill Barry, Bernard Bates, Alan Bowen, H. James Clifford, Michael Crowe, Thatcher Deane, Owen Gingerich, Thomas L. Hankins, John Heilbron, Ronald Lawson, Paul Loeb, Lilian McDermott, Robert Mitchell, Matthew Moelter, Brian Popp, Jamil Ragep, Mark Rosenquist, Patricia Sperry, Noel Swerdlow, and Alan Thorndike. In Canada: J. L. Berggren, Hugh Thurston, and Alexander Jones. In Great Britain: Richard Evans and Michael Hoskin. In Sweden: Jöran Friberg. In Denmark: Kristian Peder Moesgaard. In France: Suzanne Débarbat, François De Gandt, Libby Grenet and Franz Grenet, Michel Lerner, Henri Hugonnnard-Roche, Alain Segonds, René Taton, Georges Telier and Renée Telier, Jean-Pierre Verdet, and Christiane Vilain. In Germany: Gabi Hansen and Klaus Hansen. My warmest thanks to them all. This book is dedicated to Sharon, Elizabeth, and Virginia.

Seattle, Washington   J. E.
September 1997

ONE
The Birth of Astronomy

1.1 Astronomy around 700 B.C.: Texts from Two
     Cultures 3
1.2 Outline of the Western Astronomical Tradition 11
1.3 Observation: The Use of the Gnomon 27
1.4 On the Daily Motion of the Sun 27
1.5 Exercise: Interpreting a Shadow Plot 31
1.6 The Diurnal Rotation 31
1.7 Observation: The Diurnal Motion of the Stars 39
1.8 Stars and Constellations 39
1.9 Earth, Sun, and Moon 44
1.10 The Annual Motion of the Sun 53
1.11 Observation: The Motion of the Moon 58
1.12 The Uses of Shadows 59
1.13 Exercise: Using Shadow Plots 63
1.14 The Size of the Earth 63
1.15 Exercise: The Size of the Earth 66
1.16 Observation: The Angular Size of the Moon 67
1.17 Aristarchus on the Sizes and Distances 67
1.18 Exercise: The Sizes and Distances of the Sun
     and Moon 73

Contents

TWO
The Celestial Sphere

2.1 The Sphere in Greek Astronomy 75
2.2 Sphairopoiïa: A History of Sphere Making 78
2.3 Exercise: Using a Celestial Globe 85
2.4 Early Writers on the Sphere 87
2.5 Geminus: *Introduction to the Phenomena* 91
2.6 Risings of the Zodiac Constellations: Telling Time
     at Night 95
2.7 Exercise: Telling Time at Night 99
2.8 Observation: Telling Time at Night 99
2.9 Celestial Coordinates 99
2.10 Exercise: Using Celestial Coordinates 105
2.11 A Table of Obliquity 105
2.12 Exercise: Using the Table of Obliquity 109
2.13 The Risings of the Signs: A Table of Ascensions 109
2.14 Exercise: On Tables of Ascensions 120
2.15 Babylonian Arithmetical Methods in Greek Astronomy:
     Hypsicles on the Risings of the Signs 121
2.16 Exercise: Arithmetic Progressions and the Risings of the
     Signs 125
2.17 Observation: The Armillary Sphere as an Instrument of
     Observation 125

THREE
Some Applications of Spherics

3.1 Greek and Roman Sundials 129
3.2 Vitruvius on Sundials 132
3.3 Exercise: Making a Sundial 135
3.4 Exercise: Some Sleuthing with Sundials 140
3.5 The Astrolabe 141
3.6 Exercise: Using the Astrolabe 152

3.7 The Astrolabe in History 153
3.8 Exercise: Making a Latitude Plate for the Astrolabe 158

FOUR
Calendars and Time Reckoning
4.1 The Julian and Gregorian Calendars 163
4.2 Exercise: Using the Julian and Gregorian Calendars 170
4.3 Julian Day Number 171
4.4 Exercise: Using Julian Day Numbers 174
4.5 The Egyptian Calendar 175
4.6 Exercise: Using the Egyptian Calendar 181
4.7 Luni-Solar Calendars and Cycles 182
4.8 Exercise: Using the Nineteen-Year Cycle 188
4.9 The Theory of Star Phases 190
4.10 Exercise: On Star Phases 198
4.11 Some Greek Parapegmata 199
4.12 Exercise: On Parapegmata 204

FIVE
Solar Theory
5.1 Observations of the Sun 205
5.2 The Solar Theory of Hipparchus and Ptolemy 210
5.3 Realism and Instrumentalism in Greek Astronomy 216
5.4 Exercise: Finding the Solar Eccentricity 220
5.5 Rigorous Derivation of the Solar Eccentricity 221
5.6 Exercise: On the Solar Theory 223
5.7 Tables of the Sun 226
5.8 Exercise: On the Tables of the Sun 235
5.9 Corrections to Local Apparent Time 235
5.10 Exercise: Apparent, Mean, and Zone Time 243

SIX
The Fixed Stars
6.1 Precession 245
6.2 Aristotle, Hipparchus, and Ptolemy on the Fixedness of
the Stars 247
6.3 Observation: Star Alignments 250
6.4 Ancient Methods for Measuring the Longitudes
of Stars 250
6.5 Exercise: The Longitude of Spica 257
6.6 Hipparchus and Ptolemy on Precession 259
6.7 Exercise: The Precession Rate from Star
Declinations 262
6.8 The Catalog of Stars 264
6.9 Trepidation: A Medieval Theory 274
6.10 Tycho Brahe and the Demise of Trepidation 281

SEVEN
Planetary Theory
7.1 The Planets 289
7.2 The Lower Planets: The Case of Mercury 299
7.3 Observation: Observing the Planets 301
7.4 The Upper Planets: The Case of Mars 302

7.5   Exercise: On the Oppositions of Jupiter   305

7.6   The Spheres of Eudoxus   305

7.7.   The Birth of Prediction: Babylonian Goal-Year
Texts   312

7.8   Exercise: On Goal-Year Texts   316

7.9   Babylonian Planetary Theory   317

7.10   Babylonian Theories of Jupiter   321

7.11   Exercise: Using the Babylonian Planetary Theory   334

7.12   Deferent-and-Epicycle Theory, I   337

7.13   Greek Planetary Theory between Apollonius and
Ptolemy   342

7.14   Exercise: The Epicycle of Venus   347

7.15   A Cosmological Divertissement: The Order of
the Planets   347

7.16   Exercise: Testing Apollonius's Theory of Longitudes   351

7.17   Deferent-and-Epicycle Theory, II: Ptolemy's Theory of
Longitudes   355

7.18   Exercise: Testing Ptolemy's Theory of Longitudes   359

7.19   Determination of the Parameters of Mars   362

7.20   Exercise: Parameters of Jupiter   369

7.21   General Method for Planet Longitudes   369

7.22   Exercise: Calculating the Planets   372

7.23   Tables of Mars   372

7.24   Exercise: Using the Tables of Mars   384

7.25   Ptolemy's Cosmology   384

7.26   Astronomy and Cosmology in the Middle Ages   392

7.27   Planetary Equatoria   403

7.28   Exercise: Assembly and Use of Schöner's *Aequatorium
Martis*   406

7.29   Geocentric and Heliocentric Planetary Theories   410

7.30   Nicholas Copernicus: The Earth a Planet   414

7.31   Kepler and the New Astronomy   427

Frequently
Used Tables

2.1   Progress of the Sun through the Zodiac   96

2.2   The Length of the Night   97

2.3   Table of Obliquity   106

2.4   Table of Ascensions   110

4.1   Equivalent Dates in the Julian and Gregorian
Calendars   169

4.2–4.4   Julian Day Number   172

4.5   Some Important Egyptian/Julian Equivalents   178

4.6   Months and Days of the Egyptian Year   178

5.1–5.3   Tables of the Sun   228

5.4   The Equation of Time   236

7.1   Planet Longitudes at Ten-Day Intervals   290

7.2   Oppositions of Mars, 1948–1984   303

7.4   Modern Ptolemaic Parameters for Venus, Mars, Jupiter,
and Saturn   368

7.5–7.7   Tables of Mars   374

Appendix: Patterns for Models   445

Notes   453

Bibliography   465

Index   473

THE

HISTORY

& PRACTICE

OF ANCIENT

ASTRONOMY

ONE

The
Birth of
Astronomy

Astronomy among the Greeks of the Archaic Age

The oldest surviving works of Greek literature are the *Iliad* and *Odyssey* of Homer, which were put into written form probably around the end of the eighth century B.C. Only a little younger is Hesiod's *Works and Days*, which dates from about 650 B.C. When Homer and Hesiod were writing, the Greeks were just emerging from their dark age. Literacy had been gained, then lost in the convulsions of the twelfth century B.C., then regained. Historians turn to Homer and Hesiod for insight into the Greek societies about 700 B.C.—for insight into the Greeks' economic life, their social organization, and their religious practices. We can profitably inquire of Homer and Hesiod just what the Greeks knew of astronomy.

Homer   In the eighteenth book of the *Iliad*, Hephaistos makes a shield for Achilles and decorates it with images of the heaven and the Earth:

> *First of all he forged a shield that was huge and heavy. . . .*
> *He made the Earth upon it, and the sky, and the sea's water,*
> *and the tireless Sun, and the Moon waxing into her fullness,*
> *and on it all the constellations that festoon the heavens,*
> *the Pleiades and the Hyades and the strength of Orion*
> *and the Bear, whom men also give the name of the Wagon,*
> *who turns about in a fixed place and looks at Orion*
> *and she alone is never plunged in the wash of Ocean. . . .*
> *He made on it the great strength of the Ocean River*
> *which ran around the uttermost rim of the shield's strong structure.*[1]

Here, then, are a few stars and constellations mentioned by name: the Pleiades, the Hyades, Orion, and the Bear, which is also pictured as a Wagon. (The Bear or Wagon is our Ursa Major. The seven brightest stars of this constellation form the Big Dipper.) Elsewhere, Homer mentions the Dog Star and the constellation Boötes.[2] All these stars have therefore been called by the same names for nearly 3,000 years. In the passage above, Homer mentions that the Bear "turns about in a fixed place" and "is never plunged in the wash of Ocean." This is a reference to the fact that the Bear is a *circumpolar* constellation: it can be seen all night long turning about the celestial pole and never rises or sets. Homer also knows that sailors can steer by the Bear: Odysseus keeps the Bear on his left in order to sail to the east.[3]

How are we to imagine the place of the Earth? Homer nowhere makes a clear statement about the shape of the Earth, but he seems to picture it as flat, like a shield. As is clear from the passage above, the land makes a single island, surrounded by Ocean. Homer probably imagined the sky, or heaven, as solid, for in several passages he likened it to iron or bronze.[4]

Homer knows that different stars are conspicuous at different times of year. Diomedes' blazing armor is compared to the Dog Star (Sirius),

> *that star of the waning summer who beyond all stars*
> *rises bathed in the Ocean stream to glitter in brilliance.*[5]

Sirius is the brightest star in the sky. In Homer's time and place, Sirius made its *morning rising* in the summer. Then Sirius could be seen rising in the east just before sunrise. At this morning rising, Sirius reemerged from a period of invisibility of over two months. (Sirius was invisible when the Sun was too near it in the sky.) So here we have a reference to telling the time of year by the stars—a very important tradition in Greek culture.

A few of the stars exercise influences over men and women. Most striking is the case of Sirius. In the *Iliad*, Achilles, moving over the battlefield in his blazing armor, is compared to Sirius,

> the star they give the name of Orion's Dog, which is brightest
> among the stars, and yet is wrought as a sign of evil
> and brings on the great fever for unfortunate mortals.[6]

The morning rising of Sirius was associated with the summer heat. But there is no hint of an elaborate system of personal astrological forecasts. That was a development of the Hellenistic period, five or six centuries later, when the Greeks had become more "scientific."

The evening star and morning star are mentioned in several passages.[7] But Homer apparently did not know that these are one and the same, our planet Venus.

Hesiod's Works and Days   In Hesiod's poem, *Works and Days*, written a generation or two after Homer's time, we see a more systematic effort to connect astronomy with the lives of men and women. The central part of the poem is an agricultural calendar, which prescribes the work to be done at each season of the year. The farmer is to tell the time of year by the heliacal risings and settings of the stars (also called star phases). These are risings and settings of the stars that occur just before the Sun rises or just after the Sun sets. The calendar of works and days begins with the two famous lines:

> When the Pleiades, daughters of Atlas are rising,
> begin the harvest, the plowing when they set.[8]

The Pleiades made their morning rising in May. Then they could be seen, rising in the east just before sunrise. This was the time to harvest the wheat. The Pleiades made their morning setting (going down in the west just before the Sun came up in the east) in late fall. For Hesiod, this was the sign to plow the land and sow the grain. Fall is the time for planting what today is called winter wheat, the only kind grown in antiquity.

Hesiod's agricultural year begins in the fall with the morning setting of the Pleiades and the sowing of the grain. Hesiod warns that if the farmer puts off his sowing until the "turning of the Sun" (i.e., the winter solstice), he will reap sitting and gain but a thin harvest.[9]

Hesiod refers to the equinox as the time when "the days and nights are equal, and the Earth, the mother of all, bears her various fruits." This reference to the equinox is followed immediately by two other signs of spring—the evening rising of Arcturus and the return of the swallow:

> When Zeus has finished sixty wintry days
> after the turning of the Sun, then the star
> Arcturus leaves the holy stream of Ocean
> and first rises brilliant in the twilight.
> After him Pandion's twittering daughter, the swallow,
> comes into the sight of men when spring is just beginning.[10]

Hesiod's statement that the evening rising of Arcturus comes sixty days after the winter solstice gives a way of checking the era in which he lived. (See sec. 4.9 for the method of making such a dating.) Hesiod's statement is consistent with the date we have assumed for him, about 650 B.C.

Spring is also the time when the one who carries his house on his back (the snail) climbs up the plants "to flee the Pleiades."[11] This is a reference to the morning rising of the Pleiades, which, as mentioned above, signaled the time of the grain harvest.

When the harvest is over, Sirius makes its morning rising and the hottest time of the summer arrives. This is the season when the artichoke blooms and the cicada chirps, when goats are fattest and wine sweetest, when women are most full of lust but men are feeblest, because "Sirius parches head and knees, and the skin is dry from heat."[12] Here is another instance of the belief in the influences exerted by Sirius at its morning rising. The time for picking grapes arrives

When Orion and Sirius come into mid-sky,
and rosy-fingered Dawn looks upon Arcturus....[13]

The time is September, when Orion and Sirius are high in the sky at morning and Arcturus makes its morning rising.

The agricultural year ends as it began, with the morning setting of the Pleiades:

When the Pleiades and Hyades and strong Orion set,
remember it is seasonable for sowing.
And so the completed year passes beneath the earth.[14]

This completes the agricultural calendar in the *Works and Days*. A few other astronomical references are found in the following section of the poem, which treats sailing. The morning setting of the Pleiades and Orion around the end of October signals a stormy season and the end of good sailing. The best time for sailing is the fifty days following the summer solstice.

The poem ends with a list of lucky and unlucky days of the month. In his reckoning of days, Hesiod seems to assume a month of thirty days, divided into three parts of ten days each—the waxing, the midmonth, and the waning, which correspond to the phases of the Moon. A day is usually (though not always) indicated by specifying its place in one of these three decades. So, for example, the eighth and the ninth day of the waxing month are good for the works of man. The sixth of the midmonth (i.e., the sixteenth day of the month) is unfavorable for plants, good for the birth of males, and unfavorable for a girl to be born or married. These lucky and unlucky days are not taken up in any obvious order, nor is there any explanation of why one day should be good or bad for any particular job. There also is no distinction among months or years—the thirteenth of the month always is bad for sowing but good for setting plants.

Early Astronomy in Babylonia

In Babylonian astronomy of about 700 B.C. we can recognize many features that remind us of Greek astronomy of the same period. However, in many ways Babylonian astronomy was further advanced. We can form a fair impression of the state of Babylonian astronomy around 700 B.C. by looking in detail at two texts.

MUL.APIN MUL.APIN is the title of a Babylonian astronomical text that survives in a number of copies on clay tablets. The name of this work is taken from the opening words of the text: "Plow Star." The oldest extant copies date from the seventh century B.C., but the text is a compilation from several different sources, which may have been substantially older. The text continued to be copied down to Hellenistic times. That it was considered a standard compilation is apparent from the fact that the surviving copies differ very little from one another. In figure 1.1, we see a fragment of MUL.APIN now in the British Museum.

MUL.APIN begins with a list of stars and constellations:

FIGURE I.I. A fragment of a tablet bearing part of the text of MUL.APIN. By permission of the Trustees of the British Museum (BM 42277 Obv.).

The Plow, Enlil, who goes at the front of the stars of Enlil.
The Wolf, the seeder of the Plow.
The Old Man, Enmešarra.
The Crook, Gamlum.
The Great Twins, Lugalgirra and Meslamtaea. . . . [15]

(š represents the sound of English sh.) The star list is immediately followed by a list of the dates of the heliacal risings of various constellations, which begins thus:

On the 1st of Nisannu the Hired Man becomes visible.
On the 20th of Nisannu the Crook becomes visible.
On the 1st of Ajjaru the Stars become visible.
On the 20th of Ajjaru the Jaw of the Bull becomes visible.
On the 10th of Simanu the True Shepard of Anu and the Great Twins become visible. [16]

This is a star calendar, or what the Greeks called a *parapegma*. It enables the user to determine the time of year by noting the heliacal risings and settings of the stars. On the first day of the month of Nisannu, the Hired Man (our Aries) makes its morning rising and thus "becomes visible." The Hired Man would be seen rising in the east just before dawn. It marks the first reappearance of the constellation after a period of invisibility of a month or more. On the first of Ajjaru, "the Stars" (our Pleiades) make their morning rising. The calendar in MUL.APIN is reminiscent of the agricultural calendar in Hesiod's *Works and Days*, but it is far more complete and systematic.

The parapegma is followed by a list of stars and constellations that have simultaneous risings and settings:

The Stars rise and the Scorpion sets.
The Scorpion rises and the Stars set.
The Bull of Heaven rises and ŠU.PA sets.
The True Shepard of Anu rises and Pabilsag sets. . . . [17]

The opening lines of this section of MUL.APIN inform us that when the Pleiades are seen rising in the east, the Scorpion will be seen setting in the west (and vice versa). Why would anyone need to know this? The list of simultaneous risings and settings is undoubtedly connected with the parapegma. Using the parapegma, one tells the time of year by noting which constellation is rising in the east just ahead of the Sun. But suppose that the eastern horizon is obscured by clouds. Then one could look to see which constellation is setting in the west just before sunrise. From the list of simultaneous risings and settings, one could then infer which constellation was rising. It is interesting that a similar list of simultaneous risings and settings is given explicitly for this purpose by the Greek poet Aratus in his *Phenomena* (third century B.C.).[18]

The next section of MUL.APIN supplements the parapegma by giving the time intervals between the morning risings of selected constellations:

> 55 days pass from the rising of the Arrow to the rising of the star of Eridu.
> 60 days pass from the rising of the Arrow to the rising of ŠU.PA.
> 10 days pass from the rising of ŠU.PA to the rising of the Furrow.
> 20 days pass from the rising of the Furrow to the rising of the Scales.
> 30 days pass from the rising of the Scales to the rising of the She-goat. . . .[19]

We also find lists of this sort in later Greek papyri—for example, the so-called art of Eudoxus papyrus of about 190 B.C.[20]

The Babylonians, like most early Mediterranean cultures, used a luni-solar calendar. The month began with the new Moon. That is, a new month began when the crescent Moon could be seen for the first time in the west just before sunset. The year usually contained twelve months. But because twelve lunar months only amount to 354 days, a year of twelve months will steadily get out of step with the Sun and the seasons. (The solar year is about 365 days long.) Thus, the Babylonians, like the Greeks, inserted (or *intercalated*) a thirteenth month in the year from time to time.

The months mentioned in the parapegma of MUL.APIN are therefore not months of an actual calendar year, but rather the months of a sort of average or standard year. The Hired Man does not always make his morning rising on the first of Nisannu. Nisannu was traditionally the spring month. Whenever the Nisannu got too far out of step with the seasons (or with the morning risings of the fixed stars), a thirteenth month was intercalated into the calendar year to bring things back into alignment. Consequently, although the Hired Man always made his morning rising *around* the first of Nisannu, the date could actually slosh back and forth by up to a month. The list of time intervals between the risings of key stars was therefore in some ways more useful than the artificial star calendar, for the former was not tied to particular month names.

In the early period, the need for intercalating a thirteenth month was established without the aid of any theory, simply by observation. And the observations might not even be astronomical in nature. As we have seen, Hesiod uses signs taken from animals along with the astronomical signs: the return of the swallow and the first appearance of snails are used in combination with the heliacal risings and settings of the stars. It is noteworthy that two sections of MUL.APIN set out rules for determining whether a thirteenth month should be intercalated. For example, two of the many rules state that a leap month should be inserted to keep the morning rising of the Stars (our Pleiades) at the right time of year:

> \<If\> the Stars become visible \<on the 1st of Ajjaru\>, this year is normal.
> \<If\> the Stars become visible on the 1st of \<Simanu\>, this year is a leap year.[21]

Within a few centuries, the hodgepodge of rules governing the luni-solar calendar was regularized into a real system, based on a nineteen-year cycle. By contrast, the Greeks never did a institute a regular scheme of intercalation. One reason that the Babylonians eventually succeeded in regularizing their calendar, while the Greeks failed, is that the astronomer had a more important place in Babylonian civilization. The astronomers of Babylonia were civil servants who worked at religious temples, for example, at the great temple Esangila in Babylon itself. Thus, the practice of astronomy had a political and religious significance in Babylonian civilization that it did not have in the Greek world.

A good example of the political and religious significance of Babylonian astronomy is provided by the list of omens in MUL.APIN. Omens were taken both from the fixed stars and from the planets. Here are a few examples:

> If the stars of the Lion . . . , the king will be victorious wherever he goes.
> If Jupiter is bright, rain and flood.
> If the Yoke is dim when it comes out, the late flood will come.
> If the Yoke keeps flaring up like fire when it comes out, the crop will prosper.[22]

The Yoke appears here to be another name for Jupiter. There also exist portions of a vast compendium of omens, called *Enuma Anu Enlil*. This collection was considerably older than MUL.APIN. Its omens were frequently quoted and interpreted in later texts. From a surviving table of contents, it appears that *Enuma Anu Enlil* filled some seventy tablets, with thousands of individual omens. The temple astrologers would sometimes send reports to the king, citing an observation recently made together with the relevant interpretation quoted from the standard omen list in *Enuma Anu Enlil*. Some of the omens in MUL.APIN were drawn from those in *Enuma Anu Enlil*. As a rule, ancient Babylonian omens apply to the nation, or to the king, not to ordinary individuals.

Other sections of MUL.APIN contain information about the change in the length of the day between the solstices and the equinoxes, and the variation in the length of shadows in the course of the day. The numbers set down are not real observations, but represent idealized arithmetical patterns—though these must, of course, have been ultimately based on observation. Finally, the beginnings of a theory of the planets can be perceived in MUL.APIN. A portion of the text gives numerical values for the periods of visibility and invisibility of the planets. Although the numbers set down are crude and inconsistent, they do represent a beginning to the scientific study of the planets—the most difficult branch of ancient astronomy—which was to reach a highly succesful conclusion several centuries later.

A Circular Astrolabe

The Babylonians visualized the night sky as divided into three belts. These were named after three divinities and called the way of Ea, the way of Anu, and the way of Enlil. The stars of Anu were situated in a broad belt that straddled the celestial equator. The stars of Anu thus rose more or less in the east and set more or less in the west. The stars of Ea were located south of the belt of Anu. The stars of Ea thus rose well south of east and set well south of west. The stars of Enlil, located to the north of the belt of Anu, rose north of east and set north of west. Included among the stars of Enlil were the northern circumpolar stars, which do not rise or set.

In the first quotation from MUL.APIN, cited earlier, we read that the Plow Star "goes at the front of the stars of Enlil." The other stars in the same passage are all stars in the belt of Enlil. The text of MUL.APIN mentions some thirty-two stars (or star groups) of Enlil. Added to the stars of Enlil is the planet Jupiter (called the star of Marduk, who was the chief god of Babylon), even though the text explicitly states that the star of Marduk does not stay put but keeps changing its position. The next part of the constellation list is devoted to nineteen stars of Anu. Associated with the stars of Anu are the planets Venus, Mars, Saturn, and Mercury. The constellation list concludes with fifteen stars (or star groups) of Ea.

Other Babylonian texts give shorter lists of thirty-six star groups only. The organizing principle is that there should be one star group from each of the three belts, for each of the twelve months of the year. The lists give one star from each of the three belts that made its morning rising in the month of Nisannu, one from each of the three that made its morning rising in the month of Ajjaru, and so on. In the earliest such texts (from about 1100 B.C.), the three groups of twelve stars are simply written in parallel columns.[23]

But there also exist fragments of a list arranged in a circular pattern (see Fig. 1.2). This is usually called a *circular astrolabe*. However, this name is not especially apt, for the word *astrolabe* is also used for two kinds of astronomical instruments that were developed in late antiquity and the Middle Ages. *Circular star list* therefore might be more suitable. The fragment in figure 1.2 dates from the reign of Ashurbanipal, which would place it around 650 B.C.—roughly contemporary with the oldest surviving texts of MUL.APIN. A modern reconstruction, based on the more complete information taken from the rectangular star lists, is shown in figure 1.3.

The pie wedges represent months of the year. Wedge I in figure 1.3 is for Nisannu, the spring month. The three circular belts are the ways of Ea (southern stars, outer ring), Anu (equatorial stars, middle ring), and Enlil (northern stars, inner ring). The Plow Star, MUL.APIN, appears in the belt of Enlil, in wedge I, indicating that the Plow makes its morning rising in the month of Nisannu. The Pleiades (MUL.MUL, "the Stars," in the belt of Ea) make their morning rising in the month of Ajjaru, as we saw in the second extract from MUL.APIN cited earlier. But here we have an apparent problem with the Babylonian astrolabes—the Pleiades are near the celestial equator and ought rather to be placed in the way of Anu, as indeed the text of MUL.APIN confirms. This is one of many small ways in which the star list of MUL.APIN represents an improvement on the astrolabes, which probably derive from older material. The presence of planets in the circular astrolabe is also puzzling, for the planets cannot be used for telling the time of year, since they move

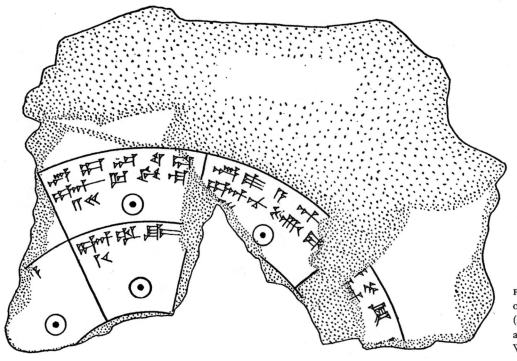

FIGURE 1.2. A fragment of a circular star list (sometimes called a circular astrolabe). From van der Waerden (1974).

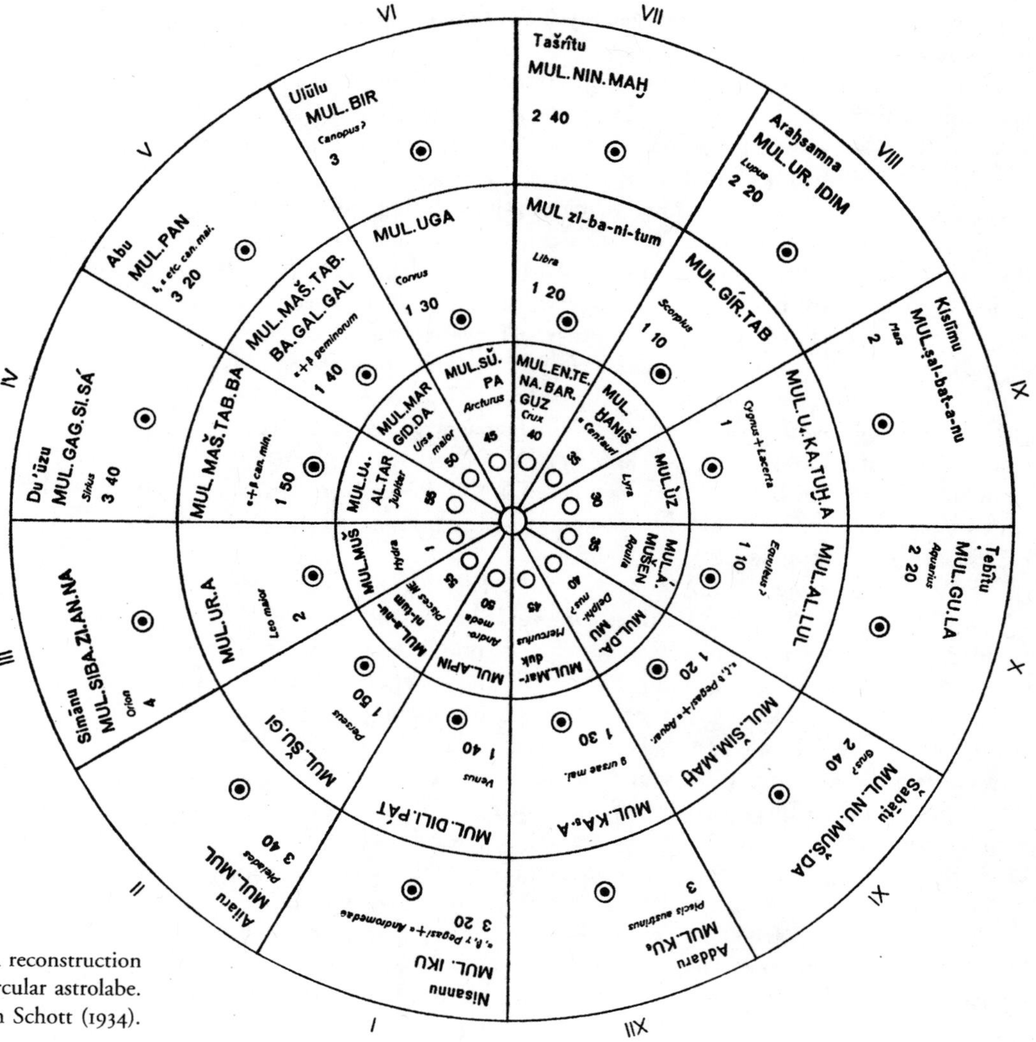

FIGURE 1.3. A reconstruction of a circular astrolabe. From Schott (1934).

around the zodiac and do not make their morning risings at the same time every year. The planet names perhaps designate some sort of "home positions" of the planets among the stars.

The Babylonian division of the night sky into the ways of Ea, Anu, and Enlil and the selection of thirty-six stars to mark the months of the year are much older than the oldest surviving astrolabes. Indeed, this organization of the sky is explicitly mentioned in the text of *Enuma Elish*, the Babylonian creation epic. (The standard title is the translation of the opening words of the text: "When above.") This long poem, which reached its definitive form by 1500 B.C., describes the births of the gods, the ascent to supremacy of Marduk, and Marduk's creation of the the world.[24] At one stage in the construction of the universe by Marduk, we read

> He [Marduk] fashioned stands for the great gods.
> As for the stars, he set up constellations corresponding to them.
> He designated the year and marked out its divisions,
> Apportioned three stars each to the twelve months.[25]

This is a clear reference to the 3 × 12 arrangement of the Babylonian astrolabes.

Each wedge in the circular astrolabe of figure 1.3 contains a number. These indicate the length of a watch. The day was divided into three watches, which were regulated by means of water clocks. The night was similarly divided into three watches. In the summer, the day watches were long and the night watches were short. In the winter, the reverse was true. The longest day watch occurs in wedge III (month of Simanu), which would be around summer solstice. The 4 in the outer segment of wedge III indicates that one should put 4 minas of water into the water clock. When this water has flowed out, a day watch is over. (The mina was a unit of weight.) The shortest day is around the winter solstice (wedge IX), when the day watch lasts for the amount of time required for 2 minas of water to flow from the water clock. Similar information is found in MUL.APIN.

The Babylonians further divided each of the three watches into four parts, which resulted in a twelve-part division of the day. The Greeks learned this twelve-part division from the Babylonians, as the Greek historian Herodotus remarked.[26] The numbers written in the inner two circles represent the lengths of half-watches and quarter-watches, respectively. Thus, in wedge III, 2 and I are one-half and one-quarter of 4. But what about the numbers in wedge II? There the sequence reads 3 40 (a day watch), I 50 (half a day watch), 55 (a quarter watch). The numbers are written in *sexagesimal* notation, that is, in base-60, after standard Babylonian practice. Thus,

$$3 \ 40 \text{ means } 3\frac{40}{60}, \quad 1 \ 50 \text{ means } 1\frac{50}{60}, \quad \text{etc.}$$

I 50/60 is half of 3 40/60. And 55/60 is half of I 50/60. Our own sixty-part divisions of the units of time and of angle derive from ancient Babylonian practice.

Let us examine the sequence of the lengths of the day watches as we go from summer to winter solstice:

Month	Watch	Change
III	4	
		0 20
IV	3 40	
		0 20
V	3 20	
		0 20
VI	3	
		0 20
VII	2 40	
		0 20
VIII	2 20	
		0 20
IX	2	

The length of the day watch decreases by steady increments of 20/60 of a mina from one month to the next. This is an example of an *arithmetic progression*. It is a characteristic feature of Babylonian mathematical astronomy. Clearly, this uniform progression is not a result of direct measurement, for the actual changes in the length of the day are smaller around the solstices and larger around the equinoxes. Rather, it represents an attempt by the Babylonian astronomers to impose an arithmetical pattern on a natural phenomenon. The application of mathematics to astronomy had already begun.

I.2 OUTLINE OF THE WESTERN ASTRONOMICAL TRADITION

By about 700 B.C. astronomy was well under way in both Greece and Mesopotamia. The texts examined in section 1.1 reveal many features in common between Greek and Babylonian astronomy. Nevertheless, these two cultures approached the subject from different perspectives and the science developed quite differently in the two regions.

Babylonian Astronomy

Early in the second millennium B.C., southern Mesopotamia was unified under the rule of Hammurapi, a king of Babylon. Marduk, the national god of Babylon, displaced competing deities and became the chief god of the Mesopotamian pantheon. The city of Babylon, at one time a minor city indistinguishable from many others, rose to become the intellectual and cultural center of the ancient Middle East. Babylonia expanded and contracted with the tides of fortune. But, apart from exceptional brief periods of military adventure, the kingdom never controlled much territory beyond the valleys of the Tigris and Euphrates. Moreover, Babylonia was repeatedly subject to conquest and occupation by foreign powers. Nevertheless, through most of the ancient period, Babylon retained a reputation for splendor, cultural brilliance, and arcane knowledge.

Cuneiform Writing   The Babylonians, who spoke a Semitic language called Akkadian, adopted the cuneiform (wedge-shaped) writing of the older civilization of their southern neighbors, the Sumerians. This style of writing was well suited to its customary medium, the clay tablet. It is easier to press an indentation into clay than to scratch it neatly. A stylus was pressed into the clay to make wedge-shaped marks. Combinations of these cuneiform marks made up the signs for words and syllables (figs. 1.1 and 1.2).

The Babylonians used Sumerian word-signs for phonetic units. Thus, an Akkadian word was broken into individual syllables, and each syllable was represented by a Sumerian sign for that syllable's sound; that is, the Sumerian signs served as *phonograms*. However, a large number of old Sumerian words were retained as *ideograms*, that is, signs that represent a meaning, rather than a sound.

The Babylonians used Sumerian word-signs in both ways when putting their spoken language into writing. For example, the Akkadian word for the constellation Libra is *zibānītu*, which means "scales" or "balance." The Sumerian word for a balance is RÍN. A Babylonian astronomer, writing in Akkadian, could write the name of the constellation Libra in two ways. He could break the word into syllables and represent it phonetically by four cuneiform signs: *zi-ba-ni-tum*. Or he could write a single cuneiform sign: RÍN. In reading aloud, he might pronounce this sign either as "rin" or as "zibanitu."[27]

The situation is very complicated, for the same sign might have multiple phonetic values as a phonogram, as well as multiple meanings as an ideogram. Consider, for example, the sign ⊬. In Sumerian, this represented the name of the the sky god, AN. But this sign also meant "god" in general. A third meaning was "sky." In Akkadian, the same sign was taken over for writing the name of the Babylonian sky god, *Anu*. It was also adopted as an ideogram for "god" in general, in which case it represented the Akkadian word *ilu*. Not surprisingly, it also served as an ideogram for *sky*, Akkadian *šamû*. Thus, as an ideogram, the sign had at least three different meanings. But the same sign also served as a phonogram for writing syllables of other Akkadian words, in which case it represented the sound *an*—the original Sumerian phonetic value of this sign. To make matters worse, the same sign also acquired the phonetic value *il*, from the Akkadian.[28]

In transliterating Babylonian texts, it is customary to distinguish Akkadian words from Sumerian words and ideograms by writing the Akkadian words in italics and the Sumerian words in Roman type. Thus, in section 1.1, we encountered MUL.APIN, the "Plow Star." Actually, the sign MUL for "star" was probably not always pronounced—it served to alert the reader to the fact that the plow intended was a star and not an ordinary plow. In modern practice, the word MUL in front of star names is sometimes omitted, and sometimes it is written in superscript: ^{mul}APIN.

In the last three centuries B.C., cuneiform writing became increasingly rare as it was displaced by Aramaic. But cuneiform continued to serve as a specialized, scholarly script for technical astronomy. Indeed, the last known cuneiform texts, from the first century A.D., are astronomical.

Numbers   In writing numbers, the Babylonians used a base-60, place-value notation. Two kinds of strokes were used, vertical and slanting. Thus, groups of from one to nine vertical strokes were used for the numbers 1 through 9:

Ⲧ	Ⲧⲧ	Ⲧⲧⲧ	Ⲧⲧⲧⲧ					
1	2	3	4	5	6	7	8	9

For 10 through 50, groups of from one to five slanting wedges were used:

⟨	⟨⟨	⟨⟨⟨		
10	20	30	40	50

Any number between 1 and 59 could be represented by combinations of these marks. For example,

16	42

The pattern starts over at 60. That is, the single vertical stroke can represent either 1, or 60, or 3,600 (= 60^2), depending on the place it holds. The lower-valued places are on the right. (This is analogous to our own practice: in the expression 111, the first 1 on the right represents a single unit, the second 1 represents 10 units, and the third 1 represents 10^2 units.) There is some ambiguity in the writing of cuneiform numerals. Thus,

can mean

$$24 \times 60 + 18 = 1{,}458,$$

or

$$24 \times 60^2 + 18 \times 60 = 87{,}480,$$

or even

$$\frac{24}{60} + \frac{18}{3600} = 0.405,$$

since fractions in base-60 were written in the same notation. The scribe would usually be able to tell the proper meaning from context.

In modern practice, the custom is to separate sexagesimal (base-60) places by commas, but to mark off the fractional part of the number by means of a semicolon. Thus,

$$5;24,36 = 5 + \frac{24}{60} + \frac{36}{3,600} = 5.41$$

but

$$5,24;36 = 5 \times 60 + 24 + \frac{36}{60} = 324.6$$

The reader should now be able to make out the numerals for 2; 20 and 1; 10 in figure 1.2.

Major Periods of Babylonian History and Astronomy   Mesopotamian civilization exhibits a great deal of continuity, even though the political situation changed through a series of military conquests. We cannot enter into a detailed history of Babylonian civilization, but it will be helpful to sketch the major periods (refer to fig. 1.4).[29] Many dates in the early part of figure 1.4 are quite uncertain.

Hammurapi's reign and the unification of southern Mesopotamia into one kingdom fall in what is called the Old Babylonian period. Epic poems describing Marduk's creation of the universe probably date from this time. As we saw in section 1.1, one of these poems, *Enuma Elish*, contains some astronomical material—references to the phases of the Moon and to the thirty-six stars used to tell the time of year.

We also have a set of observations of the planet Venus—the so-called Venus tablets of Ammi-saduqa, in whose reign the observations were made (though the copies that have come down to us were written much later). The tablets list the first and last visible risings and settings of Venus over a period of about 21 years. Although some sections of the text appear to list genuine observations, other sections contain idealized risings and settings based on a simple, but rather faulty, scheme. Thus, in the oldest significant astronomical text that we possess, both observation and some sort of theory (even if it is a crude one) are already present. Interestingly, even the apparently genuine observations are listed in omen form: "If on the 28th of Arahsamna Venus disappeared [in the west], remaining absent in the sky 3 days, and on the 1st of Kislev Venus appeared [in the east], hunger for grain and straw will be in the land; desolation will be wrought."[30] The omen form allowed the scribes to predict the status of the grain and straw supply the next time Venus went through the same pattern. While the observations in the Venus tablets are not especially remarkable, they are significant in two ways. First, they provide some help dating the reign of Ammi-saduqa, and thus in establishing the chronology of the Old Babylonian period. Second, they point out a real difference between Greek and Mesopotamian civilization. There is nothing comparable to the Venus tablets in the Greek tradition. Early Babylonian observations are not especially precise. (The remarkable accuracy of the Babylonian observers is a silly fiction that one still frequently encounters in popular writing about early astronomy.) The important thing is that there was a tradition of actually making observations and of recording them carefully and a social mechanism for preserving the records. A good deal of headway can be made with an extensive series of observations, even if the individual observations are not terribly accurate.

Part of the motivation for making the observations was religious. And part of it was practical: the stars and especially the planets were believed to provide signs of the future welfare of the king and the nation. During the long period

DATE	ASTRONOMY	GENERAL HISTORY
Old Babylonian Period 1700 BC		Reign of Hammurapi
		Enuma Elish
1600	Venus observations	
Kassite Dynasty 1500		
1400	*Enuma Anu Enlil*	
1300		
1200		
Six Dynasties 1100	Oldest rectangular astrolabe	
1000		
900		
800		
	Eclipse records	Reign of Nabonassar
700 Assyrian Rule	MUL.APIN	Reign of Ashurbanipal
600 Chaldaean Dynasty	Oldest astronomical diaries	
Persian Rule 500	Equal-sign zodiac Regularization of calendar	
400		Alexander takes Babylon
Seleucid Dynasty 300	Planetary theory	
200 BC		
100 Parthian Rule		

FIGURE 1.4. An outline of Babylonian astronomy.

from about 1570 to about 1155 B.C., Babylonia was ruled by the kings of the Kassite dynasty. The huge compilation of omens called *Enuma Anu Enlil* probably dates from the Kassite period. The Venus tablets of Ammi-saduqa were incorporated into this series.

From the middle of the twelfth century to the middle of the eighth century B.C., Babylonia was ruled by a series of unremarkable dynasties. The oldest surviving rectangular "astrolabes" (the 36-star lists discussed in sec. 1.1) date from this period. Near the end of this period, the scribes began to keep careful records of eclipses. One text, portions of which survive, reported the circumstances of successive lunar eclipses, at least for the years 731–317 B.C. In this and other such lists, the eclipses are arranged in eighteen-year groups. Because some of the circumstances of lunar eclipses repeat in an eighteen-year cycle, the scribes were soon able to use the records of past eclipses to predict future eclipses.[31]

The reign of Nabonassar (747–733 B.C.) is especially important from the viewpoint of later Greek astronomy. Both the quantity and the quality of Babylonian observations improved dramatically starting around this time. The eclipse records are only one aspect of this change. When, several centuries later, the Greek astronomers gained access to Babylonian observational records, the oldest useful material was from the eighth century. The beginning of the reign of Nabonassar was therefore used by later Greek astronomers as a fundamental reference point in their system of reckoning time.

After about 900 B.C., Babylonia came increasingly under the military and political influence of Assyria, a kingdom located to the north, farther up the valley of the Tigris. Historians of ancient Mesopotamia therefore often refer to the whole period from about 900 B.C. to the rise of the Chaldaean dynasty as the Assyrian period. The growing Assyrian intervention in Babylonian affairs came to its logical conclusion in 728 when Tiglath-Pilesar III of Assyria established direct rule over Babylonia. While the Babylonians could no longer effectively oppose the military strength of Assyria, Babylonian culture did prevail, as the conquerors adopted much of the culture of the conquered. A good example comes from the reign of the Assyrian king Ashurbanipal. This king set out to acquire a complete library of all known literature—Sumerian, Babylonian, Assyrian. Vast numbers of tablets were copied on his orders and stored at his library in Nineveh. This library, discovered by British archaeologists in 1853, is a major source of our knowledge of Babylonian literature. Ashurbanipal was also responsible for rebuilding the temple Esangila at Babylon, which had been badly damaged.

Shortly afterward, the Assyrians overextended themselves in ambitious military campaigns and their empire collapsed with astonishing rapidity. A Chaldaean, or Neo-Babylonian, dynasty was established in 625 B.C. by Nabopolassar. The Chaldaeans were originally a tribe from the southern part of Babylonia, who gradually assumed a greater importance in Babylonian affairs before finally putting a king of their own into power. During the Chaldaean dynasty, Babylonian culture underwent a renaissance which extended to astronomy. As mentioned earlier, a notable difference between early Greek and early Babylonian astronomy is that in Babylonia there was a social mechanism for making and recording astronomical observations *and for storing and preserving the records*. Scribes at the temple Esangila in Babylon had the responsibility of watching the sky every night and recording all that transpired—observations of the Moon and planets, as well as of the weather, the depth of the river, and so on. The resulting documents are called *astronomical diaries*. Large numbers of these astronomical diaries have been found. The oldest we possess come from the seventh century B.C., but they probably began a century earlier. Babylonian astronomy of the Chaldaean period and a bit later had a major influence on the development of Greek astronomy. Greek and Roman astronomical writers usually referred to the Babylonians as *Chaldaeans*. And often *Chaldaean* was used by Greek and Roman writers to mean *an astronomer or astrologer* of Babylon.

In 539 B.C., Babylon was conquered by Cyrus, the king of the Persians, an Iranian people to the east. This was during the period of rapidly rising Persian power. It was only a generation later when the Persians made the first of their attempted invasions of Greece. The crushing defeat of Xerxes' navy and army in Greece in 480/479 B.C. marked the beginning of the decline of Persian power.

Babylonian astronomy continued to develop without noticeable hindrance during the period of Persian rule. Indeed, there was a rapid increase in sophistication. Early Babylonian astronomy was fairly crude and simple, and the pace of development was very slow before the seventh century. After about 650 B.C., the pace picks up. But the most rapid advances were made starting about the middle of the Persian period. The equal-sign zodiac was developed as a rationalization of the much older zodiac constellations. In the fifth century B.C., if not a little earlier, the Babylonians regularized their calendar on the basis of the nineteen-year cycle: 19 years = 235 months. Thus, in nineteen years, one counts twelve years of twelve months each and seven years of thirteen months each. The greatest advances in Babylonian astronomy depended not so much on better observations as on better use of mathematics.[32] The scribes rapidly learned to apply elaborate arithmetical methods to astronomical problems. The use of arithmetic progressions (as in sec. 1.1) was a characteristic technique.

In 331 B.C., Alexander, called the Great, conquered the Persian empire with an army of Macedonian soldiers and Greek mercenaries. Alexander's empire lasted but eight years, for in 323 B.C. he died of a fever in Babylon. After Alexander's death, the empire broke up and his generals carved out kingdoms for themselves. It took a generation of warfare for the map to become stable. When the dust had settled, two kingdoms of considerable size and power were established in non-Greek lands.

Ptolemaios I made himself king of Egypt. A Greek-speaking Macedonian dynasty thus ruled Egypt from the end of the fourth century B.C. to the end of the first century B.C., when Egypt was finally annexed as a province of the Roman empire. The last of the Macedonian monarchs of Egypt was Queen Cleopatra. As we shall see below, the Ptolemaic dynasty was of considerable importance to the development of Greek astronomy.

In the vast lands of the old Persian empire, stretching from the borders of Egypt to the frontiers of India, and including Mesopotamia, Seleukos I set himself up as king and established the so-called Seleucid dynasty. Now a Greek-speaking ruling class administered a huge region, populated by peoples of enormous variety in language, religion, and social customs. The history of the Seleucid kingdom is quite different from that of Egypt. The central government never was able to exert the same level of direct administrative control over its far-flung provinces as could the government of Egypt. Almost as soon as the kingdom was established, the eastern provinces began to break off as the native peoples declared independence or as renegade Greek administrators rebelled and established their own kingdoms.

The Seleucid period is of enormous importance for the history of astronomy. It was during this time that Babylonian mathematical astronomy reached its full maturity. The scribes succeeded finally in devising a mathematical theory that permitted accurate numerical prediction of planetary phenomena. The Seleucid period was also the time of most intimate contact between Babylonian and Greek astronomy. However, Seleukos had made a decision that was to lead inevitably to the decline of Babylon. Rather than rebuilding the city and establishing his capital there, he built a new city, Seleucia, about thirty-five miles away on the Tigris River. Babylon never regained its former status.

On the eastern frontier of Mesopotamia, there was a resurgence of Iranian power under the Parthian or Arsacid dynasty. In 125 B.C. the Parthians under Mithradates II took Babylon and the period of Greek rule was over. The Seleucids hung on in Syria until 64 B.C., when the last of their holdings were annexed by the Roman empire.

Greek Astronomy

Hesiod's *Works and Days*, discussed in section 1.1, summarizes the status of Greek astronomy in the seventh century B.C. The subjects treated by Hesiod—phases of the Moon, the annual solar cycle, and the annual cycle of appearances and disappearances of the stars—constitute what we might call the popular and practical astronomy of the Greeks. The origins of this popular astronomy go back beyond the beginnings of writing. From about the fifth century B.C. onward, we can recognize three different astronomical traditions, all of which stemmed originally from the popular-practical astronomy of remote antiquity. These three traditions may be characterized as literary, philosophical, and scientific (see fig. 1.5).

The Literary Tradition The literary tradition involved the continuation and elaboration of Hesiod's theme. The preservers of this tradition were chiefly

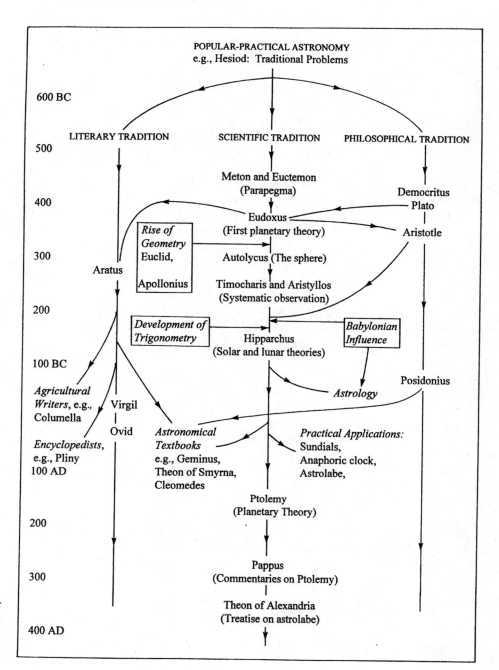

FIGURE I.5. An outline of Greek astronomy. The arrows show continuing traditions and directions of influence.

poets, who sang of the constellations, the signs of the revolving year, and the works of the farmer and the sailor. A notable poet of this genre was Aratus of Soli in Cilicia, who around 275 B.C. wrote *Phenomena*, a poem of some 1150 lines. In the *Phenomena*, Aratus treated the constellations and their risings and settings and provided a list of natural signs that might be used in making weather predictions. The poem was extraordinarily popular and was translated into Latin at least three times.

Latin poets composed original works in the same tradition. Notable among these are Ovid (*Fasti*) and Virgil (*Georgics*). When astronomy and agriculture ceased to attract the interest of the poets, the tradition was continued in modified form by prose writers on agriculture. A good example is Columella (ca. A.D. 50), whose treatise on farming was the most comprehensive of all Roman works on this subject. Columella included in the eleventh book of his treatise a farmer's calendar, which gave for each month of the year the astronomical signs of the season as well as the prescribed agricultural activity. This mix of astronomical lore and agricultural advice continued to be popular down through the Middle Ages. The same mix can be found in the grocery-store almanacs of our own day.

The popular Greek and Latin works do not reflect the level achieved by the ancient astronomers, just as today popular literature on the sciences does not give a detailed picture of our own science. But the ancient popular works are valuable because they give us an idea of the astronomical knowledge that an ordinary educated person was likely to possess.

The Philosophical Tradition   From the rise of Greek philosophy in the sixth century B.C., its practitioners concerned themselves with the fundamental causes of things. The nature of the heavenly bodies, their origin, the cause of their motion, the shape of the Earth, and its position within the cosmos—all these were subjects of intense argument and speculation. From the time of Plato onward, Greek philosophy broadened to encompass new interests (e.g., ethics and esthetics). Nevertheless, considerable effort was still devoted to physical principles. The dominant figure in physical thought was Aristotle (fourth century B.C.). His chief doctrines affecting the science of astronomy were that (1) the Earth is at rest at the center of the universe, (2) the universe is finite and (3) changeless, and (4) the motions of the celestial bodies are uniform and circular.

These doctrines were generally accepted by the Greek astronomers. The first doctrine, on the situation of the Earth, agrees with our everyday perceptions. The second, on the finiteness of the universe, has a strong commonsense appeal. The last two doctrines seemed well confirmed by observation. The Aristotelian conception of the heavens was one part of the mental equipment that every astronomer brought with him when he attacked a scientific problem. Yet it is easy to overstate its importance. In the first place, there was much greater diversity of opinion over physical matters than is commonly claimed. Second, the astronomers often showed themselves capable of questioning or even abandoning Aristotle's tenets when it seemed necessary to do so.

The diversity of physical thought can be illustrated by the controversy over a single topic—the existence of a void place. Two different possibilities must be distinguished. If the cosmos is finite, there might be a void place outside it. And there might be void places within the cosmos itself, as hollows within apparently solid objects. Three different schools of thought can be identified in Greek philosophy. The Aristotelians denied the existence of both kinds of void and gave many convincing arguments. For example, a void place would offer no resistance to the motion of objects, which would therefore rush along with infinite speed—which would be absurd.

The Stoic school, among whom Posidonius (first century B.C.) was prominent, agreed that there could be no void place within the cosmos. For, according to the Stoics, the cosmos was held together by a kind of breath or tension, which would be broken by a gap in the material of the cosmos. Without this tension to maintain it, the cosmos would fly apart. The Stoics disagreed with Aristotle, however, on the possibility of a void place outside the cosmos: they needed it to explain condensation and rarefaction. When wood is burned the smoke takes up more space than the wood did. Evidently, the cosmos must expand. Thus, in the Stoic doctrine, the cosmos alternately contracts and expands into the infinite void space beyond it.

A third distinct view was offered by the Atomists, who boldly accepted both kinds of void. The founding figures of this school were the Greek philosophers Leucippus and Democritus (fifth century B.C.). But the most comprehensive surviving exposition of their views is the Latin poem *On the Nature of Things* of Lucretius (first century B.C.). The Atomists' universe, which really is rather frightful, but which we have come ultimately to accept, consisted of atoms traveling in infinite void space. In antiquity, this was always a minority view. But the Greeks were no more unanimous in physical doctrine than in politics or religion.

The second point to bear in mind in assessing the importance of Aristotle's physics is that the astronomers were capable of abandoning it whenever it seemed expedient. It was accepted that the heavens were changeless: no shifts in the figures of the constellations had ever been detected through generations of observation. Yet Hipparchus and Ptolemy recorded many sightings of three stars that lay on the same straight line for the purpose of allowing future generations to check whether these stars might actually shift with respect to one another. The astronomers did not believe that the constellations would actually suffer any changes, but they were willing to entertain the *possibility*. And, again, when the observations seemed to demand it, the astronomers introduced nonuniform motions into their planetary theories—a serious departure from Aristotelian physics. There were even a few astronomical thinkers who asserted, against the majority opinion, that the Earth moves around the Sun. The Greek astronomers simply never were the blind slaves to Aristotle's system that they sometimes have been made out to be.

With these qualifications in mind, we may still say that Aristotelian doctrine guided the majority of Greek astronomers in their physical thought. Ptolemy, for example, mentions Aristotle by name in a prominent place: in the first chapter of the first book of the *Almagest*.

The Scientific Tradition   Scientific astronomy in Greece began in the fifth century B.C. The summer solstice of 432 B.C. was observed at Athens by Meton and Euctemon. This is the oldest dated Greek observation that has come down to us. It was used by later generations of astronomers in several successive efforts to determine a more accurate value for the length of the year.

In the early stages, much effort was devoted to traditional problems of time reckoning. Meton's name is attached to the so-called Metonic or nineteen-year cycle. The same cycle was discovered earlier in Babylonia. Whether Meton's result represents a borrowing from Babylonian sources or an independent discovery, we do not know. Euctemon is known to have devised a *parapegma* or star calendar, which listed chronologically the appearances and disappearances of the most prominent stars in the course of the year, together with associated weather predictions. The original motivation behind the star calendar must have been the desire to improve on, or to supplement, the chaotic Greek civil calendars. Each Greek city had its own calendar, with its own month names. In different cities, the year might begin with different months. Moreover, no two cities followed the same practice in the insertion of the occasional thirteenth month. Thus, a citizen of Athens and a Greek from Thessaly, for example, could not communicate a time of year to one another by mentioning a month name and a day of the month. But they could communicate unambiguously by means of star phases: the morning rising of Arcturus meant the beginning of fall for everyone. Nearly all Greek astronomers of this period concerned themselves with star calendars. In so doing, they continued and refined a tradition that dated back to Hesiod.

As their science matured, the astronomers became interested in questions with less immediate practical significance: "pure science." By the fourth century B.C. the Earth had been proven to be a sphere and its size had been estimated by means of astronomical observations. It is not known who made these observations and first deduced from them the size of the globe, nor are any details of their methods known. But Aristotle summarized their results in his treatise *On the Heavens* (ca. 350 B.C.). The famous measurement of the size of the Earth by Eratosthenes, who used the fact that the Sun shone vertically down into a well at Syene, Egypt, on the summer solstice, was made about a century later.

The first Greek attempt to explain the complex motions of the planets was made by Eudoxus, who came from Cnidus on the western coast of Asia Minor but who traveled to Athens twice and knew both Plato and Aristotle. Eudoxus is known also to have written a treatise on the celestial phenomena, which contained a description of the constellations and of the principal circles on the celestial sphere. This prose work, which has not survived, served as the inspiration and model for Aratus's verse *Phenomena*, mentioned earlier. This is a notable, and early, example of the influence of a scientific writer on a poet.

None of the works of the scientific astronomical writers mentioned thus far (Meton, Euctemon, Eudoxus) have come down to us. We know the titles of some of their works, and have a partial understanding of their contents, only because of citations made by later writers. The oldest *surviving* works of Greek mathematical astronomy are those of Autolycus of Pitane (ca. 320 B.C.). In one of his works, *On the Revolving Sphere*, Autolycus proved a number of theorems concerning the daily revolution of the sphere of the heavens. In another, *On Risings and Settings*, he provided a geometrical treatment of the old problem of the annual appearances and disappearances of the stars. It is noteworthy that one of the oldest surviving works of Greek astronomy is devoted precisely to the practical problem of telling the time of year, first sketched by Hesiod and then elaborated in the parapegma tradition. Euclid's *Phenomena*, a short, elementary treatise on geometrical matters relevant to astronomy, dates from about the same period.

During the period of early Greek astronomy (fifth and fourth centuries B.C.), Athens was the intellectual center of the Greek world. Although many intellectuals came from other parts of the Greek world—and notably Ionia, on the western coast of Asia Minor—many of them went to Athens to study or teach. Eudoxus is a good example.

Alexander's military career changed this, along with so much else. Alexander was a great founder of cities, all of which he named after himself. Most of these new cities never amounted to much. But after his conquest of Egypt, Alexander founded a city at the mouth of the Nile. This Alexandria grew rapidly in size and wealth. In a short time, it became the most important center of commerce in the eastern Mediterranean. From the late fourth to the late first century B.C., Egypt was ruled by a Greek-speaking Macedonian dynasty named for Ptolemaios I Soter, the general who made himself king of Egypt after Alexander's death.

The first two kings of the Ptolemaic dynasty, Ptolemaios I Soter and Ptolemaios II Philadelphos, were great patrons of the arts and sciences. They founded and supported the Museum and Library at Alexandria. The Museum, the first establishment ever to go by this name, was an institution of higher learning, dedicated to the muses. The members of the Museum lived on the grounds, held their property in common, and devoted themselves to literary, philosophical, and scientific studies. The Alexandrian Library became the greatest in the world.[33] Alexandria became the place to go if one wanted to study literature, mathematics, or science, as Athens once had been. Moreover, because of the fertility of the Nile valley, Egypt was a fairly wealthy country. Although the kings of Egypt carried on intermittent border warfare with their neighbors, the country's geographical situation protected it from serious military threat. Thus, the Ptolemaic kingdom of Egypt was far more stable and secure than its neighbors. As a result of all these factors—wealth, political stability, and royal patronage—the history of later Greek astronomy is largely centered on Alexandria. Many of the later astronomers and mathematicians of note lived or at least spent some time there.

Good examples are Aristyllos and Timocharis (ca. 290 B.C.), who worked at Alexandria, and who are known to us because some of their observations were used by later writers. The observations of Timocharis and Aristyllos are not all that great in number; nevertheless, they constitute the oldest surviving body of careful astronomical observations in the Greek tradition. Thus, systematic observation started much later in the Greek world than in Babylonia. Moreover, the Greeks never did develop much devotion to regular observation. For this reason, the later Greek astronomers, notably Hipparchus and Ptolemy, made as much or more use of Babylonian observations that had come into their possession as they did of observations by Greek astronomers.

The greatest successes of Greek astronomy came not so much from straight observation as from the application of geometry to the problems of astronomy and cosmology. The geometrization of the cosmos had begun with the measurement of the size of the Earth and the working out, by Eudoxus and Autolycus, of the theory of the celestial sphere. This program was continued by a calculation of the sizes and distances of the Sun and Moon by Aristarchus of Samos (ca. 270 B.C.). Aristarchus's results clearly demonstrated the power of geometrical methods in astronomy.

Indeed, one of the critical developments of this period was the rise of Greek geometry, which led rapidly to the mathematization of Greek astronomy. Notable geometers of this period were Euclid, Archimedes, and Apollonius of Perga. Apollonius (ca. 225 B.C.) seems to have been the first to experiment with combinations of deferent circles and epicycles in an attempt to provide an explanation for the motions of the planets, Sun, and Moon. The work on the solar and lunar theories was carried to a high level by Hipparchus (ca. 140 B.C.). For the first time in Greek astronomy, it became possible to make quantitative predictions of the future positions of the Sun and Moon, as in the prediction of eclipses.

Cross-Disciplinary and Cross-Cultural Fertilization   Two important events at about this time were of great benefit to Greek astronomy: first, the development of trigonometry, and second, the borrowing of astronomical results and mathematical procedures from the Babylonian tradition. In both of these developments Hipparchus played a major role.

Before the development of trigonometry, computation by geometrical methods had been laborious. For example, Aristarchus, in his treatise *On the Sizes and Distances of the Sun and Moon*, concluded that the distance of the Sun from the Earth is greater than eighteen times, but less than twenty times, the distance of the Moon from the Earth. To a modern reader, Aristarchus probably appears to be qualifying his result in accordance with the estimated sizes of the errors in his observational data. Actually, the range expressed in his final answer has nothing to do with his data, but rather reflects his methods of calculation. Aristarchus was able to prove geometrically (i.e., by the methods of Euclid) that the ratio of the Sun's distance to the Moon's distance was greater than 18, and, by another construction, that it was less than 20. He was unable to obtain the actual value of the ratio, so in proper Greek geometrical fashion, he rigorously deduced an upper and a lower limit for this ratio. A precise solution of such a problem required the methods of trigonometry, tables of sines, and so on. The rapid development of Greek mathematics greatly expanded the ability of astronomy to deal with the problems presented by the celestial motions. Indeed, astronomy was so firmly based on the methods of geometry and trigonometry that it was usually regarded *as a branch of mathematics.*

The second major stimulus in the second century B.C. came from Babylonia. By the early Seleucid period, the Babylonians had developed theoretical methods for predicting the positions of the planets. But the Babylonian theory was based on arithmetic rules rather than on geometrical models, as was the case with Greek planetary theory. Moreover, as far as we can tell from the surviving sources, the Babylonian planetary theory had no elaborate philosophical underpinning—there seems to have been no set of physical principles comparable to those that Aristotle provided for the Greek astronomers. The Greeks, therefore, could not take their physical principles or geometrical cosmology from the Babylonians, but they could, and did, borrow their observational results, as well as some techniques of calculation. The Babylonians had, for example, obtained accurate values for the tropical and synodic periods of the planets, which the Greek astronomers adopted and applied to their own geometrical planetary theory. There were also records of lunar eclipses observed in Babylon that went back to about 730 B.C.—much earlier than the oldest useful Greek observations. A number of these lunar eclipses were used by Hipparchus and Ptolemy in refining their lunar theory.

While there is evidence of earlier Greek contact with Babylonian astronomy (e.g., in the names and figures of the zodiac constellations), the great period of Babylonian influence was centered in the second century B.C. This is not surprising in view of the larger political, military, and cultural picture. Greek astronomy had already matured to the point at which it could greatly benefit from the Babylonian example. Moreover, contact was now easy, for Mesopotamia, like Egypt, was ruled by a Greek-speaking Macedonian dynasty. All over the Middle East, Greeks were thrown into contact with other peoples. It used to be common to speak of the "Greek miracle," as if the Greeks had in one swoop invented science, right along with history, poetry, and democracy. While the Greek achievement in astronomy and mathematics was truly remarkable, we can no longer regard it as without roots in other and older cultures. The debt of the Greeks to Babylonian astronomy was not recognized until our own century and was only made clear through the decipherment and study of Babylonian astronomical texts on clay tablets unearthed at the end of the nineteenth and the beginning of the twentieth century.[34] It has also become clear that the later Greek astronomers—Ptolemy, for example—did not themselves appreciate how much their own predecessors had borrowed from the Babylonians.

Ptolemy and the Culmination of Greek Astronomy   The culminating figure of Greek mathematical astronomy was Klaudios Ptolemaios—or Ptolemy, as he is usually called today. Ptolemy lived and worked at Alexandria during the first half of the second century A.D. (Ptolemy was not related to the Greek [Ptolemaic] kings of Egypt. But medieval writers sometimes made this confusion, so one can see medieval images of Ptolemy with a crown on his head.)

It was Ptolemy who brought Greek planetary theory into its final, very successful, form. Ptolemy's system was set out in a work that is known today as the *Almagest*. The original title was something like *The 13 Books of the Mathematical Composition of Claudius Ptolemy*. Later the work may simply have been known as *Megale Syntaxis*, the Great Composition. The superlative form of the Greek *megale* (great) is *megiste*. Arabic astronomers of the early Middle Ages joined to this the Arabic article *al-*, giving *al-megiste*, which was later corrupted by medieval Latin writers to *Almagest*. A thousand years of history, embracing Greek, Arabic, and Latin traditions, are thus contained in this one word. No better example could be wished of the continuity of the Western astronomical tradition.

Ptolemy's work was the definitive treatise on mathematical astronomy. The *Almagest* is one of the greatest books in the whole history of the sciences—comparable in its significance and influence to Euclid's *Elements*, Newton's *Principia*, or Darwin's *Origin of Species*. Ptolemy's *Almagest* dominated the study and practice of astronomy from the time of its composition until the sixteenth century. Just as Ptolemy influenced all who followed him, so too he tended to displace his predecessors. The technical works on mathematical astronomy by his predecessors ceased to be read and copied, since their results were included in, or superseded by, Ptolemy's work. So, for example, none of Hipparchus's technical writings have come down to us. We know their titles and their partial contents only because of quotations made by Ptolemy in the *Almagest*.

Sources for the History of Greek Astronomy   Our single most important source is Ptolemy's *Almagest*. But it was written at the very end of the historical development of Greek astronomy. The earlier works that have come down to us are mainly short treatises on specialized topics, with easy or no mathematics, that were copied because they were suitable for use in the schools. Examples are the writings of Theodosius of Bithynia *On Geographic Places* and *On Days and Nights*, Euclid's *Phenomena*, and Autolycus's writings *On the Revolving Sphere* and *On Risings and Settings*, which were mentioned earlier.

We have also three introductory Greek textbooks on astronomy that were written around the beginning of the Christian era. These are the astronomical primers by Geminus, Theon of Smyrna, and Cleomedes. Although these works are nontechnical, having been intended for beginning students, they do include a number of details that can teach us something about the development of Greek astronomy before Ptolemy. For example, Geminus provides a valuable discussion of calendrical cycles, a subject not discussed by Ptolemy. Cleomedes is our main source for the famous measurement of the size of the Earth made by Eratosthenes.

We also have some encyclopedic compositions by Latin writers. For example, the second book of Pliny's *Natural History* is devoted to astronomical matters. Pliny's treatment is nontechnical, and his understanding of astronomical matters is often defective. Nevertheless, he had access to works that now are lost and so sheds some light on the development of planetary theory during the nearly three centuries that elapsed between the activities of Hipparchus and those of Ptolemy. Several other Latin works on specialized topics are also of use. Vitruvius's work *On Architecture*, for example, is an important, if often disappointing, source of information on the theory and art of sundial construction.

Finally, we have a number of astronomical papyri from Greco-Roman Egypt. Most have been recovered by archaeological excavation of ancient garbage dumps or cemeteries—scrap papyrus was sometimes used to wrap mummies. Most astronomical papyri are of rather low intellectual quality. Many were rough notes, some probably taken by students. But they do throw light on the subject, because they are almost the only Greek astronomical documents we have that survive directly from ancient times (as opposed to medieval copies of ancient Greek texts).

Many of the later sources for the history of Greek astronomy are outside the main stream of Greek science. The primers by Geminus and others, for example, stand somewhere between the scientific and the literary traditions. These works were not, even at the time they were written, at the forefront of the science. Geminus is fond of quoting poets, such as Homer and Aratus, to illustrate astronomical points. In fact, he quotes literary men much more often than astronomers. The primers by Cleomedes and Theon of Smyrna also show marked affiliations with particular philosophical schools. Cleomedes departs from astronomical matters to expound Stoic physics. Theon of Smyrna's book has a strong Platonist flavor. In works of this kind, intended for a popular audience, or for students, we see a mixing of several traditions—literary, scientific, and philosophical.

The astronomers also devoted a good deal of time to "applied science," in works that did not break much new ground. For example, both Hipparchus and Ptolemy composed star calendars. Ptolemy's star calendar was not included in the *Almagest*, for by his time the star calendar had no place in a treatise on mathematical astronomy. Rather, he presented his star calendar in a short, separate work wholly devoted to this special topic. Ptolemy composed other works on special applications of astronomy. He wrote a short treatise, for example, on the construction of sundials, and a longer one on astrology. All of these special applications of astronomy were regarded as distinct from the pure astronomy of the *Almagest*, which was concerned with rigorous trigonometric demonstrations from accurately made observations.

After the second century A.D., Greek astronomy, and Greek science in general, went into decline. Why this happened is a great problem, bound up with the general collapse of classical culture. Some of the reasons were the rise of Christianity, which focused on the next world and had less interest in the sciences of this world; the military pressure of the tribes moving in from the Eurasian steppe; and the rigidity and weaknesses of an economic system based largely on slave labor.

Ptolemy had no successor. No Greek astronomer who followed him managed to advance the enterprise. Pappus and Theon of Alexandria (fourth century A.D.) wrote commentaries on the *Almagest*, but these had little to add to Ptolemy. After Ptolemy, astronomy marked time for six hundred years, until the Islamic revival of astronomy that began around A.D. 800.

Astronomy in Medieval Islam

During the period A.D. 800–1300, Arabic was the dominant language of science and philosophy, as Greek had been in the preceding centuries. The first flowering of Arabic astronomy occurred in Iraq and Syria. A stimulus of enormous importance was the patronage of al-Maʾmūn, the seventh Abbasid caliph (ruled 813–833). Al-Maʾmūn established at Baghdad a House of Wisdom, in which scholars, supported by the state, devoted themselves to literary, philosophical, and scientific studies, including the translation of Greek scientific and philosophical works into Arabic. Al-Maʾmūn's House of Wisdom was as significant an institution as the Museum founded at Alexandria a thousand years earlier by the Ptolemies.

Although the astronomical renaissance began in the Middle East, by the eleventh century another center of activity had emerged in Islamic Spain. The history of this development is very complex. Chronologically, we are concerned with a period of four or five centuries. Geographically, the arena stretches from the borders of India to Spain. The two unifying principles of this history are religion and language. The dominant religion was Islam, which put its imprint on every aspect of the culture, including art, literature, philosophy, and science. For this reason, some historians prefer to characterize the astronomy of this period as *Islamic astronomy*. By this, one does not mean that every astronomer was a Muslim or that the fundamental character of the astronomy came directly out of the Islamic faith. But *Islamic* does serve to characterize the cultural setting. One must keep in mind, however, that there were Christians, Jews, and followers of other faiths who practiced astronomy and wrote books about astronomy in Islamic lands. One must also keep in mind that Islamic astronomy was, in its fundamental character, a continuation of the astronomy of the Greeks.

For this reason, some historians prefer to speak of *Arabic astronomy*, referring only to the dominant language of its communication, and not meaning to imply that every astronomer was ethnically an Arab or that every astronomer wrote in Arabic. For we must keep in mind that astronomical works in other languages, such as Syriac, Hebrew, and Persian, are a part of this story.

At first, the Arabic scientists learned their astronomy by studying the classics of Greek science. The *Almagest* of Ptolemy was the standard textbook for advanced study of astronomy and was translated into Arabic several times. It was not long, however, before Arabic astronomers began to write their own astronomical treatises. Taking advantage of the long time interval that had elapsed between Ptolemy's day and their own, Arabic astronomers made discoveries that had escaped the Greeks—for example, the discovery of the decrease in the obliquity of the ecliptic. They also refined the art of making and using astronomical instruments. The solar observations used to determine the obliquity of the ecliptic and the eccentricity of the Sun's orbit are of much better quality in many medieval Arabic treatises than in Ptolemy's *Almagest*. The astrolabe, originally a Greek invention, was developed into a portable instrument of elegance and beauty. It became the characteristic astronomical instrument of the Middle Ages.

For the motions of the planets, Ptolemy's theories in the *Almagest* remained standard. Not every Islamic astronomer agreed with Ptolemy on every detail. The numerical parameters of Ptolemy's planetary theories could be, and often were, improved. Differences in philosophy sometimes led astronomers to question the geometrical models that Ptolemy had invented to explain the motions of the planets. The most frequent complaint was that Ptolemy had not stuck closely enough to the philosophy of Aristotle, especially the principle of uniform circular motion. But Arabic astronomy remained fundamentally Ptolemaic in both its basic assumptions and its methods.

Astronomy in Christian Europe

By comparison with the Islamic culture of the Mediterranean, the Christian lands of western and northern Europe were very backward. Here, in the wreckage of the Western Roman Empire, learning was protected in the monastic schools, but science and mathematics had fallen to an abysmally low level. Greek astronomical works were unknown and astronomy was studied only from a few elementary Latin works, such as Pliny's *Natural History*. The rebirth of the sciences began in the twelfth century, with the reacquisition of the classics of Greek mathematics and astronomy. At first, these were translated into Latin from Arabic translations of the Greek originals. Only somewhat later were Latin translations made directly from the Greek. An influential Latin translation of Ptolemy's *Almagest* was made from the Arabic by Gerard of Cremona, at Toledo around A.D. 1175. It was largely from this translation that Europeans learned their technical astronomy for the next three hundred years. By the middle of the thirteenth century, Europeans were writing their own introductory textbooks of astronomy. Latin versions of a number of Arabic manuals were also in circulation. But for technical astronomy, one still had to go back to Ptolemy.

The work that did the most to dismantle the universe of the Greeks was that of Copernicus, *On the Revolutions of the Heavenly Spheres* (1543). Copernicus took a radically new view of the world and asserted that the Earth is a planet moving around the Sun. This turned the old cosmology inside out. But on closer examination, Copernicus's work, great as it is, turns out to be less radical than one might suppose. The revolutionary part—the Sun-centered cosmology—is introduced in the first book and constitutes only about 5% of the text. The rest of *On the Revolutions* is a sort of rewrite of Ptolemy's *Almagest*. Theorem by theorem, chapter by chapter, table by table, these two works run parallel. Although Copernicus disagrees with Ptolemy about the arrangement of the universe, he makes use of Ptolemy's observations and methods. In the technical details, Copernicus follows Ptolemy more often than not. Copernicus may be regarded as one of the last, and one of the most accomplished, astronomers in the Ptolemaic tradition.

This book is organized by topics: chapter 2 is devoted to the theory of the celestial sphere, chapter 7 to the planets, and so on. While this arrangement is by far the best for discussion of the actual practice of astronomy, it can obscure the broad historical picture. The reader is invited to return to this survey to see how a particular writer or topic fits into the broader picture. Indeed, the reader may wish to make enlarged copies of figures 1.4 and 1.5 and add to them while working through the book.

1.3 OBSERVATION: THE USE OF THE GNOMON

The most ancient astronomical instrument is the *gnomon*—a vertical stick set up in a sunny place where it may cast a shadow. A great deal may be learned about the motion of the Sun by following the motion of the tip of the shadow. Herodotus says in his *Histories*, written around 450 B.C., that the Greeks learned the use of the gnomon from the Babylonians.[35] It may have been so, but Herodotus's remark may simply reflect the Greek fondness for assigning each advance in science and learning to a definite source. The gnomon probably was discovered independently many times in many different cultures. In any case, gnomons were used in Greece in the fifth century B.C. to observe summer and winter solstices and perhaps also to tell time.

Your gnomon can be a short nail driven perpendicularly into a flat board. Place a sheet of paper over the gnomon so that the gnomon punches a hole in the paper. Tape the paper to the board to ensure that the paper does not shift position. Label the paper in one corner with the date. Measure and record the height of the gnomon. Make sure that the board is not moved during the course of the day.

Now mark a dot at the tip of the gnomon's shadow. Next to the mark write the time of day. Mark the location of the shadow as early in the morning as possible. After that, mark the shadow's location once every half hour or so until late in the afternoon as possible. A well-made shadow plot should stretch over six to eight hours and should contain twelve to sixteen plotted points.

Make a new shadow plot once every two or three weeks to observe the changes in the Sun's behavior as the year progresses.

The interpretation of the shadow plot is the subject of section 1.4.

1.4 ON THE DAILY MOTION OF THE SUN

Interpreting a Shadow Plot

A great deal of information can be obtained from a well-made shadow plot. We shall study the shadow plot in figure 1.6, which was made at Seattle on February 19.

Local Noon *Local noon* is the time of day when the shadow of a vertical gnomon is shortest. On the example plot, local noon fell between 12:02 and

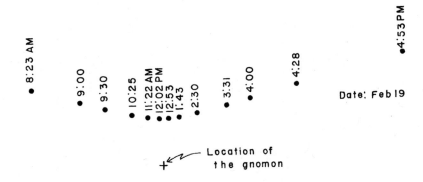

FIGURE 1.6. Shadow plot made at Seattle on February 19.

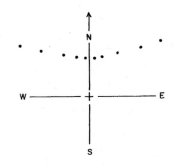

FIGURE 1.7. North is the direction of the shortest shadow cast by a vertical gnomon.

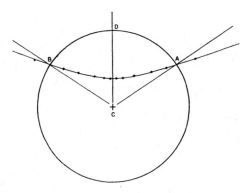

FIGURE 1.8. Alternative method for finding north.

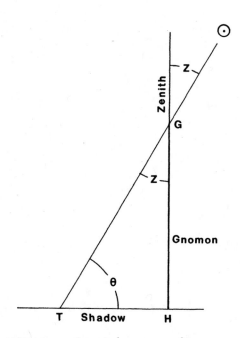

FIGURE 1.9. A vertical gnomon and its shadow. θ is the altitude of the Sun ☉. z is the Sun's zenith distance.

12:53 P.M. Local noon need not occur at twelve o'clock. There are several reasons for this. Most obviously, we set our clocks forward or backward by a whole hour when changing to or from daylight savings time. But even if we never used daylight savings time, local noon still would rarely occur at twelve o'clock, because the time of local noon wanders in the course of the year—by about half an hour from one extreme to the other. (The causes of this variation are discussed in sec. 5.9.)

The noon determined by a gnomon is called *local* because the time at which it occurs depends on the location of the observer. Local noon in New York occurs some three hours earlier than local noon in San Francisco. This is the astronomical fact that lies behind the modern use of time zones. But even within a single time zone, there is a noticeable variation in the time of local noon. A city in the extreme east of a time zone has local noon about an hour earlier than a city in the extreme west of the same zone.

Finding True North   The direction in which the shortest shadow points is called *north* (see Fig. 1.7). (On some parts of the Earth the shortest shadow points south, but everywhere in the continental United States the shortest shadow points north.) Although the clock time at which local noon occurs varies during the year, the direction of the shadow at local noon is always the same.

This is the fundamental definition of north. North definitely is not, for example, the direction in which a compass needle points. In Washington state, compass needles point about 22° to the right of north; in Maine, about 15° to the left of north. The angle by which the compass needle differs from true north is called the *compass declination* and varies with position on the Earth. Moreover, the compass declination at a particular locality may vary slightly from year to year. Every accurately determined north-south line (as in land surveying) depends on astronomical observation, usually of the Sun.

Alternative Method for Finding True North   As it is difficult to tell exactly where the shadow is shortest, let us consider an alternative procedure. Sketch a smooth curve through the points of the shadow plot (fig. 1.8). Then place the point of a drawing compass at the gnomon's base *C* and draw a circle, which intersects the shadow curve in two places, *A* and *B*. Now draw lines *CB* and *CA*. Finally, draw a line that bisects angle *ACB*. This line cuts the circle at point *D*. Line *CD* then points north.

The direction of north established in this fashion should agree with the direction of the shortest shadow. The reason this method works is that the afternoon half of the shadow plot is a mirror image of the morning half. The line that divides the shadow plot into two similar halves is the north-south line, and the shadow falls along this line at local noon.

Direction of Sunrise or Sunset   In the example shadow plot, the early morning shadows fell to the northwest (fig. 1.7). The early morning Sun had to be in the opposite direction (southeast) for the gnomon's shadow to fall so. Evidently, the Sun rose in the southeast. People often say that the Sun rises in the east and sets in the west. But this is speaking loosely. Only twice a year (at the equinoxes, March 20 and September 23) does the Sun rise exactly in the east and set exactly in the west.

Altitude of the Sun at Local Noon   The *altitude* of an object, such as the Sun, is its angular distance above the horizon. At local noon the Sun's altitude is the largest for that day. In figure 1.9, ☉ represents the Sun. *GH* is the gnomon, ☉*T* is a ray of sunlight, and *TH* is the length of the shadow. θ is the altitude of the Sun. In figure 1.9, we have

$$\tan \theta = \frac{GH}{TH}.$$

The gnomon used to make the original shadow plot of figure 1.6 was 1.0 cm tall, and the length of the noon shadow on the original plot was 1.7 cm. Using these values, we have

$$\tan \theta = \frac{1.0 \text{ cm}}{1.7 \text{ cm}} = 0.59.$$

$$\theta = \tan^{-1}(0.59) = 31°.$$

The altitude of the Sun may also be found by laying out GH and TH on a scale drawing and measuring θ with a protractor.

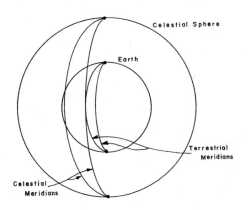

FIGURE 1.10. Terrestrial and celestial meridians.

Some Useful Terms

Several useful terms can be defined by reference to the shadow plot. A *meridian* is a line on the surface of the Earth that runs exactly north-south. The *local meridian* is the north-south line that happens to pass through the locality in question. Each meridian, extended far enough, is a great circle on the surface of the Earth. The meridians all meet at the north and south poles of the Earth.

The *zenith* is the point of the sky directly overhead (see fig. 1.9). The zenith may be defined by means of a plumb line, that is, a string from which a small weight is suspended. All plumb lines point down toward the center of the Earth.

The *zenith distance* of a celestial object is its angular distance measured down the zenith. In figure 1.9, the zenith distance of the Sun is angle z. The zenith distance is the complement of the altitude; that is, $z = 90° - \theta$.

The *celestial meridian* is a great circle on the dome of the sky. Imagine standing with your arm parallel to the ground, pointing directly north. Then swing your arm up until it points at the zenith. Then swing it on over, until it comes down behind you and points south. If you had a pencil in your hand, you could imagine drawing a semicircle on the dome of the sky. This circle, which passes through the north point, the zenith, and the south point, is called the celestial meridian.

The sky may be regarded as a great sphere that surrounds the little sphere of the Earth. When we think of the sky in this way we usually call it the *celestial sphere* (see Fig. 1.10). Directly underneath the celestial meridian is the meridian line that runs along the ground, from north to south. The meridian on the ground is sometimes called the *terrestrial meridian*. The terrestrial meridian may be regarded as a projection of the celestial one.

Historical Example of the Use of a Gnomon

Marcus Vitruvius Pollio was a Roman architectural writer who lived in the reign of Augustus (late first century B.C.). His only surviving work, commonly known as the *Ten Books on Architecture*, is an important source of information on Roman techniques of design, construction, and decoration. The work was much studied and was extremely influential during the revival of the classical style in the Renaissance. Vitruvius is a valuable source for the history of astronomy as well, mainly because of his ninth book, which includes a treatment of the design of sundials. We return to Vitruvius's discussion of sundials in section 3.2.

Here we examine only his prescription for laying out the streets of a city, which involves an interesting use of the gnomon. Vitruvius begins by remarking

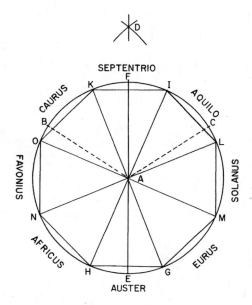

FIGURE I.II. Vitruvius's construction of the meridian and the directions of the eight winds.

that if a city is to be properly designed, some thought must be given to excluding the winds from the alleys. Lack of such foresight has, in many a case, rendered a city very unpleasant, and Vitruvius cites as an example Mytilene on the island of Lesbos. This city was built with magnificence and good taste, but it was not prudently situated. When the wind is from the south, men fall ill; when it is from the northwest, they cough. When the wind is from the north, they recover, but they cannot stand about in the streets and alleys because of the severe cold.

Vitruvius then takes up the subject of the winds themselves. Some have held that there are only four: Solanus from due east, Auster from the south, Favonius from due west, Septentrio from the north. However, more careful investigators say there are eight. To the four winds already named, Vitruvius adds Eurus from the winter sunrise point (i.e., the southeast), Africus from the winter sunset (the southwest), Caurus or Corus from the northwest, and Aquilo from the northeast. Each of these winds had its own characteristics: hot or cold, damp or dry, healthful or unhealthful. Each was most likely to blow at its own proper season. Furthermore, the wind names were often used to indicate directions. We shall encounter them, and their Greek equivalents, again in section 3.1.

Having established these preliminaries, Vitruvius tells how to determine the directions of the various winds. In the middle of the city, lay and level a marble plate, or else let the spot be made so smooth and true by means of rule and level that no plate is necessary. In the center of that spot set up a bronze gnomon to track the shadow, at a point called *A*. The following extract reproduces Vitruvius's instructions for establishing the quarters of the eight winds (see fig. 1.11).

EXTRACT FROM VITRUVIUS

Ten Books on Architecture I, 6

Let *A* be the center of a plane surface, and *B* the point to which the shadow of the gnomon reaches in the morning. Taking *A* as the center, open the compasses to the point *B*, which marks the shadow, and describe a circle. Put the gnomon back where it was before and wait for the shadow to lessen and grow again until in the afternoon it is equal to its length in the morning, touching the circumference at the point *C*. Then from the points *B* and *C* describe with the compasses two arcs intersecting at *D*. Next draw a line from the point of intersection *D* through the center of the circle to the circumference and call it *EF*. This line will show where the south and north lie.

Then find with the compasses a sixteenth part of the entire circumference; then center the compasses on the point *E* where the line to the south touches the circumference, and set off the points *G* and *H* to the right and left of *E*. Likewise on the north side, center the compasses on the circumference at the point *F* on the line to the north, and set off the points *I* and *K* to the right and left; then draw lines through the center from *G* to *K* and from *H* to *I*. Thus the space from *G* to *H* will belong to Auster and the south, and the space from *I* to *K* will be that of Septentrio. The rest of the circumference is to be divided equally into three parts on the right and three on the left, those to the east at the points *L* and *M*, those to the west at the points *N* and *O*. Finally, intersecting lines are to be drawn from *M* to *O* and from *L* to *N*. Thus we shall have the circumference divided into eight equal spaces for the winds. The figure being finished, we shall have at the eight different divisions, beginning at the south, the letter *G* between Eurus and Auster, *H* between Auster and Africus, *N* between Africus and Favonius, *O* between Favonius and Caurus, *K* between Caurus and Septentrio, *I* between Septentrio and Aquilo, *L* between Aquilo and Solanus, and *M* between Solanus and Eurus. This done, apply a gnomon to these eight divisions and thus fix the directions of the different alleys.[36]

Vitruvius recommends that the streets be laid out on lines of division between winds, for if the streets run in the direction of a wind, strong gusts will sweep through them. But if the lines of houses are set at angles to the winds, the winds will be broken up. Vitruvius's use of the gnomon to establish the meridian and other directions was not original but was a traditional technique already several centuries old. His treatment does, however, illustrate one application of elementary astronomy in antiquity.

1.5 EXERCISE: INTERPRETING A SHADOW PLOT

So, I said, it is by means of problems, as in the study of geometry, that we will pursue astronomy too. . . .

Plato, *Republic* VII, 530B–C.

Use the shadow plot that you made for section 1.3 to solve the following problems.

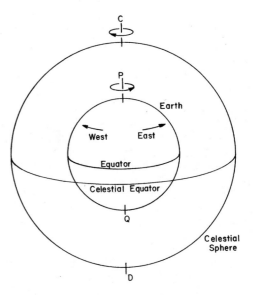

FIGURE 1.12. The Earth surrounded by the celestial sphere.

1. At what clock time did local noon occur? Connect the points of your shadow plot by a smooth curve. Try to judge exactly where the shadow would be shortest, and estimate the time accordingly.

2. Which way is north? Use the bisection-of-the-angle method to find out. Check to see that your north direction agrees fairly well with the direction of the shortest shadow. Also draw in the directions south, east, and west.

3. Which way is magnetic north? Reposition your shadow plot in exactly the place where it was made. Place a magnetic compass on the shadow plot and determine the direction in which the needle points. If your gnomon is made of iron (as are most nails, for example), be careful to keep the compass well away from the gnomon. Draw a line on your plot representing the direction in which the compass needle points. Extend this line until it crosses the true north-south line on your shadow plot, which you drew in response to question 2. Use a protractor to measure the compass declination. That is, what is the angle between true north (determined by the Sun) and magnetic north (determined by the compass)? Does the compass needle point too far east or too far west?

4. In what direction did the Sun rise? In what direction did it set?

5. At local noon, what was the altitude of the Sun?

6. Indicate on your shadow plot the directions of the eight winds mentioned by Vitruvius.

1.6 THE DIURNAL ROTATION

Some Essential Facts

In a day the Earth makes one rotation on its axis. Thus, in figure 1.12, we explain the risings and settings of the stars by supposing that the Earth rotates from west to east about axis PQ. Points P and Q are the north and south poles of the Earth.

Alternatively, we may regard the Earth as stationary and let the celestial sphere (to which the stars are fixed) revolve from east to west about axis CD. Point C is called the *north celestial pole*. Similarly, D is the *south celestial pole*. The *celestial equator* is a great circle on the celestial sphere, midway between the poles. The Earth's equator may be considered a projection of the celestial equator.

As far as *appearances* are concerned, it makes no difference which view one adopts. The ancient point of view—that the heavens revolve about a stationary

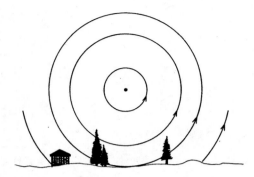

FIGURE 1.13. Paths of the circumpolar stars in the northern part of the sky.

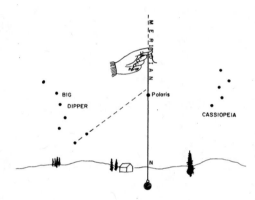

FIGURE 1.14. Using the Pointers (two stars in the Big Dipper) to locate Polaris.

Earth—is more convenient, for it pictures the world exactly as it appears to our eyes. All those whose work involves practical astronomy (surveyors and navigators as well as astronomers) habitually use the ancient point of view. We, too, shall adopt it.

In a day and a night, the stars describe circles about the celestial pole. Figure 1.13 represents a view toward the northern horizon, for an observer in the northern hemisphere. The stars nearest the pole go around in small circles, with the result that they do not rise or set but remain above the horizon for twenty-four hours every day. Such stars are called *circumpolar*. Stars that lie farther from the pole go around in larger circles, which pass beneath the horizon. These stars therefore have risings and settings.

The location of the north celestial pole is marked nearly, but not exactly, by a star of medium brightness variously called *Polaris*, the *North Star*, or the *Pole Star*. Polaris can be located in the night sky with the aid of the Big Dipper (figs. 1.14 and 1.15). Although Polaris is not especially bright, there is little risk of mistake, for it has no near neighbors of comparable brightness. Hold up a plumb line so that the string passes over Polaris. The north point *N* is then the place where the plumb line crosses the horizon.

If Polaris were located exactly at the celestial pole, it would appear never to move. Actually, Polaris is not precisely at the celestial pole; it is less than 1° from the pole. As a result, it describes a tiny circle of its own in the course of a day.

Aspects of the Celestial Sphere

Parallel Sphere   The sky takes on different aspects when observed from different places on the Earth. At the Earth's north pole the celestial pole would be seen at the zenith, as is clear from figure 1.12. Thus, at the north pole, the rotation of the celestial sphere carries all the stars around on circles parallel to the horizon, as in figure 1.16. None of the stars rise or set. Rather, half of the celestial sphere is permanently above the horizon, and half is permanently below. Medieval astronomers often referred to this arrangement as the *parallel sphere*, because the circles traced out by the stars are all parallel to the horizon.

Right Sphere   Figure 1.17 represents the situation for point *A* located on the Earth's equator. Light from the North Star shines vertically down on the north pole *P*. Because the star is very far away, compared to the size of the Earth, all rays of light that leave the star and strike the Earth are essentially parallel to one another. The ray that arrives at *A* has therefore been drawn parallel to the ray arriving at *P*. This ray coincides with the horizon at *A*. Thus, at the Earth's equator, the pole star would be seen on the horizon.

That is, at the equator, the celestial poles lie on the horizon. The diurnal rotation therefore carries the stars on circles that are perpendicular to the horizon, as in Figure 1.18. (Note that this figure is rotated by 90° with respect to fig. 1.17. It is often convenient to make the horizon horizontal!) At the equator, all the stars rise and set vertically, and every star remains for twelve hours below the horizon, and for twelve hours above. The celestial sphere, observed from a place on the Earth's equator, is said to be *right*, in the sense of upright or perpendicular, because the paths of the stars are perpendicular to the horizon.

Oblique Sphere   Figure 1.19 represents the situation for a point *A* located in the northern hemisphere. The latitude of point *A* is angle *L*. Light arriving at *A* from the North Star makes an angle α with the horizon. That is, the altitude of the star above the horizon is α. It is easy to prove that $\alpha = L$.

Draw *AB* parallel to the equator. Then $\theta = L$, since these angles are formed by the intersection of two parallel lines (the equator and *AB*) with a given line (*CA*). The rays from the North Star are perpendicular to the equator, so γ = 90° − θ. Similarly, the horizon is perpendicular to the zenith direction, so we must also have γ = 90° − α. Setting these two expressions for γ equal to each another, we obtain the result we sought, α = *L*.

That is, *the altitude of the celestial pole at a place on the Earth is equal to the latitude of that place.* This provides a way of determining the latitude of any position on the Earth's surface: measure the altitude of the pole star. (But, since the pole star is not precisely at the celestial pole, a small correction would have to be added to the measured altitude to get a perfectly accurate value for the latitude.)

At intermediate latitudes (neither at the equator nor at the poles), the axis of the celestial sphere is neither upright nor horizontal, and the sphere is said to be *oblique.* Figure 1.20 represents such a situation. The observer and the points *N, E, S,* and *W* are all on the horizon, which has been extended all the way to the celestial sphere. The observer's latitude is equal to angle *L,* the altitude of the celestial pole. The figure has been drawn with *L* = 48°, the latitude of Seattle or Paris.

As the celestial sphere revolves, the stars trace out circles parallel to the equator. The diurnal paths of five stars have been shown on the diagram: Kochab, Arcturus, Mintaka, Sirius, and Miaplacidus. These stars belong to the constellations Ursa Minor, Boötes, Orion, Canis Major, and Carina.

Stars that lie near enough to the celestial pole may be circumpolar. At a latitude of 48° N, such is the case for Kochab, which does not rise or set.

Stars that lie north of the celestial equator, but not far enough north to be circumpolar, will rise and set. These stars all spend more than twelve hours of each day above the horizon. This is the case for Arcturus. The diurnal path of Arcturus is cut by the horizon into unequal parts, with the longer part lying above the horizon. It is also evident that Arcturus rises north of east and sets north of west.

Mintaka (δ Orionis) lies almost exactly on the celestial equator. Consequently, its diurnal path is cut by the horizon into two equal parts: Mintaka spends twelve hours above and twelve hours below the horizon every day. As figure 1.20 shows, the celestial equator (which is the diurnal path of Mintaka) passes through the east and west points of the horizon. Thus, Mintaka rises exactly in the east and sets exactly in the west.

Stars that lie south of the celestial equator remain above the horizon for less than twelve hours. At 48° N, such is the case for Sirius. Sirius rises south of east and sets south of west.

Stars far enough south of the celestial equator may also be circumpolar. At 48° N latitude, such is the case for Miaplacidus (β Carinae). The diurnal path of Miaplacidus, a small circle centered on the south celestial pole, lies beneath the horizon. Thus, Miaplacidus is never seen from Seattle or Paris.

Note that a star may be circumpolar at one latitude, but rise and set at another latitude.

Methods of Demonstrating the Diurnal Rotation

The diurnal revolution of the stars is easily observed by means of a sighting tube. This is simply a hollow tube attached to a stand (fig. 1.21). The tube may be aimed at a star and clamped. After ten or twenty minutes, the rotation of the celestial sphere will carry the star out of view of the tube. For example, suppose the sighting tube is aimed toward a star that is low in the eastern part of the sky. The tube is adjusted on its stand until the star can be seen in the middle of the tube, as in figure 1.22A. After twenty minutes, the star will no longer appear to be in the tube (fig. 1.22B) but will have moved diagonally up and to the right.

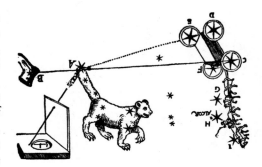

FIGURE 1.15. A sixteenth-century guide to finding the Pole Star. The Big Dipper is shown in its alternative representation as a Wagon. The Pointers (stars *D* and *E*) indicate the Pole Star, at the tip of the tail of the Little Bear (sometimes called the Little Dipper). A second means of finding the Pole Star is shown, using the left foot *B* of Cepheus (γ Cepheii) and star *C* of the Wagon.

The middle star in the team that pulls the Wagon is correctly shown as double. The principal star *H* is called Mizar. Its faint companion is Alcor. It is for this reason that the middle horse is depicted with a rider: this pair of stars is sometimes called the Horse and Rider. Anyone with good eyesight (whether corrected or naturally so) should have no trouble picking out Alcor on a clear night.

In the lower left is a folding, portable sundial of a type that was popular in the Renaissance. When the dial is opened, a string is pulled tight and serves as the gnomon. For the dial to function properly, the gnomon string must point at the celestial pole. The dial was also fitted with a magnetic compass as an aid in orienting the dial.

From the *Cosmographia* of Petrus Apianus, as reworked by Gemma Frisius: *Cosmographia . . . Petri Apiani & Gemmae Frisii* (Antwerp, 1584). Photo courtesy of the Rare Book Collection, University of Washington Libraries.

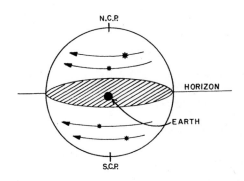

FIGURE 1.16. Parallel sphere: the sky as viewed at the Earth's north pole.

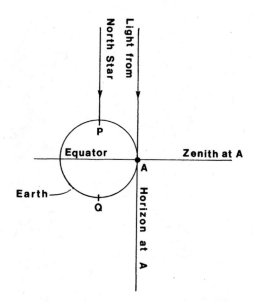

FIGURE I.17. An observer *A* at the Earth's equator sees the North Star on the northern horizon. An observer *P* at the north pole sees the North Star straight overhead.

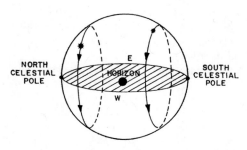

FIGURE I.18. Right sphere: the sky as viewed from a point on the Earth's equator.

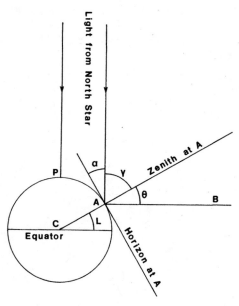

FIGURE I.19. The altitude α of the celestial pole is equal to the latitude *L* of the observer.

If, however, the sighting tube is aimed at the North Star, that star will remain in the tube all night long. Indeed, the North Star will remain in its sighting tube all day long as well, although it will not be visible.

These simple observations with a sighting tube reveal in a clear and immediate way the existence of the celestial pole and the rotation of the celestial sphere. A fine sighting tube may be made from a one- to two-foot length of aluminum or plastic tubing. For a mount, the tripod stand and clamps of the college science laboratory serve well.

An even more dramatic demonstration of the diurnal rotation may be made by means of what we shall call the compound sighting tube (refer to fig. 1.23). One sighting tube, *AB*, is directed toward the Pole Star and clamped. A second tube, *CD*, is directed toward some other star *S* and then clamped to the first tube. The apparatus is so arranged that the two-tube assembly may be rotated around axis *AB*, while the angle α between the two tubes remains unchanged. As the night goes by, star *S* revolves in a circle around the Pole Star—which is demonstrated by the fact that sighting tube *CD* may be made to follow star *S* by rotating tube *CD* about the *AB* axis. That is, if *S* has escaped from tube *CD*, the star may be recovered in that tube by a simple rotation of the apparatus about axis *AB*.

Historical Note on the Dioptra

The sighting tube was known to the ancient Greek astronomers by the name *dioptra*. In its original form it was not very different from the simple tube and stand described above (fig. 1.21). There is, of course, no need to use a closed tube; one can substitute a rod with a sight at each end (fig. 1.24), and this form is more probable. No Greek dioptra has come down to us, but the instrument is mentioned by a number of Greek writers.

Euclid, the geometer of the late fourth century B.C., mentions a dioptra in his elementary astronomical treatise, the *Phenomena*. Aim, says Euclid, a dioptra at the constellation Cancer while it is rising. Then turn around and look through the other end of the instrument and you will see Capricornus setting. Euclid's suggested observation, if carried out, would show that Cancer and Capricornus lie in diametrically opposite directions. (For Euclid's use of this fact, see sec. 2.4.)

Some version of the *compound* sighting tube was also developed in antiquity. It is impossible to be certain of the details of construction because no such ancient instrument survives and no extant Greek text provides an adequate description. However, the following extract from Geminus leaves no doubt that he was familiar with a version of the dioptra equivalent in principle to the compound sighting tube of figure 1.23. Geminus was the author of an elementary astronomy text, the *Introduction to the Phenomena*, written around A.D. 50.

EXTRACT FROM GEMINUS

Introduction to the Phenomena, XII

The cosmos moves in a circular motion from east to west. For all the stars that are observed in the east after sunset are observed, as the night advances, rising always higher and higher; then they are seen at the meridian. As the night advances, the same stars are observed declining towards the west; and at last they are seen setting. And this happens every day to all the stars. Thus, it is clear that the whole cosmos, in all its parts, moves from east to west.

That it makes a circular motion is immediately clear from the fact that all the stars rise from the same place and set in the same place [on the horizon]. Moreover, all the stars observed through the dioptras are seen to be making a circular motion during the whole rotation of the dioptras.[37]

The last quoted sentence clearly refers to an instrument more or less equivalent to the compound sighting tube of figure 1.23.

If the compound sighting tube is fitted with protractors, it can be used to measure angles. Indeed, Geminus remarks that a dioptra may be used to divide the zodiac into twelve equal parts,[38] which implies an instrument equipped with some sort of angular scale. The compound sighting tube of figure 1.23 is not suited to dividing the zodiac into equal parts (because of the obliquity, or slantedness, of the zodiac with respect to the equator), but a further modified dioptra might have done the job. However, by the time of Ptolemy (second century A.D.), and perhaps by the time of Hipparchus (second century B.C.), the dioptra was replaced by the armillary sphere as the instrument of choice for measuring the positions of the stars. From then on, the dioptra's chief role in astronomy was that of a demonstration device or teaching tool.

In surveying (as opposed to astronomy), the dioptra saw continued service. It was elaborated into a fairly sophisticated surveyor's instrument, analogous to the modern theodolite. A water level was added to the stand and the sighting tube (or rod equipped with sights) was fitted with a protractor so that it could be used to measure angles, as for example the angular height of a mountain summit. Such an elaborate dioptra was described by Hero of Alexandria (first century A.D.), the Greek mechanician best known to most modern readers for his invention of a primitive steam engine. It is doubtful that Hero's elaborate dioptra ever saw widespread use. However, it was the logical culmination of the instrument that began four or five centuries earlier as a simple sighting tube.[39]

The name *dioptra* was also applied to another instrument, a kind of cross-staff, that could be used to measure angular distances in the sky, such as the angular diameters of the Sun and Moon. This kind of dioptra is described by Ptolemy in *Almagest* V, 14.

Is the Heaven or the Earth in Motion? An Ancient Debate

As far as practical astronomy is concerned, it makes no difference whether the motion of the stars is explained by the westward rotation of the celestial sphere or by the eastward rotation of the Earth on the same axis. Observations of the heavenly bodies provide no basis for choosing. One may, however, still ask which hypothesis is *physically* true. Although opinion in antiquity was overwhelmingly in favor of a stationary Earth, there were thinkers who subscribed to the opposite view.[40]

Heraclides of Pontos   The earliest philosopher who unambiguously and undeniably taught the rotation of the Earth on its axis was Heraclides of Pontos (ca. 350 B.C.). Heraclides came from the city of Heraclea Ponticus, on the north shore of Asia Minor, the site of the modern Turkish city of Eregli. While still a young man he went to Athens, where he became a pupil of Plato and of Speusippus, Plato's successor at the Academy. Speusippus died in 347 B.C. and his place as head of the Academy was taken by Xenocrates, at which time Heraclides returned to his native city. None of Heraclides' scientific writings have come down to us. However, his opinion on the rotation of the Earth is mentioned by two later writers, Aëtius and Simplicius.

Aëtius (ca. A.D. 100) was the author of a book called *The Opinions of the Philosophers*, a guide to and history of Greek philosophy. Aëtius's handbook, extant in part, is a valuable although often disappointing source of information on the views of many writers whose works have been lost. Aëtius, on the question of the Earth's rotation, has this to say:

> Heraclides of Pontos and Ecphantus the Pythagorean move the Earth, not however, in the sense of translation, but in the sense of rotation, like a wheel fixed on an axis, from west to east, about its own center.[41]

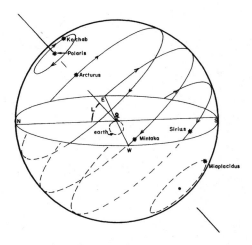

FIGURE 1.20. Oblique sphere: the sky as viewed from latitude 48° N.

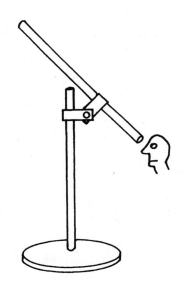

FIGURE 1.21. A dioptra, or sighting tube.

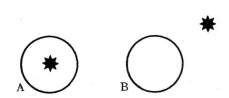

FIGURE 1.22. A star low in the east is sighted through a dioptra (A). A short time later (B) the star is seen to have moved out of the sighting tube of the dioptra.

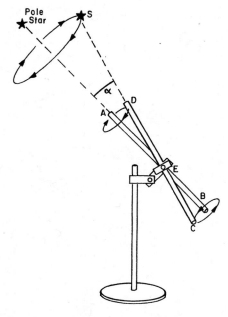

FIGURE I.23. A compound dioptra. Tube *AB* is aimed at the celestial pole. Tube *CD* makes a fixed angle α with *AB*. As the apparatus is turned about axis *AB* in the course of the night, the observer can keep star *S* in sight.

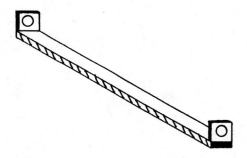

FIGURE I.24. A simple dioptra consisting of a stick with two sights.

This is as clear as one could wish. Ecphantus was a Greek who flourished about 400 B.C. at Syracuse in Sicily. He was perhaps a disciple of Hicetas of Syracuse, whose name has also been associated with the doctrine of the Earth's rotation. Hicetas and Ecphantus predate Heraclides; however, little is known of them.

The second attestation to Heraclides' opinion on the subject is provided by Simplicius, who in the sixth century A.D. wrote commentaries on the works of Aristotle. In his commentaries, Simplicius explicated difficult passages in Aristotle and sometimes supplied quotations from other writers, to compare their opinions with those of Aristotle. Simplicius was an excellent scholar and had read widely. His commentaries remain an important source of information on the interpretation of Aristotle's works in antiquity, and on the views of other writers, which are in many cases known to us only because Simplicius happened to mention them. In the course of his commentary on Aristotle's *On the Heavens*, Simplicius mentions Heraclides on the Earth's rotation:

> there have been some, like Heraclides of Pontos and Aristarchus, who supposed that the phenomena can be saved if the heaven and the stars are at rest, while the Earth moves about the poles of the equinoctial circle [i.e., the equator] from the west [to the east], completing one revolution each day. . . .

Further along, Simplicius provides a second allusion to the same fact:

> But Heraclides of Pontos, by supposing that the Earth is in the center, and rotates, while the heaven is at rest, thought in this way to save the phenomena.[42]

Aristarchus of Samos, mentioned by Simplicius in the first of these passages, flourished in the third century B.C., about two generations after Heraclides. He is known to have espoused a dual motion of the Earth: an annual revolution of the Earth about the Sun and a simultaneous rotation of the Earth on its axis. None of his writings on this subject have survived. (We do have, however, his treatise *On the Sizes and Distances of the Sun and Moon*, which is discussed in sec. I.17.)

As we have seen, there were a handful of thinkers in antiquity who asserted that the heaven was stationary and that the Earth rotated. However, none of the original writings have come down to us. Aëtius lived some 450 years after Heraclides, and Simplicius nearly 900 years after Heraclides. It is clear that by their time the original writings of Heraclides had been lost and that they were forced to rely on summaries and quotations made by other writers. In the study of the early Greek philosophers and scientists, this is far from being an unusual situation. In the case of major figures, such as Plato in philosophy and Euclid in mathematics, we have lengthy works, preserved in the whole. But in the case of many minor writers we often have only quotations or brief mentions provided by later writers. It may seem curious that the original writings espousing such a remarkable view should not have been preserved. But, in fact, once the rotation of the Earth has been asserted, and a few justificatory physical arguments made, there is little more to be said. It is unlikely that Heraclides' remarks on this subject were of any substantial length.

Aristotle on the Five Elements and Their Natural Motions　Many ancient writers who asserted that the Earth is stationary made an effort to support their claim. Because *astronomical* evidence does not bear on the case, the argument had to be *physical* in nature, that is, based on the physics or nature of the world. Aristotle (382–322 B.C.) was the most influential physical thinker of the classical age. His doctrines on the nature of the world were part of the mental furniture of every later Greek astronomer. Not everyone agreed with Aristotle on every detail, but everyone knew more or less what he had said about the nature of the cosmos.

According to Aristotle, the portion of the cosmos lower than the Moon is made up of four elements: earth, water, air, and fire. All the changes that we observe around us, the coming into being and the passing away of material objects, result from the combinations and transformations of these elements. But the celestial bodies are made of a fifth element, the ether, which is completely different in nature. It is simple and pure, and therefore incapable of any change. This is why the heavens have remained changeless for many generations.

In his theory of motion, Aristotle distinguishes between *natural* and *forced* (also called *violent*) motions. For each of the sublunar elements, the natural motion is radial motion toward or away from the center of the cosmos, in accordance with their relative heaviness or lightness. For earth, the natural motion is radially downward toward the center of the universe, as may be seen by dropping any particle of earth. A piece of earth may, of course, be given a motion that is contrary to its natural motion. It may be hurled horizontally, or even straight upward. But such forced motions do not endure: the particle eventually reverts to its natural, downward motion. In the case of fire, an element lighter than earth, the natural motion is straight up, radially away from the center. These doctrines are in keeping with a commonsense view of the world.

In the case of the fifth element, the ether, the natural motion is everlasting motion in circles about the center of the universe. This, too, anyone can see, simply by watching the nightly motion of the stars. The ether must also be extremely rare (i.e., not dense), since it lies above the four heavier elements of the sublunar world.

Aristotle's theory of the elements and of natural motions explains why the Earth is at the center of the cosmos: all the individual particles of earth strive to reach the center. It also explains why the Earth is a sphere: the center-seeking jostling of the individual particles necessarily results in this shape. Thus, the theory has a good deal of explanatory power.

The theory is also very unfavorable to the motion of the Earth. But let us suppose, nevertheless, that the Earth does move—either traveling from place to place or remaining in place and rotating on its axis. Such motion must be forced. For the natural motion of earth is in a straight line toward the center. The motion, therefore, being forced and not natural, would not endure. But the observed motion of the heavens is everlasting. So it is clear that the apparent rotation of the heavens cannot be due to a rotation of the Earth.[43]

Ptolemy on the Diurnal Rotation   Ptolemy was a Greek scientific writer who lived from about A.D. 100 to about 175. He worked at Alexandria, the intellectual center of the eastern Mediterranean and the site of the finest library in the ancient world. As is so often the case with the scientific figures of antiquity, as opposed to politicians and military men, virtually nothing is known of his life. Other scientists, both before and after his time, worked at the Alexandria Museum, either as scholars or as teachers, and received stipends from this state-supported institution. It is possible that Ptolemy had a similar appointment, although such a connection is nowhere attested. Some of his works are addressed to a certain Syrus. Whether this was a friend, colleague, or patron is not known. Ptolemy was a brilliant applied mathematician and a prodigious worker. He produced important treatises on geography and optics, as well as a highly influential handbook of astrology. But his reputation rests chiefly on his astronomical treatise, the *Almagest*.

The beginning chapters of the *Almagest* are devoted to the basic premises of astronomy. Among these are the immobility of the Earth. In the following extract, Ptolemy notes that "certain people" have argued for the rotation of the Earth. Here he may be thinking of Heraclides of Pontos and Aristarchus of Samos. Ptolemy concedes that it is impossible to refute such a theory with astronomical evidence. He therefore brings to bear *physical* arguments, based on Aristotle's doctrine of natural motions. He argues, too, that if the Earth did turn, loose objects on its surface would be left behind, thus elaborating on an argument that had been suggested by Aristotle.

EXTRACT FROM PTOLEMY

Almagest I, 7

Certain people . . . think that there would be no evidence to oppose their view if, for instance, they supposed the heavens to remain motionless, and the Earth to revolve from west to east about the same axis, [as the heavens] making approximately one revolution each day; or if they made both heaven and Earth move by any amount whatever, provided, as we said, it is about the same axis, and in such a way as to preserve the overtaking of one by the other. However, they do not realize that, although there is perhaps nothing in the celestial phenomena which would count against that hypothesis, at least from simpler considerations, nevertheless from what would occur here on Earth and in the air, one can see that such a notion is quite ridiculous.

Let us concede to them [for the sake of argument] that such an unnatural thing could happen as that the most rare and light of matter should either not move at all or should move in a way no different from that of matter with the opposite nature (although things in the air, which are less rare [than the heavens] so obviously move with a more rapid motion than any Earthy object); [let us concede that] the densest and heaviest objects have a proper motion of the quick and uniform kind which they suppose (although, again, as all agree, Earthy objects are sometimes not readily moved even by an external force). Nevertheless, they would have to admit that the revolving motion of the Earth must be the most violent of all motions associated with it, seeing that it makes one revolution in such a short time; the result would be that all objects not actually standing on the Earth would appear to have the same motion, opposite to that of the Earth: neither clouds nor other flying or thrown objects would ever be seen moving towards the east, since the Earth's motion towards the east would always outrun and overtake them, so that all other objects would seem to move in the direction of the west and the rear.

But if they said that the air is carried around in the same direction and with the same speed as the Earth, the compound objects in the air would none the less always seem to be left behind by the motion of both [Earth and air]; or if those objects too were carried around, fused, as it were, to the air, then they would never appear to have any motion either in advance or rearwards; they would always appear still, neither wandering about nor changing position, whether they were flying or thrown objects. Yet we quite plainly see that they do undergo all these kinds of motion, in such a way that they are not even slowed down or speeded up at all by any motion of the Earth.[44]

Ptolemy's physical arguments were not satisfactorily answered until the principle of inertia was understood, in the seventeenth century. The important point to be noted is that the Greek astronomers were sophisticated enough to realize that the daily motion of the heavens could be explained by a rotation of the Earth and that this view could not be refuted by astronomical observation. Nevertheless, the prevailing view was that the Earth was at rest and that the heavens really did revolve from west to east, exactly as they appeared to do.

1.7 OBSERVATION: THE DIURNAL MOTION
OF THE STARS

Contrive a few sighting tubes—four is an optimal number. Direct one tube toward a star that is rising in the east, one toward a star that is crossing the meridian in the south, and one toward a star that is setting in the west. The fourth sighting tube should be aimed at Polaris. After the tubes are set, pass half an hour reviewing the constellations. Then return to the sighting tubes to note the directions in which the first three stars have moved. As for Polaris, you will find that it has not moved perceptibly and that it remains in its sighting tube. These simple observations will reveal in a clear and immediate way the rotation of the celestial sphere.

1.8 STARS AND CONSTELLATIONS

The history of the nomenclature for the stars and constellations is complex, involving Babylonian, Greek, Arabic, and medieval Latin traditions. In many details, this history is imperfectly known. Devising constellations and naming stars are not, of course, scientific activities. But every culture in which a scientific astronomy developed did devote some effort to organizing the heaven into constellations. Perhaps this was a psychological prerequiste for scientific astronomy. And, of course, the zodiacal constellations provided a system of reference marks vital for the early investigations of the motions of the Moon and planets.

Constellations

Most of the familiar constellations have traditions going back to the Greeks. In some cases, the names are very ancient, being found in Homer and Hesiod (ca. 700 B.C.) and going back, no doubt, beyond the beginning of Greek literature. As we saw in section 1.1, Homer and Hesiod mention Orion, the Pleiades, the Hyades, the Bear, and Boötes.

While many of the constellations are undoubtedly of Greek origin, others are Hellenized versions of even older Babylonian constellations. This is particularly true of the zodiacal constellations. For example, the Babylonian GU.AN.NA, or "Bull of Heaven," corresponds to our Taurus. MAS.TAB.BA. GAL.GAL, "the Great Twins," is our Gemini, and so on. Figure 1.25 presents striking evidence of the dependence of the Greek zodiac on the Babylonian tradition. On the left are the figures of three zodiacal constellations as represented on Babylonian boundary stones of the Kassite period. (Boundary stones were used for marking out parcels of land.) On the right are the same three constellations as figured on the famous "round zodiac" from the ceiling of a temple in Dendera, Egypt. (The round zodiac is now in the Louvre in Paris.) The Egyptian figures are from the Roman period and show an incorporation of Egyptian design elements with the classical Greek imagery for these constellations. The resemblance between the Greco-Egyptian figures and the Babylonian prototypes is remarkable, extending even to such details as the positions of the forelegs of the goat-fish creature, Capricornus. The Greeks did, however, redesign some of the zodiacal constellations. For example, where the Greeks recognized Aries, a ram, the Babylonians saw a hired laborer.[45]

Among the Babylonians, the twelve-constellation zodiac emerged sometime after 900 B.C., though some of the individual constellations are considerably older. The artificial zodiacal *signs*, all of 30° length, were in use in Babylonia by end of the sixth century B.C. There is no solid evidence for a complete zodiac among the Greeks until the fifth century, when it was taken over virtually whole from the Babylonians. The earliest Greek parapegmata (star

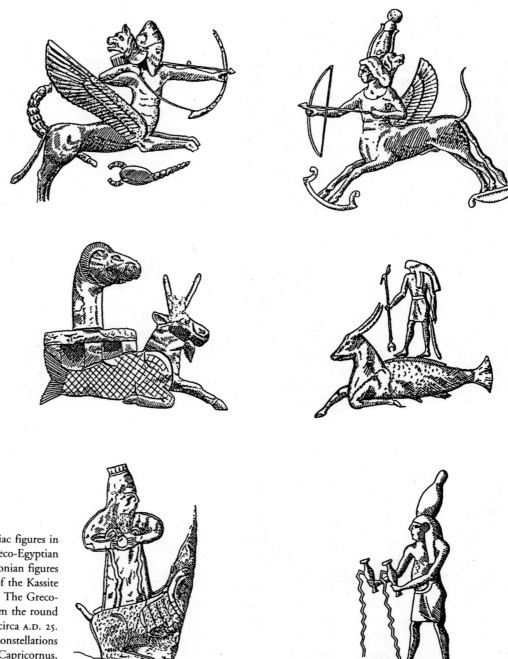

FIGURE 1.25. Zodiac figures in Babylonian (left) and Greco-Egyptian (right) styles. The Babylonian figures are from boundary stones of the Kassite period, circa 1200 B.C. The Greco-Egyptian figures are from the round zodiac of Dendera, circa A.D. 25. From top to bottom, the constellations represented are Sagittarius, Capricornus, and Aquarius. From Hinke (1907).

calendars, based on a division of the year into zodiacal signs) were those of Meton, Euctemon, and Democritus, all of the late fifth century B.C.

By the middle of the fourth century B.C., the Greek constellations were substantially complete. Eudoxus's description of the constellations probably played a great role in standardizing the Greek nomenclature. This description has not come down to us, but its essential content is preserved in Aratus's verse *Phenomena* (ca. 270 B.C.), the oldest surviving systematic account of the constellations in Greek. Aratus describes the whole celestial sphere and tells how the constellations are placed with respect to one another. But he does not give any details about the positions of individual stars within the constellations. Eudoxus had undoubtedly given more detail, but numerical data about individual star positions were rightly deemed unsuitable by Aratus for his poem. Some additional insights into Eudoxus's intentions are furnished by Hipparchus's *Commentary on Aratus and Eudoxus* (ca. 150 B.C.).

Another source of some value is the *Catasterisms* ("Constellations") of Eratosthenes.[46] This brief tract was written as a commentary on, and supplement to, the *Phenomena* of Aratus, as is evident from the fact that the constellations are treated in the same order in the two works. The work is basically a list of constellations, with a story or legend for each. Here is the entry for Cassiopeia:

> Sophocles, the tragic poet, says in his *Andromeda* that this one [Cassiopeia] came to misfortune by contending with the Nereids over beauty, and that Poseidon destroyed the region by sending a sea-monster [Cetus]. This is why her daughter [Andromeda] appropriately lies before the sea-monster. She has a bright star on the head, a dim one on the right elbow, one on the hand, one on the knee, one at the end of the foot, a dim one on the breast, a bright one on the left thigh, a bright one on the knee, one on the board, one at each angle of the seat on which she sits; in all, thirteen.

The legends of the origins of the constellations are what made the *Catasterisms* popular and guaranteed its survival. However, the *Catasterisms* does provide some genuine astronomical information, notably the mentions of how many stars are contained in each part of each constellation. This information would help a reader work out the figure of the constellation in the night sky. Aratus had provided no such detail (though Eudoxus probably had). We may regard Aratus's *Phenomena* and the *Catasterisms* of Eratosthenes as continuations and elaborations of the description of the celestial sphere set down by Eudoxus.

After Eudoxus's time, there were only a few new constellations added by later Greek astronomers, notable examples being Coma Berenices and Equuleus. The story of Coma Berenices is especially interesting. This constellation was invented by the Alexandrian astronomer Conon, and the story goes like this: Berenice was the cousin and wife of Ptolemaios III Euergetes, the third of the Macedonian kings of Egypt. (Ptolemy Euergetes ruled from 247 to 222 B.C.). When her husband departed for war in Syria, Berenice vowed to make an offering to the gods of a lock of her hair if he should return safely. When her husband did, indeed, return safely to Alexandria, she cut off a lock and placed it in a temple, from which it mysteriously disappeared. Conon, the court astronomer, consoled her by designating a new constellation Coma Berenices, the Hair of Berenice. The story has come down to us because it inspired Callimachus of Cyrene, who also worked at Alexandria at this time, to compose a poem on the subject, which survives in a fragmentary state.[47] A Latin version made two centuries later by Catullus has come down to us intact.[48] The new constellation had an unsettled history: although it appears in the *Catasterisms* of Eratosthenes (in the paragraph on the Lion), it was not included by Ptolemy among his forty-eight constellations. Ptolemy does, however, allude to the "lock" of hair in his star catalog, in his description of the unconstellated stars around Leo (*Almagest* VII, 5).

Coma Berenices, although not recognized as an independent constellation by Ptolemy, eventually won a permanent place on the sphere. Other proposed constellations, including some invented by Hipparchus, never won acceptance and have vanished from the sky. Ptolemy followed Hipparchus in recognizing Equuleus as a new constellation, but did not accept Hipparchus's Thyrsus-lance (held by the Centaur) as an independent constellation.[49] Even in the case of the old, standard constellations, there were differences among individual writers over the number of stars included in each and the manner in which one should "connect the dots" to form the figure. Ptolemy (*Almagest* VII, 4) mentions that he has departed from Hipparchus's forms to give better-proportioned figures. For example, the star Hamal (α Arietis) was said by

Hipparchus to be on the muzzle of the Ram, Aries, but Ptolemy placed this star outside of the constellation, above the head. Some stars that had been placed by Hipparchus on the shoulders of Virgo, Ptolemy described as "on her sides." It was Ptolemy's star catalog in books VII and VIII of the *Almagest* that did the most (after the *Phenomena* of Eudoxus and Aratus) to fix the names and figures of the classical forty-eight constellations. From Ptolemy's time, then, the basic forms of the constellations were settled.

Our modern astronomical tradition is not directly continuous with that of the ancient Greeks but is separated from it by the medieval period when, in Europe, Latin was the language of learning. Thus, the modern names of many constellations are Latin renderings of the Greek names. For example, *Gemini* is a literal translation of the Greek *Didymoi*, "Twins." There are many other examples: Latin *Cancer* for the Greek *Karkinos*, "Crab"; *Aries* for *Krios*, "Ram." In some cases the Latin words are etymologically the same as the Greek: Latin *Taurus* = Greek *Tauros*, "Bull"; Latin *Leo* = Greek *Leon*, "Lion."

During the Age of Exploration (fifteenth and sixteenth centuries), European navigators saw for the first time the stars around the south celestial pole. This was a new period of constellation making. Some of the new constellations (e.g., Tucana, a toucan) reflect the exotic animals encountered by the Europeans in tropical lands. In the eighteenth century, a deliberate effort was made to fill in the gaps between the classical constellations. Some of the smaller, inconspicuous constellations date from this period, for example, Microscopium and Telescopium, both devised by the French astronomer Nicolas Louis de La Caille.[50]

Stars

The Greek astronomical writers did not assign proper names to many individual *stars*. Most stars were identified simply by descriptions of their places within the constellations. Our Betelgeuse (α Ori), for example, was simply the "star on the right shoulder of Orion." Most of the stars with real names were deemed to be significant either as weather signs or as indicators of the season. For the Greeks, the most important named stars were

> Arcturus
> Sirius (Also called *Kyon*, "dog.")
> Procyon
> Antares
> Canopus
> Aix (Now called *Capella*, a Latin translation of the Greek, "goat.")
> Lyra (The "lyre." Now called *Vega*, a medieval Latin corruption of an
>   Arabic form. The Greeks also applied Lyra to the whole constellation
>   of which this star is the brightest member. This designation still
>   stands.)
> Aëtos (The "eagle." Now *Altair*, from the Arabic.)
> Stachys (Now *Spica*, a Latin translation of the Greek name, "ear of
>   wheat.")
> Basiliskos (Now *Regulus*, a Latin rendering of the Greek, the "little
>   king." For the Babylonians, this star was also called "king," LUGAL.)
> The Pleiades
> Protrygeter (Now *Vindemiatrix*, the Greek and Latin both signifying
>   "harbinger of the grape harvest.")
> Eriphoi (The "kids." These are the two dim stars η and ζ Aurigae.)
> Onoi (The "asses," γ and δ Cancri.)

This list nearly exhausts the individual star names used by the early Greeks. Some of the names are truly ancient: Arcturus, Sirius, and the Pleiades are all mentioned by Hesiod. The others are first attested somewhat later.

The first ten stars of the list are quite bright; their prominence in the night sky no doubt justified individual names. The last four items listed contain rather dim objects. They earned their names not by their brightness but by their significance. As we saw in section 1.1, the morning setting of the Pleiades signaled the beginning of winter weather and the time to sow grain (November). The morning rising of the Pleiades was the signal of the wheat harvest, in May. Similarly, the morning rising of Vindemiatrix, in September, signaled the coming of the grape harvest. And the morning setting of the Kids, in December, signaled the onset of the season of winter storms, as in these lines by Callimachus:

Flee the company of the sea,
O mariner, when the Kids are setting.[51]

Arcturus and Sirius played roles only a little less important than that of the Pleiades. The morning rising of Arcturus was a sign of autumn, while the morning rising of Sirius, the Dog Star, signaled the hot days of high summer (the "dog days"). The calendrical significance of the Pleiades, Arcturus, and Sirius probably accounts for the fact that these are the first stars to be mentioned by name in Greek literature.

Concerning the Asses: between these two stars is a small, fuzzy or nebulous patch. Through binoculars it is resolved into a tight cluster of stars, now popularly known as the Beehive. The Greeks called it a Manger (*Phatne*). According to Aratus, the visual appearance of the Asses and the Manger through thin clouds at night can be used to predict the weather:

If the Manger darkens and both stars remain unaltered, they herald rain. But if the Ass to the north of the Manger shines feebly through a faint mist, while the southern Ass is gleaming bright, expect wind from the south. But if in turn the southern Ass is cloudy and the northern bright, watch for the north wind.[52]

In the star catalog in Ptolemy's *Almagest*, more than a 1,000 stars are listed together with their coordinates and magnitudes, but no more than a dozen are given proper names. The remainder are identified in terms of their places within the constellations. Some stars, which did not fit very well into the figure of a traditional constellation, were said to be outside the constellation, but their positions were nevertheless described in terms of relationships to the constellated stars. It is still possible today to identify with absolute certainty the great bulk of Ptolemy's 1,000 stars. The identities of about ten percent of the stars are, however, not quite certain (and a few are completely uncertain) because of errors in Ptolemy's measured coordinates and lack of precision in the written descriptions.

The revival of astronomy in western Europe began in the twelfth century. The first stage in this revival required the study of the classics of ancient Greek astronomy. At first, translations were made into Latin from Arabic translations of the Greek originals. A number of Arabic treatises in astronomy and mathematics were also translated into Latin. Only somewhat later were the important Greek works translated into Latin directly from the Greek. Not surprisingly, a good deal of Arabic star nomenclature found its way into medieval and Renaissance Latin astronomy. In fact, the great majority of modern star names in the European languages are corrupt forms of the Arabic names. In some cases, the Arabic name descends from a tradition independent of the Greek. But in many cases, the Arabic name that lies behind the modern European name is merely a translation of the original Greek descriptive nomenclature. For example, our Vega is a corrupt form of the Arabic [*al-nasr*] *al-wāqiᶜ*, "the swooping [vulture]," which has no counterpart in the ancient Greek nomenclature. But our Denebola is a corrupt form of the

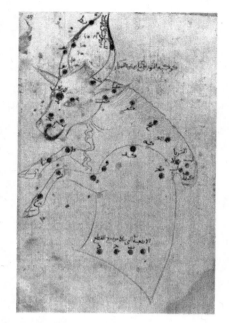

FIGURE I.26. The constellation Taurus according to al-Ṣūfī. On the top is Taurus as seen in the sky. Below this is Taurus as represented on a solid celestial globe. The five stars forming a V on the Bull's face are the Hyades. The cluster of four stars on the hump of the Bull's back are the Pleiades. Each star has been numbered to key the figure to the star list that accompanies it. The figures are photographs of a fourteenth-century Arabic manuscript copy of al-Ṣūfī's *Book on the Constellations of the Fixed Stars.* By permission of the Trustees of the British Library (Or. 5323, fols. 38v, 39).

Arabic *dhanab al-asad,* "the tail of the Lion," which was just the way the Greeks referred to this star.

One of the most important Arabic works on the stars was that of Abuʾl-Ḥusayn ʿAbd al-Raḥmān Ibn ʿUmar al-Rāzī al-Ṣūfī. Around A.D. 964, al-Ṣūfī composed his *Book on the Constellations of the Fixed Stars* (*Kitāb ṣuwar al-kawākib al-thābita*).[53] The core of al-Ṣūfī's book is Ptolemy's star catalog from the *Almagest.* Al-Ṣūfī translated the Greek descriptive nomenclature into Arabic and updated the positions of the stars by adding 12°42′ to all of the longitudes to account for the precession between Ptolemy's day and his own. But al-Ṣūfī also added a paragraph of notes for each constellation, in which he discussed problems of identification, errors in Ptolemy's coordinates, and variants for the names for individual stars, including old Arabic star names that predated Arab contact with Greek astronomy. Al-Ṣūfī's work, in the earliest extant manuscripts, is also notable for its drawings of the figures of the constellations. In fact, al-Ṣūfī included two drawings for each constellation: one as seen in the sky, and one reversed (as seen on a solid globe). The stars of each constellation were numbered on the charts and thereby keyed to the list of stars in the catalog. In figure I.26 we see the constellation of Taurus, the Bull, in a medieval Arabic manuscript of al-Ṣūfī's work. The Bull, here drawn in a fluid Arabic style, was a Greek constellation, of course. But long before that, it was a Babylonian constellation—a fact that was certainly unknown to al-Ṣūfī. Here we see one more dramatic illustration of the continuity (and complexity) of the nomenclature for the stars and constellations from the Babylonians, through the Greeks and Romans, through the Arabic and Latin astronomers of the Middle Ages, and down to our own day.

Modern astronomers frequently identify stars by means of *Bayer letters,* introduced by the German astronomer Johann Bayer in his influential celestial atlas, *Uranometria,* published in 1603. In this system, each star is labeled by a Greek letter and the Latin name (in the genitive case) of the constellation in which it is found. Thus, the stars Betelgeuse and Rigel (both in Orion) are called, respectively, α Orionis and β Orionis. But, as few twentieth-century astronomers have any Latin, a simplified system has recently arisen of using a three-letter abbreviation of the constellation name: α Ori and β Ori.

I.9 EARTH, SUN, AND MOON

A good deal of astronomy rests on an understanding of the Earth-Sun-Moon system. The fundamental questions are three: What causes the phases of the Moon? What causes eclipses? What is the shape of the Earth?

Phases of the Moon

The Moon shines by reflected sunlight. Half the Moon is always illuminated—the half that faces toward the Sun. As the Moon orbits the Earth in the course of the month, we see it from different angles. Refer to figure I.27.

> When the Moon is at position 1, the unilluminated half faces us and we cannot see the Moon at all. This is what we call new Moon. The Moon is also said to be *in conjunction* with the Sun, because they both lie in the same direction as seen from Earth.
>
> When the Moon is at position 2, we can just see a sliver of a crescent. This is what the Greeks called new Moon. The Greek new Moon came a day or two after the true conjunction.
>
> At position 3, the Moon reaches first quarter. We can see half of the illuminated portion. In the sky, the Moon looks like the letter D.

At position 4, the Moon is gibbous, or "hump-backed" in shape.

At position 5, the illuminated hemisphere of the Moon directly faces us, and the Moon is full. The Moon is also said to be *in opposition* to the Sun.

At position 6, the Moon is gibbous again.

At position 7, the Moon reaches third quarter. In the sky, it looks like a backward letter D.

At position 8, the Moon is again a crescent.

From positions 1 to 5 the Moon is said to be *waxing*, that is, growing larger. From positions 5 to 1 the Moon is said to *waning*, that is, growing smaller.

A surprisingly large number of people today believe that the phases of the Moon are due to the "shadow of the Earth." This is clearly not the case. For example, at the time of the first quarter, we can see the Sun and the Moon in the sky at the same time. Thus, the Earth cannot be between the Sun and the Moon, casting a shadow on the Moon. Phases and eclipses are due to different causes.

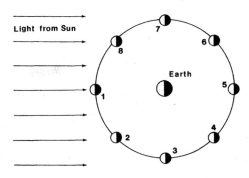

FIGURE 1.27. The phases of the Moon are due to the fact that an observer on the Earth sees different portions of the illuminated half of the Moon at different times of the month.

Eclipses

There are two kinds of eclipses, *lunar* and *solar*. Solar eclipses can occur only at the time of new Moon (position 1 in fig. 1.27). The Moon passes across the face of the Sun and blocks it out.

Lunar eclipses can occur only at full Moon (position 5 in fig. 1.27; see also Fig. 1.28). The Earth casts a long, cone-shaped shadow that stretches out in space diametrically opposite the Sun. At the time of lunar eclipse, the Moon passes through the Earth's shadow.

Although eclipses of the Moon are possible only at full Moon, a lunar eclipse does not occur *every* full Moon. The Moon's orbit is tilted a little (about 5°) with respect to the plane of the ecliptic (plane of the Sun's orbit about the Earth). Thus, most months, at the time of full Moon, the Moon misses the Earth's shadow by crossing either a little north or a little south of it. Thus, in figure 1.27, the Moon should be regarded as spending most of its time either a little above or a little below the plane of the paper.

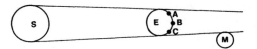

FIGURE 1.28. The Moon *M* entering the shadow of the Earth *E*, thus producing a lunar eclipse.

The Explanation of Phases and Eclipses

Parmenides of Elea (late sixth century B.C.) described the Moon in his poem as "a night-shining foreign light wandering around the Earth." In another line, Parmenides characterized the Moon as "always fixing its gaze on the beams of the Sun."[54] Some have seen in these lines a realization that the Moon shines by reflected sunlight. The second line does reveal an awareness that the bright portion of the Moon always faces toward the Sun. But this does not necessarily imply an understanding of the physical cause of the Moon's brightness.

One can know that the bright side of the Moon always faces the Sun without realizing that the Moon shines by reflected sunlight. This is abundantly clear in the "explanation" of the Moon's phases attributed by Vitruvius to Berosus. According to this view, the Moon is a ball, one half luminous and the other half of a blue color. The luminous half of the Moon always turns to face the Sun, attracted by its rays and great heat, in keeping with the sympathy between light and light. As the Moon travels through the zodiac, it gradually turns so that the luminous portion may always face the Sun. And thus we see different portions of it as the month goes by.[55] Berosus was a

Chaldaean, that is, a Babylonian astronomer and astrologer, who flourished around 300 B.C. His works were known to later Greek and Roman writers, who preserved some fragments of them. According to Vitruvius,[56] Berosus settled on the island of Cos, where he opened a school and introduced the Greeks to astrology. It is clear that Berosus did play some role in the diffusion of Babylonian astronomical knowledge among the Greeks. It is not, however, safe to ascribe Berosus's theory of the Moon's phases to early Babylonian astronomy: we simply do not know what explanation, if any, the early Babylonians offered for the phases. From the point of view of Babylonian astronomy, Berosus is rather late. Living among the Greeks, he may well have been influenced by the Greek desire for a physical explanation. If so, his view must have had little appeal, for among the Greeks the correct explanations of phases and eclipses were already several generations old.

Anaxagoras in Athens   Most of the Greek testimony attributes the discovery of the causes of the Moon's phases and eclipses to Anaxagoras of Clazomenae on the west coast of Asia Minor (just west of the modern Turkish city of Izmir). Around 480 B.C., Anaxagoras went to Athens, where he was befriended, and perhaps financially supported, by Pericles, the political leader of the democratic element in the city.

Anaxagoras correctly explained the phases of the Moon, saying that the Moon gets its light from the Sun. We have early testimony on this point. Plato mentions it as a "recent discovery" of Anaxagoras "that the Moon receives its light from the Sun."[57]

Later writers, including Hippolytus and Aëtius, say that Anaxagoras explained eclipses of the Sun by the interposition of the Moon between the Sun and the Earth, and eclipses of the Moon by the Moon's falling into the Earth's shadow. He held, too, that the Sun was larger than it appeared, that it was, indeed, "larger than the Peloponnesos."[58]

Anaxagoras is also noteworthy for his attempt to unify terrestrial physics with the physics of the heavens. In opposition to prevailing thought, he held that the celestial bodies were made of ordinary, earthy matter. He said that the Moon has plains and ravines. Perhaps inspired by the fall of a meteorite, he called the Sun a red-hot stone. These and other remarks were offensive to the religious conservatives of Athens, who believed that the Sun and Moon were gods or else were directly controlled by gods. Anaxagoras was accused of impiety and tried on that charge, among others. While there is no doubt that his views were genuinely shocking to some, the case was also used by the political conservatives as a way of discrediting Pericles. The ancient accounts vary: either Anaxagoras was tried and condemned to death in absentia, or he was, after efforts on his behalf by Pericles, merely fined and exiled. In any case, he withdrew to Lampsacus, in northwest Asia Minor, where he remained until his death around 427 B.C. The Lampsacans are said to have treated him with honor. When the rulers of the city asked him what priviledge he wished to be granted, he replied that after his death the school children should each year be given a holiday in his memory. The custom was long observed.[59]

Curiously, Anaxagoras does not appear to have pursued his astronomical ideas to their ultimate conclusion. Although he correctly explained lunar eclipses, he is said nevertheless to have maintained that the Earth is flat. He held, too, that the Moon is eclipsed not only by the Earth but sometimes also by other, unspecified bodies lying below the Moon. Thus, it appears that by about 480 B.C. the correct explanation of lunar eclipses was already current, but that this knowledge had not yet been brought to bear on the question of the Earth's shape. These related facts had not yet been integrated into a coherent world view.

The Shape of the Earth

Early Doctrines   Among the earlier philosophers, there were some who asserted that the Earth is flat or disk shaped. Moreover, in the early period there was a tendency to confound the question of the Earth's shape with the question of its support, that is, why it remained in place. Thus, Thales of Miletus (ca. 585 B.C.) declared that the Earth rests on water. His supposed student, Anaximander (ca. 570 B.C.), held the Earth to be a cylinder whose depth was one-third of its breadth. Of the two flat surfaces, one is that on which we stand and the other is opposite. According to Anaximander, the Earth is poised in emptiness, supported by nothing, and remains where it is because it is equidistant from all other things. It has therefore no predisposition to fly away in any one direction rather than in any other. Anaximenes of Miletus (ca. 540 B.C.) held the Earth to be broad and flat and supported by the air beneath it. Xenophanes of Colophon (ca. 530 B.C.) neatly disposed of the question of what supports the Earth by declaring that the Earth has its roots in infinity; that is, that the Earth reaches down forever. The cosmological doctrines of the earliest philosophers are difficult to reconstruct with certainty. For the most part, they are known only through citations and quotations by later Greek writers.[60]

Some of the Greeks attributed the discovery of the sphericity of the Earth to Parmenides (fifth century B.C.), while others gave the honor to Pythagoras (sixth century B.C.). Still others claimed, quite impossibly, that this fact had been known even to Hesiod (seventh century B.C.).[61] The statements by later Greek writers merely reflect the Greek propensity for attributing every discovery to one or another of their ancient wise men. That the idea of the sphericity of the Earth originated with the Pythagorean school of the fifth century is not, however, wholly improbable.

Aristotle on the Sphericity of Earth   The earliest writer whose work survives, who states clearly that the Earth is a sphere, and who gives adequate proof of this fact, is Aristotle, who addresses the issue in his treatise *On the Heavens* (fourth century B.C.). After refuting those who believe the Earth to be flat or drum shaped, Aristotle asserts that it is a sphere. Moreover, this spherical shape results from the natural tendency of the heavy elements to move toward the center of the universe. Thus, it is the center-seeking pressure and jostling of the separate particles of earth that brings about the spherical shape of the whole. In this way Aristotle deduces the spherical shape of the Earth from his physical doctrines. Note that in Aristotle's physics, the Earth is not the center of the universe, properly speaking. Rather, the Earth lies at the center of the universe because of the center-seeking nature of the heavy element of which it is composed. One physical principle thus explains not only the Earth's shape but also its position and its immobility.

Aristotle usually preferred an argument from physical or philosophical principles over an argument from observation. Having given first what he regarded as his best argument, he did not, however, refrain from marshaling the evidence of the senses. In the following extract, Aristotle presents three arguments in favor of the sphericity of the Earth.

EXTRACT FROM ARISTOTLE

On the Heavens II, 14

Further proof is obtained from the evidence of the senses. If the Earth were not spherical, eclipses of the Moon would not exhibit segments of the shape which they do. As it is, in its monthly phases the Moon takes on all varieties of shapes—straight-edged, gibbous and concave—but in eclipses the bound-

FIGURE 1.29. The lunar eclipse of July 16, 1981, photographed in Seattle by Brian Popp.

ary is always convex. Thus, if the eclipses are due to the interposition of the Earth, the shape must be caused by its circumference, and the Earth must be spherical.

Observation of the stars also shows not only that the Earth is spherical but that it is of no great size, since a small change of position on our part southward or northward visibly alters the circle of the horizon, so that the stars overhead change their position considerably, and we do not see the same stars as we move to the North or South. Certain stars are seen in Egypt and the neighbourhood of Cyprus, which are invisible in more northerly lands, and stars which are continously visible in the northern countries are observed to set in the others. This proves both that the Earth is spherical and that its periphery is not large, for otherwise such a small change of position could not have had such an immediate effort.

For this reason those who imagine that the region around the Pillars of Heracles joins on to the regions of India, and that in this way the ocean is one, are not, it would seem, suggesting anything utterly incredible. They produce also in support of their contention the fact that elephants are a species found at the extremities of both lands, arguing that this phenomenon at the extremes is due to communication between the two. Mathematicians who try to calculate the circumference put it at 400,000 stades.

From these arguments we must conclude not only that the Earth's mass is spherical but also that it is not large in comparison with the size of the other stars.[62]

The argument based on lunar eclipses is clear and convincing. The curved edge of the shadow may be seen on the face of the Moon, as in figure 1.29.

Aristotle's second group of arguments from sense evidence is based on observations of the stars. Aristotle does not mention particular stars, or any particular observations. In a philosphical work, particular astronomical observations would have been deemed out of place. The almost complete absence of specific observations is characteristic of most Greek astronomical writing at the *elementary* level—just as it is characteristic of elementary textbooks in our own day. Advanced astronomical treatises on specialized topics (e.g., the theory of the motion of the planets), which were produced later, naturally required the use of specific observations. Later Greek writers of elementary astronomy texts *were* fond of citing the case of Canopus, a bright star in the modern constellation Carina. Canopus is one of the brightest stars in the sky, second only to Sirius. Canopus was visible in Egypt but not in Greece. It began to peek above the horizon at about the latitude of Rhodes or Cyprus. The reports of travelers concerning this "bright star of the Egyptians" perhaps played a role in the early debate over the shape of the Earth.

We leave it to the reader to explicate the argument based on elephants. Aristotle does not claim it as his own, but attributes it to certain others left unnamed. It is clear that he regarded it as less convincing than the two astronomical arguments, but he did not refrain from using it.

Aristotle remarks that the mathematicians who calculate the circumference of the Earth put it at 400,000 stades. This is the oldest recorded calculation of the size of the Earth. The stade, or *stadion*, was a unit of length used in the Greek world. Originally the word indicated the length of a race track as, for example, that at Olympia. However, stades of several different lengths were in use. Various ancient sources give values between 7 1/2 and 9 stades to a Roman mile, the Roman mile being about 0.925 of our own.[63] Aristotle's figure thus puts the Earth's circumference between 40,000 and 50,000 statute miles. The actual circumference is about 25,000 miles, so Aristotle's value is certainly of the right order of magnitude. It is not known who the "mathematicians" were that Aristotle cites. A name that has often been proposed is that of Eudoxus of Cnidus.

Three Later Writers

Later writers added to the stock of arguments that the Earth is a sphere. The new arguments had no bearing on the outcome of the case, which had been settled by the middle of the fourth century B.C. But these new arguments for the sphericity of the Earth acquired a place in the textbooks and so played a role in education from ancient times down to the Renaissance. Three important and characteristic writers are Ptolemy, Theon of Smyrna, and Cleomedes. Ptolemy the reader already knows as the author of the *Almagest* (ca. A.D. 140).

Theon of Smyrna (an important city on the west coast of Asia Minor) was the author of a book titled *Mathematical Knowledge Useful for Reading Plato*. The book contains a number of references to Plato and is Platonic in its underlying philosophy of nature, but it is far from being a commentary on Plato's writings. In fact, Theon's book is an introductory survey of mathematics. Traditionally, the Greeks divided mathematics into four branches: arithmetic, geometry, astronomy, and music theory. A section of Theon's book probably was devoted to each of these. The section dealing with geometry and a part of that dealing with music theory have not come down to us, but we have intact the sections on astronomy and arithmetic. Theon's book is not at the same level as Ptolemy's. Ptolemy wrote a technical treatise for astronomers, containing the most advanced material of the time, while Theon's book is an introduction for beginners. In figure 1.30 we see a bust of Theon of Smyrna, found at Smyrna and now in the Musei Capitolini in Rome. The Greek inscription says that the bust of Theon, the Platonic Philosopher, was dedicated by his son, Theon the Priest. Likenesses of Greek scientific writers are extremely rare. There are, to be sure, "portraits" of Euclid and other mathematicians, but these were almost always produced centuries after the fact—they are merely symbolic figures. The bust of Theon is interesting because we know it was commissioned by his son and may actually resemble Theon. Moreover, the bust provides a means of dating Theon. Art historians date it, by its style, to the reign of Hadrian (early second century A.D.). In the *Almagest*, Ptolemy refers to observations made at Alexandria by a certain Theon. Thus, it is possible that Theon of Smyrna was an elder friend (perhaps a teacher) of Ptolemy. But Theon was a common name, so we cannot be certain that Ptolemy's Theon was the same man.

The date of Cleomedes is uncertain. Considering the other writers whom Cleomedes cites or fails to cite, we shall probably not be far wrong if we adopt a date of the first or second century A.D. Cleomedes was the author of an introduction to astronomy titled something like *On the Elementary Theory of the Heavenly Bodies*. Like the books by Geminus and Theon of Smyrna, Cleomedes' work is elementary and not mathematical. It, too, was intended for beginning students. In his physical doctrines Cleomedes was a follower of Posidonius (first century B.C.), the most famous and influential of the Stoic philosophers. Cleomedes' Stoic inclinations lend his book a flavor rather different from that of Theon's.

All three writers (Cleomedes, Theon of Smyrna, Ptolemy) repeat an argument of Aristotle's—that, as we move north or south on the Earth, we observe changes in the visibility of the stars. Both Ptolemy and Theon point out that this proves only that the Earth is curved from north to south and not necessarily from east to west. Theon cites the case of the star Canopus, which "although invisible in the parts north of Cnidus, becomes visible in more southerly regions."[64] Curiously, none of these writers mentions Aristotle's argument from the shape of the Earth's shadow. However, all add new arguments not found in Aristotle.

The Earth Is Neither Flat nor Hollow

Both Ptolemy and Cleomedes argue that the Earth could not have any of several alternative shapes: flat, hollow,

FIGURE I.30. Theon of Smyrna. This bust, found at Smyrna, was dedicated by Theon's son. Photo courtesy of Musei Capitolini, Rome.

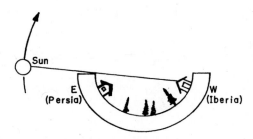

FIGURE 1.31. If the Earth were concave, observers in the west would see the morning Sun before observers in the east.

cubical, pyramidal, and so on. If the Earth were flat, the Sun and stars would rise and set simultaneously for everyone on Earth. If the Earth were concave (fig. 1.31), the Sun, rising in the east, would be seen first by those living in the west. One should not infer that these demonstrations were directed against a geographical theory in need of refutation. By the second century A.D., these demonstrations had become traditional: they were standard, if rather pedantic, fare for an introduction to astronomy.

Argument from the Delay of Dawn   All three writers (Cleomedes, Theon of Smyrna, Ptolemy) remark that the Sun, Moon, and stars do not rise simultaneously everywhere on Earth, but rise earlier for those more toward the east. As Cleomedes notes, sunrise comes four hours earlier for the Persians than for the Iberians. Since Spain is about 60° west of Persia, Cleomedes' figure is about right. However, this is a "demonstration" that never was carried out in practice. It represents an example of *backward science*, in which the supposed observation (a four-hour time difference between Spain and Persia) is in fact a deduction from an already held theory (the sphericity of the Earth). Backward science has been a common method of argument in science textbooks from antiquity to our own day.

Argument from the Local Times of Lunar Eclipses   Similar in nature is the use, made by all three writers, of the observed times at which a lunar eclipse occurs. A lunar eclipse occurs at the same instant for all who can see it. But if the Moon is eclipsed at the first hour of the day for the Iberians (to take Cleomedes' example), the same eclipse is observed at the fifth hour in Persia, and at an intermediate hour for people located between these places. Moreover, as Ptolemy says, the differences in times are proportional to the distances.

In figure 1.28, the Moon is entering the Earth's shadow, so observers at *A*, *B*, and *C* all see the eclipse beginning. For *B*, the local time is midnight, for the Sun is on the meridian below the Earth. But for *A*, it is early morning and the Sun is about to rise. For *C*, the time is early evening, that is, shortly after sunset. Thus, the three observers, who see the eclipse beginning at the same instant, report three different times of day. A difference of one hour corresponds to a difference in longitude of 15°. Such observations played no role in the original discovery of the Earth's sphericity. Timed eclipse observations were made in Mesopotamia as early as the eighth century B.C., but such observations by Greeks did not appear until Hellenistic times, when the shape of the Earth had already been decided. And *comparisons* of the observed times taken, for the same eclipse, from two different localities were rare indeed.

Several Greek geographical writers advocated the use of lunar eclipses for establishing geographical longitudes. Strabo, an Alexandrian geographer of the early first century A.D., tells us that Hipparchus advocated such a practice. In a work called *Against Eratosthenes*—a work that is now lost but was available to Strabo—Hipparchus insisted that the difference in longitude between two places could not accurately be found by any other method than that of lunar eclipses.[65] As we have seen, the *latitude* of a place on the Earth may easily be obtained by measuring the altitude of the celestial pole. The *longitude* is another matter: no simple observation made at a single locality suffices. The ancient geographers were forced to rely on distance estimates made by travelers—a notoriously unsound source.

Ptolemy, in a chapter of his *Geography* titled "That observations from the celestial phenomena ought to be preferred over those taken from the stories of travelers,"[66] took the same position as Hipparchus. But, despite the urgings of Hipparchus and Ptolemy, geographical longitude remained a very uncertain quantity almost to modern times.

There appears, indeed, to be but a single ancient lunar eclipse that was ever applied in the manner suggested by Hipparchus and Ptolemy. The eclipse of September 20, 331 B.C. occurred eleven days before the battle of Arbela, where Alexander the Great decisively defeated Darius III, the King of Persia. Ptolemy (quoting the historians of Alexander's campaign) notes that at Arbela the eclipse of the Moon occurred at the fifth hour of the night. But at Carthage in North Africa the same eclipse was recorded as occurring at the second hour. This three-hour time difference between Arbela and Carthage corresponds to a 45° difference in longitude, which somewhat overstates the case (2 1/4 hours and 34° are nearer the mark). Thus, Ptolemy considerably overestimated the distance from Arbela to Carthage.[67] Indeed, he tended to overestimate the breadth of the whole known world—a fact that, many centuries later, falsely encouraged Columbus, for it made the western ocean narrower.

Argument from Sailing Ships   Our three writers all cite an excellent argument drawn from experience with sailing. Ptolemy writes, "if we sail towards mountains or elevated places . . . , they are observed to increase gradually in size as if rising up from the sea itself in which they had previously been submerged: this is due to the curvature of the surface of the water."[68] Cleomedes adds that it is the same with the ships themselves: as they sail away from land, their hulls are seen to disappear first, while the masts and rigging may still be seen.[69] Both Cleomedes and Theon mention the fact that sailors are sometimes sent up the mast to get a longer view: "And often, during a voyage, if the land or an advancing vessel is not yet seen from the ship, those who have climbed up the mast see it, as they are in a high place and so peek over the curvature of the sea which blocked vision."[70] These arguments, although never mentioned by Aristotle, must have been old among the Greeks, who were a seafaring people.[71]

The Mountains Are Less Than Millet Seeds   How important are the irregularities produced by mountains and valleys? In the extract below, Theon of Smyrna proves that the mountains and valleys are negligible. He shows that if we represent the Earth by a one-foot sphere, the highest mountain would correspond to one-fortieth the diameter of a millet seed—truly an inconsequential irregularity. Theon uses four data in his calculation: (1) the circumference of the Earth is 252,000 stades, as shown by Eratosthenes; (2) the ratio of the circumference of a circle to its diameter is approximately 3 1/7, as shown by Archimedes; (3) approximately 12 1/2 millet seeds make a finger's breadth; and (4) the highest mountains on Earth are about 10 stades high, as measured by Eratosthenes and Dicaearchus.

EXTRACT FROM THEON OF SMYRNA

Mathematical Knowledge Useful for Reading Plato III, 24

Let not anyone believe the projection of the mountains or the depression of the plains, considered in relation to the size of the whole Earth, to be, as irregularities, sufficient cause [for doubting the Earth's sphericity]. Eratosthenes shows that the whole size of the Earth, measured by the circumference of a great circle, is approximately 252,000 stades; and Archimedes, that the circumference of a circle, stretched out in a straight line, is three times the diameter plus about one seventh of it. Thus the whole diameter of the Earth would be approximately 80,182 stades. For three times this number plus a seventh of it was the perimeter of 252,000 stades.

Just as Eratosthenes and Dicaearchus say they have found, the vertical projection of the highest mountains with respect to the lowest places of the Earth is ten stades. The projection is observed by means of instruments to be of such a size, with dioptras for measuring the height from intervals [marked on the instrument]. The height of the highest mountain is then approximately one eight-thousandth of the whole diameter of the Earth.

If we were to make a sphere a foot in diameter, since the distance of a fingerwidth is filled in length by approximately twelve and a half diameters of a millet seed, the one-foot diameter of the constructed sphere would be filled in length by two hundred millet-seed diameters, or a little less. For the foot has 16 fingers; the finger is filled by 12 diameters of a millet seed; and 16 times twelve is 192. The fortieth part of the diameter of a millet seed is therefore greater than one eight-thousandth of the one-foot diameter, for forty times two hundred is eight thousand.

It has been shown that the height of the highest mountain is approximately one eight-thousandth part of the diameter of the Earth, and thus that the fortieth part of the diameter of a millet seed has a greater ratio to the one-foot diameter of the sphere.[72]

The Dicaearchus mentioned by Theon is Dicaearchus of Messina in Sicily (ca. 320 B.C., who was a pupil of Aristotle. It was as a geographer that he made his mark, for he was among the first to grapple systematically with the arrangement of the whole known world. In this he was a predecessor of Eratosthenes, Hipparchus, Strabo, and Ptolemy. He is said to have been the first to measure the heights of mountains by triangulation, a subject on which he wrote a book. All his works are lost.

The Possibility of Circumnavigating the Globe

A common geographical view was that the outer ocean was one. That is, the known world, consisting of Europe, Africa, and Asia (although the wholes of these continents were not known to the Greeks), formed a single land mass, bathed on all sides by a single ocean. Thus, in principle, it should be possible to sail around the globe. One should be able to reach eastern Asia by sailing to the west.

We find such a possibility mentioned by Strabo. Strabo traveled widely in the Mediterranean, passing considerable time both in Rome and in Alexandria. He was the author of a long historical work that has not survived and also a *Geography* in seventeen books that has come down to us intact. Strabo's *Geography* is an important source of information on the Mediterranean nations and peoples in the first century B.C. Strabo also has a good deal to tell us about the geographical opinions of his predecessors, Eratosthenes, Hipparchus, and Posidonius, whose works are now lost.

EXTRACT FROM STRABO

Geography I, 1

We may learn both from the evidence of our senses and from experience that the inhabited world is an island; for wherever it has been possible for man to reach the limits of the Earth, sea has been found, and this sea we call "Oceanus." And wherever we have been able to learn by the evidence of our senses, there reason points the way. For example, as to the eastern (Indian) side of the inhabited Earth, and the western (Iberian and Maurusian) side, one may sail wholly around them and continue the voyage for a considerable distance along the northern and southern regions; and as for the rest of the distance around the inhabited Earth which has not been visited by us up to the present time (because of the fact that the navigators who sailed in opposite directions toward each other never met), it is not of very great extent, if we reckon from the parallel distances that have been traversed by us.

It is unlikely that the Atlantic Ocean is divided into two seas, thus being separated by isthmuses so narrow and that prevent the circumnavigation; it is more likely that it is one confluent and continuous sea. For those who undertook circumnavigation, and turned back without having achieved their purpose, say that they were made to turn back, not because of any continent that stood in their way and hindered their further advance, inasmuch as the sea still continued open as before, but because of their destitution and loneliness.

This theory accords better, too, with the behaviour of the ocean, that is, in respect of the ebb and flow of tides; everywhere, at all events, the same principle, or else one that does not vary much, accounts for the changes both of high tide and low tide, as would be the case if their movements were produced by one sea and were the result of one cause.[73]

Thus, fifteen centuries before Columbus, the possibility of reaching Asia by sailing westward from Europe was already discussed.

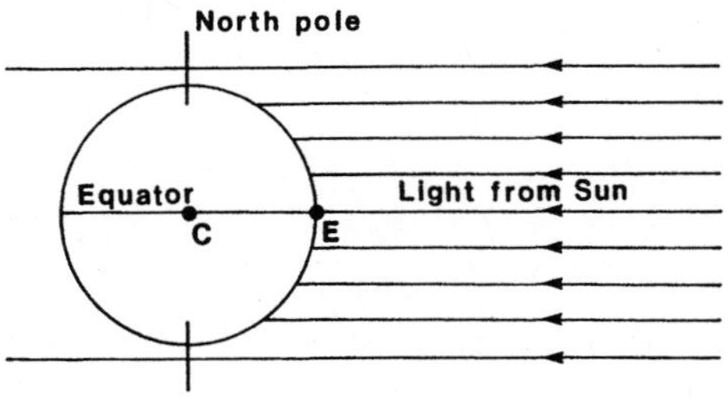

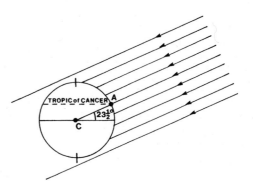

FIGURE I.32. Spring equinox (March 21) or autumnal equinox (September 23).

FIGURE I.33. Summer solstice (June 22).

I.IO THE ANNUAL MOTION OF THE SUN

The North-South Motion of the Sun

Equinoxes, Solstices, and Tropics   Let us consider the north-south motion of the Sun, starting with the *vernal* or *spring equinox* on March 21, the official beginning of spring. On this day, the Sun lies on the celestial equator. The word *equinox* refers to the fact that, on this day, the night is equal to the day: each is twelve hours long. Figure 1.32 shows a side view of the Earth on March 21. The Sun is directly above the equator, so its rays fall vertically down on point *E*. This fact determines the orientation of the central ray of light *EC*. The other rays were drawn parallel to this one. The parallelness of the rays reflects the fact that the Sun may be taken as infinitely far away.

Through March, April, and May, the Sun moves north. On June 22 it reaches its most northerly point, 23 1/2° above the equator. This day is called the *summer solstice*. It is the longest day of the year and the official beginning of summer. In Figure 1.33, the central ray has been drawn to coincide with the zenith direction at *A*, which is 23 1/2° north of the equator. The other rays were drawn parallel to this one. *A* lies on a circle on the Earth called the *tropic of Cancer*.

On September 23, the Sun, moving south, reaches the equator again. The day is twelve hours long again. This day is called the *autumnal equinox* and is the official beginning of autumn.

On December 22, the Sun reaches its most southerly point. This day is called the *winter solstice* and is the official beginning of winter. The Sun shines up from beneath the equator. At noon the Sun is straight overhead at points on the Earth's *tropic of Capricorn*, 23 1/2° south of the equator.

Seasons   Now it is easy to see why there are seasons. In figure 1.33, for June 22, five Sun rays fall on the northern hemisphere and only two on the southern hemisphere: the sunlight is distributed in such a way that the northern hemisphere receives more than the southern. (But no ancient writer ever made this argument.) Moreover, in June the Sun stays above the ground for a long time each day.

It is also easy to see that the seasons are reversed in the southern hemisphere. Note that this rather remarkable fact follows simply from the shape of the Earth and the annual motion of the Sun. There is no need to travel to Australia to find this out.

Change of the Shadow Plot through the Year   The shadow plots of figure 1.34 were all made at Seattle using the same gnomon.

March 21 (plot 4). The shadow plot on the day of the equinox is a straight line. This gives an easy way of determining the date of the equinox.

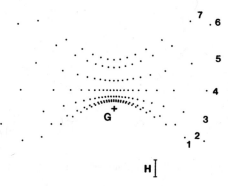

FIGURE 1.34. A sequence of shadow plots made at Seattle. On each plot, the points are half an hour apart. 1, June 22; 2, May 21 or July 23; 3, April 20 or August 24; 4, March 21 or September 23; 5, February 19 or October 24; 6, January 20 or November 23; 7, December 22. *G* labels the point at which the gnomon was set up perpendicular to the plane of the diagram. The line *H* shows the height of the gnomon that was used.

April 20 or May 21 (plot 3). In the spring months, two changes become evident in the shadow plot. First, the noon shadow becomes shorter. This reflects the fact that the Sun is now farther north on the celestial sphere and higher in the sky at noon. The second change involves the overall shape of the shadow plot. It now is curved, so as to enclose the gnomon. This change, too, reflects the fact that the Sun is north of the celestial equator: when the Sun is north of the equator, it rises north of east, crosses the meridian in the south, and sets north of east. Thus, the Sun behaves rather like Arcturus in figure 1.20.

June 22 (plot 1). On June 22, the day of the solstice, the noon shadow is its shortest, which gives a good way of determining the date of the solstice by observation.

September 23 (plot 4). On the autumnal equinox, the shadow plot is again a straight line.

October 24 or November 23 (plot 5). The shadow plot begins to curve away from the gnomon. That is, the tip of the shadow stays north of the gnomon all day long. This reflects the fact that the Sun is now south of the celestial equator and behaves rather like Sirius in figure 1.20.

December 22 (plot 7). On the winter solstice, the noon shadow is at its longest, which gives an easy way of determining the solstice.

Historical Examples of Shadow Plots   Examples of shadow tracks like those in figure 1.34 may be seen on horizontal plane sundials from Greek and Roman times. Two such dials are illustrated in figures. 3.4 and 3.5. On each of these dials, three shadow tracks are engraved: for summer solstice, equinox, and winter solstice. The upper curved track (concave upward) is for summer solstice. The equinoctial track is the horizontal straight line. The lower curved track (concave downward) is for winter solstice. The location of the gnomon, now missing, is indicated on the drawing of the Roman dial (fig. 3.4) by a small dot just above the middle of the summer shadow track. Similarly, on the drawing of the dial from Delos (fig. 3.5), the gnomon hole is indicated by a small circle above the summer shadow track. The system of eleven lines that intersect the shadow tracks was used to tell the hour of the day. These sundials were laid out by theoretical methods (as were, of course, the shadow plots of fig. 1.34). In sections 3.2 and 3.3 we show how the ancient dialers drew them.

The Eastward Motion of the Sun

The north-south motion of the Sun is responsible for the seasons. The changes produced by the Sun's eastward motion are more subtle: we see different stars at different times of the year.

The Ecliptic   The Sun's north-south and west-east motions are, of course, not separate. Rather, they are both the consequence of the Sun's yearly motion along a single circle that is oblique, or slanted, with respect to the equator. This path of the Sun's annual motion is called the *ecliptic* (see fig. 1.35). The angle between the plane of the ecliptic and the plane of the equator is called the *obliquity of the ecliptic*, which we denote ε ($\varepsilon \simeq 23\ 1/2°$).

Let us follow the Sun in its annual course about the Earth. Refer to figure 1.35. On the day of the vernal equinox, the Sun's motion along the ecliptic causes it to cross over the equator at the vernal equinoctial point *VE*. In the present age, the constellation Pisces is located at the vernal equinox. Thus, in March, Pisces cannot be seen, for it is above the horizon only when the Sun is up. The stars in the opposite part of the sky (e.g., the stars of Virgo) are visible for their longest period of time, for in March they are directly opposite the Sun and so cross the meridian at midnight.

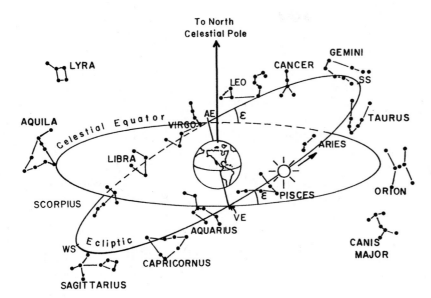

FIGURE 1.35. The Sun's path through the zodiac constellations is inclined to the plane of the equator.

In late June the Sun reaches the summer solstitial point *SS* and is as far north of the equator as it ever gets. Gemini is near the summer solstice. The stars of Gemini therefore cannot be seen in June, but conditions are favorable for viewing Sagittarius, Aquila, and Lyra, which are located in the opposite part of the sky. Sagittarius lies nearly on the ecliptic; Aquila, on the celestial equator; and Lyra, well north of the equator.

In September, the Sun again crosses the equator, this time at the autumnal equinox *AE*, while passing through the stars of Virgo.

In December, the Sun passes through Sagittarius, near the winter solstice *WS*.

The Empirical Basis of the Ecliptic   The ecliptic can be known approximately through rough observations of the stars. Consider a time of year two weeks before the Sun reaches Gemini. Let us look to the west just after sunset. Low on the horizon, near the place where the Sun went down, we would see the brighter stars of Gemini; shortly afterward, these stars would set, too.

If we repeated this operation a few weeks later, we would no longer see the stars of Gemini, for the Sun would have advanced on the ecliptic and would now be among them. But shortly after sunset, near the place on the horizon where the Sun went down, we would see the stars of Cancer. If we repeated our observations at intervals of a few weeks through an entire year, we could identify the whole path of the Sun through the constellations. The path mapped out in this fashion would not be a precisely defined great circle. Rather, it would be a broad band of constellations: the *zodiac*. We reserve the term *ecliptic* for a single circle, which is the precise path of the Sun.

The zodiac can also be picked out by observing the motions of the Moon and planets. All of these objects move more or less in the same plane. None of them travels exactly on the ecliptic, but none wanders very far north or south of it. On a night when the Moon and several of the planets are up, one sees them ranged across the sky more or less in a line—a sight that makes the orientation of the zodiac immediately sensible.

The precise location of the ecliptic circle among the stars can be defined by means of lunar eclipses. During a lunar eclipse, when the Moon falls into the Earth's shadow, the Sun, Earth, and Moon lie on a straight line. The position of the eclipsed Moon therefore identifies a part of the sky that is diametrically opposite the Sun. Indeed, this is the origin of the term *ecliptic circle*: it is on this circle alone that eclipses can occur.

FIGURE 1.36. Positions of the setting Sun on the western horizon.

Historical Origins of the Ecliptic

Knowledge of the Solstices and Equinoxes Already in Hesiod's *Works and Days* we find references to the solstices and the equinoxes (see sec. 1.1). This shows that, at least by the seventh century B.C., the solstices and equinoxes were a matter of common knowledge among the Greeks. The equinox, as a time of year, was probably known only roughly, as the season when days and nights were approximately equal. The solstices were called by Hesiod, as by later writers, *tropai helioio*, "turnings of the Sun" (from which we get our word "tropic"). This terminology reflects the manner in which the solstices were first observed. If one watches the Sun set each evening during the spring, one sees the setting point gradually work its way north along the horizon (see fig. 1.36). For several weeks around the summer solstice, the setting Sun's position scarcely changes but remains at its extreme northerly limit. Then, as the summer wears on, the Sun turns and its setting position begins to work its way south again.

Anaximander of Miletus Such practical awareness does not, however, imply knowledge of the ecliptic as a circle on the celestial sphere. The discovery of the ecliptic and of its obliquity is obscure. Anaximander (sixth century B.C.), the philosopher from Miletus, is said to have set up a gnomon at Sparta and to have used it to demonstrate the solstices and equinoxes as well as the hours of the day. Our sources are disappointingly vague, but this claim for Anaximander is not an impossible one. The summer solstice was observed, of course, by noting the day on which the noon shadow was shortest, and the winter solstice, the day on which the noon shadow was longest. The equinoxes could have been demonstrated either as the day on which the shadow plot was a straight line or as the day on which the Sun's noon altitude was midway between the two solstitial altitudes. Whether Anaximander conceived of the ecliptic as an oblique great circle we cannot say.

According to Aëtius, Anaximander taught that the Sun is a circle, twenty-seven or twenty-eight times the size of the Earth, like a chariot wheel, the rim of which is hollow and filled with fire. At one point on the rim is an opening through which the fire shines out: it is this opening that we perceive as the Sun. Eclipses of the Sun occur through the opening being stopped up. The Moon is of a similar nature; lunar eclipses and phases of the Moon are due to partial or complete closures of its vent. According to Aëtius, Anaximander also said that the circle of the Sun, like that of the Moon, is placed "obliquely." Does this rather broad cosmological speculation indicate a knowledge of the ecliptic? Perhaps. But it is also possible that the passing mention of the obliqueness of the Sun-wheel was inserted as an explanatory remark by Aëtius himself, who was describing Anaximander's philosophy to Greek readers of the first century A.D. When we consider the character of Anaximander's cosmological views, it appears a little rash to attribute to him a clear understanding of the nature of the Sun's circle and of its obliquity.[74]

Meton and Euctemon Anaximander's primitive cosmology was outmoded early in the following century, when Anaxagoras stated the true cause of eclipses (see sec. 1.9). We know also that two astronomers named Meton and Euctemon observed at Athens the summer solstice of 432 B.C. as part of an attempt to evaluate more accurately the length of the year.[75] Thus, by the latter half of the fifth century B.C. the observation of solstices was beginning to be a rather ordinary activity. It is likely that a clear conception of the ecliptic as an inclined great circle dates from around the same time.

The Babylonian Zodiac The conception of the ecliptic as an inclined path is found very clearly expressed at a much earlier date in Babylonian astronomy.

Already in MUL.APIN (seventh century B.C.), we have explicit statements that the Sun and planets follow the same path as the Moon:

> The Sun travels the [same] path the Moon travels.
> Jupiter travels the [same] path the Moon travels.
> Venus travels the [same] path the Moon travels.
> Mars travels the [same] path the Moon travels.
> Mercury, whose name is Ninurta, travels the [same] path the Moon travels.
> Saturn travels the [same] path the Moon travels.
> Together six gods who have the same positions, [and] who touch the stars of the sky and
> keep changing their positions.[76]

Moreover, the text gives a list of seventeen constellations along the path of the Moon, with the remark that the Moon "touches" them.[77] This seems to mean that the Moon can pass over, or occult, these constellations. Since the Moon is never more than 5° north or south of the ecliptic, we should find that these seventeen star groups are all within 5° of the ecliptic—which is indeed the case. The reason there are more than twelve constellations in the Moon's path is that some of the standard zodiac constellations were treated in several parts. For example, the "Stars" (the Pleiades) are listed separately from the Bull of Heaven. Thus, it appears that when MUL.APIN was written, the twelve-constellation zodiac was not yet standard.

In another part of MUL.APIN we find explicit mentions that the Sun is farther north or south on the sky at different times of the year:

> From the 1st of Addaru until the 30th of Ajjaru the Sun stands in the path of the Anu stars; wind and weather.
>
> From the 1st of Simanu until the 30th of Abu the Sun stands in the path of the Enlil stars; harvest and heat.
>
> From the 1st of Ululu until the 30th of Arahsamnu the Sun stands in the path of the Anu stars; wind and weather.
>
> From the 1st of Kislimu until the 30th of Šabaṭu the Sun stands in the path of the Ea stars; cold.[78]

(For the order of the month names, see fig. 1.3. For the ways of Enlil, Anu, and Ea, see sec. 1.1.)

During months XII, I, and II, the Sun is among the stars of Anu—or, as we would say, near the celestial equator. During months III, IV, and V, the Sun is among the stars of Enlil—the northern stars. During months VI, VII, and VIII, the Sun is again among the equatorial stars of Anu. During months IX, X, and XI, the Sun is among the southern stars—the stars of Ea. If we put all this together, we have a picture very close to figure 1.35. The Babylonians did not geometrize the world in the manner of the Greeks. There are no Babylonian equivalents of Euclid or Eudoxus, for example. But, clearly, the notion of the ecliptic as an inclined path through the stars was already well established by about 650 B.C.

Oenopides of Chios   Among the Greeks, the discovery of the ecliptic was attributed to Oenopides of Chios, who flourished about 450 B.C. (Chios is a city on a large island of the same name that lies off the west coast of Asia Minor.) Oenopides was a mathematician as well as an astronomer. He is said to be responsible for introducing the requirement that in geometric demonstrations no other instruments be allowed than straight edge and compass, and to have proved some of the compass-and-straight-edge constructions that later made their way into Euclid's *Elements.*[79]

In astronomy, Oenopides made an attempt to evaluate the lengths of the year and the month in terms of one another. According to Oenopides, 59 years contain a whole number (730) of lunar months. This period, the so-called "great year" of Oenopides, is an example of a luni-solar cycle, a subject discussed in chapter 4. In investigating the lengths of the year and month, Oenopides was a predecessor of Meton, whose work, a generation later, superseded his own.

Our authority for Oenopides' discovery of the ecliptic is Eudemus of Rhodes. Eudemus, who flourished about 325 B.C., passed some time in Athens, where he was a pupil of Aristotle. Eudemus wrote histories of the development of mathematics and astronomy, which survive only in fragmentary quotations by later writers. Eudemus may with some justice be considered the first historian of science. Eudemus is quoted by our old friend Theon of Smyrna in a short passage on early astronomical discoveries:

> Eudemus recounts in his *Astronomy* that Oenopides was the first to discover the encircling belt of zodiac and the existence of the great year . . . . Others added other discoveries to these: that the fixed stars move around the axis which passes through the poles, but that the planets move around the axis which is perpendicular to the zodiac, and that the axis of the fixed stars and the axis of the planets are separated from one another by the side of a pentadecagon, that is, by 24 degrees.[80]

This short passage presents many difficulties. Eudemus does not make the nature of Oenopides' discovery very clear. "Others" discovered the fact that all the planets move in the zodiac. Oenopides' discovery perhaps then applied only to the Sun. If he measured the obliquity of the ecliptic, he was not the source of the 24° value, for Eudemus attributes this value, again, to others. Nor could Oenopides simply have pointed out the constellations that lay along the zodiac. As discussed in section 1.8, the zodiacal constellations are of Mesopotamian origin. It is possible that Oenopides played a role in the introduction of the Babylonian zodiac to the Greeks. But this would hardly amount to a "discovery." The most likely possibility, then, is that Oenopides gave some sort of geometrical demonstration of the circular, beltlike nature of the Sun's path, that is, that he proved the ecliptic to be an oblique great circle. However, we simply do not know.

In any case, by the close of the fifth century B.C., the Babylonian zodiac was well established among the Greeks. The first Greek parapegmata (star calendars) had been composed, based on a division of the year into zodiacal signs. And the astronomers knew how to use the gnomon to observe solstices and equinoxes. A science of astronomy had begun among the Greeks, incorporating original Greek methods and discoveries, as well as borrowing from their Babylonian contemporaries.

1.11 OBSERVATION: THE MOTION OF THE MOON

In the course of a month, the Moon moves eastward all the way around the zodiac. Thus, the Moon does in a month what the Sun does in a year. Observing the zodiacal motion of the Moon is therefore a good way of learning the zodiacal constellations and of visualizing the motion of the Sun.

Begin at the time of the first visibility of the new crescent. This will be one to three days after the time of new Moon. You can find the date of the new Moon in an almanac or on a calendar. Start looking for the crescent Moon in the west just after sunset. Once every night that the weather permits, spot the Moon in the night sky and mark its location on a star chart. The Moon takes only two or three days to move through each zodiac constellation, so you can see a noticeable shift in only one day. Continue to plot the Moon's position as often as possible for the rest of the lunar month.

1.12 THE USES OF SHADOWS

A sequence of shadow plots, as in figure 1.34, can be made to yield a good deal of information.

Measuring the Obliquity of the Ecliptic

The Sun moves about 23 1/2° north or south of the equator. The arc between the two tropics is therefore about 47°. How can this angle be determined from observations of the Sun? This parameter, one of the most fundamental for the development of astronomy, is also one of the easiest to measure.

Simply measure the noon altitude of the Sun at summer solstice, perhaps with the aid of a gnomon (sec. 1.4 and fig. 1.9). Wait six months and measure the noon altitude of the Sun at winter solstice. *The angle between the tropics is equal to the difference between the Sun's noon altitudes at summer and winter solstice.* The angle between the equator and either one of the tropics is equal to half the angle between the tropics. In modern terminology, this angle is called the *obliquity of the ecliptic*, ε (ε is labeled in fig. 1.35).

Ancient and Modern Values for the Obliquity of the Ecliptic   The most ancient value for the obliquity of the ecliptic is the round figure of 24°. The degree was a Babylonian unit of measure, and it was not used in Greek astronomy until the second century B.C. Earlier Greek writers often expressed this 24° angle as *one-fifteenth of a great circle* (360/15 = 24). Or, again, the angle was described as the angle subtended by the side of a regular pentadecagon (the regular polygon with 15 sides), as in Theon of Smyrna's quotation from Eudemus at the end of section 1.10.

It is not known who first ascribed the value of 24° to the obliquity of the ecliptic. But the 24° figure was already current by the time of Euclid—about 300 B.C. In his *Elements*, Euclid shows how to construct a regular pentadecagon.[81] Proclus, the fifth-century A.D. commentator on Euclid, refers to this proposition to illustrate his statement that Euclid deliberately included a number of propositions that might be of use in astronomy.[82]

The value *one-fifteenth of a circle* for the obliquity of the ecliptic was not a measurement in the modern sense. Certainly, it was based on measurements and was, in fact, fairly close to the truth. But the adoption of this value was partly determined by the Greek propensity for neat geometrical demonstration. The idea that precise observations might be important to astronomy was slow to dawn.

The most ancient *measurement* of the arc between the tropics to have come down to us is that of Eratosthenes. According to Ptolemy (*Almagest* I, 12), Eratosthenes reckoned that the arc between the tropics was 11/83 of the whole meridian circle. Ptolemy gives no details of Eratosthenes' method. But, as Eratosthenes passed the latter half of his life at Alexandria, it is likely that the measurement was made there, toward the close of the third century B.C. The peculiar value (11/83 of a circle) probably resulted from a geometrical calculation based on gnomon measurements. Eratosthenes had to do his calculation without either trigonometry or the use of the degree, which were both later developments. According to Ptolemy, Eratosthenes' value for the arc between the tropics was also accepted by Hipparchus (ca. 140 B.C.). If we express Eratosthenes' figure for the arc between the tropics in terms of degrees, we have

$$360° \times \frac{11}{83} = 47°42'39''.$$

Ptolemy says that he himself measured the arc between the tropics several times over a period of years (ca. A.D. 140), using a meridian quadrant (see fig.

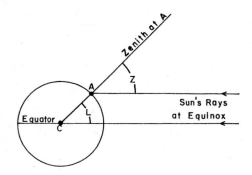

FIGURE 1.37. Sunlight incident on the Earth at equinox.

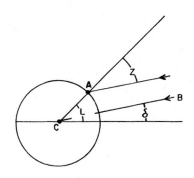

FIGURE 1.38. δ is the declination of the Sun. z is the Sun's zenith distance at noon, as measured by an observer on the Earth at latitude L. For any observer, L = δ + z.

5.2). Ptolemy found the arc always to be more than 47°40′ but less than 47°45′. He therefore adopted the value 47°42 2/3′ (i.e., 47°42′40″), which was consistent with his own measurements, as well as the work of his predecessors. Thus, according to Ptolemy, the obliquity of the ecliptic is (47°42′40″)/2, or

$$\varepsilon = 23°51′20″ \quad \text{(Ptolemy)}.$$

For the last half of the twentieth century, one should use

$$\varepsilon = 23°26′ \quad \text{(modern)},$$

which is nearly half a degree smaller than Ptolemy's value. There are two reasons for the difference. First, the ancient measurements of the obliquity of the ecliptic were all a little too high. Second, the obliquity of the ecliptic really has decreased slightly (about 1/4°) since antiquity.

Measuring the Latitude of a Place

The latitude of a place on Earth can be determined at night by measuring the altitude of the celestial pole. But observations of the Sun can also be used. Most convenient is a noon altitude measured at an equinox. However, the discussion will be simplified if it is put in terms of *zenith distance* rather than altitude. The zenith distance z of the Sun is the Sun's angular distance from the zenith; z is the complement of the altitude (see fig. 1.9).

Now, let us take up the problem of determining the latitude of a place on the Earth from solar observations. Figure 1.37 shows the situation at an equinox. The center of the Earth is C, the place of observation is A, and the latitude of this place is angle L. Since it is the time of equinox, the Sun's rays are parallel to the equator. The Sun's zenith distance is z. Evidently, $L = z$. That is, on the day of the equinox, the Sun's noon zenith distance, as measured at some place on the Earth, is equal to the latitude of that place.

Measuring the Declination of the Sun

The *declination* of the Sun is its angular distance above or below the celestial equator. Declinations north of the equator are conventionally counted as positive; those below the equator are counted as negative. Thus, we say that on June 22 the Sun's declination is about +23 1/2°, on December 22, the declination of the Sun is −23 1/2°, and on the equinoxes the Sun's declination is 0°. We will denote declinations by the letter δ. The declination of the Sun on any day can be determined from its zenith distance, provided one already knows the latitude of the place of observation.

In figure 1.38, we imagine ourselves at point A on the surface of the Earth. Our local zenith direction is line CA. The angle L that this line makes with the equator is our latitude. At local noon, the Sun's zenith distance is z, which is the angle between a ray of light arriving at A and the zenith direction. Draw a second ray of light CB, parallel to the first one, but passing through the center of the Earth C. The angle that this ray makes with the equator is the Sun's declination δ. Now, angle ACB is equal to z, so we have the following simple result:

$$L = z + \delta.$$

The relation obtained above for determining the latitude on the day of equinox (i.e., $L = z$) is a special case (for δ = 0) of this more general formula.

If L is known, and if z is measured with a gnomon or a quadrant, the Sun's declination can be calculated from $\delta = L - z$. Here we have one of the simplest procedures for measuring a *celestial coordinate*—that is, specifying the position of an object (in this case, the Sun) on the celestial sphere.

Equinoctial and Solstitial Shadows

In figure 1.9, *GH* represents a vertical gnomon, and *TH*, the length of its noon shadow. From the geometry of the figure,

$$\tan z = TH/GH.$$

Then, since $z = L - \delta$, we obtain

$$\text{length of shadow} = \text{length of gnomon} \times \tan(L - \delta).$$

This rule makes it easy to find the length of the shadow for a given place on Earth at equinox or at summer or winter solstice.

For example, take the case of Athens, latitude 38° N. At summer solstice, the Sun's declination is approximately +23 1/2°, and we have

$$\text{shadow} = \text{gnomon} \times \tan(38° - 23\tfrac{1°}{2}) = 0.26 \times \text{gnomon}.$$

Thus, at Athens on the summer solstice, the noon shadow is about one-fourth the length of the gnomon.

At equinox the Sun's declination is 0, and we find for Athens that

$$\text{shadow} = \text{gnomon} \times \tan(38° - 0) = 0.78 \times \text{gnomon}.$$

Similarly, one finds that at Athens on winter solstice the noon shadow is about 1.8 times the length of the gnomon.

Equinoctial Shadows in Ancient Sources

In antiquity the length of the equinoctial shadow was often used to specify the latitude. Shadow lengths could easily be measured by people untrained in astronomy and without elaborate instruments. Here is Vitruvius introducing his readers to the variation of the equinoctial noon shadow with geographical latitude:

> When the Sun is at the equinoxes, that is, passing through Aries or Libra, he makes the gnomon cast a shadow equal to eight ninths of its own length, in the latitude of Rome. In Athens, the shadow is equal to three fourths of the length of the gnomon; at Rhodes to five sevenths; at Tarentum, to nine elevenths; at Alexandria to three fifths; and so on at other places it is found that the shadows of the equinoctial gnomons are naturally different from one another.[83]

Pliny, the Roman encyclopedist of the first century A.D., writes in a similar vein. He first points out (as does Vitruvius) that sundials constructed for one place are not for use everywhere. The shadows change perceptibly "in three hundred stades, or five hundred at the most." Pliny continues:

> Consequently, in Egypt at midday on the day of the equinox the shadow of the pin or gnomon measures a little more than half the length of the gnomon itself, whereas in the city of Rome the shadow is 1/9 shorter than the gnomon, at the town of Ancona 1/35 longer, and in the district of Italy called Venezia the shadow is equal to the gnomon, at the same hours.[84]

Shadow lengths reported by voyagers to distant regions were an important source of information for the ancient geographers. Such observations could, in principle, establish the relative north-south positions of even very distant localities. In practice, however, the geographical writers never had access to a sufficiently large collection of data. Moreover, such shadow lengths as were reported were often defective—the ordinary traveler was not often a very good astronomer.

In the following quotation we see Strabo attempting to make use of just such data. Strabo grapples with the problem of the location of Britain: just how far north is it? Is Britain as far north as the mouth of the river Borysthenes (the modern Dnieper, which empties into the Black Sea)?

> The parallel through the mouth of the Borysthenes is conjectured by Hipparchus and others to be the same as that through Britain, from the fact that the parallel through Byzantium is the same as that through Massilia [modern Marseille]. For Pytheas found the ratio of the gnomon to its shadow in Massilia; and Hipparchus says he finds the same ratio, at the same time of year, in Byzantium.[85]

Pytheas of Massilia was a famous navigator who, about 285 B.C., explored the northwest coast of Europe. His writings are lost and are known only through quotations by later writers, such as Strabo. The dubious quality of many ancient shadow lengths, and the conclusions based on them, is demonstrated by reference to a modern map. The latitude of Marseille is about 43°18′ N, while that of Istanbul (ancient Byzantium) is about 41°02′ N. Hipparchus's use of Pytheas's doubtful measurements therefore caused a mistake of about 2 1/4° in the relative latitudes of Massilia and Byzantium. This represents a north-south displacement of some 150 miles. It should be added that Strabo considered Pytheas to be unreliable and even a great liar.

On Zones

In *Almagest* II, 6, Ptolemy discusses the characteristics of various parallels on the Earth's surface, in terms of the Sun's behavior. The region between the tropics is said by Ptolemy to be *amphiskian*, meaning that the noon shadow can point either north or south in the course of the year. The Greek adjective is a compound of *amphi* (on both sides) and *skia* (shadow).

The part of the Earth between latitudes ε and $90° - \varepsilon$ (the temperate zone) Ptolemy describes as *heteroskian*, meaning that the noon shadow always falls in the same direction. The adjective is a compound of *heteros* (to one side) and *skia*. Thus, in Greece, the noon shadow always points north.

Finally, the zone north of latitude $90° - \varepsilon$ (what we call the arctic zone) is said to be *periskian*, because, on some days of the year, the Sun is up all day long. Then the shadow goes all the way around (*peri*) the gnomon. The shadow track is a closed curve—in fact, an ellipse.

The same three terms are used by Strabo,[86] who attributes them to Posidonius, the Stoic philosopher of the first century B.C. But the terms may well be older than Posidonius.

The terrestrial zones, defined by the Sun's behavior, are thus a product of Greek astronomy. However, the Greeks often disagreed about the number of zones. Five were implied by the celestial phenomena: two frigid, two temperate, and one tropical. But Posidonius added two others, for a total of seven. The two extra zones were narrow belts straddling the tropics. In each of these, the Sun stood overhead for about half a month each year. These two narrow zones, according to Posidonius, were parched by the Sun and therefore even hotter than the region around the equator.

Polybius (third century B.C.), on the other hand, advocated six zones: two frigid, two temperate, and two tropical. (Polybius divided the zone between the tropics into two, using the equator as boundary.) It should also be noted that many Greek writers defined the arctic circle (and hence the limit of the frigid zone) differently than we do (see sec. 2.5).

But Strabo sensibly opts for five zones, defining them in terms of celestial phenomena, by means of their amphiskian, heteroskian, and periskian properties, exactly as we do today.[87]

1.13 EXERCISE: USING SHADOW PLOTS

1. Use the shadow plots in figure 1.34 to measure the arc between the tropics and the obliquity of the ecliptic.

2. Use the shadow plots in figure 1.34 to measure the latitude of Seattle. Do this three times—using the plot for summer solstice, the plot for the equinox, and the plot for winter solstice. Use the value of the obliquity of the ecliptic that you obtained in problem 1 and the general rule $L = z + \delta$. Of course, the latitude of Seattle should come out the same all three times. But you may get small differences due to errors of measurement.

3. What is the declination of the Sun on April 20? Use the appropriate shadow plot in figure 1.34, together with the value you obtained in problem 2 for the latitude of Seattle.

   Find a place on the Earth at which the Sun would be directly overhead at noon on April 20.

4. Find the latitude of your own city from an atlas or a map. Use the shadow plot that you made in the exercise of section 1.3 to measure the declination of the Sun.

   For the day of your own shadow plot, find a place on Earth at which the Sun would be directly overhead at noon.

5. In section 1.12, there are a number of equinoctial shadow lengths due to Vitruvius and Pliny. How accurate are these ancient measurements? To find out, proceed as follows. For each of the ancient equinoctial shadow lengths, compute the equivalent latitude. In an atlas find the actual latitudes of the places mentioned by Vitruvius and Pliny. You may need to use a *historical* atlas to find some of the ancient place names. Compare the actual latitudes with values you have deduced from the shadow lengths.

6. Pytheas's measurement of a shadow at Massilia [Marseille] resulted in a 2 1/4° error for the latitude of that city. Suppose that at Marseille (latitude 43° N) one attempts to measure the latitude by observing the shadow cast by a 10-cm gnomon at noon on the equinox. How great a mistake in measuring the shadow is required to produce a 2 1/4° error in the latitude?

1.14 THE SIZE OF THE EARTH

Aristotle[88] says that certain mathematicians obtained 400,000 stades for the circumference of the Earth. We have no details of the method used for this first estimate of the size of the Earth, made perhaps around 350 B.C. Another early figure for the circumference of the Earth is 300,000 stades, mentioned by Archimedes in his *Sand Reckoner*.[89] Archimedes does not tell us who made this measurement, but some scholars attribute it to Dicaearchus, who died around 285 B.C.[90] The first measurement of the size of Earth for which we have any detailed information is that made by Eratosthenes later in the third century B.C.

Eratosthenes on the Size of Earth

Eratosthenes was born around 276 B.C. in Cyrene, a Greek city on the North African coast, in what is now Libya. As a young man he studied in Athens.

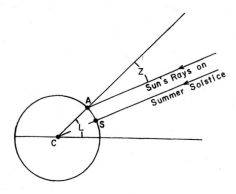

FIGURE 1.39. Sunlight incident on the Earth at summer solstice.

After making a bit of a reputation in literary studies and philosophy, he was offered a position by Ptolemaios III Euergetes, who was King of Egypt from 246 to 222 B.C. Eratosthenes spent most of his adult life at Alexandria, first as a tutor to the king's son, Philopater (who was king from 222 to 205 B.C.), then as a fellow of the Alexandria Museum, and later as head of the Library.

Eratosthenes was a man of wide interests. He was a literary critic who made studies of Homer. He wrote a philosophical study, *On the Good*, and a rhetorical treatise, *On Declamation*. But these works were criticized by Strabo for superficiality.[91] According to Strabo, Eratosthenes wanted to pass for a philosopher but did not devote himself seriously enough to this calling and vacillated among multiple interests. Eratosthenes' most significant work was his geographical treatise. He attempted to work out the arrangement of the whole known world and introduced geometrical methods into geography. None of Eratosthenes' works have survived except for his description of the constellations, the *Catasterisms*, if this indeed is really his. However, many extracts from his *Geography* are preserved by Strabo. It is likely that the attempt to measure the Earth was a part of Eratosthenes' researches in geography, part of an effort to work out the scale of the world map.

While several ancient writers mention Eratosthenes' measurement, the only one who gives much detail is Cleomedes, in his introductory textbook of astronomy and Stoic physics.[92] Eratosthenes assumed that Alexandria and Syene are on the same meridian. (Syene was a town on the Nile in upper Egypt. The modern city on the same site is Aswan.) Moreover, Eratosthenes assumed that Syene is on the tropic of Cancer.[93] Thus, at Syene, at noon on the summer solstice, the gnomons have no shadows, because the Sun is straight overhead.

But, at Alexandria at the same moment, the gnomons do cast shadows, because that city is situated to the north of Syene. According to Cleomedes, Eratosthenes measured the zenith distance of the Sun at noon in Alexandria and found it to be 1/50 of a circle. Moreover, the distance between Alexandria and Syene is 5,000 stades. From these premises, Eratosthenes worked out a figure of 250,000 stades for the circumference of Earth.

Cleomedes gives full details of the geometrical demonstration. Its essence (we simplify it a bit) is illustrated by figure 1.39. *S* is Syene, *A* is Alexandria, and *C* is the center of the Earth. On the summer solstice, the Sun was straight overhead at Syene. Thus, the ray arriving at *S*, if extended, would pass through *C*. Eratosthenes measured the Sun's zenith distance *z* at Alexandria on the same day and found it to be 1/50 of a circle. *z* is equal to angle *ACS*, which is the latitude difference between Alexandria and Syene. Thus, arc *AS* must be 1/50 of the circumference of the Earth. Eratosthenes took the distance between the two cities to be 5,000 stades. Therefore, the whole circumference of the Earth is 50 × 5,000 = 250,000 stades.

It is impossible to convert Eratosthenes' figure of 250,000 stades into modern units with any precision, because several different stades were in common use. However, whichever stade was meant, Eratosthenes' result was certainly in the right range.

How did Eratosthenes know that there were no noon shadows at Syene on the summer solstice? Probably this was common knowledge, brought back to Alexandria by travelers from upriver. Cleomedes says that there are no shadows over a region of about 300 stades in width. Strabo,[94] in his description of Egypt, gives an account of a well at Syene: at the summer solstice, the rays of the Sun reach down to the very bottom of the well. It is possible that Eratosthenes heard a similar report and realized what it meant.

What about Eratosthenes' figure of 5,000 stades for the distance between Syene and Alexandria? It is clear that this was only a rough estimate, expressed as a round number. It was probably based on reports of the time required for travelers to pass from one city to the other, rather than on any real measurement of the distance. In the same way, Eratosthenes' figure of 1/50 of a circle (we would say 7.2°) for the zenith distance of the Sun at Alexandria is clearly a round number. Thus, it might be fairer to describe Eratosthenes' assessment of the size of the Earth as an estimate rather than a real measurement.

According to some ancient authorities,[95] Eratosthenes put the circumference of the Earth, not at 250,000 but at 252,000 stades. This modification was probably introduced, not as the result of a refined measurement, but rather for the sake of arithmetical convenience. It was common practice to divide the circle into sixty equal parts. Eratosthenes' sixty-part division of the circle is attested by Strabo.[96] And we shall see other Greek astronomers and geographers using the same convention. (The degree, or 360-part division of the circle, was not adopted by the Greeks until about a century after Eratosthenes' time.) By putting the circumference of the Earth at 252,000 stades, Eratosthenes obtained an even number of stades per part: 252,000/60 = 4,200 stades per sixtieth part. It happens, happily, that this also results in an even number of stades per degree: 252,000/360 = 700 stades per degree.

Later Estimates

Posidonius (early first century B.C.) made an estimate of the size of the Earth, based on the fact that the star Canopus could be seen in Egypt but not at places farther north. Again, our most detailed source is Cleomedes (see sec. 1.15). At Rhodes, Canopus barely grazed the southern horizon. But at Alexandria, Canopus was 1/48 of a circle above the horizon. Then, taking 5,000 stades for the distance between Alexandria and Rhodes, Posidonius arrived at 5,000 × 48 = 240,000 stades as the circumference of the Earth. Again, the round numbers show that no serious effort was made to secure accurate measurements.

But, according to Strabo, Posidonius put the circumference of the Earth at 180,000 stades.[97] Thus, it appears that Posidonius changed his mind. The smaller figure was undoubtedly based on a lower value (3,750 stades) for the distance between Rhodes and Alexandria. (This figure of 3,750 stades for the distance between Rhodes and Alexandria is attributed by Strabo to Eratosthenes, who must have been Posidonius's source.) Posidonius's second value of 180,000 stades for the circumference of the Earth also results in an even number of stades per degree: 180,000/360 = 500 stades per degree.

These two figures left a considerable range of uncertainty in the scale of the world map:

700 stades/degree (Eratosthenes)
500 stades/degree (Posidonius)

The lower value of 500 stades/degree (along with the associated and circumference of 180,000 stades) was accepted by Ptolemy in his *Geography* and thus became the preferred figure. However, Eratosthenes' value never dropped completely out of sight. It reappears, for example, in the *Sphere* of Sacrobosco, a thirteenth-century introduction to astronomy that was widely used in the medieval European universities.[98]

In the early Middle Ages, a number of Arabic astronomers made measurements of the circumference of the Earth. For example, the astronomers under the patronage of al-Maʾmūn made such a measurement on the plain of Palmyra (Syria) around A.D. 830.[99] One motive for making new measurements was that the Arabic astronomers of the ninth century had no idea (any more than we have) of the length of the stade used by Eratosthenes or Ptolemy.

In the later Middle Ages, both Greek and Arabic estimates of the size of the Earth were in circulation in Europe. As the variety of estimates were compounded by uncertainties over the values of the Greek and Arabic units of measure, the European geographer was left with considerable freedom of

FIGURE I.40. Posidonius, the Stoic philosopher. This bust is a marble copy made during the early Empire of an original sculpted around 70 B.C. Museo Nazionale, Naples.

choice. When Columbus tried to convince himself and others of the practicality of his proposed voyage to Asia, he deliberately selected the smallest of the available estimates for the size of the Earth and the largest possible estimate for the width of the Eurasian continent. That made the western ocean as narrow as possible and the voyage as attractive as possible. By sheer luck, it turned out that Columbus's voyage was of about the distance he expected. He counted on a trip of under 3,000 miles between Europe and Asia.[100] In 3,000 miles he did, indeed, reach land. The true distance to Asia was more than 10,000 miles.

I.15 EXERCISE: THE SIZE OF THE EARTH

Around 100 B.C., Posidonius (see fig. 1.40), the Stoic philosopher and teacher of Cicero, calculated the circumference of the Earth from information obtained by observation of the star Canopus. Canopus is located in the steering oar of Argo (in the modern constellation Carina). Cleomedes, writing perhaps 200 years later, described the method which Posidonius used. Cleomedes' account of Posidonius's measurement immediately precedes his report of the more famous measurement of Eratosthenes.

EXTRACT FROM CLEOMEDES

On the Elementary Theory of the Heavenly Bodies I, 10, 2.

[Posidonius] says that Rhodes and Alexandria lie under the same meridian. Meridian circles are [circles] drawn through the poles of the cosmos, and through the point which is above the head of each [person] standing on the Earth. The poles of all these [meridian circles] are the same, but the point in the direction of the head is different. . . . Now Rhodes and Alexandria lie under the same meridian, and the distance between the cities is reputed to be 5,000 stades. Suppose it to be so. . . . Posidonius says next that the very bright star called Canopus lies to the south, practically on the steering oar of Argo. This [star] is not seen at all in Greece; hence Aratus does not even mention it in his *Phenomena*. But, as one goes from north to south, it begins to be visible at Rhodes and, when seen on the horizon [there], it sets immediately with the rotation of the cosmos. But when we have sailed the 5,000 stades from Rhodes and are at Alexandria, this star, when it is exactly on the meridian, is found to be at a height above the horizon of one-fourth of a sign, that is, a forty-eighth of the meridian [drawn] through Rhodes and Alexandria. It follows, therefore, that the segment of the same meridian that lies above the distance between Rhodes and Alexandria is one forty-eighth part of [the said circle], because the horizon of the Rhodians is distant from the horizon of the Alexandrians by one forty-eighth of the zodiac circle. . . . And thus the great circle of the Earth is found to be 240,000 stades, assuming that from Rhodes to Alexandria it is 5,000 stades; but, if not, [it is] in [the same] ratio to the distance. Such then is Posidonius's way of dealing with the size of the Earth.[101]

The Exercise

1. What does the writer mean when he says, "But, as one goes from north to south, it begins to be visible at Rhodes and, when seen on the horizon there, it sets immediately with the rotation of the cosmos"?
2. What does the writer mean when he says, "At Alexandria, this star, when it is exactly on the meridian, is found to be at a height above the horizon of one-fourth of a sign, that is, a forty-eighth of the meridian"?
3. Draw a diagram of the Earth showing the horizon at Rhodes and at Alexandria and the various angles mentioned by Cleomedes. Prove that

the observations do indeed lead to the stated value for the circumference. Make your geometrical arguments clear and convincing.

I.16 OBSERVATION: THE ANGULAR SIZE OF THE MOON

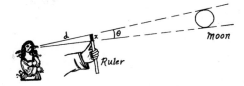

FIGURE I.41. Measuring the angular diameter of the Moon.

The angular diameter of the Moon may be determined with the aid of a millimeter scale. Hold the ruler out at arm's length, as in figure 1.41, so that the top of the ruler appears to coincide with the top limb of the Moon. Place your thumb on the part of the ruler that coincides with the bottom limb of the Moon. Let us call x the length thus marked off on the ruler. Ask a friend to measure the distance d from your eye to the ruler.

The angular diameter θ of the Moon is then given by

$$\theta = \sin^{-1}\left(\frac{x}{d}\right).$$

I.17 ARISTARCHUS ON THE SIZES AND DISTANCES

Aristarchus of Samos

Aristarchus is remembered for two remarkable achievements. He advocated the motion of the Earth around the Sun. And he was the author of a book *On the Sizes and Distances of the Sun and the Moon*, the oldest surviving geometrical treatment of this problem. Aristarchus was born around the beginning of the third century B.C. He was a native of Samos, one of the larger Greek islands in the Aegean Sea. He is said to have been a pupil of Strato of Lampsacus. This Strato was at one time tutor, then advisor, to King Ptolemaios II Philadelphos, the patron of the Museum in Alexandria. After the death of Theophrastus, Strato succeeded him as head the the Lyceum, the school of Aristotlelian philosophy at Athens. Thus, Aristarchus could have been Strato's pupil either at Alexandria or at Athens. We do not know which. Ptolemy, in *Almagest* III, 1, cites an observation of the summer solstice of 280 B.C. made by Aristarchus. This single observation is the only event of Aristarchus's life that may be dated. He is credited by Vitruvius with the invention of a type of sundial (the *scaphe*). According to Aëtius, Aristarchus also wrote a book on vision, light, and colors.[102]

The book in which Aristarchus argued for the motion of the Earth around the Sun has not survived. The best testimony is a remark by Archimedes in the *Sand Reckoner*, which is very close to Aristarchus in time.

EXTRACT FROM ARCHIMEDES

Sand Reckoner

Aristarchus of Samos brought out a book consisting of some hypotheses, in which the premises lead to the result that the cosmos is many times greater than that now so called. His hypotheses are that the fixed stars and the Sun remain unmoved, that the Earth revolves about the Sun in the circumference of a circle, the Sun lying in the middle of the orbit, and that the sphere of the fixed stars, situated about the same center as the Sun, is so great that the circle in which he supposes the Earth to revolve bears such a proportion to the distance of the fixed stars as the center of the sphere bears to its surface.[103]

As Archimedes says, Aristarchus realized that his premises implied a cosmos that was vastly larger than previously believed. If the Earth moved around the Sun, there should be a large *annual parallax*. That is, the stars should

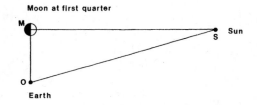

FIGURE 1.42. Aristarchus's use of the lunar quadrature.

appear to shift with respect to one another as the Earth moves. Because the constellations do not show any changes in the course of the year, Aristarchus realized that radius of the sphere of stars must be vastly greater than the radius of the Earth's orbit. (The annual parallax does, indeed, exist. But it is so small—less than a second of arc, even for the nearest visible stars—that it was not detected until the nineteenth century.)

Aristarchus's Sun-centered cosmology drew some unfavorable attention from his contemporaries. For example, the Stoic philosopher Cleanthes of Assos said that Aristarchus ought to be "indicted on a charge of impiety for putting into motion the hearth of the universe."[104] As far as we know, no formal indictment was made. (But let us recall that Anaxagoras had gotten into just such a legal scrape for calling the Sun a red-hot stone.) For the most part, the Sun-centered cosmology was simply ignored. The lack of interest in this far-fetched idea no doubt contributed to the failure of Aristarchus's book to be preserved. In any case, it is unlikely that Aristarchus worked out the consequences in detail. There is no indication, for example, that he discussed the consequences of the motion of the Earth for the apparent motions of the other planets.

Aristarchus on the Sizes and Distances

While earlier writers had speculated on the sizes and distances of the Sun and Moon, Aristarchus was the first to address this problem geometrically.

EXTRACT FROM ARISTARCHUS OF SAMOS

On the Sizes and Distances of the Sun and Moon

[Hypotheses:]

1. That the Moon receives its light from the Sun.
2. That the Earth is in the relation of a point and center to the sphere in which the Moon moves.
3. That, when the Moon appears to us halved, the great circle which divides the dark and the bright portions of the Moon is in the direction of our eye.
4. That, when the Moon appears to us halved, its distance from the Sun is less than a quadrant by one-thirtieth of a quadrant.
5. That the breadth of the [Earth's] shadow is [that] of two Moons.
6. That the Moon subtends one fifteenth part of a sign of the zodiac.

We are now in a position to prove the following propositions:

1. The distance of the Sun from the Earth is greater than eighteen times, but less than twenty times, the distance of the Moon [from the Earth]; this follows from the hypothesis about the halved Moon.
2. The diameter of the Sun has the same ratio [as aforesaid] to the diameter of the Moon.
3. The diameter of the Sun has to the diameter of the Earth a ratio greater than that which 19 has to 3, but less than that which 43 has to 6; this follows from the ratio thus discovered between the distances, the hypothesis about the shadow, and the hypothesis that the Moon subtends one fifteenth part of a sign of the zodiac.[105]

Ratio of the Distances   The first of the astronomical conclusions to be proved is that the Sun is between eighteen and twenty times farther away from us than the Moon is. Refer to figure 1.42. *O* is the Earth, which may (by hypothesis 2) be considered a mere point. *M* is the Moon at the time of quarter Moon, when we see it divided exactly in half. *S* is the Sun. From hypothesis 3, it follows that the Sun's ray *SM* must then be perpendicular to *OM*.

Now, by hypothesis 4, the angular distance between the Sun and the Moon as we observe them in the sky at the time of quarter Moon (angle *SOM*) is 87°. Aristarchus, writing before the degree was in use among the Greeks, says "less than a quadrant by one-thirtieth of a quadrant," that is, less than 90° by 3°, which is 87°. Thus, angle *OSM* must be one thirtieth of a quadrant, or 3°. Aristarchus, working before the invention of trigonometry, proves by one geometrical construction, following the methods of Euclid, that side *OS* is more than 18 times greater than side *OM*. By another construction he proves that *OS* is less than 20 times *OM*. We can apply trigonometry to solve the problem much more easily:

$$OM = OS \sin(OSM) = OS \sin(3°)$$

Thus, $OS = 19.1 \, OM$. The Sun is nineteen times farther away than the Moon.

Relative Sizes of the Sun and Moon   Aristarchus next uses the fact that the Sun and Moon have the same angular diameter, as is clear from the phenomenon of the total solar eclipse (see fig. 1.43). According to Aristarchus, the Moon exactly covers the disk of the Sun. There is no ring of uncovered Sun, which proves that the angular diameter of the Moon is not smaller than that of the Sun. Moreover, the total eclipse does not last for any appreciable time, which proves that the angular diameter of the Moon is not greater than that of the Sun. Therefore, in figure 1.43, half the Moon and half the Sun subtend the same angle α. So, if *OW* is between 18 and 20 times *OU*, then *WX* is between 18 and 20 times *UV*. In other words, the Sun is between 18 and 20 times larger than the Moon.

Absolute Sizes of the Sun and Moon   Aristarchus next works out the absolute sizes of the Sun and Moon, in terms of the size of Earth. His demonstration is based on a diagram for a lunar eclipse (fig. 1.44). We shall follow a simpler geometrical argument than the one Aristarchus gives and shall also take advantage of modern trigonometry. But the basic method and the final results are Aristarchus's.

As a preliminary, we introduce the concept of *horizontal parallax*. Refer to figure 1.45. An observer at *A* on the surface of the Earth sees a celestial body *B* (the Moon, say) on his horizon. A fictitious observer at the center *C* of the Earth would see *B* a little higher in the sky. The angular difference between the two lines of sight is called the horizontal parallax, marked *P* in the figure. The distance *d* of the object from the center of the Earth is related in a simple way to the horizontal parallax:

$$\sin P = r/d,$$

where *r* is the radius of the Earth. Thus, if the horizontal parallax of the object is small, the distance of the object is great. (The horizontal parallax should not be confused with the annual parallax mentioned earlier. The horizontal parallax involves shifts in our point of view as we move about on the Earth. The annual parallax involves shifts in our point of view as the Earth moves around the Sun.)

Now let us take up the eclipse diagram (fig. 1.44). *GH* is the path of the Moon through the shadow during a lunar eclipse. σ and τ are the angular radius of the Sun and of the shadow, respectively. For an observer at *A*, both the edge of the Sun and the edge of the shadow are on the horizon. Thus, P_S is the horizontal parallax of the Sun and P_M is the horizontal parallax of the Moon.

In figure 1.44, σ + τ = 180° − angle *XCH*. But the three angles in triangle *XCH* must add to 180°; thus, $P_M + P_S = 180° −$ angle *XCH*. Combining the two results,

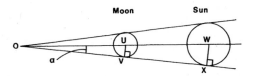

FIGURE 1.43. Aristarchus's use of the solar eclipse.

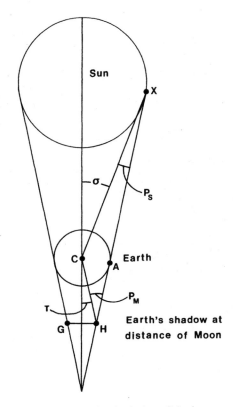

FIGURE 1.44. Aristarchus's use of the lunar eclipse.

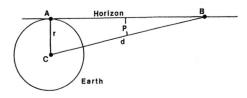

FIGURE 1.45. Horizontal parallax. Angle *P* is the horizontal parallax of a celestial object (such as the Moon) located at *B*.

$$\sigma + \tau = P_M + P_S.$$

This is the relation on which we will base the remaining work.

We know the Sun is about 19 times farther away than the Moon is. (We will dispense from now on with Aristarchus's "greater than eighteen . . . but less than twenty times.") It follows that the Moon's horizontal parallax is about 19 times greater than the Sun's. That is, $P_M = 19\ P_S$. So,

$$\sigma + \tau = 20\ P_S.$$

By hypothesis 5, the breadth of the Earth's shadow is two Moons. From figure 1.29, this seems plausible. Thus, τ = one angular diameter of the Moon. According to hypothesis 6, the Moon's angular diameter is one-fifteenth of a sign (a sign being 30°). Thus, $\tau = 2°$.

Since Aristarchus argues that the Sun and the Moon have the same angular diameter, σ (which is half the Sun) = 1°.

When these values for σ and τ are substituted into the preceding equation, we obtain a numerical value for the Sun's parallax:

$$P_S = \frac{3}{20}^{\circ}$$

The Moon's parallax is 19 times larger:

$$P_M = \frac{57^{\circ}}{20} = 2\ \frac{17^{\circ}}{20}$$

The distances of the Sun and Moon may now be calculated from the parallaxes:

$$d_S = r/\sin P_S; \quad d_M = r/\sin P_M.$$

As before, r denotes the radius of the Earth. Thus,

$$d_S = 382 \text{ Earth radii}; \quad d_M = 20.1 \text{ Earth radii}.$$

These are the absolute distances of the Sun and Moon. Aristarchus does not actually give values for these distances. But these are the values that result from his premises.

To find the actual diameter of the Sun, note that in figure 1.43,

$$\text{diameter of Sun} = 2\ (WX) = 2\ OW \sin \alpha = 2\ d_S \sin \alpha.$$

According to Aristarchus, $\alpha = 1°$, so

$$\text{diameter of Sun} = 2\ (382 \text{ Earth radii}) \sin 1° = 13.3 \text{ Earth radii, or}$$
$$= 6.67 \text{ Earth diameters.}$$

Aristarchus's actual result, quoted above, is that the diameter of the Sun is between 19/3 and 43/6 Earth diameters, that is, between 6.33 and 7.17 Earth diameters. In a similar way, we may calculate from Aristarchus's data that

$$\text{diameter of Moon} = 0.351 \text{ Earth diameters.}$$

Aristarchus's actual result is that the diameter of the Moon is between 43/108 and 19/60 Earth diameters, that is, between 0.398 and 0.317 Earth diameters.

To sum up, Aristarchus found that the Sun is about 19 times farther away than the Moon, that the diameter of the Sun is about 6.67 Earth diameters, and that the diameter of the Moon is about 0.351 Earth diameters.

Critique of Aristarchus

Aristarchus's demonstrations were a brilliant application of mathematics to a cosmological problem. The results, in their general drift, are also admirable: the Sun and Moon are very far away; the Moon is a bit smaller than the Earth; the Sun is considerably larger than the Earth. While earlier philosophers had *speculated* about the sizes of the Sun and Moon, Aristarchus showed that they could be *measured*.

However, there are some puzzles surrounding Aristarchus's data. Most glaring is his use (hypothesis 6) of 2° for the angular diameter of the Moon. In fact, the Moon is four times smaller than this—about 1/2°. It takes little effort to get a good value for the angular diameter of the Moon. It seems that Aristarchus made no measurement at all, but simply made this figure up for the purposes of demonstration. In Greek astronomy of the third century B.C., the *method* was still considered more important than the actual numbers. Interestingly, Archimedes tells us in the *Sand Reckoner* that Aristarchus "discovered that the Sun appeared to be about 1/720th part of the circle of the zodiac," that is, 1/2°.[106] Thus, Aristarchus may actually have made a measurement, presumably after having written his treatise.

The statement (hypothesis 5) that the Earth's shadow is exactly twice as wide as the Moon is, by contrast, a good round value. However, the shadow is actually a bit wider. In *Almagest* V, 14, Ptolemy says that the shadow is 2 3/5 times the width of the Moon. The reader can make an independent estimate by using figure 1.29.

Let us see what results from these two improved values. We leave all of Aristarchus's other hypotheses unchanged. σ (half the angular diameter of the Sun) is then 1/4°. The whole diameter of the shadow is 2 3/5 × 1/2° = 13/10°. Thus, τ (half the shadow) is 13/20°. If we put these values for σ and τ into the fundamental equation (and continue to assume that $P_M = 19\ P_S$), we find

Distance of Sun   = 1,273 Earth radii.
Distance of Moon =   67 Earth radii.
Diameter of Moon = 0.292 Earth diameters.
Diameter of Sun =   5.55 Earth diameters.

The distance and diameter of the Moon are now close to the truth. (The actual mean distance of the Moon is about 60 Earth radii. The actual diameter of the Moon is about 0.273 Earth diameters.) But the size and the distance of the Sun are still too small by a factor of 20. Thus, while Aristarchus's method could, with a little effort at more realistic measurements, yield good values for the size and distance of the Moon, it was incapable of yielding good values for the size and distance of the Sun.

The problem was hypothesis 4, that the angle between the Sun and the quarter Moon is 87°. This leads to the conclusion that the Sun is 19 times farther from us than the Moon is. In fact, the Sun is about 389 times farther from us than the Moon is. From this it follows that, at quarter Moon, angle *SOM* in figure. 1.42 is 89°51′—less than a right angle by only 9′. Even several centuries later, when Greek observational astronomy reached its peak, nobody could measure angles to a precision of 9′. Moreover, the required measurement was a very difficult one. Many historians have pointed out that it is impossible to judge the exact moment when the Moon reaches quadrature, and that it is also difficult to measure the angular distance between the centers of extended

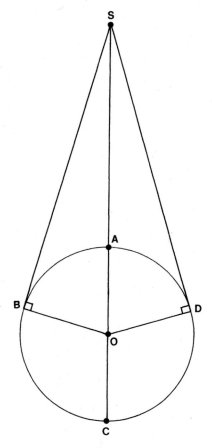

FIGURE 1.46. The Moon's path about the Earth O in the course of a synodic month. The Sun is at S.

bodies like the Sun and Moon. The brightness of the Sun posed additional problems—as well as risks for the vision of the observer.

However, hypothesis 4 was probably not the result of any measurement at all. Rather, Aristarchus simply made up the value *less than a quadrant by one-thirtieth of a quadrant*. This was probably a conjectural value based on the following considerations. The time from new Moon to new Moon is about 30 days. In figure 1.46, O is the Earth, and S is the Sun. A, B, C, and D represent the Moon's positions at new Moon, first quarter, full Moon, and third quarter, respectively. It is clear that arc BCD is greater than arc DAB, and thus that the month is divided into unequal parts by the quarter Moons. These two parts of the month cannot be as different as 16 days and 14 days, for then the excess would be apparent to us all. Let us suppose, therefore, that the time from first quarter to third quarter (arc BCD) is 15 1/2 days and that the time from third quarter to first quarter (DAB) is 14 1/2 days. The Moon then runs arc DA in one half of 14 1/2 days, that is, in 7 1/4 days. But the Moon takes 7 1/2 days to run a quadrant of the circle (since it takes 30 days for the whole circle). Thus, arc DA is less than a quadrant by 1/4 day's worth of motion. One day's worth of motion is 4/30 of a quadrant (4 quadrants in 30 days). Thus, 1/4 day's worth of motion is 1/30 of a quadrant. In short, *arc DA is less than a quadrant by 1/30 of a quadrant*. It looks as if Aristarchus arrived at hypothesis 4 by simply assuming the *largest imperceptible inequality* between the two portions of the month.

Aristarchus's successors soon arrived at accurate values for the size and distance of the Moon. But none of them made any substantial improvement on the values for the size and distance of the Sun. Why not? The answer lies in a property of the relation

$$\sigma + \tau = P_M + P_S,$$

derived from the eclipse diagram. The left side of this equation contains two quantities (angular radii of the Sun and shadow) that are relatively easy to measure. The right side contains one quantity (horizontal parallax of the Moon) that is hard to measure and one (horizontal parallax of the Sun) that actually was impossible to measure by the methods available to the Greek astronomers. Only the *sum* of the two parallaxes is easily determined: it must be equal to $\sigma + \tau$. The fundamental problem is deciding how to divide the total of the parallaxes between the Sun and Moon. Some other fact of observation must be introduced. Aristarchus used a value for the separation of the Sun and Moon at quarter Moon (fig. 1.42) to argue that, in modern langauge, $P_M = 19P_S$. But, suppose we give all of the total parallax to the Moon and nothing to the Sun. That is, suppose we assume that $P_S = 0$ and that $P_M = \sigma + \tau$. This only makes about a 5% difference in the Moon's parallax and thus brings the Moon about 5% closer to the Earth. But giving the Sun a parallax of zero pushes the Sun out to infinity. So we can make huge changes in our estimate of the Sun's distance without affecting the Moon's distance very greatly.

Later Measurements

Aristarchus's eclipse diagram remained a central feature of later efforts to improve the values for the sizes and distances of the Sun and Moon. However, the later astronomers, notably Hipparchus and Ptolemy, sensibly abandoned the method of the lunar quadrature. This meant they were obliged to introduce an additional fact of observation.

Hipparchus (second century B.C.) wrote a work on the sizes and distances, but it has not come down to us. However, some information about Hipparchus's method is preserved in Ptolemy's *Almagest* and in the commentary on the *Almagest* written by Pappus (third century A.D.). In one calculation, Hipparchus attempted a direct assessment of the Moon's parallax. He made use of information about a solar eclipse (probably that of March 14, 189 B.C.): the Sun was seen totally eclipsed near the Hellespont, but only four-fifths of its diameter was eclipsed at Alexandria. *Assuming then that the Sun's parallax was zero*, Hipparchus was able to take one-fifth of the Sun's angular diameter as the lunar parallax between the Hellespont and Alexandria. We do not know the details of his procedure, but he arrived at 71 and 83 Earth radii as the Moon's least and greatest distances.[107] (The distance of the Moon varies slightly in the course of the month.)

In another calculation, Hipparchus made use of the eclipse diagram and simply *assumed* that the horizontal parallax of the Sun was 7′, which he perhaps took to be the largest imperceptible parallax. He took the angular diameter of the Moon at mean distance to be 1/650 of a circle (33′14″), and the diameter of the shadow to be 2 1/2 times that of the Moon. This leads to a figure of 67.2 Earth radii for the Moon's mean distance. Hipparchus's methods allowed him to home in on sound values for the distance of the Moon, and in this he made a considerable improvement over Aristarchus. But it is clear that he had doubts about the possibility of obtaining a reliable value for the Sun's distance.

Ptolemy, in *Almagest* V, devotes a good deal of effort to the sizes and distances. Ptolemy attempts a direct measurement of the Moon's parallax by comparing the Moon's position as observed at Alexandria with a theoretical value for the Moon's position computed from his lunar theory. With the Moon's parallax thus in hand, together with his own values for the angular diameters of the Moon, Sun, and shadow, he uses the method of the eclipse diagram to get the distance of the Sun. Ptolemy's results are

Mean distance of Moon at new or full Moon =		59 Earth radii.
Mean distance of Sun	=	1,210 Earth radii.
Diameter of Moon	=	0.292 Earth diameters.
Diameter of Sun	=	5.5 Earth diameters.

Ptolemy shows much originality. For example, he attempts to improve on the measurement of the angular size of the shadow by means of a clever technique of comparing two lunar eclipses of different degrees of totality. Nevertheless, when he is all done, his ratio of solar to lunar distance is 1210/59 = 20.5, very close to the traditional ratio handed down by Aristarchus. Ptolemy's figures were destined to have a great influence. There were a few minor adjustments by medieval writers, but Ptolemy's values were never substantially changed. The 20-to-1 ratio between the Sun's and the Moon's distance was not called seriously into question until the seventeenth century.

1.18 EXERCISE: THE SIZES AND DISTANCES OF THE SUN AND MOON

1. Measure the angular diameter of the Moon, using the method described in section 1.16. Assume that the angular diameter of the Sun is the same as that of the Moon. This is justifiable since, at the time of a solar eclipse, the Moon often appears to just cover the Sun's disk. Aristarchus assumed the same.

2. Use the photograph of the lunar eclipse of July 16, 1981 (fig. 1.29), to determine how many times larger than the Moon the shadow is. To do this, use a drawing compass. Find, by trial and error, the center of the shadow. Measure with a ruler on the photo the diameter of the shadow and of the Moon.

3. Take Aristarchus's result, that the Sun's distance is 19 times the Moon's distance, as given. Use the method of the eclipse diagram ($\sigma + \tau = P_S + P_M$) to determine the solar parallax:

   A. First, take σ from your own measurement of the Moon's angular diameter in step 1, by assuming that the angular diameter of the Sun is the same as that of the Moon.
   B. Combine your results from steps 1 and 2 to determine τ.
   C. Note that $P_M = 19\, P_S$ according to Aristarchus.
   D. Finally, find P_S.

4. Use the result of step 3 to determine the Sun's distance, in terms of Earth radii.
5. Find the Moon's distance in terms of Earth radii.
6. Find the actual diameter of the Sun in terms of Earth diameters. To do this, combine the results of steps 1 and 4.
7. Find the actual diameter of the Moon in terms of Earth diameters.

TWO

The Celestial Sphere

Basis in Observation

To a naive observer it is by no means obvious that the sky has the shape of a dome or hemisphere. Indeed, the Egyptians in their art often represented the sky by the sky goddess, Nut, arched over the land, often, though not always, with her back flattened (see fig. 2.1). And the Egyptian hieroglyph for *sky* (⊓) is reminiscent of the flat roof of a long, low building.[1]

Although the stars themselves suggest no particular shape for the heaven, their motions do suggest a sphere: the Moon, stars, and planets are seen moving on parallel circles, climbing up together from the eastern horizon, crossing the sky, and going down together in the west. Even more suggestive are the circumpolar stars, which can be seen all night long, moving in circles about the celestial pole. Ptolemy pointed to exactly these facts of observation in trying to explain how his remote predecessors had come to the idea of a spherical heaven.[2]

"*The heaven is spherical and moves spherically.*"[3] This is the most fundamental assumption of Greek astronomy. Many Greek astronomical texts begin with it or something like it.[4] Although this view often was supported with arguments of a philosophical or even mystical nature, it is actually suggested by observation of the sky. Aristotle regarded the sphere of the fixed stars as a real, material sphere, and all later astronomical writers, including Ptolemy, followed him in this.

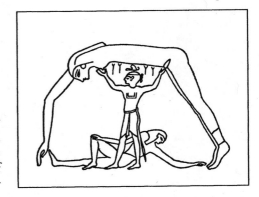

FIGURE 2.1. The Egyptian sky goddess, Nut. *Top*: Nut supported by the air god, Shu. Beneath them lies the earth god, Seb. *Bottom*: Shu supporting the boat of the sun god beneath the sky goddess, Nut. From Budge (1904).

Eudoxus and Aratus on the Sphere

The idea of a spherical cosmos can be attributed to sixth- and fifth-century B.C. philosophers such as Pythagoras and Parmenides. But the first figure in whom we see a clear and complete understanding of the celestial sphere is Eudoxus of Cnidus (ca. 370 B.C.). Eudoxus was the author of a number of astronomical works, including a star calendar and a treatise on the eight-year luni-solar cycle. He is the likely, if unproved, source of the earliest known measurement of the circumference of the Earth, mentioned in passing by Aristotle in *On the Heavens*. Among his other writings, Eudoxus is known to have composed two books on the celestial sphere, called the *Phenomena* and the *Mirror*. These books, which apparently differed little from one another, contained systematic descriptions of the constellations and their relative positions on the sphere.

Not one of Eudoxus's works has survived. But in the case of the *Phenomena* we have a paraphrase of one of them, for Eudoxus's description of the night sky inspired the poet Aratus of Soli to produce a versified version about a century later. The verse *Phenomena* of Aratus proved to be very popular: commentaries were written on it, it was on several occasions translated into Latin, and it even inspired sculptors and other artists to treat astronomical themes. Among the many commentators on Aratus was Hipparchus of Bithynia (ca. 150 B.C.), the most creative astronomer of the Hellenistic age. Hipparchus still had access to Eudoxus's original prose *Phenomena*. In his *Commentary* Hipparchus makes a painstaking examination of the works of his two predecessors. Often, he finds them inexact or mistaken about positions of the stars and constellations. Hipparchus tells us, however, that we ought not blame Aratus, who was a poet and not an astronomer, and who was in any case only following Eudoxus, but Eudoxus, as an astronomer, must be held accountable for the errors.[5] Hipparchus makes Aratus's dependence on Eudoxus clear by quoting parallel passages from the two versions of the *Phenomena*. Thus, although Eudoxus's *Phenomena* has not come down to us, we may safely assume that its astronomical content is reflected in the poem of Aratus.

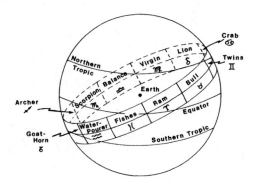

FIGURE 2.2. The celestial sphere and the signs of the zodiac.

Symbol	Greek	Latin	English
♈	Krios	Aries	Ram
♉	Tauros	Taurus	Bull
♊	Didymoi	Gemini	Twins
♋	Karkinos	Cancer	Crab
♌	Leōn	Leo	Lion
♍	Parthenos	Virgo	Virgin
♎	Zygos	Libra	Balance
♏	Skorpios	Scorpius	Scorpion
♐	Toxotēs	Sagittarius	Archer
♑	Aigokerōs	Capricornus	Goat-Horn[ed creature]
♒	Hydrochoös	Aquarius	Water-Pourer
♓	Ichthyes	Pisces	Fishes

Aratus begins with descriptions of the constellations and their positions on the sphere, along with stories and legends about them. Then he briefly describes the four principal circles of the celestial sphere: equator, zodiac, and the two tropic circles.[6] The image that emerges can be represented as in figure 2.2. Girding the celestial sphere are the three parallel circles of the equator and the two tropic circles. The fourth important circle is the slanted zodiac, which lies athwart the tropics. Although the Greek heaven of figure 2.2 still bears mythological images, it represents a radical break with traditional cosmologies, typified by the Egyptian images in figure 2.1. In introducing the theory of the celestial sphere, the Greeks took a decisive step toward geometrizing their worldview.

Aratus goes on to list the constellations that lie on each of the four circles. The northern tropic circle passes through the heads of the Twins (Gemini) and the knees of the Charioteer (Auriga), passes just below Perseus, but straight across Andromeda's right arm above the elbow. Also lying on the northern tropic are the hoofs of the Horse (Pegasus), the head and neck of the Bird (Cygnus), and the shoulders of Ophiuchus. The Virgin (Virgo) is a little south of the tropic and does not touch it, but both the Lion and the Crab (Cancer) are squarely on it. This detailed description would enable the reader to visualize the tropic of Cancer in the night sky. Aratus gives similar lists of the constellations lying on the equator and tropic of Capricorn. For the zodiac—the fourth major circle—the list consists of the familiar twelve zodiacal constellations.

A note on terminology: it is important to distinguish between *zodiac* and *ecliptic*. The Greek astronomers thought of the zodiac as a band of finite width, as in figure 2.2, rather than as a vanishingly thin circle. The circle that runs down the middle of this zodiacal band is the ecliptic (the Sun's path), which the Greeks called *the circle through the middles of the signs*. The Moon and the planets move nearly along the ecliptic, but the Moon may wander north or south of it by as much as 5°. The maximum latitudinal wanderings of the planets range from about 2° in the case of Jupiter to about 9° in the case of Venus. The zodiac was conceived of as a band wide enough to encompass these wanderings.

Fundamental Propositions of Greek Astronomy

From the time of Eudoxus on, Greek astronomy was based on five fundamental propositions:

1. The Earth is a sphere,
2. which lies at the center of the heaven,
3. and which is of negligible size in relation to the heaven.
4. The heaven, too, is spherical
5. and rotates daily about an axis that passes through the Earth.

We have discussed propositions 4 (sphericity of the heaven, in the present section) and 1 (sphericity of the Earth, in sec. 1.9). In section 1.6, we examined the ancient debate over proposition 5 (rotation of the heaven). In considering the two remaining propositions, we will examine some of the arguments offered by Ptolemy in *Almagest* I.

That Earth Is in the Middle of the Heaven Suppose, says Ptolemy, that the Earth is not at the center of the celestial sphere. Then it is either

(a) off the axis of the sphere but equidistant from the poles,
(b) on the axis but farther advanced toward one of the poles, or
(c) neither on the axis nor equidistant from the poles.

Let us examine case (a). In figure 2.3, the Earth lies off the axis of the celestial sphere, but at equal distances from the two celestial poles. In this case there will be trouble with the equinoxes. Let an observer be at *A* on the Earth's equator, with horizon *YAW*. At the time of the equinox, the Sun lies on the celestial equator and therefore runs around circle *WXYZ* in the course of one day. The observer at *A* will see the Sun above the horizon only for the short time the Sun requires to run arc *YZW*, and the Sun will be below for the long time it takes to travel arc *WXY*. But this contradicts the observed fact that, at equinox, the period of daylight is equal to the period of darkness at all places on Earth.

Now consider case (b), in which the Earth is on the axis of the universe but nearer one of the poles. Then everywhere (except at the Earth's equator) the plane of the horizon will cut the celestial sphere into unequal parts, which is contrary to observation, since one half of the sphere is always found above the horizon (fig. 2.4). And it is not possible to advance to case (c) since the objections to (a) and (b) would apply here also.

That the Earth Is a Mere Point in Comparison with the Heaven   In the first place, says Ptolemy, if the Earth had an appreciable size compared with the celestial sphere, the same two stars would appear, to observers at different latitudes, to have different angular separations. For example, in figure 2.5, observers at *D* and *E* will measure different angular separations between the stars *F* and *G*. That is, angles *FDG* and *FEG* are not the same. Further, star *G* will appear brighter to the observer at *E* than to the observer at *D*. But all of this is in contradiction to the facts, for the stars actually appear the same in the different latitudes.

Second, the tips of shadow-casting gnomons can everywhere play the role of Earth's center, which could not be the case if the Earth had any appreciable size. For example, as in figure 2.6, let gnomon *AB* be perpendicular to the terrestrial meridian *CG*. At noon on the winter solstice, the Sun is at *H* and produces the shadow *BD*; at noon on the equinox, the Sun is at *I* and produces shadow *BE*; and finally, at summer solstice, the Sun, at *J*, produces shadow *BF*. Now, at any place whatever on the Earth it is found that angle *JAI* = angle *IAH*, roughly 24°. Thus, the tip *A* of the gnomon may always be taken as the center of the sphere of the Sun's motion. But if this is true everywhere, the Earth must be very small compared to the celestial sphere. The fact that, for any place on Earth, the tip of the gnomon can be treated as the center of the cosmos would have been familiar to any of Ptolemy's readers who had studied the techniques of constructing sundials. We make use of this fact ourselves, in section 3.2, where we study the construction of Greek and Roman sundials.

We have expanded some of Ptolemy's arguments and illustrated them with figures for the sake of greater clarity. These arguments were not, however, original with Ptolemy, since some of them were used by earlier writers, for example, Euclid and Theon of Smyrna. Indeed, the essential arguments concerning the heaven and the Earth's place within it were already hundreds of years old by the second century A.D., when Ptolemy wrote. Ptolemy merely presented the case with greater thoroughness and organization. These arguments remained the common stock of all astronomers down to the Renaissance.

Critique of the Ancient Premises   In general, the Greek astronomers believed in the literal truth of all five propositions. In an introductory astronomy course, the teacher would probably have marched his students through the five propositions, giving ample proofs of each, the proofs being based not only on appeals to observation but also on physical and philosophical argument.

But from our perspective, while some of these propositions may be regarded as rigorously proved, others only reflect a point of view. In particular, propositions 1 and 3 (sphericity and smallness of the Earth) are not only provable but actually were proved in antiquity. Proposition 2, which places the Earth

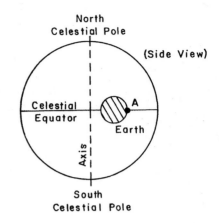

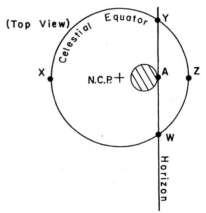

FIGURE 2.3.

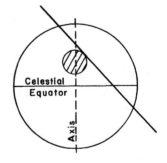

FIGURE 2.4.

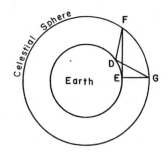

FIGURE 2.5.

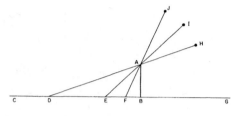

FIGURE 2.6.

at the center of the heaven, rests partly on empirical evidence and is partly conventional. Certainly, the axis of the daily rotation must pass through the Earth. But, granted this, as long as the stars are very far away, it can make no difference whether the Earth is exactly in the middle of things or not. Every observer, whether on the Earth, the Moon, or Jupiter, can legitimately treat his or her own home as the center of the universe (as far as appearances are concerned). Proposition 4, that the heaven is spherical, is wholly conventional. Because the stars are very far away from us, it makes no difference whether they all lie on a single spherical surface or not. But we will not get into trouble by assuming that they do. Proposition 5 also reflects a point of view. We may say with equal validity that the heaven rotates once a day from east to west or that the Earth rotates from west to east.

2.2 SPHAIROPOIIA: A HISTORY OF SPHERE-MAKING

Some Representative Globes and Armillary Spheres

The most ancient known celestial globe is a large stone sphere supported by a statue of Atlas, in the Museo Nazionale at Naples. This statue, transferred to its present location from the Farnese Palace in Rome, is called the Farnese Atlas. The globe is a Roman copy (first or second century A.D.) of a Hellenistic original made perhaps several centuries earlier.

The Farnese globe is shown in figure 2.7. The Earth, not represented, would be a tiny sphere located inside the celestial sphere. Part of the globe is obscured by the hand of the statue of Atlas that supports it. In figure 2.7,

FIGURE 2.7. The Farnese globe. This ancient marble celestial sphere was supported by a statue of Atlas, whose hand is visible on the globe. From G. B. Passeri, *Atlas Farnesianus . . .* , in Antonio Francesco Gori, *Thesaurus gemmarum antiquarum astriferarum*, Vol. III (Florence, 1750). By permission of the Houghton Library, Harvard University.

the zodiac belt is the triplet of rings arching across the upper part of the globe. (The ecliptic is the middle of the three rings.) On the zodiac are several familiar constellations: (1) Taurus, (2) Gemini, (3) Cancer, and (4) Leo. A number of nonzodiacal constellations may also be seen: (6) Canis Major, (7) Argo, (8) Hydra, (9) Crater, (10) Corvus, and (15) Auriga. Celestial circles represented on the Farnese globe include the celestial equator *CD*, the tropic of Capricorn *EF*, the tropic of Cancer *OK*, and the solstitial colure *AB*. (The solstitial colure is a great circle that passes through the celestial poles and the summer and winter solstitial points.)

In the placement and representation of the constellations, the globe is consistent with the descriptions of the sky in the *Phenomena* of Aratus. For example, the constellation Hercules is, in Aratus, simply called the Kneeling Man. (The identification with the hero came after Aratus's time.) Hercules is usually depicted carrying a club and a lion's skin, which is not the case with the Kneeling Man on the Farnese globe.

The Farnese globe was, of course, a display piece and not a usable globe. Figure 2.8 shows one of the oldest known portable globes, from medieval Islam.[7] Although no portable globe (of the type suitable for teaching) has come down to us from Greek times, we know that they were fairly common. Geminus, the author of an introductory astronomy textbook (*Introduction to the Phenomena*, first century A.D.), refers to celestial globes in several passages and clearly expected his readers to be familiar with them. Moreover, globes appear in Greek and Roman art, for example, on coins and on murals. (See fig. 5.13 for a coin from Roman Bithynia that shows Hipparchus seated before a small celestial globe.)

Ptolemy (*Almagest* VIII, 3) gives detailed directions for building a celestial globe. He says it is best to make the globe of a dark color, resembling the night sky, and gives directions for locating the stars on it. The stars are to be yellow, with sizes that correspond to their brightnesses. A few stars, for example, Arcturus, that appear reddish in the sky, should be painted so. The globe described by Ptolemy was of unusual sophistication, for it was fitted with a stand that allowed the user to duplicate not only the daily rotation about the poles of the equator, but also the slow precession about the poles of the ecliptic.

Similar to the celestial globe, but easier to construct, is the armillary sphere, in which the heavens are represented not by a solid ball but by a few rings or bands which form a kind of skeleton sphere. ("Armillary" from the Latin *armilla*, arm-band, bracelet.) This model emphasizes the various circles in the sky that are associated with the Sun's motion. Figure 2.9 shows a Renaissance illustration of an armillary sphere.

FIGURE 2.8. An Arabic celestial globe (Oxford, Museum of the History of Science). The stars are represented by inlaid silver disks, with sizes corresponding to the magnitudes of the stars. The globe is pierced by holes at the poles of the equator and at the poles of the ecliptic. (There is a third pair of holes whose function is not obvious. Perhaps they were drilled by mistake.) A series of holes in the stand permits adjustment of the axis of rotation for geographical latitude at increments of 10°. The inscription informs us that the globe was made in A.H. 764 (A.D. 1362/ 1363) and that the maker of the globe took the star positions from the *Book of the Constellations* of al-Ṣūfi. Ursa Major may be seen, upside down, near the middle of the globe. The Pointers point at the middle hole, which is the pole of the equator. (Compare with fig. 1.15.)

Uses of the Globe

With either a globe or an armillary sphere it is possible to reproduce a variety of astronomical events—the risings and settings of stars, the annual solar cycle, and so on. One can make apparent in a moment what would require months to observe in the sky, so the models can be used to supplement, or even replace, real observation in the teaching of astronomy.

Even after the development of spherical trigonometry (by the end of the first century A.D.), concrete models continued to serve as aids to visualization and understanding. Indeed, if one desires only numerical answers (rather than mathematical formulas), and if one does not insist that these numbers be very precise, one can perform all the trigonometric "calculations" necessary to astronomy by manipulating a concrete model. A well-made celestial globe or armillary sphere is a kind of analog computer.[8]

These models were also aids in the discovery of the world. Many facts about the Earth are read directly on the celestial globe: the existence of a

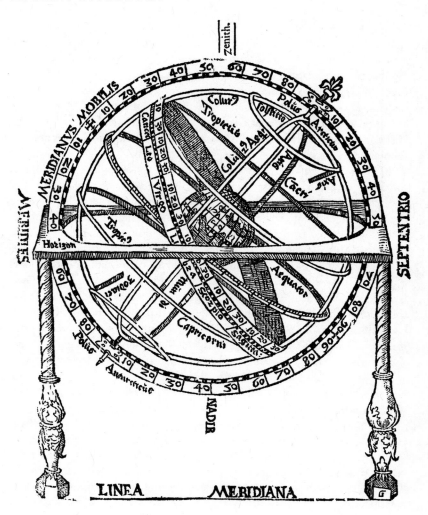

FIGURE 2.9. An armillary sphere. From *Cosmographia . . . Petri Apiani & Gemmae Frisii* (Antwerp, 1584). Courtesy of the Rare Book Collection, University of Washington Libraries.

midnight Sun in the extreme northern and southern latitudes, "days" of six months at the poles, the existence of a tropical zone in which the Sun sometimes stands at the zenith, and the reversal of the seasons in the southern hemisphere. All these facts of geography can be demonstrated with the celestial globe. These facts about the Earth were discovered through thought and not by exploration.

Armillary spheres were common teaching tools in Greek antiquity, and they are mentioned as such by Geminus. But if an armillary sphere is made well enough, and large enough, and equipped with sights, it can also function as an instrument of observation. It can be used, for example, to measure the celestial coordinates of stars or planets in the night sky. By Ptolemy's time, the armillary sphere had become the preferred instrument of the Greek astronomers. (See fig. 6.8 for the instrument described by Ptolemy in *Almagest* V, 1.)

History of Model-Making

According to Sulpicius Gallus,[9] Thales of Miletus (sixth century B.C.) was the first to represent the heavens with a sphere. This would have been a solid sphere on which stars were marked. Whether this sphere was made to turn about an axis we do not know. Indeed, because most of what we know of Thales is mere rumor and legend, it is far from certain that the celestial globe really originated with him. Among the ancients, Thales' name was a catchword for wisdom and learning and many discoveries were attributed to him that really were made much later.

In any case, by the time of Plato (fourth century B.C.) such models must have been fairly common. When Plato described the creation of the universe by the craftsman-god in his *Timaeus*, he had in mind the physical image of the universe provided by the armillary sphere. According to Plato, the craftsman-god first of all prepared a fabric from which he intended to construct the world, and this fabric was made of world-soul. The craftsman-god

> then took the whole fabric and cut it down the middle into two strips, which he placed crosswise at their middle points to form a shape like the letter X; he then bent the ends round in a circle and fastened them to each other opposite the point at which the strips crossed, to make two circles, one inner and one outer. And he endowed them with uniform motion in the same place, and named the movement of the outer circle after the nature of the Same, of the inner after the nature of the Different. The circle of the Same he caused to revolve from left to right, and the circle of the Different from right to left on an axis inclined to it; and he made the master revolution that of the Same.[10]

"Motion in the same place" means circular motion. The two intersecting circles are, of course, the equator and the ecliptic. The daily motion from east to west, shared by all the heavenly bodies, is the "master revolution," or the revolution "of the Same," and is associated with the equator. The ecliptic partakes of the nature of the Different because the Sun, Moon, and planets all tend to move in the contrary direction—from west to east—along this circle. There is no doubt, then, that Plato's conception of the universe owed something to the concrete example of the armillary sphere. This is perhaps the earliest example we have of something that has since become commonplace: a successful scientific model or theory may affect our picture of the world and cause shifts in religion and philosophy.

Farther on in the same discussion, Plato mentions the creation of the planets and the motions with which god has endowed them. But he forswears any detailed explanation of these motions, saying, "It would be useless without a visible model to talk about the figures of the dance [of the planets]," which again makes one think that models were in use by Plato's time.

Eudoxus of Cnidus sought to explain this dance of the planets by a system of nested spheres, turning about several different axes inclined to one another. He was able in this way to reproduce fairly well the variations in speed, the stationary points, and the retrogradations that are characteristic of the planets' motions. Whether he made a concrete model to illustrate his theory is not known. Eudoxus, who was a mathematician of the first order, would not have needed mechanical aids, but such a model might have made discussion with others easier. If it existed, the model of Eudoxus would have been the first orrery. Such a device, which duplicates the motions of the planets, is much more complicated than a globe or armillary sphere, which merely reproduces the daily revolution of the celestial sphere.

In any case, *sphairopoiïa* ("sphere making")—the art of making models to represent the celestial bodies and their motions—soon became an established branch of mechanics and was carried to a high level by the time of Archimedes (ca. 250 B.C.). According to Plutarch,[11] this brilliant mathematician repudiated as sordid and ignoble the whole trade of mechanics and every art that lent itself to mere use and profit. Archimedes is famous for inventing machines of all kinds—water screws, hoisting machines, and engines of war—but these he is supposed to have designed not as matters of any importance but as mere amusements in geometry. And so Archimedes did not "deign to leave behind him any commentary or writing on such subjects" but "placed his whole affection and ambition in those purer speculations where there can be no reference to the vulgar needs of life." Yet, he seems to have made an exception in the case of sphere making, perhaps because it helps one attain

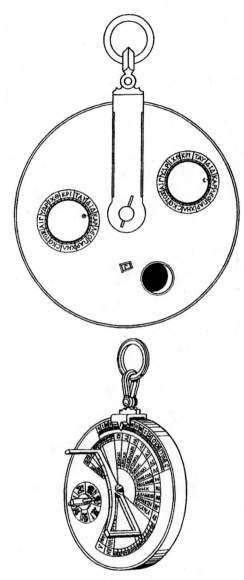

FIGURE 2.10. A. A portable sundial and gear-work calendrical calculator from the Byzantine period (circa A.D. 500). *Top*: Conjectural reconstruction of the back. A dial at the left indicates the position of the Sun in the zodiac. A dial at the right indicates the position of the Moon. The window at the bottom indicates the phase of the Moon. *Bottom*: Reconstruction of the front. The suspension ring must be positioned for the latitude of the observer. The sundial vane must be adjusted for the time of year. The user holds the dial vertically and turns it until the shadow of the gnomon falls on the scale of hours engraved on the curved part of the vane, indicating the time of day. The dial at the left is to be turned one notch a day. The gear train inside the device then advances the Sun and Moon indicators by the appropriate amounts. From Field and Wright (1984).

an understanding of divine objects. For, according to Pappus,[12] Archimedes composed a special treatise on this subject, which is now lost. This book on sphere making was the only work on mechanics that Archimedes judged worthwhile to write.

That Archimedes actually made models of the heavens is beyond doubt, for Cicero tells us that after the capture of Syracuse (212 B.C.) the Roman general Marcellus brought two of them back to Rome. One was a solid celestial globe, which Marcellus placed in the temple of Vesta, where all might go and see it. This seems to be the same globe that Ovid mentions in these lines on Vesta and her hall:

> There stands a globe hung by Syracusan art
> In closed air, a small image of the vast vault of heaven,
> And the Earth is equally distant from the top and bottom.
> That is brought about by its round shape.[13]

Ovid's description of the sphere as an image of the heavens with the Earth inside, equally distant from top and bottom, makes it sound more like a hollow armillary sphere than the solid globe described by Cicero. Ovid wrote these lines around A.D. 8, more than 200 years after the globe was brought to Rome. And, although Ovid writes as if the globe still existed in his time, it is possible that it did not and that he never saw it. Besides, Ovid's astronomical knowledge is often defective, so Cicero's description is more to be trusted.[14]

According to Cicero, the second of Archimedes' models was taken home by Marcellus, "though he took home with him nothing else out of the great store of booty captured." Years later, it was shown by Marcellus's grandson to Gaius Sulpicius Gallus, who was evidently one of the few who understood the workings of the machine. This second model,

> on which were delineated the motions of the Sun and the Moon and of those five stars which are called wanderers, or as we might say, rovers, contained more than could be shown on the solid globe, and the invention of Archimedes deserved special admiration because he had thought out a way to represent by a single device for turning the globe those various and divergent movements with their different rates of speed. And when Gallus moved the globe, it was actually true that the Moon was always as many revolutions behind the Sun on the bronze contrivance as would agree with the number of days it was behind in the sky. Thus, the same eclipse of the Sun happened on the globe as would actually happen. . . . [15]

This orrery of Archimedes must have been quite a marvel, for Cicero expresses disapproval of some who "think more highly of the achievement of Archimedes in making a model of the revolutions of the firmament than that of nature in creating them, although the perfection of the original shows a craftsmanship many times as great as does the counterfeit."[16]

Archimedes was not the only master of the art of sphere making, for Cicero also mentions "the orrery recently constructed by our friend Posidonius, which at each revolution reproduces the same motion of the Sun, the Moon and the five planets that take place in the heavens every twenty-four hours." Cicero probably saw this device himself, for as a young man he had attended Posidonius's lectures in Rhodes, and again befriended him when the philosopher came to Rome as ambassador from Rhodes in 87–86 B.C. But, alas, Cicero gives us no details of the construction of this machine.

The orreries of Archimedes and Posidonius were intended primarily to give a visual representation of the universe. But it is clear from Cicero's remarks that these two orreries also incorporated some quantitative features of the planets' motions—at least their relative speeds along the zodiac. Two related kinds of constructions can be mentioned here. One was the simple cosmological model, which did not incorporate any quantitative features, but which gave the viewer an overall visual impression of the arrangement of the universe. For example, Theon of Smyrna[17] tells us that he himself made a model of the nested spindle-whorls cosmos described by Plato in the tenth book of the *Republic*. This model would not have *done* much, but it did illustrate Plato's cosmology in a visually striking way.

Less visual, but more quantitative, was the gearwork calendrical computer. Parts of two such devices have been discovered, one dating from the first century B.C.[18] and one from the fifth or sixth century A.D.[19] The user was expected to turn a wheel through one "click" each day. A gearwork mechanism then advanced indicators showing the phase of the Moon and the position of the Sun in the zodiac (see fig. 2.10).

The Place of Sphairopoiïa among the Mathematical Arts

One should not take Archimedes' disdain for mechanics as representative of his time. By Archimedes' time, mechanics not only was a useful trade, but also had become a recognized genre of technical writing. *Sphairopoiïa*, the subdivision of mechanics devoted to models of the heavens, was also a recognized specialty. Sphairopoiïa included the construction of celestial globes, to be sure. But, as we have seen, it also included the making of other kinds of images of the heavens, such as models of the planetary system and mechanical calculating devices intended to replicate features of the motions of the Sun, Moon, and planets. Two recognized branches of astronomy proper were also devoted to concrete constructions: *gnomonics* (the making of sundials) and *dioptrics* (the design and use of sighting instruments).

The relation of these three arts to the rest of mathematical knowledge is discussed by Geminus, a Greek scientific writer of the first century A.D. Geminus wrote an elementary astronomy textbook (*Introduction to the Phenomena*) that has come down to us more or less intact. He also wrote a large book on mathematics, which contained a good deal of philosophy and history of mathematics. This book has not come down to us. But much of its content is summarized by Proclus in his *Commentary on the First Book of Euclid's Elements*. In his mathematical treatise, Geminus discussed the organization of mathematical knowledge and the relation of its various branches to one another. Geminus's outline of the mathematical sciences can be summarized thus:

Organization of the mathematical sciences according to Geminus

- Pure mathematics (concerned with mental objects only)
  - Arithmetic (study of odds, evens, primes, squares)
  - Geometry
    - Plane geometry
    - Solid geometry
- Applied mathematics (concerned with perceptible things)
  - Practical calculation (analogous to arithmetic)
  - Geodesy (analogous to geometry)
  - Theory of musical harmony (an offspring of arithmetic)
  - Optics (an offspring of geometry)
    - Optics proper (straight rays, shadows, etc.)
    - Catoptrics (theory of mirrors, etc.)
    - Scenography (perspective)
  - Mechanics
    - Military engineering
    - Wonderworking (pneumatics applied to automata)
    - Equilibrium and centers of gravity
    - Sphere making (mechanical images of the heavens)

FIGURE 2.10. B. Portable sundial and gearwork calendrical calculator from the Byzantine period. *Top:* A modern reconstruction in metal. *Bottom:* The extant portion of the gear train. At the right can be seen the ratchet (the oldest known ratchet), which prevented the user from turning the day dial in the wrong direction. Science Museum, London.

· Astronomy
 · Gnomonics (sundials)
 · Meteoroscopy (general astronomical theory)
 · Dioptrics (instruments of observation)[20]

Mathematical knowledge is divided into the pure (which deals with mental objects only) and the applied (which deals with perceptible things). Astronomy is one among six branches of applied mathematics. Geminus was not alone in making astronomy a part of mathematics, for this was the general view among the Greeks. Ptolemy, for example, always refers to himself as a *mathematician*. Two of Geminus's three subbranches of astronomy are concerned with the construction and use of instruments: *gnomonics* and *dioptrics*. As Geminus says, gnomonics is concerned with the measurement of time by means of sundials, while "dioptrics examines the positions of the Sun, Moon, and the other stars by means of just such instruments [i.e., dioptras]." As for sphere making, Geminus makes it a part of mechanics, no doubt because it involves the use of geared mechanisms and water power to activate its images of the heavens.

Some Reservations about Sphairopoiïa

The purpose of sphairopoiïa was to make immediately evident facts that could otherwise demonstrated only by difficult geometrical argument or prolonged observation of the skies. The danger of this method was that the desire to perfect the concrete mechanism would replace the taste for reflection and observation and so would lead one away from real astronomy into simple tinkering. Plato had already criticized the geometers who made use of mechanical devices to solve problems,[21] saying that this was the corruption and annihilation of the one good of geometry, which ought to concern itself with the contemplation of the unembodied objects of pure intelligence, rather than with base material things. And Ptolemy, the greatest astronomer of antiquity, objected to traditional sphere making on the grounds that, in the majority of cases, it only reproduced the appearances of things without troubling itself over causes and gave proofs of its own technical accomplishment rather than of the justice of astronomical hypotheses.[22] Ptolemy's complaint probably was justified, especially when applied to the orreries, which certainly had an air of the marvelous and extravagant. But there is no doubt that the simpler models—the armillary sphere and the celestial globe—played an important part in the teaching of astronomy and even, in the early days of this science, in fundamental research and discovery. Perhaps it was because he realized this that Ptolemy, in his *Planetary Hypotheses*, decided after all to give a summary of ideas that might be useful to those who wish to make concrete models of the cosmos.

A Renaissance Armillary Sphere

In the Renaissance, armillary spheres became enormously popular, and many examples survive in museums.[23] In figure 2.11 we see a well-made, functional model, suitable for instructional use. This armillary sphere is of sixteenth-century German workmanship. The circles are of brass. The outside diameter of the meridian is about 9 1/2 inches. This sphere has an interesting special feature: it is equipped with rotatable auxiliary rings that allow markers representing the Sun and Moon to be moved and positioned at will along the zodiac. The Sun and Moon markers may be seen on the inside of the zodiac ring.

FIGURE 2.11. A sixteenth-century brass armillary sphere. Science Museum, London.

2.3 EXERCISE: USING A CELESTIAL GLOBE

Directions for Use of the Sphere

A usable celestial globe must have the following features: (1) a fixed horizon stand, (2) a moveable meridian ring that allows the model to be adjusted for the observer's latitude, and (3) an axis of rotation. (These features are all displayed by the Renaissance model in fig. 2.9.) If you use a solid celestial globe, you should visualize the Earth as a geometric point, at rest at the center of the sphere.

The four most important circles of the model are the horizon, the meridian, the equator, and the ecliptic.

The horizon and the meridian are fixed circles that do not participate in the revolution but form a base or stand for the revolving sphere. The *horizon ring* represents the observer's own horizon. Therefore, points of the sphere that are above the horizon are visible, and those below, invisible. The horizon should be marked all around at 5° or 10° intervals. On most horizon stands, the cardinal points (north, east, south, and west) also are marked. These markings enable one to tell in just what direction a given star rises or sets.

The *meridian ring* may be turned in the stand so that the elevation of the celestial pole above the horizon may be varied. By this means the model may be adjusted to give the appearance of the sky at any desired latitude. The latitude of a place on Earth is equal to the altitude of the north celestial pole (or arctic pole) at that place.

The equator and the ecliptic both participate in the daily revolution of the heavens. The *equator* is divided into hours. These marks may be counted as they turn past the meridian ring to measure elapsed time. In other words, the celestial equator, turning past the fixed meridian, constitutes a giant clock. Thus, one may determine, for example, the time between the rising and setting of a particular star. (Technically, the stars take about four minutes less than twenty-four hours to complete a revolution. For most purposes this small difference may be ignored.)

The *ecliptic* is the path that the Sun follows in its annual motion. On your model, it may be marked in degrees of celestial longitude, or with the dates on which the Sun reaches each point, or with both kinds of information.

If you are using a celestial globe, you will see that it is marked with many stars. If you are using an armillary sphere, it may be marked with the approximate positions of a few prominent stars that happen to lie on or near one of the circles.

Example

Problem: What will an observer at 50° north latitude see the Sun do on April 20?

Solution: First set the meridian so that the arctic pole is 50° above the north point of the horizon (as in fig. 2.9).

Then place the April 20 mark of the ecliptic on the horizon and find that the Sun rises about 17° north of east. Note that the 19-hour mark of the equator is now at the meridian. (This is, in modern parlance, called the *sidereal time* of sunrise. The sidereal time is indicated by the hour mark of the equator that is on the meridian above the horizon. Sidereal time is *not* the same as ordinary clock time.)

Then turn the sphere until the April 20 mark reaches the western horizon and note that the 8 3/4 hour mark of the equator is on the meridian. (Thus, the sidereal time of sunset is 8 3/4 hours.) To find the length of the day, subtract the sidereal time of sunrise from the sidereal time of sunset: the Sun was above the horizon for 8 3/4 − 19 = 24 + 8 3/4 − 19 = 13 3/4 hours.

A note on reckoning time intervals: In computing a time interval, the rule is always (time of final event) − (time of initial event). So, to find the length of the day, we compute sunset minus sunrise. If you subtract in the wrong order, you will find the length of the night. If you cannot carry out the subtraction because the first number is smaller than the number being subtracted, you can always add 24 hours to the first number, as in ordinary clock arithmetic.

Finally, place the Sun (the April 20 mark) at the meridian, simulating local noon. Find that the Sun is 51° above the horizon.

In summary, on April 20, an observer at 50° north latitude will see the Sun rise 17° north of east, cross the meridian 51° above the horizon, and set 17° north of west, 13 3/4 hours after it rose.

The Motion of the Sun

Use a celestial globe or an armillary sphere to investigate the behavior of the Sun at different times of the year and at different latitudes. In particular:

1. Make a graph of the altitude of the Sun at noon versus the date. Plot at least one point for each thirty days over an entire year. You should make three such graphs, for latitudes 0°, 35°, and 70°. Your graphs will be more meaningful if you display them on a single sheet of graph paper.
2. Make a graph of the rising direction (number of degrees north or south of east) of the Sun versus the date for a whole year. Do this for each of the same three latitudes.
3. Make a graph of the length of the longest day of the year (in hours) versus latitude. Vary the latitude by 10° steps from 0° to 90°. (This graph has a historical as well as an astronomical interest. The Greeks often designated the latitude of a place by giving the length of the longest day there.)

Questions and Problems

1. Is there any place on Earth at which the Sun rises directly in the east every day of the year?
2. Is there any time of year at which the Sun rises directly in the east everywhere on Earth?
3. Use the celestial globe to determine the truth or falsity of the following two familiar statements: "At the equator, the Sun always rises directly in the east. Moreover, the Sun is above the horizon twelve hours every day there."
4. Suppose the Sun crosses the local meridian south of the zenith at some particular place on Earth and on some particular day. Can there be any place on Earth at which the Sun crosses the meridian north of the zenith on that same day?
5. Suppose the Sun rises south of east at some particular place on Earth and on some particular day. Can there be any place on Earth at which the Sun rises north of east on that same day?
6. The tropic of Cancer that is often marked on globes of the Earth is a projection of the celestial tropic of Cancer. Therefore, it is a circle on the Earth's surface at a latitude of about 23°. What is special about this latitude? In what way are latitudes above this different from those below? Think in terms of the apparent motion of the Sun.
7. The arctic circle is a circle on the Earth at a latitude of about 67°. In what way are latitudes above the arctic circle different from those below?
8. Suppose we divide the Earth into five zones, with boundaries formed

by the arctic circle, the tropic of Cancer, the tropic of Capricorn, and the antarctic circle. Describe characteristics of each zone as fully as possible, in terms of the Sun's behavior.

The Greeks divided the Earth into these same zones, but some writers limited the frigid zones by the arctic and antarctic circles of the Greek horizon. See section 2.5 for an explanation of the "local arctic circle." On the zones, see section 1.12.

2.4 EARLY WRITERS ON THE SPHERE

Autolycus of Pitane

The oldest surviving works of Greek mathematical astronomy are those of Autolycus, *On the Moving Sphere* and the two books called *On Risings and Settings*.[24] Autolycus (roughly 360–290 B.C.) came from the city of Pitane on the western coast of Asia Minor, opposite Mytilene. His works date from the time when Greek mathematical astronomy was just emerging. Together, the three works contain several dozen propositions, all simply and geometrically proved.

On the Moving Sphere treats twelve elementary propositions concerning a sphere that rotates about a diameter as axis. For example,

1. If a sphere rotates uniformly about its axis, all the points on the surface of the sphere which are not on the axis will trace parallel circles that have the same poles as the sphere, and that are perpendicular to the axis.

One notices here something that is common in all the elementary astronomical works: a reversal of the line of historical development. Thus, although astronomy began with observation of the circular motion of the stars, from which the spherical form of the heavens was inferred, Autolycus *assumes* a spherical universe and deduces the circular orbits of the stars.

4. If on a sphere an immobile great circle perpendicular to the axis separates the invisible from the visible hemisphere, then during the rotation of the sphere about its axis, none of the points on the surface of the sphere will set or rise. Rather, the points located on the visible hemisphere are always visible; and those on the invisible hemisphere are always invisible.

The "immobile great circle that separates the invisible from the visible hemisphere" is the horizon. Circumlocutions such as this were common in an age in which a technical vocabulary was still emerging. Our term *horizon* derives from the Greek verb *horizo*, to divide or separate. In this fourth proposition, then, Autolycus considers a situation in which the axis of the universe is perpendicular to the horizon. Such is the case at the north or south pole of the Earth, where the celestial pole stands directly overhead. Here, none of the stars rise or set.

A curious aspect of Autolycus's style in *On the Moving Sphere* is the absence of any overt reference to the astronomical applications of the theorems. The objects that rise and set are not stars but merely points (*semeia*), and the object on which these points are fixed is not the cosmos but a hypothetical revolving sphere. This was probably deemed to make the book better (because purer) geometry.

5. If a fixed circle passing through the poles of the sphere separates the visible from the invisible part, all points on the surface of the sphere will, in the course of its revolution, both set and rise. Further, they will pass the same time below the horizon and above the horizon.

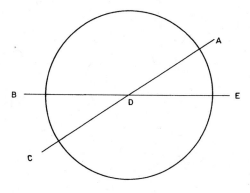

FIGURE 2.12.

Proposition 5 applies at the Earth's equator, where the celestial poles lie on the horizon.

9. If on a sphere a great circle oblique to the axis separates the visible part of the sphere from the invisible part, then, of all the points that rise at the same time, the ones closer to the visible pole will set later; and, of all the points that set at the same time, the ones closer to the visible pole will rise sooner.

If the horizon is neither perpendicular nor parallel to the axis, it is said to be oblique or inclined. So proposition 9 applies wherever 4 and 5 do not—that is, everywhere on Earth except at the poles and the equator.

The propositions quoted here give a fair idea of the subject matter and the level of difficulty of *On the Moving Sphere*. The propositions are all of the kind that could be discovered by experiment on a concrete model and then proved by elementary geometrical reasoning. In Autolycus we find nothing that could serve as a basis for a general method of calculation—there is as yet no trigonometry—but only knowledge of the sort that implies a thorough familiarity with the celestial globe.

Euclid's Phenomena

Almost contemporary with Autolycus was Euclid, whose masterwork on geometry dates from the third century B.C. The thirteen books of the *Elements* represent the culmination of classical geometry and remain among the most studied works in the history of thought. More than a 1,000 editions have appeared since the invention of printing.[25] However, the *Elements* contains little that is of special interest for astronomy. The geometry of the sphere, for example, is scarcely treated.

But Euclid did leave us a more astronomical work, the *Phenomena*, which covers in eighteen theorems some important, if elementary, features of spherical astronomy. Let us examine a few:[26]

1. The Earth is in the middle of the cosmos and occupies the position of center with respect to the cosmos.

How does Euclid prove the centrality of the Earth? Suppose, as in figure 2.12, that *ABC* is the circle of the horizon, with *C* in the eastern part and *A* in the west. The observer is at *D*. Look through a sighting tube (dioptra) at the Crab when it is rising at *C*. If you then turn around and look through the other end of the tube you will see the Goat-Horn setting at *A*. Thus, *ADC* is a diameter of the sphere of stars, for the arc between the Goat-Horn and the Crab is six zodiac signs. In the same way, aim the dioptra at *B* when the Lion is rising there. If you then look through the other end of the dioptra, you will see the Water-Pourer setting at *E*. *B* and *E* are six signs apart, so *BDE* is also a diameter of the sphere of stars. Therefore, the point *D* of intersection is the center of the sphere. This is the Earth, where the observer stands.

Like Ptolemy's demonstrations of the place of the Earth, Euclid's argument is a thought-proof rather than a true appeal to observation. No one ever became convinced of the centrality of the Earth by making such observations. The conventional nature of the proof is clear from Euclid's use of Crab and Goat-Horn as if they were points on the sphere rather than zodiacal signs each 30° long. Conventional demonstrations can have a long life. Copernicus, for example, some 1,800 years later, gave exactly the same "proof" that the Earth is as a point in comparison with the heavens.[27] Even the figure is the same. Copernicus, of course, deduces from these considerations only the smallness of the Earth: he points out, rightly, that this evidence does not prove that the Earth is at the center of the universe.

3. Of the fixed stars that rise and set, each [always] rises and sets at the same points of the horizon.

Euclid proves this from the spherical nature of the heavens, but this elementary fact was certainly known long before anyone had any conception of the celestial sphere.

11. Of [two] equal and opposite arcs of the ecliptic, while the one rises the other sets, and while the one sets the other rises.

17. Of [two] equal arcs [of the ecliptic] on either side of the equator and equidistant from the equator, in the time in which one passes across the visible hemisphere the other [passes across] the invisible hemisphere. . . .

These theorems, like those in Autolycus, contain little that would be useful to a practicing observer, nor are they the results of observation. Rather, they are simple consequences of the spherical nature of the heavens, sprung from the realm of thought.

The Little Astronomy

Euclid's *Phenomena* and Autolycus's *On the Moving Sphere* are preserved in Greek manuscripts of the medieval period—manuscripts that were copied by hand more than a thousand years after the originals were set down by their authors. In many cases, these two works are found bound together with a number of other minor works of Greek astronomy. As an example, take the manuscript *Vaticanus graecus* 204 (i.e., Greek manuscript no. 204 in the Vatican Library). This manuscript is valuable for our knowledge of Autolycus, both because of the care with which it was copied, and because of its age: it dates from the ninth or tenth century A.D., which makes it the oldest surviving copy of Autolycus's Greek text. The partial contents of this manuscript are as follows:[28]

- Theodosius of Bithynia, *Spherics*. First century B.C. The *Spherics* is a treatise on the geometry of the sphere, in the style of Euclid's *Elements*. The *Spherics* of Theodosius may be considered a continuation of, and a supplement to the *Elements*. Thus, it is more sophisticated than the *Phenomena* of Euclid.
- Autolycus, *On the Moving Sphere*.
- Euclid, *Optics*. This is an elementary geometrical treatise on various effects involving the straight-line propagation of light: shadows, perspective, parallax, and so on. Example: When one observes a sphere with both eyes, if the diameter of the sphere is equal to the distance between the pupils, one will see exactly half the sphere; if the distance between the pupils is greater, one will see more than half; if the distance between the pupils is less, one will see less than half. Second example: If several objects move at the same speed, the most distant will appear to move most slowly.
- Euclid, *Phenomena*.
- Theodosius of Bithynia, *On Geographic Places*. This little book, similar in flavor to Euclid's *Phenomena* and Autolycus's *On the Moving Sphere*, describes, in twelve propositions, the appearance of the sky as seen from various places on the Earth. Example: an inhabitant of the north pole would see always the northern hemisphere of the celestial sphere; the southern hemisphere would be forever unseen; no star would rise or set.
- Theodosius of Bithynia, *On Days and Nights*. This work presents thirty-one propositions concerning the lengths of the days and nights at different times of the year, at different latitudes on the Earth.
- Aristarchus of Samos, *The Sizes and Distances of the Sun and Moon*, third century B.C. (discussed in sec. 1.17).

- Autolycus, *On Risings and Settings* (discussed in sec. 4.9).
- Hypsicles, *On Ascensions*, second century B.C. In this short treatise, Hypsicles proves a number of propositions on arithmetical progressions and uses the results to calculate approximate values for the times required for the signs of the zodiac to rise above the horizon. Such information had practical applications, for example, in telling time at night. Hypsicles' treatise is discussed in section 2.16.
- Euclid, *Catoptrics*. This work, whose attribution to Euclid is disputed, is concerned with what we would today call optics proper. It treats the reflection of light and the formation of images by mirrors.
- Euclid, *Data* (= "given"). This treatise on elementary geometry consists of propositions proving that, if certain things in a figure are given, something else may be figured out.

This hodgepodge of short, elementary astronomical and geometrical works is found in many medieval manuscripts. Sometimes one or more are wanting. Often, other minor works are present, such as commentaries on the mathematical works of Apollonius and Euclid. The particular collection of short works listed above is sometimes called the *Little Astronomy*. Also usually included is the *Spherics* of Menelaus (first century A.D.), which no longer survives in Greek but is known through Arabic translations. Menelaus's work treats the geometry of spherical triangles. It has been said that, from the second century A.D. onward, the *Little Astronomy* served as an introductory-level textbook for students who were not yet prepared to tackle the "Big Astronomy," that is, the *Almagest* of Ptolemy.[29] It may have been so, but the evidence is slight.[30] Indeed, the chief evidence is simply the fact that many Byzantine manuscripts contain more or less the same assortment of elementary astronomical and geometrical works.

The supposed title of the collection is provided by a remark at the beginning of the sixth book of the *Mathematical Collection* of Pappus of Alexandria. Of Pappus himself we know very little. He lived and taught at Alexandria, during the last half of the third and the first half of the fourth century A.D. He had a son, Hermodoros, to whom he addressed two of his books. He had as friends two geometers, Pandrosios and Megethios, who are otherwise unknown. Pappus wrote a commentary on the *Almagest* of Ptolemy, which survives in part. But his most important work is the one that has come down to us under the title *The Mathematical Collection of Pappus of Alexandria*. This consists of a vast collection of propositions extracted from a great number of works on mathematics, astronomy, and mechanics (many of which are lost today), accompanied by Pappus's explanatory notes, alternative demonstrations, and new applications. The work does not seem to have been written according to any plan, but was probably the result of many years' reading and note taking, no doubt in connection with Pappus's teaching duties at Alexandria. The sixth book of the collection is devoted to the astronomical writers. Pappus discusses works by Theodosius, Menelaus, Aristarchus, Euclid, and Autolycus. At the beginning of the sixth book, we find the remark, written as a subtitle, "It contains the resolutions of difficulties found in the little astronomy."

Whether or not there really existed a definite collection of treatises known as the *Little Astronomy*, there is no doubt that the individual works were used by teachers from the late Hellenistic period down to Byzantine times. The tradition was continued by Arabic teachers, who made use of the same treatises in translation and added others as well. It was the schoolroom usefulness of these works that guaranteed their survival, for many works of greater scientific and historical importance have been lost, for example, most of the writings of Hipparchus and all those of Eudoxus.

Aristarchus's work, *On the Sizes and Distances of the Sun and Moon*, is quite different from the others of the collection: Aristarchus attempted to arrive at new astronomical knowledge by calculations based on astronomical data. The rest of the purely astronomical works of the *Little Astronomy* are theoretical developments of various properties of the celestial sphere, devoid of any reference to particular observations. The oldest works of the collection, those of Autolycus and Euclid (ca. 300 B.C.), represent the first attempts to grapple with the problems of spherical geometry, and therefore are endowed with a great historical interest. Some of the later works, for example, Theodosius's treatises *On Geographical Places* and *On Days and Nights* (ca. 100 A.D.), lag considerably behind the astronomical and mathematical knowledge of their own day and must actually have been written as primers for students. Their elementary nature and pedantic style would reveal them as textbooks in any age. Taken together, the treatises of the *Little Astronomy* illustrate the level of Greek mathematical astronomy around the beginning of the second century B.C., before the revolution in calculating ability brought about by the development of trigonometry. Menelaus's book was one that helped point the way to the new mathematics.

2.5 GEMINUS: *INTRODUCTION TO THE PHENOMENA*

In addition to the works of the *Little Astronomy*, we have several other elementary texts from a slightly later period. A notable example is the *Introduction to the Phenomena* by Geminus, a writer of the first century A.D.[31] This work is sometimes called the *Isagoge*, from the first word of its Greek title. This work differs markedly from most of those in the *Little Astronomy*. In the first place, it is longer. And second, it is written with grace and style. It is, in fact, a well-organized and more or less complete introduction to astronomy, intended for beginning students of this subject.

Geminus takes up the zodiac and the motion of the Sun, the constellations, the celestial sphere, days and nights, the risings and settings of the zodiacal signs, luni-solar periods and their application to calendars, phases of the Moon, eclipses, star phases, terrestrial zones and geographical places, and the foolishness of making weather predictions by the stars. From this lively and readable book we have extracted some sections devoted to the principal circles of the celestial sphere.

Italicized subheadings in the extract do not appear in the original, but have been added for the reader's convenience. Likewise, the numbering of statements is not a part of the original text, but is a practice introduced by modern scholars for their own convenience. An asterisk (*) in the text indicates that an explanatory note, keyed to the statement number, follows the extract.

EXTRACT FROM GEMINUS

Introduction to the Phenomena V

The Circles on the Sphere

1 Of the circles on the sphere, some are parallel, some are oblique, and some [pass] through the poles.

The Parallel Circles

The parallel [circles] are those that have the same poles as the cosmos. There are 5 parallel circles: arctic [circle], summer tropic, equinoctial,* winter tropic, and antarctic [circle].

2 The arctic circle* is the largest of the always-visible circles, [the circle] touching the horizon at one point and situated wholly above the Earth. The stars lying within it neither rise nor set, but are seen through the whole night turning around the pole. *3* In our *oikumene,** this circle is traced out by the forefoot of the Great Bear.*

4 The summer tropic circle is the most northern of the circles described by the Sun during the rotation of the cosmos. When the Sun is on this circle, it produces the summer solstice, on which occurs the longest of all the days of the year, and the shortest night. *5* After the summer solstice, however, the Sun is no longer seen going towards the north, but it turns towards the other parts of the cosmos, which is why [this circle] is called "tropic."*

6 The equinoctial circle is the largest of the 5 parallel circles. It is bisected by the horizon so that a semicircle is situated above the Earth, and a semicircle below the horizon. When the Sun is on this circle, it produces the equinoxes, that is, the spring equinox and the fall equinox.

7 The winter tropic circle is the southernmost of the circles described by the Sun during the rotation of the cosmos. When the Sun is on this circle it produces the winter solstice, on which occurs the longest of all the nights of the year, and the shortest day. *8* After the winter solstice, however, the Sun is no longer seen going towards the south, but it turns toward the other parts of the cosmos, for which reason this [circle] too is called "tropic."

9 The antarctic circle is equal [in size] and parallel to the arctic circle, being tangent to the horizon at one point and situated wholly beneath the Earth. The stars lying within it are forever invisible to us.

10 Of the 5 forementioned circles the equinoctial is the largest, the tropics are next in size and—for our region—the arctic circles are the smallest. *11* One must think of these circles as without thickness, perceivable [only] with the aid of reason, and delineated by the positions of the stars, by observations made with the dioptra, and by our own power of thought. For the only circle visible in the cosmos is the Milky Way; the rest are perceivable through reason. . . .

. . .

Properties of the Parallel Circles

18 Of the 5 forementioned parallel circles, the arctic circle is situated entirely above the Earth.

19 The summer tropic circle is cut by the horizon into two unequal parts: the larger part is situated above the Earth, the smaller part below the Earth. *20* But the summer tropic circle is not cut by the horizon in the same way for every land and city: rather, because of the variations in latitude, the difference between the parts is different. *21* For those who live farther north than we do, it happens that the summer <tropic> is cut by the horizon into parts that are more unequal; and the limit is a certain place where the whole summer tropic circle is above the Earth. *22* But for those who live farther south than we do, the summer tropic circle is cut by the horizon into parts more and more equal; and the limit is a certain place, lying to the south of us, where the summer tropic circle is bisected by the horizon.

23 <For the horizon in Greece, the summer tropic> is cut <by the horizon> in such a way that, if the whole circle is [considered as] divided into 8 parts, 5 parts are situated above the Earth, and 3 below the Earth. *24* And it was for this clime* that Aratus seems in fact to have composed his treatise, the *Phenomena*; for, while discussing the summer tropic circle, he says:

> *If it is measured out, as well as possible, into eight parts,*
> *five turn in the open air above the Earth,*
> *and three beneath; on it is the summer solstice.*

From this division it follows that the longest day is 15 equinoctial hours* and the night is 9 equinoctial hours.

25 For the horizon at Rhodes, the summer tropic circle is cut by the horizon in such a way that, if the whole circle is divided into 48 parts, 29 parts are situated above the horizon, and 19 below the Earth. From this division it follows that the longest day in Rhodes is 14 1/2 equinoctial hours and the night is 9 1/2 equinoctial hours.

26 The equinoctial circle, for the whole oikumene, is bisected by the horizon, so that a semicircle is situated above the Earth, and a semicircle below the Earth. For this reason, the equinoxes are on this circle.

27 The winter tropic circle is cut by the horizon in such a way that the smaller part is above the Earth, the larger below the Earth. The inequality of the parts has the same variation in all the climes as was the case with the summer tropic circle, because the opposite parts of the tropic circles are always equal to one another. For this reason, the longest day is equal to the longest night, and the shortest day is equal to the shortest night.

28 The antarctic circle is hidden wholly beneath the horizon. . . .

. . .

45 The distances of the circles from one another do not remain the same for the whole oikumene. But in the engraving of the spheres, one makes the division in declination in the following way. *46* The entire meridian circle being divided into 60 parts, the arctic [circle] is inscribed 6 sixtieths from the pole; the summer tropic is drawn 5 sixtieths from the arctic [circle]; the equinoctial 4 sixtieths from each of the tropics; the winter tropic circle 5 sixtieths from the antarctic; and the antarctic [circle] 6 sixtieths from the pole.

47 The circles do not have the same separations from one another for every land and city. The tropic circles do maintain the same separation from the equinoctial at every latitude, but the tropic circles do not keep the same separation from the arctic [circles] for all horizons; rather, the separation is less for some [horizons] and greater for others. *48* Similarly, the arctic [circles] do not maintain a distance from the poles that is equal for every latitude; rather, it is less for some and greater for others. However, all the spheres are inscribed for the horizon in Greece. . . .

. . .

The Zodiac

51 The circle of the 12 signs is an oblique circle. It is itself composed of 3 parallel circles,* two of which are said to define the width of the zodiac circle, while the other is called the circle through the middles of the signs. *52* The latter circle is tangent to two equal parallel circles: the summer tropic, at the 1st degree of the Crab, and the winter tropic, at the 1st degree of the Goat-Horn. It also cuts the equinoctial in two at the 1st degree of the Ram and the 1st degree of the Balance. *53* The width of the zodiac circle is 12 degrees. The zodiac circle is called oblique because it cuts the parallel circles. . . .

. . .

The Milky Way

68 The Milky Way* also is an oblique circle. This circle, rather great in width, is inclined to the tropic circle. It is composed of a cloud-like mass of small parts and is the only [circle] in the cosmos that is visible. *69* The width of this circle is not well defined; rather, it is wider in certain parts and narrower in others. For this reason, the Milky Way is not inscribed on most spheres.*

This also is one of the great circles. *70* Circles having the same center as the sphere are called great circles on the sphere. There are 7 great circles: the equinoctial, the zodiac with the [circle] through the middles of the signs, the [circles] through the poles, the horizon for each place, the meridian, the Milky Way.[32]

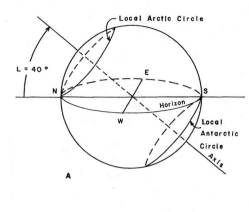

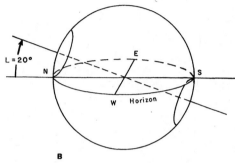

FIGURE 2.13. A (*top*). The local arctic and local antarctic circles in the sense of the Greek astronomers, shown for a latitude of 40° N. B (*bottom*). Local arctic and antarctic circles for a latitude of 20° N.

Notes to the Extract from Geminus

1. Equinoctial. The equinoctial circle is the celestial equator. It is called equinoctial because the Sun makes the day and night equal when it is on this circle. The Greek term is *isemerinos kuklos* (equal-day circle).

2. Arctic circle. For the Greeks, the arctic circle is a circle on the celestial sphere, with its center at the celestial pole, and its size chosen so that the circle grazes the horizon at the north point (see fig. 2.13A). The stars within the arctic circle are circumpolar; that is, they never rise or set. The size of the local arctic circle depends on the latitude of the observer. Figure 2.13A shows the arctic circle for latitude 40° N, and figure 2.13B, the arctic circle for latitude 20° N. The radius of the arctic circle is the angular distance of the celestial pole above the north point of the horizon. But this angular distance (the altitude of the pole) is equal to the latitude of the place of observation. Thus, the radius of the arctic circle for a particular place is equal to the latitude of that place. The modern celestial arctic circle is fixed in size: it is a circle, centered on the pole, with a radius of about 24°. The modern, celestial arctic circle is the path traced out by the pole of the ecliptic during the daily revolution of the heaven. (The arctic circle one sees marked on globes of the Earth can be regarded as a projection of the one in the sky.) Because the size of the arctic circle in the Greek style varies with the location of the observer, we shall call this the *local arctic circle*. The fixed circle of the present age will be called the *modern arctic circle*.

Oikumene. The *oikumene* is the *inhabited world*. It is used by Greek writers in two different senses. It may designate the Greeks' portion of the Earth, as opposed to barbarian lands. But the word is also used by geographical writers to mean the whole inhabited world, namely, Asia, Europe, and Africa. Geminus here appears to speak in the more restricted sense.

Great Bear. In ancient Greece the foreleg of the constellation of the Great Bear (Ursa Major) did not quite set, but grazed the horizon in the north. The foreleg of the Bear was therefore situated on the local arctic circle, and in the course of the night, it traced out this circle in the sky. Our word *arctic* derives from *arktos*, the Greek word for *bear*.

5. Tropic. Our word *tropic* derives from the Greek word *trope*, a *turn*, *turning*.

24. Clime. "Clime," from *klima* (region or zone), but originally a *slope* or *inclination*: the clime is determined by the *inclination* of the axis of the cosmos to the horizon. A clime is a zone of the Earth lying near one parallel of latitude. Often, climes were designated in terms of the length of the longest day. Thus, one might say that Seattle and Basel are in the clime of sixteen hours: at both these cities the length of the day at summer solstice is sixteen hours. The verses that Geminus quotes are from Aratus's *Phenomena*, lines 497–499.

24. The longest day is 15 equinoctial hours. In Greece five-eighths of the summer tropic circle is above the horizon. The length of the solstitial day is therefore 24 hours × 5/8 = 15 hours.

51. Composed of 3 parallel circles. See figures 2.2 and 2.7.

68. The Milky Way. The "Milky Way" is *galaktos* (milky) *kuklos* (circle), from which comes our word *galaxy*. Aristotle, in *Meteorology* I, 8 (345a11–346b15), discusses several theories of the Milky Way: (1) Some of the Pythagoreans held that the Milky Way was a former course of the Sun and that this track had been burned. (2) Anaxagoras and Democritus said that the Milky Way was the light of stars lying in the Earth's shadow. Many of the stars on which the Sun's rays fall become invisible because of the brightness of these rays. But faint stars in the Earth's shadow do not have to overcome the brightness of the Sun's rays and thus they become visible. (3) According to a third opinion, the Milky Way was a reflection of the Sun. Aristotle refutes each of these theories in turn: (1) If the Milky Way were a former, scorched track of the Sun, one would expect the zodiac also to be scorched, but it is not. (2) If the Milky Way were the light of stars lying in the Earth's shadow, the position of the Milky Way should change during the year as the Sun's motion on the ecliptic causes the shadow to move. Besides, the Sun is larger than the Earth and therefore the cone of the shadow does not extend as far as the sphere of the stars. (3) The Milky Way cannot be a reflection of the Sun, for it always cuts through the same constellations, although the Sun's position among the stars is constantly changing. But if one moves an object around in front of a mirror the location of the image is also seen to change. Aristotle's own opinion is that the Milky Way consists of the halos seen around many individual stars. These halos arise in the following way. Above and surrounding the Earth, at the upper limit of the air, is a warm, dry exhalation. This exhalation, as well as a part of the air immediately beneath it, is carried around the Earth by the circular revolution of the heavens. Moved in this manner, it bursts into flame wherever the situation happens to be favorable, namely, in the vicinity of bright stars. Aristotle points out that the stars are brighter and more numerous in the vicinity of the Milky Way than in other parts of the sky. The only objection one might make is that the dry exhalation ought also to be inflamed in the vicinity of the Sun, Moon, and planets, which are brighter than any of the stars. But, according to Aristotle, the Sun, Moon, and planets dissipate the exhalation too rapidly, before it has a chance to accumulate sufficiently to burst into flame. Note that for Aristotle the Milky Way is an atmospheric, and not a celestial, phenomenon: it is produced at the outer boundary of the air.

68. *The Milky Way is not inscribed on most spheres.*   However, Ptolemy, in his directions for constructing and marking a celestial globe (*Almagest* VIII, 3), includes the Milky Way among the objects to be represented.

2.6 RISINGS OF THE ZODIAC CONSTELLATIONS: TELLING TIME AT NIGHT

In everyday life, the Greeks kept time differently than we do. Rather than dividing the time between one midnight and the next into twenty-four equal parts, they divided the time between sunrise and sunset into twelve seasonal hours, which changed in length through the year as the day itself changed. Similarly, the night was divided into twelve seasonal hours, all equal to one another, but not equal to the day hours (except at the equinox). "Two hours after sunset" meant one-sixth of the way from sunset to sunrise. It did not matter that the time from sunset until the second seasonal hour was nearly twice as long in winter as in summer.

The seasonal hour may seem strange to a modern reader. But nature provides a means of observing the time at night, at least approximately, in terms of seasonal hours. In the course of any night, six signs of the zodiac rise. The proof of this assertion is elementary. At the beginning of night (sunset), the point of the ecliptic that is diametrically opposite the Sun will be on the eastern horizon. At the end of the night (sunrise), the point opposite the Sun will have advanced to the western horizon. The half of the ecliptic following this point is then seen above the horizon and is the very part of the ecliptic that rose in the course of the night.

The risings of six zodiacal signs every night divide the night into six roughly equal parts, of two seasonal hours each. A glance toward the eastern horizon, to see which zodiacal constellation is rising, will suffice to determine the time of night, provided that one knows which constellation the Sun is in.

This information is provided by table 2.1. From March 21 to April 20, the Sun travels from longitude 0° to longitude 30°; that is, it traverses the sign

TABLE 2.1. Progress of the Sun through the Zodiac

On these Days	The Sun Is in the Sign of the	Corresponding *Roughly* to the Constellation
Mar 21–Apr 20	Ram	Pisces
Apr 20–May 21	Bull	Aries
May 21–Jun 22	Twins	Taurus
Jun 22–Jul 23	Crab	Gemini
Jul 23–Aug 24	Lion	Cancer
Aug 24–Sep 23	Virgin	Leo
Sep 23–Oct 24	Balance	Virgo
Oct 24–Nov 23	Scorpion	Libra
Nov 23–Dec 22	Archer	Scorpius
Dec 22–Jan 20	Goat-Horn	Sagittarius
Jan 20–Feb 19	Water-Pourer	Capricornus
Feb 19–Mar 21	Fishes	Aquarius

of the Ram. In antiquity, the stars of the constellation Aries (the Ram) were in this sign. But, this is no longer the case today. Because of precession, the *sign* of the Ram (the first 30° of the zodiac) is now mostly occupied by the *constellation* Pisces. Thus, in the present era, the Sun is among the stars of the constellation Pisces between March 21 and April 20. (Precession is discussed in detail in sec. 6.1.) For a rough-and-ready method of telling time, we will rely on observations of the stars and not the signs. We will use the third column of table 2.1, not the second. (In Greek antiquity, the third column would have been the same as the second.)

Example

Problem: It is the night of July 23. We look toward the eastern horizon and see that Aries has risen completely and is well above the ground; none of the stars of Taurus are visible. Evidently, Taurus is only just beginning to rise. What time is it?

Solution: On July 23, the Sun is entering the constellation Cancer (see table 2.1). The first part of Taurus is beginning to rise. Next will rise the first part of Gemini, then the first part of Cancer, where the Sun is now located. From Taurus to Cancer is two signs, each of which takes roughly 2 seasonal hours to rise. The time is therefore *4 seasonal hours before sunrise*. Or, since 6 seasonal hours elapse between midnight and sunset, we may also say *2 seasonal hours after midnight*.

Conversion to Modern Time Reckoning   An ancient Greek would have been satisfied with either of these manners of expressing the time. A modern reader, however, is likely to be dissatisfied with a time of night expressed in terms of seasonal hours.

Conversion to equinoctial hours can be made with the aid of table 2.2, which gives the length of the night, for each of six latitudes, on the days when the Sun enters each of the zodiacal signs. For example, on July 23, the night at latitude 41°27′ lasts 9^h29^m. The peculiar values of the latitudes result from a choice made in the construction of the table, that the longest and shortest nights should be whole numbers of hours: these are the geographical *climes* of the old astronomers. The method of calculating the table is explained

TABLE 2.2. The Length of the Night

Sun's Place	Approx. Date	North Latitude					
		0°00′	16°46′	30°51′	41°27′	49°05′	54°33′
0° Crab	Jun 22	12^h00^m	11^h00^m	10^h00^m	9^h00^m	8^h00^m	7^h00^m
0° Lion or 0° Twins	Jul 23 or May 21	12 00	11 09	10 19	9 29	8 40	
0° Virgin or 0° Bull	Aug 24 or Apr 20	12 00	11 32	11 04	10 37	10 12	
0° Balance or 0° Ram	Sep 23 or Mar 21	12 00	12 00	12 00	12 00	12 00	
0° Scorpion or 0° Fishes	Oct 24 or Feb 19	12 00	12 28	12 56	13 23	13 48	
0° Archer or 0° Water-Pourer	Nov 23 or Jan 20	12 00	12 51	13 41	14 31	15 20	
0° Goat-Horn	Dec 22	12^h00^m	13^h00^m	14^h00^m	15^h00^m	16^h00^m	17^h00^m

in section 2.13. The completion of the entries under latitude 54°33′ is left for the exercise of section 2.14.

Let us take up the conversion problem. On July 23, the time is 2 seasonal hours past midnight. Suppose we are at Seattle (latitude 48° N). We wish to express the time in terms of equinoctial hours. In table 2.2 we find that, at this latitude and at this time of year, the night lasts about 8 hours 40 minutes (equinoctial hours, of course). By definition, there are 12 seasonal hours in the night. Thus,

$$12 \text{ seasonal night hours} = 8^h40^m$$

So,

$$2 \text{ seasonal night hours} = (8^h40^m) \times \frac{2}{12}$$
$$= 1^h27^m.$$

(The two seasonal night hours are short in July, because the night itself is short.) The time is thus 1^h27^m after midnight, or 1:27 A.M. In July, Seattle uses daylight savings time. If we wish to compare our result with a clock, we must add one hour to the time obtained from the stars: clocks will read 2:27 A.M.

This method of telling time is only approximate, for two reasons. First, the six signs of the zodiac that rise in the course of a night do not all take exactly two seasonal hours to rise: some take a little more, some a little less. And, second, we are not using zodiacal signs, but constellations. These constellations are not all of the same size. Virgo, for example, is much larger than Aries. Nor do they all lie exactly on the ecliptic. Some, like Leo and Gemini, are north of the ecliptic; some, like Taurus and Scorpius, are south of the ecliptic. These variations in the sizes and positions of the zodiacal constellations have an effect on the times they take to rise. Nevertheless, this rough-and-ready method should always give the time correct to the nearest hour.

*Ancient References to Telling Time by
the Zodiacal Constellations*

Aratus refers to this method of telling time at night in his *Phenomena*:

> Not useless were it for one who seeks for signs of the
> coming day to mark when each sign of the zodiac rises.
> For ever with one of them the sun himself rises.[33]

But Aratus takes the traditional method one step further. He points out that the zodiac constellation that is rising may sometimes be obscured by clouds or hills. Therefore, for each zodiac constellation, he gives a list of other constellations that rise or set while the zodiac constellation is rising:

> Not very faint are the wheeling constellations that are set
> about Ocean at East or West, when the Crab rises,
> some setting in the West and others rising in the East.
> The Crown sets and the Southern Fish as far as its back. . . .[34]

This list, which constitutes a major section (some 164 lines) of the poem, would have permitted a person to tell the time of night, if any portion of the horizon were visible. The other necessary ingredient was, of course, knowledge of the Sun's position in the zodiac. But, from the fifth century B.C. on, this was information the average person was likely to have—just as the average person today can be counted on to know the current month of the calendar. In some towns, *parapegmata* (public calendars) were set up, displaying the current place of the Sun in the zodiac, along with other information.

Hipparchus, always a stickler for precision, criticized this portion of the poem in his *Commentary on Aratus and Eudoxus*,[35] pointing out that each zodiac sign does not really take the same amount of time to rise (as we, too, have mentioned above).

Note on Computations with Base-60 Numbers

In the example above, we found that 12 seasonal hours last $8^h 40^m$. To obtain the length of 2 seasonal hours, it was necessary to multiply by 2/12:

$$(8^h 40^m) \times \frac{2}{12} = ?$$

There are several ways to perform this computation. We could express the time interval solely in terms of minutes ($8^h 40^m = 520^m$), do the arithmetic, and then regroup the minutes into whole hours. Alternatively, we could express the 40^m as a decimal fraction of an hour ($8^h 40^m = 8.67^h$) and then do the arithmetic. Both of these methods are awkward. Their awkwardness comes from pushing the calculations through base-10 forms, when the original time interval was expressed in base-60. Calculations involving time (in hours, minutes, seconds) or angle (in degrees, minutes, seconds) simplified if one exploits the base-60 nature of the numbers.

First, write 2/12 as a fraction with denominator 60:

$$(8^h 40^m) \times \frac{2}{12} = (8^h 40^m) \times \frac{10}{60}.$$

Next, perform the division by 60, *which merely changes hours to minutes and minutes to seconds* (1/60 of an hour is a minute):

$$(8^h 40^m) \times \frac{10}{60} = (8^m 40^s) \times 10.$$

Complete the arithmetic:

$$(8^m 40^s) \times 10 = 80^m + 400^s$$
$$= 87^m$$
$$= 1^h 27^m.$$

A great deal of time and trouble will be saved if computations involving base-60 numbers are performed in this way.

2.7 EXERCISE: TELLING TIME AT NIGHT

Use the method explained in section 2.6 to deduce the time of night in each of the following situations.

1. Date: December 22. Place: Columbus, Ohio (latitude 40° N). Observation of the sky: Libra has completely risen, but none of the stars of Scorpius is visible yet.
   A. What is the time of night, expressed in seasonal hours? (Answer: 4 seasonal hours after midnight.)
   B. What is the time expressed in terms of equinoctial hours? (Answer: roughly 5:00 A.M.)
2. Date: February 5. Place: Columbus, Ohio (latitude 40° N). Observation of the sky: Scorpius has risen fully. None of the stars of Sagittarius is up yet.
   A. What is the time of night, in terms of seasonal hours? (Note that on February 5, the Sun is in the middle of a zodiac sign, rather than at the beginning of one.)
   B. What is the time expressed in terms of equinoctial hours?

2.8 OBSERVATION: TELLING TIME AT NIGHT

On a clear night, go outdoors and look to see which zodiac constellation is rising. If necessary, consult a star chart as an aid in identifying the constellations. Use your observation, together with table 2.1, to figure out the time of night in terms of seasonal hours. Then use table 2.2 to convert to a time expressed in terms of equinoctial hours. Compare your result with the time given by a clock. (In summer, don't forget to allow for daylight savings time, if necessary.)

2.9 CELESTIAL COORDINATES

Coordinates of a Point on the Surface of the Earth

The reader is no doubt familiar with the common way of specifying a location on the surface of the Earth. A meridian is chosen to represent the zero of longitude. By an international agreement more than a hundred years old, this is the meridian through the old observatory at Greenwich, England. The longitudes of other meridians are are measured in degrees east or west of the Greenwich meridian. Thus, one says that the longitude of New Orleans is 90° west of Greenwich, or simply 90° W.

The latitude of a city is its angular distance north or south of the plane of the equator. For example, the latitude of New Orleans is 30° N.

Longitude and latitude are said to form an *orthogonal pair* of coordinates. The circles of constant latitude are at right angles to the meridians. Thus, the two coordinates are cleanly separated.

Because all appearances place us at the center of a celestial sphere, we may use a similar method to specify the locations of stars. There are several ways of doing this, depending on the choice one makes for the plane of reference. Four planes are in common use: the planes of the horizon, the celestial equator, the ecliptic, and the Milky Way. We need to become familiar with all but the last of these.

Horizon Coordinates

The simplest way to give the position of a star is to tell how high it is above the horizon and in which direction it lies. One might say, for example, that a star is 23° east of north and 43° above the horizon. If one pointed one's arm directly north, parallel to the horizon, then swung it horizontally 23° toward the east, then up vertically 43°, one could expect to end by pointing at the star.

The angular distance of the star above the horizon is called its *altitude*. The direction of the star in the horizontal plane is called its *azimuth*. We have already made use of altitudes (secs. 1.4 and 1.12). Azimuth is measured clockwise around the horizon, usually from the north point. So one says that a star directly in the east has an azimuth of 90°, while a star directly in the west has an azimuth of 270°. However, azimuth is often measured instead from the south point, so one must make sure of the convention being followed in any particular situation.

Like most angles, altitude and azimuth are commonly measured in terms of the degree, which is 1/360 of a complete circumference. If fractions of a degree must be specified, one makes use of the minute of arc, which is 1/60 of a degree. Similarly, the second of arc is 1/60 of a minute. The degree, minute, and second of arc are represented by the marks °, ′, ″, respectively:

$$1 \text{ circumference} = 360° \quad 1° = 60' \quad 1' = 60''$$

These units were convenient for the Babylonians who invented them more than two thousand years ago, because they did their arithmetic in a system based on the number 60, rather than on 10 as ours is. Today, this division of the circle is a little cumbersome. In the last few decades, it has become more popular to use decimal fractions of the degree, but the ancient sexagesimal division is still in use.

Celestial Equatorial Coordinates

The horizon coordinates are easy to measure and they are a natural choice. However, they have the disadvantage that observers at different places on Earth will obtain different coordinates for the same star, because each observer uses his own personal horizon as his plane of reference. Moreover, even for a single observer in a fixed place, the coordinates of all the stars will change as the diurnal revolution carries them through the sky. We can overcome the first difficulty if all observers will agree to use the same reference plane. And the second difficulty is removed if we fix the coordinates to the revolving celestial sphere rather than to the stationary Earth.

Imagine drawing on the sphere of the heavens a set of parallels and meridians like those you see on globes of the Earth. These are the basis for celestial equatorial coordinates. The reference plane is the plane of the celestial equator. In figure 2.14, draw a circular arc from the north celestial pole *G* through star *S* and extend it until it meets the celestial equator perpendicularly at *A*. The angular distance of star *S* above the plane of the equator is called its *declination*. This is angle *AOS*, which is measured at the Earth *O*. The declination in astronomy is then analogous to the latitude in geography. Declination is often denoted by the Greek letter δ. (Declination was introduced in sec. 1.12.)

The angle analogous to geographical longitude is called *right ascension*. We must choose a place on the celestial equator as the zero of right ascension—and the choice, by agreement, is the vernal equinox ♈. In figure 2.14, the right ascension of star *S* is angle ♈*OA*. Right ascension often is designated by the Greek letter α. Unlike geographical longitude, right ascension is measured only eastward from the zero point, so that right ascensions run from zero to 360°.

Conventionally, right ascensions are usually not measured in degrees, but rather in hours. So, rather than dividing the celestial equator into 360°, we divide it into 24 equal parts, and each of these parts is called an *hour*. Thus, one hour of right ascension is the same as 15°, that is, one twenty-fourth of the celestial equator. The circles through the celestial poles that play the roles of meridians, such as *G*♈*H* and *GSH*, are called *hour circles*. The hour is further divided into sixty minutes, usually denoted *m* to distinguish them from sixtieths of a degree. Similarly, the minute of right ascension is divided into sixty seconds, denoted *s*:

$$1 \text{ circumference} = 24^{h} \quad 1^{h} = 60^{m} \quad 1^{m} = 60^{s}$$

It must be emphasized that 1′ and 1^{m}, which should be read as "one minute of arc" and "one minute of right ascension," respectively, are not angles of the same size, since 1′ is a sixtieth of 1/360 of a circle, while 1^{m} is a sixtieth of 1/24 of a circle. The following relations may be useful:

$$1^{h} = 15° \quad 1^{m} = 15' \quad 1^{s} = 15''$$

The advantage of this system of celestial coordinates is that the coordinates revolve with the stars—the meridians and parallels are, as it were, painted on the celestial sphere. A given star therefore keeps the same celestial coordinates for years at a time. As an example, in 1977, the coordinates of Arcturus were right ascension 14^{h}14^{m}, declination +19°19′. A plus or minus sign with the declination indicates whether the star is north or south, respectively, of the celestial equator.

Ecliptic Coordinates

A different set of celestial coordinates is obtained if one selects the ecliptic rather than the celestial equator as the plane of reference. The ecliptic is inclined about 23° to the celestial equator, as shown in figure 2.15. The Earth *O* is the center of the celestial sphere. If at *O* we raise a line perpendicular to the plane of the ecliptic, this line will pass through the sphere at two points *J* and *K* called the *poles of the ecliptic*. The poles of the ecliptic have the same relation to the ecliptic as the celestial poles have to the celestial equator, or as the zenith (which is the pole of the horizon) has to the horizon. The north ecliptic pole *J* is therefore 23° distant from the north celestial pole *G*.

In figure 2.15 draw a circular arc through the north pole *J* of the ecliptic and star *S* so that the arc intersects the ecliptic perpendicularly at *B*. The angular distance of the star away from the ecliptic (angle *BOS*) is the star's *celestial latitude* and is positive or negative depending on whether the star is north or south of the ecliptic. Celestial latitude is often designated by the Greek letter β.

The other ecliptic coordinate is called *celestial longitude* (denoted λ) and is measured eastward along the ecliptic from the vernal equinox ♈. In figure 2.15 this is angle ♈*OB*. Note that, while in geography "latitude" and "longitude" are referred to the equator, in astronomy these terms are reserved for ecliptic coordinates.

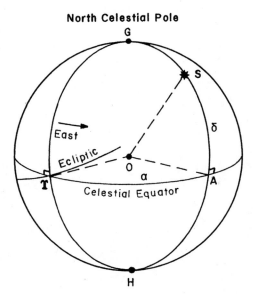

FIGURE 2.14. Celestial equatorial coordinates. α is the right ascension of star *S* (angle ♈*OA*. δ is the declination of star *S* (angle *AOS*).

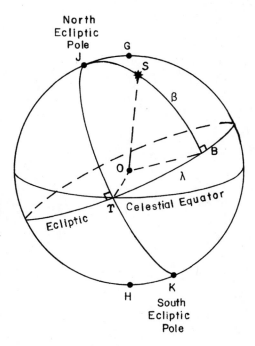

FIGURE 2.15. Ecliptic coordinates. λ is the celestial longitude of star *S*. β is the celestial latitude.

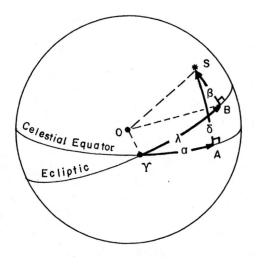

FIGURE 2.16. Equatorial coordinates (α and δ) and ecliptic corrdinates (λ and β) for a single star S.

Today, celestial longitudes are usually measured in degrees, minutes, and seconds so that they run from zero right up to 360°. But among the ancients it was common practice to divide the ecliptic into twelve signs, each 30° long, and to use the sign as a larger unit of angular measure. The vernal equinox marks the first point of the Ram, the sign of the zodiac that runs from 0° to 30° longitude. The next sign is that of the Bull, which runs from 30° to 60° longitude; then comes the Twins, from 60° to 90°, and so on. (The names and symbols of the signs are given in fig. 2.2.) So, for example, one may write the longitude of Aldebaran either as 68° or as Twins 8°. Similarly, the longitude of Spica may be written either as 202° or as Balance 22°. Also, in giving the measure of any angle, the custom was once to express it in terms of signs, degrees, and minutes. The difference in longitude between Spica and Aldebaran, which is 134°, may also be written 4 signs, 14°, or $4^s14°$.

Any star may be located on the celestial sphere by means of either the equatorial coordinates α and δ or the ecliptic coordinates λ and β. As an example, star S in figure 2.16 is located on the celestial sphere to represent Capella, whose position is completely specified by either one of the following pairs of coordinates:

Coordinates of Capella

Equatorial coordinates	Right ascension	α =	5^h15^m
	Declination	δ =	+45°59′
Ecliptic coordinates	Longitude	λ =	81°32′
	Latitude	β =	+22°52′

If either pair of coordinates is known, it is possible to obtain the other, either by calculation or by examination of a celestial globe or armillary sphere.

Why is there a need for two sets of celestial coordinates if one set will suffice? The answer is that it is a matter of convenience. If one is interested in effects that depend on the diurnal rotation, then work will be simplified by the use of equatorial coordinates. On the other hand, the study of planetary motion is simplified by the choice of ecliptic coordinates, since the planets all move nearly in the plane of the ecliptic. The nature of the particular problem under consideration determines which set of coordinates ought to be used.

Coordinates in Greek Geography and Astronomy

If we examine the geographical and astronomical work of Ptolemy, we see the modern orthogonal coordinates fully developed. In his *Geography*, Ptolemy gave a list of 8,000 cities and other localities and specified their locations in terms of longitudes and latitudes, measured in degrees and minutes, exactly in our fashion. Ptolemy's reference meridian was the meridian through the "Fortunate Islands," that is, the Canary Islands.[36] Ptolemy selected the meridian through the Fortunate Islands as his zero of longitude because these islands were the westernmost part of the known world. Ptolemy's list of cities is laid out much like a modern gazetteer.

Ptolemy's *Geography* was one of the first works to make a thoroughgoing use of longitudes and latitudes. Ptolemy makes many references to his predecessor in geography, Marinus (ca. A.D. 100). From Ptolemy's remarks, it is clear that Marinus also used longitudes and latitudes, but not as systematically as Ptolemy. For example, Ptolemy complains that in Marinus's work, one must look in one place to find the latitude of a city and in another to find the longitude.

The Greeks before Marinus's time commonly specified the latitude of a place not in our fashion but in terms of the length of the summer solstitial day. For example, a Greek of the first century would have said that Rhodes is in the clime of 14 1/2 hours (see Geminus, *Introduction to the Phenomena* V, 23–25, in sec. 2.5). Another method of specifying latitude was in terms of the lengths of equinoctial shadows (as in sec. 1.12). Longitudes were often specified in terms of the time difference separating a locality from Alexandria. Even less systematic were the handbooks then in circulation, which gave the locations of various places in terms of their distances or their travel times from one another. Thus, while there are examples of earlier uses of latitudes and longitudes in our style, it seems that this usage did not become systematized until about the beginning of the second century A.D. Ptolemy's *Geography* played a major role in popularizing this approach.

In astronomy, too, Ptolemy's systematic use of orthogonal coordinates was decisive. Most of the *Almagest* uses ecliptic coordinates, that is, celestial longitudes and latitudes. For example, Ptolemy's catalog of stars in books VII and VIII gives the longitudes, latitudes, and magnitudes of some 1,000 stars. This catalog, which was not replaced until the Renaissance, is the direct ancestor of all modern star catalogs. In his planetary work, as well, Ptolemy regularly used ecliptic coordinates. Ptolemy, like most of his successors, specifies the longitude of a body by giving its zodiac sign, the degree within the sign, and minutes of angle (if required). For Ptolemy, as for us, the first sign of the zodiac is the Ram, which begins at the vernal equinox.

When we look at what remains of the astronomical work of Ptolemy's predecessors, it seems that a systematic use of orthogonal ecliptic coordinates was late to emerge. Of all the extant works of Ptolemy's Greek predecessors, the only one that contains a substantial amount of numerical data on star positions is Hipparchus's *Commentary on the Phenomena of Aratus and Eudoxus*. In that work, Hipparchus makes use of rather peculiar (from our point of view) sets of mixed coordinates. To be sure, Hipparchus does sometimes give declinations and right ascensions. But much more frequently he gives the longitude of the ecliptic point that rises at the same time as the star in question, or the longitude of the ecliptic point that culminates with the star, and so on. These are not orthogonal pairs and are not very convenient in calculation.

In the *Almagest*, Ptolemy cites many observations of his predecessors. Notable among these is a list of declinations of eighteen stars, attributed to Timocharis and Aristyllos (third century B.C.). Thus, Greeks of the third century were beginning to use and to measure actual celestial coordinates.

Ptolemy also cites a fair number of planetary observations by his predecessors. But here the situation is rather different. In the observations attributed to Timocharis, the planet being observed is said to be next to a certain star, with no actual numerical value assigned to the position. For, example, according to Ptolemy (*Almagest* X, 4), Timocharis observed Venus during the night between Mesore 17 and 18, in year 476 of Nabonassar; the planet appeared to be exactly opposite the star η Virginis. Because Ptolemy had measured the longitude of the star (set down in his star catalog), he was able to turn Timocharis's observation into a longitude of the planet: Venus was at Virgin 4 1/6° (longitude 154°10′). Even the planet observations of Ptolemy's own contemporary at Alexandria (a certain Theon) were given as angular distances from certain stars.

As discussed in section 6.4, the measurement of absolute celestial longitudes is a delicate business. It appears that there was little effort in this direction among the Greeks before Ptolemy's time. In this, as in so much else, his work proved to be very influential. The decisive event was perhaps the clarification of the nature of precession, which meant that ecliptic coordinates should be favored over equatorial coordinates.

The division of the zodiac into signs and the measurement of star and planet places in terms of zodiacal longitudes were, of course, a Babylonian inventions. Ptolemy's thoroughgoing use of ecliptic coordinates can be viewed as a Greek systemization of a practice adapted from Babylonian astronomy.

But it is important to note that there were several different conventions regarding the beginning points of the signs. For Ptolemy, as for later European astronomers, the beginning of the Ram is at the vernal equinoctial point. But the Babylonians placed the signs in such a way that the equinox fell at the 8th degree of the Ram. That is, the Babylonian sign of the Ram is shifted by 8° with respect to the Greek sign of the Ram. All of the other signs are similarly displaced by 8°. (The Babylonians also had a convention that placed the equinox at the 10th degree of the Ram. See sec. 5.2 for more detail.)

Among the early Greek astronomers, yet another convention was popular. Eudoxus, for example, defined the signs so that the equinoctial and solstitial points fell at the midpoints of the signs—the vernal equinox at the 15th degree of the Ram, the summer solstice at the 15th degree of the Crab, and so on. The vernal equinox was the same point for all: the difference lay in the way the artificial signs were defined.

Moreover, because the Athenian calendar year began with the new Moon immediately after the summer solstice, some of the earlier Greek writers made the Crab, rather than the Ram, the first sign of the zodiac. The *Almagest* standardized practice once and for all.

Conventional Symbols for the Signs of the Zodiac

The modern symbols for the signs of the zodiac are given in figure 2.2. Some of these symbols, such as the arrow for Sagittarius, seem to be truly ancient, but most of them date only from the Middle Ages. In ancient texts the names

Modern	Handy Tables Paris Grec 2493 XVI Cent.	Geminus Paris Grec 2385 XV–XVI Cent.	Anonymous Ast. Treatise Paris Grec 2419 XV Cent.	Alfonsine Tables Paris Latin 7432 before 1488
♈				
♉				
♊				
♋				
♌				
♍				
♎				
♏				
♐				
♑				
♒				
♓				

FIGURE 2.17. Zodiac symbols in some late medieval astronomical manuscripts.

of the signs were generally written out in ordinary fashion, or sometimes in abbreviated form. Another common notation represented the signs by the numbers one through twelve, but Aries was not always counted as the first sign.[37]

Figure 2.17 presents symbols collected from several late medieval astronomical and astrological manuscripts at the Bibliothèque Nationale, Paris. No attempt has been made to be exhaustive. The purpose of the table is merely to offer a few examples of the notation employed before the modern period.

Mathematical Postscript

As mentioned above, one may always perform a conversion between equatorial and ecliptic coordinates by manipulation of a globe. But, for the sake of convenience, here are trigonometric formulas for effecting the same conversion:

$$\sin \beta = \sin \delta \cos \varepsilon - \cos \delta \sin \alpha \sin \varepsilon$$

$$\tan \lambda = \frac{\sin \alpha \cos \delta \cos \varepsilon + \sin \delta \sin \varepsilon}{\cos \alpha \cos \delta}.$$

ε is the obliquity of the ecliptic, which for the modern era has the value $23°26'$. To obtain formulas for converting from ecliptic to equatorial coordinates, interchange the symbols in the above formulas according to the scheme $\beta \leftrightarrow \delta$, $\lambda \leftrightarrow \alpha$, $\varepsilon \to -\varepsilon$. Derivations of these formulas may be found in any textbook on spherical astronomy.

2.10 EXERCISE: USING CELESTIAL COORDINATES

1. The ecliptic coordinates of Regulus are approximately $\lambda =$ Virgin $0°$, $\beta = 0°$. Express the longitude of Regulus in degrees measured from the vernal equinox.
2. Use an armillary sphere or celestial globe to determine the equatorial coordinates α and δ of Regulus.
3. Use a celestial globe to determine the ecliptic coordinates of the following stars:

Star	Equatorial coordinates	
	α	δ
γ Sag	18^h	$-30°$
Betelgeuse (α Ori)	6	$+ 7$
Menkalinan (β Aur)	6	$+45$
Caph (β Cas)	0	$+60$
Phecda (γ UMa)	12	$+54$
Hamal (α Ari)	2	$+24$

The first three stars, which all lie on the solstitial colure, should be easy, since this colure, which is their hour circle, also happens to be perpendicular to the ecliptic. The last three may be a little tricky. On the celestial globe, stretch a string from the star down to the ecliptic so that string and ecliptic meet at right angles. The place where the string cuts the ecliptic gives the longitude of the star, and the length of the string, expressed in degrees, gives the latitude.

2.11 A TABLE OF OBLIQUITY

Table 2.3 is a *table of obliquity*, which gives the declination δ of every degree of the ecliptic. For example, the vernal equinox, which is the zeroth degree of the Ram, has declination $0°00'$.

TABLE 2.3. Table of Obliquity

Ram (Scales)	Decl. + (−)		Bull (Scorpion)	Decl. + (−)		Twins (Archer)	Decl. + (−)	
0°	0°00′	30°	0°	11°28′	30°	0°	20°09′	30°
1°	0°24′	29°	1°	11°49′	29°	1°	20°21′	29°
2°	0°48′	28°	2°	12°10′	28°	2°	20°33′	28°
3°	1°12′	27°	3°	12°31′	27°	3°	20°45′	27°
4°	1°35′	26°	4°	12°51′	26°	4°	20°57′	26°
5°	1°59′	25°	5°	13°11′	25°	5°	21°08′	25°
6°	2°23′	24°	6°	13°31′	24°	6°	21°18′	24°
7°	2°47′	23°	7°	13°51′	23°	7°	21°28′	23°
8°	3°10′	22°	8°	14°10′	22°	8°	21°38′	22°
9°	3°34′	21°	9°	14°30′	21°	9°	21°48′	21°
10°	3°58′	20°	10°	14°49′	20°	10°	21°57′	20°
11°	4°21′	19°	11°	15°07′	19°	11°	22°05′	19°
12°	4°45′	18°	12°	15°26′	18°	12°	22°13′	18°
13°	5°08′	17°	13°	15°44′	17°	13°	22°21′	17°
14°	5°31′	16°	14°	16°02′	16°	14°	22°28′	16°
15°	5°54′	15°	15°	16°20′	15°	15°	22°35′	15°
16°	6°18′	14°	16°	16°37′	14°	16°	22°42′	14°
17°	6°41′	13°	17°	16°55′	13°	17°	22°48′	13°
18°	7°04′	12°	18°	11°11′	12°	18°	22°53′	12°
19°	7°26′	11°	19°	17°28′	11°	19°	22°59′	11°
20°	7°49′	10°	20°	17°44′	10°	20°	23°03′	10°
21°	8°12′	9°	21°	18°00′	9°	21°	23°08′	9°
22°	8°34′	8°	22°	18°16′	8°	22°	23°12′	8°
23°	8°56′	7°	23°	18°31′	7°	23°	23°15′	7°
24°	9°19′	6°	24°	18°46′	6°	24°	23°18′	6°
25°	9°41′	5°	25°	19°01′	5°	25°	23°20′	5°
26°	10°02′	4°	26°	19°15′	4°	26°	23°22′	4°
27°	10°24′	3°	27°	19°29′	3°	27°	23°24′	3°
28°	10°46′	2°	28°	19°43′	2°	28°	23°25′	2°
29°	11°07′	1°	29°	19°56′	1°	29°	23°26′	1°
30°	11°28′	0°	30°	20°09′	0°	30°	23°26′	0°
	(−) + Decl.	(Fishes) Virgin		(−) + Decl.	(Water-Pourer) Lion		(−) + Decl.	(Goat-Horn) Crab

The names of the southern signs are in parentheses. In these signs the declinations are negative.

As a less trivial example, consider the point on the ecliptic with longitude Ram 25°. The declination of this point is δ = 9°41′. That is, when the Sun comes to Ram 25°, it will be 9°41′ north of the equator. This information is useful for a number of applications, for instance, in computing the Sun's noon altitude.

Suppose that we are located at north latitude L = 48° on the day that the Sun reaches longitude Ram 25° (April 15). From the table of obliquity (table 2.3), we find that the Sun's declination is δ = 9°41′. Then, at noon the Sun's zenith distance will be

$$z = L - \delta$$
$$= 48° - 9°41′$$
$$= 38°19′.$$

(Refer to sec. 1.12 and fig. 1.38 if necessary.) The Sun's altitude θ is the complement of its zenith distance: θ = 90° − z = 51°41′.

Note that in table 2.3, there is one other point on the ecliptic with declination 9°41′, namely Virgin 5°. The signs of the Ram and the Virgin are situated similarly with respect to the equator, as figure 2.2 illustrates, so the first column of declinations in the table serves for both signs. Similarly, the second column of declinations serves both for the Bull and the Lion: the two points Bull 10° and Lion 20° have declination 14°49'. For the six signs below the equator, the declinations should be regarded as negative. These signs are written within parentheses in the table. For example, the declination of the tenth degree of the Archer (and of the twentieth degree of the Goat-Horn) is −21°57'.

Our table was computed for use in the last half of the twentieth century, when the obliquity of the ecliptic has the value 23°26'. It corresponds to the table given by Ptolemy in *Almagest* I, 15, which is based on an obliquity of 23°51'20".

Historical Specimen

Figure 2.18 is a photograph of a table of obliquity in a fourteenth-century manuscript of the *Alfonsine Tables*, now in the Bibliothèque Nationale, Paris. The *Alfonsine Tables* were compiled around A.D. 1270 under the patronage of Alfonso X, King of Castile and Leon (Spain). The original Spanish version

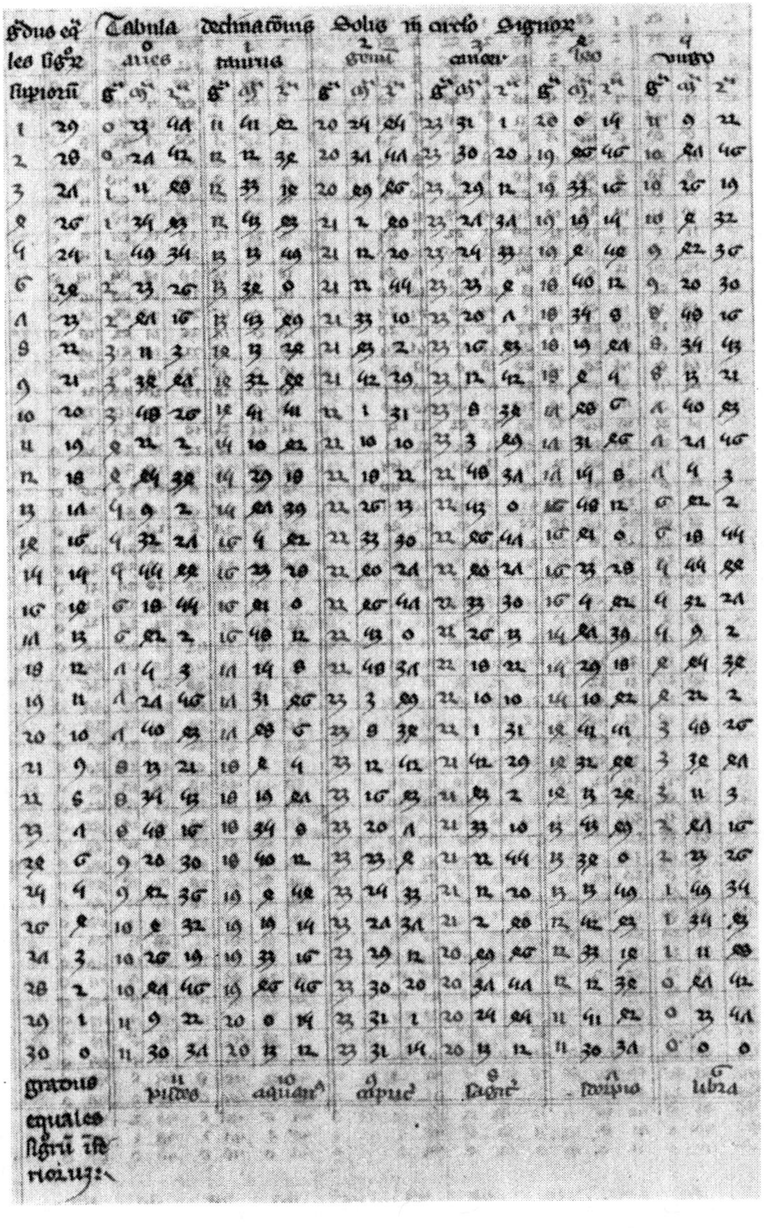

FIGURE 2.18. A table of obliquity from a fourteenth-century manuscript of the *Alfonsine Tables*. Bibliothèque Nationale, Paris (MS. Latin 7316A, fol. 114v).

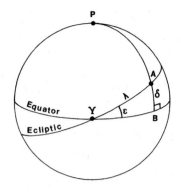

FIGURE 2.19.

of the tables is lost, but by about A.D. 1320, copies of the Spanish tables reached Paris, where they were reedited by one or several Parisian astronomers. From Paris, the revised form of the *Alfonsine Tables* was quickly disseminated throughout Latin-reading and -writing Europe. For two hundred years or more, these were the standard astronomical tables in use hroughout the Latin West.[38] The *Alfonsine Tables* exhibit a number of interesting innovations in arrangement and layout but they were, in basic principle, modeled on the ancient Greek tables of Ptolemy's *Almagest*. No more striking demonstration could be desired of the thousand-year continuity between Greek astronomical practice and the astronomy of the Middle Ages and the Renaissance.

The manuscript is neatly lettered in black and red ink. The table of obliquity is headed (if we write out fully a number of abbreviations) *Tabula declinationis Solis in circulo Signorum*: "Table of the declination of the Sun in the circle of signs." The names of the northern signs, numbered 0 through 5, run from left to right across the top of the table, beneath the main heading: Aries, Taurus, . . . Virgo. Similarly, the names of the southern signs, numbered 6 through 11, run from right to left across the bottom: Libra, Scorpio, . . . Pisces. The numerals are written in a common version of their medieval forms:

0 1 2 3 4 5 6 7 8 9

At the upper left corner is the heading for the leftmost column of the table: "Equal degrees of the upper signs." Below, the numbers run down from 1 to 30. These numbers are for use with the signs whose names appear at the top of the table. Similarly, in the lower left corner is the label for the second column of the table: "Equal degrees of the lower signs." The numbers run up from 0 to 29 and are for use with the signs whose names appear at the bottom of the table.

Within the column for each sign, the Sun's declination is given in degrees, minutes, and seconds. For example, when the Sun is at the 10th degree of Aries, its declination is 3°58′26″ north of the equator. From the declination of the Sun for the 30th degree of Gemini we see that the table is based on a value of the obliquity of the ecliptic of 23°31′15″. This is an accurate value for its time. In A.D. 1400, the obliquity of the ecliptic was indeed approximately 23°31′.

Mathematical Postscript

It is not necessary to know how the table of obliquity was computed in order to use it. However, for the sake of completeness, we present a postscript on the method of computation.

In figure 2.19, P is the north pole of the equator. A is the ecliptic point whose declination is desired. Draw the great circle arc PA and extend it until it reaches the equator at B. Arc AB is the declination that is sought. The law of sines applied to the right spherical triangle ♈AB yields

$$\frac{\sin \lambda}{\sin 90°} = \frac{\sin \delta}{\sin \varepsilon},$$

where $\lambda = ♈A$ (the Sun's longitude) and $\delta = AB$ (the Sun's declination). Thus,

$$\sin \delta = \sin \varepsilon \sin \lambda.$$

From this formula, the declination of point A is easily calculated in terms of its longitude and the obliquity of the ecliptic ε.

2.12 EXERCISE: USING THE TABLE OF OBLIQUITY

1. Find the declination of each of these points of the ecliptic: Bull 15°, Virgin 20°, Scorpion 10°, Water-Pourer 15°.

2. When the Sun is at Ram 25°, how long a shadow will a vertical gnomon cast at noon in Columbus, Ohio (latitude 39°58′ N)? Express the length of the shadow in terms of the gnomon's height.

3. Suppose you are shipwrecked on a desert island on August 24 and have no idea of your location. You do happen to remember, however, that August 24 is the date on which the Sun reaches the beginning of the sign of the Virgin. You set up a vertical gnomon 100 cm high and find that at noon it casts a horizontal shadow 26 cm long that points toward the north. What is the latitude of your island?

4. An astronomer at Seattle (latitude 47°40′ N) wanted to determine the exact time of the Sun's entry into the sign of the Bull in the year 1980. Since Bull 0° is 30° beyond the vernal equinox, and since the Sun moves roughly 1° in longitude per day, the astronomer knew that the Sun's entry into the Bull would take place within a few days of April 20. Therefore, the astronomer measured the Sun's altitude at noon on several days before and after April 20, with the following results:

Date (local noon, Seattle)	Noon altitude of Sun
April 17	53° 05′
18	53° 26′
19	53° 47′
20	54° 06′
21	54° 27′
22	54° 48′
23	55° 07′

Use these data to determine, to the nearest hour, the date of the Sun's entry into the sign of the Bull in the year 1980.

2.13 THE RISINGS OF THE SIGNS: A TABLE OF ASCENSIONS

The method of telling time by the risings of the zodiacal constellations described in section 2.6 is inexact. This inexactness has two sources: (1) the use of irregular constellations instead of the uniform zodiac signs and (2) the assumption that all the signs that rise during a given night rise in equal times. More precise time reckoning is possible if one knows the actual amount of time required for each sign to rise. A list of the rising times of the signs, called a *table of ascensions*, turns out to have many uses.

A Ptolemaic Table of Ascensions

Table 2.4 is modeled on Ptolemy's in *Almagest* II, 8. The parallels for which the rising times are given are specified in two ways: by means of the latitude, and by the length of the solstitial day. For example, the parallel of Mobile, Alabama, is that whose latitude is 30°51′ and whose solstitial day is 14 equinoctial hours.

In table 2.4 the rising times are expressed, not in hours, but in "time-degrees," where 360° represents one whole diurnal revolution. For example, at the latitude of Mobile, the first ten-degree segment of the Ram rises in 6°49′ (degrees of time), the second ten-degree segment of the Ram rises in 6°57′, and so on.

TABLE 2.4. Table of Ascensions

Signs	Tens	Right Sphere 12 hours, Lat. 0°		Parallel through Guatemala 13 hours, lat. 16°46'		Parallel through Mobile, Ala. 14 hours, lat. 30°51'	
		Time	Total Time	Time	Total Time	Time	Total Time
Ram	10	9°11'	9°11'	8°00'	8°00'	6°49'	6°49'
	20	9°17'	18°28'	8°06'	16°06'	6°57'	13°46'
	30	9°27'	27°55'	8°19'	24°25'	7°11'	20°57'
Bull	10	9°41'	37°36'	8°37'	33°02'	7°33'	28°30'
	20	9°57'	47°33'	9°00'	42°02'	8°03'	36°33'
	30	10°16'	57°49'	9°27'	51°29'	8°37'	45°10'
Twins	10	10°33'	68°22'	9°55'	61°24'	9°16'	54°26'
	20	10°45'	79°07'	10°21'	71°45'	9°58'	64°24'
	30	10°53'	90°00'	10°45'	82°30'	10°36'	75°00'
Crab	10	10°53'	100°53'	11°01'	93°31'	11°09'	86°09'
	20	10°45'	111°38'	11°09'	104°40'	11°34'	97°43'
	30	10°33'	122°11'	11°10'	115°50'	11°49'	109°32'
Lion	10	10°16'	132°27'	11°05'	126°55'	11°54'	121°26'
	20	9°57'	142°24'	10°55'	137°50'	11°53'	133°19'
	30	9°41'	152°05'	10°45'	148°35'	11°49'	145°08'
Virgin	10	9°27'	161°32'	10°35'	159°10'	11°42'	156°50'
	20	9°17'	170°49'	10°27'	169°37'	11°36'	168°26'
	30	9°11'	180°00'	10°23'	180°00'	11°34'	180°00'
Scales	10	9°11'	189°11'	10°23'	190°23'	11°34'	191°34'
	20	9°17'	198°28'	10°27'	200°50'	11°36'	203°10'
	30	9°27'	207°55'	10°35'	211°25'	11°42'	214°52'
Scorpion	10	9°41'	217°36'	10°45'	222°10'	11°49'	226°41'
	20	9°57'	227°33'	10°55'	233°05'	11°53'	238°34'
	30	10°16'	237°49'	11°05'	244°10'	11°54'	250°28'
Archer	10	10°33'	248°22'	11°10'	255°20'	11°49'	262°17'
	20	10°45'	259°07'	11°09'	266°29'	11°34'	273°51'
	30	10°53'	270°00'	11°01'	277°30'	11°09'	285°00'
Goat	10	10°53'	280°53'	10°45'	288°15'	10°36'	295°36'
	20	10°45'	291°38'	10°21'	298°36'	9°58'	305°34'
	30	10°33'	302°11'	9°55'	308°31'	9°16'	314°50'
Water-Pourer	10	10°16'	312°27'	9°27'	317°58'	8°37'	323°27'
	20	9°57'	322°24'	9°00'	326°58'	8°03'	331°30'
	30	9°41'	332°05'	8°37'	335°35'	7°33'	339°03'
Fishes	10	9°27'	341°32'	8°19'	343°54'	7°11'	346°14'
	20	9°17'	350°49'	8°06'	352°00'	6°57'	353°11'
	30	9°11'	360°00'	8°00'	360°00'	6°49'	360°00'

If we want to express this in terms of ordinary hours and minutes, we need only multiply by the conversion factor (24 hours/360°), that is, $1^h/15°$. So, the rising time of the first ten-degree segment of the Ram is $(6°49') \times (1^h/15°) = 27^m16^s$.

Another way of looking at this is in terms of a sign's co-rising segment of the celestial equator. At Mobile, the first 10° of the Ram rise in the same amount of time as it takes 6°49' of the equator to rise. Indeed, Ptolemy expresses his "rising times" in just this way, that is, in terms of the arc length of the equator that rises in the same time as the sign.

The column headed "total time" may be interpreted as the time at which each ecliptic point rises, measured from the rising of the vernal equinox. For example, at the latitude of Mobile, the 30th degree of the Ram rises 20 57/60 time-degrees after the equinoctial point rises.

The columns for the "Right Sphere" apply to places on the Earth's equator.

TABLE 2.4. (continued)

Signs	Tens	Parallel Through New London, Conn. 15 hours, lat. 41°27'		Parallel Through Vancouver, B.C. 16 hours, lat. 49°05'		Parallel Through Ketchikan, Ala. 17 hours, lat. 54°33'	
		Time	Total Time	Time	Total Time	Time	Total Time
Ram	10	5°41'	5°41'	4°37'	4°37'	3°37'	3°37'
	20	5°49'	11°30'	4°44'	9°21'	3°44'	7°21'
	30	6°06'	17°36'	5°02'	14°23'	4°00'	11°21'
Bull	10	6°30'	24°06'	5°27'	19°50'	4°27'	15°48'
	20	7°03'	31°09'	6°04'	25°54'	5°04'	20°52'
	30	7°46'	38°55'	6°53'	32°47'	5°56'	26°48'
Twins	10	8°36'	47°31'	7°53'	40°40'	7°06'	33°54'
	20	9°32'	57°03'	9°03'	49°43'	8°30'	42°24'
	30	10°27'	67°30'	10°17'	60°00'	10°06'	52°30'
Crab	10	11°18'	78°48'	11°28'	71°28'	11°40'	64°10'
	20	12°00'	90°48'	12°29'	83°57'	13°01'	77°11'
	30	12°29'	103°17'	13°12'	97°09'	13°59'	91°10'
Lion	10	12°45'	116°02'	13°39'	110°48'	14°35'	105°45'
	20	12°52'	128°54'	13°51'	124°39'	14°51'	120°36'
	30	12°52'	141°46'	13°54'	138°33'	14°56'	135°32'
Virgin	10	12°48'	154°34'	13°52'	152°25'	14°53'	150°25'
	20	12°44'	167°18'	13°49'	166°14'	14°49'	165°14'
	30	12°42'	180°00'	13°46'	180°00'	14°46'	180°00'
Scales	10	12°42'	192°42'	13°46'	193°46'	14°46'	194°46'
	20	12°44'	205°26'	13°49'	207°35'	14°49'	209°35'
	30	12°48'	218°14'	13°52'	221°27'	14°53'	224°28'
Scorpion	10	12°52'	231°06'	13°54'	235°21'	14°56'	239°24'
	20	12°52'	243°58'	13°51'	249°12'	14°51'	254°15'
	30	12°45'	256°43'	13°39'	262°51'	14°35'	268°50'
Archer	10	12°29'	269°12'	13°12'	276°03'	13°59'	282°49'
	20	12°00'	281°12'	12°29'	288°32'	13°01'	295°50'
	30	11°18'	292°30'	11°28'	300°00'	11°40'	307°30'
Goat	10	10°27'	302°57'	10°17'	310°17'	10°06'	317°36'
	20	9°32'	312°29'	9°03'	319°20'	8°30'	326°06'
	30	8°36'	321°05'	7°53'	327°13'	7°06'	333°12'
Water-Pourer	10	7°46'	328°51'	6°53'	334°06'	5°56'	339°08'
	20	7°03'	335°54'	6°04'	340°10'	5°04'	344°12'
	30	6°30'	342°24'	5°27'	345°37'	4°27'	348°39'
Fishes	10	6°06'	348°30'	5°02'	350°39'	4°00'	352°39'
	20	5°49'	354°19'	4°44'	355°23'	3°44'	356°23'
	30	5°41'	360°00'	4°37'	360°00'	3°37'	360°00'

Here, the celestial sphere is said to be "right" because the tropics and equator are perpendicular to the horizon.

Table 2.4 differs from Ptolemy's table in a number of minor ways. First, we have reduced the length of the table by eliminating the half-hour climes. Ptolemy included the climes of 12 1/2, 13 1/2, 14 1/2, 15 1/2, and 16 1/2 hours. Second, our table is founded on the modern value 23°26' for the obliquity of the ecliptic, rather than Ptolemy's value 23°51'20", which produces minor numerical differences in the table. The final and most obvious difference is in the place names associated with the parallels. Where Ptolemy has the lower part of Egypt (14 hours), we have Mobile.

The table of ascensions is a versatile tool allowing easy solution of many different kinds of problems. The advantage offered to the user of such a table is that *the compiler of the table has already done the trigonometry.* All that is required of the user is some simple arithmetic.

Directions for Using the Table of Ascensions

The use of the table of ascensions is easy but requires close attention. Rather than doggedly reading straight through the directions for all the applications, the reader is advised to read only the first one or two, and then try the analogous problems in the exercise of section 2.14. The reader may then proceed with more confidence to the other applications. The directions are modeled on Ptolemy's in *Almagest* II, 9, but are supplemented by worked examples (not to be found in the *Almagest*).

Length of the Day or Night

Example: How long is the day at latitude 49° when the Sun is in the 20th degree of the Lion (i.e., on August 13)?

The six signs following the Sun all rise in the course of the day. Therefore, we compute the rising time for the half of the zodiac starting at the Sun (Lion 20°) and extending eastward to the diametrically opposite point of the ecliptic (Water-Pourer 20°).

This may be done by adding up the rising times for the successive 10° segments of the ecliptic, starting with the last third of the Lion and extending to the first two-thirds of the Water-Pourer:

$$
\begin{array}{rll}
13° & 54' & \text{(latitude 49°)} \\
13° & 52' & \\
13° & 49' & \\
13° & 46' & \\
& \cdots & \\
6° & 53' & \\
+\;6° & 04' & \\
\hline
\text{Total}\quad 215° & 31' &
\end{array}
$$

The same result is found more easily by subtracting the total time for the arc Ram 0° → Lion 20° from the total time for the arc Ram 0° → Water-Pourer 20°:

$$340°10' - 124°39' = 215°31'.$$

This arc may be converted to equinoctial hours if desired. As 15° correspond to one hour, we divide by 15 to effect the conversion to hours:

$$215°31' = (215^h 31^m) \times \frac{1}{15}.$$

The division by 15 is most easily accomplished by the technique explained in section 2.6; that is, we write 1/15 as 4/60 and exploit the base-60 character of the ordinary units of time:

$$
\begin{aligned}
(215^h 31^m) \times \frac{1}{15} &= (215^h 31^m) \times \frac{4}{60} \\
&= (215^m 31') \times 4 \\
&= 14^h 22^m 4^s.
\end{aligned}
$$

This is the length of the day that was sought.

The length of the night may be found in a similar way, by computing the rising time of the ecliptic arc stretching from the point opposite the Sun eastward to the Sun itself, or by simply subtracting the daylight period from a whole cycle:

$$\text{length of the night} = 360° - 215°31'$$
$$= 144°29'$$
$$= 9^h38^m.$$

The result is in any case only an approximation—although a very good one. In the first place, the method assumes that the Sun remains at the same point of the ecliptic all day long, rather than moving the better part of a degree. Second, atmospheric refraction will cause the Sun to become visible a little before its geometrical rising and to remain visible a little after its geometrical setting. And, finally, daybreak really occurs when the upper limb of the Sun crosses the horizon, while the method of calculation applies to the center of the Sun's disk. However, all these effects combined will affect the length of the day by only 15^m or so.

Conversion of Times   As we have found, on August 13 at 49° N latitude, the day lasts 14^h22^m (equinoctial hours). If we divide by 12, we find how many equinoctial hours correspond to one seasonal hour:

$$\text{1 seasonal (day) hour} = 1^h12^m.$$

The length of a night hour may be computed in the same way. Dividing the length of the night, which is 9^h38^m, by 12 we obtain

$$\text{1 seasonal (night) hour} = 48^m.$$

Note that one seasonal day hour and one seasonal night hour always sum to two equinoctial hours.

Finding the Rising Point of the Ecliptic, Given the Seasonal Hour   Suppose we are given the date and the seasonal hour and are required to find the degree of the ecliptic that is rising. First, convert the seasonal hour to time-degrees. The resulting number expresses the time elapsed since sunrise (for a day hour) or since sunset (for a night hour).

Then, in the case of the day, enter the table for the appropriate latitude at the Sun's point and take out the total time. Add to this the time-degrees elapsed since sunrise, rejecting one cycle of 360° if the total exceeds this. Find in the table the degree of the ecliptic corresponding to the total. This will be the degree of the ecliptic that is rising at the given time.

In the case of the night, one proceeds similarly, but the table is entered with the point opposite the Sun rather than with the point of the Sun itself.

Example: Latitude 49°; the Sun is in Lion 20°; the time is three (seasonal) hours after sunset. Which point of the ecliptic is rising?

We found above that, for the given latitude of the observer and the given place of the Sun, the night lasts 144°29' (time-degrees). Three seasonal hours are one-fourth of this total, or 36°07'. At sunset the ecliptic point opposite the Sun (Water-Pourer 20°) is rising. We take, from the table for latitude 49°, the total time for this point and add the elapsed time:

Oblique ascension of point opposite Sun	340°	10'
Time elapsed since this point's rising	36°	07'
Total	376°	17'
Reject 360°	16°	17'

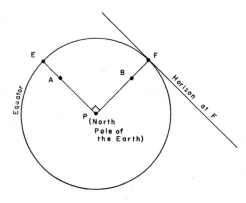

FIGURE 2.20.

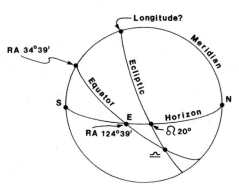

FIGURE 2.21.

Going back to the table we find that this oblique ascension corresponds to a point between Ram 30° and Bull 10°. Linear interpolation gives Bull 4°. This is the point of the ecliptic that is rising at the given time.

Finding the Culminating Point of the Ecliptic, Given the Hour   Suppose we are given the date and the seasonal hour and wish to find the degree of the ecliptic that is *culminating* (crossing the meridian in the south). It may seem that the table of ascensions cannot be used to solve this problem, since the table gives the time each sign requires to cross the horizon—not the times required to cross the meridian. However, the table can indeed be used, *because the meridian through any point on the Earth is equivalent to the horizon of some place on the Earth's equator.* In figure 2.20, the meridian plane through *A* is represented by line *PAE* and is parallel to the horizon plane at *F*, a point located 90° farther west than *E*. This means that the signs will cross the meridian anywhere on Earth in just the same way as they cross the horizon at the equator. Thus, Ptolemy directs his reader to use the table of ascensions for the right sphere in solving meridian problems.

The method: Express the seasonal hour in time-degrees, reckoned from noon for a day hour, or from midnight for a night hour, rather than from sunrise or sunset. Enter the table of ascensions for the right sphere (regardless of the actual latitude) with the point of the Sun (day) or with the point opposite the Sun (night). Add algebraically the time elapsed since noon or midnight, and find the degree of the ecliptic corresponding to the total time, again using the table for the right sphere. The result will be the degree of the ecliptic that is culminating at the given time.

Example: Latitude 49°; the Sun is in Lion 20°; the time is three seasonal hours after sunset. Which point of the ecliptic is culminating?

The time is three seasonal hours *before midnight*. As we already have shown, three night hours for the given place and date amount to 36°07' of time. We enter the table *for the right sphere* with the place opposite the Sun (Water-Pourer 20°) and take out the value 322°24'. This represents the total time at the moment of midnight, that is, the moment when the point opposite the Sun crosses the meridian. As we wish a time somewhat earlier than this, we subtract the three seasonal hours:

Total time at midnight	322°	24'
Less three seasonal hours	−36°	07'
Difference	286°	17'

286°17' is the total time at the desired moment. In the table for the right sphere, this time corresponds to Goat-Horn 15°, which is therefore the ecliptic point culminating at the given moment. Again, the essential feature is that the horizon at the equator plays the role of the meridian.

Finding the Degree Culminating, Given the Degree Rising   Since this problem does not involve the time of day, we will find it easier if we interpret the table of ascensions slightly differently than we have so far. The entry we have so far called "total time" should now be regarded as the right ascension of the equatorial point that rises at the same time as the given ecliptic point. For example, at latitude 49°, let Lion 20° be rising. The table gives 124°39': this is the right ascension (measured in degrees, rather than the usual hours) of the point of the equator that rises simultaneously with Lion 20° (see fig. 2.21).

Now we wish to find the degree of the ecliptic culminating, given the degree rising. Enter the table of ascensions for the appropriate latitude and take out the right ascension of the co-rising equatorial point. Subtract 90° to find the right ascension of the equatorial point that is simultaneously culminating (since there always is a 90° arc of the equator between the horizon and the meridian). Then go to the table of ascensions for the right sphere (used to represent the meridian) and find the point of the ecliptic that culminates with this point of the equator.

Example: At latitude 49°, when Lion 20° is rising, which point of the ecliptic is culminating?

R.A. of equatorial point that rises with Lion 20° (table for 49° latitude)	124°	39′
Less 90° of R.A.	−90°	00′
R.A. of the equatorial point on the meridian	34°	39′

(R.A. = right ascension.) We go to the table for the meridian (i.e., the table of ascensions for the right sphere) and find that this right ascension corresponds to Bull 7°. So, at latitude 49°, when Lion 20° is rising, Bull 7° is on the meridian.

Our explanation of this use of the tables is more detailed than that of Ptolemy, who gives no numerical example but only general, rather terse, directions. Nor does he explain why one subtracts the 90° for the quadrant of the equator, but simply states the rule. Perhaps he felt that this procedure would be transparent to the average reader of his work!

Equatorial Coordinates of an Ecliptic Point   Finally, let us point out an application used by Ptolemy, but not explained in so many words by him. The table of ascensions for the right sphere, together with the table of obliquity, can be used to determine the equatorial coordinates (right ascension and declination) of a point on the ecliptic. To review these coordinates, see section 2.9 and figure 2.14.

Example: What are the equatorial coordinates of Fishes 0°? (The zeroth degree of the Fishes is at longitude 330°, latitude 0°. We wish to convert these ecliptic coordinates into equatorial coordinates.)

Entering the table of ascensions (table 2.4) for the right sphere with Water-Pourer 30° (the same point as Fishes 0°), we take out $\alpha = 332°05′$. This is the right ascension of the point in question, expressed in degrees rather the usual hours. If we wish to express this quantity in the usual fashion, we divide by 15, with the result $\alpha = 22^h08^m$. To obtain the declination, we enter the table of obliquity (table 2.3) with the zeroth degree of the Fishes and find $\delta = -11°28′$.

Historical Notes

The risings of the signs were first studied because of their usefulness in telling time at night. Already in the *Phenomena* of Aratus (third century B.C.) it is noted that in any night six signs rise and six set, and that one can tell the time by looking to see which sign is rising. Commentators on Aratus often took pains to explain how it can be that in every night six signs rise, even though the nights are of unequal durations.

The *mathematical* attack on the problem began shortly after Aratus's time. In section 2.15 we shall see how, in the second century B.C., Hypsicles of Alexandria applied the Babylonian method of the arithmetic progression to obtain an approximate numerical solution. In a very short time (later in the second century B.C.), Hypsicles' work was made obsolete by the development of trigonometric methods, which for the first time made possible an exact solution of the old problem. By Ptolemy's time (second century A.D.), the problem was so completely solved that convenient tables were available to the astronomer and astrologer for practical use.

Most of the work in book II of the *Almagest* is not original with Ptolemy. The subjects treated there (gnomon problems, climes, day lengths, ascensions of the zodiac signs, etc.) were treated earlier by Hipparchus (second century B.C.) and others. In a number of example calculations that illustrate the construction of the tables, Ptolemy uses the latitude of Rhodes—a fact that Tannery[39] took as evidence that Ptolemy simply reproduced a treatise by Hipparchus, who worked at Rhodes. It may be so, but Ptolemy's use of Rhodes is no proof of it. Still, it is likely that Ptolemy was able to borrow something from the work of his predecessor.

Strabo[40] says that Hipparchus had given in tables, for all the places situated between the equator and the north pole, the various changes that the state of the sky presented. If Hipparchus really did construct such tables, Ptolemy may have had access to them. Very likely, Ptolemy's main contribution to this branch of astronomy was to refine the methods of calculation and to extend the scope of the tables. This should not be read as disparaging of Ptolemy: one does not blame the inventor of the automobile because he did not also invent the horse.

The contents of *Almagest* II, including the table of ascensions, became a standard part of astronomical knowledge. Every subsequent astronomical work that purported to be complete had to include the same material. For example, book II of Copernicus's *On the Revolutions of the Heavenly Spheres* (1543) covers this same ground. Copernicus's table of ascensions is based on a slightly smaller value of the obliquity of the ecliptic ($23°29'$) and gives the cumulative rising times for every sixth degree along the ecliptic (rather than every tenth as in Ptolemy) for latitudes running from $39°$ to $57°$ by $3°$ steps. In more recent times, the interest in horizon problems gradually died out, and the table of ascensions dropped out of the textbooks.

Finally, we should say something of the applications of the table of ascensions to Greek astrology. In making a prognosis for any person or event, it was essential to know the state of the heavens at the moment in question. For a person, this would be the moment of birth (or of conception, if known); for an event, for example, the accession of a king, the moment of the event itself. One of the most important points in the heavens was the *horoscope*—the degree of the ecliptic that was rising at the given moment. The importance of the horoscope is reflected in the fact that its name later came to signify the entire chart or method by which predictions are made. Now, why was the degree of the ecliptic that was rising called the horoscopic point? The Greek *hora* is the word for the hour of the day. *Skopos* is an object on which the eye is fixed, a mark. So the *horoskopos* is the "hour mark"—the sign one uses to tell the time during the night. This term originally had no astrological connotation, but was bound to acquire one due to its astrological applications.

The horoscope could not be determined accurately unless the time was known to within a fraction of an hour. In his work on astrology, Ptolemy[41] criticizes the majority of astrologers because of their use of sundials and water clocks. The first of these are liable to error due to shifts of their positions or of their gnomons, and the second due to irregularities in the flow of their water. Only observation *by means of horoscopic astrolabes* at the time of birth can give the minute of the hour. Ptolemy's astrolabe probably corresponds to what we would call a quadrant or a sextant, that is, a graduated circle equipped with sights by which the altitudes of stars above the horizon can be taken. Once the time is accurately known, it is possible to determine the degree of the zodiac that is rising; this is done, as Ptolemy says, *by the method of ascensions*, that is, by the use of tables like those we have discussed.

Historical Specimen

Figure 2.22 is a photograph of part of the table of ascensions in a Greek manuscript copy of Ptolemy's *Almagest*. This manuscript, copied in the ninth century and now more than a thousand years old, is one of the oldest surviving

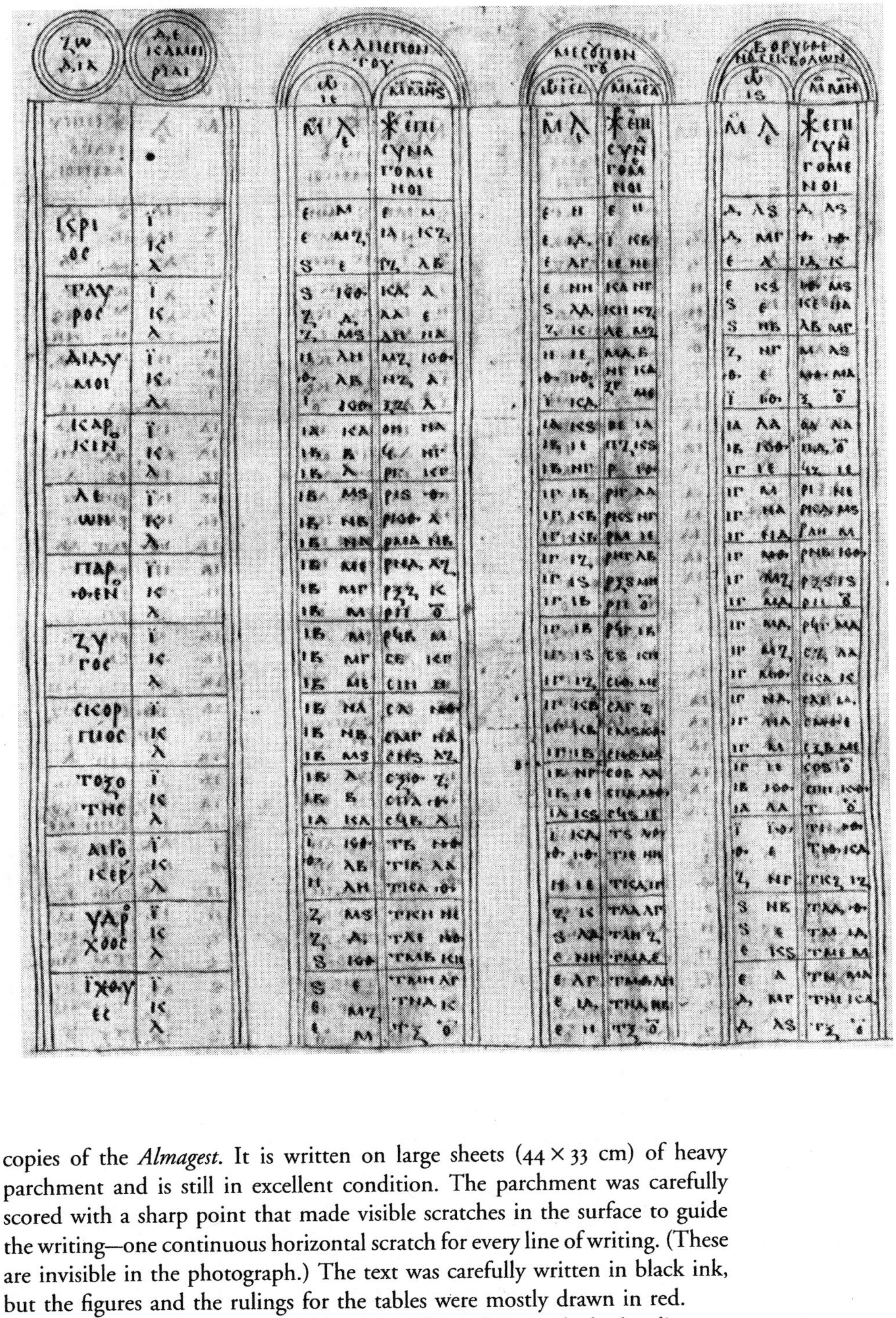

FIGURE 2.22. Part of the table of ascensions from a ninth-century parchment manuscript of the *Almagest*. Bibliothèque Nationale, Paris (MS. Grec 2389, fol. 44v).

copies of the *Almagest*. It is written on large sheets (44 × 33 cm) of heavy parchment and is still in excellent condition. The parchment was carefully scored with a sharp point that made visible scratches in the surface to guide the writing—one continuous horizontal scratch for every line of writing. (These are invisible in the photograph.) The text was carefully written in black ink, but the figures and the rulings for the tables were mostly drawn in red.

The leftmost column is headed Ζωδια, "signs." Beneath the heading are the names of the twelve signs of the zodiac: Κριος (Aries), Ταυρος (Taurus), and so on. The second column is headed "ten-degree segments." Beneath the heading run repeating cycles of the numbers 10, 20, 30 (ι, κ, λ). These columns correspond exactly to the first two columns of table 2.4.

The next pair of columns gives the rising times of the ten-degree segments of the ecliptic for the clime of 15 hours, that is, for the parallel through the Hellespont (latitude 40°56'). The first parts of these columns are translated in figure 2.23. The values of the rising times may be compared with those in table 2.4. Our values differ slightly from Ptolemy's because they are based on a slightly different value for the obliquity of the ecliptic.

The last two pairs of columns of the manuscript page are for the climes

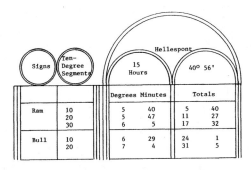

Signs	Ten-Degree Segments	Hellespont			
		15 Hours		40° 56'	
		Degrees	Minutes	Totals	
Ram	10	5	40	5	40
	20	5	47	11	27
	30	6	5	17	32
Bull	10	6	29	24	1
	20	7	4	31	5

FIGURE 2.23. Translation of the upper left corner of figure 2.22.

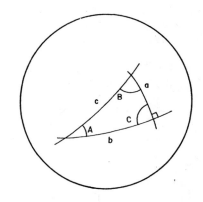

FIGURE 2.24.

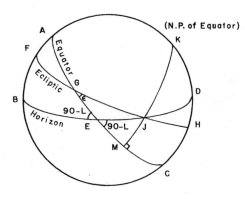

FIGURE 2.25.

of 15 1/2 hours (middle of the Pontos, or Black Sea) and 16 hours (mouth of the river Borysthenes, the modern Dnieper).

Mathematical Postscript

It is not necessary to know how tables of ascensions are calculated in order to use them. However, for the sake of completeness, a method of calculation is outlined here. Readers who are not on friendly terms with trigonometry may skip this postscript.

Oblique Ascensions We shall need one theorem from spherical trigonometry. Refer to figure 2.24. Let a, b, c denote the sides of a right spherical triangle, and A, B, C, the opposing angles. Let C be the right angle. Then

$$\sin a = \tan b \cot B.$$

For the problem of ascensions in the oblique sphere, refer to figure 2.25. *ABCD* is the celestial meridian, *BED* the horizon, *AEC* the equator, and *FGH* the ecliptic. *G* represents the vernal equinoctial point. *K* is the north pole of the equator. From *K*, we drop a great circle arc through *J*, which meets the equator perpendicularly at *M*.

The angle ε at *G* is the obliquity of the ecliptic. The angle between the equator and the horizon is the co-latitude, that is, $90° - L$ where L is the latitude of the place of observation.

At the moment represented in the figure, point *J* of the ecliptic is on the horizon. It is clear that arc *GJ* of the ecliptic rises with arc *GE* of the equator. (The first point of each arc is the same, point *G*, and the last points of the arcs, *J* and *E*, are on the horizon at the same time.) In a table of ascensions, arc *GE* goes in as the total time opposite ecliptic longitude *GJ*. The problem, then, is to calculate *GE* in terms of *GJ*.

In right spherical triangle *GJM*, we apply our theorem to obtain

$$\sin GM = \tan JM \cot \varepsilon.$$

Note that *JM* is the declination of point *J* of the ecliptic. *JM* can therefore be taken from the table of obliquity (table 2.3) for any desired *GJ*. Hence, *GM* is determined.

In right spherical triangle *EJM* we apply the same theorem to obtain

$$\sin EM = \tan JM \cot (90 - L)$$
$$= \tan JM \tan L,$$

so *EM* is also determined.

The desired arc *GE* is the difference between our two results:

$$GE = GM - EM.$$

As an example, let us compute *GE* for the clime of 14 hours ($L = 30°51'$) and the case where *J* is the 30th degree of the Ram ($GJ = 30°$). We enter the table of obliquity with Ram 30° and take out

$$JM = 11°28'.$$

Now we have

$$\sin GM = \tan (11°28') \cot (23°26')$$
$$= 0.46800.$$
$$GM = 27°54'.$$

Similarly,

$$\sin EM = \tan (11°28') \tan (30°51')$$
$$= 0.12116.$$
$$EM = 6°58'.$$

Finally,

$$GE = 27°54' - 6°58'$$
$$= 20°56',$$

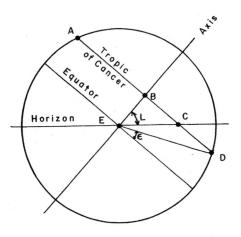

FIGURE 2.26.

which is the number tabulated in the table of ascensions (to within a 1 minute discrepancy attributable to rounding).

The method of calculating ascensions given here is more streamlined than Ptolemy's. The equivalent of the sine function was known and used in antiquity, but the tangent was not. As a result, the ancient methods of calculation are slightly more cumbersome.[42]

Latitudes and Solstitial Days   The Greeks identified parallels either by latitude or by the length of the solstitial day. We shall see here how to calculate the one if given the other.

Figure 2.26 presents a side view of the celestial sphere at the time of summer solstice. The Earth is at E. The axis of the cosmos makes an angle L with the horizon, this angle being equal to the latitude of the place of observation. The Sun is on the tropic of Cancer, north of the equator by an angle ε equal to the obliquity of the ecliptic. Line $ABCD$ is a side view of the Sun's day circle. The radius of this day circle is

$$BD = r \cos \varepsilon,$$

where $r = ED$ is the radius of the celestial sphere. Also,

$$EB = r \sin \varepsilon,$$

and thus,

$$BC = EB \tan L$$
$$= r \sin \varepsilon \tan L.$$

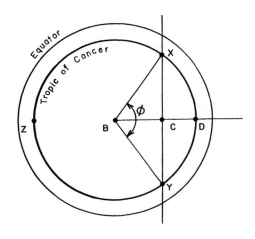

FIGURE 2.27.

Now, figure 2.27 presents a view of the sphere as seen looking down the axis. Arc XDY is the part of the day circle lying below the horizon, and YZX is the part above. Let us denote the length of the night at summer solstice by N_S. N_S is related to ϕ by

$$\phi = N_S \times (360°/24^h) = N_S \times (15°/^h).$$

We need only calculate ϕ, the "night angle":

$$\cos (\phi/2) = BC/BY.$$

But $BY (= BD)$ and BC are both known. Thus,

$$\cos (\phi/2) = r \frac{\sin \varepsilon \tan L}{r \cos \varepsilon},$$

and we obtain

$$\cos (N_S \times 7.5°/^h) = \tan \varepsilon \tan L.$$

This formula may be used to calculate the latitude L at which the shortest night has some particular value N_S.

Example: Let us calculate the latitude where the summer solstitial day is 14 hours, and the night is 10:

$$\tan L = \cos (10^h \times 7.5°/^h)/\tan (23°26')$$
$$= 0.59714$$
$$L = 30°51'.$$

This is the latitude at which the shortest night lasts 10 hours, that is, the latitude of the clime of the 14 hour solstitial day.

Finally, we should point out that our formula can be applied to find the length of *any* night, not just the solstitial night. One simply uses the Sun's declination δ for the day in question rather than the obliquity. That is, the general formula is

$$\cos (N \times 7.5°/^h) = \tan \delta \tan L.$$

As an example, let us calculate the length N of the night at latitude 30°51' when the Sun is at the zeroth degree of the Twins. Using the table of obliquity (table 2.3) we find $\delta = 20°29'$ when the Sun is at Twins 0°. So, we have

$$\cos (N \times 7.5°/^h) = \tan (20°09') \tan (30°51')$$
$$= 0.21917.$$
$$N = \cos^{-1}(0.21917)/(7.5°/^h)$$
$$= 10^h 19^m,$$

which agrees with the length of the night given in table 2.2.

2.14 EXERCISE: ON TABLES OF ASCENSIONS

Problems for the table of ascensions (table 2.4):

1. How long does the day last when the Sun is at the 20th degree of the sign of the Scorpion (Nov. 13) in the clime of 16 hours (Vancouver)? Express your answer in terms of equinoctial hours. (Answer: $9^h 7^m$.)
2. Suppose that, in the same situation as described in problem 1, a traffic accident occurs 4 seasonal hours after sunset. Express the time in terms of equinoctial hours. (Answer: 9:31 P.M.).
3. What are the equatorial coordinates of the ecliptic point at the 20th degree of the Bull? Express the right ascension in terms of the usual hours and minutes. (Answer: $\alpha = 3^h 10^m$, $\delta = +17°44'$.)
   In problems 4–7, assume a clime of 14 hours (Mobile, Alabama) and suppose the Sun is in the 30th degree of the Balance (October 24). Note that you may check your answers with an armillary sphere or celestial globe. The concrete model will not reveal very small errors, of course, but if you go very far wrong in the table of ascensions, the armillary sphere will warn you that something is amiss.
4. Find the length of the day in equinoctial hours, using the table of ascensions.
5. Suppose an observation of the Moon is made at night, three seasonal hours after sunset. Express the time in terms of equinoctial hours.

6. At the time given in problem 5, which degree of the ecliptic is rising?
7. Which degree of the ecliptic is culminating?
8. Can one use the table of ascensions to solve problems for places on the Earth south of the equator? If so, how?
9. Use the table of ascensions to compute the missing entries of table 2.2 (the length of the night).

2.15 BABYLONIAN ARITHMETICAL METHODS IN GREEK ASTRONOMY: HYPSICLES ON THE RISINGS OF THE SIGNS

Tables of Ascensions before Ptolemy

Exact calculation of the rising times of the signs requires trigonometry. The oldest known table of ascensions constructed by exact trigonometric methods is that of Ptolemy, discussed in section 2.13. However, approximate solutions of the problem were obtained earlier by purely arithmetical methods. In particular, Hypsicles of Alexandria, in a little book on the rising times (*Anaphorikos*), which dates from the first half of the second century B.C., demonstrated a plausible solution that can teach us a good deal about the state of mathematics at that time.

Hypsicles of Alexandria   In Hypsicles' time, Greek geometry was in full bloom: Euclid's *Elements* already was a century old; Apollonius's treatise on the conic sections and Archimedes' mathematical works had been around for about half that time. Hypsicles himself was an able mathematician. His book on the dodecahedron and the icosahedron extended Euclid's book XIII. But it is one thing to prove a general proposition about triangles, or about the dodecahedron, and quite another to resolve any given triangle, that is, to calculate numerical values for its unknown angles and sides. The first sort of problem belongs to geometry, for which the methods of Euclid suffice. The second belongs to trigonometry, and its solution implies the knowledge of theorems for the addition and multiplication of sines and cosines, and so on, and the existence of trigonometric tables. These tools were not available to Hypsicles but began to be developed shortly after his lifetime. His approximate calculation of the rising times of the signs represents the attempt of a highly developed mathematics to come to grips with a problem that was essentially beyond the scope of its powers.

Hypsicles' simplifying assumption is that the rising times of the signs increase in arithmetic progression from the Ram to the Virgin, and decrease in the same way from the Balance to the Water-Pourer. Let T denote the time required for either the Ram or the Fishes to rise. (These are the signs that rise most quickly.) Then, according to Hypsicles, the times required for the signs to rise are as follows:

Ram	T
Bull	$T + x$
Twins	$T + 2x$
Crab	$T + 3x$
Lion	$T + 4x$
Virgin	$T + 5x$
Balance	$T + 5x$
Scorpion	$T + 4x$
Archer	$T + 3x$
Goat-Horn	$T + 2x$
Water-Pourer	$T + x$
Fishes	T

Here x stands always for the same increment of time. Of course, the rising times do not really follow this simple pattern. Nevertheless, the assumption of an arithmetic progression was a definite step forward. The simplest possible hypothesis for the rising times would have made them all equal—two hours each—but even cursory observation would have revealed its inadequacy, since, at Alexandria for instance, the Ram rises in less than an hour and a half while the Virgin requires nearly two and a half hours. Once one recognizes differences among the rising times, the arithmetic progression becomes the simplest possible way of accounting for them.

The Table   Assuming the arithmetic progression and using the known ratio of the length of day to that of night at the summer solstice (which is 7/5 for the latitude of Alexandria), Hypsicles calculates the rising times given in the second column of table 2.5. Hypsicles' rising times are expressed, not in hours, but in "degrees of time," where 360° represents one whole diurnal revolution (15 time-degrees = 1 hour). Another way of looking at this is in terms of a sign's co-rising segment of the equator. So, for example, the Ram rises in the same amount of time as it takes 21 2/3° of the equator to rise.

The arithmetic progression is apparent in Hypsicles' results, for the rising times change regularly from one sign to the next by 3°20′, which corresponds to about 13 minutes. The actual rising times are given in the third column table 2.5, as calculated trigonometrically by Ptolemy. The real times evidently do not form an arithmetic progression; nevertheless, Hypsicles' results follow them fairly closely.

How the Table Is Constructed   At summer solstice, the Sun is at the beginning of the Crab. Six signs rise during the period from sunrise to sunset. These signs, and their assumed rising times, are:

Crab	$T + 3x$
Lion	$T + 4x$
Virgin	$T + 5x$
Balance	$T + 5x$
Scorpion	$T + 4x$
Archer	$T + 3x$

Total rising time $6T + 24x$ = length of the day.

During the night, the other six signs rise (from Goat-Horn through Twins). Adding up their rising times, we obtain

$$6T + 6x = \text{length of the night.}$$

Now, Alexandria is in the clime of 14 hours. That is, at summer solstice, the day lasts 14 equinoctial hours and the night lasts 10 equinoctial hours. If we express these in terms of degrees of time, the day is 210°, the night 150°. We thus obtain two equations in two unknowns:

$$6T + 24x = 210°$$
$$6T + 6x = 150°.$$

These equations suffice to determine T and x, with the result

$$T = 21\frac{2}{3}°$$

$$x = 3\frac{1}{3}° .$$

From these two values the whole table can be filled out.

TABLE 2.5 Hypsicles' Table of Ascensions for the Parallel of Alexandria (clime of 14 hours)

Sign	Hypsicles' Rising Times	Differences	Ptolemy's Rising Times	Differences
Ram	21°40′		20°53′	
		+3°20′		+3°19′
Bull	25°00′		24°12′	
		+3°20′		+5°43′
Twins	28°20′		29°55′	
		+3°20′		+4°42′
Crab	31°40′		34°57′	
		+3°20′		+0°59′
Lion	35°00′		35°36′	
		+3°20′		−0°49′
Virgin	38°20′		34°47′	
		0		0
Balance	38°20′		34°47′	
		−3°20′		+0°49′
Scorpion	35°00′		35°36′	
		−3°20′		−0°59′
Archer	31°40′		34°37′	
		−3°20′		−4°42′
Goat-Horn	28°20′		29°55′	
		−3°20′		−5°43′
Water-Pourer	25°00′		24°12′	
		−3°20′		−3°19′
Fishes	21°40′		20°53′	

Arrangement of Hypsicles' Book Hypsicles' *Anaphorikos* is interesting from the point of view of history of mathematics for its statement and proof of several propositions about arithmetic series. Let a_1, a_2, a_3, . . . a_n be an arithmetic progression of n terms. That is, each term differs from the preceding one by a constant difference δ. Thus, $a_2 = a_1 + \delta$, $a_3 = a_2 + \delta$, and so on.

Hypsicles proves that if the number of terms in the series is even, the sum of the series is

$$a_1 + a_2 + a_3 + \cdots + a_n = n(a_1 + a_n)/2.$$

That is, the sum of the series is equal to the number of terms times half the sum of the first and last terms.

If the number of terms is odd, and a_m is the middle term, Hypsicles proves that

$$a_1 + a_2 + \cdots + a_m + \cdots + a_n = na_m.$$

That is, the sum of the series is equal to the number of terms times the middle term.

Hypsicles also proves that, if there are an even number of terms,

$$\text{(sum of second half)} - \text{(sum of first half)} = n^2\delta/4.$$

Hypsicles does not, of course, use algebraic formulas. Moreover, he proves his theorms for a specific number of terms (six). Hypsicles then applies these theorems to the problem of determining the rising times of the signs at Alexandria, assuming 14 hours for the length of the solstitial day. The algebraic solution of the problem outlined above is a considerable simplification of Hypsicles' actual procedure.[43]

Another interesting feature of Hypsicles' book is its division of the circle into 360 parts. Hypsicles' is the earliest known Greek work to use the degree, a Babylonian unit of measure. We have seen (sec. 1.2) that the Babylonians divided the day into watches, which varied in length the course of the year, like the Greeks' seasonal hours. However, the Babylonian astronomers also divided the whole 24-hour period into 360 parts, each of which is called one UŠ (degree). Thus, Hypsicles' time-degree is of Babylonian origin.

Origin of Hyspicles' Method and Its Later History

The use of the degree is only most obvious Babylonian influence in Hypsicles' work. Indeed, the whole scheme of using an arithmetic progression to represent the rising times of the signs is of Babylonian origin. The evidence for the Babylonian origin of this method comes from cuneiform clay tablets of the Seleucid period (third century B.C. and later). In fact, the Babylonians used two slightly different versions of the system. In one system (called system A by modern historians), the rising times form a strict arithmetic progression. In the other (system B, of course), the rising times form an arithmetic progression with two exceptions: the change in rising times is twice as big as normal (a difference of $2x$ rather than x) between Twins and Crab, and also between Archer and Goat-Horn. (The values of T and x for a given clime must therefore be different in system B than in system A.)

The arithmetic progression in rising times was first deduced by Otto Neugebauer from Babylonian values for the lengths of days at different times of year. There is, after all, an intimate connection between the length of the day and the rising times of the signs: the length of any day is equal to the time it takes for six zodiac signs to rise, beginning with the Sun's position. However, the rising times of the signs also turn up explicitly on some tablets, so there is no question that the Babylonians fully understood the whole system we see discussed by Hypsicles.[44]

There are echoes of the Babylonian arithmetical scheme (both versions) in many later Greek and Roman writers. For example, Geminus (*Introduction to the Phenomena* VI, 38) says that the differences in the lengths of the days themselves form an arithmetic progression. That is, the lengths of the days form a progression of *constant second differences*. We shall see in the exercise of section 2.16 that this results directly from the use of an arithmetic progression for the rising times of the signs. Geminus thus seems to follow system A. But Cleomedes[45] gives values for the day lengths that show an anomalous jump characteristic of system B.

In many cases, it appears that later Greek and Roman writers were unaware of the Babylonian origin of their schemes for rising times and day lengths. Even after the development of trigonometry made the arithmetic methods obsolete, many Greek and Roman astrologers continued to use the old arithmetic methods because they were easier than trigonometry. Moreover, the arithmetic formulas for rising times and day lengths were taken up and used by writers who did not even understand the connection between them. Thus, Manilius, the author of a long Latin astrological poem (first century A.D.), gives a list of rising times that follows system A[46] and a list of day lengths that follows system B,[47] without realizing that these are inconsistent with one another.

Many histories of Greek astronomy have tended to overemphasize its cultural independence, its logical coherence, and its allegiance to philosophical principles. Certainly, the Greek achievement in astronomy was remarkable—one of the most remarkable in the history of science. But our brief examination of arithmetic techniques in Greek astronomy provides a necessary corrective.

The dependence on Babylonian methods is quite clear. Moreover, the slapdash use of these methods by some later writers shows that not every practitioner of Greek astronomy was a Ptolemy. Far from it![48]

2.16 EXERCISE: ARITHMETIC PROGRESSIONS AND THE RISINGS OF THE SIGNS

1. Rising times of the signs: Use Hypsicles' method (sec. 2.15) to calculate the rising times of the signs at the Earth's equator (clime of 12 hours). Calculate also the rising times for the clime of 16 hours. Compare your results with the exact rising times obtained from table 2.4. At which clime—12, 14, or 16 hours—does Hypsicles' approximation work best? How large are the errors?

2. Day lengths resulting from Hypsicles' scheme: The assumption that the rising times of the signs form an arithmetic progression leads to other consequences. For example, it turns out that the lengths of the days form a progression with *constant second differences*. The length of the day when the Sun is at the beginning of the Goat-Horn (winter solstice) can be found by adding the rising times for the six signs starting with the Goat-Horn. This we did in section 2.15, with the result $6T + 6x$. Similarly, the length of the day when the Sun is at the beginning of the Water-Pourer is found by adding the rising times for the six signs starting with the Water-Pourer; the result is $6T + 7x$. Continuing, we get

Day lengths resulting from Hypsicles' scheme

Sun at beginining of	Length of day	Differences	Second differences
Goat-Horn	$6T + 6x$		
		x	
Water-Pourer	$6T + 7x$		$2x$
		$3x$	
Fishes	$6T + 10x$		

Finish out the table and show that the day lengths form an ascending and then a descending progression with constant second difference $2x$.

For the clime of 14 hours, put in numerical values for T and x to determine the day lengths. Plot a graph of day length versus the Sun's position in the zodiac. Plot the actual day lengths (from table 2.2) as well as the day lengths resulting from Hypsicles' assumption on the same graph. The use of day lengths with constant second differences is mathematically equivalent to fitting a parabola to the day-length curve. How well does this scheme approximate the actual variation in the length of the day?

2.17 OBSERVATION: THE ARMILLARY SPHERE AS AN INSTRUMENT OF OBSERVATION

The armillary sphere is a scale model of the celestial sphere. Thus, if the model is properly aligned, it can show the actual orientation of the heavens. This is the basis for the use of the armillary sphere as an instrument of observation.

The directions below are given for the use of an armillary sphere. However, they may also be applied to many celestial globes. The globe should be of transparent plastic and should have an axis passing through it, with a miniature globe of the Earth at the center. The globe must also have a horizon stand

and a meridian ring that is adjustable for latitude. It will be helpful to make the shadows of the ecliptic and the equator more visible. This can be done by sticking narrow tape (e.g., typewriter correction tape) on the ecliptic and on the equator, all the way around the globe. The two taped rings thus turn the solid globe into an armillary sphere, if we imagine away the plastic surface of the globe!

The Armillary Sphere As a Sundial   Set up an armillary sphere (or celestial globe) in a level, sunny place so that the meridian line on the base points along the local terrestrial meridian—that is, exactly in the north-south direction. Also adjust the model for your own latitude. (The axis should make an angle with the horizon that is equal to your latitude.) With these adjustments made, the equator of the model lies parallel to the plane of the celestial equator, and the axis of the model points at the celestial pole. That is, the axis of the model is the axis of the universe.

Note that the *shadow of the axis* of the model falls on the equator ring, which is marked in hours. As the day goes by, the Sun moves in a circle around the axis, and the shadow will move along the equator ring. If it were local noon the shadow would fall on the meridian, so you need simply count the number of hour marks that separate the shadow from the meridian. This will be the time of day, expressed in hours before or after local noon. The sundial reads local Sun time, of course, and may depart from clock time.

The armillary sphere can also be made to read the time of day directly, in 24-hour military or international style. Turn the sphere until the vernal equinox (and 0-hour mark) coincides with the shadow of the axis. The time of day will then be indicated by the hour mark of the equator that is crossing the upper part of the meridian ring.

The armillary sphere is an example of an *equatorial sundial*. The simplest possible version of such a dial would consist of only these essential parts: a fixed ring in the plane of the equator, marked in hours, and a gnomon perpendicular to it.

Using the Armillary to Measure the Longitude of the Sun   In using the armillary sphere as a sundial, one must point the axis of the model at the celestial pole. This involves adjustments only to the fixed circles of the model: the base and the meridian ring. The movable part of the sphere (made up of ecliptic, equator, etc.) can be turned at will without affecting the usefulness of the dial. But if we wish to use the armillary to determine the longitude of the Sun, we must position the movable sphere so that the ecliptic ring lies in the plane of the true ecliptic.

To do this, adjust the model, as before, for the local meridian and latitude. Then slowly turn the sphere about its axis until the shadow of the ecliptic ring falls across the middle of the Earth globe in the center of the model. Take a pencil point and run it along the ecliptic ring, keeping it perpendicular to that ring, until the shadow of the pencil point also falls across the center of the Earth globe. You should then see the shadow of the ecliptic and the shadow of the pencil point intersecting in an X on the Earth globe. The pencil point marks the Sun's position on the ecliptic, so the Sun's longitude and the date may be read off.

One complication must be mentioned. On a given date, there may be two different ways to orient the ecliptic ring that will produce a shadow of that ring on the Earth globe. On April 20, for example, the Sun is 11° north of the equator. On August 24 the Sun is again 11° north of the equator. The length of the day, the rising direction of the Sun, and the lengths and directions of shadows are all the same on August 24 as on April 20. Therefore, no single observation made with sundial or armillary sphere will enable one to choose between these two dates. But if one waits a few days after the original observation, the Sun will advance in longitude and its declination will change. If the date was April 20 the Sun will move farther north. But if the date was August 24 the Sun will move south. So, a second observation will remove the ambiguity in the first one.

The armillary sphere, equipped with sights, was Ptolemy's chief instrument for measuring the positions of stars and planets (see fig. 6.8 and sec. 6.4).[49]

Our knowledge of the scientific and technological activity of the ancients is based mostly on the written testimony of the ancients themselves. The *physical* (as opposed to textual) evidence for the scientific activity of the Greeks is meager, for delicate scientific instruments tend not to survive. Objects preserved from antiquity tend to be made of relatively indestructible stuff: building stone, ceramics, marble statuary. Thus, it should come as no surprise that the corpus of some 250 Greek and Roman sundials, found at sites all over the Mediterranean, constitutes the great bulk of the physical evidence for the place of astronomy in classical civilization.[1]

Varieties of Ancient Sundials

The ancient dial makers were very inventive and designed many different kinds of dials. To judge by the number that survive, one of the most popular kinds was the *spherical dial*. In its simplest form, this consisted of a hemispherical cavity cut into a block of stone. A gnomon was set into the stone with its tip at the center of curvature of the cavity. This spherically shaped cavity was a model of the celestial sphere. The concave surface of the cavity typically was engraved with circles representing the tropics and the equator, as well as with other curves that served to indicate the hours. The principal of the spherical dial with central gnomon is illustrated in figure 3.1. This design, however, would have been impractical. It would have been laborious to cut a complete hemisphere out of stone. Besides, the cavity would have filled with rain water. Fortunately, a complete hemisphere is not needed. The shadow of the gnomon's tip cannot fall just anywhere on the spherical surface. In particular, the shadow tip can never fall outside the belt between the two tropic circles. Therefore, the entire hemisphere is not needed, and the unnecessary portions of the block can be cut away as in figure 3.2. The popularity of the spherical dial derived from its simplicity. Because the spherical dial is merely a reduced version of the celestial sphere, the theory governing the placement of the curves is very simple.

Many *conical dials* have also been found. Indeed, the known number of this type exceeds even that of the spherical dials. In a conical dial, the shadow-receiving surface is a portion of the inner surface of a cone. Typically, the conical depression was cut into the edge of a rectangular slab of stone (fig. 3.3). The stoneworking involved in making a conical depression was simpler than that required for a spherical cavity. But, by compensation, the theory was slightly more complicated: it was necessary to project the celestial sphere onto a conical surface.

About forty *plane dials* are also preserved. The theory underlying such dials is more complicated than the theory of spherical or conical dials, for the celestial sphere must be projected onto a plane surface. Some dials were designed for horizontal receiving surfaces, others for vertical surfaces.

Two Horizontal Plane Sundials of the Hellenistic Period

Among the surviving Greek and Roman sundials engraved on plane surfaces, there are fifteen that were designed to be horizontal. Figure 3.4 represents a horizontal plane sundial pieced together from three fragments found in 1814 on the Vigna Cassini near the Via Appia, Rome.[2] The extant fragments form approximately the left half of the dial, the portion bound by the wavy line in the figure. The part of the dial to the right of the wavy line is a conjectural restoration. This dial is made of white marble, and the extant portion measures approximately 35 cm wide and 54 cm high. The marble is 36 cm thick.

The engraved line that runs vertically in the figure is the meridian. The

THREE

Some Applications of Spherics

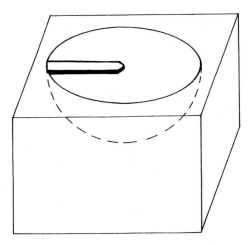

FIGURE 3.1. The principle of the spherical dial.

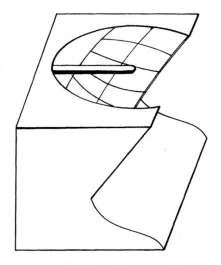

FIGURE 3.2. Spherical dial with cutaway south face. Day circles are visible for winter solstice, equinox, and summer solstice. The dial has been cut away along the circle representing the tropic of Cancer (the day circle for summer solstice). Also visible is a family of curves for the seasonal hours.

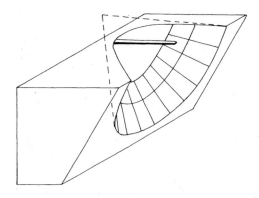

FIGURE 3.3. Principle of the conical dial.

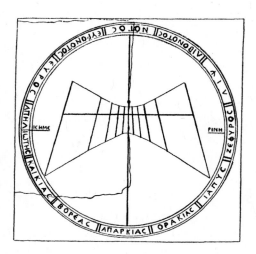

FIGURE 3.4. A horizontal, plane sundial found near Rome. From Diels (1924).

upper part is the southern end of the dial, and the lower part, the northern end. In use, the dial would lie on the ground, with its face horizontal and its meridian aligned north-south. A trace of a gnomon hole is preserved on the meridian line, at the broken right edge of the dial, just above the bat-wing shape.

This bat wing is characteristic of all Greek and Roman horizontal plane dials. The upper curve is the track of the tip of the gnomon's shadow for the day of summer solstice. The horizontal straight line below this is the shadow track for the equinox. The lowermost curve is the shadow track for the winter solstice. These three shadow tracks may be compared with figure 1.36.

The eleven more or less vertical lines in figure 3.4 are hour lines. These indicate the time of day as the tip of the shadow crosses them one by one. The hours are *seasonal*. That is, the period from sunrise to sunset consists always of twelve hours, by definition. At sunrise, the gnomon's shadow would be infinitely long and would point to the west, that is, toward the right on the figure. As the Sun rose higher, the shadow would shorten until the tip of the shadow reached the first hour line on the right. The time would then be one seasonal hour after sunrise. The shadow would continue to shorten until noon, the sixth hour, when it would fall on the meridian. In the afternoon, the shadow would lengthen, crossing the eleventh hour line (the last on the left) one seasonal hour before sunset.

This dial was engraved around its perimeter with the names of the winds, in Greek. On the preserved part of the dial, the wind names are ,

Notos	South
Euronotos	South-southeast
Euros	East-southeast
Apeliotes	East
Kaikias	East-northeast

Figure 3.5 shows another horizontal, plane sundial, found on the island of Delos in 1894.[3] The dial is engraved on a slab of white marble streaked with gray. The slab measures 37 cm × 50 cm and is about 6 cm thick. As is almost invariably the case, the gnomon has not been preserved. However, traces of iron remain in the gnomon hole, which is about 1.3 cm in diameter. The three day curves are engraved with Greek inscriptions (from top to bottom):

ΤΡΟΠΑΙ ΘΕΡΙΝΑΙ	Summer solstice
ΙϹΗΜΕΡΙΑ	Equinox
ΤΡΟΠΑΙ ΧΕΙΜΕΡΙΝΑΙ	Winter solstice

An unusual feature, found on only a few other dials, is the triangular wedge formed by two straight lines that radiate from the noon mark on the day curve for the winter solstice. These two lines may be translated

where the time of every day remains (right)
where the time of every day has passed (left)

These lines call attention to the variation in the length of the day. The amount of time the shadow spends outside the triangular wedge is the same for every day. In the case of the equinox, approximately 3 seasonal hours are cut out by the triangular wedge (1 1/2 hours on each side of the noon line). Thus, the twelve seasonal hours of the winter solstitial day are equal to 9 equinoctial hours. In the case of the summer solstice, about 4 2/3 seasonal hours are cut out by the wedge. Thus, the twelve seasonal hours at the winter solstice are equal to only 7 1/3 seasonal hours of the summer solstitial day. As already mentioned, all surviving Greek and Roman dials are marked in seasonal hours. There is not a single example of a dial that indicated equinoctial hours throughout the year. Interestingly, on the very few dials that call attention to the varying length of the day, such as this dial from Delos, the winter hour is chosen as the standard of comparison.

The triangular wedge on the Delos dial affords an easy way of determining the clime, or latitude, for which the dial was designed. The wedge indicates that 12 winter hours = 9 equinoctial hours. That is, at winter solstice, the period from sunrise to sunset (12 seasonal hours) lasts 9 equinoctial hours. This implies that at summer solstice the day would last 15 equinoctial hours. The dial seems, then, to have been designed for the clime of 15 hours. This corresponds to a latitude of about 41° (as can be seen in table 2.2 or 2.4). In section 3.4 we examine another method of determining the latitude for which the dial was designed, using the lengths of the noon shadows.

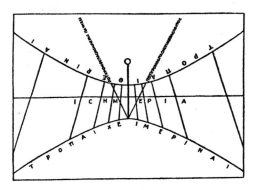

FIGURE 3.5. A horizontal, plane sundial found on Delos. From Diels (1924).

The Tower of the Winds

The most remarkable feat of dial making preserved from the ancient period is the Tower of the Winds in the Agora (marketplace) of Athens. This eight-sided marble building was constructed by a Macedonian astronomer, Andronikos of Kyrrhos, around 50 B.C. Figure 3.6 shows a view of the octagonal tower, sketched by James Stuart and Nicholas Revett, two British historians of

FIGURE 3.6. The Tower of the Winds in Athens. From Stuart and Revett (1762). Photo courtesy of Yale University Library.

architecture who published a series of engravings of Greek and Roman antiquities in 1762.[4] Each face of the tower bears a relief sculpture of a wind god. In figure 3.6, we see, from left to right, Apeliotes (the east wind), Kaikias (northeast), and Boreas (north). Below each relief is a sundial. The eight sundials, which face in eight different directions, had, of course, to be individually designed.

Vitruvius mentions the Tower of the Winds in his *Ten Books on Architecture* (I, 6). According to Vitruvius, on the top of the tower there was a bronze weather vane in the shape of a Triton (a son of Poseidon). The Triton turned to face the wind. A wand in the Triton's hand pointed to the name of the wind that was blowing. This weather vane had disappeared by the eighteenth century, but Stuart and Revett added it to their engraving (fig. 3.6), following Vitruvius's description.

The stone floor of the interior was carved with channels that received no explanation until an archaeological investigation of the 1960s suggested they were channels for conducting water to run a water clock.[5] The water clock displayed inside the Tower of the Winds was almost certainly of the type known as an *anaphoric clock*. A wheel, representing the sky and adorned with the figures of the constellations, turned around once each day. As the wheel turned, the constellations passed by a metal wire representing the horizon. Thus, the anaphoric clock would show at a glance which constellations were rising and setting—even during the daytime. The anaphoric clock is closely connected with the astrolabe in both its underlying theory and its historical origins and is discussed in section 3.7.

3.2 VITRUVIUS ON SUNDIALS

Our principal source on Greek and Roman sundials is Vitruvius, a Latin writer on architecture who lived in the age of Augustus. We have already seen (sec. 1.4) his use of the gnomon for laying out city streets. Vitruvius believed that the architect should be equipped with knowledge from many branches of study, including geometry, history, philosophy, music, medicine, and astronomy, as well as the more specialized arts of building construction. For all these sciences bear on architecture in some way. Accordingly, Vitruvius's *Ten Books on Architecture* ranges over many fields of ancient science and technology. The greatest digression from purely architectural matters is found in Vitruvius's ninth book, which consists of an elementary survey of astronomy.

Beginning of Book Nine

Book IX begins with an introduction in which Vitruvius laments the fact that authors are not accorded the same honors and riches as athletes. After some grumbling, which differs very little from the grumbling still heard today when the salaries of professional athletes are discussed, Vitruvius gives examples of several authors who have benefitted mankind with their discoveries: Plato (the doubling of the square), Pythagoras (the theorem on right triangles), Archimedes (the famous "Eureka!" story), and Archytas and Eratosthenes (the doubling of the cube). The discoveries of these men are everlasting. But the fame of athletes declines rapidly with their bodily powers.

Now Vitruvius takes up the subject of book IX: astronomy and time reckoning. The first six chapters are devoted to the zodiac, the motions of the planets, the phases of the moon, the constellations, and the prediction of the weather from the stars. Throughout, the scientific level is very low. Vitruvius had but a weak grasp of astronomy and was writing for an audience he deemed to be interested in only a superficial introduction.

The Analemma

In chapter 7 Vitruvius takes up the subject of sundials. Most of this short chapter is devoted to a description of the *analemma*, which is a two-dimensional projective drawing of the celestial sphere (see fig. 3.7). An analemma plays the same role in the construction of sundials as a lemma plays in the construction of a mathematical proof. It is a preliminary construction that permits one to reach the desired goal. The analemma described by Vitruvius was not original with him, but its inventor is unknown. A number of mathematicians devoted themselves to the theory of sundials, and many must have used analemmas of one sort or another. However, the only other ancient writer on this topic whose work has come down to us is Ptolemy, who was later than Vitruvius. The analemma treated by Ptolemy in his book *On the Analemma*[6] is not the same as Vitruvius's. Unfortunately, neither Ptolemy nor Vitruvius can teach us exactly how the ancients used their analemmas in the construction of sundials. Vitruvius, as we shall see, contents himself with explaining the construction of the analemma itself and forswears giving any example of its uses lest he should prove tiresome by writing too much. Ptolemy probably did provide examples, but this part of his treatise has been lost. Vitruvius's chapter 7 is here presented in its entirety:

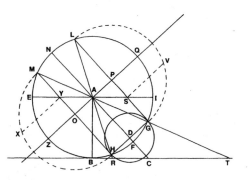

FIGURE 3.7. The analemma of Vitruvius.

EXTRACT FROM VITRUVIUS

Ten Books on Architecture IX, 7

1. In distinction from the subjects first mentioned, we must ourselves explain the principles which govern the shortening and lengthening of the day. When the Sun is at the equinoxes, that is, passing through Aries or Libra, he makes the gnomon cast a shadow equal to eight ninths of its own length, in the latitude of Rome. In Athens, the shadow is equal to three fourths of the length of the gnomon; at Rhodes to five sevenths; at Tarentum, to nine elevenths; at Alexandria, to three fifths; and so at other places it is found that the shadows of equinoctial gnomons are naturally different from one another.

2. Hence, wherever a sundial is to be constructed, we must take the equinoctial shadow of the place. If it is found to be, as in Rome, equal to eight ninths of the gnomon, let a line be drawn on a plane surface, and in the middle thereof erect a perpendicular, plumb to the line, which perpendicular is called the gnomon. Then, from the line in the plane, let the line of the gnomon be divided off by the compasses into nine parts, and take the point designating the ninth part as a center, to be marked by the letter *A*. Then, opening the compasses from that center to the line in the place at the point *B*, describe a circle. This circle is called the meridian. [See fig. 3.7.]

3. Then, of the nine parts between the plane and the center on the gnomon, take eight, and mark them off on the line in the plane to the point *C*. This will be the equinoctial shadow of the gnomon. From that point, marked by *C*, let a line be drawn through the center at the point *A*, and this will represent a ray of the Sun at the equinox. Then, extending the compasses from the center to the line of the plane, mark off the equidistant points *E* on the left and *I* on the right, on the two sides of the circumference, and let a line be drawn through the center, dividing the circle into two equal semicircles. This line is called by mathematicians the horizon.

4. Then, take a fifteenth part of the entire circumference, and, placing the center of the compasses on the circumference at the point where the equinoctial ray cuts it at the letter *F*, mark off the points *G* and *H* on the right and left. Then lines must be drawn from these [and the center] to the line of the plane at the points *T* and *R*, and thus, one will represent the ray of the Sun in winter, and the other the ray in summer. Opposite

E will be the point *I*, where the line drawn through the center at the point *A* cuts the circumference; opposite *G* and *H* will be the points *L* and *M*; and opposite *C*, *F*, and *A* will be the point *N*.

5. The, diameters are to be drawn from *G* to *L* and from *H* to *M*. The upper will denote the summer and the lower the winter portion. These diameters are to be divided equally in the middle at the points *P* and *O*, and those centers marked; then, through these marks and the center *A*, draw a line extending to the two sides of the circumference at the points *Z* and *Q*. This will be a line perpendicular to the equinoctial ray, and it is called in mathematical figures the axis. From these same centers open the compasses to the ends of the diameters, and describe semicircles, one of which will be for summer and the other for winter.

6. Then, at the points where the parallel lines cut the line called the horizon, the letter *S* is to be seen on the right, the letter *Y* on the left; and from the letter *S* draw a line parallel to the axis as far as the semi-circle on the right, which it cuts at *V*; and from *Y* to the semi-circle on the left draw in the same way a parallel which cuts it at *X*. These parallels are called . . . <Further, draw a parallel line from the point *H*, where the summer ray cuts the circumference, to the point *G*, where the winter ray cuts the circumference. This parallel is called> . . . *loxotomus*. Then, place the point of the compasses at the intersection of this line and the equinoctial ray—call this point *D*—and open them to the point where the summer ray cuts the circumference at the letter *H*. Around the equinoctial center, with a radius extending to the summer ray, describe the circumference of the circle of the months, which is called *menaeus*. Thus we shall have the figure of the analemma.

7. This having been drawn and completed, the scheme of hours is next to be drawn on the baseplates from the analemma, according to the winter lines, or those of summer, or the equinoxes, or the months, and thus many different kinds of dials may be laid down and drawn by this ingenious method. But the result of all these shapes and designs is in one respect the same: namely, the days of the equinoxes and of the winter and summer solstices are always divided into twelve equal parts. Omitting details, therefore—not for fear of the trouble, but lest I should prove tiresome by writing too much—I will state by whom the different classes and designs of dials have been invented. For I cannot invent new kinds of myself at this late day, nor do I think that I ought to display the inventions of others as my own. Hence, I will mention those that have come down to us, and by whom they were invented.[7]

The analemma is most easily understood as a side view of an armillary sphere (compare fig. 3.7 with fig. 2.9). In figure 3.7, circle *NIFE* represents the celestial meridian. The axis of the universe is line *ZQ*. The horizon plane is represented by line *EI*, and the Earth is at *A*. The two tropics and the equator are seen in an edge-on view as lines *HM* (tropic of Capricorn), *FN* (equator), and *GL* (tropic of Cancer). These three circles naturally stand perpendicular to the plane of the figure, with a semicircle above the plane and a semicircle below. In figure 3.7, a semicircle of each tropic has been folded into the plane of the diagram. These folded tropics are represented by the dashed semicircles *MZH* and *LVG*.

On the day of the summer solstice, the Sun's diurnal path through the sky coincides with the tropic of Cancer. Point *L* will be its position at noon, and *G* its position at midnight. Now, if the dashed semicircle *LVG* is imagined folded up (the fold being along the diameter *LG*) so that the tropic is in its correct position, then it is easy to see that the dashed line *SV* will lie in the plane of the horizon. Thus, *V* represents the position of the Sun at sunrise or sunset, on the day of the summer solstice. If we let the semicircle *GVL* represent the period from midnight to noon, then arc *GV* is the portion of the night from midnight till sunrise, and arc *VL* is the portion of the day from sunrise till noon. The same semicircle *LVG* can also represent the second half of the day, that is, the period from noon to midnight: in this case we interpret *V* as sunset. Similarly, on the folded-down tropic of Capricorn (semicircle *MXH*), point *X* represents either sunrise or sunset on the day of the winter solstice.

Conclusion of Book Nine

As is apparent from the extract, Vitruvius is content to describe the construction of the analemma and does not bother to explain its use in the design of sundials. (We shall see in sec. 3.3 the likely technique in the case of a horizontal plane sundial.) In the eighth and final chapter of book IX, Vitruvius lists a number of different kinds of dials, together with the names of their supposed inventors. The cut hemispherical dial, illustrated in figure 3.2, is attributed to Berosus the Chaldaean; the conical dial, to Dionysodorus; the plane disk, to Aristarchus of Samos. A number of other types of dials are named, but because Vitruvius gives no details, it is difficult to identify them all today with any certainty. Vitruvius ends his discussion of sundials with the remark that anyone who wishes to learn how to mark a dial can find out how to do so from the works of those who have written on this subject, provided that he understands the figure of the analemma. Evidently, there were other treatises on dialing that have not come down to us. Chapter 8 ends with a discussion of water clocks. Despite some gaps in the discussion, Vitruvius is fairly clear and provides valuable detail on this branch of ancient technology.

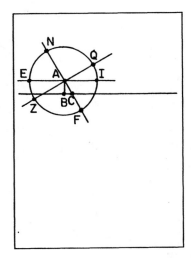

FIGURE 3.8.

3.3 EXERCISE: MAKING A SUNDIAL

Vitruvius describes the construction of the analemma but does not demonstrate its application to the design of sundials. It is therefore impossible to say exactly how the ancient dial makers used the analemma to produce the face of a sundial, but it is possible to make a good guess.

Gustav Bilfinger[8] has shown how the analemma can be used to construct horizontal and spherical sundials. Other solutions are possible, but Bilfinger's has the merit that it uses all the parts of the analemma and requires no new ones. Whether this is the actual method followed by the ancients cannot be proved, but, even if it is not, it must be close in spirit.

Step 1: Construction of the Analemma

The first step in making a sundial is the construction of the analemma according to the directions given in the extract from Vitruvius's *Ten Books on Architecture*. However, it will be convenient to modify Vitruvius's directions slightly.

Obtain a large sheet of paper, 20″ × 30″ or larger. Lay the paper on your working table with the shorter sides running left to right, as in figure 3.8. Draw the meridian circle, with at least a 4″ radius, in the upper left corner of the paper, but be careful to leave several inches of space between the circle and the edges of the paper. Through the center *A* of the circle, draw the horizontal line *EI* that represents the horizon.

Draw the axis *ZQ* of the universe through *A* so that it makes an angle with the horizon equal to the latitude of the place for which you wish to construct the sundial. The axis intersects the meridian at point *Q*, above the horizon, and at point *Z*, below. Angle *QAI* is equal to the latitude.

Choose an appropriate length for the gnomon, 1″ or 1 1/4″, and draw the gnomon AB perpendicular to the horizon. Draw through the base *B* of the

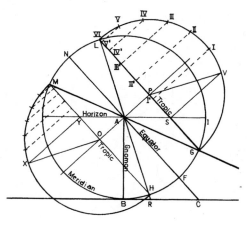

FIGURE 3.9. Division of the diurnal circle into hours (step 2).

gnomon a line parallel to the horizon and let it extend all the way across the paper. This baseline represents the ground, on which the shadows of the gnomon are to be projected. (Here we depart slightly from Vitruvius's directions. It is important to make the meridian circle large so that the pencil work can be accurate. But if we made the gnomon 4″ high—equal to the radius of the meridian—the sundial face would turn out six or seven feet wide. It is the height of the gnomon that determines the size of the finished sundial. The tip A of the gnomon must lie at the center of the celestial sphere, but the base B may be put at any convenient place. In all that follows, our diagram will be exactly the same as Vitruvius's, except that the baseline is shifted upward.)

Draw the equator FN through point A, perpendicular to the axis ZQ. The equator cuts the meridian at N, above the horizon, and at F, below. It also cuts the baseline at C. When the Sun is at N, at noon on the equinox, it will produce the shadow BC. (Note that Vitruvius begins his construction by specifying the ratio of the equinoctial noon shadow BC to the gnomon's height AB. But since the modern reader is more likely to know the latitude of the place where he or she lives than the length of the equinoctial shadow there, we begin by constructing the latitude angle QAI. The correct length of the equinoctial noon shadow then follows automatically.)

We have so far finished the parts of the construction that Vitruvius describes in paragraphs 2 and 3. We have also drawn the axis, which is described in paragraph 5. Complete the figure of the analemma according to Vitruvius's prescription in paragraphs 4, 5, and 6. Note that angle $LAN = NAM = 24°$, that is, the obliquity of the ecliptic. Also, we will have no need for line HG (loxotomus) or the small circle (menaeus) having HG as diameter, so you may leave these out.

Step 2: Division of the Diurnal Circles into Hours

Now that the analemma has been drawn, we shall prepare it for use in the construction of a horizontal sundial. The figures accompanying our directions are drawn for the latitude of Rome. To make the shadows longer, for the sake of clarity in the little diagrams, we have drawn the baseline tangent to the meridian circle as, indeed, Vitruvius prescribes. However, as explained above, you should draw your own baseline higher.

Summer Solstice On the day of the summer solstice, the Sun's diurnal motion will carry it around a circle coinciding with the tropic of Cancer. In figure 3.9 this circular path is seen edge on as line LG. The actual diurnal circle would stand up perpendicular to the plane of the diagram with LG as its diameter. Semicircle LVG represents this diurnal circle folded down into the plane of the diagram—or really one-half of the diurnal circle. Arc LV represents the part of this semidiurnal arc that is above the horizon, and VG, the part below.

According to modern practice, we should place the center of a protractor at P and divide arc LVG into 12 equal segments of 15° each, representing the twelve equal hours that we count from midnight (point G) to noon (L). However, the ancient Greeks divided the period between sunrise and sunset into twelve *seasonal hours*, which were therefore longer in summer and shorter in winter. Equivalently, the time between sunrise (V) and noon (L) is divided into six parts. Accordingly, place the center of a protractor at P, find by direct measurement that angle LPV is 115°, and divide arc LV into intervals of one-sixth this size, or about 19° each. (These particular numbers apply only to the latitude of Rome, of course.) The resulting hour marks are labeled I, II,

III, . . . VI. (Be careful not to make the mistake of dividing arc *LV* by placing the center of the protractor at *S*. The center of the protractor must be placed at the center *P* of the Sun's diurnal circle.)

Project each of the points *I, II, . . . V* onto line *LS* by means of lines drawn parallel to the axis *AP*. The resulting points *I′, II′, . . . V′* represent the position of the Sun at each of the hours, as seen in a side view of the celestial sphere. Note that point *VI′* (noon) would be the same as *VI*.

Winter Solstice   The diurnal path of the Sun at winter solstice coincides with the tropic of Capricorn. Seen from the side, this circle appears in figure 3.9 as line *MH*. When half the circle is folded down into the plane of the diagram, it appears as semicircle *MXH*. Arc *MX* represents the portion of the semidiurnal path that is above the horizon. Therefore, place the center of a protractor at *O* and divide angle *MOX* into six equal parts, marking the divisions along arc *MX*. (Be careful not to place the center of the protractor at *Y*.) The resulting points are then projected onto line *MY* by means of lines parallel to the axis.

Step 3: Construction of the Shadow Tracks

Refer to figure 3.10. Mark point *W* below the analemma exactly on the line determined by points *A* and *B*. It does not matter how far *W* lies below the analemma, but place *W* roughly in the middle of the available empty space. *W* will be the gnomon's position on the functioning sundial.

In constructing the shadow tracks, we must locate the tip of the shadow at each hour of the day. We locate the tip of the shadow by finding (1) the

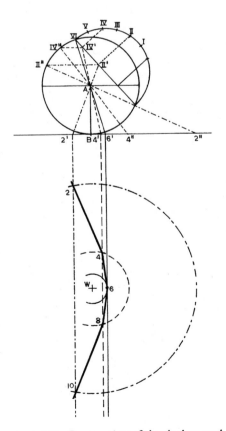

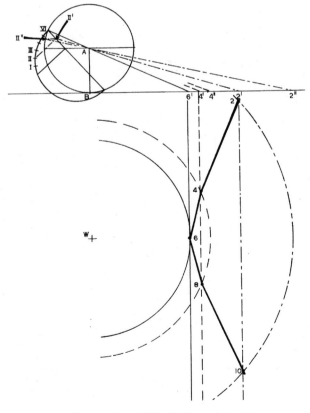

FIGURE 3.10 Construction of the shadow track, summer solstice (step 3).

FIGURE 3.11 Construction of the shadow track, winter solstice (step 3).

distance by which it lies north or south of the gnomon's base, and (2) the actual length of the shadow.

Summer Solstice To find the noon shadow, project a line from point *VI* (which represents the Sun's place on the meridian at noon) through the tip *A* of the gnomon (fig. 3.10). The point 6′ where this line crosses the baseline determines the length of the noon shadow. Therefore, draw a line down from 6′ parallel to line *ABW*. Open the compasses to make a radius equal to *B6′*, then place the point at *W* and draw an arc, which will touch the new line at the point we have marked 6. The line segment *W6* (whose length is equal to *B6′*) is the actual noon shadow on the day of the summer solstice.

To find the tip of the shadow at any other hour of the day (take the second hour as a case), proceed as follows. From point *II′*, project a line through *A* to the baseline, which it cuts at 2′. Distance *B2′* represents the distance by which the shadow tip lies south of the gnomon's base (*B* or *W*). Therefore, draw a line from 2′ straight down the page parallel to the line *ABW*. The tip of the shadow must lie somewhere along this line at the second hour of the day. To find just where, we must make another projection, which will determine the actual length of the shadow.

The projection from *II′* gave us information only about the north-south length of the shadow, because point *II′* represents the position of the Sun as viewed from the east side of the celestial sphere. All east-west information is therefore lost in this projection. That is, we cannot tell (from point *II′* alone) whether the Sun lies in the plane of the paper, or an inch below it, or two inches above. Therefore, we must somehow move the Sun into the plane of the diagram without changing the length of its shadow. To do this, project from *II′* a line parallel to the horizon that will intersect the meridian circle at a point we shall call *II″*. From *II″*, project a line through *A* to the baseline that will be cut at 2″. Distance *B2″* then is the actual length of the shadow at the second hour. (It is the fact that we moved the Sun from *II′* to *II″* keeping it always the same distance above the horizon that guarantees we have not changed the actual length of the shadow.) Then set the compasses to give a radius equal to *B2″*, place the point at *W*, and draw a circle. The place, marked 2, where this circle intersects the line drawn previously through 2′ then gives the location of the shadow tip at the second hour. There is of course another place, east of the gnomon, where the line and circle intersect. This point, marked 10, is the location of the shadow tip at the tenth hour. (As 2 gives the position of the shadow tip four hours before noon, so 10 gives the position four hours after noon.)

In the same way, find the location of the shadow tip at each of the other hours of the day. The projections for the fourth and eighth hours are also illustrated in the diagram. The odd hours have been left out, to avoid cluttering the figure, but you should include them when you make your own sundial.

Winter Solstice The construction of the shadow track for the day of winter solstice goes the same way. Figure 3.11 shows the projections for the shadows at the sixth hour (noon) and at the second hour. For clarity, we have illustrated the construction for the winter shadow track in a separate diagram, but when you set about making a real sundial you will, of course, draw the summer and winter tracks on the same sheet of paper, using the same gnomon.

To find the noon shadow, project a line from *VI* through *A* to the baseline, which it cuts at 6′. Then *B6′* represents the length of the noon shadow. This distance is reproduced on the face of the sundial as *W6*, exactly as before.

To find the tip of the shadow at the second seasonal hour, proceed as follows. Project a line from *II′* through *A* to the baseline, which it cuts at 2′. From 2′ draw a line parallel to line *ABW*. The tip of the shadow must lie somewhere along this line. Next project a line from *II′* to the left, parallel to the horizon, and cut the meridian circle at a point called II''. From II'' project a line through A to the baseline, which it cuts at $2''$. Then $B2''$ is the actual length of the shadow at the second hour. Set the compasses to make a radius equal to $B2''$, place the point at W and draw a circle. The point, marked 2, where this circle intersects the line through $2'$ is the location of the tip of the shadow at the second hour. A second intersection, labeled 10, marks the location of the shadow's tip at the tenth hour. All this is exactly like the procedure for the summer shadow. It may look a little different, however, because of the use of the different diurnal circle (MXH instead of LVG).

Equinox The equinoctial shadow track is a straight line (see fig. 3.12). From N (the Sun's position at noon on the day of the equinox), we have already projected the ray NA, which cuts the baseline at C. And BC is the length of the noon equinoctial shadow. Draw a line down from C, parallel to the line ABW. This line is the desired track of the equinoctial shadow.

Step 4: Drawing the Hour Lines

Refer to figure 3.12. Simply connect the line 2–2 between the points that mark the shadow tip at the second hour of the summer solstice and the second hour of the winter solstice. Proceed similarly for each of the other ten hours (only the even hours are shown in the simplified diagram), producing the characteristic bat-wing shape of the ancient horizontal sundial.[9] Your sundial is finished and ready for use.

Step 5: Use of the Sundial

Place your sundial on a level, sunny surface with the noon shadows pointing north. Set a gnomon at W perpendicular to the surface. The length of the gnomon must be equal to AB, and your dial will be accurate only at the latitude for which you designed it. If the shadow tip happens to fall on the line for the fourth hour, the time is four seasonal hours after sunrise, that is, two-thirds of the way from sunrise to noon. If the shadow tip falls between two shadow lines, the time can be interpolated.

Postscript on the Use of the Menaeus

The directions given above explain how to determine the lengths and directions of the shadows only for the summer solstice, the equinoxes, and the winter solstice. Since a typical Greek sundial requires shadow data only for these three days, we had no need of the menaeus circle. The menaeus is used to find the lengths and directions of the shadows at other times of the year. You will not need to use the menaeus for the dial described in section 3.3. But for the sake of completeness, we will conclude with this postscript on the use of the menaeus circle.

Refer to figure 3.13, which shows the analemma constructed in the usual fashion, except that the day circles for the summer and winter solstices have been omitted. Divide the menaeus circle into twelve equal segments, beginning at G. These twelve segments represent the twelve signs of the zodiac, as shown in figure 3.13. The beginning of the sign of the Crab coincides with G; the beginning of the sign of the Goat-Horn coincides with H. (The direction in which the signs are labeled around the menaeus is immaterial. In fig. 3.13, they run counterclockwise.)

Suppose we wish to use the analemma for the day of the Sun's entry into the sign of the Bull (April 20). Through point U on the menaeus (representing the beginning of the Bull), draw a line parallel to the equator. This line cuts

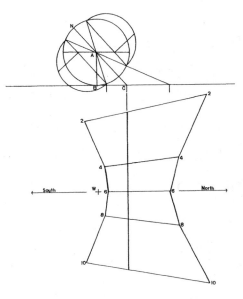

FIGURE 3.12. Construction of the shadow track, equinox (step 3) and drawing the hour lines (step 4).

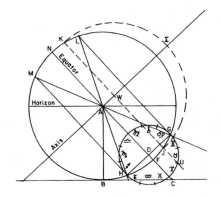

FIGURE 3.13. The use of the *menaeus* circle.

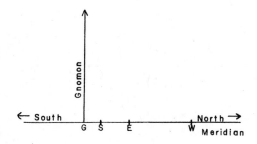

FIGURE 3.14.

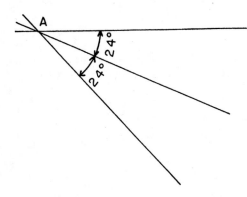

FIGURE 3.15.

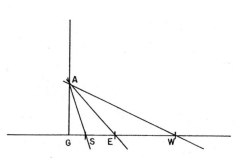

FIGURE 3.16.

the meridian at *J* and *K* and cuts the axis at *W*. Line segment *JK* is a side view of the Sun's diurnal path on the day in question. To construct a semicircle of this path, folded down into the plane of the meridian, place the point of a compass at *W* and draw semicircle *KΣJ*, as shown in figure 3.13. This semicircle may be divided in the usual way, and the usual projections may be made to determine the shadow lengths.

3.4 EXERCISE: SOME SLEUTHING WITH SUNDIALS

In this exercise, we take up a problem first raised in section 3.1—determining the latitude for which an ancient horizontal plane dial was designed. The problem is complicated by the absence of the original gnomons. Thus, two related questions must be answered simultaneously: what was the height of the missing gnomon, and for what latitude was the dial designed? We shall solve this problem for the two dials found at Rome and Delos, illustrated in figures 3.4 and 3.5.

The Rome Dial

We begin with the dial from Rome, because the problem is simpler for this dial. Construct a diagram like figure 3.14, which represents the meridian and the gnomon seen in a side view. We assume that the gnomon was set vertically into the gnomon hole *G*. The height of the gnomon is not yet determined, so the gnomon is represented as a line of indeterminate length. Along the meridian, points *S*, *E*, and *W* mark the tip of the gnomon's noon shadow at summer solstice, at equinox, and at winter solstice, respectively. The distances should be carefully drawn to actual size. For this purpose, use the following measurements taken from the original dial:[10]

$$GS = 1.5 \text{ cm}$$

$$SE = 2.3 \text{ cm}$$

$$EW = 5.6 \text{ cm}$$

On a separate sheet of transparent plastic or tracing paper, draw figure 3.15, which has three rays intersecting at a common point *A*. The angle between the central ray and each of the others should be equal to the obliquity of the ecliptic, that is, 24°. These three rays will represent rays of the noon Sun at summer solstice, equinox, and winter solstice. Point *A* will represent the tip of the gnomon.

Place the transparency over the diagram of the gnomon and the meridian. Then slide the transparency up and down on the diagram, always keeping point *A* on the gnomon line, until you can make the three rays pass through points *S*, *E*, and *W*, as shown in figure 3.16. When this is achieved, the problem will be solved. (Note that fig. 3.16 has not been drawn to scale, as your own must be.)

Once the transparency is properly oriented on the diagram, simply measure distance *AG*. This distance is the height of the gnomon for which the dial must have been designed. Place the center of a protractor at point *E* and measure angle *GEA*. This angle represents the angle between the celestial equator (line *AE*) and the horizon (line *GE*). Thus, angle *GEA* is the co-latitude for which the dial was designed. The latitude is then 90° − *GEA*.

Use the method just outlined to determine the latitude for which the Rome dial was designed and the height of its missing gnomon.

The Delos Dial

The problem for the Delos dial is a little more complicated. As we shall see, the gnomon of this dial was not perpendicular to the surface of the dial, but was bent toward the north. This was by no means an unusual situation. In all ancient sundials, only the shadow of the very tip of the gnomon played any role. As long as this tip was placed correctly over the proper spot of the dial, it made no difference where the base of the gnomon happened to be inserted. Many ancient dials make use of this freedom of placement.

The analysis should be based on the following measurements[11]:

$$GS = 3.0 \text{ cm}$$

$$SE = 2.3 \text{ cm}$$

$$EW = 4.9 \text{ cm}$$

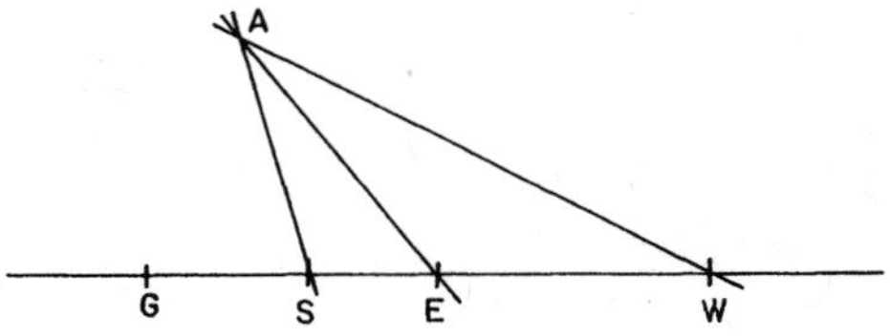

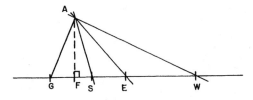

FIGURE 3.17.

In applying the method outlined above, do not assume that the tip of the gnomon must be directly above the gnomon hole G. Rather, simply adjust the position of the transparency until the three rays pass through points S, E, W, as shown in figure 3.17. The position A of the point of intersection of the rays will then indicate the proper position of the gnomon's tip. The actual gnomon then probably is represented by line GA (fig. 3.18). From A, drop a perpendicular to the meridian line. This perpendicular will cut the meridian at a point we shall call F. The proper functioning of the dial would remain undisturbed if the actual gnomon GA were replaced by a vertical gnomon FA. Use a ruler to measure the length of the actual gnomon and the length of the equivalent vertical gnomon. To determine the co-latitude at which the Delos dial would function correctly, place the center of a protractor at E and measure angle FEA.

FIGURE 3.18.

3.5 THE ASTROLABE

The astrolabe is a working model of the heavens, a kind of analog computer. In the astrolabe, the celestial sphere has been projected onto a plane surface. Thus, the astrolabe can be considered a two-dimensional version of a celestial globe or armillary sphere. The basic principle of the astrolabe was a discovery of the ancient Greeks, but the oldest surviving astrolabes are medieval. Throughout the Middle Ages, first in Islam and later in Christian Europe, the astrolabe was the most common astronomical instrument. When precise results were called for, the astronomer had recourse to specialized instruments and to tedious trigonometric computation. The beauty of the astrolabe was that approximate solutions (good to the nearest degree or so) to astronomical problems could be found by a mere glance at the instrument.

Parts of the Astrolabe

In the appendix are patterns for making an astrolabe. Photocopy fig. A.1–A.3 onto card stock, or photocopy them onto paper and glue them to cardstock using a glue stick or other glue that will not cause the paper to curl. Photocopy figure A.4 onto transparent plastic. Most copy centers can do this for you. You may, if you wish, enlarge the patterns when you photocopy. All the parts must be enlarged by the same amount. Cut out the parts of the astrolabe. Punch or cut out the eight 1/4″-diameter holes (solid black on the patterns). Glue the large disks of figures A.1 and A.2 back to back so that the holes and tabs line up and place under a stack of books until dry. You will need a 1/4″-diameter bolt and nut to hold your astrolabe together. Lightweight nylon machine screws (obtainable at most hardware stores) are a good choice.

Rete   Examine the *rete* (fig. 3.19), which is made of transparent plastic. (*Rete* is a two-syllable word; it rhymes with "treaty.") The rete represents the celestial sphere and is marked with a number of stars. A few constellations are also traced in outline on the rete: the Big Dipper, containing the stars Merak and Dubhe; the W-shaped constellation of Cassiopeia, containing the star Caph; the Great Square of Pegasus, containing Alpheratz; Orion, containing Rigel, Betelgeuse, and Bellatrix; and the Hyades (part of Taurus), containing Aldebaran.

The north celestial pole is the hole in the center of the rete. The rete is designed to turn around a screw stuck through this hole. This turning of the rete represents the daily rotation of the celestial sphere.

On the rete, the ecliptic is the off-centered circle divided into signs of the zodiac. Note that the ecliptic ring has a certain thickness. The actual ecliptic is the *fiducial edge* of the ring (the edge divided into degrees).

There are two scales around the perimeter of the rete. The outermost scale,

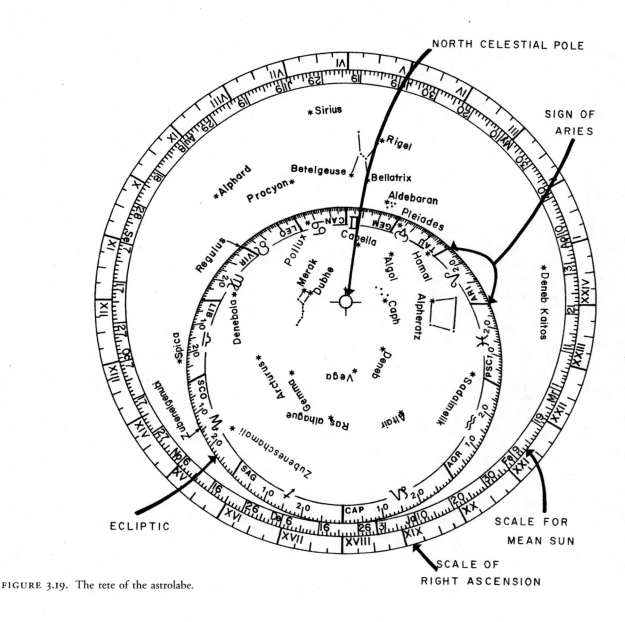

FIGURE 3.19. The rete of the astrolabe.

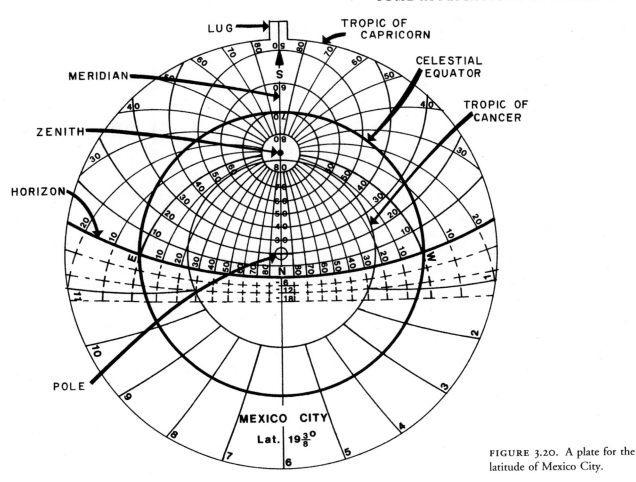

FIGURE 3.20. A plate for the latitude of Mexico City.

divided into hours, is a scale of right ascension. The other scale, marked with days of the year, is for making conversions between Sun time and clock time.

Latitude Plate Most parts of an astrolabe (including the rete) can be used at any latitude on Earth. But a *latitude plate* (fig. 3.20) must be designed for a specific latitude. Your astrolabe comes with two plates—one for the latitude of Seattle (latitude 47 2/3° N) and one for the latitude of Los Angeles (34° N). The plate for Seattle is built into the *mater* of the astrolabe. The plate for Los Angeles is separate.

On each plate, the heavy circle centered on the pole is the celestial equator. The tropic of Capricorn is the southern, or outer, boundary of the plate. The tropic of Cancer is the smallest of the three concentric circles.

The *horizon* is the heavy curve that runs off the edge of the latitude plate. Three cardinal points are marked around the horizon: east (E), north (N), and west (W). The south point does not fit on the plate, but it lies in the direction indicated by the arrow near the letter S. The fact that the plate of figure 3.20 is designed for the latitude of Mexico City shows up in a simple way: the center of the hole (the north celestial pole) is about 20° above the north point N of the horizon. Thus, the altitude of the pole on this plate is about 20°.

The *zenith* (straight overhead) is marked by a heavy dot. The *meridian* is the straight line running down the center of the plate. It passes through the north point N of the horizon and through the zenith.

Glue two small scraps of card to the mater, on either side of the meridian

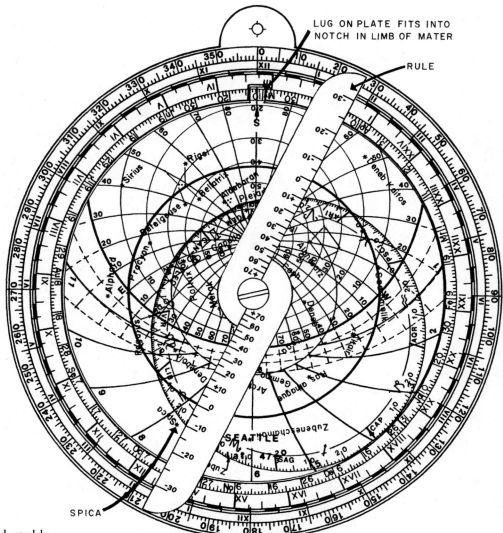

FIGURE 3.21. The assembled astrolabe.

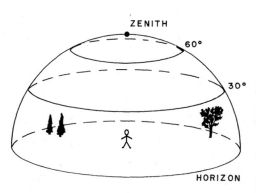

FIGURE 3.22. Almucantars.

line near the letter S (for "south"), thus forming a notch, into which the lug of your extra plate may fit (as in fig. 3.25). This will keep the Los Angeles plate from slipping around when you want to use it. When you are solving problems for Seattle, just lay the Los Angeles plate aside.

Place the rete face up on top of the plate and put the rule on top of everything (see fig. 3.21). Fasten the assembly together with the screw. Parts of the rete within the horizon circle are above the observer's horizon. Parts of the rete outside the horizon circle are below the observer's horizon. In figure 3.21, Orion (with its stars Betelgeuse and Rigel) is above the horizon and therefore visible. But Spica and Alphard are below the horizon.

The direction of a star in the sky may be specified by two angular coordinates, the altitude and the azimuth (see sec. 2.9). Imagine going outside and drawing circles on the sky, all equally spaced and parallel to the horizon, as in figure 3.22. These circles of constant altitude are called *almucantars*. On the latitude plate, the almucantars show up as a family of nonintersecting circles (as in fig. 3.24. The most important almucantar is the horizon itself. As we go up from the horizon, the almucantars become smaller until we reach the zenith point. Each almucantar is labeled with its altitude. Note that three of the almucantars are *below* the horizon. These three are indicated by dashed lines and are labeled with negative altitudes.

Imagine going outside again and drawing a second family of circles on the sky, as in figure 3.23. Begin by facing due east and drawing a circle that starts at the east point, goes straight up through the zenith, then continues straight down to the west point. Such a circle is an example of an *azimuth* circle. Every azimuth circle is perpendicular to the horizon, and they all intersect at the zenith. On the latitude plate, the azimuths show up as a family of circles that meet the almucantars at right angles (fig. 3.24). The most important azimuth is the meridian. Each azimuth is labeled with a number indicating its angular distance away from due east or due west. In figure 3.24, the position of point X is azimuth 40° south of west and altitude 20°.

The almucantars below the horizon are useful in twilight problems. Today, three kinds of twilight are distinguished. *Civil twilight* begins or ends when the Sun is 6° below the horizon. At this time only the brightest stars are visible. *Nautical twilight* begins or ends when the Sun is 12° below the horizon. Most stars of middling brightness are then visible. *Astronomical twilight* begins or ends when the Sun is 18° below the horizon and the sky becomes perfectly dark.

On the latitude plate (fig. 3.24), a system of eleven curves is used for problems involving *seasonal hours*.

Rule The *rule* (fig. 3.21) is marked with declinations, from −30° to +70°. Note that the zero of declination on the rule lines up with the celestial equator on the plate. Turn the rule until it lies beside Spica. You can read off the declination of Spica as −11°. That is, Spica is located 11° south of the celestial equator.

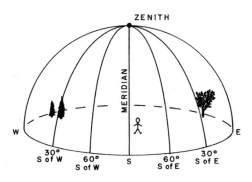

FIGURE 3.23. Azimuths.

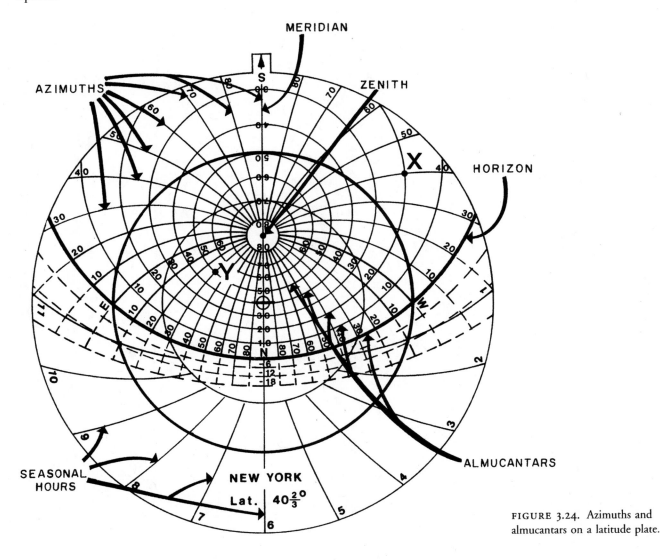

FIGURE 3.24. Azimuths and almucantars on a latitude plate.

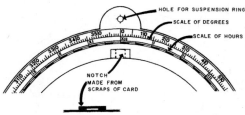

FIGURE 3.25. Upper half of the mater.

The edge of the rule, extending from Spica to the Roman numerals on the outer edge of the plastic rete, also indicates that the right ascension of Spica is 13 1/2 hours. Thus, the rule, in conjunction with the right ascension scale, can be used to read off the celestial equatorial coordinates of stars.

Mater   The *mater* (fig. 3.25) serves as a base on which the plate and rete are stacked. The notch in the limb of the mater receives the lugs on the plates and keeps the plates properly oriented. The limb of the mater is furnished with two scales. The inner *scale of hours* is used for telling time. The XII at the top represents noon and the XII at the bottom represents midnight. The hours on the left half of the limb are morning (A.M.) hours, and those on the right are afternoon hours. The outer scale of the mater is divided into degrees from 0° to 360°. It may be used for converting times or right ascensions into degrees. For example, 5 hours of time correspond to 75°.

Features of the Back of the Astrolabe   The back of the astrolabe (fig. 3.26) has three circular scales. The two innermost are a *calendar scale* and a *zodiac scale*. These are used together to determine the Sun's celestial longitude for any day of the year. The fiducial edge of the *alidade* is placed at the desired date on the calendar scale. The Sun's position is read off on the zodiac scale.

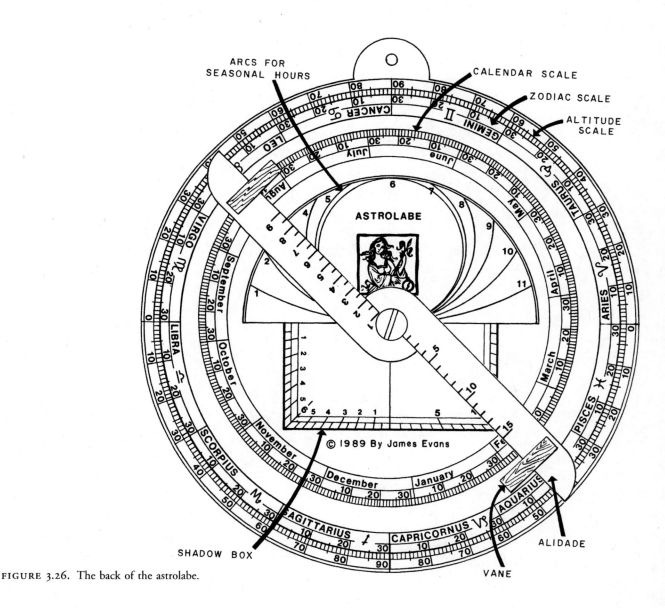

FIGURE 3.26. The back of the astrolabe.

The outermost scale (the *altitude scale*) consists of four quadrants, each graduated from 0° to 90°. This scale is used with the alidade for observing the altitude of a star or of the Sun. The alidade can be furnished with two *vanes*, each of which is pierced by a sighting hole.

The *shadow box* (fig. 3.26) is used to solve problems involving shadows. The last feature of the back of the astrolabe is a family of circular arcs used for telling the *seasonal hour* by means of the altitude of the Sun.

Many features of your astrolabe were more or less standard during the whole history of the astrolabe. But some features (e.g., the scales on the rete for mean time and for right ascension) are modern conveniences.

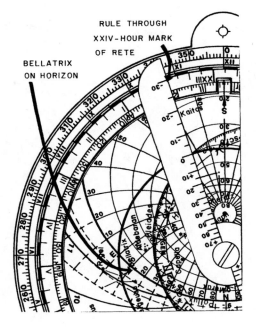

FIGURE 3.27.

Using the Astrolabe

Some of the most important applications of the astrolabe are described here in the form of worked examples. Unless otherwise noted, each example is worked for the latitude of Seattle (47 2/3° N). The secret of using the astrolabe is to *visualize* the meanings of the various circles. Once you have worked through a few problems, you should be able to solve new types of problems without instructions.

First Group: Problems Involving Stars

1. Rising position of a star:
   Example Problem: Where on the horizon does Bellatrix rise?
   Solution: Turn the plastic rete until Bellatrix appears on the eastern side of the horizon. (The horizon is the heavy circle on the plate marked with the letters E, N, and W; see fig. 3.27). Bellatrix crosses the horizon about 13° north of east.

2. Meridian altitude of a star:
   Problem: How high above the horizon is Rigel when it crosses the meridian?
   Solution: Turn the rete until Rigel comes to the meridian above the horizon. (The meridian is the straight line running through the middle of the plate; see fig. 3.28). Rigel is 33° above the horizon.

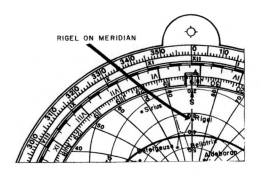

FIGURE 3.28.

3. The time a star spends above the horizon:
   Problem: How long does Bellatrix spend above the horizon each day?
   Solution: There are several ways to do this. The solution given here is the simplest. First, orient the rete so that Bellatrix is on the eastern horizon. Second, position the rule so that it passes through the XXIV-hour mark on the right ascension scale of the plastic rete (see fig. 3.27). (The XXIV-hour mark is the same thing as a zero-hour mark.) Be sure to use the hour marks *on the rete* and ignore the hour marks on the mater. Third, while holding the rule in place with your thumb, turn the rete until Bellatrix reaches the *western* horizon. You should find that the XIII-hour mark of the rete is about 1/4 hour past the rule. That is, the rete turned through 13 1/4 hours while Bellatrix went from the eastern to the western horizon. Thus, at Seattle, Bellatrix is above the horizon for 13 1/4 hours.

Second Group: Problems Involving the Sun

4. Position of the sun on the ecliptic:
   Problem: What is the Sun's position on the ecliptic on February 4?
   Solution: Turn to the back of the astrolabe. Orient the alidade so that it passes through the February 4 mark on the calendar scale, as in figure 3.26. Then read off the Sun's position on the zodiac scale. On February 4, the Sun is at the 15th degree of Aquarius.

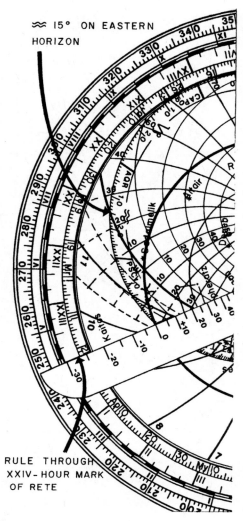

≈ 15° ON EASTERN
HORIZON

RULE THROUGH
XXIV-HOUR MARK
OF RETE

FIGURE 3.29.

5. Rising position of the sun:

Problem: For an observer in Seattle, where on the horizon does the Sun rise on February 4?

Solution: From problem 4, we know that on February 4 the Sun is at the 15th degree of Aquarius. Orient the rete so that AQR 15° is on the eastern horizon, thus simulating sunrise (see fig. 3.29). AQR 15° crosses the eastern horizon about 25° south of east. Note that this problem is essentially the same as problem 1.

6. Noon altitude of the sun:

Problem: What is the noon altitude of the Sun in Seattle on February 4?

Solution: As we know from problem 4, on February 4, the Sun is at the 15th degree of Aquarius. Turn the rete so that AQR 15° comes to the meridian, thus simulating local noon. You should find that is about 26° above the horizon. This is much like problem 2.

7. Length of the day:

Problem: How long is the Sun up at Seattle on February 4?

Solution: On February 4, the Sun is at Aquarius 15° (from problem 4). Orient the rete so that AQR 15° is on the eastern horizon; this represents sunrise. Then orient the rule so that it passes through the XXIV-hour mark on the right ascension scale of the rete (fig. 3.29). While holding the rule down with your thumb, turn the rete until AQR 15° comes to the western horizon (sunset). You should now find that the IX-hour mark on the rete is about 1/2 hour past the edge of the rule. Thus, at Seattle on February 4, the Sun is above the horizon for 9 1/2 hours. Note that this problem is essentially the same as problem 3.

8. Time of sunrise or sunset:

Problem: At what time does the Sun rise at Seattle on February 4?

Solution: There is more than one way to solve this problem. One way is to use the result of problem 7, that the Sun is above the horizon for 9 1/2 hours at Seattle on February 4. Half of this 9 1/2 hours is the length of the morning and half is the length of the afternoon. Thus, the Sun sets at 4:45 P.M. It rises 4 hours and 45 minutes before noon, at 7:15 A.M.

Alternative (and more elegant) solution: On February 4, the Sun is at the 15th degree of Aquarius. Orient the rete so that AQR 15° is on the eastern horizon, simulating sunrise. Now orient the rule so that it passes through the Sun (AQR 15°). The time of day is indicated by the rule's position on the scale of hours marked on the limb of the mater: sunrise occurs at 7:15 A.M.

The time of day obtained in this way is *Sun time* (what astronomers call *local apparent time*). There are several reasons why Sun time might differ from standard (or clock) time. Methods for obtaining clock time from the astrolabe are described in the fifth group of problems. But even with the simple procedures of Problem 8, the time obtained from the astrolabe will usually differ from clock time by less than half an hour. Only be sure not to neglect daylight saving time, when applicable.

9. Time of dawn:

Problem: In Seattle on February 4, at what time does dawn break?

Solution: Dawn breaks when the Sun is about 6° below the horizon. On February 4, the Sun is at the 15th degree of Aquarius. Place AQR 15° on the −6° almucantar on the eastern side of the astrolabe. Turn the rule so that it passes through the Sun (AQR 15°). The edge of the rule then indicates the time on the scale of hours of the mater. The time is about 6:40 A.M. Note that this solution is exactly like the alternative solution of problem 8, except that we use the −6° almucantar instead of the horizon.

Third Group: The Astrolabe as an Instrument of Observation

To determine the time of day by means of an astrolabe, one must first be able to make a relevant astronomical observation with satisfactory accuracy. The astrolabe is best suited to measuring *altitude*. Make some balsa-wood vanes and glue them to the alidade, as shown in figure 3.26. The vanes should have holes or notches in them. The holes should lie directly over the fiducial edge of the alidade.

10. Measuring the altitude of the sun:

Never sight the Sun directly by looking at it through the holes in the vanes—you could permanently damage your eyes. Sun observations should always be made indirectly, by observation of shadows.

Put a paper clip or metal key ring through the hole at the top of the mater to serve as a *suspension ring*. Hold the astrolabe by the suspension ring so that it dangles freely. Turn the astrolabe so that it is edge on toward the Sun (fig. 3.30, left). Adjust the angle of the alidade until the shadow of the upper vane falls on the lower vane, and the spot of light (coming through the hole in the upper vane) falls on the hole in the lower vane. Then read the altitude of the Sun on the altitude scale. In fig. 3.30, the altitude of the Sun is 60°.

11. Altitude of the sun: shadow box method:

Most people are about six feet tall when measured with their own feet. It follows that each person carries a standard six-foot shadow-casting gnomon wherever he or she goes. To measure the length of your shadow, note on the ground the location of a twig or stone that marks the end of your shadow. Then pace off the length of the shadow by placing one foot in front of the other, heel to toe.

Problem: Your shadow is 4 feet long (your own feet). What is the altitude of the Sun?

Solution: The shadow box on the back of the astrolabe is divided in half. One half is calibrated in sixes, the other in tens. When working with shadows cast by the human body, it is convenient to use the side calibrated in sixes.

Set the edge of the alidade on the 4 along the bottom edge of the shadow box, as in figure 3.31. Then read the altitude of the Sun on the altitude scale. The altitude of the Sun is about 56.3°. In doing shadow-box problems, it is helpful to *visualize the triangle* formed by your body, your shadow, and the Sun's ray that just grazes the top of your head.

12. Using the shadow box with long shadows:

Problem: Your shadow is 18 feet long (your own feet). What is the altitude of the Sun?

Solution: As in problem 11, we use the sixes side of the shadow box, because we are working with a shadow cast by a human body. However, the longest shadows marked on the shadow box are 6 feet long and there is no way to use an 18-foot shadow directly. This situation arises whenever the shadow is longer than the gnomon that casts it.

As the longest possible shadow is 6, we perform the following calculation involving similar triangles (see fig. 3.32). The 6-foot body casts an 18-foot shadow. We ask how tall a gnomon would be if it cast a 6-foot shadow in the same situation. That is, we solve for *x* in the second triangle, using the fact that the two triangles are similar. Thus, $6/18 = x/6$ and we find $x = 2$. That is, a 2-foot gnomon would cast a 6-foot shadow in this situation.

If the shadow is longer than the gnomon, it helps to turn the astrolabe upside down (fig. 3.33). The horizontal line (6 units long) represents the shadow. The edge of the alidade represents the Sun's ray. Set the edge of the alidade at 2 on the vertical side of the shadow box. The vertical side of the

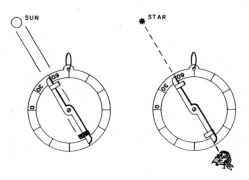

FIGURE 3.30. Observing altitudes with the astrolabe.

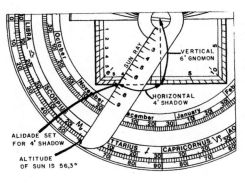

FIGURE 3.31.

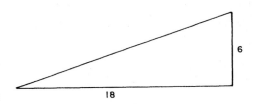

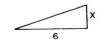

FIGURE 3.32.

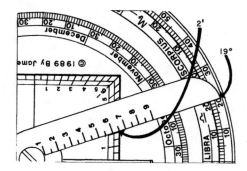

FIGURE 3.33.

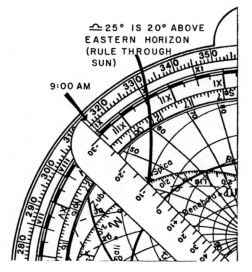

FIGURE 3.34.

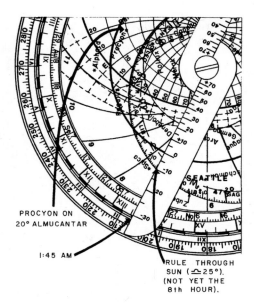

FIGURE 3.35.

triangle (2 units high) represents the human body. Read off the altitude of the Sun on the altitude scale: the Sun is 19° above the horizon.

13. Measuring the altitude of a star:

To measure the altitude of a star, hold the astrolabe by the suspension ring, so that you can look directly through the hole in the lower vane of the alidade (fig. 3.30, right). Adjust the alidade until you can see the star through the holes in both vanes. The fiducial edge of the alidade indicates the star's altitude on the altitude scale.

Fourth Group: Telling Time

14. Telling time during the day:

Problem: You are in Seattle on October 18. It is morning. After pacing off the length of your own shadow (as in problems 11 and 12), you find that the Sun is 20° above the horizon. What time is it?

Solution: First, use the zodiac and calendar scales on the back of the astrolabe to find the Sun's position on the ecliptic (as in problem 4). On October 18, the Sun is at the 25th degree of Libra (LIB 25°).

Now, turn the astrolabe over to use the front side (refer to fig. 3.34). Position the rete so that the mark for LIB 25° is on the 20° almucantar on the eastern side of the plate. (Since it is morning, the Sun is still to the east of the meridian.) Now put the rule through the Sun (LIB 25°). The end of the rule indicates the time of day by its position on the scale of hours marked on the limb of the mater. (Ignore the hours marked on the rete.) The answer: a few minutes before 9:00 A.M.

The time obtained from the astrolabe by means of Sun observations is most accurate in the early morning or the late afternoon. During the hours just before and just after local noon, the Sun's altitude changes very slowly. Solar altitudes taken near noon cannot, therefore, determine the time with precision.

15. Telling time at night:

Problem: On October 18, in Seattle, you observe that Procyon is 20° above the eastern horizon. What time is it?

Solution: Orient the rete so that Procyon is on the 20° almucantar in the eastern part of the sky (fig. 3.35). On October 18, the Sun is 25° within Libra. Let the rule pass *through the Sun* (LIB 25°), for *the Sun is the keeper of time.* Look and see where the rule hits the scale of hours on the limb of the mater. The time is about 1:45 A.M.

The altitude of a star changes slowly when the star is near the meridian. For this reason, the time will be determined most precisely if you use a star well away from the meridian.

16. Telling the time of night in seasonal hours:

Problem: Consider the situation posed in problem 15. We are in Seattle on October 18 and Procyon is 20° above the eastern horizon. What is the time, expressed in seasonal hours?

Solution: Instead of using the scale of hours on the limb of the mater (for telling time in equinoctial hours), use the *seasonal hour curves* on the latitude plate. Orient the rete so that Procyon is on the 20° almucantar in the eastern part of the sky (fig. 3.35). Then find the Sun (LIB 25°) among the set of seasonal hour curves. The Sun is two-thirds of the way between the curves for the 7th and the 8th hour. Thus, the time is 7 2/3 seasonal hours after sunset. Or, if you wish, 4 1/3 seasonal hours remain until sunrise. There is no need to use the rule. You simply find the Sun's position among the seasonal hour curves.

17. Telling the time of day in seasonal hours

Problem: In Seattle, on the morning of September 3, the Sun is 30° above the horizon. What is the time in seasonal hours?

Solution: On September 3, the Sun is at VIR 10°. Place this point of the ecliptic 30° above the eastern horizon (it is morning). No seasonal hour marks are drawn on the plate above the horizon. Therefore, we must locate the point of the ecliptic that is *diametrically opposite the Sun* and examine its position among the seasonal hour marks below the horizon. The point of the ecliptic diametrically opposite VIR 10° is PSC 10°. Now simply look and see where PSC 10° is among the seasonal hour curves. The answer: the time is almost three seasonal hours after sunrise. Note again that you need make no use of the rule.

18. Alternative method for finding the seasonal hour of day:

The seasonal hour curves on the *back* of the astrolabe give an alternative method for finding the time of day in terms of seasonal hours.

Problem: Consider again the situation of problem 17. In Seattle on the morning of September 3, the altitude of the Sun is 30°. What is the time of day, in seasonal hours?

Solution: First, we find the Sun's meridian altitude for this day and place. By the method of problem 6, find that in Seattle on September 3, the noon Sun is 50° above the horizon.

Now, on the back of the astrolabe, set the alidade to the Sun's *noon* altitude, 50° on the altitude scale (fig. 3.36). See which mark on the alidade hits the noon circle (the circle labeled 6). The 6-circle hits the edge of the alidade at about 10.5. (When used for this purpose, the marks on the alidade are simply reference marks with no deep significance.) Now rotate the alidade until it comes to the Sun's *present* altitude of 30° (fig. 3.37). Look to see where the 10.5 mark of the alidade lies among the hour curves. The answer: the time is shortly before the third seasonal hour, which agrees with the answer in problem 17.

Fifth Group: Finding Clock Time

The time determined from the altitude of the Sun is called local apparent time. In this group of problems, we explore ways to convert from local apparent time to zone time (clock time). An explanation of the various measures of time and of their relations to one another is given in section 5.9, so the reader may wish to skip the fifth group of problems until after studying section 5.9.

19. Finding the time of day directly in local mean time:

Your astrolabe is equipped with several modern advantages not found on medieval astrolabes. One of these is the scale for the *mean Sun* on the rete (Fig. 3.19). If we use this, we can directly obtain local mean time, rather than local apparent time.

Problem: In Seattle, in the afternoon of November 23, the Sun is 10° above the horizon. What is the local mean time? What is the zone time (which a clock would read)?

Solution: On November 23 the Sun is at the beginning of Sagittarius (SAG 0°). Place the Sun (SAG 0°) 10° above the western horizon. To find the local mean time directly, let the fiducial edge of the rule pass through the *mean Sun* (the Nov 23 mark of the mean Sun scale). Read the time on the scale of hours on the limb of the mater. The local mean time is about 2:50 P.M. (Note that if you put the rule through the true Sun [SAG 0°] instead, the rule will indicate the local apparent time, about 3:05 P.M.).

The final step is the conversion from local mean time to zone time (see sec. 5.9). The longitude of Seattle is 122 1/2° W. The standard meridian for

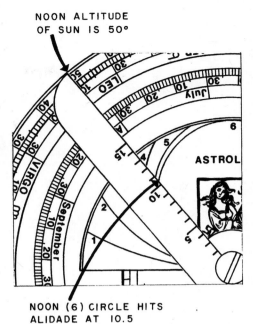

NOON ALTITUDE
OF SUN IS 50°

NOON (6) CIRCLE HITS
ALIDADE AT 10.5

FIGURE 3.36.

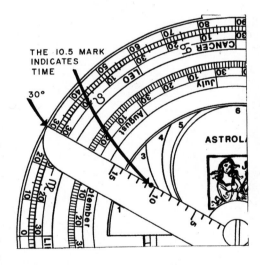

THE 10.5 MARK
INDICATES
TIME

30°

FIGURE 3.37.

Pacific Time is 120° W. Seattle is therefore 2 1/2° west of its standard meridian. The time difference between Seattle and the standard meridian is 2 1/2° × 4 min/° = 10 min. Since Seattle is *west* of the standard meridian, we *add* this amount to the local mean time. Thus, a clock should read 3:15 P.M.

20. Finding the time of night directly in mean time:

Problem: In Seattle on the evening of July 23, you ignore Schiller's warning (in *The Death of Wallenstein*): "Not everyone doth it become to question the far-off, high Arcturus." You sight Arcturus and find that it is 30° above the western horizon. What is the time by the clock (Pacific Daylight Time)?

Solution: Put Arcturus on the 30° almucantar in the west. Put the rule through the mean Sun (the July 23 mark on the scale for the mean Sun). Read the local mean time on the scale of hours of the mater: the local mean time is about 10:30 P.M. As explained in problem 19, in Seattle we must add 10 minutes to the local mean time to obtain the standard zone time. Thus, standard zone time is 10:40 P.M. In July, daylight saving time is in effect so we must add an hour. The pacific daylight time is therefore 11:40 P.M.

The problems solved above cover some applications of the astrolabe, but by no means all. As Chaucer wrote in the introduction to his *Treatise on the Astrolabe*, "Understand that all the conclusions that have been found, or possibly might be found in so noble an instrument as the astrolabe, are not known perfectly to any mortal man in this region, as I suppose."

3.6 EXERCISE: USING THE ASTROLABE

First Group: Problems Involving Stars

1. In Seattle, where on the horizon does Sirius rise? (*Answer:* 27° south of east.)
2. In Seattle, how high above the horizon is Rigel when it crosses the meridian?
3. In Seattle, how long does Arcturus stay above the horizon? (*Answer:* 14 3/4 hours.)

Second Group: Problems Involving the Sun

4. What is the longitude of the Sun (i.e., its position on the ecliptic) on June 13? (*Answer:* 21° within Gemini.)
5. In Seattle, where on the horizon does the Sun rise on summer solstice (June 22)? Note that on summer solstice the Sun is just entering the sign of Cancer. (*Answer:* 36° north of east.)
6. In Seattle, how high is the noon Sun on summer solstice (June 22)? (*Answer:* 66° above the horizon.)
7. How long does the day last at Seattle on summer solstice (June 22)?
8. At Seattle on summer solstice (June 22), how long must we wait after sunset for the Sun to be 18° below the horizon?

Third Group: Shadow Box Problems

9. Your shadow is 5 feet long, measured with your own feet. What is the altitude of the Sun? (*Answer:* 50°.)
10. When the Sun is 15° above the horizon, how long a shadow will a 10-foot pole cast? (Use the tens side of the shadow box.)

Fourth Group: Telling Time

11. At Seattle on summer solstice, in the afternoon, you get home from work when the Sun is still 20° above the horizon. What time is it? (*Answer:* 5:55 P.M., local apparent time.)

12. In the same situation as in Problem 11, what is the time expressed in seasonal hours? (*Answer:* 10 1/2 seasonal hours after sunrise, or 1 1/2 seasonal hours before sunset.)

13. In Seattle on April 1, in the early evening you see Spica 20° above the eastern horizon. What is the time, both in equinoctial hours and in seasonal hours?

3.7 THE ASTROLABE IN HISTORY

Some Representative Astrolabes

Figure 3.38 is a front view of a brass astrolabe[12] similar in design to the astrolabe of your kit. This specimen dates from the late fourteenth or early fifteenth century and is of French or Italian workmanship. It is small, about 3 3/4″ in diameter, but is fairly well made. The features of this astrolabe are typical of European astrolabes of its period.

Figure 3.39 is a photograph of the *rete* of the same astrolabe. The rete is an open, metal lacework that represents the celestial sphere. The hole in the center of the rete is the north celestial pole. The off-center circle is the ecliptic, divided into signs of the zodiac, which are labeled with abbreviations: ARI for Aries, TAU for Taurus, and so on. The eighteen small pointers represent stars, and each is labeled with a name. The reader may recognize some of the star names (e.g., *Rigil* is our "Rigel," a star in Orion). Most of the star names are in a late Gothic script. A few were added or reengraved at a later date (probably the late sixteenth or early seventeenth century) in an Italic script. For example, *Cauda Leonis* ("tail of the Lion") identifies our Denebola.

The same rete may be used for any geographical location in the northern hemisphere, but a plate must be engraved with the horizon, almucantors, and azimuths for a particular latitude. This is why a complete astrolabe normally came with a set of plates, numbering anywhere from a few to a dozen. The plate shown in figure 3.40 was engraved on both sides to save bulk—one side for latitude 40° and the other for 45°. This astrolabe has one other plate (not shown), similarly engraved on both sides, for latitudes 49° and 50°. Usually, the recess in the mater of an astrolabe was deep enough to allow storage of the plates.

Figure 3.41 shows the back of a late fifteenth-century astrolabe,[13] about a century younger than the astrolabe of figure 3.38. The main evidence for dating is the date associated with the vernal equinox on the zodiac and calendar scales. The modern forms employed for the numerals 4, 5, and 7 also point to a later date. The features shown in figure 3.41 are typical of the backs of medieval European astrolabes. Besides the zodiac and calendar scales, we see a shadow box and a set of seasonal hour curves. There is also a rotatable alidade (a crooked ruler) equipped with sights.

The fronts of astrolabes are stereographic projections of the celestial sphere and vary little from astrolabe to astrolabe, regardless of the century or the culture in which an instrument was constructed.[14] In contrast, the *back* of the astrolabe offered blank space, which the maker could fill with whatever seemed useful. Not surprisingly, the backs are much more variable than the fronts. In particular, there were different traditions for astrolabe furnishings in Islam and in Christian Europe.

FIGURE 3.38. An early fifteenth-century astrolabe of French or Italian workmanship. Photograph courtesy of the Time Museum, Rockford, Illinois.

FIGURE 3.39. The rete of the astrolabe in figure 3.38. Photo courtesy of the Time Museum.

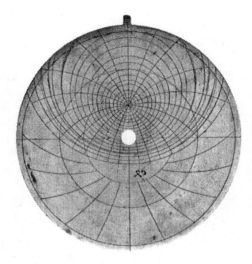

FIGURE 3.40. A plate for latitude 40° from the astrolabe in figure 3.38. Photo courtesy of the Time Museum.

FIGURE 3.41. The back of a fifteenth-century European astrolabe. National Museum of American History, Smithsonian Institution (Photo No. 77-13841).

Figure 3.42 shows the back of a seventeenth-century Arabic astrolabe and its alidade.[15] The rim of the upper half bears circular altitude scales divided into 6° arcs, with 2° subdivisions. The upper left quadrant contains a family of parallel, horizontal lines, which function, together with the altitude scale, as a table of sines of angles.[16]

The upper right quadrant contains six equally spaced circular arcs cut by two *prayer lines*. Observations of the Sun's altitude can be used with these curves to determine the proper times of day for prayer, as prescribed by the Muslim religion.

In the the lower center, a shadow box is recognizable. The right side of the box is for use with a gnomon seven units high; the left side, for use with a 12-unit gnomon. As most people are between six and seven feet tall when measured with their own feet, the right side of the shadow box is clearly intended to be used to solve problems involving the shadows cast by the human body. The left side was probably intended for use with a one-foot (12-inch) gnomon.

Around the rim of the lower half are cotangent scales, with unequally spaced marks. The user simply sets the alidade at the desired angle on the altitude scale (on the upper half of the back) and reads off the cotangent of the angle on the cotangent scale (on the lower half). The cotangent scales are useful for solving shadow problems.

Two semicircular tables of zodiacal signs and lunar mansions fill up the space between the shadow box and the cotangent scales. This astrolabe provides a nice example of efficient use of the space on the back of the astrolabe. Its features are typical of late Islamic astrolabes.

Stereographic Projection: Theory of the Astrolabe

Stereographic projection is one way (among many) of mapping a sphere onto a flat surface. It is the projection on which the astrolabe is based. The nice features of this projection are two: preservation of circles and conformality.[17] By *preservation of circles*, we mean that every circle on the celestial sphere gets mapped onto the astrolabe surface as a circle (or as a straight line, which can be regarded as a circle of infinite radius). By *conformality*, we mean that the projection preserves angles: two circles that intersect on the celestial sphere at a certain angle will intersect at the same angle on the face of the astrolabe. These properties of stereographic projection make the construction of the rete and latitude plates easy.

Figure 3.43 illustrates the principles of stereographic projection and shows the first few steps in the construction of a latitude plate for 40° north latitude. In the upper portion of the figure, we see a side view of the celestial sphere. The south celestial pole *SCP* serves as the center of projection. Points on the celestial sphere are projected from *SCP* onto the plane of the equator. For example, to project point *H* of the tropic of Capricorn onto the plane of the equator, we draw a line from *SCP* through *H* and extend this line until it crosses the plane of the equator at *H'*. Thus, *H'* is the stereographic projection of *H*. The projected tropic of Capricorn is a circle of radius *H'C*, centered at *C*. The latitude plate takes shape in the lower portion of figure 3.43. The center *C* of the latitude plate is the stereographic projection of the north celestial pole.

The projections of the celestial equator and the tropic of Cancer are made in the same way. The projections of these two circles are also circles centered on *C*. For circles parallel to the equator (such as the tropics), the more southerly will appear larger in the projection. Thus, the tropic of Capricorn is considerably larger than the tropic of Cancer on the plate of the astrolabe (lower portion of figure).

In the upper portion of figure 3.43, the horizon has been drawn for a latitude of 40°. Recall that the geographical latitude is equal to the altitude of the pole, that is, the angle between the north point N of the horizon and the north celestial pole NCP. The north point N of the horizon is projected onto the plane of the equator as N'. Similarly, the south point S of the horizon is projected as S'. Thus, on the latitude plate (lower portion of figure), the horizon will be a circle, but it will be *off-center* from C. The radius of the horizon circle will be one-half the distance $N'S'$, and the center of the horizon circle will be located halfway between N' and S'.

Often (as in the astrolabe of your kit), the plate is cut off at the radius of the projected tropic of Capricorn. In the lower part of figure 3.43, the part of the horizon curve that lies south (or outside) of the tropic Capricorn has been sketched in broken line. This part of the horizon curve would not appear on the finished latitude plate.

Figure 3.43 illustrates the essential idea of stereographic projection. Of course, a complete latitude plate would require a good deal more work: the system of almucantars and azimuths still must be drawn in. In section 3.8 we shall see how this is done.

History of the Astrolabe

The oldest surviving astrolabes are from the ninth and tenth centuries A.D. Some eleven astrolabes have been dated before the year 1000,[18] and all of these originated in different parts of eastern Islam: Syria, Egypt, Iraq, and Persia. The oldest Islamic astrolabes are rather severe in style, with simple, triangular star pointers on the rete. In most cases, the backs are not as fully furnished with supplementary scales as are the backs of later instruments. Nevertheless, the essential features of the astrolabe were already standard by the ninth century. After this period, the further development of the astrolabe consisted largely in changes of style and ornamentation and the addition of supplementary scales that simplified the solution of specialized problems. Despite these minor differences, a nineteenth-century astrolabe would have been, in most of its features, perfectly comprehensible to a ninth-century astronomer of Syria or Iraq. The stability of the astrolabe tradition through a thousand years is a striking demonstration of the continuity between ancient and modern astronomy. It also attests to the perfection already achieved by the astrolabe in its early form.

Although the astrolabe reached its definitive form in medieval Islam, the instrument is much older than one would guess from the oldest surviving examples. The astrolabe was in fact an invention of the ancient Greeks. Although far from conclusive, there is evidence that stereographic projection was invented by Hipparchus (second century B.C.).[19] In any case, stereographic projection was certainly in use by the first century B.C.

The Anaphoric Clock Vitruvius,[20] the Roman writer on architecture of the first century B.C., describes a water clock, the *anaphoric* clock, that evidently made use of stereographic projection. The moving part of the clock was a drum, around which was wrapped a chain. A float was attached to one end of the chain and a counterweight was attached to the other end. As water filled a container, the float rose and the counterweight descended, causing the drum slowly to turn at the rate of one rotation per day. Vitruvius tells us that the flat face of the drum was inscribed with an image of the heavens, including the zodiac circle. Moreover, the zodiac was drawn in such a way that some signs were larger than others, depending upon their distance from the center. This is a feature characteristic of stereographic projection. In front of the drum was a grid of wires that represented the horizon and the seasonal

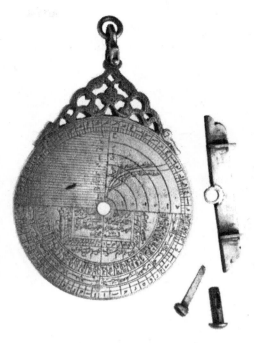

FIGURE 3.42. The back of a seventeenth-century Arabic astrolabe from Pakistan. National Museum of American History, Smithsonian Institution (Photo No. 78-5996).

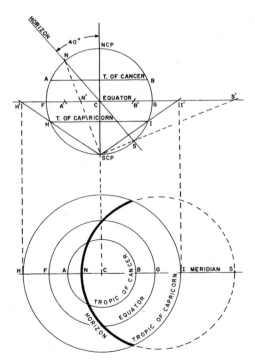

FIGURE 3.43. The principle of stereographic projection.

FIGURE 3.44. Fragment of a bronze sky plate from an anaphoric clock, first or second century A.D. The constellations represented are, from left to right across the middle of the piece, Triangulum, Andromeda, Perseus and Auriga. The names Andromeda and Auriga are engraved near the constellations. The ecliptic was represented by a series of holes, into which a marker representing the Sun could be placed. The plate has broken away along this line of holes, resulting in the serrated edge now seen. Parts of three zodiac constellations may be seen along the edge: Pisces, Aries and Taurus, from left to right. The whole disk must have been about 120 cm in diameter and must have weighed about 40 kg. Museum Carolino-Augusteum, Salzburg.

hours. The anaphoric clock was thus inside out with respect to later astrolabes. The face of the drum (corresponding to the rete of an astrolabe) was a solid sheet of metal and was placed behind an open grid of wires representing the seasonal hours. Fragments of the zodiac disks of two such clocks have actually been found—one at Salzburg in Austria and the other in the village of Grand near Neufchâteau in northeastern France. Figure 3.44 shows the fragment of the zodiac disk discovered at Salzburg.[21] Both fragments belong to the period from the first to the third century A.D. Analysis of the fragments confirms that the zodiac circle was positioned by means of stereographic projection.

Ptolemy on the Planisphere The oldest surviving mathematical treatise on stereographic projection is a short work by Ptolemy (second century A.D.), called *The Planisphere*. Ptolemy sets out the mathematical procedures for mapping the zodiac and other celestial circles onto a plane. His remarks make it clear that he intended these procedures actually to be used in making a concrete instrument. The original Greek text has not come down to us. What we have is an Arabic translation, made around A.D. 1000, and a Latin translation from the Arabic,[22] made around A.D. 1143 by Hermann the Dalmatian. It appears that the end of the treatise is missing and that Ptolemy included more information on practical construction than is now present in the text. From the text we have, it is not clear whether Ptolemy's instrument was an astrolabe in the modern sense or a type of anaphoric clock. Although the mathematics of stereographic projection were known before Ptolemy's time, his work was important, for it provided a good summary of the mathematical technique and served as a point of departure for later writers.

Theon of Alexandria The first treatise on an astrolabe in the modern sense was probably written by Theon of Alexandria (fourth century A.D.). Theon was a teacher of mathematics and a prolific writer. He worked hard at editing the classics of Greek mathematics, for example, the *Elements* of Euclid. With his daughter, Hypatia, Theon also wrote commentaries on the works of Ptolemy, including the *Almagest*. It is clear from remarks by later, Arabic writers that Theon wrote a treatise on the astrolabe and that Theon's astrolabe had all the essential features that we now associate with the instrument.

Theon's treatise has not come down to us. But we know what it contained, for there are two surviving treatises on the astrolabe that were based in great part on Theon's. One of these is a work in Greek by John Philoponus,[23] written around A.D. 530. The other is a work in Syriac by Severus Sebokht, the Bishop of Nisibis, written before A.D. 660.[24] These are the oldest *surviving* works on the astrolabe. The works by Philoponus and Sebokht describe the parts of the astrolabe and give directions for using it to solve various problems. A simple way to give a rough idea of the contents of these works is to say that section 3.5 has a fair amount of overlap with Philoponus and Sebokht.

Medieval Treatises on the Astrolabe in Arabic and Latin The tradition of treatises on the astrolabe, begun by Theon of Alexandria, flourished in Islam. In many cases, the first contact of Islamic astronomers with Greek science was through the intermediary of the Syriac language. Islamic astronomers began by making Arabic translations of Syriac astronomical works based on Greek sources. The treatise of Sebokht is an example of such a Syriac work. Shortly afterward, Arabic scholars began to make translations of Greek scientific works directly from the Greek, and then to compose original astronomical treatises directly in Arabic. The first Arabic language treatises on the astrolabe were written as early as the eighth century A.D. In the ninth and tenth centuries, the Middle East was the center of manufacture of astrolabes. By the eleventh century, astrolabes were also being made, and tracts on the astrolabe were being written, in Muslim Spain. The oldest treatise on the astrolabe that has survived from Muslim Spain was by Ibn al-Ṣaffār. His brother, Muḥammad Ibn al-Ṣaffār, made at Cordoba in A.D. 1026 the earliest dated astrolabe that has survived from this part of the world.[25]

Christian Europe received its first knowledge of the astrolabe—and its first astrolabes as well—from Muslim Spain. By the beginning of the eleventh century, astrolabes were known in southern France and Germany. By the early twelfth century, the astrolabe had become so esteemed in Paris that Abélard and Héloïse named their son Astrolabe. It is significant that there survive from this time several Arabic astrolabes to which Latin inscriptions were added. It was not long, however, before European astronomers and craftsmen were making their own astrolabes. A body of Latin treatises on the astrolabe began to accumulate, at first based on or directly translated from Arabic sources. One particularly famous and important treatise was a thirteenth-century Latin compilation based partly on the treatise by Ibn al-Ṣaffār, mentioned above. This compilation was falsely ascribed to Messahalla and circulated for centuries under that name.[26] A reader with the necessary astronomical knowledge could make an astrolabe from the directions given by pseudo-Messahalla. In the late twelfth and early thirteenth centuries, European astronomers began to write works on the use and the theory of the astrolabe that were not translations but original compositions.[27] At first these treatises were in Latin. But works on the astrolabe were soon composed in the vernacular languages of Europe.

Chaucer on the Astrolabe   The oldest surviving, moderately sophisticated scientific work in the English language is a *Treatise on the Astrolabe*, written by Geoffrey Chaucer. Chaucer was well-educated in astronomical matters, by the standards of fourteenth-century Englishmen, and the *Canterbury Tales* are studded with astronomical references.[28] He composed his treatise on the astrolabe, as he tells us, for his ten-year-old son, Lewis. In one manuscript, the treatise is subtitled "Bread and Milk for Children." Chaucer's treatise is not very original but is based in large part on pseudo-Messahalla. Moreover, Chaucer treats only the use, and not the construction of an astrolabe, as is appropriate for a work addressed to a ten-year-old boy. Nevertheless, the work is admirable for its clarity and patience. Figure 3.45 is a photograph of the drawing of the rete as it appears in one of the better manuscripts. The star at the tip of the tongue of the dog's head is Sirius, the Dog Star, which Chaucer calls *Alhabor*. The mater and rete of the instrument are labeled, in fourteenth-century English spellings, *the moder* and *thy ret*.

By Chaucer's time, the astrolabe had become a fixture in learned circles of Europe. The instrument was best known around the universities and in the courts, for many a king kept an astrologer on retainer. The most important use of the astrolabe was in time telling, especially for the purpose of casting horoscopes. But, quite apart from its practical merits, the astrolabe was prized as a mathematical tour de force that placed an image of the heavens in the human hand. Astrolabes began to appear as decorative features in illuminated manuscripts and on church facades, often merely as symbols of astronomical learning.

A Renaissance Master: Georg Hartmann   For European astrolabes, the sixteenth century was a golden age. Craftsmen produced astrolabes in greater quantity and better quality than ever before. Considerable originality was also displayed in the design of auxiliary scales and of new types of astrolabe. Georg Hartmann (1489–1564) was one of the most accomplished makers of astrolabes. From his workshop in Nuremberg he produced a steady stream of astrolabes, probably with the aid of assistants. Hartmann's metal astrolabes were expensive. Many were made for members of various European royal families. But Hartmann was also among the first makers of cheap astrolabes in kit form: he

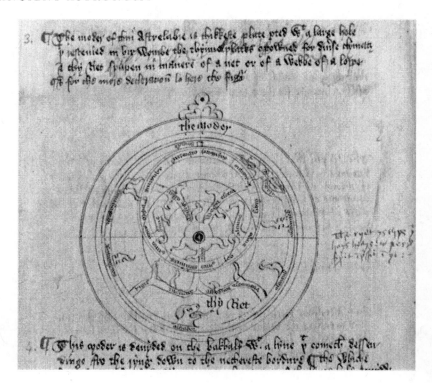

FIGURE 3.45. Drawing of an astrolabe from a manuscript copy of Chaucer's *Treatise on the Astrolabe*. Photo courtesy of the Cambridge University Library (MS. Dd.3.53).

printed parts on paper, which the purchaser could cut out, glue to wood or to heavier paper, and assemble. A number of other sixteenth-century engravers produced paper astrolabes. The astrolabe kit the reader has in hand is perfectly in keeping with a tradition that began in the Renaissance.

3.8 EXERCISE: MAKING A LATITUDE PLATE FOR THE ASTROLABE

In this exercise, you can make a latitude plate, for some city of your choice, to use with your astrolabe kit.[29]

Preliminary Drawing of the Celestial Sphere

1. On a large piece of paper (20″ × 40″), draw a circle to represent the celestial sphere, as in figure 3.46 (top). *The radius of the circle must be exactly the same as the radius of the celestial equator on your astrolabe kit.* Put in the north celestial pole *P* and the south celestial pole *Q*. At right angles to *PQ* draw a long line to represent the plane of the celestial equator. Figure 3.46 (top) represents the celestial sphere as seen from the side.

2. Draw the horizon through the center *C* of the circle. The angle between *CP* and the horizon (i.e., the altitude of the pole) should be equal to the latitude of the place for which you wish to design the plate. Figure 3.46 has been drawn for latitude 40°.

3. Draw the almucantars. The almucantars are parallel to the horizon. The method of locating the 20° almucantar is shown in figure 3.46 (top); it is 20° up from the horizon. A few other almucantars are also shown. You should draw all the almucantars from 10° to 80°, at 10° intervals. If you wish, you can put in one or more negative almucantars. The almucantar for −12° is shown (12° below the horizon).

4. Put in the zenith point *Z* and the nadir point *W*. The nadir is the

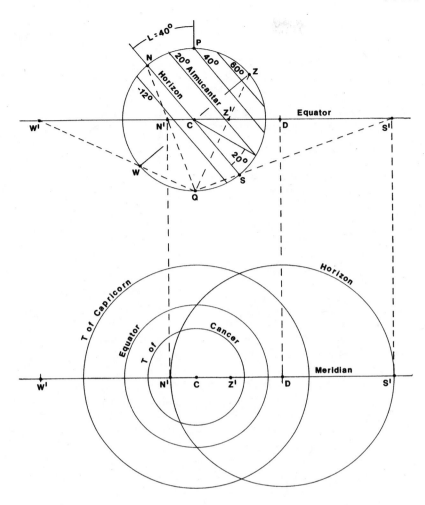

FIGURE 3.46. Construction of a latitude plate. Step 1, construction of the almucantars.

point directly "underfoot," just as the zenith is the point directly overhead. Thus, *ZW* passes through *C* and is at right angles to the horizon. This completes the preliminary construction.

Layout of the Plate: Almucantars

5. Begin the drawing of the actual latitude plate, as in figure 3.46 (bottom). First, locate the center *C* of the plate directly under line *PQ*. About *C* as center, draw circles for the celestial equator and the two tropics. These should have the same radii as on the plates that came with your kit. Thus, determine the radii by measuring on the kit astrolabe. Draw also the meridian line, through *C*, at right angles to the line *PQ* on which *C* is located.

6. Project the horizon (the most important of the almucantars). In figure 3.46 (top), *N* and *S* are the north and south points of the horizon. Their projections onto the plane of the equator are *N'* and *S'*. Find the midpoint *D* of line segment *N'S'*. Thus, in figure 3.46 (top), *D* is exactly halfway between *N'* and *S'*.

7. Draw the horizon circle on the actual plate, as in figure 3.46 (bottom). To do this, measure *CD* in figure 3.46 (top) and lay out this same distance in figure 3.46 (bottom). *D* will be the center of the horizon circle. The radius of the circle will be *DN'* (or *DS'*, which is the same. The dashed vertical lines show how *N'*, *C*, *D*, and *S'* are transferred from figure 3.46 (top) to figure 3.46 (bottom). But it is not recommended that you actually draw such lines. The best way to proceed is

to measure CD and DN' with a ruler, and to use these measurements to draw the circle.

8. Draw all the other almucantar circles in exactly the same way. Note that for each almucantar, you must find two distances in figure 3.46 (top): the radius of the circle, and the distance of its center from C. Every almucantar will have a different radius, as well as a different location for its center. (Hint for drawing circles of very large radius: make a beam compass. Tape the point of a compass to a meter stick. Tape your pencil to the meter stick at the desired distance.)

9. Project the zenith Z and nadir W onto the plane of the equator (points Z' and W'). Transfer these points to the plate by measuring CZ' and CW'.

Completion of the Plate: Azimuths

10. Refer to figure 3.47. On the latitude plate, the azimuth circles must all pass through the zenith Z' and the nadir W' in exactly the same way. It follows that the centers of the azimuth circles must all be located on the same line, the line of centers, which is the perpendicular bisector of line segment $W'Z'$. So, on your latitude plate, find the midpoint of $W'Z'$ and draw a line through this point at right angles to the meridian.

11. You can choose whether to put in azimuths at every 10° or at every 30°. Figure 3.47 shows the construction for placing them at every 30°. On the plate, at the zenith point Z', draw lines that make 30° and 60° angles with the meridian, as shown. Extend these lines until they intersect the line of centers. The intersection points, $A_1, A_2, \ldots A_5$, are the centers of the azimuth circles. Simply place the point of your compass at A_1 and open it up so that the pencil reaches the zenith point Z'. Then draw the azimuth circle, form the horizon curve, on

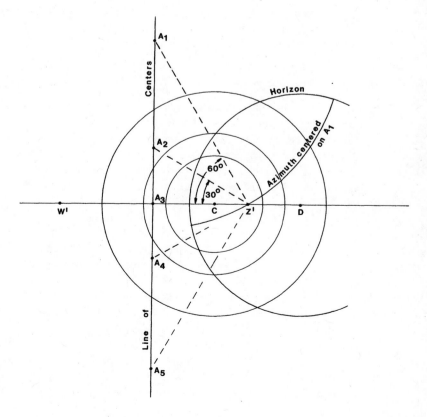

FIGURE 3.47. Construction of a latitude plate.
Step 2, construction of the azimuths.

through Z', all the way to the opposite side of the horizon curve. Draw the other azimuths in the same way.

The justification for this method lies in the conformality property of stereographic projection. Azimuth circles all pass through the zenith. Azimuth circles that make 30° angles with their neighbors on the real celestial sphere must make these same angles on the astrolabe plate.

Finishing Up

12. Erase any construction lines showing on your plate. Cut out the plate around the perimeter of the tropic of Capricorn, but be sure to leave a short tab at the south end to fit into the notch of the mater. Label your circles. Punch a small hole at C to receive the screw of your astrolabe.

Optional Construction: Seasonal Hours

13. Divide the half of the equator that is below (outside of) the horizon curve into twelve equal parts. To do this, place the center of a protractor at C, and put marks on the equator at 15° intervals between the sunset and sunrise points (the two points where the equator crosses the horizon).

Similarly, divide the portion of the tropic of Cancer that is below (outside of) the horizon curve into twelve equal parts (see fig. 3.48). On summer solstice, the Sun runs around the tropic of Cancer in the course of the day. On this day, E is the sunrise point and F is the sunset point. Place the center of a protractor at C and measure angle ECF (which is proportional to the length of the night). Divide ECF into twelve equal parts (the twelve seasonal hours of the night), placing marks on the tropic and labeling them as shown.

Do the same thing for the tropic of Capricorn.

For each seasonal hour (e.g., the second hour), you will have three points: the points labeled 2 on the tropic of Cancer, the equator, and the tropic of Capricorn. Connect these three points, from tropic to tropic, by a smooth curve. One way to do this is to find, by trial and error, using a compass, the location of the center and the necessary radius. The seasonal hour curves are not really supposed to be circles. But they are always drawn so on medieval astrolabes.

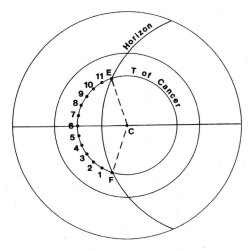

FIGURE 3.48. Division of the tropic of Cancer into seasonal hours.

One may regulate a calendar by means of the Sun alone, by means of the Moon alone, or by means of the Sun and Moon together. Thus, there are three principal types of calendar: solar, lunar, and luni-solar. At various times and in various cultures, all three types of calendars have been used. Indeed, all three types are still in use today. A good example of a lunar calendar is the Muslim calendar, which is still used in some countries of the Middle East, and which is used worldwide in Muslim religious practice. The most important luni-solar calendar still in use is the Jewish calendar. But the ancient Greek and Babylonian calendars were also of this type. The most familiar example of a solar calendar is the Gregorian calendar, which is used nearly worldwide today. However, to reckon time reliably in astronomical and historical work, one must also understand its relation to the Julian calendar that preceded it.

The Julian Calendar

Structure of the Julian Calendar The Julian calendar was instituted in Rome by Julius Caesar in the year we now call 45 B.C.. It reached its final form by A.D. 8 and continued in use without further change until A.D. 1582, when it was modified by the Gregorian reform. The Julian calendar adopts a mean length of 365 1/4 days for the year. This is in good agreement with the length of the tropical year, that is, the time from one spring equinox to the next. The Julian calendar is therefore a solar calendar and keeps good pace with the seasons. Two kinds of calendar year are distinguished: common years and leap years. Three years of every four are common years of 365 days each. One year of every four is a leap year of 366 days.

The months of the calendar year, and the number of days contained in them are

January	31	July	31
February	28 (29 in leap year)	August	31
March	31	September	30
April	30	October	31
May	31	November	30
June	30	December	31

The average length of the synodic month (the time from one new Moon to the next) is about 29 1/2 days. But, except for February, every month in the Julian calendar is longer than this. The calendar months therefore have no fixed relation to the Moon: the new Moon does not, for example, fall on a fixed day of the month.[1]

Years are customarily counted from the beginning of the Christian era. The first year of the Christian era is A.D. 1. The immediately preceding year is 1 B.C. There is no year 0. This arrangement is inconvenient for doing arithmetic. More convenient is the *astronomical way* of representing years before the beginning of the Christian era by negative numbers. In this system, the year immediately before A.D. 1 is called the year 0; the year before that, −1, and so on:

Historical way	Astronomical way
A.D. 2	+2
A.D. 1	+1
1 B.C.	0
2 B.C.	−1
3 B.C.	−2

163

The utility of the astronomical system can be made clear by an example. Let us compute the time elapsed between January 1, 23 B.C. and January 1, A.D. 47. The simplest approach is to express the B.C. date astronomically and then subtract:

$$23 \text{ B.C.} = -22.$$
$$47 - (-22) = 69.$$

Thus, 69 years elapsed between the two dates.

The leap years are those evenly divisible by four: A.D. 4, 8, 12, and so on. This rule may be extended to the years before the beginning of the Christian era, if the years are expressed astronomically: 0, −4, −8, −12 are all leap years. Note that if the years are expressed in the historical way the leap years are 1, 5, 9, 13 B.C.

The Roman manner of designating the days of the month was not the same as our own. The first day of the month was called *Kalendae*, or Kalends in English. The 5th of most months was called *Nonae* (Nones in English). The 13th day of most months was called *Idus* (Ides). However, four months had the Nones on the 7th and the Ides on the 15th (March, May, July, October). Other days of the month were specified in terms of the days remaining until the next of these three guideposts. For example,

Our way	Roman way
April 1	Kalends of April
2	4th day before the Nones of April
3	3rd day before the Nones of April
4	the day before the Nones of April
5	Nones of April
6	8th day before the Ides of April
7	7th day before the Ides of April
. . .	. . .
11	3rd day before the Ides of April
12	the day before the Ides of April
13	Ides of April
14	18th day before the Kalends of May
15	17th day before the Kalends of May
. . .	. . .
29	3rd day before the Kalends of May
30	the day before the Kalends of May
May 1	Kalends of May

The Roman way of counting the days continued in use to the end of the Middle Ages. In manuscripts of the fifteenth century, for example, one sees the Roman way and the modern way of counting used side by side. The fifteenth century was a period of transition. Note also the Roman manner of *inclusive counting*. We would say that April 11 is two days before the Ides. But the Romans called it the third day before the Ides—counting the 11th, 12th, and 13th. The Nones was so called because it came nine days before the Ides (counting inclusively). Time expressions based on inclusive counting survive in the Romance languages. For example, in French, an expression for a week is *huit jours*, literally eight days. Similarly, for two weeks, the French often say *quinze jours*, fifteen days.

The Julian calendar did not exist before 45 B.C., but that does not prevent us from using it as if it did. We say that Xerxes invaded Greece and fought the battle of Salamis in 480 B.C., or that Alexander died in 323 B.C. A Julian calendar date used in this way is always a translation into modern terms of a more ancient, and now defunct, system of chronology. An ancient Greek, for example, might have said that the battle of Salamis was fought in the year that Kalliades was archon of Athens[2] and that Alexander died during the archonship of Kephisodoros.[3]

History of the Julian Calendar   The Roman calendar that Caesar eliminated was a luni-solar calendar, consisting of twelve months.[4,5] There were four months of 31 days, Martius, Maius, Quintilis (= July), and October; seven of 29 days, Ianuarius, Aprilis, Iunius, Sextilis (= August), September, November, and December; and one of 28 days (Februarius). The length of the year was therefore 355 days, in fair agreement with the length of twelve lunar months. But as this year was some ten days shorter than the tropical or solar year, its months would not maintain a fixed relation to the seasons. Consequently, roughly every other year an intercalary month, called *Intercalaris* or *Mercedonius*, consisting of 27 or 28 days, was inserted after February 23, and the five remaining days of February were dropped. Thus, the year with an intercalated month consisted of 377 or 378 days. Some scholars suggest that the intercalary month alternated regularly between its two possible lengths,[6] so that the calendar years went through the regular four-year cycle: 355, 377, 355, 378. Four successive calendar years therefore totaled 1,465 days, and the average year amounted to 366 days, about one day longer than the tropical year. The intercalation was in the charge of the pontifices (priests of the state religion). But, through neglect, incompetence, or corruption, the necessary intercalations had not been attended to, and by 50 B.C. the calendar was some two months out of step with the seasons.

Julius Caesar, who had been elected Pontifex Maximus in 63 B.C., abandoned the old luni-solar calendar entirely and adopted a purely solar calendar. In the technical details he followed the advice of Sosigenes, a Greek astronomer from Alexandria. To bring the calendar back into step with the seasons, it was decided to apply two intercalations to the year 46 B.C.. The first was the customary insertion of a month following February 23, which was scheduled to be done in that year anyway. The second was the insertion of two additional months totaling 67 days between the end of November and the beginning of December, to make up for previous intercalations that had been neglected. The effect of this was to bring the vernal equinox back to March. After this *annus confusionis* ("year of confusion"), as it was called by Macrobius, the new calendar began to operate in 45 B.C..

The year was to consist of 365 days, ten more than in the old calendar. To make up the new total, the ten days were distributed among the old 29-day months: January, Sextilis, and December received two days each, while April, June, September, and November each gained one day. The old 31-day months (March, May, Quintilis, and October) remained unchanged,[7] as did February. An intercalary day was to be added to the month of February one year out of every four. After Caesar's assassination in 44 B.C., the Senate decided to honor his memory by renaming his birth-month (Quintilis) Iulius.

Unfortunately, owing to a mistake by the pontifices, the intercalation was actually performed once every three years so that, by 9 B.C., 12 intercalary days had been inserted, while Caesar's formula had called for only 9. The pontifices, who were inclusive counters like all Romans, had misunderstood Sosigenes' prescription. To bring the calendar back into step with the original plan, Augustus decreed in 8 B.C. that all intercalations be omitted until A.D. 8. In that year, the Roman Senate honored Augustus by renaming for him the month of Sextilis, since it was in this month that Augustus was first admitted to the consulate and thrice entered the city in triumph. From A.D. 8 the Julian calendar operated without further change until the Gregorian reform of 1582.

The week was not originally a feature of the Julian calendar. There is some evidence for an eight-day cycle of market days in Rome. The seven-day week seems to have originated from the Jewish practice: six days of work and one day of rest. The Jews had no names for the days of the week, except the Sabbath, and simply numbered them. As the week penetrated to the western Mediterranean, the practice grew up of naming the days of the week after the planets.[8] Most of these planetary names are still apparent in the French:

Planet	Latin	French	English
Saturn	Dies Saturni	Samedi	Saturday
Sun	Dies Solis	Dimanche	Sunday
Moon	Dies Lunae	Lundi	Monday
Mars	Dies Martis	Mardi	Tuesday
Mercury	Dies Mercurii	Mercredi	Wednesday
Jupiter	Dies Jovis	Jeudi	Thursday
Venus	Dies Veneris	Vendredi	Friday

In the Teutonic languages, the names of the Roman deities Mars, Mercury, Jupiter, and Venus were replaced by their counterparts Tiu, Woden, Thor, and Frigga. The seven-day planetary week was made official by the emperor Constantine in 321.

The practice of reckoning years from the beginning of the Christian era was introduced in the sixth century A.D. by the Roman abbot Dionysius Exiguus. Before this time, a year was commonly specified by the names of the consuls for that year or, later, in terms of the number of years elapsed since the beginning of the reign of some emperor, for example, Diocletian. In his tables for computing the date of Easter, Dionysius Exiguus identified A.D. 532 with year 248 of the Diocletian era. This fixed once and for all the relation of the Christian era to the Julian calendar—but not quite correctly. Modern scholarship has placed the actual year of Jesus's birth between 8 and 4 B.C.

The Gregorian Reform

The Error in the Julian Year   The *Julian year* (the average length of the Julian calendar year) is 365.25 days. But the time required for the Sun to travel from one tropic, all the way around the ecliptic, and return to the same tropic is about 365.2422 days. This is called the *tropical year*. Obviously, the tropical year can only be measured with such precision over an interval of many years. The Julian year exceeds the tropical year by 0.0078 day:

$$1 \text{ Julian year} = 1 \text{ tropical year} + 0.0078 \text{ day.}$$

In any one year, or even over a period of several years, this discrepancy would not be noticed. But over the centuries, it mounts up. In A.D. 300, to take a definite example, the vernal equinox fell on March 20. For the next several decades the equinox continued to fall on March 20 or 21. (The date of the equinox oscillated between the 20th and 21st, because of the leap day system.) But gradually, over a longer period of time, a systematic shift in the date of the equinox occurred. Consider an interval of 400 years. If we multiply the relation above by 400 we obtain

$$400 \text{ Julian years} = 400 \text{ tropical years} + 400 \times 0.0078 \text{ day}$$
$$= 400 \text{ tropical years} + 3.12 \text{ days.}$$

Therefore, the spring equinox of the year 700 did not take place on March 20, but on March 17. *Because of the difference in length between the Julian and the tropical year, the date of the equinox retrogresses through the Julian calendar by about 3 days every 400 years.* By the sixteenth century, the equinox had worked its way back to the 11th of March.

The Easter Problem   The principal motive for reform was the desire to correct the ecclesiastical calendar of the Catholic church, particularly the placement of Easter. As Easter is the festival of the resurrection, its celebration depended on the proper dating of the crucifixion and the events around it. According to the Gospels, the last supper occurred on a Thursday evening; the trial, crucifixion, and burial of Christ on Friday. On the evening of the same Friday, the Passover was celebrated by the Jews.[9] Finally, the resurrection occurred on the following Sunday.[10] The Passover, around which all these events center, is celebrated for the week beginning in the evening of the 14th day of Nisan in the Jewish calendar. Now, the Jewish calendar is of the luni-solar type, and the beginning of each month corresponds closely to a new Moon. It follows, then, that the 14th day of Nisan was the date of a full Moon. Moreover, the month of Nisan was traditionally connected with the spring equinox: a month was intercalated before Nisan whenever necessary to ensure that Passover week did not begin before the Jewish calendrical equinox. The proper time to celebrate Easter was therefore shortly after the first full Moon of spring.

In the early church, this general principle was interpreted in a number of different ways. Some Christians celebrated Easter on the third day after the full Moon, regardless of whether this was a Sunday or not. Most, however, celebrated Easter on a Sunday, although there was disagreement over which Sunday was proper. An attempt to regularize practice was made by the Council of Nicaea in 325. The rule adopted by the Council, expressed somewhat inexactly, was this: Easter is the Sunday following the full Moon that occurs on or just after the vernal equinox. The Council also decreed that if the date of Easter, so calculated, coincided with the Jewish Passover, then Easter should be celebrated one week later. This description of the Council's rule is the one commonly encountered today in nontechnical books on the subject, but it is inexact for the following reason: neither the true Sun nor the true Moon was used in the determination of Easter. For example, the Council fixed the date of the equinox at March 21. (This was correct for A.D. 325, as we have seen.) Moreover, the determination of the Easter Moon was not carried out through observation of the real Moon, but through calculation based on lunar cycles.

The Council of Nicaea does not seem to have regularized practice regarding the Moon, for different lunar cycles continued to be used in the East and the West. Thus, Easter was sometimes celebrated on different Sundays by different sects. For example, in A.D. 501, Pope Symmachus, following the cycle then used at Rome, celebrated Easter on March 25. But his political and religious opponents at Rome, the Laurentians, followed the Greek cycle and celebrated Easter that year on April 22. Moreover, they sent a delegation to the emperor at Constantinople to accuse Symmachus of anticipating the Easter festival.[11] Uniform practice between East and West was not achieved until 525, when the nineteen-year Metonic cycle was introduced at Rome. It had long been used in the East, where Greek influence predominated. Tables were prepared, based on this cycle, by means of which the date of Easter in any year could readily be determined. Again, the date of the full Moon on or next after March 21 was determined from these tables, not from astronomical observation; the Sunday following was Easter. Even after 525, other cycles continued to be used in Gaul and Britain. Feeling often ran high. The celebration of Easter on the wrong day was often deemed sufficient grounds for excommunication.[12] Completely uniform practice across Europe was not achieved until about A.D. 800.[13]

The Reform   In practice, then, Easter was celebrated on a Sunday in March or April following March 21. But by the sixteenth century the date of the equinox had retrogressed to March 11, so that Easter was steadily moving toward the summer. The need for reform had long been felt, but the state of astronomy in Europe had been inadequate for the task.[14] In 1545, the Council of Trent authorized Pope Paul III to act, but neither Paul nor his successors were able to arrive at a solution. Work by the astronomers continued, however, and when Gregory XIII was elected to the papacy in 1572 he found several proposals awaiting him and agreed to act on them. The plan finally adopted had been proposed by Aloysius Lilius (the Latinized name of Luigi Giglio, an Italian physician and astronomer, d. 1576). The final arrangement was worked out by Christopher Clavius, Jesuit astronomer and tireless explainer and defender of the new system.[15] The reformed calendar was promulgated by Gregory in a papal bull issued in February, 1582.

The most difficult part of the reform involved adjustments to the luni-solar ecclesiastical calendar used for calculating Easter. The details of this part of the reform need not concern us. New lunar tables were constructed to restore the ecclesiastical Moon to agreement with the true Moon. This reformed luni-solar calendar has never been accepted by the Orthodox churches, which still reckon Easter according to the tables that the Roman Church abandoned in 1582. As a result, the Orthodox Easter may coincide with the Roman Easter, or it may lag behind it by one, four, or five weeks.[16]

By contrast, the reform of the solar, or Julian, calendar was simple. First, to bring the vernal equinox back to the 21st of March, the day following October 4, 1582, was called October 15. That is, ten days were omitted. However, there was no break in the sequence of the days of the week: this sequence has therefore continued uninterrupted since its inception. Second, to correct the discrepancy between the lengths of the calendar year and the tropical year, it was decided that three leap days every 400 years were to be omitted. These were to be centennial years not evenly divisible by 400. Thus, in the old Julian calendar the years 1600, 1700, 1800, 1900, 2000, 2100, and so on, were all leap years. But under the new Gregorian calendar, 1700, 1800, 1900, and 2100 are not leap years.

The new calendar was immediately adopted in the Catholic countries of southern Europe, but in the Protestant north, most refused to go along. Denmark did not change over until 1700; Great Britain, not until 1752. In a few countries that had been dominated by the Eastern church, the change was not made until the twentieth century. Thus, Russia did not adopt the Gregorian calendar until 1918, after the revolution.

Using the Julian and Gregorian Calendars

In historical writing, the common practice is to use the Julian calendar for dates before 1582 and the Gregorian for dates after 1582. Consistent practice therefore requires translating many Julian calendar dates—for example, from seventeenth-century England—into their Gregorian equivalents. However, in astronomical discussion it is sometimes preferable to use the Gregorian calendar even for the remote past, since the dates of the equinoxes and solstices are nearly fixed in that calendar. *The only safe practice is to clearly specify which calendar is being used whenever there is any possibility of confusion.* Sometimes, in older writing, one comes across references to the "old style" and "new style," which refer to the Julian and the Gregorian calendar, respectively.

Table 4.1 may be used to make conversions. For example, Russia changed from the old to the new calendar on February 1, 1918 (Julian calendar). Let us express this date in terms of the Gregorian calendar. From table 4.1, we find that in 1918 there was a 13-day difference between the two calendars. The corresponding Gregorian date is therefore February 14, 1918. To put things as clearly as possible, "February 1, 1918 (Julian calendar)" and "February 14, 1918 (Gregorian calendar)" are two different names for the same day: it was a Thursday. Note that when the Gregorian calendar was promulgated in 1582 the difference between the two calendars was 10 days. But 1700, 1800, and 1900 were leap years in the Julian calendar, and not in the Gregorian; thus, by 1918 the difference had grown to 13 days. The Russian Orthodox Church uses the Julian calendar to this day. They celebrate Christmas on December

TABLE 4.1. Equivalent Dates in the Julian and Gregorian Calendars

	Time Interval	Difference
From	−500 Mar 6 Julian (= Mar 1 Gregorian)	−5 days
Through	−300 Mar 4 Julian (= Feb 28 Gregorian)	
From	−300 Mar 5 Julian (= Mar 1 Gregorian)	−4 days
Through	−200 Mar 3 Julian (= Feb 28 Gregorian)	
From	−200 Mar 4 Julian (= Mar 1 Gregorian)	−3 days
Through	−100 Mar 2 Julian (= Feb 28 Gregorian)	
From	−100 Mar 3 Julian (= Mar 1 Gregorian)	−2 days
Through	100 Mar 1 Julian (= Feb 28 Gregorian)	
From	100 Mar 2 Julian (= Mar 1 Gregorian)	−1 day
Through	200 Feb 29 Julian (= Feb 28 Gregorian)	
From	200 Mar 1 Julian (= Mar 1 Gregorian)	+0 days
Through	300 Feb 28 Julian (= Feb 28 Gregorian)	
From	300 Feb 29 Julian (= Mar 1 Gregorian)	+1 day
Through	500 Feb 28 Julian (= Mar 1 Gregorian)	
From	500 Feb 29 Julian (= Mar 2 Gregorian)	+2 days
Through	600 Feb 28 Julian (= Mar 2 Gregorian)	
From	600 Feb 29 Julian (= Mar 3 Gregorian)	+3 days
Through	700 Feb 28 Julian (=Mar 3 Gregorian)	
From	700 Feb 29 Julian (= Mar 4 Gregorian)	+4 days
Through	900 Feb 28 Julian (= Mar 4 Gregorian)	
From	900 Feb 29 Julian (= Mar 5 Gregorian)	+5 days
Through	1000 Feb 28 Julian (= Mar 5 Gregorian)	
From	1000 Feb 29 Julian (= Mar 6 Gregorian)	+6 days
Through	1100 Feb 28 Julian (= Mar 6 Gregorian)	
From	1100 Feb 29 Julian (= Mar 7 Gregorian)	+7 days
Through	1300 Feb 28 Julian (= Mar 7 Gregorian)	
From	1300 Feb 29 Julian (= Mar 8 Gregorian)	+8 days
Through	1400 Feb 28 Julian (= Mar 8 Gregorian)	
From	1400 Feb 29 Julian (= Mar 9 Gregorian)	+9 days
Through	1500 Feb 28 Julian (= Mar 9 Gregorian)	
From	1500 Feb 29 Julian (=Mar 10 Gregorian)	+10 days
Through	1700 Feb 28 Julian (= Mar 10 Gregorian)	
From	1700 Feb 29 Julian (= Mar 11 Gregorian)	+11 days
Through	1800 Feb 28 Julian (= Mar 11 Gregorian)	
From	1800 Feb 29 Julian (= Mar 12 Gregorian)	+12 days
Through	1900 Feb 28 Julian (= Mar 12 Gregorian)	
From	1900 Feb 29 Julian (= Mar 13 Gregorian)	+13 days
Through	2100 Feb 28 Julian (= Mar 13 Gregorian)	

25 of the Julian calendar, which is January 7 in the Gregorian—13 days after the Christmas of the Roman Church.

As a second example of the relation between the two calendars, consider the birth date of George Washington. In encyclopedias, this date is given as February 22, 1732. However, an entry in the Washington family Bible preserved at Mt. Vernon reads

> George Washington Son to Augustine & Mary his Wife was Born ye 11th Day of February 1731/2 about 10 in the Morning & was Baptiz'd on the 30th of April following.[17]

Two features of this entry require comment. First, the date of birth recorded by the family was the 11th of February (Julian calendar). Virginia in 1732 was an English colony and therefore used the same calendar as did the English. The colonies changed with England to the Gregorian calendar in 1752. The date that eventually became a national holiday, February 22, is the Gregorian equivalent of the date recorded in the family Bible. In 1732 there was an 11-day difference between the two calendars.

The second feature that requires comment is the designation of the year as 1731/2. There were several different practices regarding the beginning of the year. The most common initial dates were December 25, January 1, March 1, and March 25. These different reckonings of the year were known as *styles*—not to be confused with the usage *old style, new style* for designating the Julian and Gregorian calendars. In England, the Nativity style (December 25) was used until the fourteenth century, when it was superseded by the Annunciation style (March 25). This was the style still in use in the first half of the eighteenth century, when Washington was born. That is, in England and the English colonies the year officially began on March 25. However, by this time most of Europe was using the January 1 style. Therefore, to avoid ambiguity, it was common to specify both years in cases where the date fell between January 1 and March 24. The designation 1731/2 therefore means "1731 in the March 25 style, but 1732 in the January 1 style." The January 1 style was adopted in England in 1752 in connection with the change to the Gregorian calendar. The January 1 style is always used in modern historical writing.

4.2 EXERCISE: USING THE JULIAN AND GREGORIAN CALENDARS

1. Octavian assumed imperial powers and took the name Augustus in January, 27 B.C. He died in August, A.D. 14. How long did he reign?

2. The following list gives the Julian calendar dates of the vernal equinox over an interval of 3,000 years.

Year	Date of vernal equinox (Julian calendar)
A.D. 1500	11 March
1000	14 March
500	18 March
0	22 March
−500	26 March
−1000	30 March
−1500	3 April

Express these dates in terms of the Gregorian calendar. For year −500 and later, use table 4.1. For the earlier dates you will have to apply the rule for the leap years governing the centurial years in the Gregorian calendar.

3. Consider the following common remark: Isaac Newton was born in 1642, the year of Galileo's death. The popularity of this remark stems from its symbolic value. It seems to signify a passing of the torch of intellect. And it even seems to be true. Galileo died on January 9, 1642.[18] Newton was born on December 25, 1642.[19]

However, as Galileo lived in Italy, where the Gregorian reform was immediately accepted, the date of his death is naturally expressed in terms of the Gregorian calendar. Newton was born in England when that nation still used the Julian calendar. (Both dates have been expressed in the January 1 style.)

Express both dates in terms of the same calendar—first try the Julian, then the Gregorian. Do both fall in the same calendar year in one system or the other?

4. Compute the length of the *Gregorian year*, that is, the average length of the calendar year according to the Gregorian calendar. (Hint: begin by counting the number of common years and the number of leap years in the 400-year cycle.) Is the Gregorian year too long or too short in comparison with the tropical year? How much time will elapse before the Gregorian calendar loses step with the Sun by one day? The tropical year is 365.2422 days long.

4.3 JULIAN DAY NUMBER

The *Julian day number* is a count of days, widely used by modern astronomers. The day January 1, 4713 B.C. is called day zero, and for each successive day the count increases by 1.

For example, the Julian day number of December 31, A.D. 1899, is 2,415,020. The Julian day number of September 15, A.D. 1948, is 2,432,810. Knowledge of the Julian day numbers makes the calculation of time intervals simple:

$$
\begin{array}{ll}
\text{September 15, 1948} = & \text{J.D. } 2,432,810 \\
\text{December 31, 1899} = & \underline{\text{J.D. } 2,415,020} \\
\\
\text{Difference} & 17,790
\end{array}
$$

Thus, 17,790 days elapsed between the two dates. The calculation of this time interval by some other method would be much more complicated, for it would involve the reckoning of months of different lengths and the careful counting of leap days.

When the Julian day number is a whole number, as in the examples quoted so far, it signifies Greenwich mean noon of the calendar day:

$$\text{September 15, 1948, noon (at Greenwich)} = \text{J.D. } 2,432,810$$

If the time of day falls after noon, the appropriate number of hours may be added to the Julian day number:

$$\text{September 15, 1948, 6 p.m. (Greenwich)} = \text{J.D. } 2,432,810^d 6^h,$$

where ^{d} and ^{h} stand for days and hours. If the time falls before noon, the appropriate number of hours must be subtracted from the Julian day number:

$$\text{September 15, 1948, 9 A.M. (Greenwich)} = \text{J.D. } 2,432,809^d 21^h.$$

The Julian day number, although used now as a continuous count, originally specified the location of the day within a repeating period, called the *Julian period*. The length of the Julian period is 7,980 years. In principle, after 7,980 years have elapsed the Julian day numbers are supposed to start over again. (Whether the astronomers will actually consent to begin the count of days afresh at the start of the second Julian period in A.D. 3268, we shall have to wait and see!) In publications from the early part of the twentieth century, one often sees the expression "day of the Julian period," where we would now say, "Julian day number." The two expressions mean the same thing.

The Julian period and the practice of numbering the days within this period were introduced in 1583 by Joseph Justus Scaliger, the founder of modern chronology.[20] The period was formed by combination of three shorter periods. The first of these is the 19-year luni-solar (or Metonic) period, discussed in section 4.7. The second is a 28-year calendrical period: for any two

TABLE 4.2 Julian Day Number: Century Years. Days Elapsed at Greenwich Mean Noon of January 0

Julian Calendar							Gregorian Calendar	
A.D. 0	172 1057	A.D. 600	194 0207	A.D. 1200	215 9357		A.D. 1500†	226 8923
100	117 7582	700	197 6732	1300	219 5882		1600	230 5447
200	179 4107	800	201 3257	1400	223 2407		1700†	234 1972
300	183 0632	900	204 9782	1500	226 8932		1800†	237 8496
400	186 7157	1000	208 6307	1600	230 5457		1900†	241 5020
500	190 3682	1100	212 2832	1700	234 1982		2000	245 1544

†Common years.

years in the Julian calendar that are 28 years apart, all the days of the year will fall on the same days of the week. Thus, the calendars for the years 1901, 1929, 1957, 1985, and so on, are exactly the same. (Note that in the Gregorian calendar, this pattern is broken by the three century years in four that are not leap years.) The third period, called *indiction*, was a 15-year taxation period introduced in the Roman empire in the third century A.D. The Julian period is simply the product of these three: $19 \times 28 \times 15 = 7980$ years. Scaliger's starting year for the Julian period, 4713 B.C., is the most recent year in which all three periods were simultaneously at their beginnings.

Tables 4.2, 4.3, and 4.4 provide a convenient way of obtaining the Julian day number for any date.

Precepts for Use of the Tables for Julian Day Number

Dates after the Beginning of the Christian Era   For years before 1500, the date must be expressed in terms of the Julian calendar. For the year 1800 and thereafter, the date must be expressed in terms of the Gregorian calendar. Between the dates 1500 and 1800, either calendar may be used. In any case, the date must be expressed in terms of Greenwich mean time.

1. Enter the table of century years (table 4.2) with the century year immediately preceding the desired date and take out the tabular value. If the Gregorian calendar is being used and if the century year is marked with a dagger †, note this fact for use in step 2.
2. Enter the table of the years of the century (table 4.3), with the last two digits of the year in question and take out the tabular value. If the century year used in step 1 was marked with a dagger †, diminish the tabular value by one day unless the tabular value is zero.
3. Enter the table of the days of the year (table 4.4) with the day in question, and take out the tabular value. If the year in question is a leap year, and the table entry falls after February 28, add one day to the tabular value. The sum of the values obtained in steps 1, 2, and 3 then gives the Julian day number of the date desired. This Julian day number applies to noon of the calendar date.

First Example: September 15, A.D. 1948, Greenwich mean noon:

1. Century year	1900†		241 5020
2. Year of the century	48	17 532 − 1 =	1 7531
3. Day of the year	September 15	258 + 1 =	259

Julian day number   243 2810

Note that in step 2 the tabular value has been diminished by 1 because 1900 is a common year (marked with † in table 4.2). In step 3, the tabular value

TABLE 4.3. Julian Day Number: Years of the Century. Days Elapsed at Greenwich Mean Noon of January 0

Year	Days	Year	Days	Year	Days	Year	Days	Year	Days
0§	0	20*	7 305	40*	14 610	60*	21 915	80*	29 220
1	366	21	7 671	41	14 976	61	22 281	81	29 586
2	731	22	8 036	42	15 341	62	22 646	82	29 951
3	1 096	23	8 401	43	15 706	63	23 011	83	30 316
4*	1 461	24*	8 766	44*	16 071	64*	23 376	84*	30 681
5	1 827	25	9 132	45	16 437	65	23 742	85	31 047
6	2 192	26	9 497	46	16 802	66	24 107	86	31 412
7	2 557	27	9 862	47	17 167	67	24 472	87	31 777
8*	2 922	28*	10 227	48*	17 532	68*	24 837	88*	32 142
9	3 288	29	10 593	49	17 898	69	25 203	89	32 508
10	3 653	30	10 958	50	18 263	70	25 568	90	32 873
11	4 018	31	11 323	51	18 628	71	25 933	91	33 238
12*	4 383	32*	11 688	52*	18 993	72*	26 298	92*	33 603
13	4 749	33	12 054	53	19 359	73	26 664	93	33 969
14	5 114	34	12 419	54	19 724	74	27 029	94	34 334
15	5 479	35	12 784	55	20 089	75	27 394	95	34 699
16*	5 844	36*	13 149	56*	20 454	76*	27 759	96*	35 064
17	6 210	37	13 515	57	20 820	77	28 125	97	35 430
18	6 575	38	13 880	58	21 185	78	28 490	98	35 795
19	6 940	39	14 245	59	21 550	79	28 855	99	36 160

*Leap year.
§Leap year unless the century is marked †.
In Gregorian centuries marked †, subtract one day from the tabulated values for the years 1 through 99.

TABLE 4.4. Julian Day Number: Days of the Year

Day of Mo.	Jan	Feb	Mar	Apr	May	Jun	Jul	Aug	Sep	Oct	Nov	Dec
1	1	32	60	91	121	152	182	213	244	274	305	335
2	2	33	61	92	122	153	183	214	245	275	306	336
3	3	34	62	93	123	154	184	215	246	276	307	337
4	4	35	63	94	124	155	185	216	247	277	308	338
5	5	36	64	95	125	156	186	217	248	278	309	339
6	6	37	65	96	126	157	187	218	249	279	310	340
7	7	38	66	97	127	158	188	219	250	281	312	342
8	8	39	67	98	128	159	189	220	251	281	312	342
9	9	40	68	99	129	160	190	221	252	282	313	343
10	10	41	69	100	130	161	191	222	253	283	314	344
11	11	42	70	101	131	162	192	223	254	284	315	345
12	12	43	71	102	132	163	193	224	255	285	316	346
13	13	44	72	103	133	164	194	225	256	286	317	347
14	14	45	73	104	134	165	195	226	257	285	318	348
15	15	46	74	105	135	166	196	227	258	288	319	349
16	16	47	75	106	136	167	197	228	259	289	320	350
17	17	48	76	107	137	168	198	229	260	290	321	351
18	18	49	77	108	138	169	199	230	261	291	322	352
19	19	50	78	109	139	170	200	231	262	292	323	353
20	20	51	79	110	140	171	201	232	263	293	324	354
21	21	52	80	111	141	172	202	233	264	294	325	355
22	22	53	81	112	142	173	203	234	265	295	326	356
23	23	54	82	113	143	174	204	235	266	296	327	357
24	24	55	83	114	144	175	205	236	267	297	328	358
25	25	56	84	115	145	176	206	237	268	298	329	359
26	26	57	85	116	146	177	207	238	269	299	330	360
27	27	58	86	117	147	178	208	239	270	300	331	361
28	28	59	87	118	148	179	209	240	271	301	332	362
29	29	*	88	119	149	180	210	241	272	302	333	363
30	30		89	120	150	181	211	242	273	303	334	364
31	31		90		151		212	243		304		365

*In leap years, after February 28, add 1 to the tabulated value.

has been increased by 1 because 1948 was a leap year and the date fell after February 28.

Second Example: February 9, A.D. 1584 (Gregorian calendar), 10:30 A.M. Greenwich mean time:

1. 1500[†] (Gregorian)		226 8923
2. 84	3 0681 − 1 =	3 0680
3. February 9		40
Julian day number		229 9643

1 1/2 hours before noon of the 9th: $2,299,642^d 22^h 30^m$

Note that although 1584 was a leap year, the tabular value in step 3 is not changed because the date fell before the end of February.

Dates before the Beginning of the Christian Era   Express the date astronomically; add the smallest multiple (*n*) of 1,000 years that will convert the date into an A.D. date; determine the Julian day number of the A.D. date; then subtract the same multiple (*n*) of 365250. The result is the Julian day number desired.

Example: March 12, 3284 B.C. Greenwich mean noon:

$$\text{March 12, B.C. 3284} = -3283 \quad \text{March 12}$$
$$4 \times 1000 = 4000$$
$$\text{sum} = 717 \quad \text{March 12}$$

1. 700	197 6732
2. 17	6210
3. March 12	71
Julian day number, March 12, A.D. 717 noon	198 3013
Less 4 × 365250	−146 1000
Julian day number, March 12, B.C. 3284, noon	52 2013

4.4 EXERCISE: USING JULIAN DAY NUMBERS

1. Work out the Julian day numbers for the following dates. The time is Greenwich noon unless otherwise noted.

   A. June 13, 1952 (answer: 243 4177).
   B. June 10, 323 B.C. (death of Alexander).
   C. November 12, 1594, 6 A.M. Greenwich (Gregorian calendar).

2. Days of the week: The Julian day number provides a handy method of determining the day of the week on which any calendar date falls. Divide the Julian day number by 7, discard the quotient, but retain the remainder. The remainder determines the day of the week:

Remainder	Day of week
0	Monday
1	Tuesday
2	Wednesday
3	Thursday
4	Friday
5	Saturday
6	Sunday

A. Columbus, on his first voyage of discovery, first sighted land on October 12, 1492. What day of the week was this? (Answer: Friday.)

B. July 4, 1776 (Gregorian) fell on what day of the week?

3. Length of the tropical year: The vernal equinox of 1973 fell on March 20 at 6 P.M. Greenwich time. Copernicus observed the vernal equinox of the year 1516, "4 1/3 hours after midnight on the 5th day before the Ides of March"[21] That is, the vernal equinox fell at 4:20 A.M. March 11, A.D. 1516. (Is this the Julian or the Gregorian calendar?) Copernicus's time of day is referred to his own locality, that is, to the meridian through Frauenberg, on the Baltic coast of Poland. Frauenberg lies about 19° east of Greenwich, which amounts to about 1 1/4 hour of time. Expressed in terms of Greenwich time, then, Copernicus's vernal equinox fell at about 3 A.M. (We ignore the small fraction of an hour.)

Use these two equinoxes (1516 and 1973) to determine the length of the tropical year. To do this, compute the Julian day number of each observation, subtract to find the time elapsed, then divide by the number of years that passed. Compare your result with the modern figure for the tropical year, 365.2422 days.

4.5 THE EGYPTIAN CALENDAR

An understanding of the ancient Egyptian calendar is essential for every student of the history of astronomy. Because of its great regularity, the Egyptian calendar was adopted by Ptolemy as the most convenient for astronomical work, and it continued to be used by astronomers of all nations down to the beginning of the modern age. In the sixteenth century, Copernicus, for example, constructed his tables for the motion of the planets, not on the basis of the Julian year, but on the basis of the Egyptian year. When Copernicus wanted to calculate the time elapsed between one of Ptolemy's observations and one of his own, *he converted his own Julian calendar date into a date in the Egyptian calendar.*[22]

Structure

The Egyptian calendar from a very early date consisted of a year of twelve months, of thirty days each, followed by five additional days. The length of the year was therefore 365 days. Every year was the same: there were no leap years or intercalations. The names of the months are

1. Thoth	7. Phamenoth
2. Phaophi	8. Pharmuthi
3. Athyr	9. Pachon
4. Choiak	10. Payni
5. Tybi	11. Epiphi
6. Mecheir	12. Mesore
	Plus 5 additional days.

The names transcribed here, as commonly written by scholars today, represent their Greek forms. (Greeks of the Hellenistic period, living in Egypt, spelled the old Egyptian month names as well as they could in the Greek alphabet.) The additional days at the end of the year are sometimes called "epagomenal": the Greeks called them *epagomenai*, "added on."

The Egyptian year, being only 365 days, will after an interval of four years begin about one day too early with respect to the solar year. As a result, the Egyptian months retrogress through the seasons, making a complete cycle in about 1460 years (1461 Egyptian years = 1460 Julian years).[23] It therefore came about that religious festivals once celebrated in winter (on fixed calendar dates) fell in the summer. In 238 B.C., Ptolemaios III attempted to correct this supposed defect of the calendar by a plan that would insert one extra epagomenal day every four years. The reform was unsuccessful, as both the religious leaders and the populace insisted on retaining the old system.

The Astronomical Canon

The ordinary way of expressing the year in Egypt, as almost everywhere in the ancient world, was in terms of the regnal years of kings. Thus, a reliable king list is the first requirement of an accurate chronology. The *astronomical canon* is a king list that was used by the Alexandrian astronomers as the basis of their chronology. The canon is preserved in manuscript copies of Theon of Alexandria's redaction of Ptolemy's *Handy Tables*.[24] In some manuscripts, the list is titled *kanon basileion*, "Table of Reigns," and begins as follows:

Years of the reigns before Alexander, and of his reign	Years	Sums of these years
Nabonassar	14	14
Nadios	2	16
Chinzer and Poros	5	21
Ilouaios	5	26
Mardokempad	12	38
. . .	. . .	. . .
Nabonadios	17	209

These were kings of Babylonia in the eighth century B.C. The most ancient records of Babylonian astronomical observations that were available to the Alexandrian astronomers went back no farther that this. For example, the oldest observations cited by Ptolemy in the *Almagest* are three lunar eclipses that occurred during the reign of Mardokempad, in the years corresponding to 721–720 B.C. Thus, the Greek astronomers' king list went back just as far as was likely to be useful, and no farther.

The first column of numbers represents the lengths of the reigns of the individual kings. Nabonassar reigned 14 years; Nadios, only two. The second column gives the running total of all of the foregoing reigns. The 26 opposite Ilouaios signifies that his reign plus all those that went before, back to the time of Nabonassar, totaled 26 years. These cumulative totals are useful if one wishes to refer events in several different reigns to the same standard epoch, say, the beginning of the reign of Nabonassar. The first year of the reign of Mardokempad, for example, is also designated the 27th year of Nabonassar.

The years of the reigns are Egyptian years of 365 days, the year adopted by the Alexandrian astronomers for purposes of calculation. The lengths of the reigns therefore do not directly represent information recorded by the ancient Babylonians themselves, for the Babylonians used a luni-solar year of variable length. Rather, the lengths of the reigns given in the astronomical canon are the result of a translation and recalculation of the Babylonian data performed by the Greek astronomers for their own purposes. The list is also somewhat conventionalized. All regnal years are considered to begin with the 1st of Thoth, that is, the beginning of the Egyptian calendar year. Of course, kings do not generally begin their reigns on the 1st of Thoth. But, as a matter of convention, the whole Egyptian year that includes a king's assumption of power is counted as the first year of his reign. Kings who reigned less than a year are not included in the list.

Finally, the names of the Babylonian kings as given in the canon are Greek versions that are not very faithful to the Babylonian originals. More accurate transcriptions, based on Babylonian archives, are: Nabonassar, Nabunadinzri, Ukinzir and Pulu, Ulula, Mardukbaliddin, . . . Nabonidus.[25]

The list of Babylonian kings ends with Nabonadios, whose reign ended in the 209th year of Nabonassar (538 B.C.). The canon then continues with the Persian kings, the last of whom is Alexander. The astronomical canon thus reflects the political and military history of the Middle East: the Persians conquered Babylonia and were themselves eventually conquered by the Macedonians.

Persian kings

Kyros (Cyrus the Great)	9	218
Kambysos	8	226
Dareios the First (Darius)	26	262
. . .	. . .	. . .
Dareios the Third	4	416
Alexander the Macedonian	8	424

Here the list is interrupted by a new major heading:

Years of the Macedonian Kings
after the Death of Alexander

Philippos	7	431	7
The other Alexander	12	443	19
. . .	. . .	. . .	. . .
Dionysios the Younger	28	696	272
Cleopatra	22	718	294

Again, the first column gives the lengths of the individual reigns. The second continues the cumulative total since the era Nabonassar, without a break. The third column, which is new, begins a new cumulative total reckoned from the beginning of the reign of Philippos. Thus, the 22nd (and last) year of the reign of Cleopatra may also be called the 718th year of Nabonassar or the 294th year of Philippos. These years of Philippos are more often called years since the death of Alexander. For example, the last year of Cleopatra's reign was the 294th year after the death of Alexander. In many manuscripts the middle column of figures is not given. This reflects the widespread use of the era Alexander in Greek chronology.

After Cleopatra, counted as the last of the Macedonian monarchs, the canon takes up the Romans without a break:

Roman kings

Augustus	43	761	337
Tiberius	22	783	359
. . .	. . .	. . .	. . .
Trajan	19	863	439
Hadrian	21	884	460
Aelius-Antoninus	23	907	483

Each scribe generally continued the list down to his own time. In some manuscripts, the list is continued to the fall of Constantinople (A.D. 1453). We shall not need any of the Romans after Hadrian and Antoninus, whose reigns span the period of Ptolemy's astronomical work.

Calculation of Time Intervals

As an example of the use of the Egyptian calendar and the astronomical canon, we shall work out the number of days that passed between two eclipses of the moon that were used by Ptolemy in *Almagest* IV, 7, to determine the

TABLE 4.5. Some Important Egyptian/Julian Equivalents

1 Thoth, Year 1 of Nabonassar	26 February, 747 B.C.
1 Thoth, Year 1 of Philippos[*]	12 November, 324 B.C.
1 Thoth, Year 1 of Hadrian	25 July, A.D. 116
1 Thoth, Year 1 of Antoninus	20 July, A.D. 137

[*]Also known as the first year since the death of Alexander.

Moon's mean motion in longitude. The date of the first eclipse is given by Ptolemy as year 2 of Mardokempad, Thoth 18. The date of the second is year 19 of Hadrian, Choiak 2. (For simplicity, we ignore in each case the hour of the day.) The problem is to find the number of days separating these two events.

Going to the astronomical canon, we find that year 2 of Mardokempad = year 28 of Nabonassar. (Ilouaios's reign ended with the end of the 26th year of Nabonassar.) Similarly, year 19 of Hadrian = year 882 of Nabonassar. Now both years have been expressed in terms of a single standard era.

Since Thoth is the first month, Thoth 18 is the 18th day of the year. Choiak is the fourth month: three complete months, totaling 90 days, elapse before the beginning of Choiak. Therefore, Choiak 2 is the 92nd day of the year.

The time elapsed between the two lunar eclipses may now be computed:

882	years of Nabonassar,	92 days	(19 Hadrian, Choiak 2)
−28	years of Nabonassar,	18 days	(2 Mardokempad, Thoth 18)
854	years,	74 days	

The years are, of course, Egyptian years of 365 days, so the number of days elapsed is

$$854 \times 365 + 74 = 311{,}784 \text{ days,}$$

which agrees with the answer obtained by Ptolemy.

Expressed in terms of the Julian calendar, the dates of the two eclipses are March 8, 720 B.C., and October 20, A.D. 134. The calculation of the number of days elapsed directly from these Julian calendar dates would be a great deal more troublesome. There are three sources of trouble in such a calculation: the absence of a zero year at the transition between B.C. and A.D., the fact that the Julian months are not all the same length, and the necessity of counting the exact number of leap days involved.

Conversion of Dates between the Egyptian and Julian Calendars

Tables 4.5 and 4.6 provide all the information needed for converting most of the Egyptian calendar dates mentioned by Ptolemy in the *Almagest*. Table 4.5

TABLE 4.6. Months and Days of the Egyptian Year

Months	Days	Total Days	Months	Days	Total Days
Thoth	30	30	Phamenoth	30	210
Phaophi	30	60	Pharmuthi	30	240
Athyr	30	90	Pachon	30	270
Choiak	30	120	Payni	30	300
Tybi	30	150	Epiphi	30	330
Mecheir	30	180	Mesore	30	360
			Epagomenai	5	365

provides the Julian equivalents of a number of important dates in the astronomical canon. Table 4.6 gives the number of days elapsed at the end of each month of the Egyptian year.

Example   In *Almagest* X, 1, Ptolemy discusses a position measurement of Venus with respect to the Pleiades made by a certain Theon who was Ptolemy's elder contemporary. Ptolemy records the time of this observation as

In the 16th year of Hadrian,
in the evening between the 21st and 22nd of Pharmouthi.

We want to express this date in terms of the Julian calendar. According to Table 4.5,

1 Thoth, Hadrian 1 = 25 July, A.D. 116.

Starting from this date, we reckon forward to the date of Theon's observation of Venus:

From 1 Thoth, Hadrian 1 *to* 1 Thoth, Hadrian 16 is 15 Egyptian years.
From 1 Thoth *to* 1 Pharmouthi is 210 days.
From 1 Pharmouthi *to* 21 Pharmouthi is 20 days.

The elapsed time is therefore 15 Egyptian years, 230 days.

Now we break the 15 Egyptian years up into multiples of 4, plus a remainder. That is, we write $15 = 12 + 3$ (since $12 = 3 \times 4$). The 12 Egyptian years are all 365 days long. However, in the Julian calendar, one year of every four contains a leap day. Therefore, 12 Egyptian years are shorter than 12 Julian years by 3 days:

$$12 \text{ E.Y.} = 12 \text{ J.Y.} - 3 \text{ days}$$

The elapsed time may therefore be written as

$$15 \text{ E.Y.} + 230^d = 12 \text{ E.Y.} + 3 \text{ E.Y.} + 230^d$$
$$= (12 \text{ J.Y.} - 3^d) + 3 \text{ E.Y.} + 230^d$$
$$= 12 \text{ J.Y.} + 227^d + 3 \text{ E.Y.}$$

This time interval is to be added to the Julian calendar date for the beginning of the first year of Hadrian:

$$
\begin{array}{rll}
116 \text{ A.D.,} & \text{July } 25 & \\
+\quad 12 \text{ J.Y.} & + 227^d & + 3 \text{ E.Y.} \\
\hline
129 \text{ A.D.,} & \text{March } 9 & + 3 \text{ E.Y.}
\end{array}
$$

Note that the addition of 227 days to July 25 carried us forward into the next calendar year (129 A.D.). The only remaining problem is to dispose of the 3 Egyptian years. These may or may not be equivalent to 3 Julian calendar years. We will have to examine whether the addition of these 3 years causes us to roll over a leap day. The three additional Egyptian years will bring us to March, 132 A.D. That is, we will pass through the end of February, 132, when a leap day should be inserted. As the 3 Egyptian years do not contain this leap day, we will come up one day short. The final date is therefore

A.D. *132, in the evening of March 8.*

The Alexandrian Calendar

As mentioned earlier, Ptolemaios III Euergetes attempted in 238 B.C. to reform the Egyptian calendar by inserting a leap day once every four years, but the new arrangement was not accepted by his subjects. However, the same reform was reintroduced more successfully by Augustus some two centuries later, after Egypt had passed under Roman control. A sixth epagomenal day was inserted at the end of the Egyptian year 23/22 B.C., and every fourth year thereafter. The modified calendar, now usually called the *Alexandrian calendar*, is nearly equivalent to the Julian calendar: every four-year interval contains three common years of 365 days and one leap year of 366 days. As a result, the two calendars are locked in step with one another. For example, the Alexandrian month of Thoth always begins in the Julian month of August.

More precisely, the first day of Thoth in the Alexandrian calendar falls either on August 29 or August 30 of the Julian calendar, depending on the position of the year in the four-year leap day cycle. Figure 4.1 illustrates the relation of the two calendars. The leap years in each sequence are marked with asterisks. (The numbers 0, 1, 2 . . . written above the Julian calendar years indicate the positions of those years within the leap-year cycle. These numbers are the remainders that would be left if the year were divided by 4. The Julian leap years are those with remainder zero, i.e., evenly divisible by 4.) The dates written at the left and right edges of the boxes indicate the first and last days of each year. Finally, the figure indicates the Julian calendar day on which each Alexandrian year begins. Thus, Thoth 1 falls on August 30 if the August in question belongs to the Julian year preceding a leap year. Otherwise, Thoth 1 falls on August 29. Once the Julian equivalent of Thoth 1 is known, all the other days of the Alexandrian year fall into place.

The Alexandrian calendar was not uniformly and immediately accepted in Egypt. Rather, the old and the new calendar (referred to as the "Egyptian" and "Alexandrian" calendars, respectively) continued to be used side by side. The month names are the same in both calendars, so it is not always possible to decide which calendar is being used in a particular ancient document, unless there is either an explicit mention or a connection to some event that can be dated independently. Since the two calendars diverge rapidly, at the rate of one day every four years, it is usually easy to tell which calendar is being used in an astronomical text. The astronomers tended to prefer the old one because of its greater simplicity. Ptolemy, for example, used the Egyptian calendar exclusively in the *Almagest*, even though he composed it more than a century after the introduction of the new calendar.

In one of his works, however, Ptolemy did adopt the Alexandrian calendar. This was his *Phaseis*, which contained a parapegma, or star calendar, listing the day-by-day appearances and disappearances of the fixed stars in the course of the annual cycle (see sec. 4.11). For example, in the *Phaseis*, Ptolemy writes that the winter solstice occurs on the 26th of Choiak and that, for the latitude of Egypt, α Centauri "emerges" on the 6th of Choiak. (I.e., the star first becomes visible on this date as the Sun moves away from it.) It would make less sense to compose an astronomical calendar in terms of the Egyptian calendar. Neither the winter solstice nor the emergences and disappearances of the fixed stars would take place on fixed dates, since all these events advance through the months of the Egyptian calendar at the rate of one day every four years. But, in terms of the Alexandrian calendar, these annual astronomical events really do occur on about the same date every year. The winter solstice, for example, fell every year on the 26th of Choiak in the Alexandrian calen-

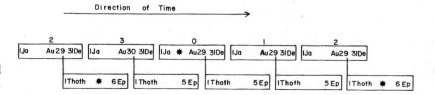

FIGURE 4.1. Relation between the Julian and the Alexandrian calendar.

dar—at least for a century or so. (Over periods of many centuries, of course, the Alexandrian calendar suffers from the same defect as the Julian, namely that the solar year is not quite exactly 365 1/4 days, the adopted length of the mean calendar year.)

One must exercise care when expressing dates in terms of Egyptian month names: it is important to state clearly whether the Egyptian or the Alexandrian calendar is meant, since the month names are the same in both. The situation is analagous to the use of identical month names for the Julian and the Gregorian calendars. In the present text, all dates using Egyptian month names are expressed in terms of the Egyptian calendar, unless explicitly stated otherwise.

Historical Specimen

Figure 4.2 is a photograph of the beginning of the table of reigns found in a Greek manuscript of the *Handy Tables*. There are two more pages to the king list, not reproduced here. The manuscript, now in the Bibliothèque Nationale in Paris, is carefully, though not elegantly, written in black ink. The ruled lines were drawn in red. Found in the same bound volume, or codex, are other astronomical works, including the *Treatise on the Astrolabe* of John Philoponos. The manuscript was written in the thirteen or fourteen century.

The first king listed in the photograph is Nabonassar, the last is Alexander the Macedonian. The name Xerxes can be seen eighth from the bottom.[26] To translate the numbers, the reader need only know that the Greeks used the letters of their alphabet as numerals,[27] with the following correspondences:

FIGURE 4.2. The beginning of the astronomical canon, from a manuscript of the *Handy Tables* that dates from about A.D. 1300. Bibliothèque Nationale, Paris (MS. Grec 2497, fol. 74).

α	1	ι	10	ρ	100
β	2	κ	20	σ	200
γ	3	λ	30	τ	300
δ	4	μ	40	υ	400
ε	5	ν	50	φ	500
ϛ	6	ξ	60	χ	600
ζ	7	ο	70	ψ	700
η	8	π	80	ω	800
θ	9	ϟ	90	ⳃ	900

4.6 EXERCISE: USING THE EGYPTIAN CALENDAR

1. Computing a time interval in the Egyptian calendar: In *Almagest* XI, 3, devoted to the determination of Jupiter's mean speeds (of the planet on the epicycle and of the epicycle around the deferent), Ptolemy makes use of an ancient observation and an observation of his own. According to the ancient observation, Jupiter occulted the Southern Ass (δ Cancri) on Epiphi 18 in the 83rd year after the death of Alexander. The second observation, made by Ptolemy himself, involved an opposition of Jupiter to the mean Sun. The date was Athyr 21, in the 1st year of Antoninus.

   Compute the exact number of days between these two observations. (Hint: First express the second date in terms of the era Alexander. Find in the extract from the astronomical canon that Hadrian's reign ended with the 460th year of Alexander, and that Antoninus's therefore began with the 461st.) (Final answer: 377 Egyptian years, 128 days = 137,733 days.)

2. Another time interval problem: Similarly, in *Almagest* XI, 7, Ptolemy makes use of an old and a recent observation to determine the rates of motion associated with Saturn. In year 519 of Nabonassar, on Tybi 14, Saturn was seen two digits below the Virgin's southern shoulder (γ

Virginis). The recent observation, made by Ptolemy himself, involved an opposition of Saturn to the mean Sun. The date (as given in *Almagest* XI, 5) was Mesore 24, in the 20th year of Hadrian. Compute the number of days between these two observations.

3. Conversion of dates, Egyptian to Julian: Ptolemy records the time of the middle of a partial lunar eclipse, which he observed at Alexandria, as follows: in year 20 of Hadrian, four equinoctial hours after midnight on the night between the 19th and 20th of Pharmouthi (Egyptian calendar). Convert this date into its equivalent in the Julian calendar. (Answer: March 6, A.D. 136, 4 A.M., Alexandria local time.)

4. Another conversion problem: In *Almagest* IV, 9, Ptolemy reports the beginning of a partial eclipse observed by him: in the 9th year of Hadrian, in the evening between the 17th and 18th of Pachon, 3 3/5 equinoctial hours before midnight. Express this date in terms of its Julian equivalent.

4.7 LUNI-SOLAR CALENDARS AND CYCLES

All luni-solar calendars contain two features. First, the months alternate between 29 and 30 days long. In this way, the calendar months closely match the synodic month (the time from new Moon to new Moon). (But because the synodic month is a little longer than 29 1/2 days, there must be a few more 30-day months than 29-day months.) Second, the calendar year contains sometimes twelve months and sometimes thirteen. Twelve synodic months amount to 354 days, which is shorter than the tropical year (365 1/4 days). Thus, if every calendar year had only twelve months, the calendar would progressively get out of step with the seasons. The occasional insertion of a thirteenth month restores the calendar to its desired relation to the seasons. In a well-regulated luni-solar calendar, the calendar months slosh back and forth a bit with respect to the seasons, but they do not continually gain or lose ground. For example, in the Jewish calendar, the month of Nisan comes always in the spring, but it does not always begin on the same date of the Gregorian calendar.

The Greek Civil Calendars

The Months of four Greek calendars[28]

Athens	Delos	Thessaly	Boeotia
1. Hekatombaion	Hekatombaion	Phyllikos	Hippodromios
Metageitnion	Metageitnion	1. Itonios	Panamos
Boedromion	Bouphonion	Panemos	Pamboiotios
Pyanepsion	Apatourion	Themistios	Damatrios
Maimakterion	Aresion	Agagylios	Alalkomenios*
Poseideon*	Poseideon	Hermaios	1. Boukatios
Gamelion	1. Lenaion	Apollonios*	Hermaios
Anthesterion	Hieros	Leschanopios	Prostaterios
Elaphebolion	Galaxion	Aphrios	Agrionios
Mounychion	Artemision	Thuios	Thiouios
Thargelion	Thargelion	Homoloios	Homoloios
Skirophorion	Panamos*	Hippodromios	Theilouthios

In the calendars of ancient Greece, the month began with the new Moon. Generally, months of 30 days, called "full" (*pleres*), alternated with months of 29 days, called "hollow" (*koiloi*). Ordinarily, the civil year consisted of twelve months, but occasionally a thirteenth month was intercalated.

Despite the simplicity of the basic calendrical scheme, Greek chronology is a difficult, even obscure, field. Most cities had their own calendars, which differed in the names of the months, the starting point of the year, and the place in the calendar where intercalary months were inserted. The list above gives the names of the months in four Greek calendars, starting from summer solstice. The first month of the year is marked 1. In the Athenian calendar, the year began with Hekatombaion, around the time of summer solstice. But in Delos the year began with Lenaion, around winter solstice. The months that were customarily doubled in leap years are marked with asterisks. For many cities, for example, for Argos and Sparta, the complete list of month names is not even known.

The most vexing complication is not, however, that each city followed its own practice, but that even in a single city the practice was not uniform. No regular pattern determined the intercalation of months. Moreover, individual days were sometimes intercalated or suppressed at will. For example, the Athenians held a theatrical presentation in connection with the cult of Dionysos on Elaphebolion 10. In 270 B.C., for some reason, the performance was postponed. Accordingly, the day following Elaphebolion 9 was counted as Elaphebolion 9 *embolimos* ("inserted"), and the next three days were counted as the second, third, and fourth "inserted" Elaphebolian 9.[29] Religious practice did not permit tampering with the *names* of days on which feasts were held, but the archons were free to intercalate days as needed, to place the feasts at a more convenient time. In a famous passage of *The Clouds* (lines 615–626), Aristophanes ridicules Athenian calendrical practice. The Moon complains that although she renders the Athenians many benefits—saving them a drachma each month in lighting costs through moonlight—nevertheless they do not reckon the days correctly, but jumble them all around. Consequently, the gods threaten her whenever they are cheated of their dinner because the sacrifices have not been held on the right days. As Samuel[30] points out, this illustrates that the festival calendar was out of step with the Moon, *and that the Athenians were aware of it*. Consequently, it is not surprising to see Athenian writers distinguish between "the new Moon according to the goddess" (Selene, the Moon) and "the new Moon according to the archon" (the head magistrate of the city).[31] We might call these the actual new Moon and the calendrical new Moon.

Because no fixed system was used to regulate the intercalation of either months or days, it is usually impossible to convert a date given in terms of the Athenian calendar into its exact Julian equivalent. Such a conversion would be possible only if we had a more or less complete record of the intercalations actually ordered by the authorities at Athens. No such record has come down to us. The same uncertainty attaches to most ancient calendars, with the notable exceptions of the Egyptian calendar and the Roman calendar after the Julian and Augustan reforms. The superiority of these two calendars derives from their regularity. In the Egyptian calendar there were no intercalations at all, while in the Julian calendar the only intercalation is the regular insertion of one day every four years.

Years were designated by the Greeks in several different manners. One practice involved the counting of Olympiads and the years (numbered one through four) within the Olympiad. The particular Olympiad was singled out both by number and by the name of the athlete who had won one of the important competitions, usually the foot race called the *stadion*.

More common was the use of the eponymous year, that is, the year named after a ruler then in power. The expression of eponymous years in terms of equivalent years of the Christian era requires a list of the rulers, and the lengths of their reigns, for the city or nation in question. We have fairly good king lists for Babylon, Persia, Egypt, Sparta, and so on, and lists of the archons of Athens and the consuls of Rome, so it usually is possible to determine at least the year to which an ancient writer refers.[32] As a rule, the farther back we go into the past, the less reliable the lists become.

A good example of these ancient manners of designating the year is provided by Diodorus of Sicily (Diodorus Siculus), a Greek historical writer who lived in Rome during the reigns of Caesar and Augustus. Diodorus completed an enormous work, of which less than half has come down to us, that treated the history of the whole known world from the time before the Trojan War down to Caesar's conquest of Gaul. Diodorus's arrangement is chronological. Each year's events are introduced by two or three equivalent designations of the year in question. For example, Diodorus begins his account of the year corresponding to 420/419 B.C. in the following way:

> When Astyphilos was archon at Athens, the Romans designated as consuls Lucius Quinctius and Aulus Sempronius, and the Greeks celebrated the 90th Olympiad, in which Hyperbios of Syracuse won the stadion. In this year, the Athenians, to abide by an oracle, restored to the Delians their island; and the Delians, who had been living at Adramyttium, returned to their homeland. . . . [33]

Luni-Solar Cycles

All ancient luni-solar calendars were originally regulated by observation, without the aid of any astronomical system. In most cultures, the month began with the first visibility of the crescent Moon—in the west, just after sunset. For this reason, in Babylonian as well as Jewish practice, the day began at sunset. A few generations of experience would suffice to show that the month varied between 29 and 30 days. Therefore, if because of unfavorable weather the new crescent could not be sighted on the 31st evening, a new month could be declared anyway.

The intercalation of months arose as a method of maintaining a roughly fixed relation between the seasons of the year and the months of the calendars. The ancient Jews inserted a thirteenth month to delay the beginning of the spring month if the lambs were still young and weak, if the winter rains had not stopped, if the roads for Passover pilgrims had not dried, if the barley had not yet ripened, and so on. Similar considerations must have governed the intercalation of months in all cultures that used a luni-solar calendar. Only later did observations of the heliacal risings and settings of the fixed stars play any part. It was much later still before any use was made of observations of solstices and equinoxes.

The Eight-Year Cycle   The lengths of the two fundamental periods are

Synodic month: 29.5306 days,
Tropical year: 365.242 days.

Their ratio is 365.242/29.5306 = 12.3683. Thus, on the average, a calendar year ought to contain 12.3683 months.

A real calendar year, however, contains a whole number of months. Suppose we let every year contain twelve months. After the first year, the calendar will be deficient by 0.3683 months. The calendrical deficit after n years will be $n \times 0.3683$ months. We simply wait until this deficit amounts to a whole month; then it will be time to intercalate a month. For example, after three years the deficit will be 3×0.3683 months = 1.1 months. If we insert a thirteenth month in the third calendar year, then at the end of that year the deficit will be nearly (although not exactly) eliminated. Unfortunately, 3×0.3683 is not very near a whole number. The central problem, then, is to find an integer n such that $n \times 0.3683$ is as close to a whole number as possible. One possible solution is $n = 8$, for then we have

$$8 \times 0.3683 = 2.946, \text{ which is pretty nearly 3.}$$

This near-equality allows us to construct a luni-solar cycle. In eight calendar years, we insert three additional months. Of the eight calendar years, five will consist of twelve months, and three will consist of thirteen months:

Eight-year cycle

	5 years of 12 months =	60 months
	3 years of 13 months =	39 months
So,	8 calendar years =	99 months

The average length of the calendar year in this system is 99 months/8 = 12.3750 months, which is close to the figure we were trying to match (12.3683 months per year). The correspondence is not perfect, however. Indeed, the calendar year is 0.0067 months too long (12.3750 − 12.3683 = 0.0067). In about 149 years, this surplus will amount to a whole month. Thus, the eight-year cycle will operate satisfactorily for about 149 years, but then one month will have to be omitted to restore the balance.

The Nineteen-Year Cycle   The eight-year cycle is tolerably accurate, but let us search for a better one. Again, the tropical year is longer than twelve synodic months by 0.3683 month. We search for an integer n such that $n \times 0.3683$ is a whole number. A very satisfactory solution is $n = 19$:

$$19 \times 0.3683 = 6.9977, \text{ which is very nearly 7.}$$

Thus, we may construct a nineteen-year luni-solar cycle. In nineteen calendar years, we insert seven additional months. Of the nineteen years, then, twelve will consist of twelve months and seven will consist of thirteen months:

Nineteen-year cycle

	12 years of 12 months =	144 months
	7 years of 13 months =	91 months
So,	19 calendar years =	235 months

The average length of the calendar year in this system is 235 months/19 = 12.3684 months, which agrees very well with the length of the solar year (12.3683 months).

Nineteen tropical years therefore contain 235 synodic months, almost exactly. The astronomical meaning of this statement is that after nineteen tropical years, both the Sun and the Moon return to the same positions on the ecliptic. The Sun returns to the same longitude after any interval containing a whole number of tropical years. The special feature of the nineteen-year period is that it also contains a whole number of synodic months. Thus, the Moon will be in the same phase on two dates that are nineteen years apart.

The explanation of the eight- and nineteen-year cycles given above is not meant to reflect the actual process of discovery: the ancient Greeks and Babylonians did not begin with a knowledge of the lengths of the year and the month. Rather, a knowledge of these cycles emerged after several generations of keeping track of the Moon.

The nineteen-year cycle was introduced at Athens in 432 B.C. by the astronomer Meton, for which reason it is also known as the *Metonic cycle*. The Greeks simply called it the *nineteen-year period*. Unfortunately, the Athenians never adopted it as the regulatory device of their calendar, although the archons may have taken it into account while pondering the need for an intercalation. Whether the Greeks discovered this cycle independently or learned it from the Babylonians, it is not possible to say. Borrowing may be considered likely in view of other demonstrated debts of Greek astronomy to Babylonian practice. On the other hand, the fundamental relation (235 months = 19 years) is very simple, and independent discovery cannot be ruled out.

Geminus on the Structure of the Nineteen-Year Cycle   In chapter VIII of the *Introduction to the Phenomena*, Geminus gives a detailed account of the nineteen-year cycle as used by the Greeks. According to Geminus, this cycle was based on the identity

$$19 \text{ years} = 235 \text{ months} = 6{,}940 \text{ days}.$$

In one nineteen-year cycle there were, of course, twelve years of twelve months and seven years of thirteen months. Geminus adds that there were 125 full months (30 days each) and 110 hollow months (29 days). Thus, 125 × 30 + 110 × 29 = 6,940 days.

Geminus asserts that the arrangement of full and hollow months should be as uniform as possible. There are 6,940 days in the nineteen-year period and 110 hollow months. If all the months were temporarily considered full, it would therefore be necessary to remove a day after every run of 63 days (6,940/110 ≃ 63). That is, every 64th day number would be removed. According to Geminus, the thirtieth day of the month is not always the one scheduled for removal. Rather, the hollow month is produced by removing whichever day falls after the running 63-day count. Such a procedure would have enormously complicated the construction of a calendar. Neugebauer[34] therefore doubts that this rule was ever followed. However, Geminus is unambiguous on this point, and both recent efforts at a reconstruction of the Metonic cycle have taken him at his word.[35]

The Callippic Cycle and the Callippic Calendar   The length of the year implied by Meton's nineteen-year cycle is

$$6{,}940 \text{ days}/19 = 365 \frac{5}{19} \text{ days.}$$

As Geminus points out, this is too long by

$$365 \frac{5}{19} - 365 \frac{1}{4} = \frac{1}{76} \text{ day.}$$

Therefore, after 76 years (which is four consecutive Metonic cycles), we will have counted one day too many in comparison with the solar year.

In the late fourth century B.C., Callippus proposed a new luni-solar cycle, the seventy-six-year or *Callippic cycle*, as it is often called. The Callippic cycle is formed from four consecutive nineteen-year cycles, but one day is dropped. The average length of the year in Callippus's cycle is therefore exactly 365 1/4 days. The cycle also preserves the good agreement with the length of the month that had already been achieved in Meton's nineteen-year cycle.

Callippus's seventy-six year cycle served as the basis of an artificial calendar used by some of the Greek astronomers. The best evidence for this comes from Ptolemy's citations of older observations in the *Almagest*. For example, Ptolemy cites an occultation of the Pleiades observed by Timocharis in the third century B.C.:

> Timocharis, who observed at Alexandria, records the following. In the 47th year of the First Callippic 76-year period, on the eighth of Anthesterion, . . . towards the end of the third hour [of the night], the southern half of the Moon was seen to cover exactly the rearmost third or half of the Pleiades.[36]

In this artificial calendar, the years were counted by their place in the seventy-six-year cycle. The month names were borrowed from the Athenian calendar. But it is important to stress that Callippus's calendar had no relation to the calendar of Athens. It was a scientific calendar used by astronomers for their own purposes. This extreme step was taken because the civil calendars of the Greeks were completely unsuitable for accurate counting of the days—for all the reasons mentioned above. Year one of the first Callippic cycle began with the summer solstice of 330 B.C. Timocharis's occultation of the Pleiades, quoted above, was observed in 283 B.C.

Ptolemy provides Egyptian calendar equivalents for the Callippic dates he cites. Thus, Ptolemy says that Anthesterion 8, year 47 of the first Callippic cycle, was equivalent to Athyr 29, year 465 of Nabonassar. All attempts to reconstruct Callippus's calendar have been based on the handful of equivalences provided by Ptolemy and the short description of the Metonic-Callippic cycle by Geminus. However, Geminus's discussion should be viewed as a pedagogical effort to explain the luni-solar cycle, rather than a serious historical account, and Ptolemy provides us very few hard facts. Thus, we cannot reconstruct the Callippic calendar with any certainty.

In the second century B.C., Hipparchus used the Callippic cycle only for specifying the year and preferred to name the day in terms of the Egyptian calendar. For example, in *Almagest* III, 1, Ptolemy cites a list of equinoxes observed by Hipparchus. According to Hipparchus, the autumnal equinox of 162 B.C. occurred in the 17th year of the third Callippic cycle, on Mesore 30, about sunset. This mixed reckoning, involving the use of a solar (Egyptian) calendar for the month and day, and a luni-solar (Callippic) calendar for the year, did not last long. In his own work, Ptolemy used the Egyptian calendar, which was the simplest, most rational option of all.

The Babylonian Calendar

The Babylonian year began with the new Moon of the spring month. Years contained either twelve or thirteen months. The thirteenth month was intercalated either by adding a second month VI or a second month XII.

Babylonian month names[37]

I	BAR	Nisannu	VII	DU_6	Tešrítu
II	GU_4	Ajjaru	VIII	APIN	Arahsamnu
III	SIG	Simānu	IX	GAN	Kislímu
IV	ŠU	Du'ūzu	X	AB	Ṭebētu
V	IZI	Abu	XI	ZÍZ	Šabāṭu
VI	KIN	Ulūlu	XII	ŠE	Addaru
VI$_2$	KIN.A		XII$_2$	DIRIG, A	

In this list, the Babylonian month name is preceded by the Sumerian ideogram often used in Babylonian astronomical texts. Thus, the name of the spring month, *Nisannu* (which would require several cuneiform signs), is usually replaced by a single ideogram, BAR. (Subscripts and accent marks on some ideograms are the Assyriologists' way of distinguishing among several cuneiform signs with the same sound.)

Originally, the intercalations were performed irregularly. Notices were sent in the king's name to the priestly officials at temples throughout Babylonia. This practice was still followed in the Chaldaean period. Later, during the Persian period, the announcements of intercalations came from the scribes at the temple Esangila, who sent notices to the officials at other temples throughout Babylonia.[38] Thus, it appears that the regulation of the calendar passed into the hands of the bureaucracy. This is what made possible the eventual adoption of a regular system of intercalation.

The few official announcements of intercalary months that have survived prove that no regular system of intercalation was in place at the beginning of the Persian period. As discussed in section 1.1, already in MUL.APIN (650 B.C.) there was an attempt at formulating some guidelines for intercalation of months, based on the heliacal risings of the stars. But the appearance of a fixed luni-solar cycle was a later development. There is some evidence that an eight-year cycle was used for the brief period from 529 to 503 B.C. (three intercalary months inserted every eight years). From 499, the nineteen-year cycle was probably in use (seven intercalary months inserted every nineteen years). However, there are some gaps in our knowledge, since we do not have records of some intercalations. Also, the scribes had not yet finalized the rules for deciding when the intercalary month should follow month VI and when it should follow month XII. A definite glitch in the pattern occurred in 385, when that year (rather than the following year) was made a leap year. But from 383 B.C. down to the first century A.D. (when the cuneiform texts cease), a regular pattern of intercalations was followed.[39]

After Alexander's conquest and the establishment of the Seleucid dynasty, the Babylonian texts use the Seleucid era, which we shall abbreviate SE. That is, the old luni-solar calendar based on the nineteen-year cycle continued to function without interruption. But the years were counted from the year that Seleukos I decided to count as the official beginning of his reign. (1 Nisannu, year 1 of Seleucid era = 3 April 311 B.C.)[40]

In terms of the Seleucid era, the leap years are those marked with asterisks in the following sequence:

$$1^* \ 2 \ 3 \ 4^* \ 5 \ 6 \ 7^* \ 8 \ 9^* \ 10 \ 11 \ 12^* \ 13 \ 14 \ 15^* \ 16 \ 17 \ 18^{**} \ 19$$

Thus, years 1, 4, and so on, of the Seleucid era were leap years. In years marked with a single asterisk, month XII was doubled. In years marked with a double asterisk (i.e., year 18), month VI was doubled. To determine whether any year of the Seleucid era was a leap year or not, divide the year number by 19, discard the quotient, but retain the remainder and compare it with the sequence above.

Features of the Babylonian calendar persist in two luni-solar calendars still in use today—the Jewish calendar and the Christian ecclesiastical calendar. After Israel and Judah were conquered by the Babylonians, in the sixth century B.C., the Babylonian calendar was adopted by the Jews. The nineteen-year cycle remains the basic operating principle of the modern Jewish calendar, which is the official calendar of Israel, and which is used worldwide for Jewish religious practice. The month names in the modern Jewish calendar clearly reflect their Babylonian origins: Nisan corresponds to Nisannu, Iyyar to Ajjaru, and so on. The Christian church, drawing on both the Jewish calendar and the Greek astronomical tradition, adopted the nineteen-year cycle as the basis of the ecclesiastical calendar that governs the date of Easter. Thus, in the twentieth century, the celebration of religious festivals such as Passover and Easter is in part regulated by decisions made by anonymous Babylonian scribes 2,500 years ago. This is another striking example of the continuity of the Western astronomical tradition.

4.8 EXERCISE: USING THE NINETEEN-YEAR CYCLE

As discussed in section 4.7, nineteen tropical years contain a whole number of synodic months. This is the basis of the Metonic cycle: 19 years = 235 months. It follows that the dates of the new Moons in the Gregorian calendar should repeat the same pattern, almost exactly, after an interval of nineteen years. The list below gives the dates of the first new Moons for the years 1960–1979. (The dates refer to Greenwich time.)

Year	Date of first new Moon	Year	Date of first new Moon
1961	January 16	1971	January 26
62	January 6	72	January 16
63	January 25	73	January 4
64	January 14	74	January 23
65	January 2	75	January 12
66	January 21	76	January 1
67	January 10	77	January 19
68	January 29	78	January 9
69	January 18	79	January 28
70	January 7		

During this nineteen-year period, the date of the first new Moon moved back and forth all over the month of January. However, after nineteen years, we find the pattern repeating, very nearly. The dates of the first new Moons for the next few years are

Year	Date of first new Moon
1980	January 17
1981	January 6
1982	January 25

We can predict the dates of all the new Moons in any desired year, by use of this list of new Moons. Suppose we want the new Moons for a year that is contained in the list, say 1963. Then, beginning with the date of the first new Moon, we add increments of 30 days and 29 days alternately (i.e., we alternate full and hollow months):

1963	First new Moon	January 25	
			+30 days
	Second new Moon	February 24	
			+29
	Third new Moon	March 25	
			+30
	Fourth new Moon	April 24	
			+29
	Fifth new Moon	May 23, etc.	

The dates obtained by this approximate scheme will rarely differ from the date of true new Moon by more than a day.

Suppose we want the new Moons for a year not in the list, say 1948. We then simply determine which year of the list occupies the same position in the nineteen-year cycle as does 1948. The answer is 1967, since 1948 + 19 = 1967. The new Moons for 1948 may therefore be written out, exactly as explained above, by use of the date of the first new Moon of 1967 as starting point:

1948	First new Moon	January 10	
			+30 days
	Second new Moon	February 9	
			+29
	Third new Moon	March 9, etc.	

Problems

1. Use the nineteen-year cycle and a pattern of alternating full and hollow months to write out the dates of the new Moons for the current year. Compare your results with the dates given by a calendar or almanac.

2. Do you see any evidence for an eight-year cycle in the list of new Moons given above?

3. The technical name for the position of a year in the nineteen-year cycle is the *golden number*. (This term originated in the Middle Ages.) The golden number may be obtained by dividing the year by 19, discarding the quotient, and adding 1 to the remainder. Thus, the golden number of 1961 is 5. (1961/19 = 103, with remainder 4. Golden number = 4 + 1.) The golden number for 1962 is 6; for 1963 it is 7, and so on. Golden numbers are usually written as Roman numerals.

   Construct a table of two columns. The first column should contain the golden numbers I through XIX. The second column should contain the date of the first new Moon of the year corresponding to each golden number.

4.9 THE THEORY OF STAR PHASES

The cycle of appearances and disappearances of the fixed stars was an important part of both early Greek and early Babylonian astronomy. As an example, take the case of the Pleiades. During the spring, the Pleiades disappeared for a month and a half when the Sun moved near them on the ecliptic. Then (in late May), the Pleiades emerged from their period of invisibility. They could be seen, for the first time in the year, rising in the east, a few minutes before dawn. This event was the *morning rising* of the Pleiades. It signaled the wheat harvest and the beginning of summer weather. In the same way, the morning rising of Arcturus was recognized everywhere in the Greek world as the beginning of autumn. The risings and settings of stars that occur just before sunrise, or just after sunset, are called *heliacal risings and settings* (because they occur in connection with the Sun). They are also called *fixed star phases*.

By the fifth century B.C., this lore was systematized into the *parapegma*, or star calendar. (The star calendar was a bit older among the Babylonians. As we have seen, the seventh-century B.C. compilation, MUL.APIN, included a star calendar.) A parapegma listed the heliacal risings and settings of the stars in chronological order. The user of the parapegma could tell the time of year by noting which stars were rising in the early morning. The parapegma served as a supplement to the chaotic civil calendars of the Greeks. Usually, but not always, the star phases were accompanied in the parapegma by weather predictions.

One could compile a list of the heliacal risings and settings of the constellations, simply by observations made at dawn and dusk over the course of a year. There is no need for any sort of theory. In this sense, the parapegma may be considered prescientific. But understanding the annual cycle of star phases was an important early goal of Greek scientific astronomy. Indeed, one of the oldest surviving works of Greek mathematical astronomy is devoted to this subject. This is the book (or really two books) written by Autolycus of Pitane around 320 B.C. and called *On Risings and Settings*. Autolycus defines the various kinds of heliacal risings and settings, then states and proves theorems concerning their sequence in time and the way the sequence depends on the star's position with respect to the ecliptic. No individual star is mentioned by name. Autolycus's goal is to provide a theory for understanding the phenomena. His style is that of Euclid.

True Star Phases

Autolycus and all the Greek scientific writers who followed him distinguished between *true* and *visible* star phases. An example of a true star phase is the *true morning rising* (TMR), which occurs when the star rises at the same moment as the Sun. At such a time the star would be invisible, owing to the brightness of the sky. The *visible morning rising* (VMR) would occur some weeks later, after the Sun had moved away from the star. The visible phases are the observable events of interest to farmers, sailors, poets, and astrologers. However, the true phases are more easily analyzed. Accordingly, Autolycus begins his treatise with a discussion of the true risings and settings. There are four true phases:

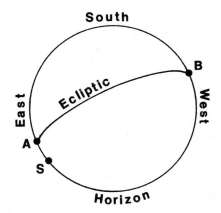

FIGURE 4.3.

TMR	True morning rising	(Star rises at sunrise.)
TMS	True morning setting	(Star sets at sunrise.)
TER	True evening rising	(Star rises at sunset.)
TES	True evening setting	(Star sets at sunset.)

Properties of True Star Phases   For any star, the TMR and the TER occur half a year apart.

For any star, the TMS and the TES occur half a year apart.

These propositions are easily proved. Let star *S* be rising in the east, as in figure 4.3. Let the Sun be rising at *A*. The star is making its TMR. The TER will occur when the star is rising at *S* and the Sun is setting at *B*. The ecliptic is bisected by the horizon; thus there are six zodiac signs between *A* and *B*. If we suppose the Sun moves uniformly on the ecliptic, it will take the Sun half a year to go from *A* to *B*. Thus, the TMR and the TER occur six signs (about six months) apart in the year. The same sort of proof is easily made for the TMS and the TES.

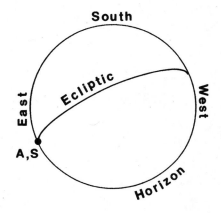

FIGURE 4.4.

The stars have their true phases in different orders according to whether they are south of the ecliptic, on the ecliptic, or north of the ecliptic.

Ecliptic Stars: If a star is exactly on the ecliptic, its TMR and TES will occur on the same day. Let star *S* be at ecliptic point *A*, as in figure 4.4. When the Sun is also at *A*, *S* and *A* rise together, thus producing the star's TMR. In the evening, *S* and *A* will set together in the west, thus producing the star's TES. (We assume that the Sun stays at the same ecliptic point for the whole day.) In the same way, one may show that for ecliptic stars, the TER and TMS occur on the same day.

Northern Stars: If a star is north of the ecliptic, the TMR will precede the TES. Let the northern star *S* be making its TMR, rising simultaneously with ecliptic point *A*, as in figure 4.3. Now, of any two points on the celestial sphere that rise simultaneously, the one that is farther north will stay up longer and set later. (We assume the observer is in the northern hemisphere.) *S* and *A* rise together. But *A* will set first. Thus, when *S* sets, the situation will resemble figure 4.5. *S* is on the western horizon. *A*, located farther south on the sphere, will already have set and will be below the horizon. The TES of star *S* occurs when the Sun is at *C*. Thus, we must wait a few weeks for the Sun to advance eastward on the ecliptic from *A* to *C*. The TES therefore follows the TMR.

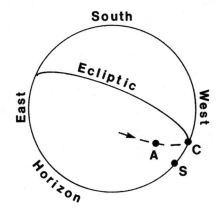

FIGURE 4.5.

Southern Stars: If a star is south of the ecliptic, the TMR will follow the TES. The proof may be made in the same way.

The proofs given above are more concise than Autolycus's proofs of the same propositions, but follow his basic method.

Example: Betelgeuse, a Southern Star   Let us examine the annual cycle of a particular star, Betelgeuse, which lies in Orion's right shoulder. We will assume

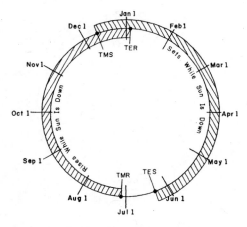

FIGURE 4.6. True phases of Betelgeuse at 40° N latitude. TMR = true morning rising; TMS = true morning setting; TER = true evening rising; TES = true evening setting.

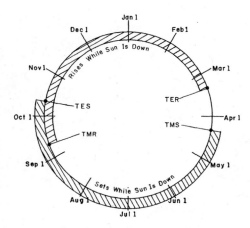

FIGURE 4.7. True phases of Denebola at 40° N latitude.

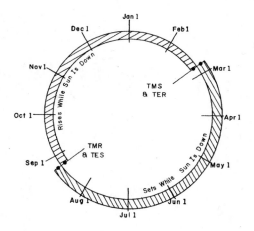

FIGURE 4.8. True phases of Regulus.

that our observations are made at 40° N latitude. The dates of the true phases are easily read off a celestial globe. The dates below are for the twentieth century.

TMR	July 4
TMS	December 8
TER	January 2
TES	June 6

If we mark these dates on a circle which represents the year[41] and if, for simplicity, we treat the months as if they all were of the same length, we get figure 4.6. The dates of the risings and settings are particular to Betelgeuse. Further, they depend on the latitude of the observer. But several more or less general features may still be noted.

First of all, the TMR and the TER are separated by six months. Similarly, the two settings are separated by six months.

Moreover, all stars south of the ecliptic have their true phases in the same order: TMR, TMS, TER, TES (assuming an observer in the northern midlatitudes).

Example: Denebola, a Northern Star　Manipulation of a celestial globe gives the following dates for the true phases of Denebola (δ Leo):

TMR	September 10
TES	October 10
TER	March 8
TMS	April 6

Again the TMR is separated by six months from the TER; similarly, the two settings are six months apart (see fig. 4.7). But there the similarity to the case of Betelgeuse ends, for the phases of Denebola occur in a different order.

All stars north of the ecliptic have their true phases in the same order as Denebola: TMR, TES, TER, TMS (assuming an observer in the northern midlatitudes).

Example: Regulus, an Ecliptic Star　Because Regulus is located on the ecliptic, its TMR and TES occur on the same day, August 24, when the Sun is at the same point of the ecliptic as Regulus itself. Similarly, the TMS and TER occur on the same day, February 20, when the Sun is 180° away from Regulus. The time chart for the phases of Regulus therefore looks like figure 4.8. Note that there is no area of overlap; that is, there is no period of time in which the star both rises and sets while the Sun is down.

Visible Star Phases

The true star phases are unobservable. If a star crosses the horizon at the same moment as the Sun, it will be lost in the general brightness of the sky. The ancient writers therefore distinguish between the true risings and settings and the visible ones. There are four visible phases:

VMR	Visible morning rising	(Before sunrise, star is seen rising for the first time.)
VMS	Visible morning setting	(Before sunrise, star is seen setting for the first time.)
VER	Visible evening rising	(After sunset, star is seen rising for the last time.)
VES	Visible evening setting	(After sunset, star is seen setting for the last time.)

Relation between the Visible and the True Phases　The visible morning phases follow the true ones. But the visible evening phases precede the true ones. The truth of these propositions follows simply from the fact that the Sun's motion on the ecliptic is from west to east, that is, opposite the diurnal revolution.

Let star S be rising while the Sun is rising at A, as in figure 4.9. This is the star's TMR, which will be invisible. But some weeks later, when the Sun has advanced from A to D on the ecliptic, the star's rising will be visible for the first time. Then the Sun will be far enough below the horizon at the rising of S for the star to be seen. Thus, when the Sun is at D, star S will make its VMR.

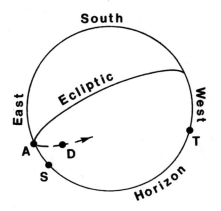

FIGURE 4.9.

Note that the same argument may be applied to the star setting at T. This star makes its TMS when the Sun rises at A. The setting will be invisible, however. The first setting of T to be visible will occur when the Sun has advanced to D.

Thus, the visible morning phases (whether risings or settings) follow the true ones. Also, the morning phases are the first events to be visible: the VMR is the first visible rising of the star in the annual cycle; the VMS is the first visible setting.

In a similar way, it may be shown that the visible evening phases (whether risings or settings) precede the true ones. Also, the evening phases are the last to be visible. That is, the VES is the last setting of the star to be visible. Similarly, the VER is the last visible rising.

A Simplifying Assumption   The number of days that separate a star's visible rising or setting from its corresponding true one depends on many factors: the brightness of the star itself, the star's exact position on the horizon (the farther from the Sun's position, the better), the steepness with which the ecliptic meets the horizon when the star is is rising or setting, as well as the observer's latitude. All these factors can at least be subjected to calculation. But there are also a number of variable conditions that affect the visibility of stars—particularly stars near the horizon—such as the clarity of the atmosphere, city lights, the observer's eyesight, and so on.

A theory of visible star phases that took all these factors into account would be very complicated—too complicated, in fact, to be very useful. And the level of Greek mathematics in the fourth century B.C. would not have permitted such a treatment even if it had been desired. Autolycus was able to dispense with all these complications by means of one simplifying assumption: a star's rising or setting will be visible if the Sun is below the horizon by at least half a zodiac sign *measured along the ecliptic.*

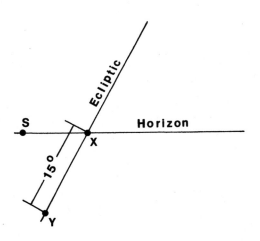

FIGURE 4.10.

In figure 4.10, let star S rise simultaneously with point X of the ecliptic. Then when the Sun is at X, star S will have its true morning rising. According to Autolycus, S will have its visible morning rising when the Sun reaches Y, which is half a zodiacal sign (15°) from X.

According to modern astronomers, the period of astronomical twilight extends from the Sun's setting until the time it reaches a position 18° (vertically) below the horizon. Then the sky becomes dark enough to permit observation of even the faintest stars. However, the brighter stars can be seen when the Sun is only 10° or 12° below the horizon, and these are precisely the stars that play the most prominent role in the ancient literature on star phases. Autolycus's use of 15° measured obliquely to the horizon is a pretty good approximation to 12° measured vertically, especially in the lower latitudes, where the ecliptic rises and sets fairly steeply.

Example: Betelgeuse, a Dock-Pathed Star   It is easy to apply Autolycus's visibility rule to the case of Betelgeuse. Let us suppose, for simplicity, that the Sun moves 1° per day along the ecliptic. Then, by Autolycus's rule, the visible morning phases occur fifteen days after the true ones, while the visible evening phases occur fifteen days before the true ones. Making the appropriate fifteen-day adjustments to the dates of the true phases (listed above), we get

TMR July 4	+ 15 days	→	VMR July 19
TMS December 8	+ 15 days	↘	VER December 18
TER January 2	− 15 days	↗	VMS December 23
TES June 6	− 15 days	→	VES May 22

If we modify the calendar diagram of figure 4.6 to show the visible phases, we obtain figure 4.11.

Note that the order of the visible phases is different from the true phases. The change of order is due to the fact that for Betelgeuse, observed at 40° N latitude, the TMS and the TER are only 25 days apart, while this period would have to be at least 30 days to prevent a reversal of the order when the 15-day visibility rule is applied.

Betelgeuse is completely invisible between the VES and VMR. This is the time when the Sun, moving on the ecliptic, reaches the general vicinity of the star. Betelgeuse rises after dawn and sets before dusk. It is thus up only in daylight and so remains invisible for nearly two months.

The period of maximum visibility of Betelgeuse is in late December, for this date is near both the evening rising and the morning setting. That is, the star rises in the evening and sets in the morning: it crosses the sky during the night. But note in figure 4.11 that there is no overlap between the rings of visibility. During the period between the VER and the VMS, neither the star's rising nor its setting is visible. Rather, the star simply appears in the eastern sky, already above the horizon, shortly after sunset. It can then be seen during most of its transit of the sky. However, the dawn arrives and makes the star disappear shortly before it has a chance to set. In his book on star phases, *Phaseis*, Ptolemy calls this kind of star *dock-pathed*.[42] That is, the

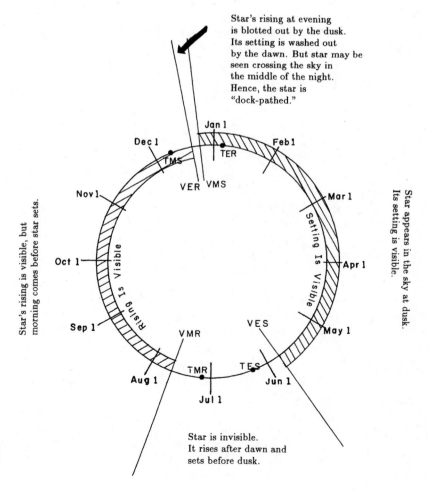

FIGURE 4.11. Visible phases of Betelgeuse at 40° N latitude. Betelgeuse, located near the ecliptic, is a "dock-pathed" star.

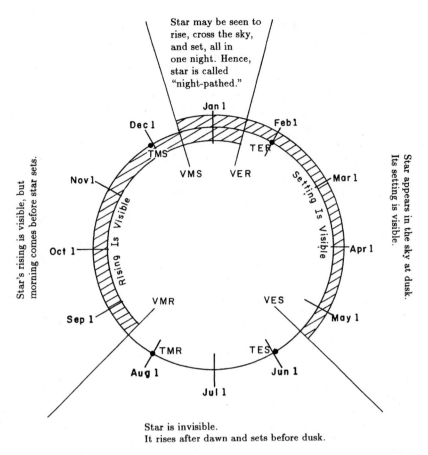

Star may be seen to rise, cross the sky, and set, all in one night. Hence, star is called "night-pathed."

Star's rising is visible, but morning comes before star sets.

Star appears in the sky at dusk. Its setting is visible.

Star is invisible. It rises after dawn and sets before dusk.

FIGURE 4.12. Visible phases of Sirius at 40° N latitude. Sirius, located well south of the ecliptic, is a "night-pathed" star.

ends of its visible path across the sky are docked, or cut off. This is a convenient term for characterizing Betelgeuse's behavior between the VER and the VMS.

All stars located on or near the ecliptic are dock-pathed and have their visible phases in the same order as Betelgeuse: VMR, VER, VMS, VES.

Example: Sirius, a Night-Pathed Star   Using a celestial globe, set for latitude 40° N, we find for the true phases of Sirius:

TMR	August 2
TMS	November 30
TER	January 31
TES	May 29

Note that, as always, the TMR and TER are separated by about six months, as are the TMS and TES. Note, too, that the true phases are in the correct order for a southern star, as established above. Applying Autolycus's fifteen-day visibility rule, we obtain the following approximate dates for the visible phases of Sirius, observed at 40° N latitude:

TMR August 2	+ 15 days	→ VMR August 17
TMS November 30	+ 15 days	→ VMS December 15
TER January 31	– 15 days	→ VER January 16
TES May 29	– 15 days	→ VES May 14

The phases of Sirius are represented in figure 4.12. Note that the visible phases occur in the same order as the true ones. This is true for stars, such as Sirius, that are far enough south on the celestial sphere. The TMR and TER are far enough apart (more than 30 days) that the 15-day rule does not result in a reversal of order for the dates of the visible phases.

Sirius, like Betelgeuse, is invisible between the VES and the VMR.

The period of the Sirius's greatest visibility is between the VMS and the VER, for then Sirius rises in the evening and sets in the morning, so it crosses the sky at night. Further, as figure 4.12 shows, both the rising and setting of the star are visible. Sirius is therefore not dock-pathed. Rather, it belongs to the class of stars that Ptolemy calls *night-pathed*. At this one time of year, the whole of Sirius's transit of the visible hemisphere takes place in the night: the star's whole path, from horizon to horizon, is visible.

All stars located far enough south of the ecliptic are night-pathed and have their visible phases in the same order as Sirius: VMR, VMS, VER, VES. (We assume the observer is in the northern midlatitudes.)

Example: Arcturus, a Doubly Visible Star   From a celestial globe, set for latitude 40° N, we take the dates for the true phases of Arcturus. To obtain rough dates for the visible phases, we apply Autolycus's fifteen-day visibility rule:

TMR October 7	+ 15 days	→ VMR October 22
TES December 4	− 15 days	→ VES November 19
TER April 5	− 15 days	→ VER March 21
TMS June 2	+ 15 days	→ VMS June 17

If we plot these dates on our visibility diagram, the result is figure 4.13. Note that the visible phases occur in the same order as the true ones. This is the case for stars, such as Arcturus, that are far enough north on the celestial sphere. The TMR and TES are far enough apart (more than 30 days), that the 15-day rule does not result in a reversal of order.

The behavior of Arcturus is different from that of either Betelgeuse or Sirius. The most interesting period for Arcturus is between the VMR and the VES. During this period of a bit less than a month, both the rising and the setting of the star are visible each night. But note that the star rises in the morning and sets in the evening: it crosses the sky in the daytime. In the

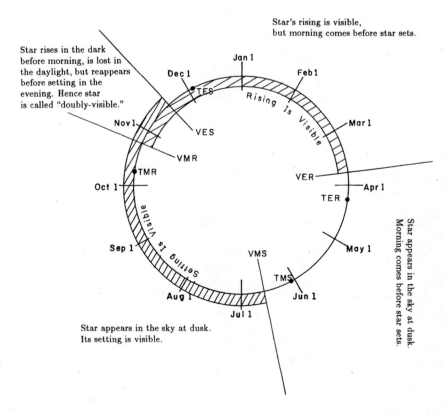

FIGURE 4.13. Visible phases of Arcturus at 40° N latutude. Arcturus, located well north of the ecliptic, is a "doubly visible" star.

early evening one therefore sees Arcturus appear well up in the western sky. Shortly afterward, it sets. But it may be seen again before the night is over, rising in the east. In the *Phaseis*, Ptolemy characterizes such a star as *doubly visible*, or *seen on both sides*. This remarkable behavior results from Arcturus's northern position on the celestial sphere: the star stays below the horizon (at latitude 40° N) for only about 9 1/2 hours each day. And so it is possible, at a particular time of year, for Arcturus to set after dark in the west and yet rise before dawn in the east.

There is another property of doubly visible stars: they never are completely obscured by the Sun. For this reason, as Ptolemy says in the *Phaseis*, they are also called *visible all year long*. These two properties are correlated: both result from the fact that the VMR precedes the VES.

All stars far enough north of the ecliptic are doubly visible and have their visible phases in the same order as Arcturus: VMR, VES, VER, VMS. (We assume the observer is in the northern midlatitudes.)

The standard terms for the phases (true morning rising, visible evening setting, etc.) were introduced by Autolycus and were universally followed. Autolycus discussed the properties of the stars that we have called dock-pathed, night-pathed, and doubly visible but did not assign these names to them. Since every star that has risings and settings may be assigned to one of these three classes, it is convenient to have names for the groups. This rigorous systemization and naming appears for the first time in Ptolemy's *Phaseis*. These terms were used by earlier writers, but less systematically—which seems to confirm the lack of standard terms for the three star classes before Ptolemy's time.

Some Inconvenient Modern Terminology   In modern writing on star phases, one may encounter these terms:

> Heliacal rising   = VMR
> Acronychal rising = VER
> Heliacal setting   = VES
> Cosmical setting = VMS

Except for "acronychal rising," none of these terms are used by the ancient Greek astronomers. And even *acronychal* poses a problem. *Akronychos = akron* (tip, extremity) + *nyktos* (of the night). This adjective is used by Greek writers on star phases and, indeed, belongs to the vocabulary of everyday speech. Usually, it means *in the evening*. But Theon of Smyrna points out that the morning is also an extremity of the night, and therefore, logically enough, he applies the same word both to evening risings and to morning settings.[43] Thus, it is better and clearer to stick to Autolycus's technical vocabulary (visible evening rising, etc.), which can hardly be improved on.

Note on the Variation of Star Phases with Time

The dates of the heliacal risings and settings of stars vary slowly with time. For a century or two, the dates of the risings and settings can be taken as fixed. But if we want to compare the date of the heliacal rising of the Pleiades, for example, in Greek antiquity with the date for the same event in our own century, we must face up to the change.

The reason for the change in the dates of the heliacal risings and settings is *precession*, a slow, progressive shift in the positions of the stars on the celestial sphere. Precession is discussed in Section 6.1. For now it suffices to say that all the stars move gradually eastward, on circles parallel to the ecliptic, at the slow rate of 1° in 72 years. Suppose that the true morning rising of the Pleiades occurs on a certain day. After 72 years, the Pleiades will have shifted eastward by 1°. So, the Sun will have to run an extra degree to reach the Pleiades, and the true morning rising will occur a little later in the year.

The rate at which the star phases shift through the calendar is easily calculated. The stars' longitudes increase by 1° in 72 years. The rate of the Sun's motion on the ecliptic is 360° in 365.25 days. Thus, the time required for a one-day shift in the dates of the star phases is

$$(72 \text{ years}/°) \times (360°/365.25 \text{ days}) = 71 \text{ years/day}.$$

Let us apply this fact to an example. The visible morning rising of Betelgeuse (at latitude 40° N) occurs on July 19. When did the VMR of Betelgeuse occur in the first century B.C.? The first century B.C. was about 2,000 years ago. Every 71 years produced a one-day shift in the star phases. The total shift was therefore 2,000/71 = 28 days. In the past, the star phases occurred earlier in the year. Thus, the VMR of Betelgeuse should have occurred around June 21 (28 days earlier than July 19). Note that it is simplest to express all dates in terms of the Gregorian calendar.

This rough-and-ready method works well for stars near the ecliptic. For stars far from the ecliptic (such as Arcturus), the situation is more complicated and the rough-and-ready method is not usable. Instead, one should replot the stars on a celestial globe in their ancient positions and read off the dates of the star phases directly.

4.10 EXERCISE: ON STAR PHASES

1. In section 4.9, it was proved that, for a star north of the ecliptic, the TMR precedes the TES. Prove that, for a star south of the ecliptic, the TMR follows the TES. (Assume an observer in the northern hemisphere.)
2. In section 4.9, it was proved that the visible morning phases follow the true ones. Prove that the visible evening phases precede the true ones.
3. The dates of a star's phases depend on the observer's latitude.

   A. Use a globe to determine the dates of the true phases of Betelgeuse at latitudes 30° N and 30° S. Draw a calendar diagram like figure 4.6 for each of these latitudes. Note a symmetry: the four dates for 30° S are the same as the four dates for 30° N, but different phases go with the dates. Can you prove the general validity of this rule?

   B. Apply the fifteen-day visibility rule to determine approximate dates for the visible phases of Betelgeuse at latitude 60° N. Draw a calendar diagram like figure 4.11. The diagram is divided into four sections by the star's visible phases. Label each section with a brief description of the star's behavior. Pay particular attention to the section between the VMS and the VER. At 60° N, is Betelgeuse dock-pathed, night-pathed, or doubly visible?

4. The following list gives the "actual dates" of the visible phases of three stars for year −300 and latitude 38° N (Athens). The dates were calculated by Ginzel[44] and are expressed in terms of the Julian calendar.

Star	VMR	VES	VER	VMS
Pleiades	May 22	April 7	Sept 27	Nov 5
Sirius	July 28	May 4	Jan 2	Nov 23
Vega	Nov 10	Jan 23	April 20	Aug 16

   A. Use the order of the phases to place each of these stars in one of the three classes: night-pathed, dock-pathed, or doubly visible. Do your assignments make sense in view of the stars' positions on the sphere?

B. Test the rough-and-ready rule that the dates of star phases shift forward by one day every 71 years. To do this, it will be enough to work with the morning phases for these stars in the year −300.

First, express Ginzel's dates for the VMR and the VMS in terms of the Gregorian calendar (use table 4.1).

Next, use a celestial globe and the fifteen-day visibility rule to estimate the dates of the VMR and the VMS for these stars in the twentieth century.

Use the twentieth-century dates and the average shift of one day in 71 years to estimate when the VMR and the VMS of these stars occurred in the year −300. Compare your estimates with Ginzel's more elaborate calculations. For which stars does the rough-and-ready method work best?

4.11 SOME GREEK PARAPEGMATA

The Geminus Parapegma

The parapegma appended to Geminus's *Introduction to the Phenomena* is one of our most important sources for reconstructing the early history of this genre among the Greeks. The Geminus parapegma is a compilation based principally on three earlier parapegmata (now lost) by Euctemon, Eudoxus, and Callippus, but it also includes a few notices drawn from other authorities. The latest writer cited is Dositheus (ca. 230 B.C.). Thus, some historians believe that the parapegma was actually compiled not by Geminus (first century A.D.), but by some unknown person early in the second century B.C. Be that as it may, this parapegma is always found in the manuscripts appended to Geminus's *Introduction to the Phenomena*.

In the Geminus parapegma, the year is divided according to zodiac signs. Each sign begins with a statement of the number of days required for the Sun to traverse the sign. Then there follow, in order of time, the risings and settings of the principal stars and constellations, together with associated weather predictions and signs of the season. Here is the portion of the parapegma for the sign of the Virgin (an asterisk indicates that an explanatory note follows the extract):

EXTRACT FROM GEMINUS

Introduction to the Phenomena (Parapegma)

The Sun passes through the Virgin in 30 days.

On the 5th day, according to Eudoxus, a great wind blows and it thunders. According to Callippus, the shoulders of the Virgin rise;* and the etesian winds cease.*

On the 10th day, according to Euctemon, the Vintager appears,* Arcturus rises, and the Bird sets at dawn; a storm at sea; south wind. According to Eudoxus, rain, thunder; a great wind blows.

On the 17th, according to Callippus, the Virgin, half risen, brings indications; and Arcturus is visible rising.

On the 18th, according to Eudoxus, Arcturus rises in the morning; <winds> blow for the following 7 days; fair weather for the most part; at the end of this time there is wind from the east.

On the 20th, according to Euctemon, Arcturus is visible: beginning of autumn.* The Goat, great star in the Charioteer,* rises <in the evening>; and afterwards, indications;* a storm at sea.

On the 24th day, according to Callippus, the Wheat-Ear of the Virgin rises; it rains.[45]

Notes to the Extract from Geminus   The time of year covered by this passage is the last week of August and the first three weeks of September. All of the star phases mentioned in a parapegma are, of course, visible phases and not true phases. Let us examine several of the entries more closely.

Day 5. *The shoulders of the Virgin rise*. When no modifier is added, the morning rising should be understood. The morning rising was the first rising to be visible in the course of the year.

The etesian winds cease. The etesian winds are annually recurring (hence their name) north winds that blow in the Mediterranean in summer, giving some relief from the heat. They were held to begin blowing about the morning rising of the Dog Star and to continue for about two months.

Day 10. *The Vintager appears*. The Vintager (our Vindemiatrix, ε Vir) is a dim star in Virgo. Its morning rising marked the beginning of the grape harvest.

Day 20. *Arcturus is visible: beginning of autumn*. The morning rising of Arcturus was widely taken as the beginning of autumn.

The Goat, great star in the Charioteer. This is our Capella, in the constellation Auriga.

Indications. This is our rendering of *episemainei*, literally, "it indicates, it signifies." Presumably, this means that a particular day's weather bore special watching, either because it was likely to undergo a sudden change, or because it would serve to indicate the weather for the immediate future. A good many days throughout the year are called significant, but the basis on which they were singled out is not clear.

In all the Greek parapegmata, we should be careful to distinguish seasonal signs from weather predictions. Thus, the cessation of the etesian winds and the beginning of autumn are best understood as seasonal changes, predictable with some security. The notices of individual rains and wind storms are, at least to a modern reader, much more dubious.

There was a debate in antiquity over the nature of the connection between star phases and weather changes. Aristotle was disposed to believe that the changes in the air were *caused* by the motions of the celestial bodies. If even philosophers believed that the weather was influenced by the stars, we can only conclude that the common people were even more strongly of this view. Geminus devotes an entire chapter (17) of his *Introduction to the Phenomena* to refuting this opinion. For Geminus, the stars *indicate*, but do not *cause* the weather. Ignorant people believe, for example, that the great heat of midsummer is brought on by the Dog Star at its rising with the Sun. But Geminus argues that the Dog Star merely happens to make its morning rising at the hottest time of the year. And he brings the whole weight of physics and astronomy to bear on the question. Geminus's refutation of the doctrine of stellar influences is the most patient and detailed that has come down to us from Greek antiquity. But the fact that he took such pains only confirms how many of his readers he suspected of harboring mistaken views.

The great historical value of the Geminus parapegma is that it cites its authorities by name. Thus, the Geminus parapegma allows us to trace the development of these calendars between the time of Euctemon (ca. 430 B.C.) and the time of Callippus (ca. 330 B.C.). Euctemon had included some fifteen stars and constellations in his parapegma. Some were traditional as seasonal markers, notably Sirius, Arcturus, the Pleiades, and Vindemiatrix. Others served as traditional weather signs, for example, Capella, the Kids, the Hyades, Aquila, Orion, and Scorpius—constellations that are all given special mention as weather signs in Aratus's *Phenomena*. Euctemon did not attempt a systematic coverage of the globe. For the stars he did select, Euctemon provided the dates of all four phases: morning rising and setting, evening rising and setting. Eudoxus's parapegma was quite similar.

Callippus's parapegma shows a number of striking differences from those of Euctemon and Eudoxus. First, Callippus introduced a systematic use of the twelve zodiac constellations in the parapegma. Second, he was much more selective in his use of nonzodiacal stars. Indeed, besides the twelve zodiac constellations, he used only Sirius, Arcturus, the Pleiades, and Orion. Third, Callippus introduced a systematic treatment of extended constellations in their parts. Thus, he tells us when the Virgin starts to make her morning rising, when she has risen as far as her shoulders, when she has risen to her middle, when the wheat ear (Spica) has risen, and when the Virgin has finished rising. To be sure, Euctemon and Eudoxus had already done a bit of this. But Callippus carried it much further. Finally, Callippus did not bother to record the dates of all four phases but confined himself to the morning risings and settings.

The purpose of a parapegma was twofold: to tell the time of year and to foretell the weather. In Callippus's parapegma, we see a shift away from weather prediction toward more precise time reckoning. This is clear in his use of zodiac constellations, in his exclusion of nonzodiacal stars except for those with traditional importance as seasonal signs, in his breaking down of extended constellations into their parts, and even in his exclusive use of morning phases. The morning rising of the Pleaides is the first rising to be visible in the year; the evening rising is the last to be visible. Thus, the morning phases are more certain: you know when you've see the Pleiades rise for the first time, but it may take a few days to be certain whether you've seen them rise for the last time.

A Stone Parapegma from Miletus

In 1902, during the excavation of the theater at Miletus conducted by the German archaeologist Wiegand, four marble fragments were found that were recognized as parts of two parapegmata. It subsequently developed that a fifth fragment, which had been found in 1899, belonged with the others.[46] These five fragments are crucial for our understanding of the public use of star calendars in ancient times. Before the turn of the century, not a single physical parapegma was known; all investigations could be based only on the literary sources (such as Geminus). Naturally, the literary sources left some questions unanswered. For example, even the origin of the name *parapegma* remained obscure. Since the turn of the century, other parapegma fragments have been discovered, but none compares with the Miletus fragments in either importance or state of preservation.[47]

Figure 4.14 is a sketch of one fragment of the so-called first Miletus parapegma. This fragment contains parts of three columns of writing, but only the left and center columns can be read. The left column is for the sign of the Archer and the middle column is for the sign of the Water-Pourer. To the left of each day's star phases, a hole is bored in the stone. Someone probably had the job of moving a peg from one hole to the next each day. Alternatively, it is possible that pegs numbered for the days of the entire month were inserted into the holes in advance. Either practice would permit a passerby to tell at a glance what was transpiring in the heavens. This is why the calendar was called a *parapegma*: the associated verb, *parapegnumi*, means "to fix beside."

A Papyrus Parapegma from Greek Egypt

In the same year, 1902, the British archaeologists Grenfell and Hunt were excavating in Egypt and were approached by a dealer who offered them a

Left Column

○ The sun in the Archer.
○ Orion sets in the morning and Procyon sets in the morning
○ The Dog sets in the morning.
○ The Archer begins rising in the morning and the whole of Perseus sets in the morning.
  ○ ○
○ The stinger of the Scorpion rises in the morning.
  ○ ○
○ The Arrow rises in the morning.
○ The Southern Fish begins to set in the evening.
○ The Eagle rises in the morning.
○ The middle parts of the Twins are setting.

Middle Column

30
○ The sun in the Water-Pourer
○ [The Lion] begins setting in the morning and the Lyre sets.
  ○ ○
○ The Bird begins setting in the evening.
  ○ ○ ○ ○ ○ ○ ○ ○ ○
○ Andromeda begins to rise in the morning.
  ○ ○
○ The middle parts of the Water-Pourer rising.
○ The Horse begins to rise in the morning.
  ○
○ The whole Centaur sets in the morning.
○ The whole Hydra sets in the morning.
○ The Great Fish begins to set in the evening.
○ The Arrow sets. A season of continual west winds.
  ○ ○ ○ ○
○ The whole Bird sets in the evening.
○ [Arcturus] rises in the evening.

FIGURE 4.14. Fragment of a stone parapegma found at Miletus. After the photograph in Diels and Rehm (1904).

large quantity of broken papyrus. The papyrus included literary fragments from the third century B.C., which made it a matter of interest. All this papyrus had been used as mummy cartonnage, that is, as wrapping for mummies. With some difficulty, Grenfell and Hunt learned that the source of the papyrus was the town of Hibeh, on the upper Nile. Subsequent excavation of the necropolis at Hibeh produced more mummies and more papyrus. Among the documents recovered was a parapegma (see fig. 4.15). This portion of the parapegma begins as follows:

EXTRACT FROM THE HIBEH PAPYRI

P. Hibeh 27

<Choiak 1,>. . . The night is 13 4/45 hours, the day 10 41/45.

16, Arcturus rises in the evening. The night is 12 34/45 hours, the day 11 11/45.

26, the Crown rises in the evening, and the north winds blow which bring the birds. The night is 12 8/15 hours and the day 11 7/15. Osiris circumnavigates, and the golden boat is brought out.

Tybi <5,> the Sun enters the Ram.

20, spring equinox. The night is 12 hours and the day 12 hours. Feast of Phitorois.

27, the Pleiades set in the evening. The night is 11 38/45 hours, the day 12 7/45.[48]

This document is of considerable historical interest for several reasons. First of all, as a document written in the third century B.C., it is the oldest surviving example of a Greek parapegma.

Second, it reveals something of the extent to which the Greek-speaking ruling class of third-century Egypt had adopted Egyptian customs. The Greeks adopted mummification of the dead. The parapegma is arranged according to months of the Egyptian year and mentions feasts of the Egyptian religious cycle along with the usual star phases and seasonal signs. Also of interest are the notices of the lengths of days and nights. According to the scheme used in this parapegma, the shortest day is 10 hours long and the longest day is 14 hours. Between winter solstice and summer solstice, the length of the day increases uniformly by 1/45 hour from one day to the next. There are 180 such steps between winter and summer solstice. Thus, the total change in the length of the day is 180/45 = 4 hours. Similarly, the length of the day decreases uniformly in 180 steps between summer and winter solstice. To make up the balance of the year, the day is assumed to remain unchanged (at 10 hours) for two days at winter solstice, and to remain unchanged at 14 hours for three days at summer solstice. This crude scheme is mentioned in Egyptian sources going back to about the twelfth century B.C.[49]

Third, the convention adopted in this parapegma for the zodiac signs is an example of the Eudoxian norm (mentioned in sec. 2.9). Note that the Sun enters the sign of the Ram on the 5th of Tybi, but equinox does not occur until the 20th of Tybi—15 days later. Thus, the equinoctial and solstitial points are at the midpoints of their signs. By the time that the Geminus parapegma was composed, this convention had been replaced by the one that became standard—the equinoctial and solstitial points are at the beginnings of their signs.

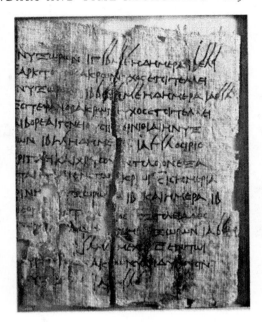

FIGURE 4.15. Fragment of a papyrus parapegma from Greek Egypt, written about 300 B.C. By permission of The Board of Trinity College Dublin. (P. Hibeh i 27. TCD Pap. F. 100 r.)

Ptolemy's Parapegma

Ptolemy's parapegma, which forms a part of his *Phaseis*, introduced some innovations into the tradition. First of all, Ptolemy carried to its logical conclusion the improvement in precision that had been begun by Callippus: Ptolemy did not give the dates of the heliacal risings and settings of constellations or parts of constellations, but only of *individual stars*. He included fifteen stars of the first magnitude and fifteen of the second. In this way, he eliminated the uncertainty in the first or last appearances of extended constellations such as Orion or Cygnus.

Moreover, Ptolemy made no use of the traditional dates of star phases due to Euctemon, Eudoxus, Callippus, and so on. Rather, Ptolemy observed for himself the heliacal risings and settings at Alexandria. He then *computed* the dates on which the stars ought to make their heliacal risings and settings in other climes (i.e., at other latitudes). The "calculations" may well have been performed with aid of a celestial globe. Thus, although Ptolemy gives a complete set of heliacal risings and settings for five different climes (from 13 1/2 to 15 1/2 hours, by half-hour steps), he does not report any star phases for the older authorities. He does, however, give an ample selection of weather predictionsattributed to specific authorities.

Ptolemy attempted to sharpen even the weather-predicting function of the parapegma. This he did by adding a list of the climes to which the weather predictions of his authorities ought to be applied. For example, according Ptolemy, Eudoxus made his observations in Asia Minor, and his weather predictions therefore apply to the climes of 14 1/2 and 15 hours.

Ptolemy's parapegma is arranged according to the months of the Alexandian calendar. The extract below is from the begining of the month of Choiak, corresponding to the end of November.

EXTRACT FROM PTOLEMY

Phaseis (Parapegma)

Choiak

1. 14 1/2 hours: The Dog sets in the morning.
   15 hours: The bright star in Perseus sets in the morning.

According to the Egyptians, south wind and rain. According to Eudoxus, unsettled weather. According to Dositheus, indications. According to Democritus, the sky is turbulent, and the sea generally also.

2. 13 1/2 hours: The star in Orion's eastern shoulder rises in the evening, and the star common to the River and Orion's foot rises in the evening.

14 hours: The star in the head of the western Twin rises in the evening, and the star in the eastern shoulder of Orion sets in the morning.

14 1/2 hours: The bright star in the Northern Crown sets in the evening . . .

. . .

5. 13 1/2 hours: The star called Goat sets in the morning, and the one in the head of the western Twin rises in the evening.

14 hours: The Dog sets in the morning.

15 1/2 hours: The star in the western shoulder of Orion rises in the evening.

According to Caesar, Euctemon, Eudoxus and Callippus, it is stormy.[50]

Note that the Dog Star makes its visible morning setting on Choiak 1 in the clime of 14 1/2 hours, but on Choiak 5 in the clime of 14 hours. Julius Caesar is mentioned as a weather authority on Day 5. This is because he published a parapegma at Rome in connection with his reform of the Roman calendar.

4.12 EXERCISE: ON PARAPEGMATA

Make a parapegma for your own latitude and for the twentieth century. Use a celestial globe to estimate the dates of the true phases of a few important stars. Apply the fifteen-day visibility rule to obtain approximate dates for the visible phases. You may decide for yourself which weather predictions and seasonal signs to include!

A solar theory is a mathematical system that can be used to calculate the position of the Sun in the zodiac at any desired date. Mastery of the solar theory is a prerequisite for any serious study of the history of astronomy. In the first place, the motion of the Sun is much simpler than that of either the Moon or the planets, yet it involves many of the same features. The solar theory therefore serves as an excellent introduction to problems and techniques that are encountered in the planetary theory.

Second, the solar theory is the foundation on which the whole of positional astronomy must be constructed. It is not possible to measure the position of any other celestial body until one has a working theory of the Sun, as Ptolemy himself says in the introduction to the third book of the *Almagest*.

Finally, the ancient solar theory is of great historical importance. It was advanced by Hipparchus in the second century B.C. and accepted by every astronomer of the Greek, Arabic, and Latin tradition down to the beginning of the seventeenth century, when it was finally displaced by the new astronomy of Kepler.

Equinoxes and Solstices: Methods of Observation

The parameters, or elements, of the ancient Greek solar theory were derived from only two kinds of observations: equinoxes and solstices. The fundamental parameter is the length of the tropical year, defined as the period from one summer solstice to the next, or from one spring equinox to the next. The success of the solar theory therefore depends on the accuracy with which these guideposts can be observed.

Gnomon   In the fifth century B.C. the gnomon was the chief, and perhaps the only, instrument available for making observations of the Sun. The solstices were determined by observing the lengths of the noon shadows. Summer solstice occurred on the day of the shortest noon shadow; the winter solstice, on the day of the longest noon shadow. Of course, the solstices do not necessarily occur at noon. The summer solstice is the moment of the Sun's greatest northward displacement from the equator, and this is just as likely to occur at the middle of the night as at noon. It is possible to interpolate between observations of noon shadows to obtain a more precise estimate of the moment of solstice. A summer solstice often cited by the ancients[1] was observed at Athens by Meton and Euctemon in the archonship of Apseudes, on the 21st day of the Egyptian month of Phamenoth, *in the morning* (June 27, 432 B.C.). This time of day for the solstice—in the morning—must have been the result of interpolation between noon observations.

However, there are severe limitations on the accuracy with which a solstice may be determined. The Sun's motion in declination is so slow around the solstices that the moment of greatest declination is very uncertain. As the *table of obliquity* (table 2.3) reveals, after the summer solstice, the Sun must run 6° or 7° along the ecliptic for its declination to decrease by only 10'. So, for a week on either side of the solstice, the length of the noon shadow scarcely changes.

The gnomon itself produces an additional uncertainty: the edges of shadows are fuzzy and indistinct, so a precise location of the shadow's tip is impossible. This fuzziness arises from the fact that the Sun is not a geometric point, but has an appreciable (1/2°) angular diameter.

Meridian Quadrant   In the *Almagest*, Ptolemy describes several instruments for observing the Sun that represent marked improvements over the gnomon of the fifth century astronomers. The instruments described by Ptolemy were

FIVE

✦ ✦ ✦

Solar
Theory

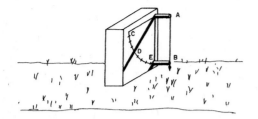

FIGURE 5.1. One of Ptolemy's quadrants for taking the altitude of the Sun.

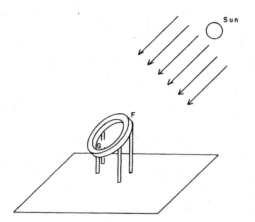

FIGURE 5.2. Ptolemy's equatorial ring.

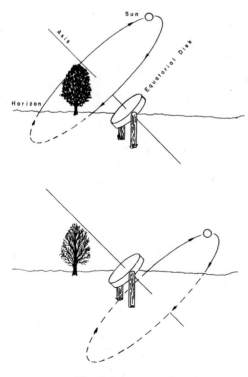

FIGURE 5.3. The daily motion of the Sun in spring or summer (top) and in fall or winter (bottom).

not all original with him. Similar instruments were used by Hipparchus and may have been used by astronomers of the third century B.C.

The most useful of the new instruments was the quadrant set in the plane of the meridian. In *Almagest* I, 12, Ptolemy describes two versions of this instrument. The simplest consisted of a block of wood or stone, with a smoothly dressed face set in the plane of the meridian (fig. 5.1). Near one of the upper corners, a cylindrical peg *A* was fixed at right angles to the face. This peg served as the center of a quarter-circle *CDE* that was divided into degrees and, if possible, into fractions of a degree. Below *A* was a second peg *B*, which served as an aid in leveling the instrument. A plumb line was suspended from *A*, and then splints were jammed under the block until the plumb line passed exactly over *B*. When, at noon, the Sun came into the plane of the meridian, the shadow cast by peg *A* would indicate the altitude of the Sun. At noon the shadow becomes rather faint. Therefore, as Ptolemy says, one may place something at the edge of the graduated scale, and perpendicular to the face of the quadrant, to show more clearly the shadow's position. Ptolemy does not say how large his quadrant was. Judging by the descriptions of similar instruments in the writings of Pappus and Theon of Alexandria, Ptolemy's quadrant most likely had a radius of from one to two cubits (18–36 inches).

One advantage of the quadrant was that it reduced the uncertainty due to the fuzziness of shadows. In using the quadrant, the observer locates the *center* of the shadow, rather than the edge. As the eye is able to spot the center of a narrow line of shadow very accurately, the new procedure represented an important advance over the old. The summer solstice was determined by taking several noon altitudes and interpolating to find the moment of the Sun's greatest declination.

Unfortunately, even a well-made and accurately aligned meridian quadrant could not determine the time of solstice very precisely, owing to the nature of the solstice itself. A precise measurement of the length of the year could not be based on the solstices. More reliable for this purpose were the times of the equinoxes. (It makes no difference whether we measure the year as the interval from summer solstice to summer solstice, or from spring equinox to spring equinox.) The time of equinox also could be determined from observations made with the meridian quadrant. To begin, one had to measure the altitude of the noon Sun at summer and at winter solstice. The altitude of the noon Sun at equinox should fall exactly midway between these two values. One therefore measured the noon altitude of the Sun for several successive days around the expected time of the equinox and interpolated to find the moment of equinox. Because the Sun's declination changes rapidly around the equinox (about 24′ in a single day), this method was capable of fixing the moment of equinox to the nearest quarter or half day. Whether this precision was actually achieved depended on the skill with which the instrument was constructed and aligned.

Equatorial Ring　A second specialized instrument was also available for the determination of the equinoxes: the equatorial ring. This consisted of a large metal ring (probably one to two cubits in diameter) placed in the plane of the equator (fig. 5.2). The operation of this instrument is illustrated in figure 5.3. During spring and summer, the Sun is north of the celestial equator. Its daily motion carries it in a circle parallel to the equator. Consequently, the Sun shines all day on the top face of the ring and never on the bottom face. In fall and winter, the Sun is below the equator and shines all day on the bottom face of the ring. Only at the moment of equinox does the Sun come into the plane of the equator. At this moment, the shadow of the upper part of the ring (*F* in fig. 5.2) will fall on the lower part (*G*) of the ring. Since the Sun has some angular size, the shadow of *F* will actually be a little thinner than the ring itself. The equinox is indicated when the shadow falls centrally on the lower part of the ring, leaving narrow illuminated strips of equal widths above and below the shadow.

The advantage of the equatorial ring is that it can indicate the actual moment of equinox: one need not restrict oneself to observations taken at noon. If the Sun comes into the plane of the equator at 9 A.M., the equatorial ring will indicate it. Of course, if the equinox should happen to occur at night, it will still be necessary to interpolate between observations taken on two successive days.

A major disadvantage of the equatorial ring is the difficulty of attaining and maintaining an accurate orientation. If the ring is tilted slightly out of the plane of the equator, it will not indicate the equinox correctly. Indeed, Ptolemy notes in *Almagest* III, 1, that one of the large rings fixed to the pavement in the palaestra of Alexandria had shifted imperceptibly, with the result that it suffered a change in lighting *twice at the same equinox*. That is, according to the ring, the equinox occurred twice, at times a few hours apart. Ptolemy seems for this reason to have mistrusted the equatorial ring and to have determined his own equinoxes with the meridian quadrant, an instrument that is easier to align. But earlier astronomers, including Hipparchus, seem to have taken at least some of their equinoxes by means of the equatorial ring.[2]

It is possible that the double equinox indicated on the ring at Alexandria was produced just as Ptolemy surmised, by a misalignment of the ring. But there are two other causes that could have produced the same effect. First, the ring might have been warped.[3] Second, false equinoxes can be produced by atmospheric refraction.[4] The effect of refraction is to make the Sun appear slightly higher than it really is. Refraction is appreciable only for objects quite near the horizon. For example, when the Sun appears to be on the horizon, it is actually below the horizon by more than half a degree. But at an altitude of 15° above the horizon, refraction amounts to only 3′. The meridian quadrant is superior to the equatorial ring because the former permits observations only at noon, when the Sun is highest and refraction is least. Ptolemy was not aware of atmospheric refraction, but he could not help noticing that the meridian quadrant gave better results than did the equatorial ring.[5]

The Length of the Year

Early Values   The oldest accurate value for the length of the year is 365 days. This figure was ancient in Egypt, where it served as the basis of the calendar. In the third century B.C., the king of Egypt, Ptolemy III Euergetes, attempted to reform the calendar by the adoption of a year of 365 1/4 days. It does not, however, follow that the Egyptians first realized that the 365-day year was too short only in the third century. Knowledge of the 365 1/4 day year arose long before, through the observed "slipping" of the agricultural year (especially the annual flooding of the Nile) and various celestial phenomena (such as the morning rising of Sirius) with respect to the 365-day calendar year. Civil calendars, which are conventional schemes for reckoning time, rarely embody the best astronomical knowledge of the age.

Among the Greeks, knowledge of the 365 1/4 day year goes back to the fifth century, if not a little earlier. When Meton introduced his nineteen-year luni-solar cycle at Athens in 432 B.C., he meant it as an improvement over the old eight-year cycle, and the eight-year cycle had been constructed on a solar year of 365 1/4 days.

The tropical year is actually a little shorter than 365 1/4 days, a fact reflected in the construction of the Gregorian calendar. In Athens of the fifth century B.C., an accurate measurement of the tropical year would have been difficult for two reasons: the lack of any earlier reliable observations, and the inadequacies of the instruments available. The first difficulty could be remedied only by time: even two crudely determined summer solstices will give a good value for the length of the year if the observations are separated by a long time interval, say, several centuries. The second difficulty was remedied by advances in the technology of instruments. By the second century B.C., the new instruments and the relatively great age of the earliest recorded solstices made possible a definite improvement in the measured length of the year. But before a value could be settled on, it was necessary to show that the year had a constant length.

Is the Length of the Tropical Year Constant or Variable?   There seems to have been a common suspicion that the length of the year might be variable.[6] The apparent variability actually resulted from errors of observation. Around 128 B.C., Hipparchus made a study of this question, as a part of his treatise *On the Change of the Tropic and Equinoctial Points*. This work has been lost, but portions of its contents were summarized by Ptolemy in the *Almagest*, so we know something of Hipparchus's procedure. Hipparchus examined the data at hand and determined that the variation in the length of the year could not be larger than about a quarter of a day. Further, he judged that the error in the observed time of an equinox or a solstice could easily amount to a quarter day. He concluded that there was no basis for accepting a variation in the length of the year: the supposed variation was no larger than the errors of observation. The most decisive evidence came from a dozen equinoxes observed carefully by Hipparchus himself over a period of nearly thirty years.

Ptolemy's discussion of this question in *Almagest* III, 2, based on Hipparchus's work, is valuable for the light it sheds on the Greek astronomers' methods of handling discordant observations. Sometimes it is hard to know whether the Greeks had too much or too little respect for observational data. On the one hand, they were capable of doctoring or adjusting data that did not seem to fit, which seems to show little faith in the possibility of accurate measurement. On the other hand, when the observation of a few solstices seemed to show that the length of the year was variable, many astronomers accepted this conclusion at face value. This seems to show too much faith in the data. In the same way, we know that as late as the second century A.D., some astronomers still believed that the Sun might wander a little north or south of the plane of the ecliptic.[7] This erroneous conclusion probably resulted from someone mistakenly noting that the Sun's most northerly rising point varied a little from one year to the next: again, a case of putting too much faith in a sloppily made observation.

There really is no such thing as the "straight facts": only careful consideration can tell us how much faith to put in an observation. In the *Almagest* Ptolemy usually avoids the problems presented by discordant or redundant data. Ptolemy generally reports no more data than he needs to determine the parameter whose value is under discussion. The planetary theories of the *Almagest* therefore appear to be based on an extremely limited number of observations. No doubt the observations actually reported by Ptolemy were selected from a larger pool of data he did not bother to report. Did he select arbitrarily one observation of a particular type from a collection of discordant observations of the same type? If he did not select arbitrarily, what criteria did he apply in making his choice? Again, Ptolemy has little to say. But his discussion of the length of the year, based on Hipparchus's, shows us that, by the second century B.C., the Greek astonomers had become much more sophisticated in handling discordant observations than we might otherwise have suspected.

Hipparchus and Ptolemy on the Length of the Tropical Year   Once Hipparchus had satisfied himself that the length of the year was constant, his next step was to assess that length as accurately as possible. This he attempted in a book *On the Length of the Year*, which also is lost but which is cited by Ptolemy in *Almagest* III, 1. Hipparchus compared a summer solstice observed by Aristarchus of Samos at the end of year 50 of the first Callippic period (the summer solstice of 280 B.C.) with one observed by himself at the end of year 43 of the third Callippic period (135 B.C.). The interval between these observations was 145 tropical years. If the length of the year were exactly 365 1/4 days, then the number of days elapsed should have been 145 × 365 1/4. Hipparchus found that the time interval was twelve hours shorter than this. He concluded that the year was shorter than 365 1/4 days by about half a day in 150 years, or a whole day in about 300 years. That is, the length of the tropical year was about

$$365 + \frac{1}{4} - \frac{1}{300} \text{ days} = 365.2467 \text{ days}.$$

This value, although still a little high, represented a distinct improvement over the old 365 1/4 day value. (The modern Gregorian calendar is based on a year of 365 + 1/4 − 3/400 days.) Indeed, it is difficult to see how Hipparchus could have done any better with the instruments and recorded observations available to him.

Some 285 years after Hipparchus, Ptolemy confirmed the length of the year measured by his predecessor. Comparing an autumnal equinox observed by himself in A.D. 139 with one observed by Hipparchus in 147 B.C., Ptolemy found that the length of the year is less than 365 1/4 days by one day in 300 years, just as Hipparchus had found. Ptolemy compared also a spring equinox observed by Hipparchus in 146 B.C. with one observed by himself in A.D. 140 and again came to the same conclusion. Finally, because of its antiquity, Ptolemy compared the summer solstice observed by Meton and Euctemon in 432 B.C. with one of his own, observed in A.D. 140. Again, the result gave a tropical year of 365 + 1/4 − 1/300 days. Ptolemy adopted this as the length of the tropical year and based his solar theory on it. In doing so, he adopted a year that was still a little too long and failed to improve on Hipparchus's value despite the advantage of the additional 285 years that separated him from Hipparchus.

For this reason, Ptolemy's solar observations have been much criticized by historians of science. Ptolemy's evaluation of the length of the year is based on the autumnal equinox of A.D. 139, the spring equinox of A.D. 140, and the summer solstice of AD 140, all observed by him. These three observations contain rather substantial errors and yet they each, when compared with the more ancient observations, give a tropical year that agrees exactly with Hipparchus's value. It has therefore been suggested that Ptolemy did not make these observations at all, but rather made them up to have "observations" in agreement with Hipparchus's solar model.[8] A less radical hypothesis is that, on examining many discordant solar observations, Ptolemy saw that he could not improve on Hipparchus's result and therefore simply selected those of his own observations that were in best agreement with Hipparchus's. Probably he also adjusted the times of his own observations slightly to produce perfect agreement. The times reported by Ptolemy for his equinoxes and solstice (one hour after sunrise, one hour after noon, two hours after midnight) are suspiciously precise, especially as they result in perfect agreement with Hipparchus's tropical year. While adjusting the observed times to produce agreement cannot be justified from our point of view, Ptolemy may have felt that small adjustments were permissible in view of the uncertainty attached to the observations themselves. Certainly, a textbook of astronomy—which is what the *Almagest* is—would be less objectionable to students and teachers alike if the numerical values reported in it were as harmonious and consistent as possible.[9]

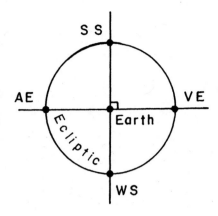

FIGURE 5.4.

5.2 THE SOLAR THEORY OF HIPPARCHUS AND PTOLEMY

In developing a theory of the Sun's motion, we have to account in the first place for the striking seasonal changes—the annual changes in the number of hours of daylight, the rising and setting directions of the Sun, and the length of the noon shadow. We have found that all of these changes can be accounted for by a model in which the Sun moves on a circle (the ecliptic) inclined to the equator. Without making too much fuss about it, we have assumed that the Earth lies at the center of the circle of the Sun's motion and that the Sun travels around the circle at a uniform rate of 360° in about 365 1/4 days.

The Solar Anomaly

This picture is about right, since it does account for the seasonal changes. However, it cannot be exactly right, because it fails to account for another observable effect—the inequality in the lengths of the seasons. In a typical year, the equinoxes and solstices fall around these dates:

Vernal equinox	Mar 21 (modern era)
Summer solstice	Jun 22
Autumnal equinox	Sep 23
Winter solstice	Dec 22

By counting days we find that the seasons have the following lengths:

Spring	93 days (modern era)
Summer	93
Autumn	90
Winter	89

The differences in the lengths of the seasons were noticed as early as 330 B.C. by Callippus, who had their lengths correct to the nearest day.[10] The definitive values for the lengths of the seasons in antiquity were those of Hipparchus, measured around 130 B.C.:

Spring	94 1/2 days (Hipparchus, ca. 130 B.C.)
Summer	92 1/2
Autumn	88 1/8
Winter	90 1/8

Note that these are not quite the same as the lengths of the seasons today. This is not a mistake on the part of Hipparchus: the seasons really have changed in length, although they still add up to the same total of 365 1/4 days. Thus, in antiquity spring was the longest season, but summer is the longest today.

In a naive model of the Sun's motion, the Sun is assumed to travel at uniform speed on a circle whose center is at the Earth, as in Figure 5.4. In this model, the equinoctial and solstitial points are equally spaced at 90° intervals around the zodiac. So, if the Sun did travel at a uniform rate around this circle, all the seasons would all be the same length, namely 365.25 days/ 4 = 91.31 days.

But, in fact, the Sun does not appear to travel at the same angular speed everywhere on its orbital circle. In the modern era, it requires 93 days to travel the 90° from summer solstice to autumnal equinox, and only 89 days for the 90° from winter solstice to spring equinox. Evidently, the Sun travels a little faster on its circle in January than in July. This apparent variation in speed is called the *solar anomaly* or the *solar inequality*.

To account for the solar inequality, the Greek astronomers had to give up one or more of their original assumptions about the Sun's annual motion.

These assumptions were three: (1) the Sun's orbit is a circle, (2) centered on the Earth, (3) along which the Sun travels at constant speed. We know today that all three assumptions are false: the orbit really is an ellipse; the Earth is not at its center but at one of the two foci; and the Sun's speed on the ellipse is not constant. But in the second century B.C. it would have been rash to reject completely a model that worked very well and evidently required only a minor modification. (The modification will be minor, because the lengths of the seasons differ from one another by only a little.) Giving up any of the three assumptions meant violating a principle of Aristotelian physics. But it would have been worse to give up either the circular path or the uniform speed, since this would have greatly complicated calculation.

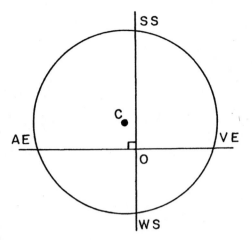

FIGURE 5.5.

Choice of a Model

Hipparchus showed that the solar anomaly can be accounted for by a much less painful change in the model. Hipparchus still let the Sun move on a circle at a uniform speed, but the center of the circle was no longer assumed to coincide with Earth. Rather, the center of the circle was slightly displaced from Earth (the center of the world). Hence, the Sun's orbital circle was said to be *eccentric*.

Figure 5.5 shows the situation for the present era. That is, we attempt to account for the present lengths of the seasons by means of Hipparchus's model. *C* is the center of the Sun's eccentric circle, and *O* is Earth. The equinoxes and solstices are still equally spaced around the sky, as viewed from Earth, but these points no longer divide the circle of the Sun's annual motion into equal intervals. Thus, we obtain seasons of unequal lengths. The placement of *C* shown in figure 5.5 produces the effect desired, for it makes summer the longest season and winter the shortest.

Let us modify figure 5.5 by drawing a line through *C* and *O*. This line cuts the circle in two places, as shown in figure 5.6. One of these intersections is the *apogee* of the eccentric circle, marked *A*. At apogee, the Sun is at its greatest distance from the Earth. The other intersection is the *perigee* of the eccentric, marked Π. At this point, the Sun is closest to the Earth in the course of the year. Either *A* or Π is called an *apse* ("arch" or "vault") of the orbit. Line *ACO*Π is called the *line of apsides*.

In Hipparchus's model, the Sun travels at a constant speed on its circle, but *appears* to travel more quickly at perigee and more slowly at apogee, because of its varying distance from the Earth.

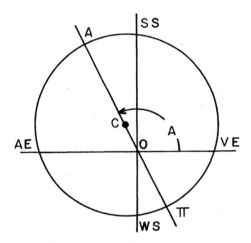

FIGURE 5.6. Eccentric-circle theory of the motion of the Sun.

Points labeled in the figure:

O, Earth	VE, vernal equinox
C, center of eccentric	SS, summer solstice
A, apogee	AE, autumnal equinox
Π, perigee	WS, winter solstice

Parameters of the model:

angle *A*, longitude of the apogee
$OC/CA = e$, eccentricity.

The Parameters of the Solar Theory

The solar theory has four *parameters*, or *elements*, that must be determined from observations before the model can be used to predict future positions of the Sun:

- the length of the tropical year, which determines the rate of the Sun's motion on the circle.
- the longitude of the apogee (i.e., the angle marked *A* in fig. 5.6). This angle specifies the direction in which *C* is located, as seen from Earth.
- the eccentricity of the eccentric circle (i.e., the ratio of *OC* to the radius of the circle). This quantity specifies the amount by which *C* is displaced from the Earth.
- the longitude of the Sun at one particular moment.

If these four quantities are known, the model can be used to predict the longitude of the Sun at any desired date.

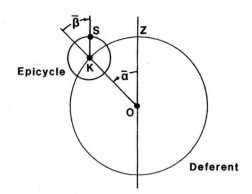

FIGURE 5.7. Concentric deferent and epicycle model for the motion of the Sun. The Sun *S* moves on an epicycle while the center *K* of the epicycle moves around a deferent circle centered on the Earth *O*. Both motions are completed in one year. Thus, angles $\bar{\beta}$ and $\bar{\alpha}$ are always equal.

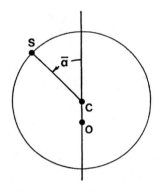

FIGURE 5.8. The eccentric-circle model for the motion of the Sun. Angle $\bar{\alpha}$ is called the mean anomaly.

Another Model

It happens that an entirely different model of the Sun's motion will produce the same result. Refer to figure 5.7. Let point *K* move uniformly and counterclockwise around a circle centered on the Earth *O*. Thus angle $\bar{\alpha}$ increases uniformly with time. The large circle on which *K* moves is called the *deferent circle*. The deferent circle is said to be concentric to the Earth (i.e., centered on the Earth).

Let the moving point *K* be the center of a small circle called the *epicycle*. Thus, the epicycle rides around the rim of the deferent. Meanwhile, let the Sun *S* move uniformly and clockwise on the epicycle. Thus, angle $\bar{\beta}$ also increases uniformly with time. Further, let these two motions occur at the same rate, so we always have $\bar{\beta} = \bar{\alpha}$. *The motion of the Sun* S *resulting from the two combined motions is nothing other than uniform motion on an eccentric circle.* That is, the concentric-deferent-plus-epicycle model of figure 5.7 is mathematically equivalent to the eccentric-circle model of figure 5.8.

The equivalence is easily demonstrated (in modern language) by means of the commutative property of vector addition. In figure 5.7, because $\bar{\beta} = \bar{\alpha}$, the turning radius *KS* in the epicycle will always remain parallel to the direction *OZ*. Now, if we regard the Earth *O* as the origin of coordinates, the position vector **OS** of the Sun can be expressed as the sum of two vectors:

$$\mathbf{OS} = \mathbf{OK} + \mathbf{KS}.$$

But we may add these vectors in either order. Thus, we must also have

$$\mathbf{OS} = \mathbf{KS} + \mathbf{OK}.$$

The model resulting from the second form of the vector addition is shown in figure 5.8. We start from the Earth *O* and lay out a vector (which we call **OC**) whose length and direction is the same as that of **KS** in figure 5.7. As already shown, **KS** points in a fixed direction. Thus, in figure 5.8, **OC** is a vector of fixed direction. To complete the vector sum in figure 5.8, we attach to **OC** a vector **CS**, equal to **OK** in figure 5.7. The result in figure 5.8 is precisely the off-center circle model. The Sun *S* moves uniformly around a circle centered on *C*, a point eccentric to the Earth *O*.

In figure 5.9, the deferent and epicycle are drawn in solid line. The resulting eccentric circle orbit is shown in dashed line. The center of the effective eccentric circle is *C*. Thus, the radius *KS* of the epicycle (in the deferent-and-epicycle version of the solar model) is equal to the eccentricity *OC* (in the eccentric circle version).

Early History of the Solar Theory

The early history of the solar theory is not very well known. Traditionally, it is ascribed to Hipparchus. But, as mentioned above, the existence of the solar anomaly was already known to Callippus (late fourth century B.C.). Moreover, it is clear from some remarks by Ptolemy in *Almagest* IX that the equivalence of the deferent-and-epicycle model to the eccentric circle model was proved by Apollonius of Perga in the third century B.C.[11] The real originator of the solar theory may therefore have been Apollonius: the existence of the solar anomaly had been demonstrated by his time, and he is known to have proved theorems concerning epicycle motion.

However, there is no evidence that Apollonius had any idea of producing a quantitative, predictive theory of the Sun. For Apollonius, as for the earlier Greek astronomical thinkers, such as Eudoxus and Callippus, the goal was broad physical explanation. The chief problem was explaining how the Sun could appear to move at a varying speed, while actually moving uniformly. Apollonius's problem was to reconcile the observed inequality in the lengths of the seasons with the universally accepted physical principle that the heavenly bodies must move in uniform circular motion.

As far as we know, Hipparchus was the first to show how to derive numerical values for the parameters of the model from observations. The very idea that a geometrical theory of the motions of the Sun and planets *ought to work in detail* was a new one in the second century B.C. It was Hipparchus who turned a broadly explanatory geometrical model into a real theory.

What was Hipparchus's motivation? On the one hand, we can see Hipparchus as continuing the geometrization of astronomy that was already well under way. The theory of the celestial sphere had subjected all the phenomena associated with daily risings and settings to a geometrical treatment. Autolycus of Pitane had shown how to explain the annual cycle of heliacal risings and settings in terms of the theory of the sphere. Apollonius had shown that deferent-and-epicycle theory could explain, at least qualitatively, the irregular motions of the Sun, Moon, and planets. All of this earlier work, however, showed only a limited interest in numerical detail. Most of the earlier numerical work concerned time periods, as in the construction and refinement of luni-solar cycles. The rest of Greek theoretical astronomy really amounted to a branch of geometry—strict geometrical proofs, but without realistic numbers, and sometimes with no numbers at all. Aristarchus of Samos was an important transitional figure. Aristarchus's derivation of the distances of the Sun and Moon was the first calculation of a cosmic *length*. In a way, Hipparchus's derivation of the eccentricity of the Sun's circle can be seen as a continuation of Aristarchus's program. But even Aristarchus's calculation was based, as we have seen, on "plausible" numerical data, rather than on real observations. Before Hipparchus, the method was always more important than the actual numbers.

If we can see Hipparchus as part of a tradition, it must also be said that there was something special about Hipparchus himself. Where his predecessors had been content with broad physical explanation, he insisted on precision. Hipparchus would have been a difficult and demanding man to have for a thesis adviser! His only surviving work is his *Commentary on the Phenomena of Aratus and Eudoxus*, in which he criticizes Aratus and Eudoxus for inaccuracies in their descriptions of the constellations. Moreover, Strabo tells us that Hipparchus wrote a book called *Against Eratosthenes*, in which he took Eratosthenes to task for sloppiness in his mathematical geography.[12] And, finally, Ptolemy tells us that Hipparchus criticized the planetary theories of his predecessors and contemporaries and showed them to be in disagreement with the phenomena.[13] Hipparchus's attitude toward observation represented a radically new way of regarding the world—at least among the Greeks.

For it is clear that Hipparchus was strongly influenced by Babylonian astronomy. The Babylonians had worked out quantitative, predictive theories of the motions of the Sun, Moon, and planets shortly before his time. In the Babylonian solar theory, the Sun moved around the zodiac at a varying pace, according to fixed, arithmetic rules. In fact, there were two different versions of the Babylonian solar theory, now called system A and system B.[14]

In the Babylonian system A, the Sun was assumed to have two different constant speeds on two portions of the zodiac. On the fast portion, from Virgin 13° to Fishes 27°, it moved at a constant rate of 30° per synodic month. On the slow portion, from Fishes 27° to Virgin 13°, it moved only 28°7'30" per synodic month (see fig. 5.10). The middle of the slow portion is at Twins 20°. This point should correspond to the apogee of the Sun's circle in the geometrical theory of Hipparchus.

Now, the Babylonians placed the beginnings of the signs differently than did the Greeks. The Greeks (whose system has become standard) defined the

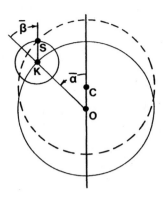

FIGURE 5.9. Equivalence of the concentric-plus-epicycle model (fig. 5.7) to the eccentric-circle model (fig. 5.8). If the radius K of the epicycle is equal to the eccentricity OC of the eccentric, and if the rates of motion are chosen so that one always has $\bar{\beta} = \bar{\alpha}$, the two models are mathematically equivalent.

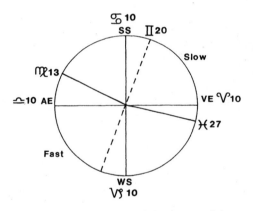

FIGURE 5.10. The fast and slow zones of the Babylonian solar theory of system A. In each zone of the ecliptic, the Sun moves at a uniform angular speed.

signs so that the spring equinoctial point fell at the beginning of the Ram (zeroth degree), the summer solstitial point fell at the beginning of the Crab, and so on. But the Babylonians put the equinoctial points at either the 10th degrees of the signs (in system A) or at the 8th degrees of the signs (in system B).[15] The center of the slow arc in the Babylonian solar theory of system A was Twins 20°, that is, the twentieth degree of the Babylonian sign of the Twins. The Greek sign of the Twins begins 10° later than the Babylonian sign. Thus, the "effective apogee" of the Babylonian theory was at Twins 10°, reckoned according to the Greek system of coordinates. This is very close to the longitude of the apogee adopted by Hipparchus (Twins 5 1/2°) in his geometrical model.

In the Babylonian solar theory of system B, the Sun's speed followed a "linear zigzag function." That is, instead of merely two values for the Sun's speed, there was a sequence of values, with the Sun's speed changing by equal increments from step to step. System A, which is simpler, seems to be older. However, both versions of the Babylonian solar theory were simultaneously in use for the whole period (roughly 250 B.C. to A.D. 50) for which evidence is preserved.[16] It was no doubt the simplicity of calculation afforded by system A that guaranteed its survival even after the introduction of the more sophisticated system B.

The goals and methods of Babylonian astronomy were very different from those of the Greeks. In particular, the Babylonians seem to have had little interest in the actual motions of the celestial bodies. Rather, their goal was the direct arithmetical calculation of the times and positions of particular celestial phenomena, for example, new and full Moons, eclipses, first and last visibilities of the planets. As far as we know, there was no underlying geometrical picture of the workings of the cosmos. It is likely that the Babylonians first became aware of the solar anomaly by noticing that times of successive full Moons were not equally spaced. (For the Greeks, as we have seen, the clue seems to have been the inequality in the lengths of the seasons.) Later Greek astronomers were well aware of the difference in the Babylonian approach. For example, Theon of Smyrna[17] says that the Babylonian astronomers, using arithmetical methods, succeeded in confirming the observed facts and in predicting future phenomena, but that, nevertheless, their methods were imperfect, *for they were not based on a sufficient understanding of nature, and one must also examine these matters physically.*

Hipparchus's use of Babylonian material is amply attested. For example, some of the numerical values that Hipparchus used for the periods of the Moon and planets (quoted by Ptolemy in the *Almagest*) were actually of Babylonian origin: they turn up on the preserved clay tablets. How did Hipparchus in particular, and the Greeks more generally, learn about Babylonian astronomy? Gerald Toomer[18] has suggested that Hipparchus went to Babylon and learned calculational astronomy from the scribes. B. L. van der Waerden[19] has suggested that the many references to the Chaldaeans (Babylonian astronomers) by Greek and Roman writers point to the existence of a Greek compendium of Babylonian astronomy, now lost, if it ever existed. Alexander Jones[20] has pointed out that we need only assume that one or several Babylonian scribes emigrated and took their skills and texts with them. These possible explanations are not mutually exclusive, of course. The details of the transmission of Babylonian astronomical knowledge to the Greek world are simply not known.

But it is clear that the second century B.C. was a period of important contact and that Hipparchus played a major role in the incorporation of Babylonian material into Greek astronomy. Hipparchus must have been forcibly struck by the Babylonian success in accurate theoretical calculation of celestial events. But he must also have been puzzled by its lack of any philosophical or physical foundation. It was Hipparchus's great accomplishment to show that the geometrical astronomy of the Greeks, which began with a concern for physical explanation, could also be made a precise tool of calculation and prediction—at least in the cases of the Sun and the Moon, for Hipparchus was unable to provide a satisfactory theory of the planets. That remained for Ptolemy to do.

A Hipparchus Coin

Figure 5.11 presents a small bronze coin from Roman Bithynia. The coin was minted during the reign of Severus Alexander, who was Emperor of Rome A.D. 222–235. The obverse of the coin bears the customary portrait of the emperor himself. But the reverse (shown here) bears an image of the astronomer Hipparchus. The Greek legend around the edge reads, "Hipparchus of Nicaea." Nicaea was Hipparchus's native town in Bithynia. In Hipparchus's day (second century B.C.), Bithynia was an independent nation. It became a Roman province in 74 B.C.

On the coin we see Hipparchus. He is bearded and he wears a himation, the familiar over-the-shoulder garb of ancient Greece. He sits at a table that supports a celestial globe. The coin was minted several centuries after Hipparchus's death and cannot, therefore, be taken as a literal likeness. It does demonstrate, however, that Hipparchus's astronomical accomplishments were remembered, if not understood, by his countrymen. It is the oldest piece of money that carries the portrait of an astronomer.

The Motion of the Apogee

From Hipparchus's lengths of the seasons, it is clear that the solar apogee lay in a different direction in antiquity than it does today. The apogee must have been in the spring quadrant of the ecliptic, as in figure 5.12, since spring was the longest season and fall the shortest. Evidently, the center of the Sun's orbit has moved since the days of the Greeks. The center C of the orbit may be regarded as traveling on a small circle about the Earth O, thus producing a slow, eastward rotation of the line of apsides.

The motion of the line of apsides is very slow, amounting to less than 2° per century. Consequently, for intervals shorter than century, the apogee may be regarded as fixed. We need take into account the motion of the apogee only when we wish to follow the motion of the Sun over a period of several centuries or more.

The motion of the apogee can also be explained in terms of the epicycle-plus-concentric version of the solar theory (fig. 5.7). We need only imagine that angles $\bar{\alpha}$ and $\bar{\beta}$ do not increase at quite the same speed. If $\bar{\beta}$ increases a bit more slowly than $\bar{\alpha}$, the effective apogee will be slowly displaced in the counterclockwise direction.

The motion of the solar apogee was unknown to the ancients. Ptolemy measured the lengths of the seasons and obtained the same values as Hipparchus had 285 years before him. He concluded that the apogee was fixed with respect to the equinoctial points as in figure 5.12. Theon of Alexandria, in his edition of Ptolemy's *Handy Tables* (late fourth century A.D.), adhered to the same principle. The motion of the solar apogee with respect to the equinoctial points was discovered by Arabic astronomers in the great revival of astronomy that began in the ninth century.

A Solar Equatorium

Striking visual evidence of the longevity of Hipparchus's and Ptolemy's solar theory is provided by figure 3.41. The device for finding the longitude of the

FIGURE 5.11. A small bronze coin from Roman Bithynia, bearing the image of Hipparchus of Nicaea. Courtesy of the Trustees of the British Museum (BMC 97 119/15A).

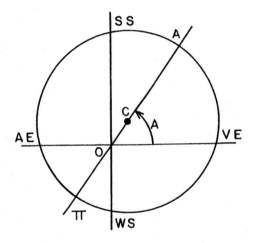

FIGURE 5.12. The eccentric-circle solar theory for the time of Hipparchus and Ptolemy. The longitude of the Sun's apogee is 65 1/2°.

Sun, commonly found on the backs of medieval European astrolabes, is nothing other than a concrete realization of the theory of the Sun. The hole in the center of the back of the astrolabe represents the Earth. The outer zodiac scale is concentric with the Earth. The eccentric calendar scale represents the Sun's eccentric circle.

This eccentric-circle device for finding the longitude of the Sun is an example of an *equatorium*. The difference between the actual position of the Sun and the position it would occupy if it moved uniformly around the zodiac is called the *equation*. An equatorium is a device for "equating" the Sun (or planets), that is, a device that supplies the equation.

There are two reasons why the equatorium on the astrolabe in figure 3.41 cannot be used for the present day. First, this fifteenth-century equatorium was designed for use under the Julian calendar, while we use the Gregorian. This explains why, according to the equatorium, the Sun reaches the vernal equinox (beginning of Aries) on March 11. Second, this equatorium places the Sun's apogee near the beginning of Cancer, which was correct for the fifteenth century. Today, the longitude of the apogee is about 103° (13° into Cancer), as on the astrolabe in figure 3.26. There are, however, examples of medieval and Renaissance equatoria that are not parts of astrolabes but are drawn separately on paper and designed with *movable* apogees, so that they could be used for any era. We shall study equatoria in more detail in section 7.27.

Conclusion

The solar theory advanced by Hipparchus in the second century B.C. was accepted by Ptolemy and virtually every other astronomer of the Greek-Arabic-Latin tradition down to the sixteenth century A.D., with occasional modifications in the numerical values of the four parameters. The solar theory of Hipparchus had several advantages that helped to ensure its long survival. First, because it was based on uniform circular motion, it was mathematically simple. Second, the theory conformed to ancient physical doctrines about the motions natural to celestial bodies. But neither mathematical convenience nor physical plausibility would have saved the model if it had been in bad agreement with the appearances. In fact, Hipparchus's model is very good. With accurately determined parameters the theory is capable of predicting the position of the Sun with an error of less than 1′—an error well below the precision of the best naked-eye observations. The ancient solar theory did not, however, achieve its full potential accuracy because of unavoidable errors in the observations of the equinoxes and solstices from which the numerical parameters were derived.

5.3 REALISM AND INSTRUMENTALISM IN GREEK ASTRONOMY

Alternative Realities: Epicycles or Eccentrics?

As we have seen, the Greek astronomers knew two versions of the solar theory: the epicycle-plus-concentric model illustrated in figure 5.7 and the eccentric-circle model illustrated in figure 5.8. That these two models are mathematically equivalent was known from the time of Apollonius of Perga. It was remarkable that two theories that seemed physically very different should turn out to be mathematically identical. There arose a debate over which model was correct.

According to Theon of Smyrna,[21] Hipparchus said that it was worth the attention of the mathematicians to investigate the explanation of the same phenomena by means of hypotheses that were so different. Theon also tells us that Hipparchus expressed a preference for the epicycle theory, saying that it was probable that the heavenly bodies were placed uniformly with respect to the center of the world.[22]

Ptolemy, on the other hand, preferred the eccentric-circle version of the solar theory, saying that it was simpler, since it involved one motion rather than two.[23] But simpler in what way? The eccentric model is not *mathematically* simpler, for the two models are mathematically equivalent, as Ptolemy himself proves. A calculation of the Sun's position would be of similar complexity in the two models. Indeed, the calculations would be virtually identical, line by line. Ptolemy was clearly thinking of *physical* simplicity. He preferred the eccentric model because it seemed physically simpler and, therefore, more likely to be true. It is clear that Hipparchus's preference for the epicycle model was also based on broad physical or cosmological principles: the heavenly bodies are all situated uniformly with respect to the Earth. Thus, even when it was recognized that the two models were mathematically equivalent, astronomers could disagree about which represented the actual motions in the universe.

"Saving the Phenomena" in Ancient Greek Astronomy

The ancient debate over eccentrics and epicycles has frequently been misinterpreted by modern writers on the history of astronomy. It is often said that the Greeks renounced any interest in finding the true arrangement of the cosmos. According to this view, they sought only to *save the phenomena*, that is, to find combinations of uniform circular motions that would reproduce the apparently irregular motions of the Sun, Moon, and planets. As long as the astronomers could accurately predict the positions of the planets, they did not trouble themselves over the truth of their models. This interpretation of the history of Greek astronomy is often called "instrumentalist": according to this view, the Greek astronomers used their theories only as instruments of calculation and prediction and did not assert that they corresponded to physical reality.

The instrumentalist interpretation of ancient Greek astronomy was popularized by the French philosopher of science Pierre Duhem in his book *To Save the Phenomena*, published in 1908.[24] Similar views have been expressed by other influential writers, including Dreyer,[25] Sambursky,[26] and Dijksterhuis.[27] An especially clear statement of the instrumentalist position was given by Arthur Koestler in his book *The Sleepwalkers*:

> The astronomer "saved" the phenomena if he succeeded in inventing a hypothesis which resolved the irregular motions of the planets along irregularly shaped orbits into regular motions along circular orbits—*regardless whether the hypothesis was true or not*, i.e., whether it was physically possible or not. Astronomy, after Aristotle, becomes an abstract sky-geometry, divorced from physical reality.... It serves a practical purpose as a method for computing tables of the motions of the sun, moon, and planets; but as to the real nature of the universe, it has nothing to say. Ptolemy himself is quite explicit about this.[28]

The instrumentalist interpretation gains some of its appeal by providing an excuse for the Greeks, who were wrongheaded about the motion of the Earth. If they didn't mean it seriously, if they only meant it as a tool for calculation, then we can more readily pardon them for their mistakes. They become heroes of positivism. Needless to say, this is an anachronism of the worst kind.

The historical evidence for the instrumentalist view comes partly from the ancient debate over eccentrics versus epicycles and partly from misinterpretation of the Greek philosophers and astronomical writers, such as Geminus, who discussed the relation between astronomy and physical thought. The Greeks were quite sophisticated in distinguishing between what could be known by observation and mathematical demonstration and what could not. But this does not mean that they renounced any interest in true nature of the cosmos. The instrumentalist view simply is not sustainable in the face of the evidence.[29]

Early Greek astronomy was far more concerned with broad physical explanation than with numerical details. The notion of saving the phenomena probably entered Greek astronomy at about the time of Eudoxus, and this actual expression is used from time to time by later writers, such as Theon of Smyrna and Simplicius. However, what counted as phenomena in need of saving evolved with time. As we shall see in section 7.6, the first Greek planetary theories had no numerical predictive power at all. Rather, they were concerned only with explaining the basic features of the planets' motions, such as retrograde motion, in terms of accepted physical principles. Eudoxus's own planetary theory was criticized, not for failing to pass some exacting numerical test, but for failing to account for the variations in brightness of Mars and Venus in the course of their synodic cycles—an obvious physical change readily perceivable without instruments or measurements. Thus, in Greek thought about the motion of the Sun, Moon, and planets, the desire for broad physical explanation is manifested long before the Greeks start to make careful observations or to devise theories that actually could be used to make accurate predictions. Plantary theories with numerical predictive power—and thus the idea of saving the phenomena *in a quantitative sense*—did not enter Greek astronomy until relatively late, around the time of Hipparchus.

Several Greek writers point out that it is impossible to know whether the eccentric-circle model of the Sun should be preferred over the epicycle, or vice versa. But this does not mean that they did not care which model was true. The opposing positions taken by Hipparchus and Ptolemy in this debate were clearly determined by their views about the physical nature of the universe. Moreover, as we shall see in section 7.25, when Ptolemy tried to calculate the size of the whole cosmos (in his *Planetary Hypotheses*), by nesting the mechanisms for the various planets one within another, he certainly took the planetary models as physically real.

Another example of Ptolemy's realist stance is his attempt to measure the variation in the Sun's angular diameter. It followed from Hipparchus's solar theory that the Sun should be farther from the Earth in the spring than in the fall. In *Almagest* V, 14, Ptolemy says that he tried to measure the variation in the angular size of the Sun using a sort of dioptra. Although Ptolemy was unable to detect the tiny change in the Sun's apparent size, his effort shows that he did take this consequence of the solar theory seriously.

Geminus on the Aims of Astronomy and of Physics

Perhaps the clearest ancient statement of the relation of mathematical astronomy to physics and philosophy is provided by Geminus. Besides his *Introduction to the Phenomena*, Geminus wrote an abridgment of, or commentary on, the lost *Meteorology* of Posidonius. This work of Geminus has not come down to us, but a fascinating fragment has been preserved because it was quoted by Simplicius in his commentary on Aristotle's *Physics*.

EXTRACT FROM GEMINUS

As preserved in Simplicius's *Commentary Aristotle's Physics*

It is the task of physical speculation to inquire into the nature of the heaven and the stars, their power and quality, their origin and destruction; and, indeed, it can even make demonstrations concerning their size, form and arrangement. Astronomy does not attempt to speak of any such thing but demonstrates the arrangement of the heaven, presenting the heaven as an orderly whole, and speaks of the shapes, sizes and distances of the Earth, Sun and Moon, of eclipses and conjunctions of the stars, and of the quality and quantity of their motions. Therefore, since it deals with the investigation into quantity, magnitude and quality in relation to form, naturally it needed arithmetic and geometry. And concerning these things, the only ones of which it undertook to give an account, astronomy has the capacity to make demonstrations by means of arithmetic and geometry.

Now in many cases the astronomer and the physicist will propose to demonstrate the same point, such as that the Sun is large or that the Earth is spherical, but they will not proceed by the same paths. The physicist will prove each point from considerations of essence or inherent power, or from its being better to have things thus, or from origin and change; but the astronomer will prove them from the properties of figures and magnitudes, or from the amount of motion and the time appropriate to it. Again, the physicist will often reach the cause by looking to creative force; but the astronomer, when he makes demonstrations from external circumstances, is not competent to perceive the cause, as when, for example, he makes the Earth and the stars spherical. Sometimes he does not even desire to take up the cause, as when he discourses about an eclipse; but at other times he invents by way of hypothesis and grants certain devices, by the assumption of which the phenomena will be saved.

For example, why do the Sun, Moon and planets appear to move irregularly? [The astronomer would answer] that, if we assume their circles are eccentric or that the stars go around on an epicycle, their apparent irregularity will be saved. And it will be necessary to go further and examine in how many ways it is possible for these phenomena to be brought about, so that the treatment of the planets may fit the causal explanation which is in accord with acceptable method. And thus a certain person, Helaclides Ponticus, coming forward, says that even if the Earth moves in a certain way and the Sun is in a certain way at rest, the apparent irregularity with regard to the Sun can be saved.

For it is certainly not for the astronomer to know what is by nature at rest and what sort of bodies are given to movement. Rather, introducing hypotheses that certain bodies are at rest and others are moving, he inquires to which hypotheses the phenomena in the heaven will correspond. But he must take from the physicist the first principles, that the motions of the stars are simple, uniform and orderly, from which he will demonstrate that the motions of them all are circular, some moving round on parallel circles, some on oblique ones.

In this manner, then, does Geminus, or rather Posidonius in Geminus, give the distinction between physics and astronomy, taking his starting point from Aristotle.[30]

The distinction is clear. We cannot know, from astronomical observation, whether the Sun goes around the Earth or the Earth goes around the Sun. Similarly, we cannot know whether the observed motion of the Sun results from an eccentric circle or from an epicycle. We must base our astronomy on physical hypotheses, which are the results of physical or philosophical enquiry. Astronomy cannot decide every question. But that does not mean that the questions are unimportant. As we have seen, Theon of Smyrna criticized the Babylonians because they had *merely* saved the phenomena, without seeking deeper for the underlying physical principles. All the Greek astronomers were realists, who thought they were grappling with the nature of the universe—as, indeed, they were. They would have been very strange people to have developed successful models of planetary motion and then refused to believe that these models had anything to do with the nature of things.

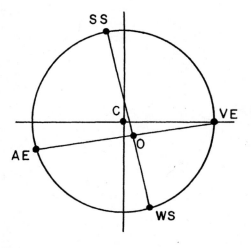

FIGURE 5.13.

5.4 EXERCISE: FINDING THE SOLAR ECCENTRICITY

Our problem is to measure the eccentricity of the Sun's circular orbit and the longitude of its apogee, starting from the lengths of the seasons.

General Directions

Rather than using the trigonometric methods of Hipparchus and Ptolemy, we shall use a simple graphical construction to solve this problem. The analysis will be based on the accurate lengths of the seasons that follow from the data given below.

On a large sheet of graph paper, draw a circle of circumference 365.25 units. Each unit of circumference will represent one day of the Sun's annual motion. Place the center C of the circle at the intersection of a vertical line and a horizontal line on the graph paper, as shown in figure 5.13. The size of the unit you choose is arbitrary, but a convenient scale will be to let 2 mm represent one day's motion of the Sun. The circumference of the circle will then be $365.25 \times 2 = 730.5$ mm. A circle of this size is large enough to give accurate results, but still small enough to be easily handled.

In figure 5.13, let point VE represent the vernal equinox. If spring were exactly one-fourth of the year, it would last $365.25/4 = 91.31$ days, and summer solstice would occur when the Sun reached X. However, spring is a bit longer than 91.31 days. Suppose spring were exactly two days longer than this. (This is *not* the correct figure. Your own work must be based on the actual season lengths.) Then summer solstice would occur when the Sun reached SS, two days (or two units of circumference) beyond X. These two units (4 mm, using the scale suggested above) can be measured off along the circle using a ruler.

Similarly, suppose summer were one day longer than the average seasonal length of 91.31 days. An autumnal equinox AE placed as shown in figure 5.13 would give a summer of the correct length. The arc from SS to AE is one day's worth of motion (one unit) longer than a quarter-circle. This one unit must be added to the original 2-unit displacement of SS. Thus AE is placed 3 units beyond Z. At a scale of 2 mm per day, this amounts to 6 mm. The winter solstice WS can be placed in a similar manner.

Now, draw a line through the two equinoxes (fig. 5.14). Draw another line through the two solstices. The two equinoxes, observed from the Earth, are directly opposite one another in the sky. The same is true of the two solstices. Therefore, the Earth must lie on the intersection of the two lines, at O.

Draw the line of apsides through the Earth O and the center C of the Sun's circular path (fig. 5.15). The point marked A is the apogee of the orbit; Π is the perigee. Angle A is the longitude of the apogee and may be measured with a protractor whose center is placed at O. Distance OC may be measured with a ruler; the ratio of this quantity to the radius is the eccentricity of the Sun's circular orbit.

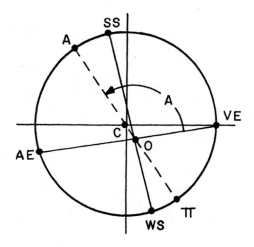

FIGURE 5.14.

FIGURE 5.15.

Problems

1. Here are the times at which the equinoxes and solstices fell in four successive years (Greenwich mean time):

Year	Month	Day	Hour	
1972	Mar	20	12	vernal equinox
	Jun	21	7	summer solstice
	Sept	22	23	autumnal equinox
	Dec	21	18	winter solstice
1973	Mar	20	18	
	Jun	21	13	
	Sept	23	4	
	Dec	22	0	

Year	Month	Day	Hour
1974	Mar	21	0
	Jun	21	19
	Sept	23	10
	Dec	22	6
1975	Mar	21	6
	Jun	22	0
	Sept	23	16
	Dec	22	12

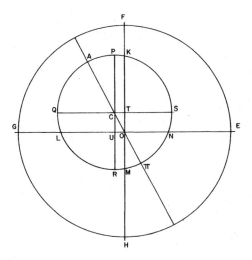

FIGURE 5.16.

Determine the length of each season above. Don't forget that one of the above years must be a leap year! You should find that the four seasons are not all of the same length, but that the lengths are very steady and do not change from one year to the next.

In contrast, the actual *times* at which the equinoxes and solstices fall are quite variable. Note the steady shift of about six hours per year in the time of the vernal equinox. Why is this? At what time do you think the vernal equinox fell in 1976? In 1971? Why?

2. Use the graphical method explained above to locate the Earth with respect to the center of the Sun's circular path. (Again, a convenient scale to use is 1 day = 2 mm.)

From your drawing, determine the eccentricity of the Sun's circular orbit and the longitude of its apogee. Use a star chart to find out what constellation the Sun is in when it is at the apogee. What constellation is the Sun in at perigee?

3. At what time of the year is the Sun closest to us? At what time is the Sun farthest away? Comment on the often-heard claim that we have winter when the Sun is farther away from us.

4. The equinoxes and the solstices are equally spaced around the ecliptic. Check to see whether this condition is satisfied by your new model of the Sun's motion. That is, as seen from the Earth, are the equinoctial and solstitial points regularly spaced at 90° intervals?

5. Use Hipparchus's values for the lengths of the seasons (given in sec. 5.2) to determine the eccentricity of the Sun's orbit and the longitude of its apogee in the second century B.C. By how much has the apogee moved between Hipparchus's era and our own? What is its motion in a single century?

5.5 RIGOROUS DERIVATION OF THE SOLAR ECCENTRICITY

We show how to calculate the magnitude and direction of the solar eccentricity from the lengths of the seasons. Our method is that of Hipparchus and Ptolemy, given in *Almagest* III, 4. But we shall base the calculation on the modern lengths of the seasons derived from the data in section 5.4:

Spring	92^d	19^h
Summer	93	15
Fall	89	20
Winter	89	0
Total	365^d	6^h

Summer is the longest season, so the center of the Sun's circle must lie toward the summer quadrant of the zodiac. In figure 5.16, *EFGH* is the circle of the zodiac, in the sphere of the fixed stars. *E* represents the spring equinox; *F*, summer solstice; *G*, fall equinox; and *H*, winter solstice. Lines *EG* and *FH* meet at right angles at the Earth *O*. *C* is the center of the Sun's eccentric circle *KLMN* with apogee *A* and perigee Π.

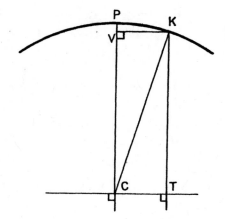

FIGURE 5.17.

We wish to determine angle *EOA* (the longitude of the apogee) and distance *OC* (the eccentricity of the solar orbit).

Draw line *PR* through *C* and parallel to *FH*. Similarly, draw *QS* through *C* and parallel to *GE*.

Now, in Hipparchus's method, we need not use all four season lengths. Two will suffice, together with the assumption that the equinoxes and solstices are spaced at 90° intervals (i.e., that *EG* and *FH* are perpendicular). Let us use the lengths of the spring and summer.

Spring: The Sun runs arc *NSK* in 92^d 19^h = 92.7917^d. The length of this arc is therefore $92.7917^d \times 360°/365.25^d = 91.4579°$.

Summer: The Sun runs the summer arc *KPL* of its eccentric circle in 93^d 15^h = 93.6250^d. The length of arc *KPL* is thus $93.625^d \times 360°/365.25^d = 92.2793°$. Thus,

$$\text{arc } NKL = \text{arc } NSK + \text{arc } KPL$$
$$= 183.7372°.$$

Now arc *SKQ* = 180°, and thus the short arcs *QL* and *SN* total 3.7372°. But *QL* and *SN* are equal, so either of them amounts to half of 3.7372°. Thus,

$$\text{arc } QL = \text{arc } SN = 1.8686°.$$

We now proceed to find arc *PK*. As already stated, arc *KPL* = 92.2793°. Furthermore, arc *PAQ* = 90°. So, we have

$$\text{arc } PK = KPL - PAQ - QL$$
$$= 92.2793° - 90° - 1.8686°$$
$$= 0.4107°.$$

We may now calculate the short line segments *CT* and *CU*. To compute *CT*, refer to figure 5.17, which shows in expanded scale the portion of the solar circle about arc *PK*. From point *K* we drop a perpendicular onto *PC* at point *V*. Now

$$CT = VK$$
$$= CK \sin KCV.$$

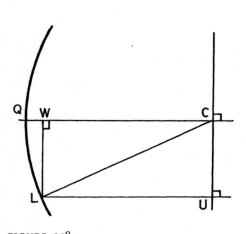

FIGURE 5.18.

But angle *KCV* = arc*PK* = 0.4107°. Thus,

$$CT = CK \sin 0.4107°$$
$$= 0.00717,$$

if we put the radius *CK* of the orbit equal to unity. For segment *CU*, refer to figure 5.18. In the same fashion we have

$$CU = WL$$
$$= CL \sin QCL$$
$$= CL \sin 1.8686°$$
$$= 0.03261,$$

since we have taken *CL* (the radius of the orbit) to be the unit of measure.

The eccentricity *e* = *OC* is now found by the rule of Pythagoras:

$$e = \sqrt{CU^2 + CT^2}$$

$$= 0.0334,$$

in units where the radius of the orbit is unity.

To determine the direction in which C lies, note that

$$\tan TOC = CT/CU$$

$$= 0.21956,$$

so

$$\text{angle } TOC = 12.40°.$$

Thus, the longitude of the apogee is

$$A = \text{angle } EOA$$

$$= EOK + TOC$$

$$= 90° + 12.40°$$

$$= 102.40°.$$

These values of e and A apply to the early 1970s—the years for which the lengths of the seasons were given. They should agree well with the values obtained by the graphical method of section 5.4. The value of the eccentricity is valid for many centuries, since the eccentricity scarely changes. The longitude of the apogee, however, increases at a slow, steady rate, as mentioned in section 5.2.

The method of calculating e and A explained here probably was invented by Hipparchus in the second century B.C. It remained standard from his day until the sixteenth century. The only criticism that can be made of it is that, since the exact moment of summer solstice is difficult to determine precisely, the measured lengths of spring and summer are subject to rather large errors—up to half a day or even more.

This difficulty can be avoided by using, instead of the equinoxes and solstices, four other reference points on the zodiac spaced at 90° intervals. For example, one could use the points placed halfway between the equinoxes and solstices. That is, rather than trying to observe the moments when the Sun reaches the zeroth degree of the Ram, Crab, and Balance, one could observe instead the moments when the Sun reaches the 45th degree of the Bull, Lion, and Scorpion. The Sun's declination changes rapidly enough at these points that the uncertainty is greatly reduced in comparison with the uncertainty associated with solstices. The moment when the Sun reaches the midpoints of each of these signs could be determined by noon altitudes taken with a meridian quadrant. The calculation of e and A from the observed times then proceeds in exactly the same manner as was used with the equinoxes and solstices. Copernicus used this modification of Hipparchus's method in his own calculation of the solar eccentricity, based on observations made by himself in the early part of the sixteenth century. Although Copernicus's method represents an improvement over that of Hipparchus, it is in essence the same. The observational and theoretical methods of astronomy had remained unchanged for 1,800 years.

5.6 EXERCISE: ON THE SOLAR THEORY

1. Trigonometric determination of the eccentricity: As remarked in section 5.2, Hipparchus found the length of the spring to be 94 1/2 days; and that

of the summer, 92 1/2 days. Use these values and the trigonometric method explained in section 5.5 to determine the solar eccentricity and the longitude of the solar apogee in the second century B.C. You should be able to confirm Hipparchus's results:

$$e = 0.0415,$$

$$A = 65.5°.$$

Hipparchus's longitude of the apogee is very good. (It is within 1° of the actual longitude of the solar apogee in 140 B.C.) His value for the eccentricity, however, is a little high. In fact, the solar eccentricity was nearly the same in antiquity as it is today.[31]

2. Relations among the lengths of the seasons: It follows from the solar theory of Hipparchus that the lengths of the four seasons are not all independent. If two are known, it is possible to calculate the other two. This is why, in section 5.5, we needed only the lengths of two seasons to determine the parameters of the orbit.

Refer to figure 5.16. Circle KLMN is the solar orbit, with center at C. The Earth is at O. N, K, L, and M mark the Sun's position at spring equinox, summer solstice, fall equinox, and winter solstice, respectively.

Arcs KP and RM are equal. The Sun takes equal times to run these small arcs; let us call this time a. Similarly, QL and NS are equal. Let us call b the time the Sun takes to run either QL or NS. Thus,

$$\text{time on } PK = \text{time on } RM = a,$$

$$\text{time on } QL = \text{time on } NS = b.$$

Summer lasts for the time required for the Sun to run arc KPL:

$$\text{arc } KPL = PQ + KP + QL.$$

Now, arc PQ is a quadrant of the circle. If we let T denote the tropical year, the Sun runs arc PQ in a time T/4. Then our equation can be expressed in terms of times instead of arcs as follows:

$$\text{length of summer} = T/4 + a + b.$$

Similarly, winter lasts for the time required for the Sun to run arc MN. This arc is shorter than a quadrant of the circle by arc RM and arc NS. Thus, we have

$$\text{length of winter} = T/4 - a - b.$$

Similar arguments lead to the following results for spring and fall:

$$\text{length of spring} = T/4 - a + b.$$

$$\text{length of fall} = T/4 + a - b.$$

Adding the lengths of summer and winter gives

$$\text{summer} + \text{winter} = T/2.$$

Similarly,

$$\text{spring} + \text{fall} = T/2.$$

That is, the longest season plus the shortest should be exactly half a year. Similarly, the remaining two seasons should make exactly half a year.

If we regard the lengths of spring, of summer, and of the year as known, then we may calculate the lengths of fall and winter:

$$\text{(1) winter} = T/2 - \text{summer}$$

$$\text{(2) fall} = T/2 - \text{spring}.$$

It is not obvious that the actual seasons obey equations (1) and (2). But these are consequences of our solar theory.

For the modern era, the lengths of the seasons are

Spring	92^d	19^h
Summer	93	15
Fall	89	20
Winter	89	0

Testing equation (2), we have

$$\text{fall} = \frac{1}{2} \times (365^d 6^h) - 92^d 19^h \ (?)$$

$$= 89^d 20^h,$$

so the relation is indeed satisfied. Testing equation (1), we have

$$\text{winter} = \frac{1}{2} \times (365^d 6^h) - 93^d 15^h \ (?)$$

$$= 89^d 0^h.$$

Thus, the lengths of the seasons actually are related to one another as the theory predicts. The Sun arrives at each of the points K, L, M, N (fig. 5.16) at exactly the moments predicted by the theory. That is, at four different times during the year, the theoretical and the actual positions of the Sun are in exact agreement. These are the only points we can check with our limited data, which consist only of the lengths of the seasons. Nevertheless, if the model represents the Sun well at these four points, it cannot be very far wrong at other places on the orbit.

The Exercise:

A. In antiquity, the solar apogee was located in the spring quadrant of the ecliptic. Redraw figure 5.16 for ancient times. Use your figure to rederive equations (1) and (2), relating the lengths of the seasons. Your derivation for ancient times will be similar in method to the derivation given above for the modern age, but different in some details. The differences will reflect the fact that in antiquity spring, and not summer, was the longest season.

B. Hipparchus measured the length of spring and of summer, obtaining 94 1/ days and 92 1/2 days, respectively. From these values and the 365 1/4 day length of the year he deduced the lengths of fall and winter. There was no need to measure the fall and winter directly. Use the relations derived by you in problem A and Hipparchus's lengths for the spring and summer to deduce the lengths of fall and winter in antiquity. To check your work, see Hipparchus's own results for the fall and winter, given in section 5.2.

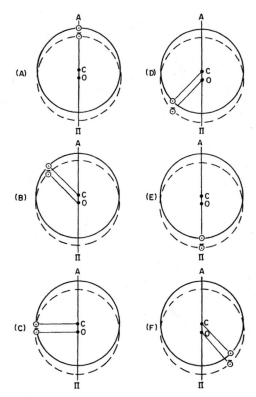

FIGURE 5.19. Relation between the mean Sun and the true Sun.

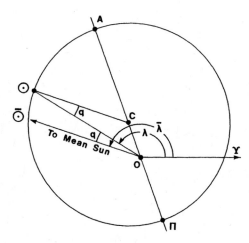

FIGURE 5.20. The equation of center (q).

C. Season lengths attributed to Euctemon (fifth century B.C.) and to Callippus (fourth century B.C.) are given in note 10 of this chapter. Do you think either Euctemon or Callippus could have anticipated Hipparchus in advocating an eccentric-circle theory of the Sun's motion? Base your argument on their season lengths and relations (1) and (2), which must apply to an eccentric-circle model.

5.7 TABLES OF THE SUN

A solar theory permits one to answer, for any date, the question, *what is the longitude of the Sun?* This problem can be solved approximately using a concrete model, such as an equatorium. If more precision is needed, the longitude of the Sun can be calculated trigonometrically. Such calculations tend to be laborious. If the longitude of the Sun is to be calculated often, it is convenient to have tables that minimize the labor. It is possible that Hipparchus constructed solar tables in the second century B.C., but if so, they have not come down to us. The solar tables of Ptolemy (second century A.D.), in the *Almagest* and the *Handy Tables*, served as a model for all those constructed later. If one learns how to use Ptolemy's tables, one has little difficulty using the solar tables of any medieval or Renaissance astronomer, whether Arabic or Latin.

Some Concepts Useful in the Solar Theory

The Mean Sun   The *mean Sun* is a fictitious body that moves uniformly on a circle centered at the Earth. In figure 5.19 the orbits of the mean and the true Sun are both shown. The true Sun ☉ moves uniformly on the solid circle, whose center is C. The mean Sun $\bar{☉}$ moves uniformly on the broken circle, whose center is the Earth O. Figure. 5.19A–F shows the positions of the mean and the true Sun at equal intervals of time. The time between two successive figures is one-eighth of a year. Note that $O\bar{☉}$ remains parallel to $C☉$.

The mean Sun lies in the same direction as the true whenever the true Sun is in the apogee A or the perigee Π of its eccentric circle. (In the modern era, these times fall in July and January, respectively.) At all other times of year, the true Sun, as seen from the Earth, is a little ahead of or a little behind the mean Sun. The mean Sun represents the position that the Sun would have if the eccentricity of its orbit were zero, that is, if the center of the orbit were the Earth.

Alternatively, we may think in terms of the concentric-plus-epicycle version of the solar theory. Then the mean Sun is the center of the epicycle. The true Sun can be a little ahead of or a little behind the mean Sun, depending on the true Sun's position on the epicycle.

Equation of Center   The *equation of center* is the angular distance between the true Sun and the mean Sun. In figure 5.20, the equation of center is angle $☉O\bar{☉}$, marked q. The mean Sun $\bar{☉}$ moves about O at a uniform rate, so the *mean longitude* $\bar{\lambda}$ (i.e., the longitude of the mean Sun) increases at a steady rate. The actual longitude λ of the Sun differs from $\bar{\lambda}$ by the small correction q. That is, $\lambda = \bar{\lambda} - q$. Note that in this situation (with the Sun lying between A and Π), the equation of center is a subtractive correction to the mean longitude. When the Sun lies between Π and A, as in figure 5.19F, the equation of center is an additive correction to the mean longitude.

Mean and True Anomaly   The equation of center varies with the Sun's position on its eccentric circle. The equation of center is zero when the Sun is at apogee or perigee (figs. 5.19, A and E) and reaches a maximum value when the mean Sun is approximately halfway between apogee and perigee (fig. 5.19C). The magnitude of the equation of center therefore depends not on the mean longitude but on the mean Sun's angular distance from the apogee.

The angular distance of the mean Sun from the apogee (angle $AO\bar{\odot}$) is called the *mean anomaly* and is denoted $\bar{\alpha}$ (see fig. 5.21). The mean anomaly is related in a simple way to the mean longitude. In figure 5.21 we see that

$$\bar{\alpha} = \bar{\lambda} - A,$$

where A is the longitude of the apogee.

Similarly, the angular distance of the true Sun from the apogee (angle $AO\odot$) is called the *true anomaly* and is denoted α. And, clearly,

$$\alpha = \lambda - A.$$

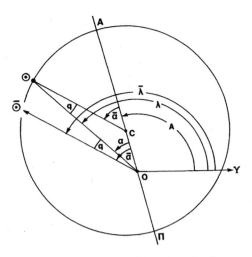

FIGURE 5.21. Angles useful in the solar theory. Longitude of apogee, A. Mean longitude, $\bar{\lambda}$. True longitude, λ. Mean anomaly, $\bar{\alpha}$. True anomaly, α. Equation of center, q.

The name *equation of center* for the small correction q requires an explanation. In astronomy an *equation* is the difference between the actual value and the mean value of some quantity. The equation of center is the difference between the true longitude and the mean longitude of the Sun. In medieval Latin, the regular name for the anomaly was *centrum*, the angle "at the center." Since the equation q depends on the value of this angle, q is still today called the "equation of center."

Maximum Solar Equation The largest value q_{max} of the equation of center occurs when the Sun is $90°$ from apogee, that is, when $\alpha = 90°$ or $270°$. The situation for $\alpha = 90°$ is shown in figure 5.22. From the figure,

$$\sin q_{max} = OC/C\odot = e,$$

where e is the solar eccentricity. From section 5.5, $e = 0.0334$, and thus

$$q_{max} = \sin^{-1}(0.0334) = 1°55'.$$

That is, the true Sun is never more than $1°55'$ from the mean Sun.

The Tables of the Sun

A modern version of Ptolemy's solar theory is embodied in tables 5.1–5.3, (tables of the sun). Table 5.1 gives the amount by which the mean longitude $\bar{\lambda}$ changes in 1, 2, 3, . . . days, in 10, 20, 30, . . . days. For example, in one day $\bar{\lambda}$ increases by $59.1'$, and in 20 days, by $19°42.8'$. In 10,000 days, $\bar{\lambda}$ increases by $136°28.4'$, over and above complete circles. The blanks in the table are left for the exercise of section 5.8. Table 5.1 also gives the amount by which the mean longitude $\bar{\lambda}$ changes in hours and minutes. At the bottom of table 5.1 is given the value of the mean longitude for one particular date. On January 0.5 (Greenwich mean time) 1900, the mean longitude of the Sun was $279°$ $42'$ (The notation January 0.5, 1900, means noon of January 0, 1900, i.e., noon of December 31, 1899.)

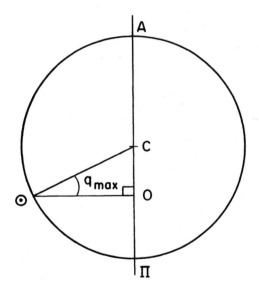

FIGURE 5.22.

Table 5.2 permits the determination of the longitude of the solar apogee A for any desired date. For example, for 1900, $A = 101°06'$. In 1940, A was greater than this by $42'$, which is forty years' motion.

Table 5.3 permits the determination of the equation of the center q if the mean anomaly $\bar{\alpha}$ is known.

The advantage of tables such as these is that they permit rapid, precise computation of the Sun's longitude on any desired date. For their use, the tables require only addition and subtraction. All the more complicated mathematical procedures—multiplication, division, extraction of square roots, and trigonom-

TABLE 5.1. The Sun's Mean Motion.

Days	Motion	Days	Motion	Days	Motion
100,000	284°44.0′	10,000	136°28.4′	1,000	———
200,000	209°28.0′	20,000	272°56.8′	2,000	———
300,000	134°12.0′	30,000	49°25.2′	3,000	———
400,000	58°56.0′	40,000	185°53.6′	4,000	———
500,000	343°40.1′	50,000	322°22.0′	5,000	———
600,000	268°24.1′	60,000	98°50.4′	6,000	153°53.0′
700,000	193°08.1′	70,000	235°18.8′	7,000	59°31.9′
800,000	117°52.1′	80,000	11°47.2′	8,000	325°10.7′
900,000	42°36.1′	90,000	148°15.6′	9,000	230°49.6′
100	———	10	9°51.4′	1	0°59.1′
200	———	20	19°42.8′	2	1°58.3′
300	———	30	29°34.2′	3	2°57.4′
400	———	40	39°25.6′	4	3°56.6′
500	———	50	49°16.9′	5	4°55.7′
600	231°23.3′	60	59°08.3′	6	5°54.8′
700	329°57.2′	70	68°59.7′	7	6°54.0
800	68°31.1′	80	78°51.1′	8	7°53.1′
900	167°05.0′	90	88°42.5′	9	8°52.2′

Hours	Motion	Hours	Motion	Minutes	Motion
1	0°02.5′	13	0°32.0′	10	0°0.4′
2	0°04.9′	14	0°34.5′	20	0°0.8′
3	0°07.4′	15	0°37.0′	30	0°1.2′
4	0°09.8′	16	0°39.4′	40	0°1.6′
5	0°12.3′	17	0°41.9′	50	0°2.1′
6	0°14.8′	18	0°44.4′	60	0°2.5′
7	0°17.2′	19	0°46.8′		
8	0°19.7′	20	0°49.3′		
9	0°22.2′	21	0°51.7′		
10	0°24.6′	22	0°54.2′		
11	0°27.1′	23	0°56.7′		
12	0°29.6′	24	0°59.1′		

Epoch 1900 Jan 0.5 ET = J.D. 241 5020.0 (noon at Greenwich).
Mean longitude at epoch = 279°42′.

etry—have been done by the compiler of the tables. In the days before the hand calculator, such tables offered the user great savings in labor.

Precepts for the Use of the Tables of the Sun

1. Determine the Julian day number of the moment for which the Sun's longitude is desired. Subtract from this the Julian day number of the epoch,

TABLE 5.2. Longitude of the Solar Apogee

Year	Longitude	Year	Longitude	Year	Longitude	Ten-Year Intervals	Motion
801 B.C.	53°57′	200 A.D.	71°25′	1200 A.D.	———	10	0°10′
701	55°42′	300	73°10′	1300	———	20	0°21′
601	57°27′	400	74°55′	1400	———	30	0°31′
501	59°12′	500	76°40′	1500	———	40	0°42′
401	60°57′	600	78°24′	1600	———	50	0°52′
301	62°41′	700	80°09′	1700	97°37′	60	1°03′
201	64°26′	800	81°54′	1800	99°21′	70	1°13′
101	66°11	900	83°39′	1900	101°06′	80	1°24′
1 B.C.	67°56′	1000	85°23′	2000	102°51′	90	1°34′
100 A.D.	69°40′	1100	87°08′	2100	104°36′		

TABLE 5.3. Equation of Center of the Sun

Mean Anomaly	Equation of Center	Mean Anomaly	Equation of Center
0° (360)	−(+) 0° 0′	90° (270)	−(+) 1°55′
5° (355)	0°10′	95° (265)	1°55′
10° (350)	0°19′	100° (260)	1°54′
15° (345)	0°29′	105° (255)	1°52′
20° (340)	0°38′	110° (250)	1°49′
25° (335)	0°47′	115° (245)	1°46′
30° (330)	0°56′	120° (240)	1°41′
35° (325)	1°04′	125° (235)	1°36′
40° (320)	1°12′	130° (230)	1°30′
45° (315)	1°19′	135° (225)	1°23′
50° (310)	1°26′	140° (220)	1°16′
55° (305)	1°32′	145° (215)	1°08′
60° (300)	1°38′	150° (210)	0°59′
65° (295)	1°43′	155° (205)	——
70° (290)	1°47′	160° (200)	——
75° (285)	1°50′	165° (195)	——
80° (280)	1°52′	170° (190)	——
85° (275)	1°54′	175° (185)	——
90° (270)	1°55′	180° (180)	——

1900 January 0.5 Greenwich mean time (= J.D. 241 5020.0). The result is Δt, the number of days elapsed since epoch.

2. Finding the mean longitude: Enter table 5.1 with the digit for each power of 10 in Δt and take out the corresponding motion. Take out also the motion for the hours and minutes, if required. The total mean motion is the sum of all. The total mean motion is positive if the date is after the epoch and negative if it is before. Add the mean motion to the mean longitude at epoch (279°42′) and subtract as many multiples of 360° as required to render the quantity less than 360°. Round to the nearest minute of arc. The result is the Sun's mean longitude $\bar{\lambda}$ at the required date.

3. Longitude of the apogee and mean anomaly: Enter table 5.2 with the century year immediately before the required year. For example, for A.D. 1583, use 1500; for 183 B.C., use 201 B.C. Then correct this longitude by the motion of the apogee during the interval from the century year to the required year. It is sufficient to work to the nearest decade. For example, for A.D. 1583, add 80 years' motion. If the table is handled in this way, the motion for the decades elapsed will always be added positively to the value for the century. The sum is the longitude A of the solar apogee. Calculate the mean anomaly $\bar{\alpha}$ by subtracting A from the mean longitude:

$$\bar{\alpha} = \bar{\lambda} - A.$$

If $\bar{\alpha}$ should turn out negative, add 360°.

4. Equation of center: Enter table 5.3 with the mean anomaly and take out the equation of center q. Here, the interpolation should be done with care to determine the equation to the nearest minute of arc. Note that the equation is negative if the anomaly is between 0° and 180° and positive if the anomaly is between 180° and 360°.

5. Add the equation of center to the mean longitude. (The tables have been set up so that one always adds. But the sign of q may be either positive or negative, as listed in table 5.3.) The result is the longitude of the Sun that was sought:

$$\lambda = \bar{\lambda} + q.$$

Example: Calculate the longitude of the Sun on November 4, 1973, at 10:30 A.M., Greenwich time.

1. From the tables for Julian day number (tables 4.2–4.4) we have

1900†	2,415,020	
73	26,663	(= 26,664 − 1)
Nov 4	308	
	2,441,991	

This is the Julian day number for November 4, 1973, Greenwich mean noon. We want 10:30 A.M., so we must subtract 1 1/2 hours, obtaining 244 1990^d 22.5^h. Next, compute Δt, the time elapsed since epoch:

1973 Nov 4, 10:30 A.M.	J.D. 244	1990^d	22^h	30^m
Less Julian day number of epoch	−241	5020	0	0
$\Delta t =$	2	6970^d	22^h	30^m

2. From table 5.1 we have

Time		Motion	
20,000 days		272°	56.8'
6,000		153	53.0
900		167	05.0
70		68	59.7
0		0	0.0
22 hr		0	54.2
30 min		+0	1.2

	660°	229.9'	= 663° 50'
Plus mean longitude at epoch		+ 279	42
		942	92
		= 943	32
Reject 720°		− 720	
$\bar{\lambda}$		= 223°	32',

which is the mean longitude at the required date.

3. From table 5.2 (longitude of the solar apogee):

1900	101°	06'
70 years	1	13
A	102°	19'

This was the longitude of the Sun's apogee in 1973. The mean anomaly at the required date is calculated thus:

Mean longitude	223°	32'
Less longitude of apogee	−102	19
$\bar{\alpha}$, the mean anomaly	121°	13'

4. In table 5.3 find

$$q = -1°40'.$$

5. Add this equation to the mean longitude:

$$\begin{array}{llr} \bar{\lambda} & 223° & 32' \\ +q & -1 & 40 \\ \hline \lambda & 221° & 52' \end{array}$$

This, according to our modern Ptolemaic theory, was the longitude of the Sun on November 4, 1973, at 10:30 A.M., Greenwich time. The Sun's actual longitude at this date, as calculated from modern celestial mechanics, was the same. Our modern Ptolemaic model (Ptolemy's model, but with improved numerical parameters) will never be wrong by more than 1' or 2' for any date within a few centuries of our epoch date of 1900.

Ptolemy's Solar Tables

Our tables of the sun (tables 5.1–5.3) are modeled on those of Ptolemy but differ from them in a number of minor ways.

Table of the Sun's Mean Motion   In table 5.1, the mean motion is given for 1, 2, 3 days, for 10, 20, 30 days, and so on, up to multiples of 100,000 days. This is a convenient arrangement but is not the only one imaginable. In the solar tables of the *Almagest* Ptolemy gives the mean motion for hours from 1 to 24, for days from 1 to 30, for 1 to 12 complete Egyptian months of 30 days each, for 1 to 18 Egyptian years of 365 days each, and for 18-year periods up to 810 years (= 45 eighteen-year periods). This arrangement was convenient for use with the Egyptian calendar. It demanded slightly less labor from the user than does our own arrangement, for there was no need to reduce the time interval to days. Rather, complete years and months could be dealt with as they stood. The Gregorian calendar is not quite as convenient for such a purpose as was the Egyptian calendar, because of the variable lengths of our months and years. Ptolemy's table of mean motion is based on his value (365 + 1/4 − 1/300 days) for the tropical year.

Longitude of the Apogee   Ptolemy believed that the apogee was fixed with respect to the equinoxes, at longitude 65 1/2°, because he found the same lengths for the seasons as Hipparchus had found nearly three hundred years before. Consequently, Ptolemy has nothing like our table for the longitude of the solar apogee. In this, of course, Ptolemy was mistaken, and he was eventually corrected by the Arabic astronomers of the medieval period. From the ninth century onward, the longitude of the Sun's apogee was recognized as an increasing quantity. Usually the rate of motion of the apogee was identified with the rate of the precession. That is, the Sun's apogee was considered to be fixed with respect to the stars, rather than with respect to the equinoctial point. While this represented an improvement over Ptolemy's theory, it still somewhat underestimated the true rate of motion.

Equation of Center   Ptolemy calculated the equation of center just as we have (although we have simplified things by using modern algebraic and trigonometric notation). Ptolemy's table is based on a slightly larger value for the solar eccentricity (0.0415, as compared with our 0.0334). Consequently, his maximum value for the equation of center is larger than ours (2°23', as compared to our 1°55' in table 5.3).

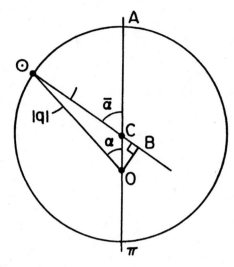

FIGURE 5.23.

Mathematical Postscript: Construction of the Tables

Let us see how the *tables of the sun* (tables 5.1–5.3) were constructed.

Mean Motion   The table of the sun's mean motion (table 5.1) is based on the following length for the tropical year:

$$1 \text{ tropical year} = 365.242199 \text{ days}.$$

As the Sun completes 360° of motion with respect to the equinoctial point in a tropical year, the mean motion in longitude is

$$360°/365.242199^d = 0.985647335 \text{ °/d}.$$

All entries in table 5.1 are multiples of this figure, with whole circles discarded.

Longitude of the Solar Apogee   Table 5.2 is based on the following two positions of the solar apogee:

$$A = 102.4° \text{ in A.D. 1974}$$

$$A = 65.5° \text{ in about 140 B.C.}$$

The first result is our own, from section 5.5. The second is due to Hipparchus. The rate (assumed constant) at which the solar apogee advances is

$$(102.4° - 65.5°)/(1974 + 139 \text{ years}) = 0.017463°/\text{year}$$

$$= 1.7463°/\text{century}.$$

The column giving the motion for ten-year intervals is based on this rate.

The longitude of the apogee in the year 1900 is obtained by subtracting 74 years' motion from the longitude of the apogee in 1974:

$$A_{1900} = A_{1974} - 74 \text{ years' motion}$$

$$= 102.4° - 74 \text{ years} \times 0.017463°/\text{year}$$

$$= 101.1°, \text{ or } 101°06'.$$

The rest of the table is easily completed by successive additions or subtractions of the motion for a single century.

Equation of Center   Table 5.3 is based on the value for the eccentricity of the Sun's circle calculated in section 5.5:

$$e = 0.0334,$$

the radius being taken as unity.

It is now required to calculate the equation of center q as a function of the mean anomaly $\bar{\alpha}$. Note that in figure 5.20, angle $C\odot O$ and angle $\bar{\odot}O\odot$ are always equal since $C\odot$ is parallel to $O\bar{\odot}$. Thus, angle $C\odot O$ is equal to the equation of center.

Our calculation will be based on figure 5.23. Line $\odot C$ has been extended beyond C, and a perpendicular has been dropped from O to meet this line at B. Now $OCB = \bar{\alpha}$. *In triangle OCB* we have

$$CB = e \cos \bar{\alpha}$$

$$OB = e \sin \bar{\alpha},$$

where e is the eccentricity OC. Therefore, the whole line segment $\odot B$ is

$$\odot B = \odot C + CB$$

$$= 1 + e \cos \bar{\alpha}.$$

In right triangle $O \odot B$, $O \odot$ is the hypoteneuse. By the rule of Pythagoras,

$$O\odot = \sqrt{\odot B^2 + OB^2}$$

$$= \sqrt{(1 + e \cos \bar{\alpha})^2 + (e \sin \bar{\alpha})^2}$$

$$= \sqrt{1 + 2e \cos \bar{\alpha} + e^2}.$$

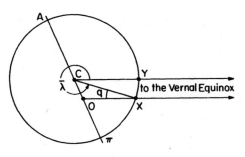

FIGURE 5.24.

Then, in triangle $O\odot B$ again, we have

$$\sin q = OB/O\odot$$

$$= \frac{e \sin \bar{\alpha}}{\sqrt{1 + 2e \cos \bar{\alpha} + e^2}}.$$

This is the result that was sought. It allows us to compute the equation of center q for any value of the mean anomaly $\bar{\alpha}$.

Actually, this expression should be regarded as giving the *magnitude* of q only. Figure 5.23 shows that q (defined by $q = \alpha - \bar{\alpha}$) should be negative for $0 < \bar{\alpha} < 180°$. q is positive for $180° < \bar{\alpha} < 360°$, as in figure 5.19F. Both the correct sign and magnitude will be obtained if q is calculated from

$$\sin q = \frac{-e \sin \bar{\alpha}}{\sqrt{1 + 2e \cos \bar{\alpha} + e^2}}.$$

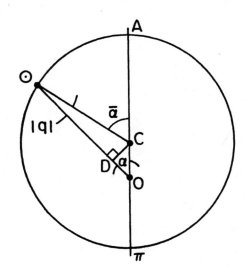

FIGURE 5.25.

Mean Longitude at Epoch   The last parameter that must be specified is the mean longitude of the Sun for some date. As epoch, we have chosen 1900 January 0.5 (noon at Greenwich). The problem, then, is to determine the Sun's mean longitude on this date. We do not yet know the mean longitude of the Sun on any date. But we do know the *true* longitude of the Sun on several dates: the equinoxes and solstices of the years 1972–1975 on which our work has been based. The first step in our procedure, then, is to determine the mean longitude of the Sun at one equinox or solstice.

Let us choose the vernal equinox of 1973. According to the data in section 5.4, this vernal equinox fell on March 20 at 18^h Greenwich time. Figure 5.24 shows the solar model at the moment of a vernal equinox. The parallel lines OX and CY point to the infinitely distant equinoctial point. At the moment of equinox, the Sun is at X, so its true longitude, as measured from the Earth O, is zero. The mean longitude $\bar{\lambda}$, as measured at the center C of the eccentric circle, has not quite reached zero but is shy of being zero by angle YCX, which is equal to the equation of center q. Therefore, let us compute q at the moment in question.

We know the Sun's true longitude λ (0°, or 360°, at vernal equinox), and we know the longitude of the apogee A. Therefore, we know the true anomaly $\alpha = \lambda - A$. The problem thus requires calculating q in terms of α (rather than in terms of $\bar{\alpha}$, as in our construction of the table for the equation of center).

The derivation of a general formula for calculating q in terms of α will be based on figure 5.25. In triangle OCD,

$$DC = OC \sin \alpha$$

$$= e \sin \alpha,$$

if the radius of the circle is taken as unity. Then, in triangle $D\odot C$, $\sin q = DC/C\odot$, or

$$\sin q = e \sin \alpha,$$

since $C\odot = 1$. This formula gives the correct magnitude, but not the correct sign, for q. To obtain the correct sign, note that in figure 5.25, q should be negative for a between 0 and 180°, since $\bar{\alpha} > \alpha$, and q is defined by $q = \alpha - \bar{\alpha}$. Thus, the correct magnitude and sign of q will be obtained if q is calculated from

$$\sin q = - e \sin \alpha.$$

Now we may proceed with the calculation of the Sun's mean longitude at the vernal equinox of 1973.

Sun's true longitude: $\lambda = 0$ (vernal equinox)
Longitude of apogee in 1973: $A = 102°19'$ (from tables)

Therefore, the true anomaly was

$$\alpha = \lambda - A = -102°19',$$

or

$$\alpha = 257°41' \text{ (adding 360°)}.$$

From this we calculate q:

$$\sin q = -e \sin \alpha$$
$$= -0.0334 \sin(257°41')$$
$$= +0.03263.$$
$$q = + 1°52'.$$

The mean longitude was less than the true by this amount:

$$\bar{\lambda} = \lambda - q = -1°52',$$

or

$$\bar{\lambda} = 358°08' \text{ (adding 360°)}.$$

This was the Sun's mean longitude at March 20, 1973, 18^h Greenwich mean time.

We have obtained the Sun's mean longitude at a particular moment in the year 1973. What we actually want is the mean longitude at our standard epoch, the beginning of the year 1900. Therefore, proceed as follows:

Mar 20, 1973, 6 p.m.	J.D.	244 1762^d	6^h
Jan 0, 1900, noon	J.D.	241 5020	0
Time elapsed since epoch	Δt	2 6742^d	6^h

During this time interval, the Sun's mean longitude increased by 78°26′, over and above complete cycles. (This result may be obtained either by multiplying Δt by the mean daily motion of 0.985 6473°/d, or by entering table 5.1 with

Δt.) To obtain the mean longitude at epoch, we set back the mean longitude at the vernal equinox of 1973 by this amount:

Mean longitude, Mar 20, 1973, 6 P.M.	358°	08′
Less the motion	−78	26
Mean longitude, Jan 0, 1900, noon	279°	42′

This is the result that was sought, the Sun's mean longitude at epoch. This figure is given at the bottom of table 5.1.

5.8 EXERCISE: ON THE TABLES OF THE SUN

1. Use tables 5.1–5.3 to compute the longitude of the Sun on March 15, A.D. 1979, Greenwich mean noon. Show all your work in a clear, orderly fashion. (Answer: 354° 19′. If your answer disagrees with this by more than 1′ or 2′ check your work to see where you went wrong.)
2. Work out the missing entries and fill in the blanks in tables 5.1, 5.2, and 5.3.
3. Using the solar tables, compute the longitude of the Sun on December 25, A.D. 1960, Greenwich mean noon.
4. Choose a date in the current year and calculate the longitude of the Sun using tables 5.1–5.3. Compare with the longitude tabulated in, for example, the current year's *Astronomical Almanac*.

5.9 CORRECTIONS TO LOCAL APPARENT TIME

The Equation of Time

An ideal clock runs at a steady rate, but the Sun does not. Indeed, the time from one local noon to the next is slightly variable; that is, the length of the solar day is not constant. The variation is not large, however, and it passes through the same cycle each year. Clocks run at a rate chosen to match the length of the *mean solar day*, that is, the average of the lengths of all the days of the year. It is the mean solar day that amounts to twenty-four hours. Any particular solar day can be a little longer or shorter than this.

Suppose that a clock's hands are adjusted so that local noon comes *on the average* when the clock reads 12:00. In such a case the clock keeps what is called *local mean time*. This kind of time is "mean" because it runs at a steady rate (unlike the time that the Sun keeps). It is "local" because it depends on the longitude of the timekeeper's position on the Earth. For example, local mean noon at New York occurs about three hours before local mean noon in San Francisco.

The time kept by the Sun and indicated by a sundial is called *local apparent time*. The difference between local apparent and local mean time is called the *equation of time*:

Equation of time = local apparent time (L.A.T.) − local mean time (L.M.T.).

Table 5.4 gives the values of the equation of time at one-month intervals throughout the year. (The missing entries are the subject of sec. 5.10.)

Example of the Use of the Table   Suppose that on July 23 a sundial reads 4:45 P.M. (i.e., local apparent time is 4:45 P.M.). What is the local mean time?

L.M.T. = L.A.T. − equation of time.

TABLE 5.4. The Equation of Time (local apparent minus mean time)

Date	Equation	Date	Equation
Jan 20	−11 min	Jul 23	−6 min
Feb 19	−14	Aug 24	−3
Mar 21	−7	Sep 23	+7
Apr 20	+1	Oct 24	+16
May 21	+4	Nov 23	_____
Jun 22	−2	Dec 22	_____

From table 5.4, we find that on July 23 the equation of time is −6 minutes. Thus, the local mean time when the sundial was consulted was

$$\text{L.M.T.} = 4:45 \text{ P.M.} - (-6 \text{ min})$$

$$= 4:51 \text{ P.M.}$$

Second Example   How much time (in mean solar days) elapses between local noon of March 21 and local noon of October 24 of the same year?

October 24 is the 297th day of the year (see table 4.4); March 21 is the 80th day. Thus, between the two noons the number of days elapsed is 297 − 80 = 217. These 217 days are not, however, mean solar days. That is, these 217 days are not units of equal length. To discover the exact amount of time elapsed, it is necessary to express the two dates in terms of mean time:

$$\text{L.M.T. of 2nd date} = \text{L.A.T. of 2nd date} - \text{equation of time}$$

$$= 12:00 \text{ P.M.} - (16 \text{ min})$$

$$= 11:44 \text{ A.M.}$$

$$\text{L.M.T. of 1st date} = 12:00 \text{ P.M.} - (-7 \text{ min})$$

$$= 12:07 \text{ P.M.}$$

The time interval runs from 12:07 P.M. March 21 to 11:44 A.M. October 24. The total time elapsed, *expressed in mean solar days*, is *216 days, 23 hours, 37 minutes*.

Zone Time

Clocks ordinarily keep, not local mean time, but *zone time*. The Earth is divided into twenty-four time zones, each approximately 15° wide. The exact boundaries of the time zones are irregular and are subject to occasional change, as they reflect political decisions. For example, it is inconvenient to let a zone boundary pass through a large city. The placement of time zone boundaries often expresses a sense of political or cultural identity. For example, all of western Europe (except the United Kingdom) lies in one time zone, although the continent is much wider than 15° of longitude.

Associated with each time zone is a *standard meridian*. There are twenty-four standard meridians, spaced at 15° intervals eastward and westward from the Greenwich meridian. The time zones are arranged, usually but not always, so that the standard meridian for each zone runs roughly down the middle of that zone. The clocks in a single time zone all keep the same time. The zone time in a particular time zone is the local mean time on the standard meridian of that zone. That is, all the clocks in a zone are set to agree with the clocks on that zone's standard meridian.

Here are the standard meridians for the time zones of the continental United States:

Eastern	75° W
Central	90° W
Mountain	105° W
Pacific	120° W

Conversion from local mean time to zone time involves a correction for the longitudinal distance of the locality from the standard meridian: 360° of longitude represents a 24-hour time difference, so the time difference associated with a single degree of longitude is 24 hours/360 = 1/15 hour, or 4 minutes. For a location *west* of the standard meridian, the correction for the longitude must be *added* to the local mean time to obtain zone time. Suppose, for example, that it is 12:00, local mean noon, at Baltimore. Baltimore, at longitude 77° W, is 2° west of the standard meridian for its zone. At the standard meridian, local mean noon will already have occurred. Thus, the zone time must be later than 12:00.

For a location *east* of the standard meridian, the correction for the longitude must be *subtracted* from the local mean time to obtain zone time.

Complete Example of Time Conversion A sundial in Boston reads 10:13 A.M. on September 23. What is the Eastern standard time?

First Step. Obtain the local mean time by applying the equation of time to the local apparent time. The equation of time for September 23 is 7 minutes. Thus,

$$L.M.T. = L.A.T. - \text{Equation of time}$$

$$= 10:13 - (7 \text{ min})$$

$$= 10:06 \text{ A.M.}$$

Second Step. Obtain the zone time (Z.T.) by applying the correction for the position of the city in its time zone: Boston, at 71° W longitude, is 4° east of the zone's standard meridian. The correction is therefore subtractive:

$$Z.T. = 10:06 - 4° \times 4 \text{ min/}°$$

$$= 10:06 - 16 \text{ min}$$

$$= 9:50 \text{ A.M.}$$

Thus, at the moment the sundial was consulted, Eastern standard time was 9:50 A.M. That is, Sun time was 10:13, but clock time was 9:50.

Summary Conversion from local apparent time to zone time involves two corrections. One of these, the equation of time, depends only on the time of year. The second correction depends only on the locality's position in its time zone. A third correction, for daylight savings time, must be applied in the summer months in localities that use this convention.

Cause of the Equation of Time

The equation of time arises from two causes. First, the ecliptic is inclined to the plane of the equator. And, second, the Sun's motion along the ecliptic is not uniform, but is sometimes faster and sometimes slower.

It is convenient to introduce a fictitious object that suffers from neither of these complications. The *equatorial mean Sun* is defined to travel around

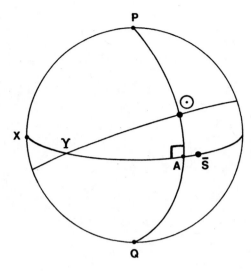

FIGURE 5.26.

the celestial equator, from west to east, at a uniform angular speed, completing one circuit of the celestial sphere in a tropical year. If the Sun traveled at a uniform angular speed around the ecliptic, and if the ecliptic coincided with the equator, the equatorial mean Sun would coincide with the true Sun. (The *equatorial mean Sun* should not be confused with the mean Sun of sec. 5.7. The mean Sun travels in the plane of the ecliptic, but the equatorial mean Sun, introduced here for the first time, travels in the plane of the equator.)[32]

In figure 5.26, the vernal equinoctial point is Υ. P and Q are the north and south celestial poles. The true Sun $\odot$ moves eastward along the ecliptic at a variable speed. Thus, the Sun's longitude (ecliptic arc $\Upsilon\odot$) increases at a variable rate. Pass a great circle through P and $\odot$. This great circle cuts the equator at A. The right ascension of the Sun is arc ΥA of the equator. As $\odot$ moves nonuniformly on the ecliptic, A will move nonuniformly on the equator. That is, the Sun's right ascension ΥA increases at a variable rate.

Moreover, even if the Sun did move along the ecliptic at a steady rate (i.e., if arc $\Upsilon\odot$ increased uniformly with time), the Sun's right ascension ΥA would still increase at a variable rate. When the Sun is near a solstitial point, as in figure 5.26, its motion along the ecliptic carries it on a path that is nearly parallel to the equator. Thus, the Sun's projection onto the equator (point A) moves along at a healthy speed. But when the Sun is near an equinoctial point, its path along the ecliptic is significantly inclined to the equator, and so point A tends to move more slowly.

The equatorial mean Sun, $\bar{S}$ in figure 5.26, is defined to move along the equator at a steady rate. The speed of $\bar{S}$ is the same as the average speed of A. Thus, A will be sometimes a little ahead of $\bar{S}$, and sometimes a little behind.

The equatorial mean Sun is the keeper of mean time. Once each day, the revolution of the cosmos carries the equatorial mean Sun across our meridian and produces local mean noon. When the true Sun crosses the meridian, either a little earlier or a little later than this, local apparent noon (= *local noon*) occurs. In figure 5.26, suppose that PXQ is the local celestial meridian, regarded as fixed. In the course of a day, the celestial sphere revolves to the west about axis PQ and all the points of the sphere are carried past meridian PXQ. The figure, drawn for a time in May, shows that the true Sun $\odot$ will cross the meridian before the equatorial mean Sun $\bar{S}$. So, apparent noon occurs a few minutes before mean noon.

The equation of time is equal to arc $\bar{S}A$ in figure 5.26. *The equation of time is the difference between the right ascension of the equatorial mean Sun and the right ascension of the Sun*:

$$\text{Equation of time} = \Upsilon\bar{S} - \Upsilon A$$

Corrections to Local Apparent Time in Antiquity

Change of Meridian There was no such thing as zone time in antiquity. Each astronomer used the local time of his own meridian. When observations made at two different locations were compared, however, the astronomer had to take into account the difference between the longitudes of the places.

Ptolemy, for example, in the construction of his lunar theory, made use of lunar eclipses observed by himself and his near contemporaries in Alexandria, but he also used records of eclipses that had been observed in Babylon eight centuries earlier. One of the Babylonian lunar eclipses used by Ptolemy occurred in the first year of the reign of Mardokempad, in the night between Thoth 29 and Thoth 30 (721 B.C., March 19/20, the most ancient dated observation in the *Almagest*). From the ancient record, Ptolemy determines that the middle of the eclipse came 2 1/2 equinoctial hours before midnight in Babylon. He continues in this vein:

Now we take as the standard meridian for all time determinations the meridian through Alexandria, which is about 5/6 of an equinoctial hour to the west of the meridian through Babylon. So at Alexandria, the middle of the eclipse in question was 3 1/3 equinoctial hours before midnight . . . .[33]

The accuracy of such a correction depended on the precision with which the longitudes of the cities could be determined. Before the seventeenth century, the measurement of longitude was a highly imperfect art. Ptolemy's 5/6 hour difference between Babylon time and Alexandria time corresponds to a 12°30′ longitude difference between the two cities. Their actual separation is about 15°, or about one hour.

Ptolemy's tables for working out the positions of the Sun, Moon, and planets were based on Alexandria time, just as our own tables of the Sun (tables 5.1–5.3) are based on Greenwich time. The choice of a standard meridian for the tables, however, is a trivial matter, as it influences only the initial values of the time-varying quantities. For example, at the foot of table 5.1 (table of the sun's mean motion), the mean longitude of the Sun is given as 279°42′ at Greenwich mean noon of December 31, 1899 (= Jan. 0.5, 1900). If we wished to adapt the tables for New York local time, it would be convenient to know the Sun's mean longitude at New York mean noon of the same day. New York is at 74° W longitude. The 74° longitude difference between Greenwich and New York corresponds to a difference of 4^h56^m between the local times. Now in 4^h56^m the mean Sun moves about 12′, as can be found from table 5.1. Thus, at local mean noon in New York, December 31, 1899, the longitude of the mean Sun was 279°42′ + 12′ = 279°54′. If this information were noted at the foot of the table, the table would contain all that was necessary for convenient calculation for the meridian of New York.

Ptolemy's tables enjoyed a long life. They were copied in but slightly modified forms during the whole medieval period by Arabic- and, later, by Latin-writing astronomers. Astronomers often adapted the tables to their own meridians by means of the simple transformation we have just illustrated. In many medieval manuscripts the strictly astronomical tables are accompanied by subsidiary tables, including geographical tables giving the longitudes and latitudes of major cities. The longitudes were required, of course, for making a change of meridian.

Equation of Time   Geminus says in his *Introduction to the Phenomena* (VI, 1–4) that the *nychthemeron* (a day and night together) is not of constant length. Moreover, Geminus gives a clear explanation of one of the causes of this inequality—the one that depends on the obliquity of the ecliptic. Geminus defines the *nychthemeron* as the time from sunrise to sunrise, rather than as the time from local noon to local noon as we (and Ptolemy) do. Nevertheless, the essential phenomena are the same. The variability in the length of the solar day was too small to be measured directly, by means of a water clock or some other device. Rather, this variability was deduced from theory. From Geminus's remark, it is clear that the Greek astronomers knew by the first century A.D. that the *nychthemeron* must vary in length.[34] However, the oldest surviving detailed, mathematical discussion of this subject is that of Ptolemy.

In antiquity, the equation of time had few important consequences, simply because of the smallness of this equation. From table 5.4, the mean time elapsed between February 19 and October 24 is some 30 minutes less than the apparent time elapsed. This 30-minute correction to the length of a time interval is about the largest ever required.

The Sun and planets move along the ecliptic at rather slow rates. The Sun moves only about 1°/day. In half an hour the Sun's motion then amounts to 0.02°, or slightly more than 1′. Thus, unless one can measure the longitudes of the Sun and planets to a precision of a minute of arc, it makes no difference whether one works with mean or apparent time. There is no need to apply the equation of time.

Only in the case of the Moon, which moves much more rapidly, need this proposition be modified. The Moon completes one revolution about the Earth, with respect to the fixed stars, in 27.3 days, which works out to 13°/day. In half an hour, therefore, the Moon moves about 16′, a distance easily detected by naked-eye observation, since it amounts to half the Moon's apparent diameter. Thus, strict handling of the time interval between two lunar observations requires that the equation of time be taken into account—as Ptolemy himself says in *Almagest* III, 9.

Ptolemy's treatment of the equation of time follows immediately on the solar theory and is his first practical application of that theory. His treatment is, in every important respect, equivalent to the modern one. Ptolemy defines the apparent solar day as the interval between two successive meridian crossings by the Sun. He distinguishes between the apparent or *anomalistic day* (*nychthemeron anomalon*) and the mean or *uniform day* (*nychthemeron homalon*). He identifies the two causes that necessitate this distinction: the obliquity of the ecliptic and the solar eccentricity. Finally, he explains how to reduce any time interval expressed in apparent solar days to mean solar days.

In one respect, his approach differs from our own. There was not in antiquity any quantity corresponding to our local mean time. Thus, Ptolemy never uses the equation of time to convert the time of a given astronomical observation from local apparent to local mean time, as we did in our first example above. Before the invention of accurate mechanical clocks, which provide its concrete realization, the notion of mean time was of little utility. Thus, Ptolemy always treats the equation as a correction to be applied to a *time interval*, as in our second example.

For example, when Ptolemy takes up the lunar theory in *Almagest* IV, 6, he considers three lunar eclipses taken from the Babylonian records. He reckons that the middles of two of these eclipses occurred at the following apparent times in Babylon:

- Year 2 of Mardokempad, Thoth 18/19, midnight
- Year 2 of Mardokempad, Phamenoth 15/16, 3 1/2 hours before midnight

By a simple count of the calendar days and hours, Ptolemy determines the apparent time interval separating these two eclipses:

$$176^d \ 20\frac{1^h}{2}.$$

He then applies the correction for the equation of time to obtain the actual length of the time interval:

$$176^d \ 20\frac{1^h}{5}, \text{ in mean solar days.}$$

Although Ptolemy explains the calculation of the equation of time in *Almagest* III, 9, he does not provide a table of this equation to simplify its application by users of his book. Nor does he offer a single worked-out numerical example. This shortcoming was rectified in Ptolemy's *Handy Tables*, composed some time after the *Almagest*.

The tables of the *Almagest* are somewhat inconvenient to use, as they are scattered throughout the text. In the *Handy Tables*, Ptolemy grouped all of the tables into one compact package. Ptolemy's original version of the *Handy Tables* has not come down to us. What we now have is a revised version composed by Theon of Alexandria around A.D. 395.[35] It does not appear, however, that Theon greatly modified Ptolemy's work. Many of the tables in the *Handy Tables* are considerably expanded in comparison to their counterparts in the *Almagest*. For example, the table of ascensions in the *Almagest* gives the rising time for each 10° segment of the ecliptic, but the corresponding table in the *Handy Tables* gives the cumulative rising time for each single degree.

The *Handy Tables* also contain material that has no counterpart in the *Almagest*, including a table of reigns (see sec. 4.5), a list of cities with their geographical coordinates—and a table for the equation of time. In the *Handy Tables* this new table appears as an extra column in the table of ascensions for the right sphere. The first part of Theon's table is translated thus:

Excerpt from the Table of Ascensions for the Right Sphere
as Found in the *Handy Tables*

| Longitude of Sun (degrees) | Goat-Horn | | Water-Pourer | |
	Ascensions	Differences of the Hours (sixtieths)	Ascensions	Differences of the Hours (sixtieths)
1	1 6	18 40	33 18	31 16
2	2 12	19 11	34 20	31 28
3	3 18	19 42	35 22	31 40
4	4 24	20 13	36 24	31 52
5	5 30	20 44	37 26	32 4
6	6 35	21 15	38 28	32 15

In the left column run the degrees from 1 to 30 for the longitude within each sign. The second and third columns are devoted to the sign of the Goat-Horn. The second column gives the right ascension of each of the thirty points in the sign of the Goat-Horn. (This part of the table may be compared with the part of table 2.4 for the right sphere.) Interestingly, in the *Handy Tables*, the zero of longitude is taken to be the beginning of the Goat-Horn, rather than the beginning of the Ram as in the *Almagest*. The third column gives the equation of time for each of the tabulated positions of the Sun.

In the *Handy Tables*, Ptolemy uses the era Philippos (see sec. 4.5), as opposed to the era Nabonassar that he used in the *Almagest*. That is, the initial positions of the Sun, Moon, and planets are given for Alexandria noon, Thoth 1, Year 1 of Philippos. The values of the equation of time in Ptolemy's (or Theon's) table are also referred to this epoch. So, for example, when the Sun is 5° within the sign of the Goat-Horn, the equation of time is $20^m 44^s$. This signifies that the mean time elapsed between the epoch (beginning of the reign of Philippos) and the date in question (the Sun being 5° within the Goat-Horn) is $20^m 44^s$ longer than the time apparently elapsed.

Because of a happenstance, the equation of time in the *Handy Tables* is always an additive, and never a subtractive, correction to the time apparently elapsed since epoch. That is, at the beginning of the reign of Philippos, the equation of time in the modern sense was near its extreme positive value (corresponding to Oct 24 in table 5.4). Thus, given any date after this epoch, the mean time elapsed can only be greater than the time apparently elapsed. The maximum additive correction given in the table for the equation of time in the *Handy Tables* is some 33 minutes and occurs when the Sun is in the sign of the Water-Pourer. This compares well with the maximum correction of 30 minutes we obtained above.

Ptolemy's treatment of the equation of time is a remarkable testimonial to the sophistication of late Greek astronomy. This effect, too small to be detected observationally, was deduced as a logical and necessary consequence of the solar theory. Ptolemy's treatment of the equation of time—like his treatment of so many other topics—proved to be definitive. His tabulation of this equation in the *Handy Tables* became the model for many similar tables during the Middle Ages.

Mathematical Postscript: Computation of the Equation of Time

Refer once again to figure 5.26. As demonstrated earlier, the equation of time is the difference between the right ascension of the equatorial mean Sun $\bar{S}$ and the right ascension of the Sun $\odot$. That is,

$$\text{Equation of time} = \Upsilon\bar{S} - \Upsilon A.$$

The position of the equatorial mean Sun $\bar{S}$ is easily specified: we begin with the solar theory, eliminate the equation of center to obtain the mean Sun, which moves uniformly on the ecliptic, then fold the ecliptic down so that it coincides with the celestial equator. In other words, *the right ascension of the equatorial mean Sun is equal to the longitude of the mean Sun.* Arc $\Upsilon\bar{S}$ is thus determined: $\Upsilon\bar{S} = \bar{\lambda}$. And ΥA is the right ascension of the true Sun, as found from the solar theory and a table of right ascensions. The equation of time is the difference between $\Upsilon\bar{S}$ and ΥA.

Example: Equation of Time on February 19   On February 19, the Sun enters the Fishes (see table 2.1 . The Sun's true longitude λ is Fishes 0°, or 330°. This is arc $\Upsilon\odot$ in figure 5.26. The right ascension of the true Sun is found from (table 2.4 for the right sphere with this longitude. Entering the table Water-Pourer 30° (= Fishes 0°), we find the right ascension 332°05′. This is arc ΥA.

It remains to find $\Upsilon\bar{S}$, the right ascension of the equatorial mean Sun. Again, this is equal to the longitude of the mean Sun. We may find this from the true longitude of the Sun, already obtained, by subtracting the equation of center q:

$$\bar{\lambda} = \lambda - q.$$

The equation of center may be computed from the true anomaly α (Sec. 5.7):

$$\sin q = - e \sin \alpha.$$

Now, $\alpha = \lambda - A$, where A is the longitude of the Sun's apogee. In the 1980s, $A = 101°30′$. Thus, on February 19,

$$\alpha = \lambda - A$$
$$= 330° - 101°30′$$
$$= 228°30′.$$

So,

$$\sin q = -e \sin \alpha$$
$$= -0.0334 \sin(228°30′)$$
$$= +0.0250.$$
$$q = +1°26′.$$

Thus,

$$\bar{\lambda} = \lambda - q$$
$$= 330° - 1°26'$$
$$= 328°34'.$$

This is the longitude of the mean Sun. It may also be interpreted as the right ascension of the equatorial mean Sun, arc $\Upsilon\bar{S}$.

The equation of time is

$$\text{Equation of time} = \Upsilon\bar{S} - \Upsilon A$$
$$= 328°34' - 332°05'$$
$$= -3°31'.$$

Finally, we convert from angular measure to time:

$$\text{Equation of time} = -(3\ 31/60)° \times 4\ \text{min}/°$$
$$= -14\ \text{min},$$

in agreement with the entry in Table 5.4.

5.10 EXERCISE: APPARENT, MEAN, AND ZONE TIME

1. Suppose that an accurately constructed and aligned sundial reads 10:30 A.M. The date is January 20. What is the local mean time?

2. How much time, in mean solar days, elapses between October 24 of one year and March 21 of the following year? (Suppose, for simplicity, that there are no leap years involved.)

3. A sundial in Seattle (122° W longitude) reads 1:45 on March 21. What would a clock read at this instant? (I.e., what is the Pacific standard time?)

4. In the eighteenth century, the development of accurate, transportable clocks made possible for the first time the reliable determination of the longitude at sea. A marine chronometer, or clock, set to Greenwich mean time, was carried onboard the ship. By observation of the Sun the navigator could determine the local apparent time on the meridian where his ship lay; this, corrected for the equation of time, gave the local mean time. When the local mean time so obtained was compared with the Greenwich mean time read by the chronometer, the navigator could deduce his longitude east or west of the Greenwich meridian. This method remained standard from the late eighteenth century until the twentieth, when it was replaced by the methods of radio navigation.

   Suppose that on October 24, a navigator onboard a ship in the Atlantic finds, by observation of the Sun, that local noon occurs when his chronometer reads 3:32 P.M. Greenwich mean time.

   A. What is the local apparent time? (Answer: 12:00 P.M.)
   B. What is the local mean time? (Answer: 11:44 A.M.)
   C. What is the longitude of the ship? (Answer: 3^h48^m west of Greenwich, i.e., 57° W longitude.)

5. Repeat problem 4 for a ship in the Pacific. Suppose that the date is May 21 and that, by taking an altitude of the Sun, the navigator finds that the local apparent time is 10:49 A.M. The chronometer, set to Greenwich mean time, reads at that instant 8:35 P.M. What is the longitude of the ship?

6. Supply the missing entries in table 5.4.

Precession is a slow revolution of the whole field of stars from west to east about the poles of the ecliptic. Figure 6.1 shows the positions of three constellations (Delphinus, Auriga, and Orion) on the celestial sphere, both for the present day and for 2,000 years ago. Each star has moved along a circular arc parallel to the ecliptic. Because the motion is parallel to the ecliptic, the stars' *ecliptic coordinates* change in a simple way: the latitudes remain unchanged, while the longitudes increase at a steady rate. Although this rate is slow (1° in 72 years, or 50″ per year) it adds up over long periods of time. In 2,000 years, the precession comes to some 28°, nearly a whole zodiac sign. Every star in the sky suffers this same change in longitude.

The *equatorial coordinates* change in a much more complicated way. The right ascension does not change at a steady rate: the rate of change depends on the star's position on the sphere. Moreover, the declination does not remain unaffected. Two thousand years ago, Orion was almost wholly south of the equator. But precession, parallel to the ecliptic, has carried Orion northward and its middle now lies on the equator.

Precession causes the Pole Star to move. Polaris is now quite near the pole of the equator and is still getting nearer. But two thousand years ago Polaris was more than 10° south of the pole. Two thousand years from now, Polaris will have moved beyond the pole of the equator and will cease to serve as a good north star. About that same time (around A.D. 4000), the star Errai (γ Cephei) will come near the pole of the equator and will serve as pole star. The set of future and past pole stars forms a small circle centered on the pole of the ecliptic, as shown in figure 6.1.

Because of precession, it is vital to distinguish between the zodiacal *signs* and the *constellations* that have the same names. The first 30° of the zodiac, going eastward from the vernal equinox, is called the sign of Aries today just as in ancient times. The next 30° of the zodiac is the sign of Taurus, and so on. Two thousand years ago, the constellation Aries was within the sign of Aries, but it is now in the sign of Taurus (fig. 6.2). The astrological charts that appear in newspapers are based on signs, not constellations. A modern person whose astrological sign is Taurus was born when the Sun was in the sign of Taurus—that is, when the Sun was among the stars of the constellation Aries.

Just as we can account for the daily motion of the heavens either by letting the stars revolve from east to west about the poles of the equator, or by letting the Earth revolve in the opposite direction about the same poles, so too there are two possible explanations of precession. The one we have given is the ancient view of the situation.

For the modern view, see figure 6.3. The Earth has a rapid rotation (once every twenty-four hours) from west to east about the poles of the equator, which produces day and night. Meanwhile, the Earth suffers a slow rotation (once every 26,000 years) from east to west about the poles of the ecliptic, which drags the Earth's axis around such that the axis points to different pole stars in different eras. From this point of view, the star field remains fixed, but the celestial equator changes. In figure 6.3, the equinoctial point ♈ moves westward through the stars. This means, of course, that the stars will appear to move eastward with respect to the equinox.

The two pictures of precession are indistinguishable as far as appearances are concerned. But the ancient theory is easier to use when trying to solve a problem concerning the appearance of the sky in the distant past or future. It is easy to use because it looks at the sky in the same way that a real observer does—from the Earth, which for all appearances is at rest in the center of the world.

SIX

The
Fixed
Stars

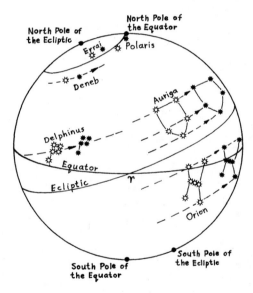

FIGURE 6.1. Precession. The sphere of stars rotates about the poles of the ecliptic from west to east. In this figure, the stars are seen as on a globe. That is, the celestial sphere is seen from the outside. The open symbols show the positions of the stars 2,000 years ago; the solid symbols, their positions today.

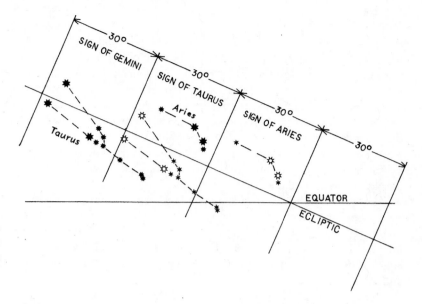

FIGURE 6.2. Precession. The open symbols show the positions of the stars 2,000 years ago; the solid symbols, their positions today. In this figure, the stars are viewed as on a chart. That is, the celestial sphere is seen from the inside.

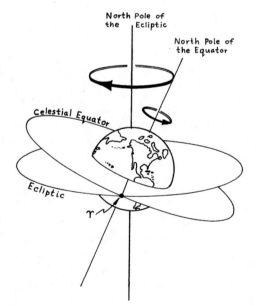

FIGURE 6.3. An alternative view of precession. The Earth's axis rotates about the poles of the ecliptic from east to west. Consequently, the equinoctial point ♈ also moves to the west.

Precession was discovered by Hipparchus in the second century B.C. but remained unexplained until Newton deduced it from his laws of motion and theory of gravity. In the Newtonian theory, the rapid rotation of the Earth on its axis produces a bulge at the equator. The precession is caused by the gravitational action of the Sun and Moon on this bulge.[1]

Dealing with Precession in Historical Studies

Whenever one wishes to study the appearance of the sky in ancient times, one must take precession into account: the constellations occupied different places on the celestial sphere than they do now. Most problems involving the appearance of the sky in ancient times may be solved to adequate precision by means of a celestial globe. Suppose we wish to determine where on the horizon Procyon rose 2,000 years ago at some particular latitude. The precession in this period comes to about 28°. Therefore, simply place a mark on the globe to represent the ancient position of Procyon. This mark will be at the same latitude as the mark for the star's modern position but will be set back in longitude by 28°. Once the star's ancient position is marked on the celestial globe, the globe may be used in the usual way to solve the given problem.

A second convenient method of investigating the disposition of the sky in ancient times involves a *precession globe*. This is a celestial globe that differs from the ordinary kind in one way only. Instead of having just one pair of holes for the axis to pass through, it has several, and these other holes correspond to the positions of the equator's poles at different epochs. By inserting the axis through the correct pair of holes, one ensures that the stars' daily rotation will take place about the correct axis for that epoch. The oldest known design for a precession globe, slightly different from the one described here, was given by Ptolemy in *Almagest* VIII, 3.

More precise mathematical methods for reducing modern star positions to an ancient epoch are discussed in textbooks of spherical astronomy.[2]

Two Kinds of Year

Because the tropic and equinoctial points are moving with respect to the stars, the word *year*, defined as "the period required for the Sun to complete one orbit around the Earth," is ambiguous. Two different years must be distinguished. The *tropical year* is the period required for the Sun to travel from the vernal equinox back to the same equinox. The *sidereal year*, speaking loosely, is the period required for the Sun to travel from one fixed star back to the same fixed star.

Because the stars advance in longitude by about 50″ in the course of a year, the Sun must travel a little farther to return to the same star than to return to the equinoctial point. Thus, because of precession, the sidereal year is a little longer than the tropical year:

$$1 \text{ sidereal year} = 1.000039 \text{ tropical year.}$$

The sidereal year is the *physically* significant period. This is the period that must be used in orbital mechanics involving Newton's laws. However, the tropical year is more fundamental *astronomically* since it is accessible to direct measurement. The calendar we use is, of course, based on the tropical year.

6.2 ARISTOTLE, HIPPARCHUS, AND PTOLEMY ON THE FIXEDNESS OF THE STARS

The Physics of Aristotle's Cosmos

The very idea of a constellation presupposes that these figures remain the same for long periods. By the fourth century B.C., this fact of observation had been promoted to the rank of a first principle: the heaven was changeless because it was physically impossible for it to be otherwise. According to Aristotle, the heaven is made of a fifth element, the ether, different in nature from the four elements that make up our world of growth and decay.[3] The essence of this fifth element is absolute changelessness, and its natural motion is circular revolution about the center.

Aristotle's method is mainly deductive and theoretical rather than empirical. But he often reinforces his physical arguments with appeals to experience. For example, after attempting to prove logically that the fifth element is neither heavy nor light, that it is ungenerated and indestructible, that it can neither grow nor diminish and is unalterable in every way, Aristotle adds that the truth of these things "is also clear from the evidence of the senses, enough at least to warrant the assent of human faith; for throughout all past time, according to the records handed down from generation to generation, we find no trace of change either in the whole of the outermost heaven or in any one of its parts."[4] Finally, Aristotle supposes that the substance of the heavens got its name from the very changelessness of its motion: he derives *aither* (ether) from *aei thein* (always runs).

This conception of the heavens, often characterized as "Aristotelian," was widely accepted in antiquity. Moreover, it dominated cosmological thinking through the Arabic and Latin Middle Ages. It was one part of the mental equipment that every ancient and medieval astronomer brought with him when he attacked a scientific problem. And yet the Greek astronomers often proved themselves capable of departing from Aristotle's physics when astronomy seemed to require it.

Hipparchus and Ptolemy on the Stars

The New Star A case in point is the hypothesis of Hipparchus that the stars might shift their positions with respect to one another. According to Pliny, Hipparchus had observed at least one startling change in the heavens: he is supposed to have seen a new star.[5] Pliny says that it was the appearance of this *nova stella* that moved Hipparchus to compile his catalog, giving the positions and magnitudes of the many stars, so that it might be possible to discern not only whether stars perish and are born, but also whether they are in motion and whether they increase or decrease in magnitude. Here we have a Greek astronomer, and therefore supposedly an intellectual stepchild of Aristotle, entertaining decidedly non-Aristotelian ideas.

Hipparchus's Discovery of Precession   Ptolemy does not repeat Pliny's story about the new star, but he does give an even more pressing reason why Hipparchus thought it necessary to determine whether the constellations really are changeless or not. Hipparchus had discovered the precessional motion by comparing his own observations of the stars with earlier ones by Timocharis and Aristyllos. The principal test star seems to have been Spica, which Hipparchus found to lie 6° west of the autumnal equinox in his own time, but 8° in the time of Timocharis. And Hipparchus found this same 2° eastward shift in the longitudes of the few other fixed stars for which he was able to make comparisons. However, the nature of Hipparchus's method (discussed below) limited him to stars near the ecliptic, such as Spica. For this reason he could not be sure whether precession was a motion shared by all the stars, or whether it belonged solely to those in the zodiac. These zodiacal stars, with their slow eastward motion, might be planets after all. The title of Hipparchus's book on the subject, *On the Change of the Tropic and Equinoctial Points*, suggests that he believed the motion was a general one. Unfortunately, this work is lost. Everything that we know of it we owe to Ptolemy's summary and quotations in *Almagest* VII. Ptolemy tells us that Hipparchus suspected the truth but was hampered in demonstrating it by the paucity of observations that had been made before him, which were only those recorded by Timocharis and Aristyllos and which were not completely trustworthy. According to Ptolemy, Hipparchus left behind many more, and better, observations than he ever got from his predecessors. It was these observations by Hipparchus that enabled Ptolemy to settle the issue—for the time being—260 years later.

Star Alignments   One class of observations handed down by Hipparchus consisted of star alignments. Ptolemy quotes more than twenty of these as affording the simplest proof that the stars maintain always the same figures. For example,

> In the case of the stars in the Bull [Hipparchus writes] that the eastern ones of the Hyades [α and ε Tauri] and the sixth star, counting from the south, of the hide that Orion holds in his left hand [π′ Orionis] are on a straight line.

This was in fact a very good alignment. Calculations based on the actual positions of these stars in 130 B.C. show that the middle star (α Tauri) lay off the line through the other two by only 6′, toward the southwest.[6]

Here we have a statement by Hipparchus that two stars in a zodiacal constellation, the Bull, are on a straight line with a third star belonging to a constellation, Orion, not in the zodiac. If precession were a motion in which only the zodiacal stars participated, then this alignment would be destroyed in a fairly short time. But if, as Hipparchus suspected, the precession were common to all the stars, then the alignment would be preserved. Later astronomers would have to decide the issue, but in either case, an extensive list of alignments would make the decision easy and sure. The Hipparchian alignments preserved by Ptolemy are all of this sort, connecting zodiacal stars with stars that lie outside the zodiac. The list begins with the Crab and takes up each sign in order, with the single exception of the Goat-Horn, for which no alignments are recorded.

In a few situations, Hipparchus attempted to specify by just how much an alignment fell short of perfection:

> In the case of the stars in the Ram [Hipparchus writes] that the western
> star of the Triangle's base [β Trianguli] deviates by one finger to the eastward
> from the straight line drawn through the star in the Ram's jaw [α Arietis]
> and Andromeda's left foot [γ Andromedae].

The "finger" (*daktylos*) was a unit of angular measure common in Babylonian
astronomy, where it generally represented 5′. Was this Hipparchus's meaning
here? β Trianguli was fully 50′ east of the line through the other two stars.
An error of 45′ seems rather large if Hipparchus were really attempting to
measure to the nearest 5′. In any case, it appears likely that these alignments
were all roughly made, without the aid of any sort of instrument.

Some historians have conjectured that Hipparchus made these alignments
by holding a taut string out at arm's length, and in this way confirmed that
three stars lay along the same straight line.[7] Actually, the string is more trouble
than it is worth. If the sky is dark, the string cannot be seen. But if one can
see the string, and if one does look at it, then one cannot focus sharply on
the stars at the same time. And, anyway a string held in one's hands at arm's
length wobbles around too much to be of any use. One can actually do much
better with the unassisted eye than with the aid of a handheld string. Whether
Hipparchus employed a string we do not know.

After listing more than twenty of Hipparchus's alignments, Ptolemy de-
clares that in these and similar configurations affording a comparison for
nearly the whole celestial sphere, nothing has changed down to his own time.
But a change would certainly have been sensible in the intervening 260 years
if only the zodiac stars participated in the eastward precessional motion. Thus,
Hipparchus's method of alignments, crude as it was, sufficed to settle this
important question.

So that others might make comparisons over a longer time from more
configurations of the same kind, Ptolemy appends a list of his own alignments,
starting with the Ram and going through all the signs in order. We will quote
several of these, in case the reader should like to go outside and, taking
advantage of the long time elapsed since Ptolemy's day, confirm with greater
certainty that the precession is common to all the stars.

> The straight line joined through the one called Goat [α Aurigae] and the
> bright star of the Hyades [α Tauri] leaves a little to the east the one in the
> Charioteer's eastern foot [ι Aurigae].

In Ptolemy's day, ι Aurigae was 23′ to the east of the line through the other
two. This alignment has scarcely changed down to the present day. This
deviation of ι Aurigae from the line is small but easily perceptible to the
unaided eye.

> The star called Goat [α Aurigae], the one in common to the Charioteer's
> eastern foot and the tip of the Bull's northern horn [β Tauri], and the one
> in Orion's western shoulder [γ Orionis] are on a straight line.

This is a poor alignment. In 100 A.D., the line through α Aurigae and γ
Orionis left β Tauri to the east by more than a degree. This is an easy
alignment to check, the stars all being very conspicuous.

> The bright one in the southern Claw [α Librae], Arcturus, and the middle
> one of the three in the Great Bear's tail [ζ Ursae Majoris] are on a straight
> line.

In Ptolemy's day, this alignment was almost perfect. Arcturus was east of the
line by only 1′. *This seems to be the only one of Ptolemy's alignments that can
be used today to reveal a proper motion:* Arcturus is now 44′ west of the line.
(*Proper motion* is the motion of a star with respect to its neighbors. As we
shall see in sec. 6.10, proper motions were not discovered until the eighteenth
century.) This shift of 45′ is almost entirely due to the strong proper motion

of Arcturus toward the southwest. The arc from ζ Ursae Majoris to α Librae is very long (some 73°), which makes the observation somewhat difficult, but Arcturus is so far west of the line now that the careful observer should be able to see this effect of proper motion.[8]

Ptolemy gives some twenty-three of his own alignments, then concludes by remarking that if one should compare these configurations with those on Hipparchus's solid sphere, he would find that the positions are very nearly the same. This could imply that Hipparchus's celestial globe, or a copy of it, still survived in Alexandria during Ptolemy's day. Or, perhaps, someone had made a globe and placed the stars on it in accordance with the positions specified by Hipparchus in his *Commentary on Aratus and Eudoxus.*

If we compare the alignments of Hipparchus with those of Ptolemy, several differences in style are apparent. Hipparchus's alignments seem to have been more carefully made: he generally used short arcs, which facilitates accurate judgment, and he was content to use faint stars if these happened to give good alignments. Ptolemy, on the other hand, seems to have sought out alignments involving three bright stars, even if these involved arcs of 60° or 70°, and even if the alignments were somewhat sloppy. This difference in care is no doubt due to the differing motives of the two astronomers in making their alignments. For Hipparchus it was still an open question whether the precession was confined to the zodiac stars or shared by the whole sphere. He therefore took some care with his alignments. By Ptolemy's time this issue had been settled. Ptolemy's aim was simply to give his readers a number of easily checked rough alignments involving bright and familiar stars.

Hipparchus and Ptolemy showed that the stars really are fixed in their constellations, as nearly as they could tell, which was what everyone had believed all along. But the fact that Hipparchus and Ptolemy entertained the notion that the stars *might* shift their relative positions, and that they set down observations explicitly for the purpose of allowing later generations to confirm or reject this possibility, demonstrates a freedom of thought and a willingness to depart from Aristotle when necessary. The star alignments also nicely illustrate the cumulative nature of the science of astronomy. Some motions in the heavens are so slow that one generation may be forced to leave certain questions for later generations to answer.

6.3 OBSERVATION: STAR ALIGNMENTS

Find on a star chart the stars involved in the alignments of Hipparchus and Ptolemy that were quoted in section 6.2. (You can find others in *Almagest* VII, 1.) Taking into account the time of year and the hour at which you wish to look, decide which stars will be visible. Then go outdoors and check for yourself whether the figures of the constellations really have remained unchanged since the days of the Greek astronomers.

6.4 ANCIENT METHODS FOR MEASURING
THE LONGITUDES OF STARS

Two methods were used by Hellenistic astronomers to measure star longitudes. The zero point for the measurement of longitudes is the vernal equinox, the point where the Sun's motion on the ecliptic carries it across and above the equator: *the zero point is defined by the motion of the Sun.* Now, the longitude of the Sun at any moment can be calculated *from theory* (Hipparchus's solar theory, in sec. 5.7). If we could measure the angular distance between some ecliptic star—Spica, for instance—and the Sun at some moment, we could simply add this to the Sun's calculated longitude to obtain Spica's longitude.

The trouble is that we cannot directly measure the angular distance between Spica and the Sun because we never can see them at the same time. The two ancient methods employed different techniques for getting around this difficulty. But both made use of the Moon.

Method Using a Lunar Eclipse (Method of Hipparchus)

In *Almagest* VII, 2, Ptolemy remarks that Hipparchus was able to deduce the longitude of Spica by using lunar eclipses. In fact, Hipparchus was able to do this not only for his own day but also for the time of Timocharis, who preceded him by several generations, since he had access to some of Timocharis's data. Ptolemy gives no details of Hipparchus's method, but it must have been very much as follows.

The general idea is to employ the relation

$$\text{longitude} \atop \text{of Spica} = {\text{longitude} \atop \text{of Sun}} + {\text{longitudinal arc} \atop \text{from Sun to Spica.}}$$

The first term on the right Hipparchus computed from his solar theory. The use of a lunar eclipse helped him get around the difficulty of measuring the second term, the longitudinal arc from the Sun to Spica.

During a lunar eclipse, the middle of the Earth's shadow marks a point in the sky that is directly opposite the Sun. Midway through the eclipse, the Moon's longitude is almost exactly 180° different from the Sun's. The longitudinal arc from the Moon to Spica at this moment can be measured. The longitudinal arc from the Sun to Spica must be 180° greater. In this method of measuring star longitudes, the Moon provides a screen on which the Earth's shadow may fall, permitting us to "see" the Sun indirectly. Once a small correction is made for the Moon's parallax, the longitude of the star can be determined:

$$\text{longitude} \atop \text{of star} = {\text{longitude} \atop \text{of Sun at} \atop \text{eclipse middle}} + {\text{longitudinal arc} \atop \text{from Moon to star} \atop \text{at eclipse middle}} + 180°.$$

In section 6.5, the reader will have the opportunity to apply this method.

Method Using the Moon as Connecting Link (Method of Ptolemy)

General Explanation of the Method   Ptolemy describes in detail his own method of measuring star longitudes. This second method still relies on the Moon, although in a different way.

Shortly before sunset, Ptolemy measures the longitudinal arc between the Sun and the Moon, using an armillary sphere equipped with sights. A little while later, the Sun has set and the star is visible. Ptolemy then measures the longitudinal arc between the Moon and the star using the same instrument. The longitude of the star is then

$$\text{longitude} \atop \text{of star} = {\text{longitude} \atop \text{of Sun}} + {\text{longitudinal} \atop \text{arc from} \atop \text{Sun to Moon}} + {\text{longitudinal} \atop \text{arc from} \atop \text{Moon to star}} + \text{correction.}$$

The first term on the right side of this equation is calculated from theory, using the solar tables. The second and third terms are the results of measurements with the armillary sphere. The correction term requires some comment.

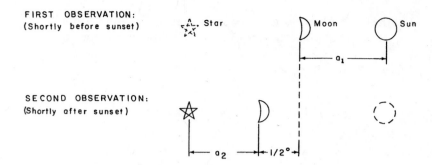

FIRST OBSERVATION:
(Shortly before sunset)

SECOND OBSERVATION:
(Shortly after sunset)

FIGURE 6.4.

It actually consists of two parts, a correction for the Moon's motion and a correction for the change in the Moon's parallax.

Correction for the Moon's Motion   The Moon moves eastward around the ecliptic in (very roughly) 30 days, so its average motion is roughly 360°/30 days = 12°/day. In a single hour, then, the Moon moves eastward through the stars by 1/2°. This effect is too large to be neglected. Suppose the three bodies are arranged in the sky in the order (going from east to west) star, Moon, Sun, as in figure 6.4. Suppose we measure the arc Moon-Sun and get a value a_1, as shown in the top part of figure 6.4. The star is drawn in broken line to remind us that it is invisible at the first observation. Suppose that *one hour later* we measure the arc star-Moon and obtain a_2, as shown in the lower half of the figure. The Sun is invisible at the time of the second observation. The actual length of the arc star-Sun at the moment of the first observation is not simply $a_1 + a_2$. Rather, it is $a_1 + a_2 + 1/2°$. The star's longitude, as deduced from the raw measurements, must be increased by the Moon's motion in the interval between the two observations.

Correction for the Change in the Moon's Parallax   The second effect for which a correction must be made is the change in the Moon's parallax between the first observation and the second. Before discussing this correction, however, we must first say a little about the lunar parallax in general.

Because of the Moon's proximity to us, an observer on the surface of the Earth will not generally see the Moon at the same position among the stars as would an observer at the center of Earth. The effect of parallax is always to make the Moon look lower in the sky than it would be if seen from the center of the Earth, as shown in figure 6.5. An observer at O would see the Moon M below the (infinitely distant) reference star T, but an observer at the Earth's center E would see the Moon above the star. The Moon's zenith distance z as measured at O is greater than its zenith distance z' measured at E. The amount by which these angles differ is the Moon's parallax $P = z - z'$.

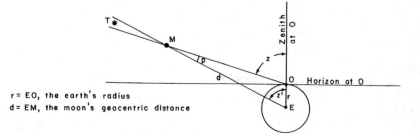

FIGURE 6.5. Parallax. The Moon's parallax is angle p. $r = EO$, the earth's radius; $d = EM$, the moon's geocentric distance.

$r = EO$, the earth's radius
$d = EM$, the moon's geocentric distance

Applying the law of sines to triangle EMO, we have

$$\frac{\sin P}{r} = \frac{\sin(180 - z)}{d}.$$

So,

$$P = \sin^{-1}\left(\frac{r \sin z}{d}\right).$$

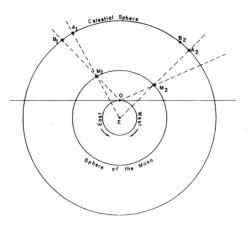

FIGURE 6.6.

Note that if the Moon is at the zenith ($z = 0°$), its parallax is zero: in this case observers at O and at E would see the Moon in the same direction. The largest possible value of the parallax, which obtains when the Moon is at the horizon ($z = 90°$), is called the *horizontal parallax*. (Horizontal parallax was introduced in sec. 1.17 and is illustrated in fig. 1.45.) The actual parallax in any situation is less than or equal to the horizontal parallax. In almanacs one often finds the Moon's horizontal parallax tabulated rather than its distance d from the Earth. These two pieces of information are equivalent, since the one is easily calculated from the other. The Moon's horizontal parallax is on the order of 1°. It varies somewhat in the course of a month, as the Moon's distance varies between about 56 and 64 Earth radii.

Because of the Moon's parallax, it is convenient to distinguish between the Moon's *apparent place* and its *true place*. In figure 6.6, E and O represent observers at the Earth's center and on the Earth's equator, respectively. The observer at E sees the Moon M_1 against the star labeled A_1, but observer O sees the Moon lower and farther east in the sky, against star B_1. We say that at this moment the Moon's apparent place on the celestial sphere (as seen from O) is farther east than its true place (as seen from E).

With this background established, we are now prepared to discuss the correction for lunar parallax in Ptolemy's method of measuring star longitudes. In this method—as opposed to the eclipse method—the Moon's parallax does not directly enter. After all, it makes no difference whether we observe the Moon's true place or its apparent place. We are only using the Moon as a convenient marker in the sky. The rule

arc from = arc from + arc from
Sun to star   Sun to Moon   Moon to star

is perfectly correct whether the true or the apparent Moon is used, provided that the two arcs involving the Moon are measured at the same instant, that is, provided that the Moon's place on the celestial sphere is the same for each measurement. The parallax would be irrelevant if it remained the same for the hour or so between the first observation and the second. The problem, of course, is that it does not.

Let us return to figure 6.6 and suppose that a few hours have elapsed since the Moon and stars were at their first position. Now the stars have been carried by the diurnal rotation to their new positions A_2 and B_2, and the Moon has come to M_2. Here we ignore the Moon's motion on the ecliptic and consider only the effect of the diurnal rotation of the heavens. At the second moment, as well as the first, the Moon lies on the line between E and A. An observer at the Earth's center would detect no parallactic shift in the Moon's position among the stars. But the real observer at O does see a shift: when the Moon was at M_1, it was seen east of star A, but when the Moon comes to M_2 it is seen west of A. In the course of its transit through the sky, the Moon suffers an apparent, continuous westward motion through the stars

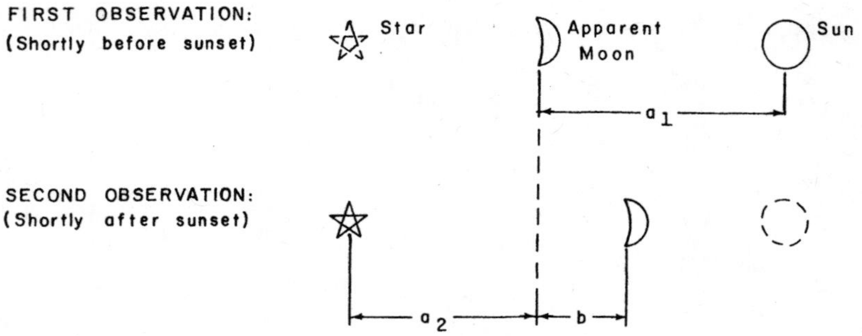

FIRST OBSERVATION:
(Shortly before sunset)

SECOND OBSERVATION:
(Shortly after sunset)

FIGURE 6.7.

due to the continuously changing parallax. This apparent westward motion is in addition to, and smaller than, the genuine eastward motion on the ecliptic.

Suppose the three bodies that we are to sight with the armillary sphere are arranged in the sky in the order (going from east to west) star, Moon, Sun, as in figure 6.7. Suppose we measure the arc Moon-Sun and get the value a_1, as in the top half of figure 6.7. The star is drawn in broken line because it is invisible at the time of the first observation. Suppose that after sunset we measure the longitudinal arc between the Moon and the star and obtain the value a_2, as in the bottom half of the figure. The Sun is drawn in broken line here because it is out of sight at the time of the second observation. We will not get the right answer for the arc Sun-star if we simply add a_1 and a_2, because the apparent Moon has shifted westward during the time between the two observations. The correct value for the arc Sun-star is $a_1 + a_2 - b$, where b is the change in the Moon's longitudinal parallax between the one observation and the next. Note that in this figure we have ignored the genuine motion on the ecliptic of the Sun and the Moon: arc b is due entirely to the change in the Moon's parallax.

Now, what is the size of b? A rigorous calculation would be complicated. But we can get a rough idea of the size of b from the following simple considerations. Suppose our observation station is at the equator and that the Moon is on the celestial equator. At moonrise, the Moon's parallax will be the full value of the horizontal parallax, about 1°. That is, the Moon's apparent place will be 1° farther east than its true place. About six hours later, the Moon will reach the zenith and the parallax will be zero. Finally, about twelve hours after its rising, the Moon will set. At this time the apparent place will be 1° west of its true place. Evidently, the apparent Moon will suffer a 2° parallactic shift toward the west during its twelve-hour transit of the visible hemisphere. The average rate of the Moon's parallactic motion is therefore 1/6° per hour. Now, in fact this motion is not uniform: it is slower near the horizon and more rapid near the zenith. Even so, the estimate we have made here illustrates the size of the effect: if our two observations are made an hour apart, we will have to subtract something on the order of 1/6° (i.e., 10′ of arc) from the measured Sun-star arc if we expect to get the true angular distance between the star and the Sun.

Example: Ptolemy's Measurement of the Longitude of Regulus   In *Almagest* VII, 2, Ptolemy gives a detailed description of this method in his report of a measurement of the longitude of Regulus.

1. Place and date of the observations: Alexandria, in Year 2 of the reign of Antoninus, on the 9th day of the Egyptian month Pharmouthi (February 23, A.D. 139, Julian Calendar).

2. First observation: As the Sun was about to set, the apparent Moon was sighted 92 1/8° east of the Sun.

3. Second observation: Half an equinoctial hour after the Sun had set, Regulus appeared to be 57 1/6° east of the Moon along the ecliptic.

The above information is all that is necessary for calculating the longitude of Regulus. However, it is first necessary to determine the time of day, so that the Sun's longitude can be calculated from the solar theory.

4. Ptolemy says that the first observation was made (as the Sun was about to set), *when the last section of the Bull was culminating, which was 5 1/2 equinoctial hours after noon.* The reader can easily confirm with a celestial globe that in Alexandria (latitude 30° N) on February 23, the Sun really does set 5 1/2 hours after noon, as the last part of the sign of the Bull is crossing the meridian. The armillary sphere is more than accurate enough for this application, so Ptolemy may have determined the time of day in just this fashion—or he may have used a table of ascensions.

5. Longitude of the Sun: Ptolemy says that at the time of the first observation, the Sun's position was 3 1/20° within the Fishes, that is, at 333°03' of longitude. This figure was taken from his solar tables, using the known date and time of year.

6. The longitude of Regulus is then calculated as follows:

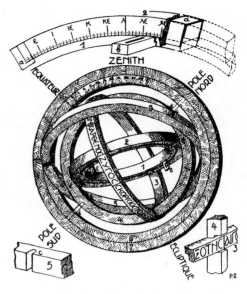

FIGURE 6.8. A modern reconstruction of the armillary sphere described by Ptolemy in *Almagest* V, 1. From Rome (1927). Photo courtesy of the University of Cincinnati Library.

Longitude of Sun at first observation (from tables)		$333\frac{1}{2}°$
Arc from Sun to Moon at first observation	+	$92\frac{1}{8}°$
Arc from Moon to Regulus at second observation	+	$57\frac{1}{6}°$
Plus Moon's eastward motion for $\frac{1}{2}$ hour	+	$\frac{1}{4}°$
Less Moon's westward parallactic motion for $\frac{1}{2}$ hour	−	$\frac{1}{12}°$
Sum		482° 30.5'
or		122° 30.5'

Ptolemy rounds this to 122 1/2°. That is, Regulus "was situated 2 1/2° within the Lion, and was a distance of 32 1/2° from the summer tropic."

The correction of 1/4° for the Moon's eastward motion corresponds to a mean rate of 12° per day, as discussed above. This is a rough value but is adequate for the short time interval involved. As for the correction for the parallactic motion, Ptolemy says nothing of how he arrived at his particular figure. In book V of the *Almagest*, Ptolemy worked out tables for computing the Moon's parallax. However, it is unlikely that the tables were used here. Rather, it looks as if the parallactic shift was estimated offhandedly, according to the scheme we suggested above—by assuming a parallactic motion of 2° per twelve hours. Since the correction is small, only 1/12°, this figure could hardly be wrong by more than 2' or 3'. So, again, Ptolemy's rough estimate is perfectly adequate in the given situation.

Ptolemy's Armillary Sphere

The armillary sphere that Ptolemy used for measuring the positions of the fixed stars is described in *Almagest* V, 1. figure 6.8 is a modern reconstruction based on Ptolemy's description.[9]

The fixed ring 7 is placed in the plane of the meridian. Ring 6 is turned within ring 7 to adjust the instrument for the geographical latitude: axis *dd* must make an angle with the horizon that is equal to the latitude. The inner nest of rings (1, 2, 3, 4, 5) may be turned as a unit about axis *dd*, thus simulating the diurnal rotation. Ring 1, equipped with sights, turns within ring 2 and is used for measuring the latitudes of the stars. Rings 2 and 5 may be turned independently of one another about ecliptic axis *ee*, which pierces the solstitial colure (ring 4). Ring 4 is rigidly attached to the ecliptic ring 3. Rings 3, 2, and 5 are used for measuring the longitudes of the stars.

How to Measure the Longitude of a Fundamental Star   Ptolemy sets up the instrument so that it is adjusted for the correct latitude and is aligned along the meridian.

Next, ring 5 is turned to the longitude of the Sun along ecliptic ring 3. This can be done either by direct observation or by calculating the Sun's longitude from theory. A direct sighting is performed as follows. One turns the inner nest of rings (1, 2, 3, 4, 5) about axis *dd* until the shadow of the ecliptic ring 3 falls on the ecliptic ring itself. This guarantees that the armillary ecliptic is oriented in the same way as the real ecliptic. One then turns ring 5 until the shadow of the upper part of ring 5 falls exactly on the lower part of ring 5. Then ring 5 will indicate the true longitude of the Sun.

Next, without disturbing the positions of the other rings, one turns ring 2 until one can sight the Moon touching its limbs. That is, when one places an eye near ring 2, one sees the Moon touch both the near arc of the ring and the opposite arc of the same ring. Ring 2 then indicates the longitude of the Moon along the scale on ring 3.

Finally, after sunset, one uses rings 2 and 5 in a similar way to measure the longitudinal distance between the Moon and the star that has been selected.

Measuring the Positions of Secondary Stars   The most difficult problem in determining the positions of the stars on the celestial sphere is measuring the *absolute longitude* of one or a few fundamental reference stars. This requires measuring the position of the star with respect to the Sun, using one of the two methods explained above. The procedure is delicate and time-consuming. However, once one has the longitudes of a few reference stars, the positions of all the other stars can be found much more easily, *by measuring their positions with respect to the reference stars.*

Assume that the longitude of one star (the fundamental star) is known. The longitude of some other star (the secondary star) can be measured as follows. One first orients the inner nest of rings of the armillary sphere so that the fundamental star is sighted at its proper longitude on the ecliptic ring 3. To do this, one turns ring 5 on the ecliptic axis *ee* until the ring is set at the proper longitude of the fundamental star along ecliptic ring 3. One then turns the whole inner nest of rings until the fundamental star appears to graze both the near and the far limb of ring 5. The armillary sphere is then, in principle, aligned exactly in the same way as the celestial sphere.

To sight a secondary star, one then turns ring 2 on axis *ee* until the secondary star appears to graze the near and the far sides of that ring. The longitudinal distance between the secondary and the fundamental star may then be read by noting on ring 3 the angular distance that separates rings 2 and 5. Alternatively, one could simply read off the absolute longitude of the secondary star on ring 3.

To measure the latitude of a secondary star, one sights it through the pinnules *bb* on ring 1, which is free to turn within the graduated ring 2.[10]

6.5 EXERCISE: THE LONGITUDE OF SPICA

We shall measure the longitude of Spica using observations of a recent eclipse of the Moon. Our method will be close to the method probably used by Hipparchus.

Figure 6.9 shows the southeastern sky as observed in Spokane, Washington, during the partial lunar eclipse of April, 1977. Figure 6.10 is an enlarged view of the part of the sky contained by the box in broken lines. The box represents a small portion of the sky, roughly what would be covered by one's hand, held at arm's length. The Moon was in the constellation Virgo, some of whose stars are visible. The bright star in the lower left corner of the box is Spica, α Virginis; the other star in the box is θ Virginis. The remaining stars in figure 6.9, beginning with the lowest and going toward the higher altitudes, are κ, ι, ζ, γ, δ, and η Virginis. Compare the figure with a star chart.

The beginning of the eclipse (entry of the Moon into the umbra of the Earth's shadow) occurred on April 3, 1977, at 7:30 P.M. Pacific Time, shortly after sunset. The eclipse ended at 9:06 P.M. Pacific Time. The middle between these two times, 8:18 P.M., is what we shall call the middle of the eclipse. It is the moment for which figures 6.9 and 6.10 are drawn. At this moment, the Moon was covered by the shadow to the maximum extent. This moment, April 3, 20^h18^m Pacific time, corresponds to April 4, 04^h18^m Greenwich mean time, as indicated on the figure.

Now, α, θ, and η Virginis lie roughly on the ecliptic, as a star chart will show. If we use these three stars to roughly sketch in the ecliptic on figure 6.9, two facts become evident. First, the Moon is a little below the ecliptic; this is why the eclipse was only partial. Second, the ecliptic makes a steep angle with the ground at this moment. It is worthwhile to confirm on a globe or armillary sphere that, indeed, this part of the ecliptic rises very steeply at Spokane (latitude 47 2/3° N). This, coupled with the fact that the Moon is at a fairly low altitude, means that it will have a large parallax in longitude, which we shall have to take into account.

The Problem

Use figure 6.9 and figure 6.10 to determine the longitude of Spica in 1977. Use your result, together with Hipparchus's longitude of the same star, to calculate the precession rate. Detailed directions follow.

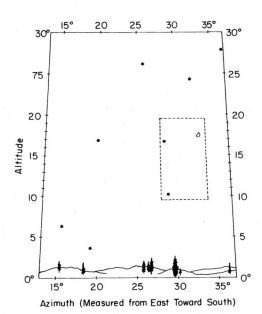

FIGURE 6.9. Partial eclipse of the Moon as observed in Spokane, Washington, April 4, 1977, 04^h18^m Greenwich mean time.

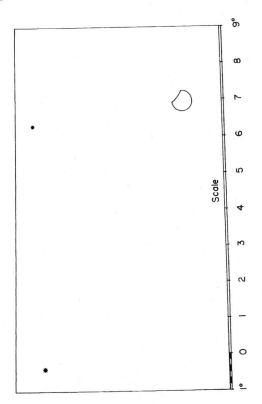

FIGURE 6.10. An enlarged view of a portion of figure 6.9.

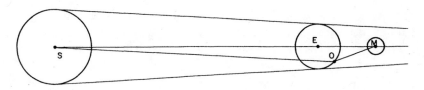

FIGURE 6.11. Sun, Earth, and Moon at the middle of a lunar eclipse.

1. How to take account of the Moon's parallax: figure 6.11 shows the situation at the middle of a lunar eclipse. The centers of the Sun, Earth, and Moon are labeled *S*, *E*, and *M*, respectively. At the middle of the eclipse an observer at the Earth's center *E* would see *S* and *M* in diametrically opposite directions, but for an observer at *O* on the surface of the Earth, this will not be so. Owing to parallax, observers at *E* and at *O* see the Moon in different directions: lines of sight *EM* and *OM* are not parallel. (The Sun, by contrast, is so far away that it has no appreciable parallax: *ES* and *OS* can be regarded as parallel.) Thus, at mideclipse, the Moon's true longitude (as measured at *E*) is equal to 180° plus the Sun's, but its apparent longitude (as measured at *O*) is not.

Now we are in a position to correct figure 6.10 for parallax. The figure shows the Moon among the stars as it was seen from Spokane. We wish now to travel to the center of the Earth, in imagination, from where we will see the Moon in a different position among the stars. A formula for calculating the parallax *P* of a celestial object derived in section 6.4 is

$$P = \sin^{-1}\left(\frac{r \sin z}{d}\right),$$

where *z* is the measured zenith distance of the object, *d* is the distance of the object from the center of the Earth, and *r* is the radius of the Earth. Measure the Moon's altitude using figure 6.9 and calculate its parallax, assuming that the Moon's distance *d* from the center of the Earth was 57 Earth radii. Then, on figure 6.10 shift the Moon vertically upward by the amount of the parallax. Then you will have the Moon's position among the stars as it was seen from the center of the Earth.

2. How to locate the ecliptic on figure 6.10: The latitude of Spica is −2°03′. Draw a circle of this radius with the star as center. (Use the scale of degrees on the right edge of the figure.) The ecliptic must be tangent to this circle somewhere on its upper part.

The latitude of θ Virginis is +1°45′. Draw a circle of this radius with this star as center. The ecliptic must be tangent to this circle somewhere on its lower part. Use the two circles to draw a line representing the ecliptic.

3. Measure the longitudinal arc from the true Moon to Spica: Project Spica and the center of the Moon (as seen from the Earth's center) perpendicularly onto the ecliptic. Measure the longitudinal difference between these projected points.

4. Calculate the Sun's longitude: Use the solar theory to calculate the Sun's longitude at the instant in question.

5. Find the longitude of Spica: At the eclipse middle, the true (not the apparent) longitude of the Moon's center is very nearly 180° different from that of the Sun's. From the result of step 4 you can now get the Moon's true longitude very nearly (within 5′ or 6′). With this and the result of step 3, you can get the longitude of Spica.

6. Calculate the precession: Hipparchus measured the longitude of Spica by a method similar to the one used here and obtained a position about 6° west of the autumnal equinox for an epoch about the year 141 B.C. (Actually, these numbers represent the average of two separate determinations made by Hipparchus using two different lunar eclipses that occurred near Spica at dates

eleven years apart. See *Almagest* III, 1.) Use your longitude for Spica together with Hipparchus's to deduce the precession rate. Express your answer both in terms of seconds of arc per year and in terms of the number of years required for 1° of motion.

6.6 HIPPACHUS AND PTOLEMY ON PRECESSION

Hipparchus's Discovery

In *Almagest* VII, 2, Ptolemy preserves the title of Hipparchus's lost work on precession and a brief summary of Hipparchus's method. All that Ptolemy says very definitely is that Hipparchus calculated that Spica was 6° west of the autumnal equinox in his own time but very nearly 8° in the time of Timocharis, and that Hipparchus came to this conclusion by using lunar eclipses observed carefully by himself and lunar eclipses observed earlier by Timocharis. About Timocharis we know very little. Ptolemy cites some of his observations in the *Almagest*, from which it is clear that Timocharis lived at Alexandria and worked in the 290s and 280s B.C. Except for the famous summer solstice of 432 B.C. (observed at Athens by Meton and Euctemon), Timocharis's are the oldest dated observations in Greek astronomy. Timocharis may be considered the founder of careful and systematic observation among the Greeks.[11]

To calculate the longitude of Spica in bygone times, Hipparchus would have needed an old eclipse record of the following form: on such a day, at such an hour of the night, the Moon was seen eclipsed, and at the middle of the eclipse, the Moon was so many degrees east of Spica. This is just the sort of information likely to have been recorded by Timocharis or any other early observer. Of course, Timocharis, in making the observation, must have thought he was recording a fact that would be useful in working out *a theory of the Moon*. He could not have guessed that Hipparchus, 150 years later, would turn the observation around and deduce from it the motion of Spica with respect to the equinoxes. From the recorded date and time, Hipparchus could calculate the longitude of the Sun. At the eclipse middle, the Moon's longitude could be assumed to be 180° different from this. The recorded distance of Spica from the Moon then gave the longitude of Spica (as discussed in secs. 6.4 and 6.5). Hipparchus needed only to compare the result with his own more recent observations of Spica to see that the star had moved.

Hipparchus's conclusion is preserved by Ptolemy: in another lost work, *On the Length of the Year*, Hipparchus wrote that the solstices and equinoxes shift westward with respect to the stars "not less than 1/100°" in a year. This rate is a bit low, the actual rate being 1° in 72 years. However, it is possible that Hipparchus intended this figure only as a lower bound and that he realized the actual motion might be faster.[12]

In any case, the evaluation of the precession rate was a chancy thing. Hipparchus himself measured the longitude of Spica on two different occasions, only eleven years apart, and got values that differed by 1 1/4°.[13] This shows the size of the possible errors in ancient measurements of absolute star longitudes. Hipparchus's estimate of the shift in Spica's longitude between the time of Timocharis and his own day was 2°, only a little greater than the size of the errors of measurement.

Ptolemy on Precession

Ptolemy, comparing Hipparchus's observations with his own, deduced a precession rate of 1° in 100 years, almost exactly. Ptolemy deduces the precession rate by comparing his own longitude of Regulus (discussed in sec. 6.4) with a similar measurement made earlier by Hipparchus. According to Ptolemy, Hipparchus observed this star in the 50th year of the third Callippic period (129/128 B.C.) and found it to be 29°50' east of the summer solstitial point. Since Ptolemy found Regulus 32°30' east of the same solstice, the star must have shifted eastward by 2°40' during the period between Hipparchus's measurement and his own. The time between Hipparchus's observation and the beginning of Antoninus's reign, when Ptolemy made most of his star observations, was about 265 years. The rate of precession was therefore 2°40'/265 years, or very nearly 1° in a 100 years, which amounts to 36" per year.

Ptolemy goes on to say that he has also measured the longitudes of Spica and other bright ecliptic stars in this same way. Then, from these fundamental stars, he has sighted others also. And he finds their distances with respect to each other very nearly the same as observed by Hipparchus, but their distances with respect to the solstitial and equinoctial points changed by about 2 2/3°.

The Axis of the Precession   It was clear, then, that Regulus, Spica, and several other stars had shifted eastward by more than 2° since the days of Hipparchus. Indeed, it looked as if the whole sphere of the fixed stars was revolving slowly from west to east. But the axis of this revolution was by no means obvious: did it pass through the poles of the ecliptic or through the poles of the equator?

The fact that the motion was easterly (like that of the Sun and planets) and not westerly (like the diurnal revolution) must have suggested to Hipparchus that this was a motion about the poles of the ecliptic. At any rate, according to Ptolemy, Hipparchus finally asserted that the motion was parallel to the ecliptic and supported this claim with observations of Spica made by Timocharis around 280 B.C. and by himself around 130 B.C.: Spica's latitude (2° south of the ecliptic) had remained unchanged, but its distance from the equator had not. But (according to Ptolemy) Hipparchus still had doubts because of inadequacies in Timocharis's observations and because the intervening time had not been long enough.

Ptolemy therefore marshals evidence that the axis of the precession does indeed pass through the poles of the ecliptic. This can be proved most readily from changes in the declinations of the stars. Because the motion is parallel to the *ecliptic*, the stars do shift northward or southward with respect to the *equator*. In particular, all the stars on the half of the sphere containing the vernal equinox, from the winter solstice eastward to the summer solstice, have increases in their declinations. That is, they all move northward with respect to the equator. (This effect is illustrated by fig. 6.1.) Those near the solstices increase their declinations by only a little, and those near the equinox by a considerable amount. To prove his point, Ptolemy quotes a number of declinations of stars in this part of the sky measured by Timocharis, Aristyllos, Hipparchus, and himself. Some of these stars are listed in table 6.1. To help

TABLE 6.1. Declinations from *Almagest* VII, 3 (Stars near the spring equinox)

Stars	Their Longitudes According to Ptolemy	Their Declinations			Changes in Declination (Ptolemy minus Hipparchus)
		Timocharis (c. 270 BC) or Aristyllos*	Hipparchus (c. 128 BC)	Ptolemy (c. 138 AD)	
Altair	♑ 3 5/6°	+ 5°48'	+ 5°48'	+ 5°50'	+0°02'
The Pleiades	♉ 3°	+14°30'	+15°10'	+16°15'	+1°05'
Aldebaran	♉ 12 2/3°	+ 8°45'	+ 9°45'	+11°00'	+1°15'
Bellatrix	♉ 24°	+ 1°12'	+ 1°48'	+ 2°30'	+0°42'
Betelgeuse	♊ 2°	+ 3°50'	+ 4°20'	+ 5°15'	+0°55'
Sirius	♊ 17 2/3°	−16°20'	−16°00'	−15°45'	+0°15'
Pollux	♊ 26 2/3°	+30°00'*	+30°00'	+30°10'*	+0°10'

TABLE 6.2. Declinations from *Almagest* VII, 3 (Stars near fall equinox)

Stars	Their Longitudes According to Ptolemy	Their Declinations			Changes in Declination (Ptolemy minus Hipparchus)
		Timocharis (c. 270 BC)	Hipparchus (c. 128 BC)	Ptolemy (c. 138 AD)	
Regulus	♌ 2 1/2°	+21°20′	+20°40′	+19°50′	−0°50′
Spica	♍ 26 2/3°	+ 1°24′	+ 0°36′	− 0°30′	−1°06′
Arcturus	♍ 27°	+31°30′	+31°00′	+29°50′	−1°10′
α Librae	♎ 18°	− 5°00′	− 5°36′	− 7°10′	−1°34′
Antares	♏ 12 2/3°	−18°20′	−19°00′	−20°15′	−1°15′

the reader visualize the positions of these stars on the celestial sphere in antiquity, we also give their longitudes according to Ptolemy's star catalog. The stars in the signs near the solstices underwent relatively small changes in declination. These are Altair, which is in the sign of the Goat-Horn and therefore near the winter solstice, and Sirius and Pollux, which are in the Twins and therefore near the summer solstice. On the other hand, the Pleiades and Aldebaran (in the sign of the Bull) had large changes in declination.

Similarly, Ptolemy examines a number of stars in the other hemisphere—the one that includes the autumnal equinox and stretches from the summer to the winter solstice. In Table 6.2 we quote a part of his evidence. The stars in this half of the sky all moved south with respect to the equator.

Confirmation of the Precession Rate by Occultations Ptolemy gives the impression that his figure for the precession is based chiefly on the measured longitude of Regulus, and perhaps also of Spica. However, he confirms his result by means of several *occultations* recorded by his predecessors. For example, Menelaus the geometer observed an occultation of Spica by the Moon, in Rome, in year 1 of Trajan, at the end of the tenth hour of the night between the 15th and 16th of Mechir (January 11, A.D. 98, Julian calendar). Using his lunar theory, Ptolemy calculates the Moon's theoretical position at this date, and corrects it for parallax. Then, since the Moon covered Spica, this gives the longitude of Spica at the time of Menelaus's observation (176 1/4°). To get the precession rate, Ptolemy uses an older near-occultation of Spica by the Moon, observed by Timocharis in 283 B.C. Again, he uses his lunar theory to calculate the Moon's theoretical position at the earlier date (172 1/2°). By subtraction, Spica appears to have advanced 3 3/4° during the intervening 379 years, so that the rate is, again, very nearly 1° in 100 years. This result is confirmed by similar pairs of occultations of the Pleiades (Timocharis in 283 B.C. and Agrippa in A.D. 92) and of β Scorpii (Timocharis in 295 B.C. and Menelaus in A.D. 98).

This method has one great strength and one great weakness. The occultation of a star is a very reliable observation, since it does not involve the measurement of an arc in the sky and therefore does not require any kind of instrument. So, the occultations observed by an ancient astronomer like Timocharis potentially afford much greater precision than any other conceivable kind of star observations that might have been made at this early date. Unfortunately, the defects of Ptolemy's lunar theory pretty well wipe out whatever advantage is offered by the observations themselves.

An Unfortunate Consequence Ptolemy's low rate for the precession (1° in 100 years, instead of 1° in 72 years) was little worse than could be expected, given the defects of both his solar theory and his observations. However, this measurement by Ptolemy had some unfortunate and far-reaching consequences for astronomy. Arabic astronomers in the ninth century, trying to reconcile Ptolemy's low value with better and more recent determinations, concluded that the rate of precession must be *variable*. An elaborate machinery was invented to account for this supposed fact of observation, which was not finally dismantled until the sixteenth century. We shall explore this medieval theory of the "trepidation of the equinoxes" in section 6.9.

A Modern Controversy   Few developments in science have so exercised the historians of science as Ptolemy's measurement of the precession rate. At stake is Ptolemy's reputation as an astronomer; at issue are his honesty and reliability as an observer. Did Ptolemy really, as he says he did, make all of the observations upon which his system is based? If so, why is his value for the precession so low? And how is it that several different methods used by Ptolemy all led to exactly the same value for the precession? Perhaps, as Delambre[14] suggested at the beginning of the nineteenth century, Ptolemy's so-called observations are actually examples calculated from tables to better illustrate his theories. Some twentieth-century scholars, notably Robert R. Newton,[15] have gone much farther. Newton claims that all of Ptolemy's "observations" were simply made up as a swindle to prove the validity of his theories. This position leaves unanswered the question of where the theories (which are generally pretty good) could have come from in the first place. In fact, there is no doubt that Ptolemy's theories were all based on observation. However, it is also clear that Ptolemy must have smoothed out or adjusted discordant observations to produce a set of data that was internally consistent. Ptolemy's many independent measurements of the precession rate, which all lead to exactly the same value, are certainly not the straightforward results of unadjusted observations. There is a large modern literature on this question.[16] We return to this issue in the context of Ptolemy's star catalog, in section 6.8.

The Acceptance of Precession in Antiquity

In modern assessments, the discovery of precession is often regarded as Hipparchus's greatest single achievement. This evaluation is colored by the connection between the precession and Isaac Newton's system of the world, as well as by the fact that the theory of precession is virtually the only part of the Greek machinery of the heavens that is still regarded today as "correct." One should keep in mind that the ancients seem not to have invested precession with the same importance. In fact, precession appears never to have become very widely known in antiquity. It is never alluded to by Geminus, Cleomedes, Theon of Smyrna, Manilius, Pliny, Censorinus, Achilles, Chalcidius, Macrobius, or Martianus Capella.[17] Pliny had an unquenchable thirst for astonishing facts, and he professed the highest admiration for Hipparchus. Had Pliny heard of precession he certainly would have mentioned it. The reader already knows Geminus as a competent writer on astronomy familiar with other parts of Hipparchus's work. A discussion, or at least a mention, of precession would have fit naturally into Geminus's lengthy discussion of the zodiac signs—if he ever had heard of it. The only ancient writers who mention precession besides Ptolemy are Proclus, who denies its existence, and Theon of Alexandria, who in his redaction of Ptolemy's *Handy Tables* accepts Ptolemy's value of 1° in 100 years.

6.7 EXERCISE: THE PRECESSION RATE
FROM STAR DECLINATIONS

In his demonstration that the precession is parallel to the ecliptic, Ptolemy uses the declinations of eighteen stars, observed by Timocharis, Aristyllos, Hipparchus, and himself. Some of these data are displayed in tables 6.1 and 6.2. Ptolemy also uses the changes in the declinations of six of these stars to confirm his numerical value for the precession rate. For these six, Ptolemy asserts, without giving details of his calculations, that the observed changes in declination are consistent with his value of the precession constant (1° per 100 years, or 36″ per year). We shall see how the precession rate can be deduced from the changes in the stars' declinations.

Method   The method is simple for a star near the ecliptic. Take Spica as an example. According to table 6.2, Hipparchus measured the declination of Spica in 128 B.C. and found +0°36′; 265 years later, Ptolemy measured the declination of the same star and found −0°30′. Thus, the star had shifted south with respect to the equator by 66′. The question is, how large a displacement along the ecliptic is required to bring about such a change in the declination of Spica?

Since Spica lies nearly on the ecliptic, we can use the table of obliquity (table 2.3) to find out. In Ptolemy's day the longitude of Spica was Virgin 26°40′. (This value, printed in table 6.2, comes from the star catalog in the *Almagest*.) From table 2.3, (the table of obliquity) we take out the declinations of the two ecliptic points Virgin 25° and Virgin 26°:

Longitude	Declination	Difference
Virgin 25°	1°59′	
		−24′ = −0.40°
Virgin 26°	1°35′	

Thus, if a star exactly on the ecliptic traveled from Virgin 25° to Virgin 26°, its declination would diminish by 0.40°.

Now we compute the ratio

1° of longitude/0.40° of declination = 2.50° of longitude/° of declination.

We can compute the precession between Hipparchus's day and Ptolemy's by multiplying this ratio by the observed 66′ change in the declination of Spica:

$$\frac{2.50° \text{ of long.}}{1° \text{ of dec.}} \times \frac{66°}{60} \text{ change in declination } = 2.75° \text{ change in longitude}$$

That is, the observed 66′ change in Spica's declination implies a precession of 2.75° between 128 B.C. and A.D. 138. The precession rate is therefore

$$2.75°/265 \text{ years} = 0.0104°/\text{year} = 37″/\text{year},$$

which is very close to Ptolemy's adopted rate of 36″ per year.

To be perfectly consistent, we should have used Ptolemy's table of obliquity (*Almagest* I, 15), which is based on a slightly different value for the obliquity of the ecliptic than is our own table. However, this would make only a tiny difference in the results.

Problem

Use this method to compute the precession rate from the data given in tables 6.1 and 6.2 for the following five stars: Aldebaran, Pollux, Regulus, α Librae, Antares. These stars are near enough to the ecliptic to justify the direct use of the table of obliquity (table 2.3) as in our Spica example. These five stars are *not* among the six that Ptolemy used to confirm his precession rate.

What value for the precession rate do you get if you average the results for these five? Why do you suppose Ptolemy left them out of his account?

6.8 THE CATALOG OF STARS

Ptolemy's Star Catalog

In *Almagest* VII and VIII, Ptolemy presents a catalog of slightly more than one thousand stars. Figure 6.12 is a photograph of the first page of the star catalog in a ninth-century parchment copy of the *Almagest*. The first part of this page may be translated thus:

Tabular layout of the constellations in the northern hemisphere

Constellation of the Little Bear	Degrees of longitude	Degrees of latitude	Magnitude
The star at the end of the tail	Twins 0 1/6	North 66	3
The one next to it on the tail	Twins 2 1/2	North 70	4
The one next to that, before the place where the tail attaches	Twins 10 1/6	North 74 1/3	4
The southernmost of the stars in the leading side of the quadrilateral	Twins 29 2/3	North 75 2/3	4
The northernmost in the same side	Crab 3 2/3	North 77 2/3	4
The southern star in the following side	Crab 17 1/2	North 72 5/6	2
The northern one in the same side	Crab 26 1/6	North 74 5/6	2

For each star, Ptolemy gives a descriptive identification of the star, then the star's longitude, latitude, and magnitude. For example, the first star, the one on the end of the tail of the Little Bear (the star we call Polaris), has a longitude of 1/6° within the Twins. Its latitude is 66° north of the ecliptic. Ptolemy divides the stars into six magnitude (or size) groups. The largest stars are magnitude 1. The faintest visible stars are magnitude 6. Today, we distinguish a star's brightness from its size. But it is a property of human vision that brighter stars do appear larger to us. Ptolemy thought of his magnitudes as measures of the stars' relative sizes. This view was not altered until the seventeenth century.

Several features of the ninth-century manuscript (fig. 6.12) are worthy of notice. The Greeks used the letters of their alphabet as numerals. (For the correspondences, see sec. 4.5.) Fractions were sometimes signaled by means of extra strokes over the numerals. Thus, $\Gamma = 3$ but $\acute{\Gamma} = 1/3$. So, for example, the longitude of the last star listed in figure 6.12 is καρ ΙΓ $\acute{\Gamma}$, that is, Crab 13 1/3°. This ninth-century manuscript also uses some special symbols for common fractions. In the longitude of the fourth star, we see Γ (the symbol for 1/3) with a dot under it, which represents 2/3. The symbol rather like *L* is the common Byzantine sign for 1/2. This occurs, for example, in the longitude of star 10. Note that this system of indicating fractions works best for *unit fractions*, that is, for fractions with a unit numerator—1/3, 1/6, and so on. There is a special symbol for 2/3, but other nonunit fractions must be indicated by addition. Thus, the fractional part of the longitude of star 10 is *L* $\acute{\Gamma}$, that is, 1/2 + 1/3, or, as we would write it, 5/6.

Also, note that there are no spaces between the words. Ancient Greek was written in imitation of speech, as a continuous stream of sounds. Finally, note the occurrence of zerolike symbols, such as in the longitude of the first star in the table: zero degrees and 1/6° within the Twins. The symbol for zero is o, the first letter of the word *ouden*, "nothing."

Ptolemy's catalog, with its thousand-odd stars, represented a fair amount of work. For the next thousand years, this was the standard star catalog everywhere in the Greek-Arabic-Latin tradition. Later astronomers who needed the positions of stars had simply to increase all the longitudes equally for the precession between Ptolemy's time and their own. The latitudes were believed to remain forever the same—which was, indeed, pretty close to the truth.

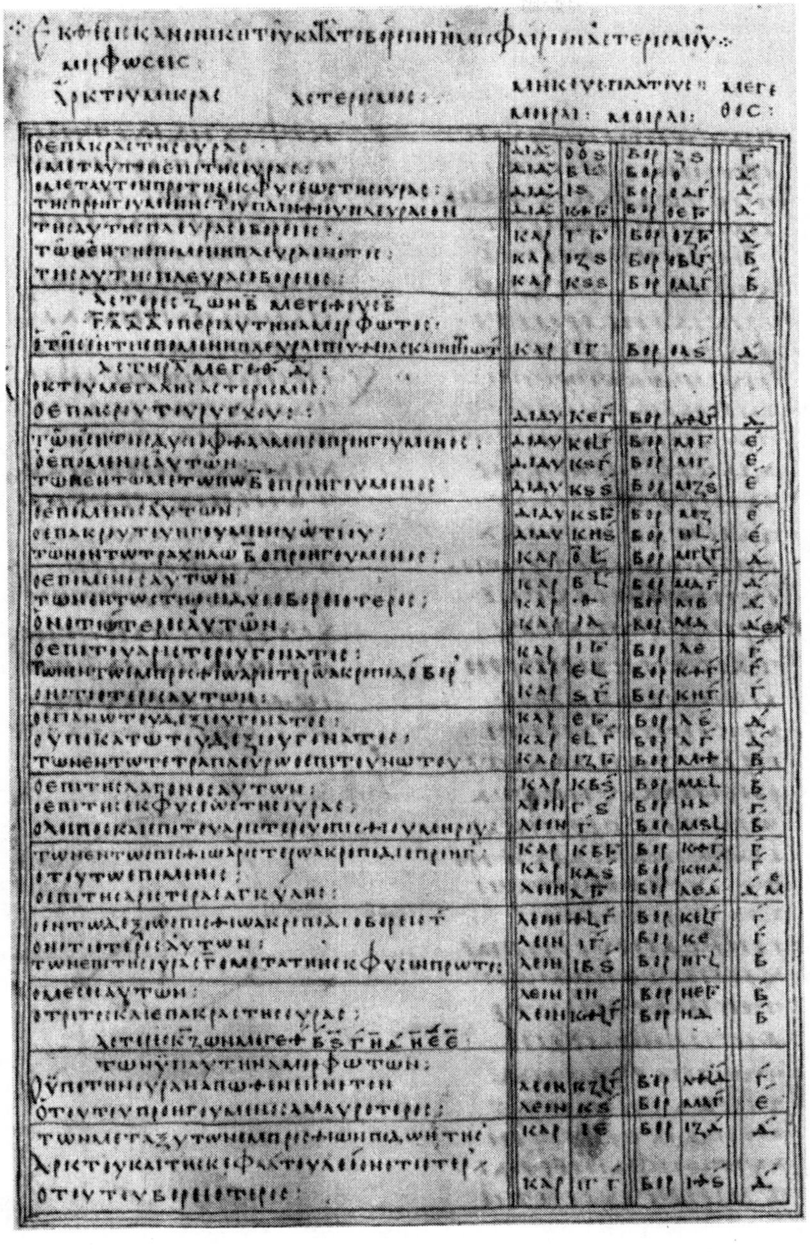

FIGURE 6.12. The first page of the star catalog in a ninth-century parchment copy of Ptolemy's *Almagest*. Bibliothèque Nationale, Paris (MS. Grec 2389, fol. 215v).

Thus, even Copernicus's catalog of stars in *De revolutionibus* (1543) is just Ptolemy's, with the longitudes all shifted by a constant increment. There are, to be sure, examples of Greek star observations that precede Ptolemy. There are also many examples of medieval observations of the stars. But Ptolemy's was the only systematic catalog covering a large number of stars until the fifteenth century. The first substantial catalogs independent of it were those of Ulugh Beg (fifteenth century) and Tycho Brahe (sixteenth century).

A Modern Controversy in the Form of a Detective Story

A dispute has raged around origin of the star catalog in the *Almagest*. Did Ptolemy, as he says he did, really compile the catalog himself? Or was Ptolemy a mere textbook writer who simply borrowed a previously existing catalog compiled by Hipparchus?

Ptolemy Discredited   In the nineteenth century, it was a passion among historians of ancient Greek thought to discover the *precursors* of scientific and philosphical figures. A common presumption was that most well-known figures of Hellenistic philosophy and science had merely developed the ideas of earlier thinkers whose works are now lost. The game, then, was to reconstruct these lost works by attributing to them various passages mined from the surviving works of later writers. It was in keeping with nineteenth-century attitudes that Ptolemy should be rated low and Hipparchus high. The prevailing opinion was that Ptolemy, whose works we possess nearly in their entirety, was not an original thinker. Rather, he compiled and systematized the works of his predecessors. In the view of many nineteenth-century authorities, the truly creative figure of Greek astronomy was Hipparchus, about whom we know very little. In particular, the most influential historians of astronomy held that the star catalog of the *Almagest* was the work of Hipparchus and not of Ptolemy.

What sort of evidence could be produced in support of this view? The star longitudes recorded in the catalog suffer from a systematic error that makes them about a degree too small, on the average. We can explain this error if we suppose that Ptolemy simply took the coordinates over from an earlier catalog by Hipparchus and added $2°40'$ to all the longitudes to account for the precession between Hipparchus's day and his own. The argument is short and sweet: Hipparchus's longitudes were accurately measured around the year 130 B.C. Ptolemy wanted to bring these longitudes down to his own epoch, A.D. 137, which was 266 years later. Since he believed that the precession rate was $1°$ in 100 years, or $36''$ per year, he added to all the longitudes the value $36'' \times 266 = 2°40'$. However, the true precession rate is $50''$ per year, so the true motion of the stars in this interval was $50'' \times 266 = 3°42'$. Ptolemy's adopted longitudes for the year A.D. 137 were therefore too small by $3°42' - 2°40' = 1°02'$. Thus, the $1°$ systematic error in the star catalog was explained in terms of Ptolemy's low value for the precession rate.

Throughout the ninteenth century the prevailing opinion was that Ptolemy's catalog had originated in exactly this way. The case could have been proved only by direct comparison of Hipparchus's catalog with Ptolemy's. Unfortunately, this was impossible, as Hipparchus's catalog (assuming he made one) has not come down to us. Early in our own century, Peters and Knobel made an exhaustive study of Ptolemy's catalog and held to the conclusion that Ptolemy had merely added $2°40'$ to Hipparchus's longitudes.[18] The case seemed to be settled beyond reasonable doubt.

Ptolemy Rehabilitated   However, the case proved to be not as simple as it seemed. In 1901, Franz Boll[19] had demonstrated that, while Ptolemy's catalog in the *Almagest* gave the coordinates of 1025 stars, Hipparchus's catalog could have contained at most 850. Boll's argument was based on recently rediscovered Greek manuscripts that listed the constellations according to Hipparchus. For example, one of the manuscripts, dated to the fourteenth century, carried the title *From the Stars of Hipparchus*. These manuscripts did not contain any star coordinates, but merely listed the names of the constellations, together with the number of stars in each, for example, "The Great Bear, 24 stars; The Little Bear 7; Draco between the Bears 15; Boötes 19," and so on. These lists seemed to give a way of comparing Ptolemy's catalog with the lost catalog of Hipparchus, not star by star, but at least constellation by constellation. There were a few minor difficulties. A couple of constellations were missing from the list, and in the cases of three other constellations the number of stars was not given. Nevertheless, the evidence seemed to point to the conclusion already mentioned, that Hipparchus's catalog could not have listed more than about 850 stars and that, if Ptolemy plagiarized it, he must have added at least 175 stars of his own.

In 1917–1918, J. L. E. Dreyer[20] published two articles on the origin of Ptolemy's star catalog, in which he considered several possible causes of the systematic 1° error in longitude. Dreyer argued that the major source of error was a defect in Ptolemy's solar theory, which made the Sun's mean longitude too small by about a degree for the epoch A.D. 137. This defect of the solar theory arose from Ptolemy's error of about one day in his observed time of the autumnal equinox of the year A.D. 132—the equinox that Ptolemy used to establish the epoch value of the Sun's mean longitude. As we have seen, the solar theory is fundamental to the measurement of star longitudes. A star's longitude is obtained by measuring, in several steps, the longitudinal arc between the Sun and the star and adding to this the absolute longitude of the Sun *as calculated from the solar theory*. Any systematic error in the Sun's longitude will therefore be transmitted to the longitudes of the stars. Thus, if Ptolemy measured the positions of the stars exactly as he says he did, we would, indeed, expect his star longitudes to be about 1° too small. Dreyer concluded that there was no reason to disbelieve Ptolemy's statement that he had made extensive observations of the fixed stars.

The final step in Ptolemy's rehabilitation was taken by Heinrich Vogt[21] in an article published in 1925. Although Hipparchus's catalog has not survived, we do possess a large number of star coordinates measured by him. By far the largest source of these is his *Commentary on the Phenomena of Aratus and Eudoxus*—the only one of Hipparchus's works that has come down to us. This *Commentary* contains 859 numerical data on some 374 stars. An additional 22 data attributed to Hipparchus are preserved by Ptolemy and Strabo. Curiously, only 2 of these 881 data are longitudes, and there are no latitudes at all. Most of the data consist of the longitudes of the ecliptic points that rise, set, or culminate simultaneously with the given stars. For example, Hipparchus says that the head of the Little Bear reaches the meridian at the same time as does the 3rd degree of the Archer, and at the same time as the 17th degree of the Water-Pourer rises. All the risings and settings refer to the clime of 14 1/2 hours. A much smaller but still substantial body of data consists of equatorial coordinates, that is, declinations and right ascensions. In the case of 122 stars, the available information was equivalent to two coordinates, and Vogt used these to compute longitudes and latitudes, which he then compared to the longitudes and latitudes in Ptolemy's catalog. If Ptolemy had plagiarized Hipparchus's data, then his latitudes should all be the same as the latitudes computed for Hipparchus, while the longitudes should all show a constant difference. The comparison demonstrated that Ptolemy's coordinates could not have been derived from Hipparchus's in this fashion. For several decades after the publication of Vogt's article, the prevailing opinion was that the star catalog in the *Almagest* was compiled by Ptolemy himself and that he did exactly what he says he did—observed as many stars as he could, down to those of the sixth magnitude.

Ptolemy Reindicted   In the writing of history, there is no rule against double jeopardy. In the 1970s, Ptolemy was reindicted for plagiarism and theft. Robert R. Newton was a geophysicist studying the secular accelerations of the Earth and Moon. These slow accelerations show up in the gradual increase in the length of the day and in the Moon's slow recession from the Earth, which produces a gradual increase in the length of the month. The increases in the length of the day and of the month are best studied by means of eclipse observations. Modern astronomical observations that are precise enough for such delicate studies really only go back to the eighteenth century. But Newton hoped to use ancient Greek observations (mostly in the *Almagest*) to extend his study. The greater time span would more than compensate for the lower precision of the ancient observations. But Newton discovered that observations in the *Almagest* were not usable. Moreover, he strongly suspected that they had been doctored, or even made up, by Ptolemy to agree with Ptolemy's own theories. Newton began a long study of all the observations cited in the

Almagest—not just eclipses. The longer he worked, the angrier he got, and he finished by being a bitter personal enemy of Ptolemy—a man who had been dead for nearly two thousand years. Newton's conclusions were published in 1977 in a book called *The Crime of Claudius Ptolemy*. Newton alleged that Ptolemy was liar, a plagiarist, and a thief and that every single observation cited and used by Ptolemy in the *Almagest* was fabricated. Newton concluded that astronomy would have been much better off if Ptolemy had never lived.

Newton's book produced a sensation. His methods and conclusions were attacked by historians of ancient astronomy and the debate sometimes spilled over from scholarly journals into the popular press. While few historians have accepted Newton's conclusions, his book did stimulate a lot of new work, reexamining Ptolemy's methods of observation and the relationship between theory and observation in Greek astronomy.

In particular, Newton brought forward new evidence that Ptolemy had stolen Hipparchus's star catalog and updated the longitudes by 2°40′ to account for the precession between Hipparchus's time and his own. If we look at the extract from the star catalog at the beginning of this section, we see that the star coordinates typically involve a whole number of degrees, plus a fraction of a degree. The most commonly occurring fractions are 1/6, 1/3, 1/2, 2/3, 5/6, and, of course, zero (which corresponds to a whole number of degrees). Expressed in terms of minutes of angle, these fractions are 0, 10, 20, 30, 40, and 50. Newton had the clever idea of simply counting up the number of times that each fraction occurs in the catalog.[22]

In the case of the *latitudes*, the most frequently occurring fraction is 0′ and the next most frequent is 30′. This is what we might expect to result if Ptolemy observed the stars with an armillary sphere divided merely to whole degrees, or perhaps to half degrees. The other fractions, presumably the results of interpolation when the stars fell between the marks, appear a good deal less often.

In the case of the *longitudes*, the situation is quite different. The most commonly occurring fraction is 2/3°, that is, 40′, which seems strange. Newton argued that this proved Ptolemy's theft of Hipparchus's catalog. The basic argument is very simple: when Hipparchus observed the stars' longitudes, the most frequently occurring fraction would have been 0′. Ptolemy added 2°40′ to all the longitudes, which shifted the peak in the distribution to 40′. Thus, the odd distribution of the fractional degrees is neatly explained.

Newton's argument convinced a number of historians. It was soon reinforced by Dennis Rawlins.[23] Borrowing an idea from Delambre,[24] Rawlins noted that the southernmost stars of the catalog crossed the meridan in Alexandria about 6° above the horizon. Now, Rawlins argued that the ancient observer—whoever he was—must have tried to observe stars right down to his southern horizon. Using heavy statistical machinery, Rawlins calculated the probability that an observer, trying to observe right down to the ground, wound up, by chance, with a star catalog that stopped 6° short of the ground. Of course, the chances came out fantastically small. Like Delambre before him, Rawlins pointed out that at Rhodes, some 5° north of Alexandria, the southernmost stars of the catalog would have crossed the meridian only 1° above the horizon. Applying a similar analysis to this situation, Rawlins found that the southern boundary of the catalog appeared to be statistically consistent with the catalog having been compiled at Rhodes. Since Rhodes was Hipparchus's presumed place of observation, Rawlins argued that his results supported Hipparchus's authorship of the catalog and made Ptolemy's impossible.

Testing the Argument Based on Fractional Degrees   One way to tell how strong these arguments are is to apply them to other star catalogs about which we know more and to see whether the arguments lead to sensible conclusions in these test cases. The first two star catalogs of substantial length in the Western tradition that are independent of Ptolemy's are those of Ulugh Beg and Tycho Brahe.

Ulugh Beg (1394–1449), the grandson of Tamerlane, ruled Maverannakr from its capital city of Samarkand from 1409 until his death by assassination. Ulugh Beg was a patron of the sciences and especially of astronomy. At his direction, an observatory was constructed at Samarkand and equipped with instruments. The astronomers in his employ undertook a program of observation and the construction of astronomical tables. The result of this activity was a book called the *Zīj* of Ulugh Beg, or the *Zīj-i Gurgāni*. A good way to give an idea of what a typical *zīj* contains is to say that it continues the tradition of the *Handy Tables* of Ptolemy and Theon of Alexandria. Indeed, Ulugh Beg's *Zīj* is a more or less complete manual of astronomy, with trigonometric and astronomical tables and instructions for their use—as well as a star catalog.[25]

Let us apply Robert R. Newton's idea of examining the fractional degrees in the longitudes and latitudes. If we examine the *latitudes* in Ulugh Beg's star catalog, we find that the fractions that occur are all multiples of 3'—that is, 0, 3, 6, 9, . . . , 51, 54, 57. The most commonly occurring fraction is 0 but there are also strong peaks at 15', 30', and 45'. The other fractions occur much less often. This is just what we might expect if Ulugh Beg's astronomers used a zodiacal armillary sphere graduated to quarter-degrees.

When we examine the *longitudes* in Ulugh Beg's star catalog, we discover an anomaly. The minutes of a degree that occur in the longitudes are not 0, 3, 6, 9, and so on, but rather 1, 4, 7, 10, and so on, in steps of 3', up to 58. Clearly, these are not the direct results of measurement. This conclusion is reinforced when we count up how often each fraction occurs. The most frequently occurring fraction is 55', and the other peaks are at 10, 25, and 40. As before, the peaks are separated by 1/4° intervals. But the shifted locations of the peaks show us again that these longitudes cannot be the direct results of measurement. Rather, it looks as if 55' plus a whole number of degrees has been added to all the longitudes. (Alternatively, 5' plus a whole number of degrees may have been subtracted.) This is how the major peak, which should have been at 0', wound up at 55'.

Were we to leap to Newton's conclusion regarding the star catalog of the *Almagest*, we would say that, contrary to Ulugh Beg's express statement, the stars were not observed by him or his astronomers and that the catalog was plagiarized from an earlier, lost version by the addition of a constant to the longitudes. But in this case there is no known catalog that might have been the source of Ulugh Beg's coordinates. All earlier catalogs are simply versions of Ptolemy's—which Ulugh Beg's certainly is not. Moreover, we know a fair amount about the observatory at Samarkand—the date of its construction, the names of the principal astronomers, some of the instruments employed, and so on—enough that it is impossible to doubt the reality of the enterprise there.[26] Archaeological excavation has uncovered the remains of the observatory itself, including the great segment of a circle used for taking noon altitudes of the Sun.

Thus, we must conclude that Newton's argument based on the fractional degrees in the star coordinates leads to an incorrect conclusion when it is applied to the test case of Ulugh Beg's catalog. This does not prove, of course, that the argument is wrong when applied to Ptolemy's catalog, but it does show that we have to ponder it with care. In particular, we will need to inquire whether there might be other possible causes for a shift in the peak of the distribution of fractional degrees.

In the case of Ulugh Beg's catalog, the most likely explanation of the shift in the peak of the longitude fractions is a change of epoch. The canons accompanying Ulugh Beg's *Zīj* say that the star catalog is set down for year 841 of the Hegira (A.D. 1437). We know also that the construction of the observatory began in A.H. 823 or 824 and that astronomical work was carried on for thirty years, until the assassination of Ulugh Beg, and even for a few years after. If the bulk of the star places were observed about A.D. 1443, say, then it would have been necessary to subract six tropical years of precession from the longitudes to reduce them to the adopted epoch of A.D. 1437. Given Ulugh Beg's precession of 1° in 70 tropical years, that amounts to 5'. If 5' were subtracted from all the measured longitudes, this would, indeed, produce the distribution that we find in the star catalog.

It is unlikely that this same process could explain the distribution of fractional degrees in Ptolemy's longitudes, for the peak would have to be displaced by adding 40' or by subtracting 20'. At Ptolemy's precession rate (1° in 100 years), these amount to corrections for 66 or for 33 years. It is unlikely that Ptolemy's own observations of the stars extended over so long a period.

Possible Origins of the Fractional Longitudes of 2/3°   There are several ways in which 40' could have become the most frequently occurring fraction in Ptolemy's longitudes. It could have happened just as Newton claimed, by Ptolemy stealing Hipparchus's catalog, adding 2°40' to all of the longitudes, and then claiming the catalog as his own.

But 40' could also become the most commonly occurring fraction if Ptolemy's fundamental star simply happened to have a longitude of a whole number of degrees plus 40'. Let us recall that the Ptolemy made a measurement of the absolute longitudes of only a small number of fundamental stars, by the laborious method described in section 6.4. For each of the other thousand stars in the catalog, he probably measured the longitudinal distance of the star from a fundamental star, then added the absolute longitude of the fundamental star. If the longitude of the fundamental star contained the fraction 40', we would naturally expect this to be the most commonly occurring fraction in the star longitudes.

Is there any way to tell which star (or stars) Ptolemy used as fundamental? One way is to examine closely all of Ptolemy's observations of the planets' positions. If we examine all the planet observations that Ptolemy made with the armillary sphere and note which reference stars were used for orienting the instrument, we find that only four stars were ever used as fundamental:

Star	Almagest longitude
Aldebaran	42°40'
Regulus	122°30'
Spica	176°40'
Antares	222°40'

These four bright ecliptic stars are just the ones we might expect him to have used. It is noteworthy that three of the four have a fractional longitude of 2/3°. This being the case, it is perhaps not surprising that 2/3° is the most common fraction in Ptolemy's catalog.

There is another plausible way in which 2/3° might have become the most commonly occurring fraction in Ptolemy's longitudes. He might, indeed, have added 2°40' to all of the longitudes—but to longitudes that he himself had measured. Suppose Ptolemy simply adopted, as a working hypothesis, Hipparchus's longitude for some fundamental reference star. Ptolemy then measured the longitudes of all the other stars using this fundamental star for orienting the armillary sphere. The result would have been a catalog compiled by Ptolemy, but with longitudes valid for the time of Hipparchus. To adjust the absolute longitudes for his own epoch, Ptolemy would then have added the amount of precession between Hipparchus's time and his own, which he took to be 2°40′.

Moreover, there are some reasons why Ptolemy might have proceeded in such a fashion. Most importantly, Ptolemy probably had not yet worked out the final version of his solar theory when he measured the coordinates of the stars for his catalog. We reach this conclusion by looking at the dates of some of Ptolemy's activities. His own observations of the stars were probably made around A.D. 137 (beginning of the reign of Antoninus). But the latest observation of the Sun, used by Ptolemy in deriving the elements of his solar theory, was of the summer solstice of A.D. 140. Thus, Ptolemy could not have been in a position to use his own solar theory to fix the absolute longitudes of the stars. Might he not have simply adopted Hipparchus's values for the longitudes of one or more reference stars so that his work could proceed?

There is good reason to suppose that Ptolemy may have worked in just this way. Ptolemy's earliest planet observation in the *Almagest* is a position of Mercury in relation to Aldebaran taken in A.D. 132. Similar measurements by a certain Theon, who apparently was an acquaintance of Ptolemy's at Alexandria, go back to A.D. 127. Thus, five or ten years before the bulk of the star places were measured, at least a few stars were being used by Ptolemy and Theon for position measurement of the planets. Again, this requires either that Ptolemy simply adopted the longitudes of one or a few reference stars as handed down by his predecessors, or that he measured at least a few star longitudes himself long before completing his solar observations (perhaps using Hipparchus's solar theory on a provisional basis).

If Ptolemy accepted the longitude of Spica, say, or of Regulus, as derived from Hipparchus's observations, he would then have been free to continue with both stellar and planetary work. A possibility, then, is that he did just that and that subsequently he applied a correction of 2°40′ to all his star longitudes.

As we have seen, Robert R. Newton's argument based on the distribution of the fractional degrees in the longitudes breaks down when it is applied to the test case of Ulugh Beg's catalog. When plausible explanations of the distribution of the fractional degrees are available that are less drastic than imputations of plagiarism and deception, some temperance of judgment may be called for.

Testing the Argument Based on the Southern Limit of the Catalog The best case for testing Rawlins's argument about the southern limit of the star catalog is provided by Tycho Brahe's catalog of stars. Brahe's catalog was based on observations made in the 1580s and 1590s at his observatory on the island of Hven (latitude 55.9°), just north of Copenhagen. There are two different versions of the catalog. The earlier version, containing 777 stars, was printed in Brahe's *Progymnasmata*. The *Progymnasmata* was not published until 1602, after Brahe's death, but the bulk of the observations for the 777-star catalog were made before the end of 1592. In 1595, the observations of the fixed stars were resumed, bringing the total number up to 1,000, so that Brahe should not be inferior to Ptolemy. The 1000 stars were completed in great haste at the beginning of 1597, immediately before Brahe's departure from Hven. According to Dreyer, the quality of the later star places is greatly inferior to that of the original 777. Because the *Progymnasmata* was still not complete, Brahe decided to circulate a limited number of manuscript copies of the 1,000-star catalog, in 1598. The 1,000-star catalog was first published by Kepler in the *Tabulae Rudolphinae* of 1627.[27] Both versions of the catalog are available in Dreyer's edition of Brahe's *Opera*.[28]

Let us recall that the southernmost stars of the *Almagest* catalog culminate about 6° above the horizon. When we examine Brahe's 777-star catalog, we find something quite similar. At Hven, only one star in the catalog was at an altitude lower than 4.3°. This was Fomalhaut (α Piscis Austrini), at an altitude of 2.6°. Fomalhaut is a first-magnitude star. In the *Almagest*, Fomalhaut appears twice: as the first star in Piscis Austrinus ("on the mouth" of the fish), and as the last and most southerly star of Aquarius ("at the end of the water"). Brahe's catalog does not include Piscis Austrinus—most of the constellation was too far south to be visible at Hven. Thus, in Brahe's catalog, Fomalhaut appears only once, as the last and only bright star of Aquarius.

The next most southerly stars in Brahe's 777-star catalog are three stars of Canis Major (α, η, and ε) with altitudes from 4.3° to 5.8°. Interestingly, Tycho did not bother to include several important stars of Sagittarius (ζ, δ, and γ). These stars are important parts of the figure of the constellation and were well above the horizon at Hven, culminating at altitudes from 3.9° to 4.3°. Were it not for the presence of Fomalhaut, a devil's advocate could argue from this omission that Brahe's catalog is really due to an unknown astronomer of Bergen, Oslo, or Helsinki, cities some 4° north of Hven.

When we turn to Brahe's 1,000-star catalog, we find that he has included four stars in the head of Centaurus (g, h, i, and k Centuari). The most southerly of these stars, g Centuari, crossed the meridian at Hven at an altitude of about 2°. Interestingly, these stars of Centuarus appear in Brahe's catalog in the same order as in Ptolemy's. (This tends to be true within each constellation, especially near the beginning of the constellation.) Thus, it appears that when Brahe was searching for the additional stars he needed to make up his 1,000, he scanned Ptolemy's catalog to see what he might have overlooked.

Our examination of Brahe's catalogs suggests some plausible conclusions. An astronomer working in the ancient tradition did not scan the sky systematically, as in a modern grid search. Rather, he surveyed the stars constellation by constellation. It is unlikely that he would try to observe below about 5° altitude (where the visibility is poor), except for particularly bright stars, or unless he were seeking to include or complete an already defined, classical constellation. Ptolemy, as an Alexandrian, had the distinction of being among the most southerly of all the astronomers in the Greek tradition. The constellations he had inherited from his predecessors did not reach down to his southern horizon. It looks as if the ragged 6° empty band at the southern edge of Ptolemy's catalog is entirely normal.

What the Magnitudes Tell Us   The arguments of Newton and Rawlins do not have enough weight to settle the issue. There is, then, no compelling reason to doubt that Ptolemy actually observed the bulk of the stars in his catalog, just as he says he did. But is there any positive evidence to be offered in support of Ptolemy's authorship of the catalog?

In fact, the magnitudes assigned to the southernmost stars of the catalog appear to support an Alexandrian origin. Star light is dimmed as it passes through the atmosphere. This phenomenon is called *atmospheric extinction*. The extinction becomes progressively more important for stars at lower altitudes. (Light from a low star must pass through much greater quantities of air, which absorbs or scatters more of the light.)

If Hipparchus compiled the catalog, and if he observed at Rhodes as is believed, he would have seen the most southerly stars about 5° lower in the sky than Ptolemy would have at Alexandria. For the most southerly stars of the catalog, the atmospheric extinction would have been much more pronounced at Rhodes than at Alexandria. That is, Hipparchus would have assigned much dimmer magnitudes to these stars than would Ptolemy. Detailed calculations show that the magnitudes set down in the catalog are consistent with these stars having been observed at Alexandria, and not with their having been observed at Rhodes.[29] These results are summarized in the following table:

Star	Magnitude in the catalog	Magnitude as seen in Alexandria	Magnitude as seen in Rhodes
α Crucis	2	2.3–2.6	3.8–4.3
α Carinae	1	0.7–1.0	3.4–4.1
α Centauri	1	1.1–1.4	2.2–2.7
γ Arae	4+	4.6–4.9	5.8–6.3
β Arae	4	4.0–4.2	4.9–5.3
τ Puppis	3+	3.9–4.1	4.8–5.2

These are the six most southerly usable stars of Ptolemy's catalog. (Other stars are not usable because there are uncertainties in the manuscript tradition about the magnitude of the star, or the identity of Ptolemy's star is uncertain, or the star is not far enough south for the difference in the extinction between Alexandria and Rhodes to be appreciable.)

The numbers in the second column are the magnitudes found in Ptolemy's star catalog.[30] The numbers in the third column are the magnitudes that would have been observed at Alexandria in the second century A.D. These numbers are the results of calculation. The stars have been reduced to their second-century places using modern theories of precession and proper motion, and the magnitude of each star has been adjusted for atmospheric extinction. The apparent magnitudes have a range, reflecting the likely lowest and highest values for the coefficient of atmospheric extinction, a parameter that represents the clarity of the atmosphere. Similarly, the numbers in the last column are the magnitudes that would have been observed at Rhodes in the second century B.C. These numbers are also the results of modern calculations.

It is clear that the magnitudes of these stars in the *Almagest* catalog are not consistent with their having been observed at Rhodes. But they appear quite consistent with observation at Alexandria.

Some Conclusions   If we have devoted more space to this controversy than it seems to deserve, it is for two reasons. First, not only historians but also practicing astronomers from medieval times down to our own day have debated Ptolemy's reliability. The argument over the origin of the star catalog is thus a part of a much larger debate about Ptolemy's use of observations and his believability as a witness. Second, although it may be of only minor importance that the right astronomer should get credit for some piece of work, it is very important that we at least get the timeline right. It makes a big difference in reconstructing the history of Greek astronomy whether we place the origin of the star catalog in the second century B.C. or in the second century A.D.

It appears quite likely that the catalog we have in the *Almagest* is the work of Ptolemy. An important lesson is the difficulty of deciding this issue only on the basis of evidence internal to the catalog. As we have seen, most aspects of the catalog are susceptible to multiple interpretations. When we look at a broader class of evidence, Hipparchus's authorship of the catalog becomes much less likely. Most compelling is the complete absence of orthogonal ecliptic coordinates from Hipparchus's only surviving work, the *Commentary on the Phenomena of Aratus and Eudoxus*. The magnitudes assigned to the southernmost stars also speak in favor of Ptolemy.

It is interesting that the errors in Ptolemy's star coordinates in the *Almagest* catalog do show correlations with the errors in Hipparchus's data in the *Commentary*.[31] This seems to imply what we have suggested already on other grounds, that Ptolemy did make some use of Hipparchus's work on the stars—perhaps by adopting the coordinates of a number of reference stars. But since the random errors in Ptolemy's star positions are significantly smaller than the random errors in Hipparchus's, it is abundantly clear that Ptolemy could not have simply copied out a large number of Hipparchus's star coor-

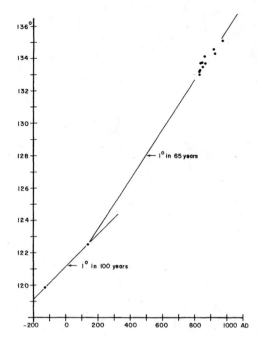

FIGURE 6.13. Some longitudes of Regulus. The points in the lower left are from Hipparchus and Ptolemy. The points in the upper right are from various Arabic astronomers, as reported by Ibn Yūnus.

dinates.[32] Moreover, as Shevchenko has pointed out,[33] the excess fractions of 2/3° in Ptolemy's longitudes are not distributed uniformly across the catalog, as Newton implicitly assumed. The *northern* constellations indeed show a pronounced excess of 40′ fractions. But among the southern constellations, this effect is almost completely absent—40′ occurs much less frequently than 0′, and also less frequently than 10′, 20′, and 50′.

Among the zodiacal constellations, the seven constellations from the Archer to the Twins show a strong excess in 40′; but the five constellations from the Crab to the Scorpion do not. Thus it appears very likely that the ancient compiler of the catalog (whoever he was) changed some part of his procedure (or his reference star) part way through the work on the catalog. While this fact does not logically weigh either for or against Ptolemy, it does show that the structure of the star catalog is more complicated than Newton supposed in his analysis.

A final conclusion is the mutability of history. Each generation needs to reevaluate the evidence, and we cannot completely prevent our prejudices from affecting the way we look at the evidence. The nineteenth-century seekers of precursors saw what they expected to see: that Hipparchus, whose works were lost, was a figure of high originality and that Ptolemy was a mere textbook writer. In our own century, Hipparchus remains a figure of great importance. But we see more clearly what he borrowed from his contemporaries and his predecessors. For example, because we know much more now about the astronomy of the Babylonians, we can see how deeply indebted Hipparchus was to Babylonian astronomy in his work on the Sun, Moon, and planets. And the more we learn about the character of Greek astronomy between Hipparchus and Ptolemy, the more Ptolemy stands out as a figure of major significance and originality, though, of course, he too made ample use of what had been accomplished before his time.

6.9 TREPIDATION: A MEDIEVAL THEORY

Greek astronomy virtually ceased after Ptolemy. Astronomy was still studied and taught in Greek, but there was little original work. The commentaries on the *Almagest* written in the third and fourth centuries A.D. by Pappus and by Theon of Alexandria did not advance the science. A revival of astronomy began in the Islamic renaissance of the ninth century. When Arabic astronomers compared their own observations with those of Ptolemy, they made two striking discoveries—one true and one illusory. First, the obliquity of the ecliptic was smaller than it had been in Greek antiquity. And, second, the rate of precession was faster than it had been.

The obliquity of the ecliptic is one of the most fundamental parameters of astronomy, and one of the easiest to measure. One need only measure the noon altitude of the Sun at summer and at winter solstice, as explained in section 1.12. Ptolemy's value for the obliquity of the ecliptic was 23°51′20″. By the ninth century, it was easy to see that the obliquity was closer to 23°33′, the value measured by the astronomers of al-Maʾmūn.[34] (23°35′ would have been accurate for the ninth century). So, the obliquity of the ecliptic appeared to have decreased by 18′ since Ptolemy's time. In fact, Ptolemy's value was too high by about 10′. The discovery of the decrease of the obliquity of the ecliptic was made much easier by Ptolemy's error of measurement. The obliquity really is decreasing, but not by as much as the astronomers of the ninth century believed.

When the ninth-century astronomers investigated precession, there they were confronted with another change. Figure 6.13 shows some longitudes of Regulus measured by the Greeks and by the Arabs. The point in the lower left is Hipparchus's longitude of Regulus. The point about A.D. 139 is from

Ptolemy. The cluster of points in the upper right is from measurements by various Arabic astronomers in the period A.D. 830–975, as reported by Ibn Yūnus around A.D. 1007.[35]

The ninth-century Arabs could not determine the precession rate with any exactness from their own measurements alone. The only hope lay in a comparison with the data handed down by Ptolemy. The straight line connecting Ptolemy's observation of A.D. 139 with the cluster of Arabic observations has a slope of 1° in 65 years. But the slope of the line connecting Ptolemy's observation with Hipparchus's is 1° in 100 years. It therefore looked as if the precession rate had increased since the days of the Greeks. Actually, the rate had been constant at 1° in 72 years during the whole period, and the apparent variation was due solely to errors of observation. If the Arabs had ignored Ptolemy and simply calculated the precession between Hipparchus's time and their own, they would have obtained a rate very close to the true one. But of course they had no way of knowing this.

Some astronomers were quite aware of the delicate nature of precession studies. Al-Battānī, the greatest of all ninth-century astronomers, after making his own investigation, adopted a uniform rate of 1° in 66 years for his tables.[36] But he expressed some doubt that the rate really had stayed the same since the time of Ptolemy. Al-Battānī remarked that if there exists any motion that we do not know and that we do not understand, those who come after us will observe and verify it, and improve our theories, just as we did with the theories of our predecessors.[37]

Thābit ibn Qurra and the Theory of Trepidation

In the ninth century a way was found to explain both these changes (decrease in the obliquity of the ecliptic and the variability of the precession rate) in terms of a single mechanism. The theory of trepidation is usually attributed to Thābit ibn Qurra, though some scholars doubt this attribution.[38] The system was described in a medieval Arabic treatise that no longer survives. We do possess a medieval Latin translation, De motu octave spere, "On the Motion of the Eighth Sphere."[39] Thābit's doctrine of the trepidation of the equinoxes had a profound influence on medieval and early Renaissance astronomy. Indeed, one can hardly understand the medieval astronomical literature without a familiarity with Thābit's system.

Al-Ṣābiʾ al-Ḥarrānī Thābit ibn Qurra (ca. A.D. 824–901) was born in Harran, in what is now southeastern Turkey, but passed most of his professional life in Baghdad. He was not a Muslim, however, but a Sabian. The Sabians, whose religious practice included star worship, preserved aspects of Babylonian religion. Most of Thābit's scientific works were written in Arabic, but some were composed in Syriac. Thābit was a talented and well-educated man, who wrote on mathematics and made competent Arabic translations of Greek mathematical works. Besides this he found time to practice as a physician and to write on physics, philosophy, logic, politics, and grammar. And he grappled with the problem of precession.[40]

A great problem was posed by the lack of data for the seven-century gap between Ptolemy and Thābit's own time. In a letter to Isḥāq ibn Ḥunayn, a physician and Arabic translator of Greek scientific works, Thābit mentions precession and says that he cannot be sure of the phenomena. "But we could if someone found between us and Ptolemy a solar observation long enough before our time, fully corrected; and if you find between us and Ptolemy a suitably distant observation according to the books of the Greeks, bring the matter to my knowledge so that I may decide the question."[41]

Although there were no ancient observations that could settle the matter, there was a bit of ancient lore that worked on Thābit's imagination. Theon of Alexandria, the fourth-century commentator on Ptolemy, mentioned in his introduction to the *Handy Tables* a strange doctrine that had been held by certain ancient astrologers.[42] According to these unnamed astrologers, the precession was not a steady, everlasting motion. Rather, they taught that the stars moved to the east at the rate of 1° in 80 years only until they had traveled 8°, *when they suddenly reversed direction* and traveled back to the west at the same rate over the same 8° arc. This to-and-fro motion continued forever. Furthermore, they fixed the year 158 B.C. (128 years before the reign of Augustus) as a year in which the precession had reversed direction. Ptolemy never alludes to this theory. And Theon points out that Ptolemy's tables, which are based on a uniform precession, are always in accord with the observed positions of the stars. For this reason, Theon advises his readers not to depart from Ptolemy.

What was the origin of this strange doctrine, mentioned by no Greek writer except Theon of Alexandria? It seems likely that it was all a mistake. There were in antiquity several different conventions for the boundaries of the zodiac signs (sec. 5.2). The later Greeks placed the equinoctial and solstitial points at the beginnings of their signs. But, in one of the Babylonian conventions, the equinoctial and solstitial points were placed at the 8th degrees of their signs. It seems likely that the astrologers mentioned by Theon mistook a difference in convention for a real 8° shift of the stars. Why they settled on 158 B.C. as a year in which the motion reversed direction it is impossible to say. But since these astrologers had heard of precession, they must have known something of Hipparchus's works. Perhaps, as Dreyer[43] has suggested, they misinterpreted Hipparchus's adoption of the Greek rather than the Chaldaean division of the signs and chose 158 B.C. as the time of a shift in the heavens because this was the time of Hipparchus's astronomical activity. The rate, 1° in 80 years, remains unexplained, but there may have been something in Hipparchus's writings that supported this. We do know that the rate adopted by Ptolemy, 1° in 100 years, was considered by Hipparchus only to be lower limit. If all this is right, then the doctrine of the trepidation of the eighth sphere, as it came later to be called, and which was to disfigure astronomy for more than a thousand years, was founded on a simple misunderstanding.

But for Thābit ibn Qurra, this ancient doctrine seemed to provide a way of explaining the apparent variation in the rate of precession. In his letter to Ishāq ibn Hunayn, Thābit mentions Theon by name in this very connection. Of course, the numerical parameters cited by Theon could not be right. For an 8° displacement at the rate of 1° in 80 years requires only 640 years. So if the motion of the stars had switched to the east in 158 B.C., it ought therefore to have reversed again in A.D. 483. The stars ought then to be traveling *westward* in Thābit's day, which of course they were not. Moreover, figure 6.13 shows that the longitude of Regulus had increased by 13° or 14° since the time of Hipparchus—considerably more than the 8° maximum claimed by the ancient astrologers.

An Overview of Thābit's System   Thābit's system is couched in terms of medieval spherical cosmology, which, of course, derived ultimately from Eudoxus, Aristotle, and Ptolemy. The sphere of the fixed stars is called the *eighth sphere*. The Sun, Moon, and the five planets (which all lie closer to the Earth) are assigned the first seven spheres. This is why the Latin translation of Thābit's (or whoever else's it was) book was called *On the Motion of the Eighth Sphere*. The ecliptic is a circle inscribed on the eighth sphere, along with the stars.

Thābit's goal was to explain two "facts" at once: the real decrease in the obliquity of the ecliptic and the supposed variation in the rate of precession. For this purpose, he surrounds the eighth sphere with a new sphere, which in medieval cosmology came to be called the ninth sphere, or the prime movable. This ninth sphere in Thābit's system is responsible for the daily rotation of the cosmos, which motion it communicates to the eight lower-lying orbs. Because the ninth sphere is the sphere of the diurnal rotation, it naturally bears the celestial equator. The principal features of both the eighth and ninth spheres are illustrated in figure 6.14.

The celestial equator *AJ* is intersected at *A* by the fixed ecliptic *AI*. This is not the real, observable ecliptic, but merely a fictitious reference circle inscribed in the ninth sphere. We call it "fixed" because it is fixed with respect to the celestial equator, with which it always makes the angle $\varepsilon_0 = 23°33'$, according to Thābit. The pole of the celestial equator is *P*, and the fixed solstitial colure (not the real, observable solstitial colure) *PIJ* intersects the fixed ecliptic and the equator at right angles. Furthermore, centered on *A* is a small circle whose radius, according to Thābit, is $4°18'43''$. There is another small circle (not shown) on the opposite side of the sphere, centered on the point opposite *A*. All these are features of the ninth sphere, absolutely fixed with respect to one another.

Moving around the small circle at a uniform rate is point *C*, which in medieval Latin astronomy is called the moving *caput Arietis*, the "head [or beginning] of Aries." Thus, angle β increases uniformly with time. According to Thābit, β goes through 360° in about 4,182 Arabic years (about 4,057 Julian years).[44] Diametrically opposite *C* is a point called the moving *caput Librae*, which similarly moves in a small circle. These two points are always opposite one another, so that when caput Arietis is north of the equator, caput Librae is south of the equator.

The true, actual ecliptic is labeled "movable" in the diagram because it moves with respect to the equator. This movable ecliptic passes through the moving caput Arietis *C* and through the unseen caput Librae on the other side of the sphere. The mechanism resembles the drive system for an old-fashioned steam locomotive. The actual, observable spring equinoctial point ♈ is the intersection of the movable ecliptic with the equator.

Point *D*, the moving *caput Cancri*, or beginning of Cancer, is 90° from *C* along the movable ecliptic. Point D of the movable ecliptic always stays on the fixed ecliptic. Its motion is therefore a simple to-and-fro vibration along the fixed ecliptic. *D* is not the true, observable solstitial point. It, too, is merely a fictitious reference point. The true solstitial point, at which the movable ecliptic is at its greatest distance from the equator, is obviously 90° from ♈ along the movable ecliptic and is labeled ♋ in the figure.

Finally, it is important to remember that the stars (and also the apogees of the planets) are embedded in the sphere of the movable ecliptic. Thus, the stars gyrate around with motion of *C* and the movable ecliptic. Consequently, the latitudes of the stars (their angular distances above or below the plane of the movable ecliptic) remain invariable. The eighth sphere in the title of Thābit's work is the sphere of the movable ecliptic, to which the stars are attached. It is the wobbling of this sphere with respect to the ninth sphere, or sphere of the equator, that causes the stars to advance and recede with respect to the equinoxes. This to-and-fro motion is called accession and recession. Moreover, the same mechanism causes the obliquity of the ecliptic to change.

Accession and Recession   To see how the motion of accession and recession arises, let us use figure 6.15. This figure shows a close-up of the region about point *A* at four different epochs according to Thābit's system. The four views are separated by equal intervals of time, each interval corresponding to 45° of motion in β. Point *C* and a star *S* both lie in the eighth sphere. Thus, the distance of *S* from *C* does not change. This particular star happens to lie exactly on the movable ecliptic (and thus has latitude zero forever). The longitude of *S* is the angular distance from ♈ to *S*, which now varies in a nonsteady way: we obtain a variable rate of precession. Note that ♈*S* increased

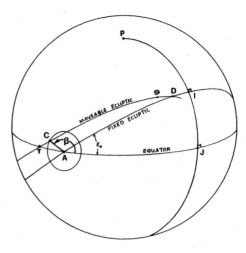

FIGURE 6.14. The system for the motion of the eighth sphere attributed to Thābit ibn Qurra.

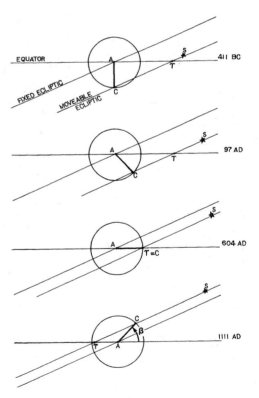

FIGURE 6.15. The motion of accession and recession according to Thābit ibn Qurra.

by a larger amount between A.D. 97 and A.D. 604 than between 411 B.C. and A.D. 97.

At any time (say, A.D. IIII in fig. 6.15), the star's longitude ΥS is its distance CS from caput Arietis plus the distance ΥC. For each star, the distance CS never changes. Thus, in a star catalog, the longitudes of all the stars could most conveniently be put down as measured from the moving caput Arietis. The distance ΥC is called the *equation in longitude*. The equation of longitude can be positive or negative. But at any moment its value is the same for all the stars. To calculate star longitudes for a particular year, one applies the rule

star's longitude = star's distance from caput Arietis + ΥC.

The first term on the right would be taken from the perpetual star catalog. The second term on the right would be calculated for the desired date and applied equally to all the stars. The equation ΥC is positive if caput Arietis is north of the equator (as in A.D. IIII) and negative if caput Arietis is south of the equator (as in A.D. 97). The *latitudes* of the stars were of course regarded by Thābit as absolutely fixed: these could be taken directly from Ptolemy's catalog.

The exact trigonometric formula for ΥC that would result from Thābit's model is rather complicated.[45] Thābit does not work out the trigonometry in detail. But he does give a table for the equation in longitude ΥC, for values of angle β running from 0 to 90° in 5° steps. Thābit's table agrees rather well with a simple sinusoidal rule:

$$\sin \Upsilon C = \sin 10°45' \sin \beta,$$

which is, indeed, a reasonably good approximation to the actual model. Nearly all practical calculation of longitudes in trepidation theory in the later Middle Ages was based on some such simple rule.

The maximum equation in longitude in Thābit's table is 10°45'. An exact calculation from Thābit's premises, using his values for r and ε_o, gives 10°40' for the maximum equation in longitude.

Thābit, or whoever it was that wrote *De motu*, is clearly aware that the system also produces a variation in the obliquity of the ecliptic, although this aspect of the system is not discussed in much detail and no table is provided. Thābit mentions that the maximum value of the obliquity is, according to the Hindus, 24°, that Ptolemy found 23°51', and that the astronomers of al-Maʾmūn found 23°33'. An exact calculation from Thābit's model shows that the obliquity varies between about 23°29' and about 23°59'.[46] The obliquity is a minimum when $\beta = 0$ or 180° and a maximum when $\beta = 90$ or 270°. Thus, the obliquity goes through two complete cycles while the equation in longitude goes through one.

How well does Thābit's system work? According to Thābit, at the beginning of the first year of the Hegira (A.D. 622), $\beta = 1°34'02''$. If we work in Julian years rather than Arabic years, the equation in longitude in Thābit's system is given approximately by

$$\sin \Upsilon C = \sin 10°45' \sin \left[\frac{360° (t - 622)}{4057 \text{ years}} + 1°34'2'' \right],$$

where t is the year expressed in the Julian calendar. For example, if we put $t = -127$ (Hipparchus), we get $\Upsilon C = -9°43$. Calculating also the equation in longitude for A.D. 138 (Ptolemy) and for A.D. 830 (al-Maʾmūn), we get the third column of the following table:

Some numerical consequences of Thābit ibn Qurra's theory of the motion of the eighth sphere

Date	Difference (years)	♈C equation in longitude	Change in the equation	Precession rate	Obliquity
−127		−9°43′			23°54′
	265		2°38′	1° in 101 years	
A.D. 138		−7°05′			23°44′
	692		10°45′	1° in 64 years	
A.D. 830		+3°40′			23°32′

The precession between Hipparchus's day and Ptolemy's is obtained by subtracting the values of ♈C at these two epochs, giving 2°38′. The precession rate is then obtained by dividing by the time interval. The resulting figure, 1° in 101 years, is very close to the value Thābit thought he had to match. In the same way, Thābit's theory results in a precession rate of 1° in 64 years for the period between Ptolemy and al-Maʾmūn—very close to the rate of 1° in 66 years that Thābit cites for this period.

The last column displays values of the obliquity of the ecliptic calculated from Thābit's premises for the same three years. Thābit's system does a good job of accounting for the phenomena he thought he had to explain.

Later History of Trepidation

Thabit's system had a mixed reception in Islamic astronomy. In the East, it was not greeted with enthusiasm. Thabit's younger contemporary, al-Battānī, seems to have rejected the theory and to have adopted a uniform precession of 1° in 66 years. However, Thābit says that al-Battānī later reversed his opinion and accepted the theory. Ibn Yūnus (d. 1009) adopted a uniform rate of 1° in 70 years, one of the most accurate of medieval values for the precession. And there were others who held back, too.

In the West, and especially in Islamic Spain, trepidation had a warmer reception. Thābit's tables for trepidation were borrowed and included as a part of the enormously influential *Toledan Tables*, compiled in Spain by a group of Muslim and Jewish astronomers and put into final form by al-Zarqālī around the year 1080.[47] The tables and their canons (directions for their use) were soon translated into Latin. The presence of trepidation tables in a manual of practical astronomy such as the *Toledan Tables* probably did more to popularize the theory than did Thābit's own treatise, which was much more difficult to understand and which was couched in speculative language.

By the thirteenth century, the total precession since Greek times had accumulated to a large enough value to make Thābit's purely back-and-forth motion impossible to retain. One had to accept the reality of a steady forward precession. But now it was believed that a back-and-forth motion was superimposed on a steady forward motion. This gave rise to a variable precession rate, without any actual backing up by the stars. A very influential version of this theory was built into the *Alfonsine Tables* (see sec. 2.11). The makers of the *Alfonsine Tables* introduced a steady and uniform precession that was supposed to carry the stars and the apogees of the planets all the way around the ecliptic in 49,000 years. Superimposed on this steady motion is an oscillation over an arc of ±9°, which is completed in 7,000 years. The equation in longitude q (i.e., the contribution of the oscillation) is calculated according to the simple rule

$$\sin q = \sin 9° \sin A,$$

where the angle A runs through 360° in 7,000 years and had the value 359°12′34″ at the Incarnation of Christ (A.D. 1). Everywhere in Latin Europe

FIGURE 6.16. The nested celestial spheres, including mechanisms for trepidation and precession. From Petrus Apianus, *Cosmographicus liber* (1524). Lilly Library, Indiana University, Bloomington, Indiana.

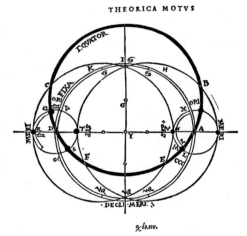

FIGURE 6.17. Thābit ibn Qurra's system for trepidation in a sixteenth-century edition of a fifteenth-century textbook. Georg Peurbach, *Theoricae novae planetarum* (Paris, 1553). Courtesy of Special Collections Division, University of Washington Libraries.

this was the standard theory of trepidation and precession from the middle of the fourteenth to the middle of the sixteenth century.[48]

In Aristotelian physics, each simple body can possess but a single simple motion. Thus, the addition of a steady precession to the trepidation required the introduction of a tenth sphere in the cosmos. Figure 6.16 is a woodcut from an early sixteenth-century textbook of cosmography written and printed by Peter Apian. *Cosmography* encompassed elements of geography, astronomy, and cosmology. The figure is meant to illustrate the general construction of the universe. We see the central Earth surrounded by seven "heavens" or spherical shells—one each for the Moon, Sun, and five planets. The eighth sphere is called the "firmament" and is spangled with stars. The ninth sphere is called "crystalline" and carries the little circle on which revolves the moving caput Arietis. This little circle may be seen centered at the beginning of the sign ♈ of Aries in the ninth sphere. Note that the beginning of Aries ♈ in the eighth sphere is offset a little, reflecting the motion of trepidation. Next comes the tenth sphere, which is called the "first movable." The offset between the beginning of Aries in the ninth sphere and the beginning of Aries in the tenth sphere represents the effect of the steady component of precession. Thus, the motion of the eighth sphere is the trepidation, the motion of the ninth is precession, and the motion of the tenth is the daily rotation. Finally, like many other European writers, Apian has Christianized his cosmos, by surrounding the whole with the "empyrean" sphere, which is the "habitation of God and of all the elect."

One often hears it said that during the Middle Ages, the old astronomy of the Greeks became more and more complicated, as epicycles were added to epicycles, until the system collapsed of its own intellectual weight. This is quite untrue. Almost everywhere in both Islam and Christendom computation of planet positions continued to follow standard Ptolemaic models. The planetary theories underlying the *Alfonsine Tables*, for example, are standard Ptolemaic models. The one big exception to this rule is trepidation theory. The addition of this oscillatory motion to the sphere of stars did make the universe more complicated—but not all that much more complicated.

In 1543, Copernicus published a book, *On the Revolutions of the Heavenly Spheres*, which claimed that the Earth was a planet in orbit about the Sun. But, as radical as this hypothesis was, Copernicus was quite conservative in the technical details of his astronomy. Some celestial motions were transferred by Copernicus from the sphere of the stars to the Earth, but most of the essential features of the standard astronomy were retained. For example, Copernicus let the Earth spin on its axis and revolve about the Sun to explain the daily and annual motions. And he used a gradual displacement of the Earth's axis to explain precession. But, like all his contemporaries, Copernicus believed that the precession rate was variable—he, too, was a believer in trepidation.[49] Copernicus introduced a rather complicated motion of the Earth's axis to explain (in one system) both the decrease of the obliquity and the variable precession rate. So, although Copernicus now attributed these motions to the Earth, he constructed a fairly traditional system to explain them. The Alfonsine compromise of an oscillatory motion superimposed on a steady precession was basically unchanged, apart from some adjustments to the numerical parameters.

In figure 6.17 we see an illustration of Thābit's system from the Paris, 1553, edition of Georg Peurbach's *Theoricae novae planetarum* (New theories of the planets). Peurbach's book, written around 1460 and first printed in 1472, was widely used as a university text.[50] The fact that we still see Thābit's original theory of trepidation (complete with a maximum equation in longitude of 10°45′!) in an edition published ten years after Copernicus's book demonstrates how thoroughly integrated into Western astronomy the theory of trepidation had become.

6.10 TYCHO BRAHE AND THE DEMISE
OF TREPIDATION

Tycho Brahe

Tycho Brahe (1546–1601) was a Danish nobleman who devoted the greatest part of his adult life to astronomy. King Frederick II of Denmark gave him an island, Hven, in fief and there Brahe lived like a feudal lord. The residents of the island contributed their rents to his support and were also obliged to supply labor for his construction projects. Brahe, the most famous astronomer of his generation, reflected glory on the Danish court—and also provided some practical services by casting horoscopes at the births of the royal children.

When Frederick died in 1588, his eldest son, Christian, was just a boy and the government passed into the hands of a board of regents. Christian IV began to rule in his own right in 1596. Brahe's relations with the court had gradually deteriorated. Brahe's own haughty manner and his stormy relations with the islanders were contributing causes. But it is also a fact that the new king placed less value on Brahe's astronomical accomplishments than had his father. When the court needed to economize, it made a series of cuts in Brahe's support. This was the last straw and Brahe left Hven for good in 1597. Eventually he found a new patron in Emperor Rudolph II of Germany and he spent the last few years of his life at Prague.[51]

But it was on Hven that the great bulk of Brahe's work was carried out. It was on Hven that Brahe constructed an observatory and place of residence that he called Uraniborg—the celestial castle. As Brahe's fame as an astronomer grew, he gathered about him a circle of assistants, students, and visiting astronomers. In many ways, Uraniborg can be considered the first modern astronomical observatory in Europe and the model for much that followed. Similar state-supported astronomical research programs had, of course, appeared earlier in Islamic lands, that of Ulugh Beg at Samarkand being a good example. But, because of their lack of contact with the West, they did not influence the development of the national European observatories of the seventeenth century.

Brahe's Uraniborg was notable for its sustained program of observation. From the mid-1570s until near the end of the century Brahe and his assistants engaged in regular position measurements of the planets, Moon, and stars. The solar observations, similarly carried out over many years, resulted in the best possible values for two fundamental solar parameters—the obliquity of the ecliptic and the eccentricity of the Sun's circle. Brahe also investigated the refraction of starlight by the Earth's atmosphere and prepared tables of refraction. Although others had pointed out the existence of atmospheric refraction, Brahe was the first to take it systematically into account.

Brahe was a capable theoretician, but he is best remembered as a painstaking observer. He aspired to bring about no less than the *reform of astronomy*. And for this, he insisted, it was first necessary to acquire a great body of observations made as accurately and as systematically as possible. Although Brahe was the most significant observer of the sixteenth century, he was not the only one to take this new stance on the importance of regular observation. For example, Wilhelm, the Landgrave of Hesse, who was about fifteen years older than Brahe, had established an observatory at Cassel (Germany). Brahe had visited Cassel before he settled at Uraniborg. And, over the years, his correspondence with Wilhelm and Wilhelm's court astronomer, Christoph Rothmann, provided not only encouragement but also a source of competition that helped to sustain Brahe's efforts.

Brahe was also notable for the great care he took in the design, construction, and alignment of his instruments. Brahe's instruments and his techniques of observation were squarely in the tradition that descended from Ptolemy.

For example, among Brahe's chief instruments were armillary spheres and quadrants. But Brahe paid greater attention to their design than had anyone before him. He made them of more rigid materials, gave them more symmetrical forms so that they would remain in balance and sag less under their own weight, and invented new kinds of sights and better methods of dividing the measuring scales on the limbs of the instruments. Brahe took comparable pains with the observations, repeating them over and over again and doing things in as many different ways as possible to provide checks for consistency.[52]

Brahe on Precession

One of the fruits of Brahe's work was a new star catalog, the first in the West that could replace Ptolemy's star catalog in the *Almagest*. (On Brahe's catalog, see sec. 6.8.) Naturally, Brahe needed to be able to include tables of precession, which required that he undertake a study of the subject. In his study of precession, Brahe drew on longitudes of Spica and Regulus measured by Timocharis, Hipparchus, Ptolemy, al-Battānī, and Copernicus, supplemented by new measurements of his own. Brahe argued that the precession had always proceeded at the rate of 51″ per year, the rate that he adopted for the tables of precession that accompanied his star catalog. The variation in the rate of precession, widely accepted from the ninth to the sixteenth century, was only due to errors of observation.

EXTRACT FROM TYCHO BRAHE

Astronomiae instauratae mechanica

I have noticed that the irregularity of the rate of change of the longitudes [of the stars] is not so considerable as Copernicus assumed. His erroneous ideas on this matter are a consequence of the incorrect observations of the ancients, as well as those of more recent times. Consequently the precession of the equinoctial point during those years is not so slow as he asserted. For in our times the fixed stars do not take a hundred years to move one degree, as indicated in his table, but only 71 1/2 years. This has practically always been the case, as appears when the observations of our predecessors are carefully checked. In fact, only a small irregularity appears, which is due to accidental causes. This we shall, God willing, explain in more detail in due course.[53]

There were two reasons why Brahe was capable of arriving at this conclusion. First, he knew better than anyone else what great care was required for accurate longitude measurements. Thus, he held in low esteem the measurements of all his predecessors and contemporaries. Trepidation theory had arisen in the first place because the medieval astronomers had put too much faith in Ptolemy's observations. This is somewhat ironic, since premodern scientists are often castigated for attaching too little importance to observation. But the medieval astronomers, right up to the time of Copernicus, had felt obliged to seek a theory that would explain all the observations.

A second reason why Brahe may have been more prepared than many to abandon trepidation theory is that he did not believe in the traditional celestial machinery—in particular, he doubted the existence of the nested planetary spheres. His own planetary theory was of the partially heliocentric type. That is, the Earth was at rest in the center of the universe. But the planets all circled the Sun, which, in turn, went around the Earth (see sec. 7.29.[54]) This is a perfectly satisfactory way of accounting for the planetary motions. But it turns out that, when one works out the details, the orbit of Mars must cross the orbit of the Sun. So, if Brahe were to maintain this system, it was important for him to insist that the orbits were mere geometrical figures and not physically real, crystalline spheres. Although Brahe's planetary theory has no essential logical connection to trepidation theory, the traditional physical mechanism (Thābit's) for trepidation involved gyrating spheres. Brahe's disbelief in the traditional celestial machinery probably made it easier for him to doubt trepidation, too.

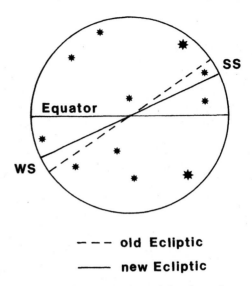

FIGURE 6.18. The rotation of the plane of the ecliptic produces changes in the observed latitudes of the stars.

Changes in the Latitudes of the Stars

It was also Tycho Brahe who set right a 700-year-old mistake about the nature of the decrease in the obliquity of the ecliptic. Almost everyone from Thābit on had supposed that this was due to a shift of the eighth sphere (bearing both the stars and the ecliptic) with respect to the equator. Thus, the latitudes of the stars should remain forever unchanged.

In fact, the decrease of the obliquity of the ecliptic is due to the rotation of the plane of the ecliptic, as in figure 6.18. From the point of view of modern celestial mechanics, this is due to a shift of the plane of the Earth's orbit around the Sun. This rotation of the plane of the orbit is caused by the weak gravitational actions of the other planets on the Earth. Over a long enough time, the motion is oscillatory, but over the whole of recorded history, the obliquity of the ecliptic has been steadily decreasing—by about $1/4°$ since Ptolemy's time.

Now if the decrease in the obliquity is due to a motion of the plane of the ecliptic, we should expect systematic shifts in the latitudes of the stars. Thus, as in figure 6.18, the stars near the summer solstice *SS* should appear to shift north with respect to the ecliptic. The stars near the winter solstice *WS* should shift south.

It was Brahe who first pointed out that the latitudes of the stars had shifted in exactly the manner to be expected if the decrease in the obliquity of the ecliptic were due to a motion of the ecliptic itself. But Brahe did not feel secure in using the latitudes set down in Ptolemy's catalog of stars. As the star catalog was simply a list of numbers, there was no sense of internal coherence to guide the copyist. One might therefore expect the entries in the manuscripts to be corrupted by copying errors. Therefore, Brahe turned to the declinations in *Almagest* VII, 3. (These were discussed in sec. 6.6. See also tables 6.1 and 6.2.) In the *Almagest*, these declinations were not simply set down in a table but were embedded in a prose passage. For example, Ptolemy had written, "Timocharis records the star in the western shoulder of Orion as being $1\ 1/5°$ north of the equator, Hipparchus $1\ 4/5$, and we find it to be $2\ 1/2°$." In a case like this, where the three numbers were related to one another in a single sentence, one could feel more confident that the successive copyists of the *Almagest* had preserved the author's original version. Ptolemy gave such declinations for eighteen stars, then, in a more extensive analysis a page or two farther on in the text, he repeated the data for six of these.

Brahe chose these declinations as the starting point for investigating whether the latitudes of the stars might suffer some change. Thus, it was necessary to convert Ptolemy's declinations to latitudes. For these, he required one other coordinate for each star. Brahe, like all his predecessors, was convinced that the distances of the stars from one another do not change. The stars are fixed in their constellations without proper motions. Thus, because the shifts in latitude (if they existed) were very small, the longitudinal distances between the stars should be practically invariable. If Brahe could then establish the longitude of just one star in Ptolemy's day, he could calculate the longitudes of all the others, simply by using the longitudinal distances that he himself had measured.

For the reference star, Brahe selected Spica. In a complex analysis based on his own accurately measured longitude of Spica, his own uniform precession theory, and the declinations of Spica cited by Ptolemy and attributed to Timocharis, Hipparchus, and Ptolemy himself, Brahe worked out values for

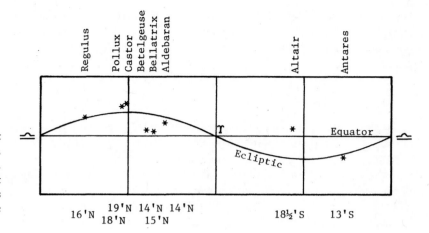

FIGURE 6.19. Shifts in the latitudes of eight test stars, as computed by Tycho Brahe, between antiquity and the sixteenth century. (The stars are located for A.D. 735, the midpoint of the time interval.) The latitude shifts calculated by Brahe (given below the stars) are the result of judicious analysis.

the longitude of Spica in the days of these three observers. Brahe concluded that the longitude of Spica in Ptolemy's time was Virgin 26°38′. But this can hardly have been worth the trouble, since Ptolemy's catalog gave practically the same thing, Virgin 26°40′. Similarly, Brahe calculated that the longitude of Spica in the days of Hipparchus was Virgin 23°53′, and in the days of Timocharis, Virgin 21°40′. Through this procedure, Brahe attempted to put the ancient longitude of a single star on a secure footing.

It was then an easy matter to get the longitude of any other star in the days of Timocharis, Hipparchus, or Ptolemy, simply by adding the longitudinal distances from Spica to the absolute longitude of Spica, just established. Brahe does this for about half of the eighteen stars that Ptolemy had mentioned in *Almagest* VII, 3. Finally, from these longitudes, so painstakingly established, and the declinations in *Almagest* VII, 3, Brahe calculated what the latitudes of these stars must have been in the days of the three ancient observers. And, finally, he compared these calculated latitudes with the latitudes he himself had measured.

The results of the comparison are displayed in figure 6.19, which shows the positions of Brahe's test stars in A.D. 735, a date roughly halfway between the activities of Hipparchus and those of Brahe. Written underneath the stars are the values for the changes in their latitudes, from ancient times to the sixteenth century, as deduced by Brahe. The stars on the part of the chart from the spring equinox eastward to the fall equinox (Aldebaran, Bellatrix, Betelgeuse, Castor, Pollux, and Regulus) have all moved north with respect to the ecliptic, while those on the other side of the chart (Antares and Altair) have moved south. Moreover, the stars nearest the solstices (Castor and Pollux near the summer solstice and Altair near the winter solstice) show the largest latitudinal shifts. The latitude shifts are consistent with the same motion of the ecliptic that would be required to explain the decrease in the obliquity of the ecliptic. It looks, then, as if this decrease is due to the shift of the ecliptic itself with respect to the star field, exactly as in figure 6.18.

The fact that the obliquity of the ecliptic was decreasing had been known since the ninth century, but everyone had mistakenly believed that was due to a shift of the eighth sphere (carrying stars and ecliptic) with respect to the equator, rather than a shift of the ecliptic with respect to the equator and the star field. Until Brahe, no one had measured the latitudes of the stars with sufficient accuracy to uncover the truth. The declinations of eighteen stars that Ptolemy had recorded in *Almagest* VII, 3, had lain like seeds, patiently waiting.

The changes in the latitudes displayed in figure 6.19 establish the shift of the ecliptic as a fact beyond dispute. The pattern in the numbers is so clear that it leaves no room for doubt. But this fine pattern is in large measure the result of careful tailoring by Tycho Brahe. The raw numbers, calculated by him according to the method described above, give a picture that is a good deal more clouded. This is partly the result of the errors in the declinations observed by Timocharis, Hipparchus, and Ptolemy.

Because the three observers were separated by only a few hundred years, the latitudes deduced from their observations should be sensibly the same. For example, in the case of Castor, Brahe calculates latitudes of 9°42 3/4', 42', and 44 3/4', respectively—fine agreement. Brahe adopts 9°43' as a good middle value, compares this to his own measured latitude for the star (10°02'), and deduces a shift of 19', as indicated in figure 6.19.

In the case of Pollux, the three astronomers' observations lead to latitudes of 6°26 1/3', 19 1/2', and 22 1/3'. Here Brahe chooses to ignore Timocharis's value as too high and adopts a value of 6°20', which is intermediate between those of Hipparchus and Ptolemy. This procedure is a little cavalier, but perhaps justifiable.

In the case of Aldebaran, Brahe allows himself even greater freedom. The three observers yield latitudes of 5°56 1/4', 33', and 7 3/4'—which is a considerable range, much greater than the size of the shift being investigated. Brahe adopts 5°45', for no very good reason except that it leads to a shift of 14' when compared with his own latitude for this star and is therefore in fine agreement with his hypothesis.

In the case of Regulus, the three ancients' declinations lead to latitudes of 0°23', 9', and 4 1/2'. Brahe chooses to reject all three and, contrary to his procedure, adopts the latitude 0°10' directly out of Ptolemy's catalog on the grounds that Ptolemy had "observed this star diligently." (Regulus, the reader will recall, was the chief star used by Ptolemy in his investigation of precession.)

Suppose we simply compare Brahe's measured latitudes with the latitudes deduced by him from Ptolemy's declinations alone, without any doctoring. The result is displayed in figure 6.20. The handsome pattern that characterized figure 6.19 is gone. Only the most judicious eye might still deduce that the shifts in the latitudes "are contingent upon the change in the obliquity of the ecliptic." In particular, note the large anomalous southward shift of Aldebaran. A hundred years later, Edmund Halley was to see this as evidence for a proper motion of this star toward the south. But Brahe, convinced as he was that the stars were fixed, smoothed away this irregularity to prove his own point. It is important to add here that there is nothing dishonest in Brahe's procedure: he displays all his calculations and even the doctoring is openly done.[55]

Why was Brahe more prepared than his predecessors to accept shifts in the latitudes of the stars? Again, a major influence can be found in the post-Copernican worldview. Copernicus had made the variation of the obliquity of the ecliptic into a motion of the Earth. Thus, the old view of an ecliptic embedded in the sphere of the stars no longer made any sense.

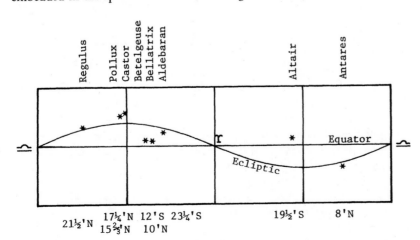

FIGURE 6.20. Shifts in the latitudes of eight test stars obtained by direct comparision of Brahe's observations with data handed down by Ptolemy in *Almagest* VII, 3.

Second, Brahe was spurred on by his correspondence with Christoph Rothmann, who also had speculated that the latitudes of the stars might change. Interestingly, Rothmann was prepared to go even further than Brahe, for Rothmann asked whether the stars might actually have proper motions— that is, whether they might not shift with respect to one another.[56] Brahe still believed in a sphere of stars, which would be inconsistent with shifts of the stars with respect to one another. Moreover, he needed the assumption that the stars' distances from one another do not change in order to carry out the painstaking analysis described above. Thus, he argued that the fixity of the stars was proved by the list of alignments set down by Hipparchus and Ptolemy (sec. 6.2). This was ironic, since Brahe's assault on the theory of trepidation was predicated on the unreliability of the ancient observations—and the alignments are the crudest observations recorded in the *Almagest*.

An Eighteenth-Century Postscript: Edmund Halley on the Proper Motions of the Stars

There was no possibility of detecting shifts in the relative positions of the stars until quite recent times, when the improved status of observational astronomy and the slowly accumulating displacements of the stars themselves finally combined to give a fair chance of success. The necessary investigation was first made by Edmund Halley, then secretary of the Royal Society, and announced by him in the *Philosophical Transactions* for the year 1718.[57] Halley had begun an investigation of precession and the decrease in the obliquity of the ecliptic in conjunction with his work as a cataloger of stars. To this purpose he compared the positions of the stars set down in the *Almagest* with those determined by Brahe and other more recent observers.

To guard against possible errors of transcription in Ptolemy's catalog, Halley followed Brahe's example and employed the declinations of eighteen stars discussed by Ptolemy in *Almagest* VII, 3. Just as Brahe had before him, Halley deduced from these declinations the latitudes that the stars must have had in ancient times and compared them with the modern latitudes. And, again like Brahe, he sought in the latitudinal shifts a confirmation of the decrease of the obliquity of the ecliptic: the stars near the solstitial points ought to have suffered a large change in latitude, while those near the equinoxes ought to have changed very little. Moreover, the stars around the summer solstice should have shifted north, while those near the winter solstice should have shifted south.

But Halley also noticed that the shifts of three or four prominent stars directly contradicted the general trend. He concluded that these stars *must have moved with respect to their neighbors*. For example, the latitudes given in Ptolemy's catalog for Palilicium (i.e., Aldebaran) and Betelgeuse revealed striking anomalies. Aldebaran and Betelgeuse, both reasonably near the summer solstitial point, should have shifted northward with respect to the ecliptic, by virtue of the decrease of the obliquity of the ecliptic. But they had actually shifted southward. To assure himself that these anomalies were not due to errors of transcription, Halley turned again to *Almagest* VII, 3, and in particular, to "the declinations . . . set down by Ptolemy, as observed by Timocharis, Hipparchus and himself, which shew that those Latitudes are the same as those Authors intended." The anomalous shifts of Aldebaran and Betelgeuse are quite plain in figure 6.20. Tycho Brahe, convinced that the stars are fixed on their sphere, had smoothed these anomalies away. But Halley, accommodated to Newton's universe, in which the stars were distributed through a vast void space through which they moved under their mutual gravitation, was prepared to accept these shifts as real. Moreover, Halley argued that as these stars are very bright, it is likely that they are near us. Thus, their motions would be among the most easily perceived.

Halley made claims of proper motion for two other bright stars—Arcturus and Sirius. Arcturus was too near the equinoctial colure to make an argument based on the obliquity. But the direct examination of the latitude of Arcturus set down in Ptolemy's star catalog showed that the star had shifted southward since Ptolemy's time by 33'. Similarly, Sirius seemed to have shifted south by about 42'.

Halley's discovery was confirmed for Arcturus by J. Cassini in 1738. Evidence for the proper motions of a large number of stars was obtained by Tobias Mayer from the comparison of Roemer's observations of 1706 with observations by himself and LaCaille from 1750 and 1756. This demonstrates the rapid improvement of positional astronomy in the seventeenth and eighteenth centuries. Halley, who compared modern observations with observations made some 2,000 years earlier, was barely able to distinguish proper motions in four stars. And for two of these—Aldebaran and Betelgeuse—he was in fact mistaken, being misled by errors in the ancient positions. So, it turns out, after all, that Tycho Brahe had been right to doctor away the anomalous shift in the positions of Aldebaran and Betelgeuse: they had resulted, not from real motions, but only from imperfections in the ancient observations. In the case of the other two stars—Sirius and Arcturus—Halley was right, and the proper motions he deduced are of about the right size. By contrast, by the middle of the eighteenth century, position measurement was so refined that proper motions were discernible with far greater certainty after intervals of only fifty years.

The history of the declinations in *Almagest* VII, 3, is quite remarkable. Timocharis undoubtedly thought that he was setting down positions for eternity. Ptolemy used these data to confirm Hipparchus's assertion about the nature of the precession—that it was parallel to the plane of the ecliptic and that it was shared by all the stars. More than a thousand years later, Brahe used the same data to show that his predecessors had all been mistaken about the nature of the decrease of the obliquity of the ecliptic. And Edmund Halley used the same data to demonstrate the motions of the stars through a vast empty space. There is no parallel to this anywhere else in the history of science.

The declinations in *Almagest* VII, 3, have had an exceptionally long and useful life. This partly was because they contained more than any one generation of astronomers dreamed of. But this longevity also owed a great deal to limitations of vision. In each age, astronomers, looking at the same numbers, have seen what their age prepared them to see or, perhaps, only what they most wished to see.

Figure 7.1 shows the constellation Leo on January 14, 1980. The two bright objects near the center of the photo are planets—Jupiter in right center, Mars in left center. To an untrained eye, observing on a single night, these planets are indistinguishable from stars. Thus, the two planets in figure 7.1 might seem to be a permanent part of the constellation Leo. Only if we carefully note the positions of the stars of Leo with respect to one another and watch them over a period of weeks or months will we notice anything strange. In figure 7.2, we see Mars and Jupiter in Leo about three months later, on April 10, 1980. In figure 7.2, Jupiter and Mars are again the brightest objects in the photo, Jupiter being a little lower than Mars. In the course of three months, Mars and Jupiter dramatically shifted their places with respect to the background stars.

The Motions of the Planets

The planets move almost on the ecliptic, the annual path of the Sun. Indeed, they are never away from the ecliptic by more than a few degrees. The maximum possible latitudes range from about 9° for Venus to about 2° for Jupiter. In our study of planetary motion, we will ignore the planets' latitudes and focus on the motion in longitude.

A Table of Planet Longitudes   Most of our study of the planets will be based on the data in table 7.1.[1] The first two columns of table 7.1 specify the date (year and day); the third duplicates this information in the form of the Julian day number. (The Julian day numbers, being integers, are for Greenwich noon.) The remaining columns give the longitudes of the Sun and the five naked-eye planets. The longitudes are measured, in the usual way, eastward from the spring equinoctial point. The longitudes in the table were obtained from computer calculations; however, we shall treat them as if they were the results of observation.

Prograde and Retrograde Motion   The planets travel generally eastward around the zodiac, just as the Sun and Moon do. But the planets occasionally *reverse direction and go backward* for a while. A planet that is traveling eastward, like the Sun, is said to be in *direct* or *prograde* motion. Or, it is said to be traveling *in the order of the signs*—for example, from Aries, to Taurus, to Gemini. In table 7.1, note that in February, 1971, Mars was in prograde motion, for the longitude steadily increases: 256°, 262°, 268°.

A planet that is traveling westward through the stars is said to be in *retrograde* motion, or traveling *contrary to the order of the signs*—from Gemini, to Taurus, to Aries. In August, 1971, Mars was in retrograde motion, for the longitude was decreasing: 318°, 315°, 313°.

When a planet is in the process of reversing its direction in the zodiac, it may appear to stand still for several days or even several weeks, depending on the planet. These standings are called *stations*. In 1971, Mars was in its normal prograde motion, until it reached its first station, at 322°, in July. Mars remained at this position for ten days or more. All through August, Mars was in retrograde motion. Mars's retrograde motion ceased at the second station, at longitude 312° (September 5–15), and Mars then reverted to its normal eastward motion.

Planetary Periods   In the case of the Sun, one period suffices to determine the mean motion. This is the *tropical year*, which is the time for the Sun to make one complete trip around the ecliptic. The motions of the planets are more complicated, and we must define two different periods.

Planetary Theory

FIGURE 7.1. Mars and Jupiter among the stars of Leo on January 14, 1980. The two bright objects near the center of the photo are planets. Jupiter is in the right center, Mars in the left center. The star at the center of the right edge of the frame is Regulus. Photo courtesy of Robert C. Mitchell.

FIGURE 7.2. Mars and Jupiter in the constellation Leo on April 10, 1980. Jupiter and Mars are the brightest objects in the photo; Jupiter is a little lower than Mars. The star very close to Jupiter is Regulus. The planets have shifted dramatically since the time of Figure 7.1. Photo courtesy of Robert C. Mitchell.

TABLE 7.1. Planet Longitudes at Ten-Day Intervals

Year	Date	J.D. 244	Sun	Mer	Ven	Mar	Jup	Sat	Year	Date	J.D. 244	Sun	Mer	Ven	Mar	Jup	Sat
1971	Feb 17	1000	328	315	283	256	244	46	1972	Jul 1	1500	99	124	77	122	273	74
1971	Feb 27	1010	338	332	294	262	245	46	1972	Jul 11	1510	109	136	76	128	271	75
1971	Mar 9	1020	348	351	306	268	246	48	1972	Jul 21	1520	119	142	80	134	270	76
1971	Mar 19	1030	358	10	318	274	246	49	1972	Jul 31	1530	128	140	85	141	269	77
1971	Mar 29	1040	8	27	330	280	247	50	1972	Aug 10	1540	138	133	93	147	269	78
1971	Apr 8	1050	18	34	341	286	246	51	1972	Aug 20	1550	147	130	101	153	268	79
1971	Apr 18	1060	28	31	353	292	246	52	1972	Aug 30	1560	157	139	111	160	268	79
1971	Apr 28	1070	37	24	5	297	245	53	1972	Sep 9	1570	167	157	121	166	269	80
1971	May 8	1080	47	23	17	302	244	55	1972	Sep 19	1580	176	176	132	173	269	80
1971	May 18	1090	57	31	29	307	242	56	1972	Sep 29	1590	186	194	143	179	270	80
1971	May 28	1100	66	43	41	312	241	58	1972	Oct 9	1600	196	210	154	186	271	80
1971	Jun 7	1110	76	60	53	316	240	59	1972	Oct 19	1610	206	225	166	192	273	80
1971	Jun 17	1120	85	80	66	319	239	60	1972	Oct 29	1620	216	239	178	198	274	81
1971	Jun 27	1130	95	103	78	321	238	61	1972	Nov 8	1630	226	249	190	205	276	80
1971	Jul 7	1140	104	122	90	322	237	62	1972	Nov 18	1640	236	253	202	212	278	79
1971	Jul 17	1150	114	138	103	322	237	63	1972	Nov 28	1650	246	241	214	218	280	78
1971	Jul 27	1160	123	151	115	320	236	64	1972	Dec 8	1660	257	237	227	225	282	77
1971	Aug 6	1170	133	159	127	318	237	65	1972	Dec 18	1670	267	246	239	232	284	76
1971	Aug 16	1180	143	160	140	315	237	65	1972	Dec 28	1680	277	259	252	238	287	75
1971	Aug 26	1190	152	153	152	313	238	66	1973	Jan 7	1690	287	274	264	245	289	74
1971	Sep 5	1200	162	147	164	312	239	66	1973	Jan 17	1700	297	290	277	252	291	73
1971	Sep 15	1210	172	154	177	312	240	66	1973	Jan 27	1710	307	306	289	259	294	73
1971	Sep 25	1220	182	170	189	314	242	67	1973	Feb 6	1720	317	324	302	266	296	73
1971	Oct 5	1230	191	189	202	316	243	67	1973	Feb 16	1730	328	342	315	273	298	73
1971	Oct 15	1240	201	206	214	318	245	66	1973	Feb 26	1740	338	356	327	280	300	73
1971	Oct 25	1250	211	222	227	322	247	66	1973	Mar 8	1750	348	358	340	287	303	74
1971	Nov 4	1260	221	238	239	328	249	65	1973	Mar 18	1760	358	349	352	294	305	74
1971	Nov 14	1270	231	252	252	333	251	64	1973	Mar 28	1770	8	345	4	301	306	75
1971	Nov 24	1280	241	264	264	340	253	63	1973	Apr 7	1780	17	350	17	308	308	76
1971	Dec 4	1290	252	269	277	346	256	61	1973	Apr 17	1790	27	0	29	315	309	77
1971	Dec 14	1300	262	259	289	352	258	60	1973	Apr 27	1800	37	14	41	322	310	78
1971	Dec 24	1310	272	252	302	358	261	59	1973	May 7	1810	47	32	54	329	311	79
1972	Jan 3	1320	282	259	314	5	263	59	1973	May 17	1820	56	52	66	336	312	81
1972	Jan 13	1330	292	272	326	12	265	59	1973	May 27	1830	66	75	79	343	312	82
1972	Jan 23	1340	303	286	338	18	267	58	1973	Jun 6	1840	75	94	91	350	312	83
1972	Jan 2	1350	313	302	350	24	269	58	1973	Jun 16	1850	85	109	103	357	312	85
1972	Jan 12	1360	323	319	2	31	271	59	1973	Jun 26	1860	94	119	116	4	312	86
1972	Feb 22	1370	333	337	14	38	273	59	1973	Jul 6	1870	104	123	128	10	311	87
1972	Mar 3	1380	343	356	26	44	274	60	1973	Jul 16	1680	114	120	140	16	310	89
1972	Mar 13	1390	353	12	37	51	276	61	1973	Jul 26	1890	123	113	152	22	308	89
1972	Mar 23	1400	3	16	48	57	277	61	1973	Aug 5	1900	133	114	164	27	306	90
1972	Apr 2	1410	13	10	59	64	278	62	1973	Aug 15	1910	142	125	176	31	305	91
1972	Apr 12	1420	23	4	69	70	278	64	1973	Aug 25	1920	152	143	188	35	304	92
1972	Apr 22	1430	32	6	78	77	279	65	1973	Sep 4	1930	162	163	200	38	303	93
1972	May 2	1440	42	15	86	84	279	66	1973	Sep 14	1940	171	181	211	39	302	94
1972	May 12	1450	52	29	91	90	278	67	1973	Sep 24	1950	181	198	223	39	302	94
1972	May 22	1460	61	46	95	97	278	69	1973	Oct 4	1960	191	213	235	38	302	94
1972	Jun 1	1470	71	66	94	103	277	70	1973	Oct 14	1970	201	226	246	35	302	95
1972	Jun 11	1480	80	89	89	109	276	72	1973	Oct 24	1980	211	235	257	32	303	94
1972	Jun 21	1490	90	108	83	115	274	73	1973	Nov 3	1990	221	236	268	29	304	94

TABLE 7.1. (continued)

Year	Date	J.D. 244	Sun	Mer	Ven	Mar	Jup	Sat	Year	Date	J.D. 244	Sun	Mer	Ven	Mar	Jup	Sat
1973	Nov 13	2000	231	224	278	26	305	94	1975	Mar 28	2500	7	347	41	319	2	102
1973	Nov 23	2010	241	222	288	25	306	93	1975	Apr 7	2510	17	5	53	327	5	103
1973	Dec 3	2020	251	232	297	26	308	93	1975	Apr 17	2520	27	25	65	334	7	103
1973	Dec 13	2030	261	246	304	27	310	92	1975	Apr 27	2530	36	46	76	342	9	104
1973	Dec 23	2040	272	262	309	30	312	91	1975	May 7	2540	46	65	88	349	12	105
1974	Jan 2	2050	282	279	313	33	314	90	1975	May 17	2550	56	78	99	357	14	106
1974	Jan 12	2060	292	294	311	37	317	89	1975	May 27	2560	65	83	110	4	17	107
1974	Jan 22	2070	302	311	306	41	319	88	1975	Jun 6	2570	75	81	120	12	18	108
1974	Feb 1	2080	312	328	300	46	321	87	1975	Jun 16	2580	84	75	131	19	20	110
1974	Feb 11	2090	322	341	296	51	324	87	1975	Jun 26	2590	94	74	140	27	21	111
1974	Feb 21	2100	332	340	297	57	326	87	1975	Jul 6	2600	103	82	148	33	22	112
1974	Mar 3	2110	342	330	301	63	329	87	1975	Jul 16	2610	113	96	155	40	24	114
1974	Mar 13	2120	352	327	308	68	331	88	1975	Jul 26	2620	123	115	160	47	24	115
1974	Mar 23	2130	2	335	316	73	333	88	1975	Aug 5	2630	132	137	162	54	24	116
1974	Apr 2	2140	12	346	325	79	335	89	1975	Aug 15	2640	142	156	160	60	25	117
1974	Apr 12	2150	22	1	335	85	337	89	1975	Aug 25	2650	151	172	154	66	25	118
1974	Apr 22	2160	32	18	346	91	340	90	1975	Sep 4	2660	161	187	149	72	24	119
1974	Mar 2	2170	42	38	357	97	342	91	1975	Sep 14	2670	171	198	145	77	23	120
1974	May 12	2180	51	61	8	103	343	93	1975	Sep 24	2680	181	205	146	82	22	121
1974	May 22	2190	61	80	19	109	344	94	1975	Oct 4	2690	190	202	150	86	21	122
1974	Jun 1	2200	70	94	31	115	345	95	1975	Oct 14	2700	200	190	156	89	19	122
1974	Jun 11	2210	80	102	42	121	346	96	1975	Oct 24	2710	210	192	164	91	18	123
1974	Jun 21	2220	89	103	54	128	347	97	1975	Nov 3	2720	220	205	174	93	16	123
1974	Jul 1	2230	99	97	66	134	348	99	1975	Nov 13	2730	230	221	184	92	15	123
1974	Jul 11	2240	109	94	77	140	348	100	1975	Nov 23	2740	240	237	194	91	14	123
1974	Jul 21	2250	118	97	89	146	347	102	1975	Dec 3	2750	251	253	206	88	14	123
1974	Jul 31	2260	128	110	101	152	347	103	1975	Dec 13	2760	261	269	217	84	14	123
1974	Aug 10	2270	137	129	114	159	346	104	1975	Dec 23	2770	271	285	229	80	14	122
1974	Aug 20	2280	147	150	126	165	345	105	1976	Jan 2	2780	281	300	241	77	15	122
1974	Aug 30	2290	157	169	139	171	343	106	1976	Jan 12	2790	291	310	253	75	16	121
1974	Sep 9	2300	166	185	151	177	342	106	1976	Jan 22	2800	302	304	265	75	18	120
1974	Sep 19	2310	176	200	163	184	341	107	1976	Feb 1	2810	312	294	277	76	19	119
1974	Sep 29	2320	186	212	176	191	339	108	1976	Feb 11	2820	322	296	289	77	21	118
1974	Oct 9	2330	196	220	188	197	338	108	1976	Feb 21	2830	332	306	302	80	22	117
1974	Oct 19	2340	206	219	201	204	338	109	1976	Mar 2	2840	342	319	314	83	24	117
1974	Oct 29	2350	216	207	213	211	337	109	1976	Mar 12	2850	352	334	326	87	27	116
1974	Nov 8	2360	226	207	226	218	338	109	1976	Mar 22	2860	2	352	339	92	29	116
1974	Nov 18	2370	236	218	238	224	338	109	1976	Apr 1	2870	12	11	351	97	31	116
1974	Nov 28	2380	246	234	251	231	339	109	1976	Apr 11	2880	22	32	4	102	33	116
1974	Dec 8	2390	256	250	264	238	340	108	1976	Apr 21	2890	31	51	16	107	36	117
1974	Dec 18	2400	266	266	276	245	341	108	1976	May 1	2900	41	61	28	112	39	118
1974	Dec 28	2410	276	281	289	253	342	107	1976	May 11	2910	51	63	41	117	41	118
1975	Jan 7	2420	287	298	301	260	344	105	1976	May 21	2920	60	58	53	123	44	119
1975	Jan 17	2430	297	314	314	267	346	104	1976	May 31	2930	70	54	65	128	46	120
1975	Jan 27	2440	307	325	326	274	348	103	1976	Jun 10	2940	79	56	78	134	48	121
1975	Feb 6	2450	317	322	339	281	350	103	1976	Jun 20	2950	89	66	90	140	50	122
1975	Feb 16	2460	327	312	351	289	353	102	1976	Jun 30	2960	99	82	102	146	52	123
1975	Feb 26	2470	337	312	4	296	355	102	1976	Jul 10	2970	108	101	114	152	54	124
1975	Mar 8	2480	347	320	16	304	357	102	1976	Jul 20	2980	118	123	127	158	56	126
1975	Mar 18	2490	357	332	28	312	0	102	1976	Jul 30	2990	127	143	139	164	58	127

(Continued)

TABLE 7.1. (continued)

Year	Date	J.D. 244	Sun	Mer	Ven	Mar	Jup	Sat	Year	Date	J.D. 244	Sun	Mer	Ven	Mar	Jup	Sat
1976	Aug 9	3000	137	159	152	171	59	129	1977	Dec 22	3500	271	269	263	131	91	151
1976	Aug 19	3010	146	173	164	177	60	130	1978	Jan 1	3510	281	261	276	129	89	151
1976	Aug 29	3020	156	183	176	183	61	131	1978	Jan 11	3520	291	267	288	126	88	151
1976	Sep 8	3030	166	188	189	190	61	132	1978	Jan 21	3530	301	279	301	122	87	150
1976	Sep 18	3040	175	183	201	196	61	133	1978	Jan 31	3540	311	293	313	118	86	149
1976	Sep 28	3050	185	174	213	203	61	134	1978	Feb 10	3550	321	309	326	115	86	148
1976	Oct 8	3060	195	177	225	210	60	135	1978	Feb 20	3560	331	326	338	113	86	147
1976	Oct 18	3070	205	191	237	216	60	136	1978	Mar 2	3570	341	344	351	112	86	146
1976	Oct 28	3080	215	208	249	223	58	137	1978	Mar 12	3580	351	4	3	113	86	146
1976	Nov 7	3090	225	225	262	230	58	137	1978	Mar 22	3590	1	20	16	115	87	145
1976	Nov 17	3100	235	241	274	237	56	137	1978	Apr 1	3600	11	27	28	117	88	144
1976	Nov 27	3110	245	257	286	244	54	138	1978	Apr 11	3610	21	22	41	120	90	144
1976	Dec 7	3120	256	272	298	252	53	137	1978	Apr 21	3620	31	15	53	124	91	144
1976	Dec 17	3130	266	286	309	259	52	137	1978	May 1	3630	41	16	66	128	93	144
1976	Dec 27	3140	276	294	321	267	51	137	1978	May 11	3640	50	24	78	133	95	144
1977	Jan 6	3150	286	287	332	274	50	136	1978	May 21	3650	60	36	90	138	97	145
1977	Jan 16	3160	296	277	343	281	50	135	1978	May 31	3660	69	53	102	143	99	146
1977	Jan 26	3170	306	282	354	289	51	134	1978	Jun 10	3670	79	74	114	148	101	146
1977	Feb 5	3180	317	292	3	297	51	134	1978	Jun 20	3680	89	96	126	153	103	147
1977	Feb 15	3190	327	306	12	305	53	133	1978	Jun 30	3690	98	116	137	159	106	148
1977	Feb 25	3200	337	321	19	312	54	132	1978	Jul 10	3700	108	132	149	165	108	149
1977	Mar 7	3210	347	339	23	320	55	131	1978	Jul 20	3710	117	144	160	171	110	150
1977	Mar 17	3220	357	358	25	328	57	131	1978	Jul 30	3720	127	152	171	177	112	151
1977	Mar 27	3230	7	18	22	335	59	130	1978	Aug 9	3730	136	152	182	183	115	152
1977	Apr 6	3240	16	36	16	343	60	130	1978	Aug 19	3740	146	144	192	189	117	153
1977	Apr 16	3250	26	45	10	351	63	130	1978	Aug 29	3750	156	140	202	196	119	155
1977	Apr 26	3260	36	43	7	359	65	131	1978	Sep 8	3760	165	147	211	202	121	156
1977	May 6	3270	46	36	9	7	67	131	1978	Sep 18	3770	175	164	219	209	122	157
1977	May 16	3280	55	34	13	15	69	132	1978	Sep 28	3780	185	183	227	216	124	159
1977	May 26	3290	65	40	20	22	72	133	1978	Oct 8	3790	195	200	231	222	125	160
1977	Jun 5	3300	74	52	28	30	74	133	1978	Oct 18	3800	205	217	234	229	127	161
1977	Jun 15	3310	84	68	38	37	77	134	1978	Oct 28	3810	215	232	232	237	128	162
1977	Jun 25	3320	94	87	48	44	79	135	1978	Nov 7	3820	225	246	227	244	128	162
1977	Jul 5	3330	103	110	58	51	81	136	1978	Nov 17	3830	235	258	221	251	129	163
1977	Jul 15	3340	113	129	69	58	83	137	1978	Nov 27	3840	245	262	218	258	129	163
1977	Jul 25	3350	122	146	80	65	85	138	1978	Dec 7	3850	255	252	220	266	129	164
1977	Aug 4	3360	132	159	91	72	87	140	1978	Dec 17	3860	265	245	224	273	128	164
1977	Aug 14	3370	141	168	103	79	89	141	1978	Dec 27	3870	275	253	231	281	128	164
1977	Aug 24	3380	151	171	114	85	91	143	1979	Jan 6	3880	286	266	240	289	126	164
1977	Sep 3	3390	161	164	126	92	93	144	1979	Jan 16	3890	296	281	249	297	125	164
1977	Sep 13	3400	170	157	138	97	94	145	1979	Jan 26	3900	306	297	259	304	123	163
1977	Sep 23	3410	180	162	150	103	95	146	1979	Feb 5	3910	316	313	270	312	122	163
1977	Oct 3	3420	190	178	162	108	95	147	1979	Feb 15	3920	326	331	281	320	121	162
1977	Oct 13	3430	200	196	175	113	96	148	1979	Feb 25	3930	336	349	293	328	120	162
1977	Oct 23	3440	210	212	187	118	96	149	1979	Mar 7	3940	346	5	304	336	119	161
1977	Nov 2	3450	220	229	200	122	96	150	1979	Mar 17	3950	356	9	316	344	119	159
1977	Nov 12	3460	230	244	213	126	95	150	1979	Mar 27	3960	6	1	328	352	119	159
1977	Nov 22	3470	240	259	225	129	95	151	1979	Apr 6	3970	16	355	340	359	119	158
1977	Dec 2	3480	250	272	238	131	93	151	1979	Apr 16	3980	26	359	352	7	120	158
1977	Dec 12	3490	260	278	251	132	92	151	1979	Apr 26	3990	36	9	3	15	121	158

TABLE 7.1. (continued)

Year	Date	J.D. 244	Sun	Mer	Ven	Mar	Jup	Sat	Year	Date	J.D. 244	Sun	Mer	Ven	Mar	Jup	Sat
1979	May 6	4000	45	22	15	23	121	158	1980	Sep 17	4500	175	192	130	223	171	180
1979	May 16	4010	55	39	27	30	123	158	1980	Sep 27	4510	184	207	141	229	174	182
1979	Mar 26	4020	64	60	40	38	124	158	1980	Oct 7	4520	194	220	152	236	176	183
1979	Jun 5	4030	74	82	52	45	126	159	1980	Oct 17	4530	204	228	164	244	178	184
1979	Jun 15	4040	84	102	64	52	128	159	1980	Oct 27	4540	214	229	176	251	180	185
1979	Jun 25	4050	93	117	76	59	130	160	1980	Nov 6	4550	224	217	188	258	182	186
1979	Jul 5	4060	103	129	89	66	132	161	1980	Nov 16	4560	234	215	200	266	184	187
1979	Jul 15	4070	112	134	101	74	134	162	1980	Nov 26	4570	244	226	212	273	185	187
1979	Jul 25	4080	122	132	113	81	136	163	1980	Dec 6	4580	255	241	225	281	187	188
1979	Aug 4	4090	131	125	126	87	138	164	1980	Dec 16	4590	265	256	237	288	187	189
1979	Aug 14	4100	141	123	138	94	140	165	1980	Dec 26	4600	275	272	250	296	188	189
1979	Aug 24	4110	151	133	150	100	143	166	1981	Jan 5	4610	285	288	262	304	189	190
1979	Sep 3	4120	160	150	163	107	145	167	1981	Jan 15	4620	295	305	275	313	190	190
1979	Sep 13	4130	170	170	175	113	147	168	1981	Jan 25	4630	305	321	288	320	190	190
1979	Sep 23	4140	180	188	188	119	149	170	1981	Feb 4	4640	316	334	300	328	190	190
1979	Oct 3	4150	190	205	200	125	151	171	1981	Feb 14	1650	326	333	313	336	190	190
1979	Oct 13	4160	199	220	212	131	153	172	1981	Feb 24	4660	336	323	325	344	189	190
1979	Oct 23	4170	209	233	225	136	154	173	1981	Mar 6	4670	346	321	338	352	188	189
1979	Nov 2	4180	219	243	237	141	156	174	1981	Mar 16	4680	356	328	350	0	187	188
1979	Nov 12	4190	229	246	250	146	157	175	1981	Mar 26	4690	6	340	3	7	185	187
1979	Nov 22	4200	240	234	262	151	158	175	1981	Apr 5	4700	16	355	15	15	184	187
1979	Dec 2	4210	250	230	275	155	159	176	1981	Apr 15	4710	25	12	27	23	183	186
1979	Dec 12	4220	260	239	288	158	159	176	1981	Apr 25	4720	35	32	40	30	182	185
1979	Dec 22	4230	270	253	300	161	160	177	1981	Mar 5	4730	45	54	52	38	181	185
1980	Jan 1	4240	280	269	312	165	160	177	1981	May 15	4740	54	73	64	45	181	184
1980	Jan 11	4250	290	284	324	165	160	178	1981	Mar 25	4750	64	87	77	52	181	184
1980	Jan 21	4260	301	301	336	165	159	177	1981	Jun 4	4760	74	94	89	59	181	184
1980	Jan 31	4270	311	318	349	164	158	177	1981	Jun 14	4770	83	94	102	66	181	185
1980	Feb 10	4280	321	335	1	161	157	177	1981	Jun 24	4780	93	88	114	73	182	185
1980	Feb 20	4290	331	349	12	158	155	176	1981	Jul 4	4790	102	85	126	81	182	185
1980	Mar 1	4300	341	351	24	154	154	176	1981	Jul 14	4800	112	91	139	88	183	186
1980	Mar 11	4310	351	341	36	150	153	175	1981	Jul 24	4810	121	104	151	94	185	187
1980	Mar 21	4320	1	337	46	148	152	174	1981	Aug 3	4820	131	123	162	101	186	187
1980	Mar 31	4330	11	343	57	146	151	173	1981	Aug 13	4830	140	144	174	107	188	188
1980	Apr 10	4340	21	354	67	146	150	172	1981	Aug 23	4840	150	163	186	114	190	189
1980	Apr 20	4350	30	8	76	147	150	172	1981	Sep 2	4850	160	179	198	120	192	190
1980	Apr 30	4360	40	26	84	149	150	171	1981	Sep 12	4860	169	194	210	127	194	191
1980	Mar 10	4370	50	46	89	152	151	171	1981	Sep 22	4870	179	206	221	133	196	192
1980	Mar 20	4380	59	68	92	156	151	171	1981	Oct 2	4880	189	213	233	139	198	193
1980	May 30	4390	69	88	92	160	152	172	1981	Oct 12	4890	199	212	244	145	200	194
1980	Jun 9	4400	79	102	87	164	153	172	1981	Oct 22	4900	209	200	255	151	202	195
1980	Jun 19	4410	88	112	80	169	154	173	1982	Nov 1	4910	219	200	266	156	204	197
1980	Jun 29	4420	98	115	75	174	156	173	1981	Nov 11	4920	229	212	276	162	207	197
1980	Jul 9	4430	107	111	74	179	157	174	1981	Nov 21	4930	239	228	286	168	209	198
1980	Jul 19	4440	117	105	77	185	159	175	1981	Dec 1	4940	249	244	295	172	211	199
1980	Jul 29	4450	126	107	83	191	161	175	1981	Dec 11	4950	259	260	302	177	212	200
1980	Aug 8	4460	136	118	91	197	163	176	1981	Dec 21	4960	270	275	307	182	214	201
1980	Aug 18	4470	145	136	99	203	165	177	1981	Dec 31	4970	280	292	310	187	216	201
1980	Aug 28	4480	155	157	109	210	167	178	1982	Jan 10	4980	290	307	308	191	217	202
1980	Sep 7	4490	165	175	119	216	169	179	1982	Jan 20	4990	300	318	303	194	218	202

(*Continued*)

TABLE 7.1. (continued)

Year	Date	J.D. 244	Sun	Mer	Ven	Mar	Jup	Sat	Year	Date	J.D. 244	Sun	Mer	Ven	Mar	Jup	Sat
1982	Jan 30	5000	310	315	297	197	219	203	1983	Jun 14	5500	83	60	129	80	244	209
1982	Feb 9	5010	320	304	293	198	219	203	1983	Jun 24	5510	92	75	138	87	243	209
1982	Feb 19	5020	330	305	295	199	220	203	1983	Jul 4	5520	102	95	146	94	242	209
1982	Mar 1	5030	341	314	299	199	220	203	1983	Jul 14	5530	111	117	153	101	242	209
1982	Mar 11	5040	351	326	306	197	220	202	1983	Jul 24	5540	121	136	158	107	241	210
1982	Mar 21	5050	0	341	314	194	220	201	1983	Aug 3	5550	130	153	160	113	241	210
1982	Mar 31	5060	10	359	323	190	219	200	1983	Aug 13	5560	140	167	157	120	242	211
1982	Apr 10	5070	20	18	334	187	218	199	1983	Aug 23	5570	150	177	152	127	242	211
1982	Apr 20	5080	30	40	344	183	217	199	1983	Sep 2	5580	159	181	146	133	243	212
1982	Apr 30	5090	40	59	355	181	214	198	1983	Sep 12	5590	169	176	142	139	244	212
1982	May 10	5100	49	71	6	180	213	197	1983	Sep 22	5600	179	167	144	145	245	213
1982	May 20	5110	59	75	17	181	212	197	1983	Oct 2	5610	189	170	148	151	246	214
1982	May 30	5120	69	72	29	182	211	197	1983	Oct 12	5620	198	185	154	158	248	215
1982	Jun 9	5130	78	66	40	185	211	197	1983	Oct 22	5630	208	203	163	164	250	216
1982	Jun 19	5140	88	67	52	189	210	197	1983	Nov 1	5640	218	219	172	170	252	218
1982	Jun 29	5150	97	75	64	193	211	197	1983	Nov 11	5650	228	235	182	176	254	219
1982	Jul 9	5160	107	89	76	197	211	197	1983	Nov 21	5660	239	251	193	182	256	220
1982	Jul 19	5170	116	109	88	202	211	198	1983	Dec 1	5670	249	266	204	188	258	221
1982	Jul 29	5180	126	131	100	208	212	198	1983	Dec 11	5680	259	280	215	193	261	222
1982	Aug 8	5190	135	150	112	214	212	199	1983	Dec 21	5690	269	287	227	199	263	223
1982	Aug 18	5200	145	166	124	219	214	199	1983	Dec 31	5700	279	280	239	204	266	224
1982	Aug 28	5210	155	181	137	225	215	200	1984	Jan 10	5710	289	270	251	209	268	225
1982	Sep 7	5220	164	191	149	232	217	201	1984	Jan 20	5720	300	275	263	214	270	225
1982	Sep 17	5230	174	198	162	239	218	202	1984	Jan 30	5730	310	286	275	219	272	226
1982	Sep 27	5240	184	194	174	245	220	203	1984	Feb 9	5740	320	300	288	224	274	226
1982	Oct 7	5250	194	184	187	252	222	204	1984	Feb 19	5750	330	316	300	228	276	227
1982	Oct 17	5260	204	185	199	259	224	205	1984	Feb 29	5760	340	333	312	232	278	227
1982	Oct 27	5270	214	199	212	267	227	207	1984	Mar 10	5770	350	351	324	235	279	227
1982	Nov 6	5280	224	215	224	274	229	208	1984	Mar 20	5780	0	12	337	237	281	227
1982	Nov 16	5290	234	232	237	282	231	209	1984	Mar 30	5790	10	29	349	238	282	226
1982	Nov 26	5300	244	248	269	289	233	210	1984	Apr 9	5800	20	37	2	238	282	226
1982	Dec 6	5310	254	263	262	297	235	211	1984	Apr 19	5810	29	34	14	237	283	225
1982	Dec 16	5320	264	279	274	305	238	212	1984	Apr 29	5820	39	28	26	235	283	224
1982	Dec 26	5330	274	294	287	313	240	213	1984	May 9	5830	49	26	39	232	283	223
1983	Jan 5	5340	285	303	299	321	242	213	1984	May 19	5840	58	32	51	228	283	222
1983	Jan 15	5350	295	297	312	329	243	214	1984	May 29	5850	68	45	63	225	282	222
1983	Jan 25	5360	305	287	325	336	245	214	1984	Jun 8	5860	78	61	76	223	281	221
1983	Feb 4	5370	315	290	337	344	247	215	1984	Jun 18	5870	87	81	88	222	280	221
1983	Feb 14	5380	325	300	350	352	248	215	1984	Jun 28	5880	97	103	100	222	279	221
1983	Feb 24	5390	335	313	2	359	249	215	1984	Jul 8	5890	106	123	113	223	277	220
1983	Mar 6	5400	345	328	15	7	250	215	1984	Jul 18	5900	116	140	125	226	276	221
1983	Mar 16	5410	355	346	27	15	251	215	1984	Jul 28	5910	125	152	137	231	274	221
1983	Mar 26	5420	5	5	39	23	251	214	1984	Aug 7	5920	135	161	150	235	274	221
1983	Apr 5	5430	15	26	51	30	251	213	1984	Aug 17	5930	144	163	162	240	273	222
1983	Apr 15	5440	25	44	63	37	251	213	1984	Aug 27	5940	154	156	174	246	273	223
1983	Apr 25	5450	35	54	75	45	250	212	1984	Sep 6	5950	164	149	187	252	273	223
1983	May 5	5460	44	55	86	52	249	211	1984	Sep 16	5960	174	155	199	258	273	224
1983	May 15	5470	54	49	97	59	248	210	1984	Sep 26	5970	183	171	211	265	274	225
1983	May 25	5480	64	46	108	66	247	210	1984	Oct 6	5980	193	190	223	271	275	225
1983	Jun 4	5490	73	49	119	73	245	209	1984	Oct 16	5990	203	207	236	278	276	226

Adapted from Stahlman and Gingerich (1963); used by permission (see n. 1).

The *tropical period* of a planet is the average amount of time it takes to go all the way around the ecliptic. Mars takes a bit less than 2 years to go all the way around the ecliptic, Jupiter about 12 years, Saturn about 30.

The *synodic period* of a planet is the average amount of time between one retrograde motion and the next. In figure 7.3, it is clear that Jupiter's synodic period is much shorter than Jupiter's tropical period, for Jupiter makes only a little progress around the ecliptic (about one zodiac sign) between retrogradations.

From figure 7.4, it is clear that, in the case of Mars, the synodic period is a little longer than the tropical period. Mars goes all the way around the ecliptic, plus a bit more, between retrograde motions.

The typical motion of Mercury is shown in fig. 7.5. The tropical period is about three times the synodic period.

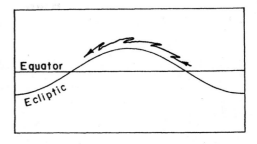

FIGURE 7.3. Typical motion in longitude of Jupiter. The planet travels only a little way forward between retrogradations. (No attempt has been made to depict the planet's latitudes.)

Two Kinds of Planets

The Inferior Planets   The inferior, or lower, planets are Mercury and Venus. In modern parlance they are called inferior because they are closer to the Sun than the Earth is. In the geocentric cosmology of Ptolemy, they were believed to be the closest planets to Earth, situated below the Sun. So, from this point of view, "inferior" is again an appropriate designation. However, as we shall see, there were in antiquity conflicting opinions about the order of the planets. Fortunately, it is not necessary to adopt any cosmological ordering of the planets to see that Mercury and Venus are different from the other three planets.

Mercury and Venus are always close companions of the Sun. They are characterized by *limited elongations* from the Sun. (The angular distance of a planet from the Sun is called its elongation.) Mercury can never be found more than about 28° from the Sun. The greatest possible elongation of Venus from the Sun is about 48°.

The best way to visualize the motion of Mercury and Venus is as follows. As the Sun marches around the ecliptic, Mercury and Venus accompany it. But they alternately dart out in front of the Sun and lag behind it. Consequently, the best time to look for Venus is in the early evening, just after sunset, or in the early morning, just before sunrise. When Venus is to the right of the Sun, it can be seen as a *morning star* in the east. When Venus is to the left of the Sun it can be seen as an *evening star* in the west. The behavior of Mercury is similar, but Mercury is harder to see, because it is fainter and because its greatest elongations are much smaller.

Because Mercury and Venus always accompany the Sun, it follows that, on the average, they take just as long to go around the ecliptic as the Sun does. *The tropical period of an inferior planet is exactly one year.* The synodic periods (the time between retrogradations) are independent. Mercury cycles through a synodic period (passing out in front of, then dropping back behind the Sun) in only 116 days. Venus takes 585 days.

The Superior Planets   Three superior, or upper, planets were known to the ancients: Mars, Jupiter, and Saturn. These planets are characterized by unlimited elongations. That is, Mars can be found at any angular distance from the Sun, right up to 180°. When a planet is 180° from the Sun, it is said to be *in opposition*.

The inferior planets manifest a connection with the Sun, by accompanying it around the ecliptic. The superior planets also manifest a connection with the Sun though it is a bit more subtle: *the superior planets undergo retrograde motion when they are in opposition to the Sun.* Consider the following excerpt from table 7.1:

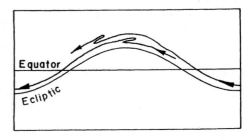

FIGURE 7.4. Typical motion in longitude of Mars. The planet travels all the way around the ecliptic, plus a bit more, between retrogradations.

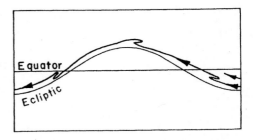

FIGURE 7.5. Typical motion in longitude of Mercury.

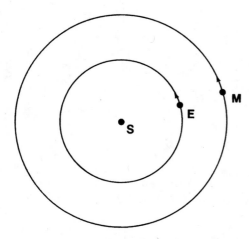

FIGURE 7.6. The Sun, Earth, and Mars in a modern heliocentric theory.

Excerpt from table 7.1

Year	Date	Longitude of Mars	Longitude of Sun
1977	Nov 12	126	230
1977	Nov 22	129	240
1977	Dec 2	131	250
1977	Dec 12	⌐132⌐	260
1977	Dec 22	131	271
1978	Jan 1	129	281
1978	Jan 11	126	291
1978	Jan 21	122	301
1978	Jan 31	118	311
1978	Feb 10	115	321
1978	Feb 20	113	331
1978	Mar 2	⌐112⌐	341
1978	Mar 12	113	351
1978	Mar 22	115	1

The longitudes of Mars and of the Sun are taken directly from table 7.1. As the entries in the excerpt begin, Mars is in prograde motion. The retrograde arc (marked with a bracket) stretches from longitude 132° to 112°. The middle of this arc is 122°, which corresponds to January 21, 1978. The Sun's longitude on this date was 301°, which is almost exactly 180° different from the longitude of Mars. Thus, at the middle of its retrograde arc, the planet was in opposition to the Sun. For the superior planets this is always the case. The connections between the Sun and planets find a ready explanation in Sun-centered cosmology. The Earth's orbit is smaller than that of Mars. Also, the Earth *E* takes only a year to orbit the Sun *S*, while Mars *M* takes two (see fig. 7.6). Thus, Mars appears to retrograde when the Earth, traveling faster, passes by Mars on the inside track. At the middle of the apparent retrograde motion, Mars is indeed in the diametrically opposite direction from the Sun. In the old astronomy of the Greeks and Babylonians, the connections between the Sun and the planets were well known as facts of nature, but there was no such simple explanation of these connections.

The Planets in Early Greek and Babylonian Astronomy

The Greeks called these objects "wandering stars." Our word *planet* comes come a Greek verb meaning *to wander*. To the early Greeks the planets were puzzling objects, with their complicated motions. The Greeks of the seventh century B.C. were not even sure how many planets there were. The Greeks had two different names for Venus: it was called Hesperos (evening [star]) in its guise as evening star and Phosphoros (light bringer) when it appeared as morning star. Some asserted that Pythagoras (sixth century B.C.) was the to first to realize that the morning and the evening stars are one, while others gave the credit to Parmenides (fifth century).[2]

Most of the Greeks considered the planets to be divine, living beings who moved by their own wills. Each planet had a proper name but was also called the star of a certain god.

Greek name	Translation	Greek god	Roman god
Phainon	shiner	Kronos	Saturn
Phaëthon	bright one	Zeus	Jupiter
Pyroëis	fiery one	Ares	Mars
Phosphoros	light bringer	Aphrodite	Venus
Stilbon	gleamer	Hermes	Mercury

For example, Saturn was called Phainon, *the shiner*, but was also known as the star of Kronos. The modern names for the five naked-eye planets are the names of the Roman divinities who were more or less equivalent to each of the Greek gods. The Babylonians, like the Greeks, associated the planets with gods. Marduk was the most important god of Babylon. His star is the planet Jupiter. The fact that the Babylonians associated the planet Jupiter with the chief god of their pantheon is an interesting parallel to Greek practice. Moreover, Venus was associated with Ishtar, the goddess of love and fertility, and Mars with Nergal, the god of war and pestilence. These parallels are too striking to be due to chance. The Greek associations are probably the result of Hellenization of earlier Mesopotamian associations. The divine associations came into use by the time of Plato.[3]

For the early Greeks, the Sun, Moon, and fixed stars were far more important than were the planets. The motion of the Sun was intimately connected with the annual cycle of agricultural labors. The phases of the Moon governed the reckoning of months. And the heliacal risings and settings of the stars told the time of year. The irregular and nonrepeating motions of the planets had no such direct utility. So, it is not surprising that Hesiod's *Works and Days* (ca. 650 B.C.), which contains a good deal of practical lore about the Sun, Moon, and stars, makes no mention of the planets. There was very little scientific activity among the Greeks concerning the planets before the fourth century B.C. Gradually, as the problems of early Greek astronomy were solved, the planets became more important. Explaining the bizarre retrogradation of the planets in terms of accepted physical principles was one of the most difficult problems of Greek astronomy. Planetary theory was the dominant problem of Greek astronomy from the time of Hipparchus to that of Ptolemy, roughly the second century B.C. to the second century A.D. Only with Ptolemy's work in the *Almagest* did it become possible for Greek astronomers accurately to predict the positions and motions of the planets from a geometrical theory.

Babylonian interest in the planets far exceeded that of the early Greeks—perhaps because the planetary gods played a greater role in Mesopotamian religion. Already in MUL.APIN (ca. 650 B.C.) there is a compendium of information about the motions of the planets, though some of it is confused or a bit inaccurate. MUL.APIN explicitly mentions that the planets change their positions among the stars but travel on the same path as the Sun and the Moon. The text also gives some detailed information about the planets' cycles. Consider the following notice of the behavior of Mars:

> Mars becomes visible in the East, stands in the sky for one year and 6 months, or for one year and 10 months, or for 2 years, and disappears in the West. This star shows either redness and is bright, or is . . . and small.[4]

This is a fair description of Mars's synodic cycle. Mars is invisible for a few months, when the Sun is too near it. Mars first reappears when it makes its morning rising, in the east just before sunrise. The planet remains visible in the night sky for a long time—nearly two years, but this is somewhat variable. Mars finally disappears again when the Sun comes too near it. The last visibility of Mars is at the time of its evening setting: Mars sets in the west shortly after sunset, and the next night it is not seen at all. The same passage also explicitly mentions that Mars usually looks red in color, but that it appears much brighter in some parts of its cycle than at others.

Because a given planet does not behave in exactly the same way from one cycle to the next, generalized rules such as those cited in MUL.APIN are not sufficient for constructing a theory of planetary motion with predictive power. Of course, one does need this sort of broad understanding before more detailed work can even begin. But it must be supplemented by specific, accurately made individual observations. Regular observation of the planets—and, what is just as important, careful record keeping of the observations—commenced in Babylonia at an early date. There are two important reasons why this happened in Babylonia but not in Greece.

In Babylonia the planets were believed to provide important signs for the future of the state. In MUL.APIN, for example, there is a list of omens associated with the star of Marduk (Jupiter).

> If the star of Marduk becomes visible at the beginning of the year, in this year the crop will prosper.
>
> If the star of Marduk reaches the Pleiades, in this year the Storm god will devastate.
>
> . . .
>
> If the star of Marduk is dark when it becomes visible, in this year there will be *asakku*-disease.[5]

MUL.APIN and other Babylonian material contain many other kinds of omens, based on the fixed stars, the winds, and the behavior of animals. The Greeks believed in many of these same kinds of omens. One need only look at the rules of personal behavior and the list of lucky and unlucky days of the month in Hesiod's *Works and Days*, or the rules for predicting the weather from the behavior of dogs and geese in the second part of Aratus's *Phenomena*. But among the early Greeks, there was no tradition of taking signs from the planets.

Of course, the Greeks *did* eventually produce a complex system of planetary astrology, but this was a development of the Hellenistic period, after Greek contact with Babylonian astronomy and astrology. The rise of interest in astrology served as a motivation among the Greeks to find better methods of predicting the motions of the planets—just as the importance of planetary omens had earlier served as a stimulus to the development of Babylonian planetary theory. Thus, planetary astronomy matured at an earlier date in Babylonia than in Greece at least partly because of its greater perceived significance for the welfare of the state. For, in early Babylonia, planetary omens foretold matters of significance to the king or to the whole nation. (Only much later do we find horoscopes for ordinary people.)[6]

A second reason planetary astronomy prospered earlier in Babylonia than in Greece is that the organization of Babylonian society favored the keeping of astronomical records. From a very early date southern Mesopotamia was centralized under one government, in which there was a fairly complete fusion of civil and religious authority. Writing was a skill not widely diffused through the populace. Rather, it was a special function of the priestly scribes at the temples in Babylon and other major cities. In Babylon, at the temple Esangila, some scribes were assigned the duty of watching the sky at night and recording everything of significance that transpired. This gave rise to a kind of document called the *astronomical diary*.[7]

The oldest astronomical diary discovered so far is for 652 B.C., but there is no doubt that regular sky watching was even older. There are Babylonian compendia of lunar eclipses that go back to the middle of the eighth century B.C., and these records were probably extracted from diaries. Some of the astronomical diaries that have been found are clearly the night-by-night records of the the sky watch: as the clay gradually dried in the course of a month of record taking, the impressions made by the writing stylus gradually became shallower and shallower. Other tablets are clearly the final, "fair copies" summarizing perhaps half a year's worth of material. On the fair copies, the writing is typically neater, and the depth is more uniform, indicating that the whole tablet was written at one time.

Here are a few lines from a diary for the year 419 B.C.:

Year 5 of Umakus, month I, night of the 7th, first part of the night, Venus was 8 fingers below β Tauri, Venus having passed four fingers to the east.

Night of the 9th, middle part of the night, the Moon was 3 fingers in front of Mars, the Moon being a little low to the south. Around the 12th or 13th, Saturn's first appearance in Pisces.

Night of the 21st, a "fall of fire" occurred in the district of Suanna. That month a fox appeared in a broad street of the city.[8]

In the first entry, after the notice of the date, we have a mention of Venus passing by one of the Babylonian standard reference stars (called *normal stars* by modern scholars). The "finger" was a Babylonian unit of angular measure—usually 1/12°. In the second entry is the notice of a conjunction of the Moon with Mars, as well as a notice of the first appearance of Saturn. Saturn made its morning rising and was in the sign of Pisces when it did so. Other planetary phenomena regularly listed in the diaries include the beginning and the end of retrograde motion (i.e., the stations). The diaries also include much nonastronomical information: reports of the weather, changes in the level of the river Euphrates, monthly reports of the prices of essential commodities such as wool, barley, and sesame, reports of monstrous births, and so on.

There is no parallel to these astronomical diaries in Greek history. In our culture, it is popular to disparage bureaucrats. But it was the existence of a centralized government and of a stable bureaucracy that made scientific astronomy possible in Babylonia. The absence of these features from Greek civilization put planetary astronomy outside of the realm of possibility for the early Greeks. To this must be added the early Greek propensity for mere philosophizing rather than making careful observations. When, a few centuries later, the Babylonian astronomers turned to the task of constructing a theory of the planets with quantitative, predictive power, they had plenty of observational data to work from.

7.2 THE LOWER PLANETS: THE CASE OF MERCURY

In the next several sections we examine the behavior of the planets in more detail, staying close to the phenomena, without imposing any particular theoretical view. Then, beginning with section 7.6, we turn to the history of planetary theory among the Greeks and the Babylonians. In the present section we examine the motion in longitude of the inferior planets, using Mercury as an example. (Again, we ignore the motion in latitude.) We begin with the following excerpt from table 7.1:

Excerpt from table 7.1

Year	Date	Sun	Mercury	Elongation
1976	Apr 1	12°	*11	1° W
1976	Apr 11	22	32	10° E
1976	Apr 21	31	51	20° E
1976	May 1	41	61	20° E
1976	May 11	51	⌈63⌉	12° E
1976	May 21	60	*58	2° W
1976	May 31	70	⌊54⌋	16° W
1976	Jun 10	79	56	23° W
1976	Jun 20	89	66	23° W
1976	Jun 30	99	82	17° W
1976	Jul 10	108	*101	7° W
1976	Jul 20	118	123	5° E
1976	Jul 30	127	143	16° E

1976	Aug 9	137	159	22° E
1976	Aug 19	146	173	27° E
1976	Aug 29	156	183	27° E
1976	Sep 8	166	188	22° E
1976	Sep 18	175	183	8° E

The first four columns give the year, the date, the longitude of the Sun, and the longitude of Mercury, all copied directly for table 7.1. We have marked a retrogradation of the planet (in May, 1976) with square brackets, as before. The planet's conjunctions with the Sun are marked with asterisks. Thus, on dates near April 1, May 21, and July 15, Mercury and the Sun had the same longitude.

Two Kinds of Conjunctions

Examination of the excerpt shows that Mercury has two different kinds of conjunctions; we shall call them *prograde* and *retrograde conjunctions*. Around April 1 and again around July 15, 1976, Mercury was in conjunction with the Sun and in prograde motion. (Thus, Mercury's longitude was increasing.) But during the conjunction of May 21, the planet was in retrograde motion (longitude decreasing). Evidently, corresponding to the conjunctions and oppositions of the superior planets, Mercury has conjunctions only, but compensates by having two different kinds.

Limited Elongations

The fifth column gives Mercury's *elongation from the Sun*, that is, the difference between the longitudes of the two bodies. Column five is obtained by subtracting columns three and four. The elongation is marked W when the planet lies west of the Sun and E when it lies east. Clearly, Mercury cannot be found at just any angular distance from the Sun: Mercury's elongations are limited in size. Mercury's greatest elongations from the Sun are not always exactly the same. In our example, the greatest (eastward) elongation of April, 1976, was 20°, while the greatest (westward) elongation of June, 1976, was 23°. (The greatest elongations are marked in the excerpt by braces.) Mercury's greatest elongations vary between 18° and 28°.

The Synodic Cycle

With this terminology established, let us now examine in detail the motion of Mercury between April 1 and July 20. The principal events in the synodic cycle are the planet's *conjunctions* with the Sun, the *greatest elongations* from the Sun (which are the times of best visibility), and the *stations* (which mark the beginnings and ends of the retrograde motion). We can see that they occur in the following order:

Prograde conjunction	April 1
Greatest eastward elongation	April 26
First station	May 11
Retrograde conjunction	May 21
Second station	May 31
Greatest westward elongation	June 15
Prograde conjunction	July 15

Four other events in the synodic cycle must be mentioned, which are not as easy to read directly off table 7.1 but which played a very important role in Babylonian astronomy. These are the *planetary phases*. Around the time of a conjunction, the planet is invisible, for it is too near the Sun. A few days after the retrograde conjunction of May 21, Mercury emerged from the Sun's rays, to the right (or west) of the Sun, and could be seen rising in the east just before sunrise. This event is called the *first morning rising* and marks Mercury's first appearance as a morning star. The planet was visible as morning star until a few days before the prograde conjunction of July 15. Then the star made its *last morning rising* and disappeared into the rays of the Sun.

A few days after the the prograde conjunction of July 15, the planet emerged to the left (east) of the Sun and could be seen in the west shortly before sunset. This is the *first evening setting*. The planet remained visible as an evening star until shortly before the next retrograde conjunction, around September 23. Then it made its *last evening setting*. The importance of first and last visible risings and settings in Babylonian astronomy can hardly be overstressed. The oldest planet observations we possess from Babylonia are notices of the first and last visible risings and settings of Venus from the reign of Ammi-saduqa (mentioned in sec. 1.2).

7.3 OBSERVATION: OBSERVING THE PLANETS

Of the five planets visible to the naked eye, three repay close watching: Venus, Mars, and Jupiter. The remaining two are less suitable for a program of observation—Mercury because it is so rarely visible, and Saturn because its motion is so slow.

The best way to get started is to have someone who knows the planets point them out to you at night. A number of magazines for amateur astronomers provide monthly notices on the state of the heavens, including charts of the positions of the planets among the constellations. Two such magazines are *Astronomy* and *Sky and Telescope*.

After you become acquainted with the planets, you will find that you can recognize them, even after long disappearances. First, the planets are always found in the zodiac constellations. So, if you see a "star" in Taurus which is not on your star chart, it must be a planet. Second, Venus, Mars, and Jupiter are all bright, usually brighter than any stars in their vicinity. (Mercury and Saturn are dimmer.) Third, the planets do not twinkle as the fixed stars do, but shine with a steady light. Finally, the colors of the planets are helpful in distinguishing them from one another. In the tenth book of the *Republic*, Plato describes their colors in the following terms: Saturn and Mercury are yellow; Venus and Jupiter are very white; Mars is a little red.

Your observations must be good enough, and extend over a long enough period of time, to reveal the motion of a planet through the star field. Especially, you will want to observe the planet reverse direction on entering or leaving retrograde motion. This requires making at least one observation a week for one or two months. You will need a good star chart on which to record the positions of the planets as you see them.[9] The observations should be made carefully, by the method of alignments. Find in the sky a line between two identifiable stars on which the planet lies. Relative distances may be estimated by counting fists or finger widths. Suppose that you sight Mars on a straight line between two stars with which you are familiar. Holding your hands out at arm's length, you find that the planet is two fingers from the upper star and four fingers from the lower star (i.e., twice as close to the upper star). This information will allow you to mark the planet's position on your star chart rather accurately. Verify your first alignment by other alignments using different stars.

After marking the planet's position on your chart, be sure to label the mark with the name of the planet and the date of the observation. In many cases, an observation made only a week or two later will reveal a shift in the planet's position.

7.4 THE UPPER PLANETS: THE CASE OF MARS

The Synodic Cycle of a Superior Planet

Consider the synodic cycle of Mars. The key events are conjunction, opposition, and the two stations. From the portion of table 7.1 for 1979–1981, we see that they occur in the following order:

Event	Date	Mars	Sun
Conjunction	1979 Jan 16	297	296
First station	1980 Jan 11	165	290
Opposition	1980 Feb 25	336	156
Second station	1980 Apr 5	146	16
Conjunction	1981 Mar 31	11	11

To these must be added the first and last visible risings and settings, which were so important in Babylonian astronomy. These *planetary phases* for a superior planet are different from those of an inferior planet (discussed in sec. 7.2). The superior planets can be at any angular distance from the Sun. Moreover, because the superior planets move more slowly than the Sun, the Sun overtakes them. Therefore, the phases of the superior planets are rather similar to those of the fixed stars (the so-called dock-pathed stars that lie near the ecliptic; see sec. 4.9).

Two phases are particularly important: the morning rising and the evening setting. Around the conjunction of January, 1979, Mars was invisible. A couple of weeks later, the Sun moved on, leaving Mars behind. Now Mars could be seen rising in the east, just before sunrise. This is Mars's *morning rising*. A few weeks before the conjunction of March, 1981, the Sun was again close to Mars. Thus, Mars could be seen setting in the west, just after sunset. The next night, Mars was no longer visible. The last observable setting was Mars's *evening setting*. The planet remained invisible between its evening setting and its next morning rising.

Constructing a Table of Oppositions

For the superior planets, Babylonian astronomy focused on predicting the first appearance (morning rising) and the disappearance (evening setting), as well as the beginning and end of retrograde motion. There was also some attention devoted to the oppositions, but these were not as important.

Ptolemy complained about just this circumstance in *Almagest* IX, 2. Most of the older observations available to Ptolemy were of stations and phases (first and last visibilities). These were undoubtedly Babylonian records, extracted from the astronomical diaries, that had come to Ptolemy through Hipparchus. Ptolemy rightly complains that stations and phases are difficult to observe with any precision. Even the *day* on which a station occurs is uncertain, since the planet may scarely move for a week or more. The dates of first morning risings and last evening settings are uncertain because they are affected by atmospheric conditions and differences in the eyesight of the observers. In his own investigations of the superior planets, Ptolemy therefore relied almost exclusively on the oppositions.

Here we will examine a list of some oppositions of Mars, learn how it was compiled, and see what it can be used for. Table 7.2 lists the oppositions of Mars that occurred between 1948 and 1984. The first column gives the date of each opposition, the second gives the Julian day number, and the third gives the longitude of the planet at the moment of opposition. The data for the years 1971 and later were taken from table 7.1. The earlier oppositions are included as extra information.

Admittedly we are being a trifle sloppy with the term "opposition": the dates and longitudes listed in table 7.2 apply strictly to the *centers of the planet's*

TABLE 7.2. Oppositions of Mars, 1948–1984

Date	Julian Day	Longitude
1948 Feb 17	243 2599	147 1/2°
1950 Mar 25	3366	181
1952 May 5	4138	220
1954 Jun 25	4919	273 1/2
1956 Sep 11	5728	349
1958 Nov 14	6522	54
1960 Dec 27	7296	99
1963 Feb 1	8062	135
1965 Mar 10	8830	168
1967 Apr 18	243 9599	204 1/2
1969 Jun 5	244 0378	250
1971 Aug 9	1173	317
1973 Oct 24	1980	32
1975 Dec 13	2760	84
1978 Jan 21	3530	122
1980 Feb 26	4296	155 1/2
1982 Apr 2	5062	189 1/2
1984 May 14	244 5835	230

retrograde arcs and not to its actual oppositions. (We will ignore the small errors entailed by blurring this distinction in order to streamline the analysis.)[10] The manner by which they were obtained is best explained by example. Let us consider the retrogradation of 1971. The entries from table 7.1 are reproduced in the excerpt below.

Excerpt from table 7.1

	Date	J.D. 244	Sun	Mars
1971	Jun 17	1120	85	319
	Jun 27	1130	95	321
	Jul 7	1140	104	322
	Jul 17	1150	114	322
	Jul 27	1160	123	320
	Aug 6	1170	133	318
	Aug 16	1180	143	315
	Aug 26	1190	152	313
	Sep 5	1200	162	312
	Sep 15	1210	172	312
	Sep 25	1220	182	314
	Oct 5	1230	191	316

Mars's retrograde arc stretches from longitude 322° to 312°. The midpoint of this arc is 317° (317 lies halfway between 312 and 322). This is the figure we have put into table 7.2 for the retrogradation of 1971. The planet's longitude was 318° on August 6 and 315° on August 16. The date when Mars reached longitude 317° therefore fell sometime between the sixth and the sixteenth of August. Linear interpolation yields August 9, which was J.D. 244 1173, and this is the date we have listed in table 7.2.

Using the Table of Oppositions

Let us see what use can be made of table 7.2. First of all, let us note its general features. The oppositions follow one another at intervals of roughly two years and a month, or two years and two months. Furthermore, the longitude increases from one opposition to the next. The *average* increase in longitude is about 50°, but there is considerable variability. (The shifts range from 34° to 75°.) Of course, Mars does not simply travel forward 50° between one opposition and the next. Rather, it goes forward all the way around the ecliptic, *plus about 50° more.*

Period Relations   The table of oppositions reveals a connection among the planet's tropical motion, its synodic motion, and the motion of the Sun. Consider two oppositions of Mars that occurred at the same longitude. Inspection of the table shows that no two oppositions satisfy this condition exactly; however, the oppositions of 1965 and 1980 occurred at nearly the same part of the sky—the former at longitude 168°, and the latter at 155 1/2°, so the discrepancy is only 12 1/2°. Now, between any two entries in table 7.2, the planet completed a whole number of synodic cycles, since the synodic cycle is nothing other than the period from one opposition to the next. Between the oppositions of 1965 and 1980, seven synodic cycles elapsed. But since these two oppositions occurred at nearly the same longitude, the time interval between them also contained very nearly a whole number of *tropical* cycles. The planet made a whole number of trips around the ecliptic.

How many trips? Between the opposition of 1965 and the next one in 1967, the planet completed one tropical revolution, plus the advance from 168° to 204 1/2°. There is one complete tropical revolution between each pair of successive oppositions, plus a bit. By 1980, the bits have added up to an additional complete revolution, so that the opposition of that year falls again at the same place as the opposition of 1965. Between 1965 and 1980, Mars therefore completed 7 + 1 = 8 tropical cycles, if we ignore the 12 1/2° discrepancy.

How much time elapsed between these two oppositions? The first opposition occurred near the beginning of March, 1965, and the second, near the end of February, 1980. The interval was therefore 15 years, almost exactly, the discrepancy being only about two weeks. Indeed, if the two oppositions really did occur at the same longitude, a whole number of years logically must separate them. At opposition the Sun is diametrically opposite the planet. But if at the two oppositions the Sun has the same longitude, then the two oppositions must occur at the same time of year.

To summarize, during the fifteen complete years that elapsed between the oppositions of Mars in 1965 and 1980, we find that seven complete synodic cycles and eight complete tropical cycles elapsed. Note the relation:

$$8 + 7 = 15 \quad \text{(Mars)}.$$

Number of tropical + number of synodic = number of years
    cycles elapsed   cycles elapsed   elapsed.

At first sight it may seem odd that the tropical and synodic motions of Mars should have any connection with the number of years gone by, which after all is determined by the motion of the Sun. But let us recall that the occurrence of retrograde motion seems somehow to be determined by the Sun: Mars retrogrades while in opposition to the Sun. The period relation above is another manifestation of this fact.

Further examination of the table of oppositions (table 7.1) will remove all suspicion that our equality 8 + 7 = 15 is a fluke. Actually, this relation is not perfectly accurate, because of the minor discrepancies of 12 1/2° in the longitude and two weeks in the time. We can do better if we use instead the oppositions of 1948 and 1980. Here the longitudes are 147 1/2° and 155 1/2°, so the discrepancy is only 8°. The dates of the oppositions, February 17 and February 26, disagree by only nine days. In the almost exactly 32 years between the two oppositions, we find 15 synodic cycles and 17 tropical cycles completed. (Between successive oppositions, the planet completes one tropical revolution, plus a bit more. In this case the bits add up to two extra complete revolutions, as one can easily see by following the progress of the longitude numbers between 1948 and 1980.) We have

17 tropical and 15 synodic cycles completed in 32 years;

$$17 + 15 = 32 \quad \text{(Mars, better result).}$$

This *period relation* is not peculiar to Mars. Similar rules apply to Jupiter and Saturn, that is, to all the superior planets. Of course, in the cases of these other planets different numbers apply.

This rule does not, however, apply to the inferior planets, Mercury and Venus. Since these planets have the same tropical period as the Sun, the period relation is obviously

$$\begin{array}{cc} \text{number of tropical} = \text{number of years} & \text{(Venus or} \\ \text{cycles elapsed} \qquad \text{elapsed} & \text{Mercury).} \end{array}$$

The number of synodic cycles elapsed is independent.

Evaluating the Lengths of the Tropical and Synodic Periods   The period relation gives us a way of measuring the tropical and synodic periods. According to our first form of the relation, Mars completes 8 tropical cycles in 15 years. One tropical period must therefore be $15/8 = 1.875$ years. Similarly, the synodic period is $15/7 = 2.143$ years. Better values for the periods can be obtained from our second, more accurate version of the period relation. Thus, the tropical period of Mars is $32/17 = 1.882$ years and the synodic period is $32/15 = 2.133$ years. These results agree with the best modern measurements to within a fraction of a day. Later, we shall see how they can be improved even further.

7.5 EXERCISE: ON THE OPPOSITIONS OF JUPITER

1. Mark all the retrograde arcs of Jupiter with brackets in table 7.1.
2. Compile a list of the oppositions of Jupiter over the whole period for which you have data available. Use as your model table 7.2. It will be sufficiently accurate to use the centers of the retrograde arcs, just as we did in section 7.4. Interpolate between the data points, if necessary, to obtain the day on which Jupiter reached the middle of each retrograde arc.
3. Use your table of oppositions of Jupiter to investigate the planet's periods. Find two oppositions that occurred at about the same longitude. Check to see whether the motions that took place between these oppositions obey the rule

$$\begin{array}{ccc} \text{number of tropical} + \text{number of synodic} = \text{number of years} \\ \text{cycles elapsed} \qquad \text{cycles elapsed} \qquad \text{gone by.} \end{array}$$

   What are the numerical values that apply to this period relation for Jupiter?
4. Use your period relation for Jupiter to make estimates of the lengths of the planet's tropical and synodic periods.

7.6 THE SPHERES OF EUDOXUS

Eudoxus on the Planets

Eudoxus of Cnidus was of the generation intermediate between those of Plato and Aristotle. He spent some time at Athens, where he undoubtedly knew both men. Eudoxus's astronomical works had a profound influence on the development of Greek astronomy and cosmology. As discussed in section 2.1, Eudoxus's *Phenomena*, the first systematic description of the Greek constellations and the celestial sphere, was the inspiration for the long poem of the same title by Aratus of Soli. Eudoxus was also the first of the Greeks to devise a geometrical-mechanical theory of the motions of the planets. This planetary theory was described in a book, *On Speeds*, which has not survived. Although Eudoxus's planetary theory was soon rejected, it was of tremendous importance, for two reasons. First, it put the Greeks on the road to a geometrical planetary theory. And second, its basic principles set the pattern for the development of Greek cosmology.

Our knowledge of Eudoxus's planetary theory is based entirely on accounts by Aristotle and Simplicius.[11] Aristotle, in his *Metaphysics*, discusses Eudoxus's system briefly, mentions some modifications of it proposed by Callippus, and finally offers a modification of his own, intended to make the system physically more workable. As Aristotle probably knew Eudoxus personally and probably had access to his writings, his account, though disappointingly brief, may be relied on with some confidence.

Simplicius was a sixth-century A.D. commentator on Aristotle. He discusses Eudoxus's system in considerable detail in his commentary on Aristotle's *On the Heavens*. Simplicius wrote nearly 900 years after Eudoxus and he makes it clear that he did not have a copy of Eudoxus's work, which was undoubtedly already lost. Rather, Simplicius drew on Aristotle, as well as on a book by Sosigenes, the Peripatetic philosopher (second century A.D., not to be confused with the Sosigenes who helped Julius Caesar reform the calendar). This Sosigenes, whose works are also lost, drew on Eudemus (a younger contemporary of Aristotle), who wrote a *History of Astronomy* (also lost), which discussed Eudoxus's system. Thus, Simplicius's *Commentary* is many steps removed from Eudoxus. Although Simplicius provides more detail than Aristotle, it is clear that he must be used with some caution.

Historians of astronomy largely ignored Eudoxus's system as a bizarre contrivance until the nineteenth century, when the details of the mechanism were reconstructed by Ideler, Apelt, and, especially, the Italian astronomer Schiaparelli.[12] Most modern accounts of Eudoxus's system follow Schiaparelli.[13] Let us begin by looking at what Aristotle had to say of his contemporary's system:

EXTRACT FROM ARISTOTLE

Metaphysics XII, 8, 1073b17–1074a15

Eudoxus assumed that the Sun and Moon are moved by three spheres in each case; the first of these is that of the fixed stars, the second moves about the circle which passes through the middle of the signs of the zodiac, while the third moves about a circle latitudinally inclined to the zodiac circle; and, of the oblique circles, that in which the Moon moves has a greater latitudinal inclination than that in which the Sun moves.

The planets are moved by four spheres in each case; the first and second of these are the same as for the Sun and Moon, the first being the sphere of the fixed stars which carries all the spheres with it, and the second, next in order to it, being the sphere about the circle through the middle of the signs of the zodiac which is common to all the planets; the third is, in all cases, a sphere with its poles on the circle through the middle of the signs; the fourth moves about a circle inclined to the middle circle [the equator] of the third sphere; the poles of the third sphere are different for all the planets except Aphrodite and Hermes, but for these two the poles are the same.

Callippus agreed with Eudoxus in the position he assigned to the spheres, that is to say, in their arrangement in respect of distances, and he also assigned the same number of spheres as Eudoxus did to Zeus and Kronos respectively, but he thought it necessary to add two more spheres in each case to the Sun and Moon respectively, if one wishes to account for the phenomena, and one more to each of the other planets.

But it is necessary, if the phenomena are to be produced by all the spheres acting in combination, to assume in the case of each of the planets other spheres fewer by one [than the spheres assigned to it by Eudoxus and Callippus]; these latter spheres are those which unroll, or react on, the others in such a way as to replace the first sphere of the next lower planet in the same position [as if the spheres assigned to the respective planets above it did not exist], for only in this way is it possible for a combined system to produce the motions of the planets. Now the deferent spheres are, first, eight [for Saturn and Jupiter], then twenty-five more [for the Sun, the Moon, and the three other planets]; and of these, only the last set [of five] which carry the planet placed lowest [the Moon] do not require any reacting spheres. Thus the reacting spheres for the first two bodies will be six, and for the next four will be sixteen; and the total number of spheres, including the deferent spheres and those which react on them, will be fifty-five. If, however, we choose not to add to the Sun and Moon the [additional deferent] spheres we mentioned [i.e., the two added by Callippus], the total number of the spheres will be forty-seven. So much for the number of the spheres.[14]

Sun and Moon   According to Aristotle, Eudoxus introduced three spheres each for the Sun and Moon. Let us consider the Moon. Refer to figure 7.7, which represents a two-sphere simplification of the system. (We shall take up Eudoxus's complete system in a moment.) The Earth (not shown) is a point at the center of the system. Sphere 1 is the sphere of the fixed stars, which rotates westward about axis PQ in a day. The Moon M rides on the ecliptic, which is the "equator circle" of sphere 2. The axles of sphere 2 are set into the surface of sphere 1 at A and B. The angle between A and P is equal to the obliquity of the ecliptic—about 24°. Sphere 2 rotates eastward about axis AB in a month. In this way, the Moon is carried eastward around the zodiac each month, while the whole sky (including the Moon) is carried westward about axis PQ in a single day.

But, of course, the Moon does not travel exactly on the ecliptic. Rather, the Moon's path is inclined (by about 5°) with respect to the ecliptic. This is why there are not eclipses of the Moon every month. Thus, Eudoxus's system for the Moon, as described by Aristotle, requires a third sphere. Refer to figure 7.8. In the complete system, the Moon M rides on the "equator circle" of sphere 3, which rotates eastward about axis CD in a month. The Moon's path is inclined (by about 5°) with respect to the ecliptic. Thus, the axles of sphere 3 are set into the zodiac sphere 2 with a slight inclination: the angle between C and A is about 5°. Point N is a *node* of the Moon's orbit: it is one of the two places where the Moon crosses over the plane of the ecliptic in the course of its monthly journey. If the Moon happens to be full when it reaches N, there will be a lunar eclipse.

Now, successive eclipses of the Moon at the same node do not occur in the same zodiac sign. Rather, the eclipses gradually work their way westward around the zodiac. Thus, if there is an eclipse when the Moon is in the Twins, later there will be an eclipse with the Moon in the Bull, and still later, an eclipse with the Moon in the Ram. The eclipses will return to the Twins after an interval of about 18.6 years. Thus, the nodes of the Moon's orbit must work their way westward around the zodiac in 18.6 years. So, in figure 7.8, sphere 2 must rotate westward about axis AD in 18.6 years.

Eudoxus's system thus explains a good deal: it accounts for the daily motion of the Moon, the Moon's motion in longitude around the ecliptic, and its motion in latitude. It explains, as well, the displacement of successive eclipses

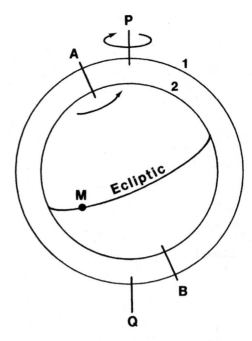

FIGURE 7.7. Simplified two-sphere model for the motion of the Moon. Sphere 1 rotates westward once a day. Sphere 2 rotates eastward once a month.

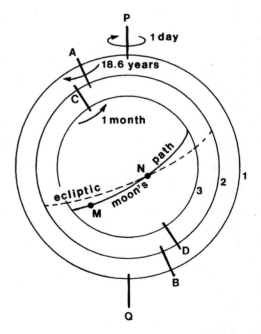

FIGURE 7.8. Eudoxus's model for the motion of the Moon. Sphere 1 produces the daily motion. Sphere 3 produces the monthly motion around a path slightly inclined to the ecliptic. Sphere 2 produces the motion of the nodes of the Moon's orbit and explains why eclipses do not occur always in the same zodiac sign. A similar model was applied to the motion of the Sun.

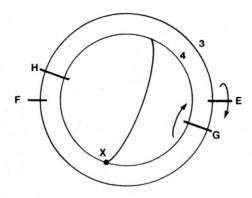

FIGURE 7.9. Eudoxus's device for producing the retrograde motion of a planet. Spheres 3 and 4 turn at the same rate about axes slightly inclined to one another. The planet X rides on the "equator" of the inner sphere.

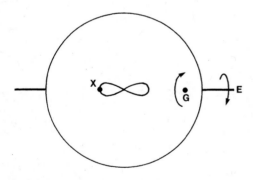

FIGURE 7.10. The figure-eight path (called a *hippopede*) of point X that results from the two motions shown in Figure 7.9. The width of the figure eight is greatly exaggerated.

westward around the zodiac. In our discussion of the system, we have added numerical values and details not present in Aristotle's discussion. But the essential features of the model are not in doubt. It is interesting that Simplicius, in his own account, botches the discussion of the Moon's system by reversing the order of spheres 2 and 3. In Simplicius's version, the Moon would stay north of the ecliptic for nine years, then remain south of the ecliptic for nine years! This is an example of why we cannot rely on Simplicius—even though it might be tempting to do so, for he provides more detail.

As is clear from the first paragraph of the extract from Aristotle, Eudoxus applied a similar three-sphere system to the motion of the Sun. That is, figure 7.8 can also represent the motion of the Sun. In this case, M will represent the Sun. Sphere 1 must still rotate to the west once a day. But now, of course, sphere 3 rotates to the east in the Sun's own tropical period—one year. The strange thing is sphere 2. According to Aristotle, angle AC is less for the Sun than for the Moon. But Aristotle says clearly that the Sun, too, has a motion in latitude—that it does not ride exactly on the ecliptic. This, of course, was quite mistaken. Probably the idea that the Sun has a motion in latitude arose from sloppy observations of the Sun's rising point at summer solstice. If one year the solstitial Sun appeared to rise a little farther north than it had in some previous year, one would infer that the Sun has a motion in latitude. This mistaken idea was still current in the early second century A.D., for Theon of Smyrna says that the Moon's motion in latitude is ±6° and the Sun's ±1/2°.[15] Even Simplicius, in his account of Eudoxus's system, still seems to accept the Sun's motion in latitude. This is somewhat strange since, after Ptolemy's time, there was no excuse for an astronomer to hold such a view. This is yet another example of Simplicius's inadequate understanding of astronomy.

The Planets   In the case of the planets, we must account not only for the daily westward motion and the tropical motion around the zodiac, but also for retrograde motion. In the extract above, we learn that Eudoxus used four spheres for each of the planets. The two outer spheres are essentially the same as in figure 7.7. Sphere 1 produces the daily westward motion. Sphere 2 carries the planet eastward around the zodiac in the planet's tropical period. Thus, for Mars, sphere 2 would complete one revolution in about 2 years; for Jupiter, in about 12 years.

To produce retrograde motion, Eudoxus inserted a two-sphere assembly inside sphere 2. For the two-sphere assembly, see figure 7.9. Sphere 3 and sphere 4 both execute one rotation in a time equal to the planet's synodic period. The axles of sphere 4 are inserted into sphere 3 at G and H. The angle between the two axes of rotation is small, and the two spheres rotate in opposite directions. The planet is a point X located on the "equator" of the inner sphere. The question is: what sort of motion of X results from the combination of two rotational motions? In fact, X executes a sort of figure-eight motion, as shown in figure 7.10. The resulting figure eight is quite narrow. If the angle between axes EF and GH is 5°, the figure eight will be 10° long, but only 2/10° wide at the two widest spots.

If axles E and F of the two-sphere assembly are inserted into the ecliptic of sphere 2 in figure 7.7, the result is Eudoxus's complete model, as shown in figure 7.11. The planet rides around in a figure-eight pattern (produced by spheres 3 and 4), while the figure eight is carried eastward about the ecliptic by the motion of sphere 2. The combination of motions therefore produces a steady eastward motion of the planet around the ecliptic with a superimposed back-and-forth motion. If the motion backward on the figure eight is fast enough, it can more than make up for the steady eastward motion of sphere 2, and retrograde motion will result.

Eudoxus's model explains in at least a rough way the basic properties of planetary motion: the westward daily motion, the eastward zodiacal motion, and the occasional retrogradations. And it does this in a way that was in keeping with the principles of Greek celestial physics. All the motions are uniform and circular. And the whole system has one single center—the center of the cosmos, where the Earth lies.

What about the numerical values of the important parameters—the rotation rates and the inclinations of the axes? The rotation periods are simple: sphere 1 rotates once a day; sphere 2, in the planet's tropical period; spheres 3 and 4, in the planet's synodic period. Moreover, Simplicius explicitly connects the tropical and synodic periods to the rotations of the spheres in just this way. According to Simplicius, Eudoxus assigned the following values to the tropical and synodic periods:

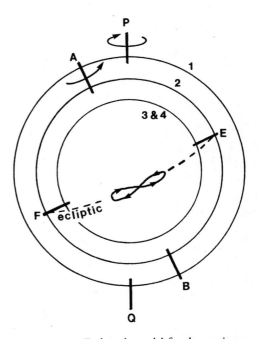

FIGURE 7.11. Eudoxus's model for the motions a planet. Sphere 1 produces the daily westward motion. Sphere 2 produces the eastward motion around the ecliptic. Spheres 3 and 4 together produce the back-and-forth motion required for retrogradation.

Periods of the planets according to Eudoxus

Planet	Tropical period	Synodic period
Mercury	1 year	110 days
Venus	1 year	19 months
Mars	2 years	8 months 20 days
Jupiter	12 years	13 months
Saturn	30 years	13 months

Actual periods of the planets

Planet	Tropical period	Synodic period
Mercury	1 year	116 days
Venus	1 year	584 days
Mars	1.88 years	780 days
Jupiter	11.86 years	399 days
Saturn	29.42 years	378 days

Thus, most of Eudoxus's periods are reasonably good approximations. The only glaring problem is the value of Mars's synodic period. It is scarcely possible that Eudoxus could have made such an error, for it takes only the most casual observation to realize that the synodic period of Mars is over two years. So we have here either a misunderstanding on the part of Simplicius or a corruption of his text.

The most delicate question is then the inclination of the axes of the two spheres (numbers 3 and 4) responsible for producing the figure eight Some modern commentators have attempted to deduce the values of these angles of inclination that would produce the best agreement with the actual planetary motions. But, in fact, we have no idea what values Eudoxus assigned to these angles—or even if he assigned any numerical values at all.

Eudoxus's Intentions   It is unlikely that Eudoxus gave numerical values for the angle between the axes of spheres 3 and 4. In the case of Venus and Mars, he could not possibly have done so, for it turns out that, for these two planets, Eudoxus's model is not actually capable of producing retrogradation at all! Let T denote the length of the tropical period and S the length of the synodic period. Let i denote the angle of inclination between the axes of spheres 3 and 4 (the two spheres responsible for producing the figure eight). It turns out that retrograde motion is possible only if

$$T \sin i > S.$$

Since $S > T$ for Mars and Venus, retrograde motion will not occur, no matter what value of i is chosen.

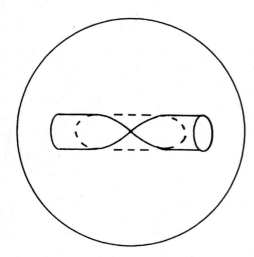

FIGURE 7.12. The hippopede, regarded as an abstract mathematical curve, can be produced by the intersection of two surfaces. A cylinder pierces a sphere and is tangent to the sphere from the inside. The hippopede is the curve of intersection.

Even in the cases of the other planets, it is unlikely that Eudoxus gave values for *i*. There was no tradition of accurate planetary observation among the Greeks that might have supplied the raw data (such as observed lengths of the retrograde arcs) required for fixing the parameters. Moreover, such an endeavor would have been alien to the spirit of Greek astronomy in the fourth century B.C.

Eudoxus probably meant his system to serve two functions. First, it did explain the basic facts of planetary motion in way that was consistent with accepted physical principles. This did not mean the model was supposed to be literally true, or that it was believed to represent the motions in quantitative detail. Rather, the model was intended as a sort of physical allegory: the universe might work more or less like this. Eudoxus's models can be seen, then, as a continuation of the tradition of physical speculation characteristic of Ionian philosophy of the preceding century.

Thus, one should not attempt to read too much significance into the details of the model. Simplicius was one of the first to go astray in this way. For example, Simplicius says that Eudoxus used the width of the figure eight to account for the planets' latitudes (i.e., their departures from the plane of the ecliptic), and a number of modern writers have accepted this explanation. But, as remarked above, to produce retrograde arcs of modest length, only very narrow figure eights are required. The model can almost be visualized as producing a back-and-forth motion in the plane of the ecliptic (superimposed on the steady eastward motion due to sphere 2). Moreover, the whole pattern of latitudinal motion is wrong: the model would require a planet to reach both the northern and the southern extremes in latitude twice in each synodic cycle—instead of reaching each limit once, as is the actual case. So it is pretty clear that Eudoxus only wanted to account for the three motions: the daily motion, the tropical motion around the ecliptic, and retrogradation.

Second, Eudoxus intended his model to serve as an arena for proving difficult and interesting geometrical theorems. The essential problem posed by Eudoxus's system is of this sort: given that a point (*X* in fig. 7.9) moves with a certain combination of motions, deduce the figure traced out. This belongs to a traditional class of problems in Greek geometry.[16]

According to Simplicius, Eudoxus called the figure eight traced out by the motions of spheres 3 and 4 a *hippopede*—that is, a horse fetter or hobble. Thus, it is clear that Eudoxus understood the general character of the curve. In modern times, the geometrical properties of the hippopede were worked out by Schiaparelli. It turns out, for example, that the hippopede is the intersection of a sphere with a cylinder that pierces it and touches it from inside (see fig. 7.12). Problems of this type also were a part of Greek mathematics. For example, Eudoxus's teacher, Archytas, is said to have solved the problem of the duplication of the cube by means of the intersection of three surfaces of revolution—a cone, a cylinder, and a torus.[17] Thus, the demonstration of the geometrical properties of the hippopede, including its equivalence to the intersection of a sphere with a cylinder, was well within the powers of Greek mathematics of Eudoxus's time.

The Modifications of Callippus and Aristotle

Aristotle tells us in the extract above that Eudoxus's planetary system was modified by Callippus (ca. 330 B.C.). Callippus needed one more sphere each for Mercury, Venus, and Mars. And he added two more spheres each to the systems for the Sun and the Moon.

According to Simplicius, the changes in the systems for the Sun and Moon were required to explain the solar and lunar anomaly—the fact that the Sun and the Moon do not appear to move at a uniform speed around the zodiac. Simplicius says explicitly that he did not have any work by Callippus to go by, but was relying on Eudemus (as quoted by Sosigenes). Nevertheless, this explanation seems quite plausible. Let us recall that by the time of Callippus, the inequality in the lengths of the seasons was well established (see n. 10 in chap. 5). The lunar anomaly is even more striking. In twenty-four hours, the Moon can move by as little as 11.7° or by as much as 14.6°. We do not know exactly how Callippus proposed to achieve a variable speed by the addition of two spheres.

Why Callippus added one sphere each to the systems for Mercury, Venus, and Mars we cannot say. Simplicius says that Eudemus explained this addition quite clearly—but then he does not tell us what Eudemus had to say! Perhaps Callippus meant to correct the obvious defect of the systems for Venus and Mars—that is, the fact that Eudoxus's models for these planets did not actually produce retrograde motion.

Aristotle's own modifications were motivated by completely different concerns. Aristotle wanted, above all else, to make the whole system into a workable mechanism. Thus, he proposed to insert the reacting or *unrolling* spheres. The point of this modification was to prevent the spheres of the outer planets from distorting the motions of the inner planets.

Consider the four-sphere system for Saturn in figures 7.11 and 7.9. In the extract from Aristotle, these are called the *deferent* or "carrying" spheres, because they carry the planet and produce its three motions. A similar four-sphere system for Jupiter is to be inserted inside the system for Saturn. But then the spheres for Jupiter will be slung wildly about by the motions of the Saturnian spheres.

Aristotle's solution is as follows. Sphere 4 for Saturn carries Saturn itself. Inside sphere 4 let there be a sphere 4′, which rotates about the same axis as sphere 4, but in the opposite direction. This sphere will unroll, or cancel out, the rotation of sphere 4. Sphere 4′ will therefore have the same motion as sphere 3. Inside sphere 4′ let there be a sphere 3′, which rotates about the same axis as but in the opposite direction to sphere 3. This will cancel out the motion of sphere 3. Similarly, inside sphere 3′, let there be a sphere 2′, which rotates about the same axis as, but in the opposite direction to, sphere 2. This will cancel out the motion of sphere 2. The result is that the innermost unrolling sphere (sphere 2′) is at rest with respect to sphere 1. That is, sphere 2′ rotates once a day about the poles of the equator. It can therefore serve as the receptacle for the system of Jupiter.

So, the sequence is (the sphere marked * actually carries the planet):

Spheres for Saturn		Spheres for Jupiter		Spheres for Mars
Deferent	Unrolling	Deferent	Unrolling	
1 2 3 4*	4′ 3′ 2′	1 2 3 4*	4′ 3′ 2′	etc.

In the same way, the system for Mars can be plugged into the innermost of the unrolling spheres of Jupiter. It is clear that Aristotle was far more concerned with making the system physically plausible than in accounting for some technical detail of planetary motion. This is perfectly consistent with the character of fourth-century Greek physics and astronomy.

Ancient Criticisms of Eudoxus

Simplicius says that the system of Eudoxus did not account for the phenomena—and not only phenomena that were discovered later, but also phenomena that were known in Eudoxus's time. As an example he mentions the fact that the planets appear sometimes to be closer to us and sometimes farther away. As Simplicius points out, this is especially clear in the cases of Venus and Mars. These planets appear much larger during retrogradation than at other times.

The Moon, too, varies noticeably in size during the course of the month. As Simplicius points out, this is clear not only from direct measurement of the Moon's angular size with instruments but also from the phenomenon of the annular eclipse. During some solar eclipses, even though the Moon covers the Sun centrally, a total eclipse is not produced; rather, an uncovered ring of Sun may be seen around the perimeter of the Moon.

According to Simplicius, "no one before Autolycus of Pitane" tried to account for this obvious variation in the distances of the celestial bodies, and even Autolycus did not succeed. Unfortunately, we have no idea of what sort of theory Autolycus proposed to explain the variation in distance. Simplicius also mentions the variation in the daily motion of the planets. He points out, quite rightly, that "the ancients" (meaning Eudoxus) did not even attempt to save these phenomena.

According to Simplicius, "the ancients" were not sufficiently acquainted with the phenomena, and the Greeks only learned enough about the phenomena when, at Aristotle's request, Callisthenes got hold of Babylonian planetary observations stretching over 31,000 years, down to the time of Alexander the Great, and sent them back to Greece! Callisthenes was the nephew of Aristotle. He went along on Alexander's campaign as a historian and specimen collector. He was executed by Alexander in 327 B.C. for disloyalty.

Simplicius's remark about the importance of Alexander's campaigns and about the role of Babylonian observations in the development of Greek astronomy is a fascinating mixture of naive gullibility and shrewd insight. The Babylonian astronomical records covering 31,000 years are of course, an absurd fiction. This story shows once again the aura of arcane knowledge that, to Greek eyes, surrounded Babylonian civilization. Simplicius's remark also shows his considerable tendency to overestimate the role of philosophers—especially Plato and Aristotle—in the development of Greek astronomy. We need not put any stock in the roll of Aristotle and Callisthenes in acquainting the Greeks with the phenomena. In fact, the period of greatest Babylonian influence on Greek astronomy came two centuries later. And yet, Simplicius is perfectly correct about the importance of Babylon to the development of Greek astronomy.

Finally, let us note that what counted as phenomena in need of explanation changed as Greek astronomy matured. For Eudoxus, the goal was to provide a philosophically and geometrically satisfactory explanation of the broad features of planetary motion. As we have seen, he ignored not only the planets' obvious variation in distance (revealed by changes in brightness), but also the anomaly of motion. It is significant that his theory was criticized most strongly, not for failing some precise numerical test, but for failing to explain the variations in brightness that were widely known and easily perceived without the aid of instruments.

7.7 THE BIRTH OF PREDICTION: BABYLONIAN GOAL-YEAR TEXTS

As we see from the example of Eudoxus, Greek thought about the planets in the fourth century was dominated by physical speculation and by the application of geometry to cosmological models with broad explanatory power but with no predictive capability. Babylonian thought at about the same time reveals completely different concerns. In Mesopotamia, a primary goal was achieving a predictive capacity. Remarkably, this may be done without much theoretical apparatus, as long as one has access to long sequences of planetary observations. The first successes in predicting the behavior of the planets came from the recognition that, over long enough time intervals, the patterns repeat. Somewhat later, the Babylonian scribes did achieve a planetary theory with an elaborate theoretical structure, which we shall study in section 7.10.

Great Cycles of the Planets

The secret to predicting the future behavior of the planets from their past behavior lies in making use of the *period relations* discussed in section 7.5. For each planet there are two things going on at once—the tropical motion eastward around the zodiac and the superimposed, back-and-forth synodic motion that is responsible for retrogradation. These two motions are snarled together so that the planet's behavior is not the same from one retrogradation to the next. Thus, a planet does not retrograde in the same part of the zodiac from one time to the next (figs. 7.3, 7.4, and 7.5). But, if we wait long enough, after a whole sequence of retrograde motions, the pattern will more or less repeat. The planet may need to go several times around the zodiac before anything approximating a repetition occurs.

For Venus, a very good period relation is

5 synodic periods = 8 tropical periods = 8 years.

(Since Venus is an inferior planet, the number of tropical cycles elapsed is equal to the number of years gone by.) So, after 8 years, everything about the motion of Venus must repeat—not exactly, but very nearly. For example, the planet must retrograde in the same part of the ecliptic, and at the same time of year, as it did eight years earlier. We shall call this 8-year period a *great cycle* of Venus. The Babylonians used the 8-year great cycle for predicting the behavior of Venus by the beginning of the Seleucid era.

In section 7.4, we discovered the following period relation for Mars:

7 synodic cycles = 8 tropical cycles = 15 years.

(Since Mars is a superior planet, the number of tropical cycles elapsed plus the number of synodic cycles elapsed is equal to the number of years gone by.) Thus, we could expect everything about the motion of Mars approximately to repeat after 15 years. We call this 15-year period a great cycle of Mars.

Of course, as table 7.2 shows, the 15-year great cycle for Mars is not terribly accurate. After 15 years, Mars does not reach opposition again at exactly the same part of the zodiac. One way to improve the predictive power of the great cycle scheme is to take this into account in making predictions. Another way is to use a more accurate great cycle. As discussed in section 7.4, a 32-year great cycle for Mars is a better approximation (15 synodic periods = 17 tropical periods = 32 years).

Predicting the Behavior of Venus Let us examine the behavior of Venus in 1972, using table 7.1. From table 7.1 we pick out the dates and longitudes of two notable events in the synodic cycle—the beginning and ending of retrograde motion (i.e., the first and second stations). We shall also pick out the days when Venus passed by some reference stars—the Pleiades (at longitude 59°) and Spica (203°).

Behavior of Venus in 1975

Passes by Pleiades	Apr 14	longitude 59°
First station	Aug 5	longitude 162°
Second station	Sep 14	longitude 145°
Passes by Spica	Dec 1	longitude 203°

If we wished to predict the behavior of Venus eight years later, in 1983, a good guess would be that everything would occur in just the same way. The beginnings and endings of retrograde motion would occur at the same places and on the same days. Venus would pass by important reference stars on the same day of the year. Let us see how well this works out by extracting from table 7.1 the Venus data for 1983:

Behavior of Venus in 1983

Passes by Pleiades	Apr 12	longitude 59°
First station	Aug 3	longitude 160°
Second station	Sep 12	longitude 142°
Passes by Spica	Nov 30	longitude 203°

The correspondence between 1983 and 1975 is amazingly good. Thus, we would rarely be off by more than a few days and a few degrees in trying to predict the behavior of Venus from one great cycle to the next. Furthermore, the next time we wanted to make a prediction, we could improve our accuracy by taking this slight defect into account and noting that after an 8-year cycle the events repeat about 2 days early.

Goal-Year Texts

The repetition of planetary patterns after each great cycle formed the basis of the first successful predictions of planetary phenomena by the Babylonian scribes. Indeed, there exists a whole category of cuneiform texts that make use of this method of prediction. These are called *goal-year texts*, a term introduced by Sachs.[18]

The great cycles attested in the cuneiform goal-year texts are these:

Jupiter	71 years	(= 6 tropical = 65 synodic periods)
Jupiter	83 years	(= 7 tropical = 76 synodic periods)
Venus	8 years	(= 8 tropical = 5 synodic periods)
Mercury	46 years	(= 46 tropical = 145 synodic periods)
Saturn	59 years	(= 2 tropical = 57 synodic periods)
Mars	47 years	(= 25 tropical = 22 synodic periods)
Mars	79 years	(= 42 tropical = 37 synodic periods)
Moon	18 years	

Most of these are considerably longer (and more accurate) than the rough great cycles mentioned above for Mars.

Suppose we wanted to predict the behaviour of all the planets during the year 1983, which would then be our *goal year*. One way to make the prediction would be to write out all the important phenomena that occurred for each planet one great cycle earlier. Thus, we would write out what Jupiter did in 1912 (one 71-year Jupiter great cycle earlier), what Venus did in 1975 (one 8-year Venus great cycle earlier), what Mercury did in 1937 (one 46-year Mercury cycle earlier), and so on. We would then have a goal-year text for 1983. And we would not be far off in any of our predictions for 1983.

In the cuneiform goal-year texts, the planets are always listed in the order given above, one paragraph of data being listed for each planet. Whether this order represented the Babylonian idea of the planets' distances from Earth, or some sort of order of importance, it is difficult to say. But it is probably significant that Jupiter (the star of Marduk) is always listed first.

A typical goal-year text usually lists for the planets both synodic phenomena and normal-star passings, as in our example above. The most important synodic phenomena are the dates of the planetary phases. For the superior planets, the goal-year texts also give the dates of the beginning and end of retrograde motion and of oppositions. Besides the dates of the synodic phenomena, the goal-year texts give the zodiac signs within which they occur.

The normal-star data are notices of when the planets passed by the most important of the Babylonian reference stars along the ecliptic (called *normal stars* by modern scholars). About thirty stars were used as normal stars, all within 10° of the ecliptic.[19] Besides listing the dates at which a planet passed by each of the normal stars, a typical goal-year text also told how far in angular measure above or below the star the planet passed. The prediction of the dates of the phenomena is, of course, a bit more complicated in the Babylonian luni-solar calendar than in our calendar. Otherwise, everything proceeds more or less as in the example above.

Note that for two planets (Jupiter and Mars) two different great cycles are used, sometimes in the same goal-year text. This may be because one cycle for each planet gave somewhat better dates for the normal-star passings and one gave somewhat better planetary phenomena.[20]

Finally, the Babylonian scribes probably also took into account the imperfections of their great cycles in making predictions. One short, fragmentary text, which is assigned to pre-Seleucid times on the basis of its use of older versions of the planet names, gives directions for applying the great cycles. For example, this text specifically says that to get the right results for Venus, you must apply an 8-year cycle, but then subtract four days.[21] We saw above that the Venus phenomena repeat about two days early after 8 of *our* (Julian) years. These two results are in good agreement.

Let us see how this works out. Eight Julian years come to $365.25 \times 8 = 2,922$ days. But the Babylonians used a luni-solar calendar. For short time intervals, it is well approximated by the eight-year luni-solar cycle (see sec. 4.7). One eight-year luni-solar cycle consists of 99 lunar months (five years of 12 months and 3 years of 13 months). The average length of the synodic month is about 29.531 days. Thus, eight successive Babylonian years should amount to approximately $99 \times 29.531 \simeq 2,924$ days—about two days more than eight Julian years. So, if the Venus phenomena repeat 2 days early after 8 years in our calendar, they will fall about 4 days early in the Babylonian calendar.

All the known goal-year texts are from the Seleucid period. Among the oldest is a text for 81 S.E. (231/230 B.C.). The goal-year texts continue well into the first century B.C. Although the oldest surviving examples happen to be from the third century, similar texts were probably produced much earlier. The main requirement for predicting the behavior of the planets in this way is the possession of a long series of continuous observations of planetary phenomena. Exactly the right sort of observational data was collected in the *astronomical diaries* discussed in section 7.1. The oldest diaries we have go back to about 650 B.C., but they may have started as early as the reign of Nabonassar (747–734 B.C.). The longest great cycles used in the goal-year texts (for Jupiter and Mars) could easily have emerged after only a century of continuous observation.

The Babylonian great cycles for the planets were eventually adopted by the Greeks. The period relations quoted by Ptolemy as the basis of his planetary theory in *Almagest* IX, 2, were of Babylonian origin, though Ptolemy himself may not have fully appreciated this fact—for he ascribes them to Hipparchus.

Lunar Phenomena

We will not deal in any detail with lunar theory in this book, but we must say enough about lunar phenomena to explain why the goal-year texts use an 18-year great cycle for the Moon.

The mean time required for the Moon to travel from one equinoctial point, all the way around the zodiac, and return to the same point is called the *tropical month*. It is about 27.3216 days.

The mean time between full Moons is called the *synodic month*, roughly 29.5306 days. The synodic month is longer than the tropical month, because the Sun advances on the ecliptic in the course of the month and the Moon must travel a bit farther than 360° to again reach opposition to the Sun. Eclipses of the Moon can occur only at full Moon; thus, if we know that an eclipse occurred on a certain day, we might look for another eclipse a whole number of synodic months later.

But, of course, eclipses do not occur every month. The Moon's orbit is inclined by about 5° to the ecliptic. Thus, at most full Moons, the Moon is located a few degrees north or south of the plane of the ecliptic and escapes falling into the Earth's shadow. A lunar eclipse is possible only if the Sun happens to be located near one of the two *nodes* of the orbit (the two places where the Moon's orbit crosses through the plane of the ecliptic). If the Sun is at one node, the full Moon will be at the opposite node and therefore in the plane of the ecliptic. This is why lunar eclipses do not occur every full Moon, but rather only a bit more often than twice a year, when the Sun and Moon are simultaneously located at opposite nodes. Now, it happens that the nodes of the Moon's orbit shift gradually westward around the ecliptic, making a complete circuit in about 18.6 years. Thus, eclipses of the Moon at the same node do not constantly recur in the same zodiac sign, but work their way westward through about two signs in three years. The time it takes the Moon to travel from one node of the orbit, all the way around the zodiac, and return to the same node is called a *draconitic month*, about 27.2122 days. (The modern term derives from the nomenclature of medieval astronomers who referred to the two nodes as the "head and tail of the dragon.") If we know the date of an eclipse of the Moon, we might look for another one a whole number of draconitic months later.

The Moon moves at a variable speed around the zodiac. It moves fastest when it is at perigee (nearest Earth) and slowest at apogee. Because of this, an eclipse might fail to occur even if the other cirumstances are favorable. Thus, suppose the mean Moon reaches a node of the orbit when the Sun is located at the opposite node. The eclipse might fail to occur because the actual Moon is a few degrees ahead of or behind the mean Moon and thus misses falling into the shadow. The Moon's perigee is not fixed at one point in the zodiac but works its way gradually eastward around the zodiac, making a complete circuit in about 9 years. The time it takes the Moon to travel from perigee (or fastest motion) all the way around its orbit and return to perigee is called the *anomalistic month*, about 27.5546 days.

For these two reasons (forward motion of the perigee and regression of the nodes) the circumstances of eclipses do not repeat from one year to the next. But we can form a longer period after which the circumstances do more or less repeat. Suppose that a lunar eclipse occurs on a certain day. If, some time later, a whole number of synodic months has elapsed, a whole number of draconitic months has elapsed, and a whole number of anomalistic months has elapsed, the circumstances will again be perfect: the Moon will again be full, it will have returned to the same node, and it will again be at the same distance from perigee as before.

A very satisfactory lunar cycle is sometimes called by modern writers the *saros*:[22]

$$223 \text{ synodic} = 242 \text{ draconitic} = 239 \text{ anomalistic months}$$

The reader can multiply out the month lengths given above and see that this equality holds very nearly. The saros amounts to roughly 6,585 1/3 days. Now 6,585.33/365.25 = 18.029 years.

This is why the Babylonian goal-year texts use an 18-year great cycle for the Moon. We would expect most lunar phenomena to repeat very closely after an interval of 18 years. The lunar phenomena listed in the goal-year texts include not only eclipse data but also information about the time separating moonset from sunset, and so on, at various key times of month.

7.8 EXERCISE: ON GOAL-YEAR TEXTS

1. The results of section 7.5 suggest that 12 years might function as a reasonably good great cycle for Jupiter. Use table 7.1 to see how well

this works. In particular, how well can you use it to predict the dates and places where retrograde motion begins and ends and the dates when Jupiter passes by Regulus (longitude 150°) and Spica (203°)?

2. Investigate table 7.1 to find an excellent great cycle for Mercury that is much shorter than (though not quite as exact as) the Babylonian great cycle for Mercury.

3. Make a short goal-year text for the current year (or for some other year you are interested in). Use the 12-year great cycle for Jupiter, the 15-year cycle for Mars (if the dates work out), the 8-year cycle for Venus, and the great cycle for Mercury that you discovered in problem 2. For each planet, we want to predict

· the date and longitude of the beginning of retrograde motion,
· the date and longitude of the end of retrograde motion,
· the date when the planet passes by Regulus, and
· the date when the planet passes by Spica.

Set up your goal-year text in the following way. List the planets in the standard Babylonian order. Thus, Jupiter comes first. For Jupiter, use table 7.1 to find when and where the phenomena of interest occurred in the year that was one 12-year great cycle before the year of interest. (Note that, if necessary, you can use multiples of great cycles, e.g., 24 or 36 years. Or, if the dates work out, you can even use the longer Babylonian periods given in sec. 7.7.)

Then write out all the phenomena for Venus as they occurred some number of 8-year great cycles before the year of interest. Continue for Mercury and Mars.

Note that a typical Babylonian goal-year text would have much more information in it. For example, the dates of planetary phases were very important. Also, most goal-year texts included data on the Moon. And more normal stars would have been included.

Finally, consult this year's issue of the *Astronomical Almanac*, or some other source giving similar information, to see how well you did. Note that for some of the poorer great cycles (e.g., the 15-year cycle for Mars) it may be helpful to take into account the amount by which the cycle falls short of perfection.

7.9 BABYLONIAN PLANETARY THEORY

The method of prediction on which the goal-year texts is based requires no elaborate theory. Predictions are made simply on the basis of repeating patterns. The price one pays is the necessity of compiling a complete set of observational data over an entire great cycle for each planet—up to 83 years in the case of Jupiter.

By contrast, the Babylonian planetary theory that emerged somewhat later is a very clever mathematical construction. The mathematical planetary theory is unlike the goal-year method in that it does not require giant compendia of data. Rather, the mathematical planetary theory is based on a small set of numerical parameters for each planet. It therefore represents a large advance in sophistication and convenience over the method of the goal-year texts.

The mathematical planetary theory of the Babylonians reached its final, successful form shortly after the beginning of the Seleucid period. Our knowledge of Babylonian planetary theory is based on about 300 tablets, almost all of which came from two sites, Uruk and Babylon.[23] Most of the material from Babylon was unearthed in the late nineteenth century by local diggers who sold it to British representatives. A smaller portion of it was turned up by British archaeological excavations. The great bulk of all this material is now in the British Museum in London. Because of the haphazard way in which it was acquired, it is often impossible to know just exactly where any individual tablet came from—whether a tablet was really from Babylon, or, if so, just where in the ancient city it was found. Some of the material from Uruk came from a German dig conducted in 1912–1913. The Uruk tablets are divided among museums in Istanbul, Berlin, Paris, Chicago, and Baghdad. Most tablets are broken. Sometimes, fragments of the same tablet may be found in different museums, on different continents, which makes it all the more difficult to establish *joins*. The dates of the tablets range from about 300 B.C. to about A.D. 50, but the great bulk fall into the short span between 200 and 50 B.C.

The astronomers of Uruk seem to have been most active from about 220 to about 160 B.C. Most of the material from Babylon belongs to the period 170 to 50 B.C. Thus, the astronomers of Babylon became most active just as activity was falling off at Uruk. However, the very oldest tablets (ca. 300 B.C.), though few in number, are from Babylon itself. So, it is not clear why Uruk was so important a center for astronomy during this one brief stretch of time. It is interesting that Pliny claimed in his *Natural History* that there were three schools of Babylonian astronomy associated with Babylon, Uruk, and Sippar.[24] So far, there is no archaeological evidence for the school at Sippar, but Pliny was certainly correct about Babylon and Uruk. The tablets we possess probably represent accidents of preservation and excavation. This material gives us snapshots of Babylonian planetary theory at two different locations, at a time when it had already achieved full maturity. We cannot say what it looked like in the process of formation—say, around 400 B.C.—for we have little to go by.

The first understanding of these tablets came through the efforts of three Jesuit priests, J. N. Strassmaier, J. Epping, and F. X. Kugler, whose studies spanned the period from the 1880s to the 1920s. Their work was based on the tablets in the collection of the British Museum. Tens of thousands of cuneiform tablets were acquired by the British Museum in the late nineteenth century. Strassmaier, an Assyriologist, worked to bring order to the collection and to make the texts available to other scholars by publishing transcriptions—a task of many years. In the course of this work, he identified a substantial number of tablets apparently of astronomical significance. He recognized the astronomical material from its extensive displays of numbers and its frequent use of month names, but had no understanding of its content. Strassmaier persuaded Epping, a Jesuit professor of mathematics and astronomy, to undertake a study. The first results were published in 1881, in an obscure Catholic theological journal. In his first article, Epping succeeded in understanding a good deal of the Babylonian lunar theory, including the use made of arithmetic progressions. He correctly identified the names of the planets and the zodiac constellations, and correctly determined the starting point of the Seleucid era. Strassmaier continued sending material to Epping and, after Epping's death, to Kugler. Kugler's work opened the way to understanding the remarkable achievement of the Babylonian astronomers.[25] A second wave of scholars carried the investigation forward in the 1930s, 1940s, and 1950s, notably Schaumberger and Neugebauer. As a result, our understanding of Babylonian planetary theory at its maturity is quite complete and detailed. There are many remarks in Greek and Roman literature about the arcane knowledge of the Chaldaeans (Babylonian astronomers and astrologers), but concrete details are few. Nothing in Greek and Roman literature could have prepared us to understand the level of sophistication and success achieved in Babylon and Uruk.

Classes of Texts

In Babylonian mathematical astronomy, there are two important classes of texts: ephemerides and procedure texts. An *ephemeris* is a text that lists planet positions or events connected with the motions of the planets (e.g., the beginnings and ends of retrograde motion), calculated from theory, in an orderly time sequence. (Tables 7.1 and 7.2 are modern examples ephemerides.) The *procedure texts* describe the methods to be followed in computing an ephemeris. If the procedure texts were clear and complete enough, one could hope to reconstruct the details of Babylonian planetary theory simply by following the directions written down by the scribes. Unfortunately, many tablets are broken, and the rules of computation are very condensed. It is unlikely that even a Babylonian scribe would have been able to compute an ephemeris from the rules in a procedure text without the benefit of face-to-face instruction by a senior scribe. Thus, most of the progress in understanding the theory has come from close study of the ephemerides. The rules of computation inferred from an emphemeris can then be checked against the relevant procedure text, if it exists.

Social Setting of Babylonian Mathematical Astronomy

Many of the tablets from Uruk have colophons. Typically, a colophon includes the name of the scribe who wrote the tablet, the name of the owner of the tablet, and the date on which the tablet was written. Sometimes the colophon includes an invocation of the gods—Anu and Antu in the case of tablets from Uruk, Bēl and Bēltī in the case of those from Babylon. In late Babylonian times, the title Bēl ("lord") became synonymous with Marduk. It is interesting that even Herodotus (ca. 446 B.C.) knew that the Chaldaeans were priests of Bēl.[26] Pliny equates Bēl with Jupiter, which shows that he, too, understood the place of Marduk in Babylonian religion, and goes on to say that Bēl was the "discoverer of the science of the stars"—another reflection of the practice of astronomy by the priests of Marduk.[27] Thus, in some cases, the remarks of Greek and Roman writers are confirmed by what we find on the Babylonian tablets.

A number of colophons include prayers for the preservation of the tablet or harm to anyone who may steal it. Some tablets demand secrecy, the informed being forbidden from showing the tablet to the uninformed. Because the scribe usually signs his name in the form "X, son of Y, son of Z, descendent of Q," it has proved possible to work out family trees for the scribes and owners of tablets. It turns out that, in the case of the Uruk tablets, all these people belonged to two scribal families, the family of Ekur-zākir and the family of Sin-leqē-unninnī. Whether these families represent real family relationships or merely the relationship of apprentices and students to master teachers is not certain. However, the passing on of a specialized craft within a family tradition is not improbable. Moreover, these family names are known also from cuneiform legal contracts. Many of the scribes indicate that they are priests, or that their ancestors were priests. So, the picture that emerges—at least for Uruk, since colophons are much rarer in the case of tablets from Babylon—is that the technical mathematical astronomy was the work of a small number of people, often related by family ties, whose astronomical endeavors were a part of the work ordinarily carried out in the temples.[28]

Mesopotamian society is often described as one in which individuals sank their personal identities in the interests of the broader community. King and temple so dominated life that the ordinary individual lost all importance. The collective nature of Babylonian society is usually compared, unfavorably, to the individualism of Greek society. There is an element of truth in this, of course. The Greek case presents us with the spectacle of egocentric poets, philosophers, and mathematicians criticizing their rivals by name, and boldly signing their names to their own works to guarantee that they receive proper credit. By contrast, we know very little about the originators of Babylonian mathematical astronomy.

However, it is important not to insist too strongly on these differences, for the Babylonian astronomical tablets do present us with a vivid picture of intellectual life in which the efforts of individual people counted and were recognized. At least at Uruk, the scribe frequently signed his tablet and often noted the name of another scribe who was the tablet's owner. Sometimes we come across information about the tablets themselves, such as "recently broken" or "checked." Thus, we can picture a lively bureaucracy in which great care was taken with organizational matters. Many tablets include a title describing the contents of the tablet. These titles were often inscribed on an edge, so that they could be read without dragging out the tablet when the tablets were stored in rows, like books on a shelf.

A trained scribe could produce a list of the upcoming retrogradations of Jupiter by applying well-established arithmetical rules. But it is important to note that there were several different versions of the theory of Jupiter. The same is true for all the other celestial bodies. Thus, a good deal of experimentation went on with the theories over the whole period for which we have evidence. Many scribes were not, therefore, mere drones but also creative theoretical astronomers.

Also, we do seem to have the names of two individual theoreticians of great significance. One tablet for the new and full Moons of year s.e. 263, in system A of the lunar theory, bears the title "<tersit>u of Nabū-(ri)-man-nu. . . ."[29] Another text of new and full Moons for two years in system B of the Babylonian lunar theory includes the title "*tersitu* of Kidinnu."[30] The same name appears in the form Kidin on another tablet for new and full Moons in system B, for years s.e. 208 to 210.[31] Now, *tersitu* may mean "tools," "apparatus," or "equipment." The reading of the name of Naburimannu is not quite certain. But the conclusion that Naburimannu and Kidinnu were the originators of systems A and B of the lunar theory is not implausible. This appears all the more likely in view of the fact that these two names were known to Greek and Roman writers. In the course of his description of Babylonia, Strabo says that "the mathematicians" (meaning the Greek astronomers) make mention of some of the Chaldaeans (meaning the Babylonian astronomers), "such as Kidenas and Naburianos and Sudines."[32]

The Character of Babylonian Planetary Theory

The Babylonians took a completely different approach to the planets than did the Greeks. As far as we know, the Babylonians did not visualize the motions of the planets in terms of geometrical or mechanical models. Thus, there is nothing analogous to the Eudoxus's theory of nested spheres or to the later deferent-and-epicycle theory of Apollonius, Hipparchus, and Ptolemy. Also, the Babylonians did not base their astronomy on an elaborate philosophy of nature. Thus, there is no Babylonian equivalent of Aristotle.

Rather, the Babylonian planetary theory was based on arithmetical methods. Moreover, rather than following the planet's motion around the zodiac, the Babylonian theory emphasized direct computation of the important events in the synodic cycle: first and last visibility, beginning and end of retrograde motion, and opposition. Consider one of these synodic events—say, the first station (when retrograde motion begins). A Babylonian ephemeris for Jupiter's first station could be constructed without worrying about any of the other events in the synodic cycle. The first station could almost be thought of as an object in its own right, which worked its way around the zodiac at a variable pace. In contrast, the actual position of a planet at any moment is not something that is immediately obtainable from the theory. Rather, one must interpolate between the events of the synodic cycle.

Finally, while the early Greeks ignored the zodiacal anomaly, Babylonian astronomy confronted it directly. Thus, the earliest workable Babylonian planetary theories already take account of the fact that the planets, Sun, and Moon do not move at a steady speed around the zodiac. The first stations of Jupiter (to continue our example) are not equally spaced around the zodiac. In the next section, we study Babylonian planetary theory by looking in detail at several versions of the theory of Jupiter.

7.10 BABYLONIAN THEORIES OF JUPITER

We will study Babylonian planetary theory by looking in detail at several versions of the theory of Jupiter. This planet is well represented in the surviving material. Indeed, there are more tablets devoted to Jupiter than to all of the other planets combined. This may only reflect accidents of preservation, but it may also be connected with the importance of Jupiter (the star of Marduk) in Babylonian omens.

Theory of Jupiter in System A

Several versions of the Jupiter theory are preserved. The simplest version is called system A by modern scholars. All synodic events work their way around the zodiac in a similar fasion. Let us focus on a single synodic event—the first station. In system A, the first station moves along the zodiac at a uniform speed, until it reaches a jump point, where the speed abruptly changes to a new constant value. There are just two zones on the zodiac, a fast zone and a slow zone. (This was also the case with system A of the solar theory, discussed in sec. 5.2.)

An Ephemeris for Jupiter in System A   As usual, things become much clearer when we study particular texts. In figure 7.13 we see a portion of an ephemeris for Jupiter. This tablet was among a large group of tablets from Uruk that were acquired by the Louvre in 1913. The sketch was made by François Thureau-Dangin, one of the leading Assyriologists of the day. We shall refer to this tablet as ACT 600, since it is number 600 in Otto Neugebauer's *Astronomical Cuneiform Texts*, which contains translations of and commentaries on all known tablets relating to mathematical astronomy. A transliteration of the first twenty lines of this tablet is printed below.[33] The line numbers in the transliteration correspond to those in Thureau-Dangin's copy.

In column III the word MUL.BABBAR ("white star") appears in the first two lines. This is one of the names for Jupiter. The words for "first station" appear in the first three lines of column IV. Thus, the ephemeris of figure 7.13 is a list of first stations of Jupiter.

Let us examine the extract column by column. Column I is a list of years. The first line tells us that we are dealing with the year s.e. 113. Line 20 is for year 133. In figure 7.13, each new line begins with a single vertical wedge ("one"). The sign ME ⊩ for a "hundred" follows. The rest of the year number is written as explained in section 1.2. Writing 113 as 100 + 13 is a departure from strict base-60 notation. So, we see that the scribes sometimes used a mixed base-10 and base-60 notation. The leap years in column I are marked "KIN.A" (when month VI is doubled) or "A" (when month XII is doubled). Counting in column I we see that there are seven leap years in a sequence of nineteen years, as we would expect. We can check to see whether the leap years fall at the right places:

$$113/19 = 5, \text{ with remainder } 18.$$

From the sequence of leap years given in section 4.7, we see that s.e. 113 should, indeed, be a leap year, with a second month VI. So everything fits.

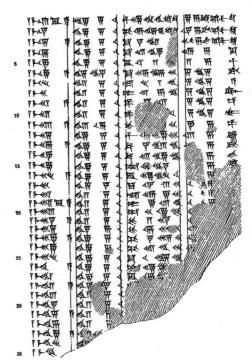

FIGURE 7.13. Portion of an ephemeris of first stations of Jupiter according to system A. This tablet, now in the Louvre, is from Uruk, second century B.C. From Thureau-Dangin (1922).

Extract from ACT 600

	I	II	III		IV	
1	113 KIN.A	48; 5,10	BAR	28;41,40 Jupiter	8; 6	MÁŠ first station
	114	48; 5,10	GU₄	16;46,50 Jupiter	14; 6	GU first station
	115 A	48; 5,10	ŠU	4;52	20; 6	*zib* first station
	116	48; 5,10	ŠU	22;57,10	26; 6	LU
5	117	48; 5,10	KIN	11; 2,20	2; 6	MAŠ
	118 A	45;54,10	DU₆	26;56,30	5;55	KUŠÚ
	119	42; 5,10	APIN	9; 1,40	5;55	A
	120	42; 5,10	GAN	21; 6,50	5;55	ABSIN
	121 A	42; 5,10	ZÍZ	3;12	5;55	RÍN
10	122	42; 5,10	ZÍZ	15;17,10	5;55	GÍR.TAB
	123 A	43;16,10	ŠE	28;33,20	7; 6	PA
	125	48; 5,10	BAR	16;38,30	13; 6	MÁŠ
	126 A	48; 5,10	SIG	4;43,40	19; 6	GU
	127	48; 5,10	SIG	22;48,50	25, 6	*zib*
15	128	48; 5,10	IZI	10;54	1; 6	MÚL
	129 A	48; 5,10	KIN	28;59,10	7; 6	MAŠ
	130	45; 4,10	DU₆	14; 3,20	10; 5	KŬSÚ
	131	42; 5,10	APIN	26; 8,30	10; 5	A
	132 KIN.A	42; 5,10	GAN	8;13,40	10; 5	ABSIN
20	133	42; 5,10	AB	20;18,50	10; 5	RÍN

Column III contains month names, which are written in terms of the Sumerian logograms that provided a compact technical vocabulary. (For the month names, see sec. 4.7.) The entries of column III, with month names and numbers that range from 1 to under 30, tell us the calendar dates on which the first stations of Jupiter occurred. For example, the date of the first station listed in line 1 is year 113, month BAR, day 28 (and 41/60 + 40/3600 of a day). In the extract, the usual semicolon notation (see sec. 1.2) has been used to separate the integral part of each number from the fractional parts.

Column II, used in the construction of column III, consists of time intervals—the times, over and above twelve complete months, that separate successive dates in column III. The dates of the stations on lines 4 and 5 are separated by one whole year plus 48 days (and 5/60 + 10/3600 of a day.) Thus,

Year 116	ŠU	22;57,10	(date of first event)
+ 1		48;05,10	(time between events)
Year 117	KIN	11;02,20	(date of second event)

Note that the time intervals in column II range from about 42 to about 48 days. An interval of twelve months plus 45 days (or so) usually causes the date of the station to advance one in year number, as well as to jump ahead to the next month. But in the case of a leap year, the thirteenth month absorbs 30 of these days. This is why, on lines 3 and 4, the station stayed in the month ŠU two years in a row.

We must mention that the "days" used in columns II and III are not exactly days in the usual sense. Recall that a Babylonian month could be either 29 or 30 days long. It would be a difficult to handle this complication in the calculation of a planetary ephemeris. Thus, the unit of time used by the scribes is actually 1/30 of a mean synodic month. Modern historians call this a *tithi*. (This is not a term used by the Babylonians. Rather it represents a borrowing from the terminology of Indian astronomy, where the same convenient idea turns up.) Since the mean synodic month is about 29.53 days, 1 tithi is 29.53/30 = 0.9843 day. The advantage of tithis is that when 30 of them have accumulated in an ephemeris, the scribe can reckon that a new month has started. For practical dating, the differences between tithis and ordinary days can be ignored. ŠU 22 can be taken to be the 22nd actual day of the month ŠU. The difference between 22 tithis and 22 days is only a fraction of a day, and we could hardly expect the ephemeris to be accurate enough for this difference to matter in making predictions.

Column IV contains the longitudes of the first stations of Jupiter, expressed in terms of degrees and minutes within zodiac signs. The Babylonian names for the zodiac signs in the forms used in this ephemeris are displayed in figure 7.14: KUŠÚ = Crab, A = Lion, and so on. Thus, in column IV, line 6, the entry 5;55 KUŠÚ means that a first station of Jupiter occurred at longitude Crab 5°55′. However, it must be kept in mind that the signs are not defined in the same manner as the Greeks defined them (see sec. 5.2). The beginning of KUŠÚ may occur 8° or 10° before the beginning of the Greek sign of the Crab. All the other Babylonian signs are offset by the same amount from the Greek signs.

The reader may enjoy comparing the cuneiform numerals in figure 7.13 with the extract. To facilitate comparison, we should mention one other detail not discussed in section 1.2. In column IV, line 17, of figure 7.13 and the extract, the longitude of the station is written as 10;5 [KUŠÚ]. The first digit ⟨ of the cuneiform numer is 10, with no units. The second digit 𝍦 is 5, with no tens. Thus, ⟨ 𝍦 could easily be mistaken for 15 rather than for the intended 10 5/60. The scribe therefore inserted a *separation mark* (two small diagonal strokes on line 17 of fig. 7.13) between the digits. The separation mark, which essentially plays the role of a zero, removes the ambiguity.

We have succeded in understanding the basic meaning of all four columns of figure 7.13. Let us now examine the system that was used to calculate the entries in this ephemeris.

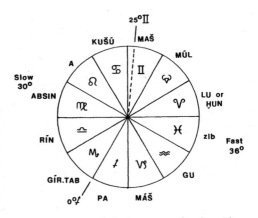

FIGURE 7.14. Babylonian names for the zodiac signs used in the cuneiform texts, with the symbols for the Greek equivalents. Thus, LU corresponds to our Ram (Aries). The figure also shows the fast and slow zones of the Jupiter theory of system A.

Calculating an Ephemeris in System A: Positions   In the Jupiter theory of system A, the zodiac is divided into two zones. Let w_f stand for the spacing between successive first stations of Jupiter in the fast zone. (The spacing between successive occurrences of the same event is called the *synodic arc*.) Let w_s stand for the spacing between successive first stations in the slow zone. The two zones are as follows:

Fast	$w_f = 36°$	Archer 0° to Twins 25°
Slow	$w_s = 30°$	Twins 25° to Archer 0°

What does this mean? If a first station of Jupiter occurs at a certain place in the fast portion of the zodiac, the next time a first station occurs, it will be 36° farther along in longitude. Similarly, if a first station occurs at a certain place in the slow portion of the zodiac, the next time a first station occurs, it will be 30° farther along in longitude. We can see this in the extract from ACT 600. Column IV gives the longitudes of successive first stations of Jupiter. Consider line 6. There a first station is listed at 5;55 KUŠÚ, that is, Crab 5°55′. KUŠÚ is in the slow zone. So, the next time a first station occurs (a bit more than a year later), it will be 30° farther along in longitude, namely, at 5;55 A (= Lion 5°55′), just as we find on line 7.

The next few first stations occur at intervals of 30°—at 5;55 ABSIN, 5;55 RÍN, and 5;55 GÍR.TAB. However, the next first station will fall past the jump point that separates the slow from the fast zone. Some part of the motion will occur in each zone, so we must perform an interpolation to find out just how far the first station will move. The interpolation works like this.

The last station we considered (line 10) was at Scorpion (GÍR.TAB) 5°55′, in the slow zone. If we add 30° that brings us to Archer (PA) 5°55′. But the slow zone ends at Archer 0°. Thus, we have an extra 5°55′ of motion into the fast zone. Let us call this 5°55′ the *overshoot* past the jump point.

At the jump point, the speed increases by the fraction

$$c_1 = \frac{w_f - w_s}{w_s}$$

$$= \frac{36 - 30}{30}$$

$$= \frac{1}{5},$$

expressed as an ordinary fraction, or as 0;12 (= 12/60) in base-60. We shall call c_1 the *interpolation coefficient* for going from the slow to the fast zone. Therefore, we expect the 5°55′ overshoot to be increased by the amount

$$5°55′ \times c_1 = 5°55′ \times \frac{12}{60}$$

$$= 1°11′.$$

The actual distance traveled into the sign of the Archer is therefore the overshoot plus the increase in the overshoot due to the change in speed:

$$5°55′ + 1°11′ = 7°06′.$$

Thus, the next first station of Jupiter occurs at Archer (PA) 7°6′, just as we find on line 11.

Note that the total synodic arc (distance between the two stations) is $w =$ PA 7;6 − GÍR.TAB 5;55 = 31°11′, which does, indeed, fall between the extreme values of 30° and 36°.

Further first stations will be separated by equal intervals of 36°, since we are now in the fast zone. Thus, we find stations at Goat-Horn (MÁŠ) 13°6′, at Water-Pourer (GU) 19°6′, and so on. Everything proceeds like this until we leave the station at Twins (MǍŠ) 7°6′. Another 36° would take us over the jump point and into the slow zone.

The interpolation coefficient for passing from the fast to the slow zone is

$$c_2 = \frac{w_s - w_f}{w_f}$$

$$= \frac{30 - 36}{36}$$

$$= -\frac{10}{60}.$$

The number of degrees by which the station would overshoot the jump point on the way into the slow zone must be multiplied by c_2.

In our example (ACT 600, line 16), we have a station at Twins (MAŠ) 7°6′. We add 36°, then see how far this takes us past the jump point at Twins 25°:

$$
\begin{array}{r r r}
\text{Twins} & 7° & 6′ \\
+ & 36° & 00′ \\
\hline
= \text{Crab} & 13° & 6′ \\
- \text{Twins} & 25° & 00′ \\
\hline
& 18° & 6′
\end{array}
$$

The overshoot past the jump point must therefore be reduced by

$$18°6' \times c_2 = 18°6' \times -\frac{10}{60}$$

$$= -3°1'.$$

The actual distance traveled past the jump point is then

$$18°6' - 3°1' = 15°5'.$$

Therefore, the position of the next station is 15°5′ past the jump point (Twins 25°), or at Crab (KUŠÚ) 10°5′, just as we find in line 17 of the extract.

The generation of a series of successive first stations in system A is therefore quite simple. The stations are equally spaced in each zone. We need take extra care only when the planet passes from the one zone to the other. The only remaining task is to explain how the dates of the stations in column III are obtained.

Calculating an Ephemeris in System A: Dates   The basic assumptions are that the same synodic event (e.g., a first station of Jupiter) occurs always when Jupiter is the same angular distance from the Sun, and that the Sun moves around the zodiac at a uniform speed. In the Babylonian solar theory, the Sun moves at a variable speed, of course. But this complication is ignored when using the Sun to analyze the motions of the planets. Therefore, the method of calculating the amount of time between successive first stations of Jupiter reduces to figuring out how far the Sun the moved. The Sun is the keeper of time.

We will need three numerical parameters for our discussion—one pertaining to Jupiter, one pertaining to the number of months in the year, and one pertaining to the Sun.

Mean Synodic Arc.   For Jupiter, the synodic arc (spacing between two successive first stations) varies between $w_s = 30°$ and $w_f = 36°$. The mean synodic arc $\bar{w}$ is the average value of the synodic arc, taken over an entire great cycle of Jupiter. System A is based on the identity

$$391 \text{ synodic periods} = 36 \text{ tropical periods.}$$

Therefore,

$$1 \text{ synodic period} = \frac{36}{391} \text{ tropical period.}$$

Between two successive synodic events, the planet therefore advances around the zodiac only a fraction of 360°:

$$\bar{w} = \frac{36}{391} \times 360°$$

$$= 33;8,45°$$

in sexagesimal notation (or 33.14578° if written as a decimal fraction). $\bar{w}$ is the mean synodic arc for Jupiter.

Epact.   The epact E is the amount by which the solar year exceeds twelve lunar months. The Babylonian lunar theory is based on the identity

$$1 \text{ year} = 12;22,8 \text{ months.}$$

Thus, the epact is

$$E = 1 \text{ year} - 12 \text{ months}$$

$$= 0;22,8 \text{ months}$$

$$= 30 \times (0;22,8) \text{ tithis,}$$

since there are by definition thirty tithis in a mean month. Carrying out the base-60 arithmetic, we get

$$E = 30 \ (0;22,8)$$

$$= \frac{60}{2} \ (0;22,8)$$

$$= \frac{1}{2} \ (22;8)$$

$$= 11;4 \text{ tithis.}$$

Mean Solar Speed.   The mean solar speed $\bar{v}$ is the number of degrees the Sun moves per tithi, the average being taken over a whole year. Thus, we need to work out how many tithis there are in a year: 1 year = 12 lunar months + 11;4 tithis. But each of the twelve lunar months consists of 30 tithis by definition. Thus,

$$1 \text{ year} = (360 + 11;4) \text{ tithis.}$$

So, the mean solar speed is

$$\bar{v} = \frac{360}{360 + 11;4}.$$

The unit of measure for $\bar{v}$ is °/^{t}, degrees per tithi.

Now we are ready to explain how the dates of the first stations of Jupiter in the Babylonian ephemeris were obtained. Between successive first stations of Jupiter, the Sun travels all the way around the zodiac plus the synodic arc w, that is, the extra distance by which Jupiter has advanced on the zodiac. w can be as much as 36° or as little as 30°. The time ΔT between successive first stations is obtained by dividing the distance the Sun has moved by the mean solar speed:

$$\Delta T = \frac{360° + w}{\bar{v}}$$

$$= \frac{(360 + w)(360 + 11;4)}{360}$$

$$= \left(1 + \frac{w}{360}\right) (360 + 11;4)$$

$$= 360 + 11;4 + w + w \, \frac{11;4}{360},$$

where everything is measured in tithis. Now, the last term on the right side of the equation is clearly much smaller than any of the others. Thus, in this term we (and the Babylonian scribes) will commit a negligible error if we replace the synodic arc w by its mean value $\bar{w}$. Thus, we put

$$w \frac{11;4}{360} \simeq \bar{w} \frac{11;4}{360}$$

$$= \frac{33;8,45 \times 11;4}{360}.$$

Carrying out the arithmetic, we get

$$w \frac{11;4}{360} \simeq 1;01,8^t,$$

where ^{t} denotes *tithis*. In the Babylonian calculations this is rounded up to $1;1,10^t$, the value that we, too, shall use.

In our expression for ΔT, the largest term was 360^t. But since a mean month ^{M} contains 30^t, this amounts to 12 months. Thus, we have finally

$$\Delta T = 12^M + (11;4 + 1;1,10 + w)^t$$
$$= 12^M + (12;5,10 + w)^t.$$

Nothing could be simpler! Let us see how the dates in column III of the extract now follow. When $w = 30°$, that is, when successive stations are separated by $30°$, we will have the minimum time difference

$$\Delta T_{min} = 12^M + 42;5,10^t.$$

This is just the amount of time separating successive stations in lines 9 and 10 of the extract.

But in the fast zone, when $w = 36°$, successive stations are separated in time by the maximum time difference

$$\Delta T_{max} = 12^M + 48;5,10^t,$$

just as we find in lines 1 and 2.

If the two stations are in different zones, we use the actual length of the synodic arc as already calculated. For example, we found above that in crossing over the jump point from the slow to the fast zone (lines 10 and 11), the station moved from 5;55 GÍR.TAB to 7;6 PA—a distance of 31°11′. We insert this value for the synodic arc into the general formula for the time difference:

$$\Delta T = 12^M + (12;5,10 + w)^t$$
$$= 12^M + (12;5,10 + 31;11)^t$$
$$= 12^M + 43;16,10^t,$$

just as we find on line 11, column II.

Relations among the Parameters   In system A for Jupiter, the slow arc is $155°$ wide (stretching from Twins $25°$ to Archer $0°$). In this zone, the synodic arc (spacing between successive synodic events of the same type) is $30°$. Thus, on the average, the number of events that occur during one trip of the planet through the slow zone is 155/30. In the same way, the width of the fast zone is $205°$ and the synodic arc in the fast zone is $36°$. Thus, on the average, the number of synodic events occuring during one trip of the planet through the fast zone is 205/36. Thus, we have

$$\frac{155}{30} + \frac{205}{36} \text{ synodic events in one tropical cycle,}$$

which is

$$\frac{155}{30} + \frac{205}{36} = 10\frac{31}{36}.$$

Thus, we have the following relation:

$$10\frac{31}{36} \text{ synodic periods} = 1 \text{ tropical period}$$

If we multiply this relation by 36 to eliminate fractions, we obtain the period relation

$$391 \text{ synodic periods} = 36 \text{ tropical periods} = 427 \text{ years},$$

where the last equality holds because Jupiter is a superior planet (and so the number of synodic periods elapsed plus the number of tropical periods elapsed must be equal to the number of years gone by).

Thus, the four fundamental parameters of the theory (width of the slow and of the fast zone, synodic arc in the slow and in the fast zone) must be carefully chosen to ensure that they accord with the period relation that has been selected as the basis of the theory. System A of the Jupiter theory is based on a period relation that is substantially longer than those used in the goal-year texts. Period relations of considerable length (up to seventy or more years) probably were discovered by the accumulation of several generations' worth of data in the astronomical diaries. But when we encounter a period relation involving a period of 427 years, we are probably dealing with a parameter that has been derived by a process of tinkering with the shorter period relations, perhaps with the idea of correcting for the slight inaccuracies of the shorter relations.

In section 7.5, we found that for Jupiter there is a tolerably good period relation of 12 years. As we saw in section 7.7, the Babylonian goal-year texts use periods of 71 years and 83 years for Jupiter. Some cuneiform procedure texts mention periods for Jupiter of 12, 71, 83, 95, and 261 years.[34]

System A'

In system A', the zodiac is divided into several zones of constant speed (rather than merely two).

A Procedure Text for System A'   A surviving procedure text for Jupiter in system A' is quite clear:

EXTRACT FROM ACT 810

Jupiter. From 9 Crab to 9 Scorpion add 30. The amount in excess of 9 Scorpion multiply by 1;7,30.

From 9 Scorpion to 2 Goat-Horn add 33;45. The amount in excess of 2 Goat-Horn multiply by 1;4.

From 2 Goat-Horn to 17 Bull add 36. The amount in excess of 17 Bull multiply by 0;56,15.

From 17 Bull to 9 Crab add 33;45. The amount in excess of 9 Crab multiply by 0;53,20.

From 9 Crab to 9 Scorpion slow. From 9 Scorpion to 2 Goat-Horn medium. From 2 Goat-Horn to 17 Bull fast.

From 17 Bull to 9 Crab medium.[35]

The synodic arcs in the slow and in the fast zone are 30° and 36°, as in system A. But now there are two intermediate zones, in which the synodic arc is 33°45'. The text says that the slow zone stretches from Crab 9° to

Scorpion 9° and that the interpolation coefficient for passing from the slow to the intermediate zone is 1;7,30. The procedure for calculating an ephemeris in system A' is therefore clear—one proceeds exactly as in system A, but one will need to perform the interpolation procedure a bit more often.

Let us check a few features of this theory. To begin with, let us check the interpolation coefficient c_1. As before, it should be given by

$$c_1 = \frac{w_i - w_s}{w_s},$$

where w_s is the synodic arc in the slow zone and w_i is the synodic arc in the intermediate zone. Thus,

$$c_1 = \frac{33\frac{45}{60} - 30}{30}$$

$$= \frac{2}{60} \times 3\frac{45}{60}$$

$$= \frac{7}{60} + \frac{30}{60 \cdot 60}$$

$$= 0;7,30.$$

Note that the procedure text does not say 0;7,30, but rather 1;7,30. That is, the text does not give c_1 as we have defined it but rather $1 + c_1$. This is actually a bit more convenient. In the example above of calculating an ephemeris in system A, we multiplied the overshoot (5°55' in the example) by c_1, then added this product to the overshoot. We could have saved a step by simply multiplying the overshoot by $(1 + c_1)$. In the same way, one may show that the other three interpolation coefficients given in our procedure text are correct, but that they represent 1 + the interpolation coefficient defined in the old way. Here we see a minor difference in procedure favored by a particular scribe. From now on, we shall stick to our original definition of the interpolation coefficient.

The Implied Period Relation   Finally, let us determine the period relation on which system A' is based. The four zones have the following properties:

Zone	Width	Synodic arc
Slow	120°	30°
Intermediate	53°	33°45'
Fast	135°	36°
Intermediate	52°	33°45'

Thus, the number of synodic events in one tropical cycle of Jupiter is

$$\frac{120}{30} + \frac{53}{33;45} + \frac{135}{36} + \frac{52}{33;45} = \frac{391}{36}.$$

Thus, we have

391 synodic periods = 36 tropical periods,

the very same period relation on which system A was based. So, it is clear that system A' was developed later, as a way of improving on system A by smoothing out the passage between the slow and the fast zone. But the scribe who developed system A' had to do some rather sophisticated arithmetic to ensure that the new system would not violate the previously adopted period relation.

Accuracies of Systems A and A'

How well do the Babylonian theories of Jupiter work? Table 7.3 compares the longitudes of some second stations of Jupiter, taken from cuneiform tablets for systems A and A', with the longitudes of the same stations calculated from modern celestial mechanics.

The first column on the left gives the year in which Jupiter's second station occurred, in terms of the Seleucid era. The first entry in this table (year 188) gives a second station of Jupiter that occurred in January, 123 B.C. The last entry (for S.E. 202) gives the second station of March, 109 B.C.

Column 2 gives the longitudes of the second stations of Jupiter taken from the cuneiform tablet ACT 605.[36] This tablet, which was calculated according to system A, gives second stations and last visibilities of Jupiter for years S.E. 188 to 222. However, to facilitate comparison with modern data, the longitudes from the cuneiform tablet have been expressed in decimal fractions of a degree, reckoned from the beginning of the first sign of the zodiac. For example, for the longitude of the second station of year 188, the tablet actually gives 15;42 MÚL, that is, Bull 15°42'. In table 7.3 the 42' is expressed as a decimal: 42/60 = 0.70. And since MÚL is the second sign of the Babylonian zodiac, the longitude is written out as 45.70°. Thus, in table 7.3 all the Babylonian longitudes are reckoned continuously from the beginning of the first sign.

If we subtract successive entries in column 2, we obtain the distances (synodic arcs) between neighboring stations. (This is our procedure for analyzing the tablet—we are not following a Babylonian procedure here.) These synodic arcs are listed in column 3, which uncovers the essential structure of system A: we see four successive synodic arcs of 30° and four successive synodic arcs of 36°. These are bridged by the interpolation scheme. The synodic arcs

TABLE 7.3. Second Stations of Jupiter in Systems A and A'

Year (S.E.)	System A		Modern		System A'	
	ACT 605	Diff.	Longitude	Diff.	ACT 612	Diff.
188	45.70		41.14		47.35	
		36.00		33.53		33.75
189	81.70		74.67		81.10	
		30.55		31.74		31.99
190	112.25		106.41		113.09	
		30.00		30.59		30.00
191	142.25		137.00		143.09	
		30.00		30.14		30.00
193	172.25		167.14		173.09	
		30.00		30.51		30.00
194	202.25		197.65		203.09	
		30.00		31.73		31.76
195	232.25		229.38		243.85	
		34.45		33.43		33.75
196	266.70		262.81		268.60	
		36.00		35.30		35.78
197	302.70		298.11		304.38	
		36.00		36.47		36.00
198	338.70		334.58		340.38	
		36.00		36.37		36.00
199	14.70		10.95		16.38	
		36.00		35.09		35.66
200	50.70		46.04		52.04	
		35.72		33.26		33.75
201	86.42		79.30		85.79	
		30.00		31.57		31.47
202	116.42		110.87		117.26	

of intermediate size take on various values, depending on just where the station falls in relation to the jump point.

Column 4 gives the actual longitudes of Jupiter's second stations for the same years, as calculated from modern celestial mechanics.[37] Column 5 gives the differences between successive entries in column 4. Thus, column 5 gives the actual values of the synodic arcs.

The longitudes of the stations in system A run consistently several degrees larger than the modern longitudes. In part, this reflects the fact that Babylonian longitudes are measured from a different reference point than are the modern longitudes. The modern longitudes are measured from the spring equinoctial point. Presumably, in system A, the longitudes are reckoned from the beginning of the first Babylonian zodiac sign—which starts 10° earlier than does the modern sign. Thus, we should expect the Babylonian longitudes to be, on average, 10° larger than the modern ones. The fact that they are only from 4° to 7° larger shows that the whole list of longitudes in tablet ACT 605 suffers from a constant shift of perhaps 5°—the longitudes all being too small by this amount. This perhaps reflects the selection of a somewhat defective starting value. We know very little about how the scribes determined their initial values. They probably compared their computed ephemerides with the actual stations of Jupiter by noting how far Jupiter was from one of the normal stars when it reached its station. (Absolute longitudes, measured from the beginning of the first zodiac sign, were not directly measurable.) Of course, the longitudes of the stars increase slowly with time, because of precession. It is still an open question whether the Babylonians were aware of precession, but there is no direct proof that they were. Lists of the longitudes of normal stars that were a few centuries out of date would have longitudes too small by several degrees. Thus, if the scribes compared their planetary ephemerides against the stars, we might expect the theoretical planet longitudes to be systematically smaller than the modern computed values. The computed *dates* of the synodic events were more easily checked and may have been deemed more significant.

What about the *spacing* between events? The pattern of modern synodic arcs shows that the minimum is, indeed, around 30° and the maximum is a bit more than 36°. But the system A synodic arcs definitely stay at their minimum and maximum values for too long.

Column 6 in table 7.3 gives the longitudes for the second stations of Jupiter, for the same run of years, taken from the cuneiform tablet ACT 612,[38] which was calculated according to the rules of system A′. Column 7, obtained by subtracting neighboring entries of column 6, gives the resulting values for the synodic arc. We still see three successive arcs of 30° in the slow zone and two successive arcs of 36° in the fast zone. But now the transition between the slow and fast zones is smoothed out and extends over several entries. This is, of course, characteristic of system A′.

Note that the system A′ longitudes of the stations (from ACT 612) are a degree or two higher than in the example from system A (ACT 605). This reflects a somewhat better initial value. The system A′ longitudes are, however, still a little too small on the average, since they should be 10° greater than the modern values. A better starting value is not, however, a necessary signature of system A′. It is interesting that two other preserved tablets[39] give longitudes of second stations of Jupiter, also in system A′, for spans of years that overlap with the tablet under consideration here. These other tablets give longitudes that are 2° or 3° lower than the longitudes in ACT 612. Thus, multiple ephemerides (calculated according to the same system) might give somewhat different answers to the same question: when and where did Jupiter stand still in a given year?

What about the spacings? When we look at the synodic arcs in system A′, we see a truly impressive accomplishment. The theoretical synodic arcs never disagree with the actual ones by more than about 6/10°. System A′ represents a substantial improvement on system A.

System B

In system A and its variants, the zodiac is divided into zones, in each of which the synodic arc is constant. The synodic arc jumps abruptly from one zone to the next. In system B, the synodic arc steadily increases by equal increments until it reaches its maximum value, after which it decreases by equal increments. In ephemerides prepared according to system B, the synodic arcs therefore form an arithmetic progression with constant differences. The longitudes of the stations show constant second differences. Let us examine a particular text.

The cuneiform tablet ACT 620 is from Uruk. The text is an ephemeris of oppositions of Jupiter for s.e. 127–194. The longitudes and dates of the oppositions were computed according to the rules of system B. Here we print an extract (for years 167–183) from the reverse of the tablet:

EXTRACT FROM ACT 620

(Reverse)[40]

	I	II	III		IV	V
1	167 A	41;45	ZÍZ	18;40	29;40	18;22 A
	168	40;44,30	ZÍZ	29;24,30	28;39	17;01 ABSIN
	170 KIN.A	42;32,30	BAR	11;57	30;27	17;28 RÍN
	171	44;20,30	BAR	26;17,30	32;15	19;43 GÍR.TAB
5	172 A	46;08,30	SIG	12;26	34;03	23;46 PA
	173	47;56,30	SIG	30;22,30	35;51	29;37 MÁŠ
	174	49;44,30	IZI	20;07	37;39	7;16 zib
	175 A	48;42	DU₆	8;49	36;37	13;53 HUN
	176	46;54	DU₆	25;43	34;49	18;42 MÚL
10	177	45;06	GAN	10;49	33;01	21;43 MAŠ
	178 A	43;18	AB	24;07	31;13	22;56 KUŠÚ
	179	41;30	ZÍZ	5;37	29;25	22;21 A
	180 A	40;59,30	ŠE	16;36,30	28;54	21;15 ABSIN
	181	42;47,30	ŠE	29;24	30;42	21;57 RÍN
15	183 A	44;35,30	GU₄	13;59,30	32;30	24;27 GÍR.TAB

Column I identifies the years, with the leap years marked A or KIN.A, as before. Column III gives the month and day of each opposition, with the "days" measured in tithis, as before. Column II is the time, over and above twelve complete months, separating successive entries in column III, as before. Column V gives the longitude of Jupiter at the time of its opposition. The longitudes are expressed, as before, in terms of degrees and minutes within a zodiac sign. Column IV (which is new) contains the *synodic arc*, that is, the distance between successive oppositions. Thus, consider lines 1 and 2:

$$\begin{array}{ll} \text{First opposition} & 18°22′ \text{ A} \\ \text{Plus synodic arc} & \underline{+28°39′} \\ \\ \text{Second opposition} & 17°01′ \text{ ABSIN} \end{array}$$

The essential character of system B is revealed by column IV. These synodic arcs form an arithmetic progression with constant differences of 1°48′. Thus, $30;27 - 28;39 = 1;48$. And $32;15 - 30;27 = 1;48$, and so on. The synodic arcs get bigger and bigger as we move down column IV from line 2 to line 7.

After that, as we move from line 8 to line 12, the synodic arcs get smaller and smaller, again by equal increments of 1°48′. In the extract, we have drawn a horizontal line (between lines 7 and 8) to separate the rising and falling sequences.

If we graph the synodic arcs we get figure 7.15. On the vertical axis we have plotted the synodic arcs (measured in degrees) of column IV. The steps on the horizontal axis are simply the line numbers from the extract. The synodic arcs form what Neugebauer has called a "linear zig-zag function."

It should be noted that, over this particular sequence of years, the synodic arcs did not happen to hit their maximum or minimum possible values. That is, none of the points happen to fall exactly on the peaks of the zigzag function. The peaks can be found by extrapolation (the dashed lines on fig. 7.15). It turns out that the maximum and minimum possible synodic arcs in the Jupiter theory of system B are

$$w_{max} = 38°02′$$

$$w_{min} = 28°15′30″.$$

The mean synodic arc $\bar{w}$ is therefore given by

$$\bar{w} = \frac{w_{max} + w_{min}}{2}$$

$$= 33°08′45″,$$

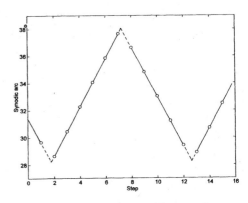

FIGURE 7.15. A linear zigzag function for Jupiter in system B. The synodic arc (longitudinal distance between successive events of the same kind) increases then decreases by equal increments.

which is exactly the same as the mean synodic arc for Jupiter that we have encountered in system A and system A′.

So, here is another example of the sophistication of Babylonian applied mathematics. System B will give the same average spacing of synodic events around the zodiac as do systems A and A′. System B therefore rests on the same fundamental period relation as do the other two Jupiter theories. But in system B the successive oppositions are spread (and then compressed) by equal increments as we move around the zodiac. System B is thus even easier to use in calculating positions than either A or A′.

Here we must pause to consider a detail we have not yet addressed. How does one bridge the breaks between the rising and falling sequences? The rule is very simple: the total change in the synodic arc (taken as the sum of the increase on the rising section and the decrease on the falling section) must still total 1°48′. For example, consider the transition from line 7 to line 8. The last synodic arc on the rising sequence was 37;39. If we add 1;48, that would take us to 39;27. But the maximum possible synodic arc is (as mentioned above) 38;02. The excess is therefore 39;27 − 38;02 = 1;25. This 1°25′ of change must be used up in the decline from the peak value. Therefore, the synodic arc will be 38;02 − 1;25 = 36;37, just as we find on line 8 of the extract.

The time intervals between successive oppositions (listed in column II) also form a linear zigzag function, with constant differences of 1;48 (i.e., 1 48/60 tithis). This follows from the expression we derived above (in the discussion of system A) for the time interval ΔT:

$$\Delta T = 12 \text{ months} + (w + 12;5,10) \text{ tithis}.$$

In system B, for some reason (probably just to have a more convenient round number), the constant 12;5,10 is rounded up to 12;5,30. Thus, the connection between the time intervals (column II) and the synodic arcs (column IV) is

$$\Delta T = 12 \text{ months} + (w + 12;5,30) \text{ tithis.}$$

For example, in line 2 (column IV), $w = 28;39$. Thus, we find for ΔT (in column II) $28;39 + 12;5,30 = 40;44,30$.

System B is an admirable accomplishment in that it manages to give a good accounting of the synodic phenomena with a much smaller number of adjustable parameters than are required in system A′. System B is also much easier to use. It is not, however, very different from system A′ in terms of accuracy.

The most remarkable Babylonian accomplishment is certainly the theory of the Moon, which is of much greater complexity than the theories of the planets. In the lunar theory, the scribes had to take account of the inclination of the Moon's orbit to the plane of the ecliptic. The complete lunar theory allowed not only the prediction of chief events during the synodic month, but also reasonably accurate prediction of eclipses.

Despite its accomplishments—or perhaps because of them—Babylonian planetary theory failed to develop further. The major variants of the theory were probably developed by the beginning of the Seleucid period. After that—that is, for the whole period for which we have evidence—the theories remained static. The last cuneiform tablets are from the first century A.D.

7.11 EXERCISE: USING THE BABYLONIAN PLANETARY THEORY

1. System A′: Let us update the Babylonian Jupiter theory of system A′ to the twentieth century. We do this by rotating the jump points forward, mostly to account for precession. Also, we shall use the modern convention for the definition of the zodiac signs. (Thus, the spring equinoctial point is at Ram 0°.) Finally, let us reckon longitudes continuously from 0 to 360°, rather than using the names of signs. The zones then look like this:

Zone	Boundaries	Synodic arc
Slow	130° to 250°	30°
Intermediate	250° to 303°	33.75°
Fast	303° to 78°	36°
Intermediate	78° to 130°	33.75°

We leave all the other parameters unchanged. Thus, the widths of the four zones are the same as in the Babylonian theory, as are the lengths of the synodic arcs and the interpolation coefficients.

A. Starting from the lengths of the synodic arcs, calculate the interpolation coefficients for the Jupiter theory in system A′. You should find

c_1 (slow to intermediate)	$= 0;07,30$	$= 0.1250$
c_2 (intermediate to fast)	$= 0;04$	$= 0.0667$
c_3 (fast to intermediate)	$= -0;03,45$	$= -0.0625$
c_4 (intermediate to slow)	$= -0;06,40$	$= -0.1111$

Here, we have given the interpolation coefficients both in base-60 and in base-10, which will be more convenient for computation.

Confirm that the base-60 forms of the interpolation coefficients are consistent with the procedure text ACT 810, quoted in section 7.10. (Note that in ACT 810, the interpolation coefficients are defined a bit differently.)

B. In table 7.1, we see that a first station of Jupiter occurred in 1971 at longitude 247°. The date of this event was J.D. 244 1040. Let us use this station as the initial event in our ephemeris.

Step	T	ΔT	w	λ
1	244 1040*			247*
		399.11	33.38	
2	244 1439.11			280.38
			34.49	
3				314.87

In this table, the column labeled λ gives the longitudes of the stations. Column w gives the synodic arcs (the differences between successive values of λ). ΔT is the time difference separating successive stations. T is the actual date of the station, expressed in terms of Julian day number. The data marked * are the initial values: the date and longitude of the initial first station.

Generate a series of eleven first stations (2 through 12) of Jupiter using the updated version of system A'. Work carefully. A mistake early on will corrupt all your later stations. (The answers for stations 2 and 3 are given so that you may check your work.) Compare the longitudes of the stations you obtain with the actual longitudes of the stations given in table 7.1 to see how well the theory works.

C. In computing the *dates* on which the stations occur, it will be more convenient to work in terms of Julian day numbers, rather than in terms of the Babylonian luni-solar calendar. We can modify the Babylonian procedure very simply as follows.

As in section 7.10, we assume the following value for the mean synodic arc $\bar{w}$ of Jupiter:

$$\bar{w} = 33;8,45° = 33.1458°$$

We shall use the decimal form for ease of computation.

For the mean solar speed $\bar{v}$ we adopt the value

$$\bar{v} = 360°/365.25^d = 0.985627°/^d.$$

Let w represent the value of Jupiter's synodic arc (which will be some number between 30° and 36°). Then, as in Babylonian practice, the time ΔT between successive first stations is given by dividing the distance the Sun moves by the mean solar speed:

$$\Delta T = \frac{360° + w}{0.985627°/^d}$$

$$= 365.25^d + 1.014583w$$

$$= 365.25^d + w + 0.014583w$$

In the small last term we can replace w by its mean value $\bar{w}$ without introducing appreciable error. Since $0.014583 \times \bar{w} = 0.014583 \times 33.1458 = 0.4834$, we obtain finally

$$\Delta T = 365.7334 + w.$$

Let us apply this formula in practice. From table 7.1, the date of the first station of Jupiter in 971 (the one at longitude 247°) was J.D. 244 1040. The synodic arc between vents 1 and 2 is obtained by subtracting the longitudes: $w = 280.38 - 247 = 33.38°$. Putting this into our general formula, we get

$$\Delta T = 365.73 + w$$
$$= 365.73 + 33.38$$
$$= 399.11^{d}$$

And thus the date of station 2 should be

$$244\ 1040.00$$
$$+\ 399.11$$
$$\overline{244\ 1439.11}$$

We see in table 7.1 that this station of Jupiter actually occurred at longitude 279°, around J.D. 244 1435. So we have not done too badly!

Finish out the table started already by computing the dates of stations 2 through 12 according to our updated system A'. Compare your results with table 7.1 to see how well you did.

2. System B: Lay out a table for using system B to compute the first stations of Jupiter from 1971 to 1983:

Step	T	ΔT	w	λ
1	244 1040*			247*
		397.73	32*	
2	244 1437.73			279*
			33.8	
3				312.8

As initial data we shall use only the following facts, taken from table 7.1. There was a first station of Jupiter at longitude 247° on J.D. 244 1040. The next first station of Jupiter was at longitude 279°. The synodic arc (distance between the two first stations) was therefore 279 − 247 = 32°. We shall assume that w was then in the rising part of the zigzag function. The initial data taken from table 7.1 are marked *. Everything else in the table will be worked out from the rules of system B.

For example, if 32° separate stations 1 and 2, the time interval (in days) separating them can be calculated from the formula obtained above:

$$\Delta T = 365.73 + w.$$

Thus, the date of station 2 must be 244 1040 + 397.73 = 244 1437.73.

The synodic arcs form a rising arithmetic progression with constant differences of 1°48′ = 1.8°. (Let us work in terms of decimal fractions.) Then the next synodic arc must be 32 + 1.8 = 33.8°. From this it follows that station 3 will occur at longitude 279 + 33.8 = 312.8°.

Work out the rest of the table, up through the first station of 1984. Thus, there should be a total of thirteen stations in your ephemeris. Note that if w should exceed w_{max} or fall below w_{min}, you must follow the procedure explained in section 7.10 for passing from a rising to a falling segment of the zigzag function. Use the Babylonian values (expressed in terms of decimal fractions):

$$w_{max} = 38.0333°$$
$$w_{min} = 28.2583°$$

Compare your completed table with table 7.1 to see how well you and system B have done.

7.12 DEFERENT-AND-EPICYCLE THEORY, I

Between the third century B.C. and the second century A.D., Greek astronomers developed a new geometrical planetary theory based on uniform circular motion. However, now the circles no longer turned about one single center. Thus, it became possible to avoid the difficulties that had plagued Eudoxus's system. The new system, based on deferent circles and epicycles, originated in the third century B.C. with Apollonius of Perga. At the beginning, Apollonius's system, like Eudoxus's before it, was intended only to be broadly explanatory. But in the next century, the Greeks' contact with Babylonian planetary theory alerted them to the possibility of a quantitative theory. The deferent-and-epicycle theory of the planets was brought into its highly successful final form by Ptolemy in the second century A.D. In the next few sections, we follow the historical evolution of the theory.

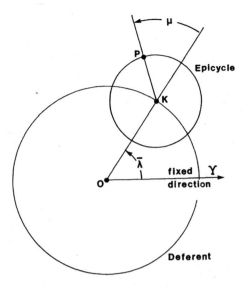

FIGURE 7.16. Apollonius's deferent-and-epicycle model. The Earth is at O. ♈ marks the direction of the vernal equinox.

General Features of Apollonius's Theory

Each planet participates in two motions. There is a steady eastward motion around the ecliptic. Superimposed on this steady progress in longitude is a back-and-forth motion that produces occasional retrogradations.

Apollonius's model is illustrated in figure 7.16. The figure lies in the plane of the ecliptic and is observed from the ecliptic's north pole. The *deferent circle* is centered on the Earth O. Along this circle, point K moves eastward at constant speed. K serves as the moving center of a second circle called the *epicycle*. The planet P travels around the epicycle at constant speed.

The motion of K around the deferent is designed to reproduce the planet's circuits around the ecliptic. K must therefore complete one revolution in one tropical period. The angular distance between K and the vernal equinox ♈ is called the *mean longitude*, denoted $\bar{\lambda}$. Thus, $\bar{\lambda}$ increases by 360° in one tropical period.

The motion of the planet P on the epicycle is designed to reproduce the planet's retrogradations. The planet's position on the epicycle is defined by the *epicyclic anomaly*, μ, which increases by 360° in one synodic period.

Let us examine the model in more detail (see fig. 7.17). The point π of the epicycle that is nearest the Earth is called the *perigee of the epicycle*. The point a farthest from the Earth is called the *apogee of the epicycle*. The planet's *actual longitude* at any moment is λ. The planet appears to be backing up (retrogressing) when it is at π, for then the motion of P on the epicycle is westward and opposed to the motion of K on the deferent.

When the two motions are put together, the motion that results is a series of loops, shown in figure 7.18. This figure is drawn with loops properly sized for Mars. Between one retrograde loop and the next, Mars makes a complete trip around the ecliptic, plus a bit more. But in the figure we have illustrated only the parts of the motion around each retrogradation.

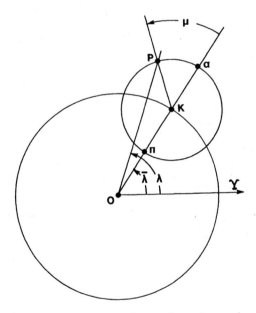

FIGURE 7.17. Terminology and notation used for Apollonius's theory.

Points labeled in figure:	Angles and radii of circles:
O, Earth	$\bar{\lambda}$, the mean longitude
K, center of epicycle	μ, the epicyclic anomaly
π, perigee of epicycle	λ, longitude of the planet
a, apogee of epicycle	$R = OK$ radius of the deferent
P, the planet	$r = KP$ radius of the epicycle

Some Technical Detail

Connection with the Sun: Superior Planet Superior planets retrograde when they are in opposition to the Sun. Further, there is a period relation that connects the planet's tropical and synodic motions to the motion of the Sun. Any planetary theory must be able to account for these facts, but the model we have been describing has so far taken no notice of them. Fortunately, it is possible to produce the necessary results by means of only a slight addition to the theory.

In figure 7.19, about the Earth O as center, we have drawn the circular orbit of the mean Sun and the deferent circle of Mars. The change we must make in our theory is to add the following stipulation: *in the case of a superior*

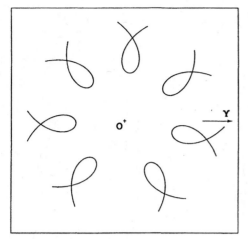

FIGURE 7.18. Retrograde loops of Mars generated by the model of Figure 7.16. The Earth is at O. ♈ marks the direction of the vernal equinox.

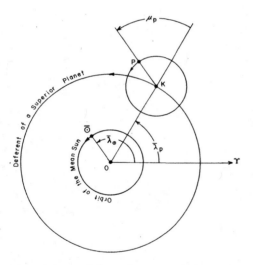

FIGURE 7.19. Relation between the mean Sun and a superior planet: $\bar{\lambda}_p + \mu_p = \bar{\lambda}_\odot$.

planet, the radius of the epicycle always remains parallel to the line from the Earth to the mean Sun. Thus, KP is parallel to $O\odot$.

From the figure we may deduce a simple relation between the planet's mean longitude $\bar{\lambda}_p$ and epicyclic anomaly μ_p and the longitude of the mean Sun $\bar{\lambda}_\odot$. First, note that the three angles with vertex at O satisfy the relation $\bar{\lambda}_p + KO\odot = \bar{\lambda}_\odot$. But since $O\odot$ is parallel to KP, angle $KO\odot$ must be equal to μ_p. Thus, we have

$$\bar{\lambda}_p + \mu_p = \bar{\lambda}_\odot.$$

In words, the planet's mean longitude plus its epicyclic anomaly equals the longitude of the mean Sun. This equation reflects the period relation for a superior planet from section 7.4:

Number of tropical + number of synodic = number of years
cycles elapsed   cycles elapsed   elapsed.

Figure 7.20, A and B, show the situation shortly before and exactly at the time of a mean opposition. In figure 7.20A, the planet is approaching the perigee of its epicycle. As always, KP and $O\odot$ are parallel. In figure 7.20B, the planet has reached the perigee of the epicycle and the middle of its retrograde motion. The parallelism of KP and $O\odot$ guarantees that, at this moment, an observer at O will see the planet P and the mean Sun in diametrically opposite directions.

(Note the importance of the mean Sun. The mean Sun moves at constant angular speed around the Earth, while the true Sun does not. Thus, we cannot require that KP remain parallel to the line from the Earth to the true Sun. In fig. 7.19, we may regard the true Sun as moving on a tiny epicycle centered on the mean Sun, as in sec. 5.2.)

Connection with the Sun: Inferior Planet   Inferior planets (Mercury and Venus) have the same tropical period as the Sun. They move alternately ahead of and behind the Sun, but they always remain its close companions. These facts may be accounted for in Apollonius's theory by the addition of the following stipulation: *in the case of an inferior planet, the direction from the Earth to the epicycle's center always coincides with the direction from the Earth to the mean Sun.* In figure 7.21, O, K, and $\odot$ all lie on one line. The planet's mean longitude $\bar{\lambda}_p$ is always equal to the mean longitude of the Sun $\bar{\lambda}_\odot$. This explains why an inferior planet has limited elongations from the mean Sun.

Direction of Revolution on the Epicycle: Inferior Planet   We have asserted that the planet revolves on the epicycle in the same direction as the epicycle's center revolves on the deferent. In the case of an inferior planet it is easy to prove that this is so. An inferior planet reaches its greatest elongation from the mean Sun when the line of sight from the Earth to the planet becomes tangent to the epicycle. In figure 7.22, OK points to the mean Sun. The planet has its greatest eastward elongation when it reaches e and its greatest westward elongation when it reaches w.

Suppose that the planet revolves clockwise on the epicycle, in the order $eaw\pi$. In this case, the time consumed going from e through a to w will be more than the time consumed going from w through π to e.

Suppose instead that the planet revolves counterclockwise on the epicycle, $e\pi wa$. Then the time from e through π to w will be less than the time from w through a to e.

This suggests a test that is easy to make. We take Mercury as our example. From section 7.2, we extract the following information:

Three successive greatest elongations of Mercury

Year	Date	Sun	Mercury	Elongation
1976	Apr 26	36	56	20° E
1976	Jun 15	84	61	23° W
1976	Aug 24	151	178	27° E

The time from the greatest eastward to the greatest westward elongation (April 26 to June 15) is 50 days. And the time from the westward to the eastward (June 15 to August 24) is 70 days. Thus, Mercury must travel counterclockwise on its epicycle.

A similar test can be made for Venus, with a similar result. It is not possible to apply this method to the superior planets, since they do not have greatest elongations. However, it turns out that they, too, travel on their epicycles in the same direction as Mercury and Venus.

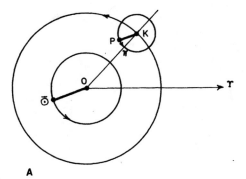

Rough Estimate of the Epicycle's Radius: Inferior Planet   For an inferior planet, one can use the greatest elongations to make a quick estimate of the size of the epicycle (see fig. 7.23.) P marks the planet's position at a greatest elongation, and θ is the angular measure of the elongation. From the figure,

$$r = R \sin \theta.$$

The greatest elongations of Mercury show considerable variability: the three used above have the values 20°, 23°, 27°. Let us put $\theta = 23.3°$, which is the average of the three. The result is

$$r = 0.40 \ R.$$

The radius of Mercury's epicycle is four-tenths the radius of its deferent.

However, two cautionary notices must be inserted here. For simplicity, we have used the elongations of Mercury from the true Sun, rather than from the mean Sun. The true elongations can differ from the mean ones by up to 2°. Thus, our result for the epicycle's radius is only a rough value. Moreover, the size of the greatest elongation varies. We used an average value to get around their variability, but this is only ducking a serious issue. Since the size of the greatest elongation varies, it almost seems that the epicycle is sometimes closer to, and sometimes farther from, the Earth. This is an issue we shall have to address eventually.

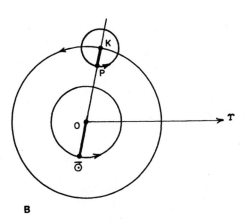

FIGURE 7.20. Relation between the mean Sun and a superior planet shortly before (A) and exactly at (B) a mean opposition.

Successes and Failures of Apollonius's Model

Apollonius's model provides a simple explanation of retrograde motion that is consistent with the principle of Aristotelian physics that celestial bodies must move on circles at uniform speed. Moreover, according to the model, Mars is closest to the Earth during retrograde motion. This is in agreement with the observed fact that Mars is brighter during retrograde motion than at other times. Apollonius's model thus represents an improvement over the homocentric spheres of Eudoxus. It is also far simpler than Eudoxus's model from a mathematical point of view.

However, Apollonius's model is not capable of predicting the motions of the planets with any real accuracy. Apollonius's model generates retrograde loops that are all of the same size and shape and that are equally spaced around the zodiac, as in figure 7.18. However, the actual retrograde arcs of Mars vary considerably in size and spacing. figure 7.24 shows the actual retrograde arcs of Mars for the years A.D. 109–122.

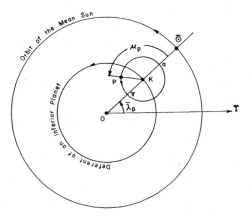

FIGURE 7.21. Relation between the mean Sun and an inferior planet: $\bar{\lambda}_P = \bar{\lambda}_\odot$.

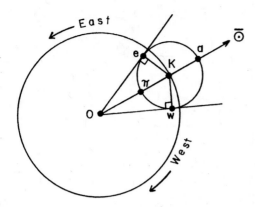

FIGURE 7.22. The time between greatest elongations allows us to prove that the inferior planets move counterclockwise around their epicycles.

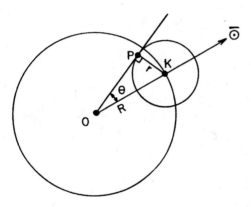

FIGURE 7.23. The size of an inferior planet's greatest elongations allows us to estimate the size of the epicycle.

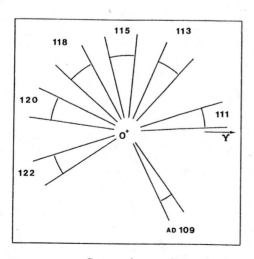

FIGURE 7.24. Retrograde arcs of Mars for the years A.D. 109–122.

If we superimpose figures 7.24 and 7.18, we obtain figure 7.25. We have started off the theoretical model to produce the retrograde loop of A.D. 109 in the right part of the zodiac. But the very next theoretical loop (for A.D. 111) falls well short of the part of the sky where the real retrogradation took place. Apollonius's model clearly has no numerical predictive power.

Two Inequalities

As figure 7.24 reveals, the retrogradations of Mars show great variability. Mars backs up over an arc whose length varies from about 10° (as in A.D. 109) to about 20° (A.D. 118). Moreover, the distance that the planet travels between one retrogradation and the next is quite variable. Thus, the centers of the retrograde arcs of A.D. 109 and 111 are 75° apart, but the centers of the A.D. 115 and 118 retrograde arcs are only 34° apart.

Inequality with Respect to the Sun Any departure of a planet from uniform angular motion is called an *anomaly* or an *inequality*. Mars has two separate inequalities. One of these is very striking and produces the reversals of direction known as retrograde motion. In Apollonius's theory, this inequality is produced by the epicycle. As we have seen, retrograde motion is intimately connected with the Sun: the superior planets retrogress when they are in opposition to the Sun. For this reason, the inequality of a planet associated with retrograde motion is sometimes called the *inequality with respect to the Sun*.

Zodiacal Inequality In the solar theory, we saw an example of a different kind of inequality, the *zodiacal inequality*. The Sun appears to move faster in some parts of the zodiac and slower in other parts. In the solar theory, this inequality can be produced by an eccentric (off-center) deferent circle. It is clear from figure 7.24 that Mars also has a zodiacal inequality. The epicycle's center appears to travel more slowly around the position of the A.D. 118 retrogradation and more quickly around the position of the A.D. 109 retrogradation. This is why the retrograde arcs are closely bunched in the first part of the sky but widely separated in the other. The zodiacal inequality is also known as the *first inequality*. This terminology has its origins in the solar theory: the Sun has only one inequality. The additional inequality displayed by all the planets, which causes retrograde motion, is logically called the *second inequality*. Apollonius's theory of longitudes accounted for Mars's second inequality, but failed to reproduce the first.

Status of the Epicycle Model in the Third Century B.C.

We know little of Apollonius's methods, for none of his writing on the planets has come down to us. All we really have is a few remarks by Ptolemy that make it clear that Apollonius proved some mathematical theorems involving epicycle motion. In particular, Apollonius proved the equivalence of an epicycle-plus-concentric to an eccentric circle (two forms of the later solar theory).

Apollonius also proved a theorem that established the conditions necessary for retrograde motion.[41] Refer to figure 7.26. The Earth is at O. The planet F travels on an epicycle, whose center K moves on a circle about the Earth. Let f_λ denote the angular speed of the epicycle's center (the rate at which OK turns). Let f_μ denote the angular speed of the planet on the epicycle (the rate at which angle FKO changes). Apollonius proved that the planet, at F, will appear stationary, as seen from O, whenever

$$\frac{FG}{OF} = \frac{2 f_\lambda}{f_\mu}.$$

This theorem permits one to calculate the length of the retrograde arc if the two angular speeds and the radius of the epicycle are known.

Apollonius's theorem directly applies only to a zero-eccentricity model, in which the center of the deferent circle is located precisely at the Earth. The first practitioners of deferent-and-epicycle astronomy evidently concerned themselves only with the inequality with respect to the Sun and took no account of the zodiacal inequality. However, Apollonius must certainly have been aware of the zodiacal inequality. As we have seen, in the case of Mars, this inequality is very striking. One would need only the roughest observations to make it apparent—it would be enough to note only the *constellation* in which each retrogradation occurred. Moreover, Ptolemy tells us in *Almagest* IX, 2, that the stations were one of two classes of planetary phenomena of greatest interest to his predecessors (the other being heliacal risings and settings).

It seems, then, that Apollonius knew of the zodiacal inequality but deliberately neglected it. His model was therefore incapable of predicting planetary positions. What, then, was the purpose of his astronomical work? First, it was an exercise in geometry. The study of curves generated by points moving in some complicated way constituted a traditional class of geometrical problems. Second, it was a response to Eudoxus. The idea of *genre* was very important in Greek mathematical writing.[42] Apollonius was writing in a genre of mathematics established by Eudoxus's book *On Speeds*. Third, it is clear that Apollonius also meant his models to apply to the world. The epicycle model explained retrograde motion, while also accounting for the variation in the brightness of the planets in the course of their synodic cycles. That is, it explained the most general and readily perceived features of planetary motion. The model was intended only to be qualitative and broadly explanatory in nature. The idea that one could demand a quantitative and predictive model must have dawned very slowly.

Whether Apollonius went so far as to work out numerical values for planetary parameters, we do not know. Nor do we know whether he discussed the relations between the motions of the planets and that of the Sun. But there is no evidence that he did. The elaboration of the theory—and, in particular, the deduction of numerical values for such parameters as the radius of the epicycle—was a later development.

An Intermediate Model

To produce a better model, suppose we take a hint from the solar theory and allow the deferent circle to be eccentric to the Earth. This gives rise to what we shall call the intermediate model. It would make sense to displace the center of the deferent one way or the other along the line of symmetry in the pattern of retrograde arcs shown in figure 7.24. Thus, we might consider displacing the center *D* of Mars's deferent in the direction of the A.D. 118 retrogradation, as in figure 7.27A. Alternatively, we might displace the deferent's center in the direction of the A.D. 109 retrogradation, which produces the model of figure 7.27B.

It is easy to predict what will result from the intermediate model: the retrograde loops will still be uniformly spaced and all of the same size, but the center of the pattern of loops will lie at *D* and not at *O*. Figure 7.28A compares the model of figure 7.27A with the real retrograde arcs of Mars. We have adjusted the position of *D* (center of the loop pattern) to obtain the best agreement with the actual positions of the retrograde arcs, as seen from the Earth *O*. As far as the *positions* go, version A of the intermediate model is pretty good: every one of the theoretical loops falls on top of the corresponding observed retrograde arc. However, the *widths* of the retrograde

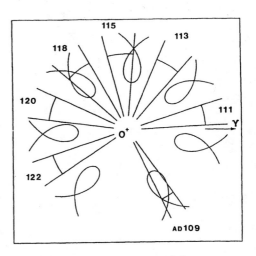

FIGURE 7.25. Retrograde arcs of the zero-eccentricity model compared with the actual retrograde arcs of Mars, A.D. 109–122.

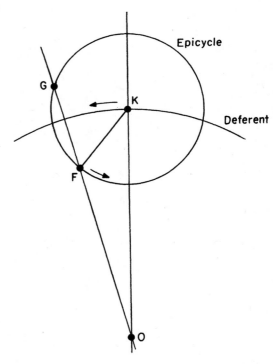

FIGURE 7.26. Illustrating Apollonius's theorem.

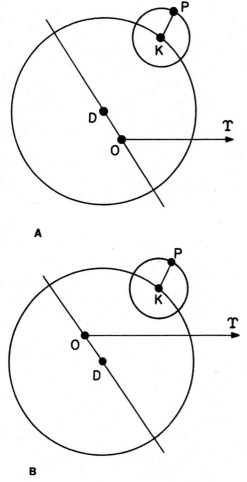

FIGURE 7.27. Two versions (A and B) of an intermediate model.

arcs are terrible. In fact, they are now even worse than in the zero-eccentricity model.

Let us therefore examine the behavior of version B of the intermediate model (fig. 7.27B). Figure 7.28B shows the theoretical loops (centered on *D*) of this model superimposed on the real retrograde arcs (centered on *O*) of Mars. Now we have done a good job with the *widths* of the retrograde arcs: in two diametrically opposite parts of the zodiac (the 109 and the 118 retrogradations) the theoretical loops closely match the observed widths. But now the *positions* of the retrogradations are terrible. Thus, the intermediate model cannot simultaneously account for both the positions and the widths of the retrogradations.

7.13 GREEK PLANETARY THEORY BETWEEN APOLLONIUS AND PTOLEMY

Hipparchus on the Planets

In the second century B.C., Greek astronomers began to grapple with the planets' zodiacal inequality. Our chief source of information is Ptolemy's brief summary, in *Almagest* IX, 2, of Hipparchus's work on the planets. As Ptolemy remarks, Hipparchus made notable contributions to the theories of the Sun and the Moon. However, according to Ptolemy, Hipparchus did not give a theory of the planets but only arranged the observations in a more useful way and showed the appearances to be inconsistent with the hypotheses of the mathematicians.

In particular, Hipparchus noted that, owing to the two separate inequalities, the retrogradations of each planet are not uniform, but the mathematicians gave their geometrical demonstrations as if there were a single inequality and as if all the retrograde arcs were of the same length. From this remark by Ptolemy, it appears that one part of Hipparchus's contribution was a demonstration that the zero-eccentricity model of his predecessors was inconsistent with the motions of the planets. But Hipparchus's predecessors could hardly have been unaware that the planets' retrogradations are unevenly spaced around the zodiac, for this is reflected quite clearly in the unequal times that separate successive retrogradations. And, at least in the case of Mars, the unequal widths of the retrograde arcs must also have been known. Hipparchus's insistence that *a planetary theory ought to work in detail* was far more significant for the future of astronomy than was his simple notice that the Greek planetary theories were not terribly accurate. As we have suggested in section 5.2, Hipparchus's insistence on models that worked in detail probably was a consequence of his contact with Babylonian astronomy.

The central problem of Greek astronomy, from the time of Eudoxus on, was to save the phenomena in terms of accepted physical principles. But what counted as the phenomena—the class of details to be explained—changed dramatically over time. While Eudoxus and Apollonius saw their job as merely giving a physically plausible, geometrical explanation of retrograde motion, Hipparchus insisted on a planetary theory that could also explain the zodiacal inequality. Now, for the first time, a geometrical planetary theory was also required to have numerical predictive power. This Hipparchus was unable to provide.

Ptolemy points out that the appearances cannot be saved either by eccentric circles, or by circles concentric with the Earth but bearing epicycles, or even by eccentrics and epicycles together. A model with an eccentric and an epicycle would be something like the intermediate model illustrated in figure 7.27, a model that was investigated by Greek astronomers between the times of Hipparchus and Ptolemy.

It is even possible to say which version of the intermediate model was preferred. Pliny, in book II of his *Natural History* (first century A.D.), has a little to say about the planets. Pliny was not an astronomer, and much of his discussion is both confused and confusing. However, his writing predates Ptolemy by two generations and he had access to pre-Ptolemaic works that are now lost. According to Pliny, the apogees of the superior planets' deferents are as follows: Saturn in Scorpio, Jupiter in Virgo, Mars in Leo.[43] These are consistent with version A of the intermediate model. That is, the apogees were placed correctly to account for the *spacing* of the retrogradations around the zodiac. No account was taken, therefore, of the *widths* of the retrograde arcs.

Version A was, indeed, the more reasonable choice. Except in the case of Mars, the variation in the widths of the retrograde arcs is not very dramatic. For most of the planets, this variation could be ignored, while the uneven spacing of the arcs could not. Besides, version A of the intermediate model had a close parallel in the solar theory.[44] As we saw in section 7.12, version A of the intermediate model is not really a satisfactory representation of planetary motion. However, as no one before Ptolemy had anything better to propose, this model continued in use down to his time.

Astrology as a Motive

The philosophically based geometrical astronomy in the tradition of Eudoxus, Apollonius, and Hipparchus was inadequate for the calculation of planetary phenomena. However, there were pressing practical reasons for the Greeks to be able to calculate planetary phenomena from theory. Babylonian astrological ideas were introduced to the Greeks in the third century B.C. and grew to have enormous currency and popularity.

Although belief in planetary influences was ancient in Babylonia, planetary omens were first interpreted as applying only to the king or to the nation as a whole. It is only at the close of the fifth century B.C. that horoscopic astrology emerges. By horoscopic astrology we mean the prediction of an ordinary person's future or disposition by examination of the positions of the Sun, Moon, and planets at the moment of his birth or conception. Only a handful of Babylonian horoscopes have been dated to the third century B.C. or earlier. The oldest known is for 410 B.C.[45]

While horoscopic astrology was certainly of Babylonian origin (as, indeed, the Greek and Roman writers always claimed[46]), it was elaborated into a complex system by the Greeks. Thus, the familiar and fantastically complicated system of horoscopic astrology with dozens of conflicting rules does not descend from remote antiquity. Rather it is a product of Hellenistic and Roman times. This fact comes rather as a blow to modern apologists for astrology who are fond of claiming ancient wisdom as a justification for their art—and the older the wiser.

Greek interest in horoscopic astrology grew rapidly starting in the first century B.C. References to astrology begin to appear in Greek and Roman literature. We have also nearly 200 Greek and Roman horoscopes in the astrological writers, as well as on papyri discovered in archaeological excavations. The oldest known Greek horoscopes are from the first century B.C. but the great bulk of them come from the first five centuries of our own era.[47]

Systematic treatises on horoscopic astrology were written in Greek and Latin. We shall mention here only three texts of considerable historical importance. The oldest surviving complete manual of horoscopic astrology is the long Latin poem *Astronomica* by Manilius, a Roman writer of the first century A.D. In the second century A.D., Vettius Valens, a Greek from Antioch who settled in Alexandria, wrote a large *Anthology* of astrology, to which we shall refer below. But the definitive Greek treatise on astrology was written by none other than Ptolemy. Ptolemy's *Tetrabiblos* came to serve as a standard manual

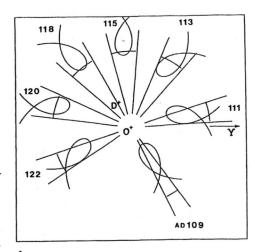

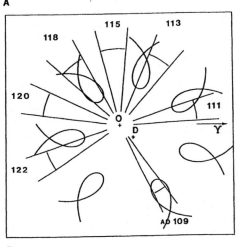

FIGURE 7.28. A, Retrograde loops of the intermediate model (version A) compared with the actual retrograde arcs of Mars, A.D. 109–122. B. Retrograde loops of the intermediate model (version B) compared with the actual retrograde arcs of Mars.

of astrology, much as his *Almagest* served to define planetary theory. The commonly used title of Ptolemy's astrological work, *Tetrabiblos*, reflects its division into four parts or books. Ptolemy addresses this work to Syrus, the same friend or patron to whom he addressed the *Almagest*. In the introduction, Ptolemy remarks that there are two kinds of prediction through astronomy. The first kind (i.e., the kind treated in the *Almagest*) deals with the motions of the Sun, Moon, and planets and ranks first both in primacy and in effectiveness. The second kind of prediction in astronomy deals with the changes produced on Earth by the planets. Ptolemy admits that this second (astrological) kind of prediction is far less certain. But he attempts to provide some physical justification for this science.[48]

To practice horoscopic astrology, a Greek of the Hellenistic or Roman period needed to be able to calculate the positions of the planets in the zodiac with ease and rapidity for the moment in question. And he needed to be able to calculate the horoscopic point—the point of the ecliptic that · as rising on the eastern horizon. Hypsicles' arithmetical methods for dealing with the risings of the zodiac signs date from the second century B.C. But, as we saw in section 2.15, even after the development of trigonometry made exact solutions possible, Greek and Roman astrological writers contined to favor arithmetical methods for ascensions, because they were easier. In the case of planetary theory, the situation was even worse, for the geometrical models were not only hard to use, but also unsatisfactory. Astrologers, who needed numerical answers rather than philosophical generalities, had no choice but to fall back on arithmetical schemes for calculating planet positions.

Many writers on the history of Greek planetary theory have overstressed the continuity of its development, the purity of its allegience to philosophical principles, and its cultural independence. The motivations underlying Greek planetary theory were complex and they also evolved with time. The Greek names of the planets reveal connections with religion. Even Ptolemy still regarded the planets as divine. Moreover, as we have seen, planetary theory had deep roots in philosophy as well as in geometry. A philosophically based geometrical theory of planetary motion, such as that of Eudoxus or Apollonius, was supposed to explain in a qualitative way how the world might work and to provide a field of play for the geometer. But it was incapable of yielding numerically accurate positions. In part, the emergence of a geometrical planetary theory with quantitative predictive power represented a continuation of a process already begun—the geometrization of the universe. Thus, we can see the work of Hipparchus and Ptolemy as a continuation of the tradition of Eudoxus, Aristarchus, and Apollonius. But, in part, the emergence of quantitative planetary theory among the Greeks also depended on the Babylonian example—which showed that such a thing was possible—and on the sense of urgency imposed by the astrological motive.

Arithmetic Methods in Greek Planetary Theory

Between the time of Hipparchus and Ptolemy, Greek astronomers largely turned away from geometrical planetary theory to arithmetical methods for calculating positions of the planets. The basic idea of using arithmetic methods came from Babylonia. However, the Greeks were not always able to take over Babylonian methods directly. As we have seen, Babylonian planetary theory focused on direct computation of the times and places of important synodic events—first and last visibility, beginning and end of retrograde motion, and so on. The position in the zodiac of a planet on a given day can be obtained from Babylonian procedures, but not easily or directly. Rather, one must interpolate between the directly computed synodic events. For the practice of astrology among the Greeks of the Roman period, the most important things to calculate were the dates of the entry of a planet into successive zodiac signs. Although there is emerging evidence of more direct use of Babylonian methods, the Greek astronomers and astrologers in the period between Hipparchus and Ptolemy largely went their own way and worked out a number of original arithmetic procedures for calculating planetary positions. Ptolemy perhaps refers to these methods in *Almagest* IX, 2, when he speaks of the unsatisfactory character of the "aeon-tables" (we might call them perpetual tables) used by his predecessors.

The most complete extant Greek text on arithmetic methods for the longitudes of the planets is that of Vettius Valens (second century A.D.), in book I, chapter 20, of his astrological compendium known as the *Anthology*.[49] Actual planetary tables based on arithmetic methods survive on papyrus and on wooden tablets from Greco-Roman Egypt of the first and second century A.D. Some of this material is written in Greek and some in demotic (the later, simplified Egyptian script). Many of the papyri are devoted to tables for the dates of entry of planets into zodiac signs, but the theories on which they are based cannot always be reconstructed in detail.[50]

While most of the papyri seem to rely on methods that are not Babylonian in origin, there is also solid evidence of direct use of Babylonian arithmetical methods by Greek astronomers. One of the most detailed Greek references to Babylonian astronomy is the final chapter of Geminus's *Introduction to the Phenomena*, which is devoted to the lunar theory. In Geminus's account, the Moon's daily motion follows a linear zigzag function. According to Geminus, the maximum amount that the Moon can move in one day is $15;14,35°$ (sexagesimal notation; see sec. 1.2). The smallest amount the Moon can move in one day is $11;6,35°$, and the mean daily motion is $13;14,35°$. Moreover, according to Geminus, the daily motion changes by equal increments of $0;18°$ from one day to the next. Geminus attributes the figure for the mean daily motion to the "Chaldaeans." But the other parameters are of Babylonian origin, too. Indeed, Geminus is describing system B of the Babylonian lunar theory.

Geminus's description of the lunar theory is not detailed enough to permit the reader to use it in practice. For example, Geminus does not provide epoch values; that is, he does not bother to tell either where the Moon was or where the Moon's line of apsides was on a particular starting date. Thus, one could still wonder whether Greek astronomers really mastered the details of Babylonian science or just gained a passing familiarity with its basic concepts.

This question was resolved beyond all doubt by the discovery of a papyrus of the second century A.D., written with Greek numerals, in which lunar phenomena were computed on the basis of system B of the Babylonian lunar theory.[51] Another Greek papyrus (of the third century) uses a theory of Mars that is related to the Babylonian theory of Mars of system A. However, the numerical parameters have been modified—and in fact made worse. Alexander Jones has suggested that the modifications were introduced to make the arithmetical scheme more consistent with a deferent-and-epicycle model for the motion of Mars. That is, the Babylonian calculating scheme may have been modified by a Greek astronomer, not to make it agree better with observations, but to make it agree better with a physical theory of the motion of the planets.[52]

Very recently, the direct evidence for Greek knowledge of Babylonian planetary theory has been vastly expanded through the study of astronomical material in the Oxyrhyncus papyri. Oxyrhyncus was a town in Greco-Roman Egypt. Its garbage dumps, excavated 1897–1934, were the richest source of papyri ever found in Egypt. More than 70% of the surviving literary papyri have come from Oxyrhyncus, but until quite recently the astronomical papyri were ignored. The astronomical material is being edited by Alexander Jones and will soon be published. According to Jones, the Oxyrhyncus papyri include Greek versions of typical ACT-style Babylonian planetary tables, with nearly exact replicas of system A and system B schemes for every planet except Venus. We do not know when the main period of transmission occurred since the Oxyrhynchus papyri are all Roman period, first century A.D. and later. (The pre-Roman levels of the Oxyrhynchus garbage dumps lay below the water table, so no papyri from these levels have survived.) But it is now abundantly clear that Greeks living in Egypt had mastered the Babylonian planetary theory in pretty full detail at least a few generations before Ptolemy's time.

We do not know in any detail how Babylonian astronomy and astrology jumped the cultural gap. But the time of transmission was probably the Seleucid period. Van der Waerden has argued that the many references by later Greek and Roman writers to "Chaldaean" practice point to the existence of a compendium of Babylonian astronomy and astrology, written in Greek.[53] The possibility of a compendium of Babylonian astronomy in Greek cannot be discounted, for we do know of an interesting parallel. The Chaldaean Berosus wrote, in Greek, around 280 B.C., a history of Babylonia, called *Babylonica*, for his patron, Antiochus I Soter, the second king of the Seleucid dynasty. This history has not survived, but many citations of it are preserved by Josephus and Eusebius. Berosus treated the history of the world from the creation down to the time of Alexander. The first portions of his book were therefore mythological, but the later portions must have been based on Babylonian chronicles.

As we saw in section 1.9, Berosus also had a reputation as an astrologer and astronomer. Vitruvius claimed that Berosus settled at Cos and played a role in introducing astrology to the Greeks. Pliny[54] says that the Athenians were so impressed with his marvellous predictions that they erected at the exercise ground a statue of Berosus with a gilded tongue. We can find in the preserved fragments of Berosus's work very little to convince us that he was an accomplished astronomer. Therefore, we need not take very seriously Berosus's purported role in introducing the Greeks to Babylonian astronomy. The important thing is the example: Berosus was a priest of Marduk who did write some sort of book in Greek for a Seleucid royal patron. If it was not Berosus, we can imagine another priest or scribe of Babylon writing some sort of astronomical compendium in Greek for some other Greek patron. But all this remains conjecture. Other writers have pointed out that it is sufficient to assume that Babylonian scribes emigrated and took their skills with them, or that Greeks who talked to the priests in Babylon picked up the essentials of Babylonian astronomy.[55]

Thus, Greek planetary theory in the period just before Ptolemy presents a very complex picture. This remains a lively area of historical research and we can expect the picture to change a bit in the next few years.

But this much is now clear: if you were a Greek steeped in Aristotelian physics and Euclidean geometry, you couldn't understand what was going on unless you thought in terms of deferents and epicycles. Thus, philosophically oriented writers expounded geometrical systems based on deferents and epicycles. A good example of such a text is Theon of Smyrna's *Mathematical Knowledge Useful for Reading Plato*, which dates from the early second century A.D. Theon was not the only one writing in this genre, for he makes it clear that he draws most of his astronomical detail from earlier writers, especially Adrastus, a Peripatetic philosopher who was a generation or two older. For Theon, the Chaldaean planetary theory was unacceptable because it was not based on a proper understanding of nature.[56] On the other hand, if you were a Greek astrologer in Roman Egypt who needed to obtain planet positions (even if they were not very accurate), you had to fall back on arithmetic methods. The philosophically based geometrical planetary theory and the arithmetically based calculating schemes still existed side by side in the second century A.D.

Despite Hipparchus's insistence that it should be able to do so, the deferent-and-epicycle theory was at this stage of its development incapable of providing satisfactory answers. The Greeks therefore experimented with a number of numerical predictive schemes with only limited success. The new double goal of providing a quantitative planetary theory based on accepted physical principles had not yet been realized. That remained for Ptolemy to do.

7.14 EXERCISE: THE EPICYCLE OF VENUS

1. Using data from table 7.1, prove that Venus travels in the same direction on its epicycle as Mercury does (counterclockwise as viewed from the north pole of the ecliptic).
2. Make a rough estimate of the radius of Venus's epicycle.

7.15 A COSMOLOGICAL DIVERTISSEMENT: THE ORDER OF THE PLANETS

Aristotle points out that the Moon is sometimes seen not only to eclipse the Sun but also to pass in front of (or *to occult*) stars and planets.[57] Thus, all ancient writers agreed in placing the Moon nearer to us than any other celestial body. Moreover, the Moon's parallax is large enough to allow a measurement of its distance (sec. 1.17). The parallaxes of the planets, however, are very small. No measurements of the planets' distances were possible with the methods of the ancient astronomers.

The deferent-and-epicycle arrangement for each planet therefore constituted an independent system. There was no *astronomical* way to tell which planets were closest to the Earth and which were farthest away. That is, there was no way to measure the absolute sizes of the deferent circles. All that was astronomically determinable was, for each planet, the *ratio* of the epicycle's radius to the radius of the deferent. But this did not prevent the Greek astronomers from speculating about the order of the planets.

In ancient science, the order of the planets was a *cosmological* question. Cosmology is the effort to understand the arrangement of the whole universe. Astronomy is one part of this endeavor. But astronomy, based on observation and calculation, cannot answer every question—a point stated emphatically by Geminus (see sec. 5.3). The selection of a model for the motion of the planets (such as Apollonius's deferent-and-epicycle model) had therefore to be made partly on the basis of nonastronomical criteria. The most important of these was, of course, the principle of ancient physics that the planets must move in circles at constant speed.

An Organizing Principle for the Cosmos

Another physical principle accepted at an early date (long before the invention of deferent-and-epicycle theory) was that the planets ought to be arranged according to their tropical periods. The slowest planet (Saturn) should be farthest away from us and closest to the fixed stars. This principle was enunciated by Aristotle and justified with physical arguments.[58] According to Aristotle, the eastward motion of the planets in the zodiac is partially retarded or restricted by the primary westward motion of the whole cosmos. It makes sense that the planet closest to the sphere of stars (Saturn) should be restricted the most. This is why it has only a feeble eastward motion.

On the basis of this principle, all the Greek writers agreed in placing Saturn (tropical period = 30 years) nearest the fixed stars, Jupiter (12 years) next below, and Mars (2 years) next. The lowest place, just above the Earth, was assigned by all writers to the Moon, which completes a trip around the zodiac in a month. However, there was a difference of opinion about the Sun, Mercury, and Venus. These three bodies must be placed between the Moon and Mars. But since all three have a tropical period of exactly one year, their order cannot be deduced from the principle that connects distances to times.

Among the early writers, several different doctrines arose. Plato chose the order Moon, Sun, Venus, Mercury, Mars, Jupiter, Saturn.[59] Theon of Smyrna remarks that some of the "mathematicians" (i.e., technical astronomers) also adopted this order but that others inverted the positions of Venus and Mercury. In either arrangement, all the planets were placed above the Sun.

The Standard Order

Pythagorean Sun Mysticism   Theon of Smyrna tells us that "certain Pythagoreans" chose the order Moon, Mercury, Venus, Sun, Mars, Jupiter, Saturn. These Pythagoreans wanted the middle circle among the planets to be that of the Sun, which was the "heart of the universe" and the most fit for command.[60] Theon distinguishes between the center of activity and the center of volume. For example, in the case of man, the center of the living creature, considered as a man and an animal, is the heart, which is warm and always in motion. The heart is the origin of all faculties of the soul, such as life, movement from place to place, desires, imagination, and intelligence. But the center of volume is different; in us, it is situated near the navel. In the same way, the center of volume of the cosmos is the Earth, cold and motionless. But the center of the cosmos, considered as cosmos and animal, is the Sun, which is the heart of the universe, and from which the soul arises to fill the universe and to spread through the whole body to the farthest limits.[61]

Another example of this Sun mysticism is provided by Pliny:

EXTRACT FROM PLINY

Natural History II, 12–13

In the middle [of the planets] moves the Sun, whose magnitude and power are the greatest, and who is ruler not only of the seasons and of the lands, but even of the stars themselves and of the heaven. Taking into account all that he effects, we must believe him to be the soul, or more precisely the mind, of the whole world, the supreme ruling principle and divinity of nature. He furnishes the world with light and removes darkness, he obscures and illuminates the rest of the stars, he regulates in accord with nature's precedent the changes of the seasons and the continuous re-birth of the year, he dissipates the gloom of heaven and even calms the storm-clouds of the mind of man, he lends his light to the rest of the stars also; he is glorious and preeminent, all seeing and even all hearing.[62]

Ptolemy on the Order of Planets   A less mystical approach is taken by Ptolemy, who begins book IX of the *Almagest* with a discussion of the order of the planets. Ptolemy remarks that some mathematicians placed Venus and Mercury higher than the Sun, because the Sun had never been seen eclipsed by the planets.[63] However, Ptolemy points out that these planets might lie a little north or south of the ecliptic at their conjunctions with the Sun and therefore fail to produce an eclipse, just as the Moon fails in the majority of cases to eclipse the Sun at the time of new Moon. In the *Planetary Hypotheses* Ptolemy adds that the occultation of the Sun by a small body might not even be perceptible, just as small, grazing eclipses of the Sun by the Moon are often not perceptible.[64] Ptolemy himself adopts the order Moon, Mercury, Venus, Sun, Mars, Jupiter, Saturn, saying that it is reasonable to place the Sun in the middle, as a division between the planets that can be at any elongation from the Sun and those that always move near it.

In the *Planetary Hypotheses*, Ptolemy supplies a number of physical arguments to justify this order. He argues that the further removed a planet's astronomical hypotheses are from those of the Sun, the farther the planet must lie, in real distance, from the Sun.[65] Thus, Mercury must lie below the Sun and close to the Moon, because the rather complex theoretical system for Mercury resembles that of the Moon. One might also expect the lowest planets to have the most complex motions because they are nearest to the air and their movement resembles the turbulent motion of the element adjacent to them. So, again, it makes sense for the Moon and Mercury, which have the most complex theories, to be lowest. This argument is Ptolemy's elaboration of an idea of Aristotle's.[66]

Ptolemy's ordering of the planets (Moon, Mercury, Venus, Sun, Mars, Jupiter, Saturn) had become standard somewhat before his time. This ordering was confirmed by Ptolemy's great authority and was almost universally accepted down to the sixteenth century.

A Partially Heliocentric System

But one other possible arrangement of Mercury and Venus must be mentioned. We know that the centers of these planets' epicycles always lie in the same direction as the Sun. It is impossible to say whether these centers are closer to us or farther from us than the Sun. Might it not be the case that the centers of Venus's and Mercury's epicycles actually *coincide* with the Sun? Then the two planets would execute circular orbits around the Sun while the Sun travels on its own circle around the Earth. More precisely, the epicycles of Venus and Mercury would be centered on the mean Sun, around which the true Sun would also revolve on its own tiny epicycle. This arrangement is a plausible extension of the principle that connects distances to times, for it explains why the tropical periods of Mercury, Venus, and the Sun are all the same: they all share the same deferent circle.

Moreover, this system was actually advocated in antiquity. Theon of Smyrna remarks that it is possible that the three bodies have three separate deferents that revolve in the same time, the Sun's being smallest, Mercury's larger, and Venus's larger yet. But, says Theon, there could also be only a single deferent common to the three stars, whose epicycles would then turn about a single center. The smallest epicycle would be the Sun's, Mercury's next larger, then Venus's. This would explain, says Theon, why these three stars are always neighbors, Mercury never being more than 20° from the Sun, and Venus never more than 50°. "One might suspect that the truer position and order is this, in order that this might be the seat of the life principle of the cosmos, considered as cosmos and living creature, as if the Sun were the heart of the universe by virtue of its motion, its size, and the common course of the planets round it."[67]

This heliocentric arrangement for Venus and Mercury is also mentioned by three late (fourth and fifth centuries A.D.) Latin writers, Chalcidius, Macrobius, and Martianus Capella. Chalcidius attributes the system to Heraclides of Pontos (fourth century B.C.)—but he is clearly mistaken, as Heraclides lived before the invention of the epicycle theory.[68] Perhaps Chalcidius attributed this view to Heraclides because he was known to have held another unorthodox astronomical opinion: the daily rotation of Earth on its axis (see sec. 1.6).

A Grand View of the Cosmos

Figure 7.29, A and B, present a grand view of the cosmos according to deferent-and-epicycle theory. The epicycle of each planet is drawn in correct proportion

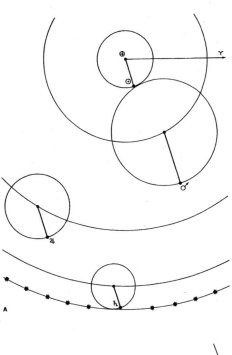

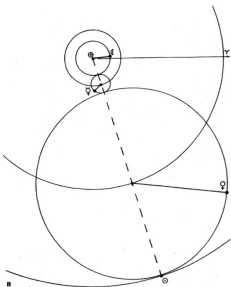

FIGURE 7.29. A grand view of the cosmos: disposition of the Sun and superior planets (A) and of the Sun and inferior planets (B) on January 7, 1900.

to its deferent. We have adopted the standard order of the planets (Ptolemy's order). We have also adopted another of Ptolemy's cosmological principles (see sec. 7.25): the universe should contain no empty or wasted space. Thus, one side of Mars's epicycle is tangent to the Sun's circle. Similarly, the other side of Mars's epicycle has just enough space to squeeze by the epicycle of Jupiter. We have, however, ignored the eccentricities of the deferent circles, so the figure somewhat simplifies Ptolemy's system. The figure has been drawn to represent the arrangement of the heavens on a particular date: January 7, A.D. 1900.

Figure 7.29A shows the outer part of the system—Saturn, Jupiter, Mars, and the Sun. A striking feature of the figure is the fact that the radii of the epicycles of the superior planets are parallel to one another and to the line from the Earth to the Sun. This is, of course, a necessary condition of the theory. Figure 7.29B shows the inner part of the system: Sun, Venus, Mercury, and Moon. This figure has been drawn on a scale eight times larger than figure 7.29A. So, if one could shrink figure 7.29B by a factor of eight, the solar circles of the two figures would be the same size, and all of figure 7.29B could be placed inside the solar circle of A. The striking feature of figure 7.29B is the fact that the centers of the epicycles of Mercury and Venus lie on the line from the Earth to the Sun.

As Ptolemy says in the *Planetary Hypotheses*, each planet has one free motion and one constrained motion. As one ponders these diagrams and remembers that the Moon gets its light from the Sun, that its phases are determined by its elongation from the Sun—then slowly one begins to appreciate the force of the old Pythagorean doctrine that the mind and heart of the universe is the Sun.

Note on Planetary Symbols   The conventional signs for the planets introduced in figures 7.29 and 7.30 provide a useful shorthand notation. To judge by the papyrus planetary tables and other texts preserved from the Hellenistic period, the ancient Greek astronomers did not use such symbols. Rather, the names of the planets were simply written out or, often, abbreviated. The modern planetary symbols first appear in Byzantine Greek manuscripts of the late Middle Ages. Figure 7.30 presents variants of the planetary symbols found in a few medieval astronomical and astrological manuscripts in the Bibliothèque

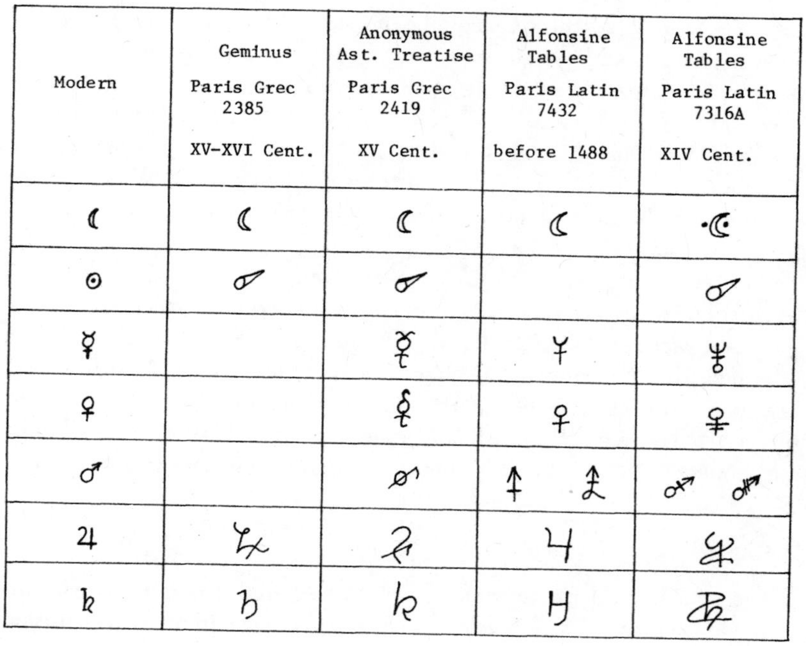

Modern	Geminus Paris Grec 2385 XV–XVI Cent.	Anonymous Ast. Treatise Paris Grec 2419 XV Cent.	Alfonsine Tables Paris Latin 7432 before 1488	Alfonsine Tables Paris Latin 7316A XIV Cent.
☾	☾	☾	☾	·☾
☉	♂	♂		♂
☿		☿	♀	☿
♀		♀	♀	♀
♂		♂	↑ ⚹	♂ ♂
♃	♃	♃	♃	♃
♄	♄	♄	♄	♄

FIGURE 7.30. Examples of planetary symbols in some late medieval manuscripts.

Nationale, Paris. The table is not intended to be exhaustive but simply to illustrate the notation actually employed by medieval astronomers. The lunar symbol is practically invariable, no doubt because it is a direct pictorial representation. Similarly, the medieval astronomical writers are more or less unanimous in their use of the symbol for the Sun. Interestingly, the modern symbol ⊙ does appear in medieval Greek manuscripts, but never as a mark for the Sun; rather, it is employed always as a sign for "circle."

Levels of Certainty in Greek Science

The Greeks wanted to find the structure and arrangement of the universe as a whole. This was a daunting task. The parts of the task that required only *astronomical* observation and geometry led to results that were certain and reliable. Examples of these include the measurement of the size of the Earth and the distance of the Moon.

Some parts of the task could not proceed unless astronomical observation and geometry were supplemented by a set of *physical* assumptions. A good example is Apollonius's theory for the motions of the planets. If we assume uniform circular motion, we may then construct a model to account for the observed behavior of the planets. The assumption of circular motion was not arbitrary or ad hoc. Rather, this was a universally agreed physical principle. Of course, it could still be asked just how well Apollonius's model agreed with the actual details of planetary motion. As we have seen, the answer is not very well. The response of the Greek astronomers to this difficulty was a gradual refinement of the model. This required bending the rules of Aristotelian physics and introducing departures from uniform circular motion. By the time of Ptolemy (second century A.D.), the deferent-and-epicycle theory was brought into a highly satisfactory and accurate form.

Some *cosmological* questions, including the order of the planets, could be decided only by means of ad hoc assumptions. These assumptions could be supported by philosophical argument. But they could not be tested against observation. Thus, the order of the planets had a completely different epistemological status than the deferent-and-epicycle theory. It was not possible to refine the order of the planets by incremental progress in observation or by adding new details to the theory. Either you agreed with Plato, or you agreed with Ptolemy, or you suggested some order of your own.

Modern writers on Greek astronomy often fail to distinguish very clearly between these three levels of certainty and three varieties of proof. The best of the Greek astronomical writers—Ptolemy and Geminus, for example—were themselves quite clear about what was demonstrable and what was based on physical assumptions. Other writers, for example, Cleomedes, are more dogmatic and show less sensitivity to the differences among astronomical, physical, and cosmological premises. The Latin encyclopedists—Pliny and Vitruvius, for example—are the least sophisticated. For them, the astronomy of the Greeks is all of a piece and no part of it is any more fundamental than any other.

7.16 EXERCISE: TESTING APOLLONIUS'S THEORY OF LONGITUDES

In this exercise we test the theory of longitudes introduced in section 7.12, using Mars as example. How well does Apollonius's theory account for the actual pattern of retrogradations? If we needed answers accurate to the minute of arc, this would involve tedious trigonometric computations. Fortunately, we can learn almost as much from numbers that are accurate only to the

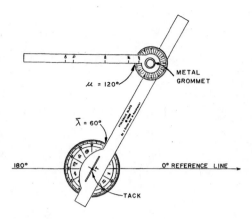

FIGURE 7.31. Using the Ptolemaic slats.

nearest degree or two. We can attain this precision and avoid trigonometry by employing a mechanical calculating tool, the *Ptolemaic slats*.

1. Assembling the Ptolemaic slats: Photocopy the Ptolemaic slats in the back of this book (fig. A.5.). If you wish, you may enlarge the figure slightly, so that the longest part is about 11 inches long. Photocopy onto stiff card stock, or glue a paper photocopy to card stock, using a good-quality glue stick. Cut out the three parts.

Attach the shorter slat (marked with planetary symbols) to the larger slat in the following way. Obtain a grommet-fastening kit at a hardware store. The smallest-sized grommets will be fine. They should be hollow, so that you can see through the hole after you have fastened the grommet. Use the tool provided in the kit to punch a hole at the point marked *H* on the larger slat, and at the point marked *H* on the smaller slat. Fasten the two parts together, as in figure 7.31. *A vital point: be careful to fasten the grommet loosely so that the smaller slat can be turned freely.*

2. Preparing the ground: Obtain a sheet of paper, about 20″ × 20″. Draw a line through the middle. Place a dot at the middle of the line to represent the Earth. The sheet represents the plane of the ecliptic. Label one end of the reference line 0° (longitude of the vernal equinoctial point) and the other end 180°. Poke a thumb tack through the Earth dot from below.

Place the large paper protractor (marked with the signs of the zodiac) from the Ptolemaic slats over the tack so that the tack sticks through the center of the protractor. Push a small eraser or a piece of balsa wood over the point of the tack so you will not stick yourself accidentally. Turn the protractor until the 0° direction coincides with the 0° reference line drawn on the paper. Stick a curl of tape under the protractor so that it remains fixed in this position.

Locate some prominent ecliptic stars around the edges of the paper to provide a frame of reference. Use a star chart or the rete of the astrolabe kit in the appendix of this book to obtain rough longitudes for Hamal (α Arietis), the Pleiades, Aldebaran, Pollux, Regulus, Spica, Zubenelgenubi (α Librae), Antares, and λ Sagittarii. For example, on a star chart, the Pleiades can be seen near the ecliptic at 58° longitude. Put a mark for the Pleiades 58° counterclockwise from the 0° reference line and at the edge of the paper.

3. How to get started: Place the Ptolemaic slats over the thumb tack so that the tack sticks through the center *T* of the cross hairs near one end of the long slat, as shown in figure 7.31. The tack will act as a pivot for this slat. The long slat will act as the revolving radius of the deferent circle. As the deferent slat is turned, the deferent circle is swept out by the grommet. The smaller slat, which is free to turn about this grommet, represents the revolving radius of the epicycle. The Ptolemaic slats may be used for any of the five naked-eye planets. The same deferent radius is used for all, but the epicycle slat is marked to show the appropriate epicycle radius for each planet. In this exercise we shall work with Mars.

As a planet moves past the fixed stars, its mean longitude $\bar{\lambda}$ and epicyclic anomaly μ both increase. Refer to figure 7.31 to see how these angles are measured on the Ptolemaic slats.

In section 7.4 we determined the tropical and synodic periods of Mars:

Tropical period: 1.88 years

Synodic period: 2.13 years

The angular speed at which the mean longitude $\bar{\lambda}$ increases will be denoted f_λ. Since $\bar{\lambda}$ must go through 360° in one tropical period,

$$f_\lambda = 360°/1.88 \text{ years}$$

$$= 0.524°/\text{day}.$$

Similarly, the epicyclic anomaly μ increases by 360° in one synodic period. The daily motion in μ, denoted f_μ, is therefore

$$f_\mu = 360°/2.13 \text{ years}$$

$$= 0.462°/\text{day}.$$

We shall make a time-lapse picture of Mars by taking a "snapshot" every time the epicycle slat has turned through 10°. Now, $\bar{\lambda}$ changes 0.524/0.462 (= 1.14) times more quickly than μ changes. Therefore, whenever μ increases by 10°, $\bar{\lambda}$ will increase by 11.4°.

The only remaining issues are the initial values of $\bar{\lambda}$ and μ. The easy way to begin is at a mean opposition, that is, at the center of a retrograde arc, for then the value of μ is known. Since the epicycle's radius points directly at the Earth at mean opposition, *the epicyclic anomaly must be 180°*. Also, since the planet is then seen in the same direction as the epicycle's center, *the mean longitude is the same as the longitude of the planet.*

Let us begin our study at the opposition of 1971. In section 7.4, we found that Mars reached its opposition to the Sun at longitude 317° on August 9, 1971. Therefore, we know that on August 9, 1971, Mars's mean longitude was 317° and its epicyclic anomaly was 180°. We can deduce these facts from the observation only because the observation was a very special one—an opposition.

Fill in the values of $\bar{\lambda}$ and μ by repeated additions or subtractions:

Step	$\bar{\lambda}$	μ	(Mars)
0	——	——	
1	——	——	
2	——	——	
3	——	——	
4	——	——	
5	——	——	
6	——	——	
7	——	——	
8	——	——	
9	317.0°	180°	(August 9, 1971)
10	328.4	190	
11	339.8	200	
12	351.2	210	
13	362.6 = 2.6	220	
14	14.0	230	
15	25.4	240	

Complete the table backward to the 0th step and forward to the 54th step. Be careful: one addition error will corrupt all the entries below it.

4. Turning the slats: Use the slats to plot the positions of Mars by adjusting the slats so that angles $\bar{\lambda}$ and μ have the values listed in your table. For each step, put a dot on the paper next to the Mars symbol on the epicycle slat. Carefully plot all 55 points. Lightly sketch a smooth path through the points.

5. Examining the plot: Your plot should contain two retrograde loops—the retrogradation of the summer of 1971 and that of the fall of 1973.

A. According to your plot, how wide were these two retrograde arcs, measured in degrees, as seen from the Earth? How far apart were the centers of the two arcs? Consult table 7.1 (planet longitudes at ten-day intervals) to see how well your plot agrees with actual data.

B. You plotted positions of Mars at 10° intervals in the epicyclic anomaly. What time interval does this represent? That is, how many days apart are two successive positions in your plot? Remember that $f_\mu = 0.4616°/\text{day}$. (Keep several decimal places in your answer for use below.)

C. To what date does the 54th point of your plot correspond? (The 9th point corresponds to August 9, 1971 = J.D. 244 1173.) According to your plot, what was the longitude of Mars at on this date? Does this agree with the position in table 7.1?

6. Completing the plot: We want to examine retrograde arcs of Mars occurring all the way around the ecliptic. (We will focus on the retrograde loops and not pay any more attention to the rest of the planet's motion.) It is clear that the loops will all be equally spaced (at 49° intervals) and they will all be of exactly the same size and shape. We need to produce retrograde loops all the way around the ecliptic. We *could* do this by turning the slats, but we will take advantage of the uniform size and spacing of the loops to simplify things.

Using your original two retrograde loops as guides, carefully trace in five more on your sheet. These five should be arranged in counterclockwise order following your second loop. The new loops should be 49° apart and exactly the same distance from the Earth as are the original loops. (You can trace a loop and then use the tracing as a master from which to trace the new loops onto the sheet. Or you can make photocopies of one of your loops and paste these on in the right positions.) When you are finished you should have seven loops. The five added loops represent the retrogradations of December 1975, January 1978, February 1980, April 1982, and May 1984.

7. Making an overlay—general test of Apollonius's theory: In table 7.1 bracket all the retrograde arcs of Mars from 1971 to 1984 inclusively.

Obtain a sheet of transparent plastic and felt-tip pen, or else a sheet of tracing paper and an ordinary pencil. Near the center of the transparency, mark a dot *O* to represent the Earth. Draw a reference line from *O* toward one edge of the transparency to represent the zero of longitude. A short mark near the edge will suffice, as shown in figure 7.24. Label the reference line with the symbol ♈ for the vernal equinox.

From the Earth *O*, draw lines of sight to the ends of the retrograde arcs of Mars that occurred in the years 1971–1984. The lines of sight should be at least four inches long. Your transparency should resemble figure 7.24, but the dates and longitudes of the planet's stations will be different, since they are to be taken from table 7.1. Your finished transparency will include seven retrograde arcs.

When you have finished the overlay, place it on top of the plot of retrograde loops. Match up the Earth dots on the two drawings and line up the zero-degree directions. How well does the deferent-and-epicycle model agree with the actual pattern of Mars's retrogradations?

The model puts the 1971 retrograde loop in the right part of the sky, of course—it had to, since we started the Ptolemaic slats rotating in August 1971, in the middle of the retrograde motion. What about the positions of the other retrograde loops? The model predicts equally spaced retrograde loops. But, as the overlay shows, the actual spacing of the retrogradations is far from uniform.

The model also predicts retrograde arcs of uniform width, but the actual widths, on the overlay, vary considerably. Note that the loop for 1971 is too wide and spills over the actual lines of sight to the planet's stations. On the other hand, the loop for 1980 is too small and does not nearly fill the space between the lines of sight. Again, the simple deferent-and-epicycle model must be judged a failure.

8. Testing the intermediate models: Remove the overlay from the plot and examine it carefully. Note that there is a definite pattern. Around longitude 320° (i.e., around the 1971 retrogradation) the retrograde arcs are small and far apart. Around longitude 140° (between the retrograde arcs of 1978 and 1980), the arcs are at their widest and most densely packed. The whole pattern is roughly symmetrical about a line drawn through the Earth toward longitudes 140° and 320°.

In the intermediate version of the deferent-epicycle theory (sec. 7.12), the center of the deferent circle is shifted away from the Earth. This model is easy to test with the plot and overlay you have already made.

The center O of the transparency, which is the origin of our lines of sight, will continue to represent the Earth. The center of the retrograde loop plot, which held the thumb tack, represents the center of Mars's deferent circle. Label this tack hole D, for the center of the deferent. Note that the Earth O is marked on the transparency, while the center D of Mars's deferent circle is marked on the paper plot. By sliding the paper plot underneath the transparency, we can move the center of the deferent away from the Earth in any direction we please.

In which direction ought we to displace D? Only two directions would make any sense, namely, along longitudes 320° or 140°, that is, in one direction or the other along the line of symmetry in the pattern of retrograde arcs. We shall try each of these directions in turn.

First, shift the paper plot underneath the transparency so that D moves an inch or more toward longitude 140°—that is, toward the 1978 and 1980 retrograde arcs. You should find that you can make the retrograde loops on the plot all fall in the right parts of the sky: they all agree pretty well with the actually observed lines of sight on the transparency. As far as the *spacing* of the retrograde loops is concerned, an off-center deferent circle seems to be what we need. But what about the *widths* of the retrograde loops? By shifting D in the 140° direction, we have actually made the widths worse. The 1971 loop is much too wide, while the 1978 and 1980 loops are much too narrow.

Let us put D back at O and try something else. This time, shift the paper plot underneath the transparency so that D moves an inch or so toward longitude 320° (toward the 1971 retrograde arc). You should find that you can make both the 1971 arc and the 1980 arc fill the space between their lines of sight, very nearly. That is, by shifting the center of the deferent toward longitude 320°, we can make the model produce retrograde arcs of about the right width in two opposite parts of the sky. Unfortunately, we have in the process made the spacing of the arcs worse.

It seems, then, that a simple shift of the deferent's center cannot save our model. To make the model reproduce the observed spacing of Mars's retrogradations, we must put the center of the deferent at longitude 140°, as seen from the Earth. But to produce retrograde arcs of the right widths, we must place the center of the deferent at longitude 320°. There is no way to produce the correct spacing and the correct widths simultaneously by a simple shift of the deferent's center. This can be made apparent by the following simple argument. Imagine a row of trees. If we back away from the trees they will appear to become smaller and closer together. But around longitude 320° we need the retrograde loops of Mars to appear small and far apart. There is no way to produce this appearance by simply shifting the position of the Earth with respect to the uniform loops of our first theory of longitudes. Some wholly new theoretical device is called for.

7.17 DEFERENT-AND-EPICYCLE THEORY, II: PTOLEMY'S THEORY OF LONGITUDES

Overview of Ptolemy's Theory of Longitudes

Figure 7.32 illustrates the theory of longitudes adopted by Ptolemy for Venus, Mars, Jupiter, and Saturn. (The Mercury theory has an extra complication.) About C as center the deferent circle $AK\Pi$ is drawn. The Earth is at O. Thus, the deferent is off-center from the Earth. For this reason the deferent circle is also called the *eccentric*: the two terms are interchangeable. The line through O and C cuts the eccentric at A, the apogee of the eccentric, and at Π, the

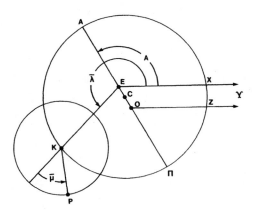

FIGURE 7.32. Ptolemy's final theory of longitudes for Venus and the three superior planets. The Earth is at O. C is the center of the deferent circle. But the epicycle's center moves at uniform angular speed as viewed from the equant point E.

perigee. (*A* and Π are the two apsides of the deferent, so *A*Π is sometimes called the *line of apsides*.) *OZ* points toward the (infinitely distant) spring equinoctial point—the zero direction for measuring longitudes. The angle marked *A* is the *longitude of the apogee* of the eccentric. The longitude of the apogee is different for each planet. For Mars in the twentieth century, it is approximately 150°. The line of apsides is the line of symmetry in the pattern of the planet's retrogradations.

The epicycle's center *K* moves eastward on the eccentric circle, but its motion is not uniform either as seen from the Earth *O* or as seen from the center *C* of the deferent. Rather, the motion is uniform *as seen from a third center E*, the center of uniform motion or *equant point*. That is, an imaginary observer at *E* would see *K* travel through equal angles in equal times, while observers at *C* or *O* would not. Ptolemy's introduction of the equant point into the planetary theory means that point *K* must physically speed up and slow down. *K* travels most slowly at the apogee and most rapidly at the perigee. Needless to say, this is a serious bending of the rules of Aristotle's physics. However, the rule governing the variation in speed is very simple, since the angular motion appears uniform from *E*. Draw line *EX* parallel to *OZ*. *EX* is the zero-degree reference line for angles measured at the equant. Thus, the *mean longitude* $\bar{\lambda}$ increases at a uniform rate. In the case of Mars, the motion in $\bar{\lambda}$ is about 0.524°/day. This rate is determined by the planet's tropical period.

The planet *P* travels on the epicycle in the same sense as *K* does on the eccentric, counterclockwise as viewed from the north pole of the ecliptic. The position of the planet on the epicycle is specified by angle $\bar{\mu}$, the *mean epicyclic anomaly*. The planet travels uniformly on its epicycle. Uniform motion must, of course, be measured with respect to the uniformly revolving line *EK*. Therefore, the uniform motion of *P* on the epicycle means that the mean epicyclic anomaly $\bar{\mu}$ increases at a steady rate. In the case of Mars the motion in $\bar{\mu}$ is about 0.462°/day. This rate is determined by the planet's synodic period.

The radius *CK* of the eccentric is arbitrary in Greek astronomy, except that it must be much greater than the radius of the Earth, since the planets have negligible parallax. We shall denote the radius of the eccentric by the letter *R*. The radius *KP* of the epicycle, denoted *r*, is fixed in terms of the eccentric's radius. For Mars, $r/R = 0.656$. Similarly, *CE* and *CO*, the distances of the equant point and of the Earth from the eccentric's center, can only be expressed in units of the eccentric's radius. In Ptolemy's theory $CE = CO$, so that the equant and the Earth are equidistant from the center of the circle. The ratio *CO/R* (or *CE/R*) is called the *eccentricity*, which we shall denote *e*. The eccentricity is different for each planet. For Mars in the twentieth century, a good value is $e = 0.103$.

Empirical Necessity of Ptolemy's Theory of Longitudes

Popular writers on the history of astronomy have often been unsympathetic toward Ptolemy and his planetary theory. Often, one reads complaints that the theory was complicated, or unnatural, or arbitrary. Such complaints usually stem from inadequate understanding. The theory is as simple as the planets themselves will allow. The deferent, with its eternal revolution from west to east, produces the steady progress in longitude associated with a planet's tropical revolution. The epicycle accounts for the second inequality, which is manifested most spectacularly in retrograde motion. But, as figure 7.24 shows, the planets also have a zodiacal inequality. The combination of equant and off-centered deferent is Ptolemy's manner of accounting for this inequality. It is important to understand how these features are forced on the model by the planets themselves. To a modern reader, the strangest feature of the theory is undoubtedly the equant point. Let us see how Mars forces this device on us.

Let us begin by examining once again the A version of the intermediate model (figs. 7.27A and 7.28A). The theoretical loops are, of course, equally spaced as seen from the center D of the deferent. Since the actual retrograde arcs coincide with them, it follows that D is the center of uniform motion. That is, *point D is acting as an equant point.* We have no choice in the matter: the planet insists on it. Of course, the A version of the intermediate model fails to agree with the widths of the retrograde arcs. The only apparent solution is precisely that adopted by Ptolemy, that is, to separate the center of the deferent from the center of uniform motion. In figure 7.27A, D represents both the equant point and the center of the deferent. We must leave the equant point at D to get the correct spacing between the retrogradations. But if we move the deferent's center closer to O, this will cause the retrograde loop of A.D. 118 to draw nearer the Earth and, therefore, to look larger. The A.D. 109 loop will recede from the Earth and, therefore, look smaller. In this way, we will have a chance of producing loops of correct and variable width, while preserving the correct spacing already achieved by the A version of the intermediate model. The final model, produced by the separation of point D into points E and C, is precisely that adopted by Ptolemy and illustrated by figure 7.32.

Just how well the final theory of longitudes will agree with the observations is, however, by no means clear. Once the equant point is separated from the center of the deferent, the retrograde loops cease to be of uniform size and shape. There is no recourse but to plot it out and see what happens. This we have done in figure 7.33. Ptolemy's final model can only be judged a stunning success and a huge improvement over all that preceded it.

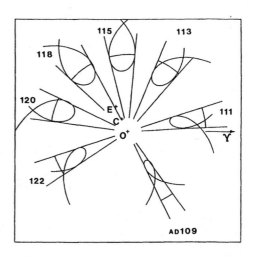

FIGURE 7.33. Splendid agreement between the retrograde loops generated by Ptolemy's theory of Mars and the actual retrograde arcs of Mars, A.D. 109–122.

Discovery of the Equant

As far as we know, the equant was Ptolemy's own discovery. Ptolemy's style in the *Almagest* is the style of most scientific writing. It is lean, elegant, and efficient and discloses very little of the original process of discovery.[69] Ptolemy presents the equant in *Almagest* IX, 5, but he offers no justification for this radical innovation, which introduces nonuniform motion in the heavens, and which therefore constitutes a serious violation of the principles of Aristotelian physics. This is uncharacteristic of Ptolemy, who usually explains the reasons that lie behind his choice of a model.

In *Almagest* IX, 2, Ptolemy apologized for the fact that he might seem to presuppose things without immediate foundation in the phenomena. He justified himself by saying that things supposed without proof cannot be without some logic if they are found to be consistent with the appearances, even though the way of arriving at them might be hard to explain. This seems to be a veiled reference to his manner of introducing the equant, which follows shortly afterward.

Thus, Ptolemy nowhere says explicitly how he arrived at the idea of the equant, which was perhaps his most important personal contribution to planetary theory. But it seems likely that he was experimenting with two versions of the intermediate model, which he found irreconcilable. In *Almagest* IX, 6, where Ptolemy takes up the derivation of the parameters for the superior planets, we find some evidence that this was the case. Ptolemy asserts, without proof, that for these planets, as for Venus, the center of the deferent lies exactly halfway between the Earth and the equant point. And then he says something, by way of justification, that is extremely interesting. Ptolemy says that *the eccentricity calculated from the zodiacal anomaly is about twice the eccentricity calculated from the lengths of the retrograde arcs at greatest and least distances.*

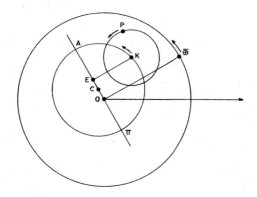

FIGURE 7.34. Connection between the mean Sun and an inferior planet: *EK* remains parallel to *O*☉.

It appears, then, that for one or more of the superior planets, Ptolemy calculated the value of the eccentricity required to save the zodiacal anomaly as manifested in the motion of the epicycle's center. That is, he found the distance *OD* in figure 7.27A that would give the correct *spacing* of the retrograde arcs, as in figure 7.28A. This was, of course, a part of the procedure he had inherited from his immediate predecessors who subscribed to version A of the intermediate model.

But Ptolemy also calculated the eccentricity *OD* (in fig. 7.27B) that gives the right widths of the retrograde arcs, as in figure 7.28B. This involved methods based on Apollonius's theorem—another bit of traditional planetary geometry. Ptolemy compared the result with the eccentricity required to explain the spacings. He found that *the two results were not the same*. As shown clearly by a comparison of figures 7.28A and 7.28B, the eccentricity *OD* required to save the spacing is substantially greater than the eccentricity required to save the widths of the retrograde arcs. Ptolemy says the one is twice the other, but this is only approximately so.

Ptolemy's insight then consisted in realizing that he might preserve the correct spacing by leaving the center of uniform motion at the required distance from the Earth and yet obtain correct regressions by placing the deferent's center at half that distance. This resulted in the model of figure 7.32, which, in a sense, splits the difference between those of figures 7.27A and 7.27B.[70]

Finally, we might ask which planet was occupying Ptolemy's attention when he came upon the equant. As we have seen, in his introduction to the theory of the superior planets, Ptolemy remarks that two different methods of determining the eccentricity in the intermediate model lead to two different results, the one being twice the other. Now, in the case of Jupiter, the retrograde arcs at apogee and at perigee are almost identical, so a calculation of the eccentricity from these data is not actually possible. For Saturn, the case is hardly better. Only in the case of Mars is the difference between the longest and shortest retrograde arcs so large that it would immediately suggest the use of these lengths in a derivation of the eccentricity. It seems most likely, then, that Ptolemy was grappling with Mars, the planet that, fourteen centuries later, was to occupy the attention of Kepler.

Some Technical Detail: Connection with the Sun

In section 7.12 we examined the connections between the Sun and the planets in deferent-and-epicycle theory. In the case of an inferior planet, the center of the epicycle lies on the line of sight from the Earth to the mean Sun. In the case of a superior planet, the radius of the epicycle remains parallel to the line of sight from the Earth to the mean Sun.

Actually, the first statement is strictly true only if the planet's orbit has no eccentricity, that is, if the equant and the center of the deferent both coincide with the Earth. For our final theory of longitudes it is necessary to restate the connections more precisely. Figure 7.34 illustrates the connection between the mean Sun and an inferior planet. The mean Sun ☉ travels at uniform speed around a circle centered on the Earth *O*. The planet *P* travels on an epicycle whose center *K* travels on an eccentric deferent: the center of the deferent is at *C* and the equant point is at *E*. In Ptolemy's theory of longitudes, *EK* remains parallel to *O*☉. Figure 7.34, although labeled as a figure for "an inferior planet," strictly applies only to Venus, since Ptolemy's model for Mercury contains an extra complication.

Figure 7.35 illustrates the connection between the mean Sun and a superior planet (Mars, Jupiter, or Saturn). The radius of the epicycle *KP* remains parallel to the line of sight from the Earth to the mean Sun. Ptolemy states this relationship very clearly in *Almagest* X, 9. It will still be the case that

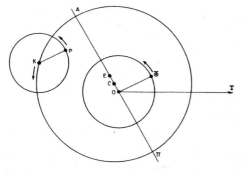

FIGURE 7.35. Connection between the mean Sun and a superior planet: *KP* remains parallel to *O*☉.

when the planet is in opposition to the mean Sun, *KP* will point directly at the Earth. However, since *E* and *C* do not coincide with *O*, the center of the retrograde arc will not in general correspond exactly to the mean opposition.

The peculiar role of the Sun (or, more precisely, the mean Sun) in the ancient planetary theory provided a clue that the Sun deserved a more important role in the world picture. But it was not until the sixteenth century that anyone saw the consequences clearly.

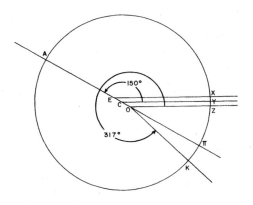

FIGURE 7.36.

7.18 EXERCISE: TESTING PTOLEMY'S THEORY OF LONGITUDES

The purpose of the exercise is to test Ptolemy's theory, using the method we employed in section 7.16 (in which we tested Apollonius's theory of longitudes and found it wanting). We shall use Ptolemy's theory of longitudes to generate a series of retrograde loops for Mars for the years 1971–1984. This theoretical prediction will be compared with the actual behavior of the planet by means of the transparent overlay of Mars's retrograde arcs that you made in section 7.16. The general method of producing the theoretical retrograde loops will be similar to that used in section 7.16. However, the change of the underlying model—that is, the separation of the equant from the center of the deferent—will entail a few modifications.

1. Preparing the ground: Obtain a large sheet of paper, about 20″ × 20″. Near the center of the paper, place a point *C*, to serve as the center of Mars's deferent circle. About *C* draw a circle with a radius equal to the radius of the deferent of the Ptolemaic slats. That is, the radius of your circle should be equal to the distance between the tack hole *T* and the center *H* of the grommet hole on the deferent slat. Then, as in figure 7.36, draw a line through *C* to represent the zero of longitude. This line cuts the circle at *Y*.

Place the center of a protractor at *C* and lay out the line of apsides along direction 150°, cutting the deferent at the apogee *A* and perigee Π, as in figure 7.36. (Measure counterclockwise from the zero-degree direction.) This longitude of the apogee (150°) is valid for Mars in the 1970s.

Along the line of apsides *A*Π, mark the location of the equant point *E* and the Earth *O*. These must be placed so that *CE* = *CO* = the eccentricity *times* the radius of the deferent. For Mars, the eccentricity is 0.103. Suppose the radius of your deferent slat (the distance between the tack hole *T* and the center *H* of the grommet hole) is 15 cm. Then you should draw your diagram with *CE* = *CO* = 0.103 × 15 cm = 1.55 cm. (If the radius of your deferent slat is different, use the actual radius.)

Draw lines from *E* and *O* parallel to *CY*. These new lines will cut the circle at *X* and *Z* and will serve as the zeros of longitude for angles measured at *E* or at *O*.

Poke a thumb tack through the equant point *E* from underneath the paper. Place the large paper protractor from the Ptolemaic slats kit over the equant so that the tack sticks through the center of the protractor. Turn the protractor until the zero-degree direction coincides with line *EX*. (Eventually, you may wish to place a curl of tape under the protractor to hold it in position. However, since you will have to remove the protractor for step 2, below, you may wish to wait until after then to apply the tape.)

Using a sharp knife, cut out the long, narrow slot near the base of the deferent slat of the Ptolemaic slats. (In cutting out the slot, you will eliminate the tack hole you made in the slat for sec. 7.16.)

Place the Ptolemaic slats on the paper so that the equant tack sticks through the slot on the deferent tack. (Push a small eraser or a cube of balsa wood over the tack so that you do not stick yourself accidentally.) As usual, the deferent slat will rotate uniformly about the tack. But, since the tack is at *E* and not at the center *C* of the circle, it will be necessary to move the deferent slat in and out in order to keep the center of the epicycle on the deferent circle. The epicycle's center is represented in the Ptolemaic slats by the metal grommet. Carefully slide the deferent slat in or out as required until you can see the deferent circle through the center of the hole in the grommet, as shown in figure 7.42.

The value of the mean longitude $\bar{\lambda}$ may be read on the large paper protractor at the edge of the deferent slat. The value of the mean epicyclic anomaly $\bar{\mu}$ is indicated on the small protractor by the edge of the epicycle slat. Figure 7.42 shows the slats oriented for $\bar{\lambda} = 197.8°$ and $\bar{\mu} = 231.0°$. The position of Mars is indicated by point *P*.

2. Initial values of the mean longitude and mean epicyclic anomaly: To start generating retrograde loops for Mars, we must know the values of $\bar{\lambda}$ and $\bar{\mu}$ at one moment. As always, the time most convenient to use will be a mean opposition. Let us choose the mean opposition of 1971. In section 7.4, we determined that Mars had an opposition to the mean Sun on August 9, 1971, and that the longitude of the planet at this moment was 317°.

From the Earth *O*, draw a line of sight at longitude 317°, as in figure 7.36. (You will have to remove the paper protractor from the equant point to draw this line.) On August 9, 1971, Mars lay somewhere on this line. Because this was a mean opposition, the radius of the planet's epicycle was pointing directly at the Earth. It follows that the center of the epicycle also lay on this same line of sight. The epicycle's center therefore lay at *K*, where the line of sight intersects the deferent circle (fig. 7.36). Again, we are able to identify the location of the epicycle's center in this simple fashion only because the date in question is a mean opposition. Replace the large paper protractor over the equant tack and put the Ptolemaic slats into position to represent the state of affairs for Mars on August 9, 1971. The center of the rivet should lie directly over point *K*, and the epicycle slat should point directly at *O*, along the 317° line of sight.

Read off the values of the mean longitude $\bar{\lambda}$ and the mean epicyclic anomaly $\bar{\mu}$ directly from the two protractors. These were the values of these two angles for Mars on August 9, 1971. You should find that $\bar{\lambda}$ is slightly greater than 317°, say, 319° or 320°. You should also find that $\bar{\mu}$ is slightly less than 180°, say, 177° or 178°. Note carefully whatever values you get for $\bar{\lambda}$ and $\bar{\mu}$ on your particular slats.

3. Preparing a table of values to be plotted: For every 10° motion in $\bar{\mu}$ Mars experiences a motion of 11.354° in $\bar{\lambda}$. (This is more precise than the value 11.4° we used in sec. 7.16.) Using this information, we can prepare a table of values of $\bar{\lambda}$ and $\bar{\mu}$ for times covering the retrograde loops of 1971–1984.

A. Retrograde loop of 1971: Suppose we found that at the mean opposition of August 9, 1971, Mars's mean longitude was 319.6° and its mean epicyclic anomaly was 177.8°. By repeated addition or subtraction we generate the following table of values for five positions before and five positions after the mean opposition.

Step	$\bar{\lambda}$	$\bar{\mu}$	
−5			
−4			
−3			
−2	296.8°	157.8°	
−1	308.2	167.8	
0	319.6	177.8	(retrograde loop of 1971)
1	331.0	187.8	
2	342.4	197.8	
3			
4			
5			

As before, $\bar{\mu}$ increases by regular intervals of 10°, and $\bar{\lambda}$ by regular intervals of 11.4°. For this purpose, the tenth of a degree is adequate precision.

B. Later retrograde loops: We are interested in plotting only the retrograde loops. Five points before and five points after the middle of the retrograde movement will be sufficient. Therefore, we use the following shortcut procedure to skip ahead to points in the immediate vicinity of the desired oppositions. The 1973 opposition was the next to occur after that of 1971. The mean epicyclic anomaly $\bar{\mu}$ must therefore have increased by approximately 360° during the interval between the two oppositions. (The 360° change in $\bar{\mu}$ from one mean opposition to the next is only approximate because of the effect of the eccentricity. Nevertheless, if we skip ahead by 360° in $\bar{\mu}$ from the 1971 mean opposition, we will be somewhere in the retrograde loop for 1973.) The change in $\bar{\lambda}$ corresponding to a 360° change in $\bar{\mu}$ is easily calculated:

$$\Delta\bar{\lambda} = 1.1354 \times \Delta\bar{\mu}$$
$$= 1.1354 \times 360°$$
$$= 408.744°$$

On August 9, 1971, we had $\bar{\lambda} = 319.6°$. Applying the motion in this angle to the initial value, we get

	$\bar{\lambda}$
August 9, 1971	319.6°
Plus motion	408.7
Sum	728.3°
Less complete circles	−720.0
	8.3°

That is, the next time that $\bar{\mu}$ had the value 177.8°, the value of $\bar{\lambda}$ was 8.3°. This point lies somewhere on the 1973 retrograde loop. It is an easy matter to fill in a table using this position as a beginning entry:

Step	$\bar{\lambda}$	$\bar{\mu}$
−5		
−4		
−3		
−2		
−1		
0	8.3°	177.8° (retrograde loop of 1973)
1	19.7	187.8
2	31.1	197.8
3	42.5	207.8
4		
5		

For the step-by-step addition it is sufficient to use 11.4° as the motion in $\bar{\lambda}$ for 10° motion in $\bar{\mu}$. However, in skipping ahead from one opposition to the next, it was important to use the more precise figure of 11.354°.

You should now prepare tables giving the values of $\bar{\lambda}$ and $\bar{\mu}$ for times around the six oppositions following that of 1971. Eleven points on each retrograde loop will be sufficient. Be sure to check as you proceed at least some of the numbers computed for each opposition, for arithmetical errors in one opposition may be propagated into those that follow.

4. Plotting the retrograde loops: Use the Ptolemaic slats to plot the retrograde loops of Mars for 1971–1984. Remember that the large paper protractor goes over the tack at the equant and that its zero-degree mark must coincide with line EX (fig. 7.42). In plotting each point, remember to slide the deferent slat in or out as necessary to keep the center of the grommet exactly on the

deferent circle. When you are finished, you should have seven retrograde arcs of rather different sizes and shapes, spaced unevenly about the zodiac.

5. Comparison with the actual data: Remove the slats, protractor, and thumb tack from your plot. Place over the plot the transparency of Mars's retrograde arcs that you drew for section 7.16. Note that you have no freedom in positioning the transparency. The Earth point on the transparency must coincide with point O on the plot, and the direction of the vernal equinox on the transparency must coincide with line OZ on the plot. What is your judgment of Ptolemy's theory of longitudes?

7.19 DETERMINATION OF THE PARAMETERS OF MARS

How can we know how large a planet's epicycle is? How can we know how large to make the eccentricity? In this section we demonstrate how the parameters for a superior planet can be determined from observations. We use Mars as an example, but the same procedures could be applied to Jupiter or Saturn. Although the methods demonstrated here do not exactly follow those of the *Almagest*, they show, clearly and simply, the connection of each parameter with the observed motion of the planet. And who knows? It is more than likely that some such rougher method preceded the elegant perfection of Ptolemy.[71]

There are seven parameters to be determined:

1. The mean angular speed of the epicycle's center around the deferent circle—in other words, the rate of change of the mean longitude $\bar{\lambda}$ (see fig. 7.32). This angular speed we denote f_λ.
2. The angular speed of the planet on the epicycle. This speed, denoted f_μ, is the rate at which the mean epicyclic anomaly $\bar{\mu}$ changes.
3. The longitude of the apogee of the deferent, denoted A.
4. The eccentricity of the deferent, denoted e. This is the ratio OC/R, or CE/R, where R is the radius of the deferent.
5. The initial value of $\bar{\lambda}$ for some specific date. This initial value will be denoted $\bar{\lambda}_o$.
6. The initial value of $\bar{\mu}$, which we will denote $\bar{\mu}_o$.
7. The radius of the epicycle, denoted r. All that matters in Greek astronomy is the size of the epicycle in relation to the deferent, that is, the ratio r/R.

1 and 2. The Angular Speeds

The best method of determining the two periods is to count the time between one opposition and another *that occurs at the same place in the sky*. This means that the epicycle's center will have returned to the same position on the deferent, and the planet will have returned to the same position on the epicycle. Only in this way can we be sure that a whole number of cycles in each motion have been completed.

We found in section 7.4 that the oppositions of March, 1965, and February, 1980, fit these conditions fairly well:

Date of opposition	Longitude of planet
1965 Mar 10 (J.D. 243 8830)	168°
1980 Feb 26 (J.D. 244 4296)	155 1/2

These two oppositions did not take place exactly at the same longitude, but they are only 12 1/2° apart. Using these oppositions we have already obtained the period relation:

8 tropical revolutions = 7 synodic revolutions = 15 years.

A rough estimate of the two angular speeds is then

$$f_\lambda = \frac{8 \times 360°}{15 \times 365\frac{1}{4} \text{ days}} = 0.5257°/^d,$$

$$f_\mu = \frac{7 \times 360°}{15 \times 365\frac{1}{4} \text{ days}} = 0.4600°/^d.$$

We can do better by taking into account our known 12 1/2° error. Thus, the planet did complete 7 retrograde cycles, but it moved through 12 1/2° less than 8 complete tropical revolutions. Further, the elapsed time should be counted exactly, to the day. This can be done by subtracting the Julian day numbers for the two dates, which gives 5,466 days for the time interval (some 13 days less than 15 whole years). Our slightly more sophisticated guess at the two parameters, using our own data, is

$$f_\lambda = \frac{8 \times 360° - 12\frac{1}{2}°}{5,466 \text{ days}} = 0.5246°/^d,$$

$$f_\mu = \frac{7 \times 360°}{5,466 \text{ days}} = 0.4610°/^d.$$

Even better values could be obtained by using a longer great cycle of Mars—for example, the Babylonian 79-year period, which Ptolemy adopted with a minor change. The Babylonian cycle leads to $f_\lambda = 0.5240°/^d$, only slightly different from our second value for f_λ.

3. Longitude of the Apogee

The longitude of the deferent's apogee can be determined from the pattern of the planet's retrograde arcs around the ecliptic. For Mars, these arcs tend to be wide and closely spaced toward the signs of the Lion and the Virgin, as illustrated in the transparency you made for section 7.16. Mars's apogee must lie somewhere in this part of the zodiac.

A more systematic approach is to plot a graph of the angular distance between neighboring oppositions as a function of the longitude at which the oppositions occurred. To this purpose we begin by reproducing the first few entries from the table of oppositions for Mars (table 7.2, from sec. 7.4):

Date of the opposition	Longitude	Difference	Average difference
1948 Feb 17	147.5°		
		33.5°	
1950 Mar 25	181.0		36.25°
		39.0	
1952 May 5	220.0		46.25
		53.5	
1954 Jun 25	273.0		64.50
		75.5	
1956 Sep 11	349.0		

The first two columns are taken directly from table 7.2. In the third column, the differences in longitude are listed for neighboring pairs of oppositions. For example, the opposition of 1950 took place at a longitude 33.5° greater

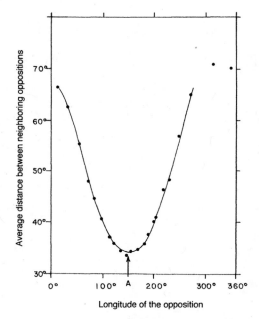

FIGURE 7.37. Determining the longitude of the apogee of Mars.

than that of 1948 (181.0 − 147.5 = 33.5). In the fourth column, we give the average distance of each opposition from its two neighbors. For example, the opposition of 1950 occurred at a place 33.5° beyond the place of the opposition of 1948, and 39.0° before the place of the opposition of 1952; the average of these numbers is 36.25° and represents the average separation of neighboring oppositions at longitude 181°. In a graph of average separation versus the longitude of the oppositions, we would plot the separation 36.25° against the longitude 181°. Such a graph is shown in figure 7.37. On the graph, the minimum separation falls at about 150° longitude, which must be the longitude of the apogee.

Conclusion: longitude of the apogee of the Martian deferent = 150°.

4. Eccentricity of the Deferent

Choose, from the table of oppositions (table 7.2), one opposition that occurred very near either the apogee or the perigee. The opposition that is nearest to one of the apsides seems to be that of February 26, 1980, which occurred at longitude 155 1/2°, only 5 1/2° from the apogee. For the purpose of determining the eccentricity, let us now suppose that the longitude of the apogee is actually 155 1/2°; that is, let us treat the opposition of 1980 as if it fell exactly on the line of apsides. Since our planetary positions are accurate only to the nearest degree, and since our value for the longitude of the apogee is uncertain by at least a few degrees, there can be no harm in making such an approximation.

Now choose a second opposition, either a neighbor or a near-neighbor of the first one. In the graphical method that we will use, the result will be more accurate if the angle between the two oppositions is fairly large. Therefore, we choose the opposition of December, 1975, whose longitude of 84° places it some 71 1/2° away from the first one. Any angular distance between 60° and 120° would have been acceptable.

Now we draw a line, as in figure 7.38, to represent the line of apsides. We may put the Earth *O* and the equant *E* wherever we please on this line—an inch apart will be convenient. The problem now is to use our two oppositions to determine the *scale* of the diagram we have begun to draw, and thus to establish the length of distance *OE compared to* the deferent's radius.

The end of the line toward the equant is labeled 155 1/2°, which is the longitude of the apogee. Now, at an opposition the radius of the epicycle

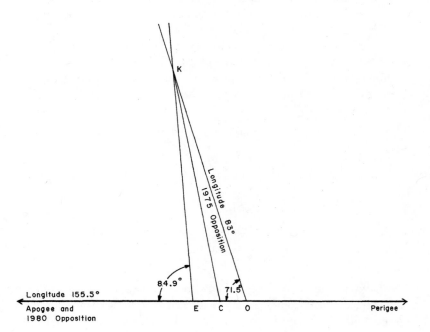

FIGURE 7.38. Determining the Martian eccentricity using an opposition in the apogee and one other opposition.

points directly at the Earth, so that the center of the epicycle, the planet, and the Earth all lie in a line: at an opposition, observers at the Earth can "see" the center of the epicycle, for the planet itself lies in exactly the same direction and marks the spot, so to speak. And since the opposition of 1980 occurred exactly on the line of apsides, at that one moment an observer on the Earth and an imaginary observer at the equant would both see the center of the epicycle in the same direction, namely, along the 155 1/2° line.

The opposition of December, 1975, occurred at longitude 84°. Therefore, we draw a line of sight from the Earth in this direction, which makes a 71 1/2° (155 1/2 − 84 = 71 1/2) angle with the direction to the first opposition. The center of Mars's epicycle must lie somewhere on this line on December 13, 1975. (Remember, at an opposition the center of the epicycle lies in the same direction as the planet itself.) To find just where on this line the center of the epicycle lies, we need to establish a line of sight from the equant as well.

During the period from December 13, 1975, to February 26, 1980, the center of the epicycle moved at a constant angular speed, as observed from the equant. So, it is an easy matter to calculate the angle through which it moved. First, we calculate the elapsed time:

Julian day numbers of oppositions 244 4296 (Feb 26, 1980)
                                 244 2760 (Dec 13, 1975)

Difference   1,536 days

The epicycle's center travels at the rate of the mean motion in longitude, 0.5240°/day, so the total motion was

$$0.5240°/^d \times 1,536^d = 804.9°, \text{ or } 84.9°$$

after eliminating two complete cycles of 360°. *This is the angular distance between the places occupied by the epicycle's center at the two oppositions, as observed from the equant.* The 1980 opposition took place two complete circuits plus 84.9° farther along in longitude than the 1975 opposition.

We draw a line of sight from the equant making an angle of 84.9° with the original line of sight. The center of the epicycle must lie on this line on December 13, 1975. Therefore, on this date the epicycle's center was at the point marked K. Since the epicycle's center always rides on the deferent circle, we have succeeded in finding a point that lies on this circle.

Now we have an easy way to establish the size of the deferent. Mark C, the center of the deferent, on the line of apsides midway between E and O. Measure CK and CO with a ruler. On the original diagram (reduced for printing in fig. 7.38), the distances were $CK = 4.5''$ and $CO = 0.5''$. Thus, $e = CO/CK = 0.11$.

Conclusion: The eccentricity of Mars's deferent is 0.11.

5 and 6. Initial Values of $\bar{\lambda}$ and $\bar{\mu}$

The values of the mean longitude $\bar{\lambda}$ and the mean epicyclic anomaly $\bar{\mu}$ are most easily found at the time of an opposition. From the table of oppositions of Mars (table 7.2), we choose the opposition of December 13, 1975, which occurred at 84° longitude.

As in figure 7.39, draw the deferent circle about center C. Draw a line through C to represent the direction of the vernal equinox. Draw the line of apsides so that it makes an angle of 150° (as determined above) with the direction to the equinox. The 150° end of the line of apsides cuts the deferent at the apogee A. Mark on the line of apsides the equant point E and the Earth O. These two points should be placed according to the value of the ec-

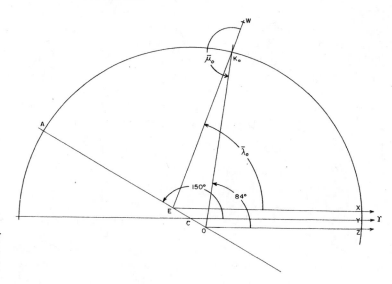

FIGURE 7.39. Determination of the initial values of the mean longitude $\bar{\lambda}_0$ and the mean epicyclic anomaly $\bar{\mu}_0$ using the opposition of December 13, 1975.

centricity already determined: $EC = OC = 0.11$ times the radius of the deferent. From O and E draw reference lines parallel to the equinoctial line already drawn through C. These three parallel lines are all to be regarded as pointing toward the (infinitely distant) vernal equinoctial point. They will serve as the zero-degree lines for any angles to be measured at E, C, or O. The reference lines cut the deferent at X, Y, and Z.

From the Earth, lay out a line in the direction of 84° of longitude. This represents the line of sight to Mars on December 13, 1975, and cuts the deferent at the point K_0. Because the date in question is the date of an opposition, we know that the center of the epicycle lies in the same direction as the planet itself. That is, the radius of the epicycle points directly at the Earth. Point K_0 is therefore the actual position of the epicycle's center on December 13, 1975. The planet itself lies somewhere on the line between O and K_0, but we cannot say exactly where, since we do not yet know the radius of the epicycle.

Finally, draw line EK_0 and mark on it a point W somewhere beyond K_0. Place the center of a protractor at E and measure the mean longitude $\bar{\lambda}$ (angle XEK_0), obtaining about 71°.

Then place the center of the protractor at K_0 and measure the mean epicyclic anomaly $\bar{\mu}$ (angle WK_0O), counterclockwise from W. The result is about 190°.

Conclusion: On December 13, 1975 (J.D. 244 2760),

$$\bar{\lambda} = 71°,$$

$$\bar{\mu} = 190°.$$

Reduction to Standard Epoch We know the values of $\bar{\lambda}$ and $\bar{\mu}$ for Mars on December 13, 1973. This particular date came up because it was the date of a mean opposition of Mars. In working out theories for the other planets, we would use other dates, as the circumstances required. It is convenient, however, to choose one standard epoch for *all* the planets. We select the date that served as epoch of our solar theory: A.D. 1900, January 0.5 (Greenwich mean noon), which was J.D. 241 5020.0. Accordingly, we calculate the values of $\bar{\lambda}$ and $\bar{\mu}$ at this epoch, starting from their values, just determined, for December 13, 1975:

December 13, 1975 J.D.	244 2760
January 0, 1900 J.D.	241 5020
Difference Δt	2 7740 days

Because the desired date (1900) falls before the original date (1975), the motions in the angles are subtractive. The values of the mean longitude and the mean epicyclic anomaly at our standard epoch are then

$$\bar{\lambda}_o = 71° - f_\lambda \times \Delta t$$

$$= 71° - 0.524072°/\text{day} \times 27{,}740^d$$

$$= -14{,}467°$$

$$= 293° \text{ (with addition of 41 complete circles)},$$

$$\bar{\mu}_o = 190° - f_\mu \times \Delta t$$

$$= 190° - 0.461576°/\text{day} \times 27{,}740^d$$

$$= -12{,}614°$$

$$= 346° \text{ (with addition of 36 complete circles)}.$$

Conclusion: At epoch 1900, Jan 0.5 (J.D. 241 5020)

$$\bar{\lambda}_o = 293°$$

$$\bar{\mu}_o = 346°.$$

7. Radius of the Epicycle

All the parameters of Mars established so far have been based on our table of oppositions. But the final parameter of the theory, the radius of the epicycle, cannot be fixed by means of oppositions. The reason is simple: since at an opposition the radius of the epicycle points directly at the Earth, one then has no means of fixing its size.

Therefore, we need one additional observation of Mars not at an opposition. Our graphical method will be most accurate if we choose an observation in which the planet is roughly halfway between opposition and conjunction. The exact location of the planet does not matter; we simply want it to lie well away from line *OK*.

From the table of longitudes at ten-day intervals (table 7.1), we choose the following position of Mars:

1976 April 11 (J.D. 244 2880) longitude 102°

This date follows the opposition of December 13, 1975, by 120 days, which means that the radius of the epicycle will have had sufficient time to turn away from line *OK*. The first step is to compute the mean longitude and the mean epicyclic anomaly on this date, by starting from their known values at the preceding opposition:

Final date	J.D.	244 2880	(1976 Apr 11)
Initial date	J.D.	244 2760	(1975 Dec 13)
Difference, Δt		120 days	

$$\bar{\lambda} = \bar{\lambda}_o + \Delta t \times f_\lambda$$

$$= 71° + 120 \text{ days} \times 0.5240°/\text{day}$$

$$= 133.9°$$

$$\bar{\mu} = \bar{\lambda}_o + \Delta t \times f_\mu$$

$$= 190° + 120 \text{ days} \times 0.4616°/\text{day}$$

$$= 245.4°$$

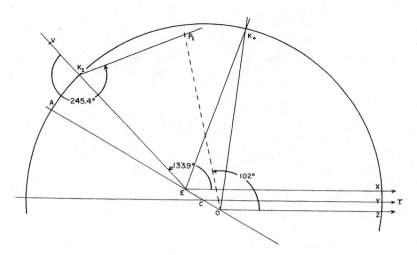

FIGURE 7.40. Determination of the radius of the epicycle.

These were the values of the mean longitude and the epicyclic anomaly on April 11, 1976.

The second step is to lay these directions out on a diagram. Place the center of the protractor at E (fig. 7.40), with the zero-degree direction along line EX. Lay out a line from E in the direction 133.9°, which will cut the deferent at K_1. Somewhere beyond K_1 on the line EK_1, mark point V. Point K_1 represents the new location of the epicycle's center, which has moved forward from K_0 during the 120 days.

Now place the center of the protractor at K_1, with the zero-degree direction along line VK_1 and lay out a line in the direction 245.4°, which is the new epicyclic anomaly. This line represents the direction of the radius of the epicycle on April 11, 1976. The planet must lie somewhere on this line at the given date.

The third and final stage of the procedure is to find just where on this line the planet lies. Mars was seen at 102° longitude on April 11, 1976. Therefore, lay out this line of sight from Earth O. The 102° line of sight cuts the epicycle's radius line at P_1. This is the actual location of Mars on April 11, 1976. Measure K_1P_1 with a ruler. In the original drawing (reduced in fig. 7.40), K_1P_1 was 2.88 inches. The radius CA of the deferent was 4.5 inches. The ratio of these numbers is 2.88/4.5 = 0.64.

Conclusion: The radius of the Martian epicycle is 0.64, where the radius of the deferent is 1.

A Table of Planetary Parameters

Table 7.4, gives the modern Ptolemaic parameters for Venus, Mars, Jupiter, and Saturn and contains all the information necessary for calculating the

TABLE 7.4. Modern Ptolemaic Parameters for Venus, Mars, Jupiter, and Saturn

Planet	Mean Motion in Longitude f_λ (°/day)	Mean Motion in Epicyclic Anomaly f_μ (°/day)	Radius of Epicycle r	Eccentricity e	At epoch January 0.5 GMT 1990 = J.D. 241 5020.0		
					Longitude of Apogee A_0	Mean Longitude $\bar\lambda_0$	Mean Epicyclic Anomaly $\bar\mu_0$
Venus ♀	0.985 647 34	0.616 521 36	0.72294	0.01450	98°10′	279°42′	63°23′
Mars ♂	0.524 071 16	0.461 576 18	0.65630	0.10284	148°37′	293°33′	346°09′
Jupiter ♃	0.083 129 44	0.902 517 90	0.19220	0.04817	188°58′	238°10′	41°32′
Saturn ♄	0.033 497 95	0.952 149 39	0.10483	0.05318	270°46′	266°15′	13°27′

General precession f_p = 0.000 038 22°/day = 1°23′45″ per Julian Century = 0.838′ per year.

longitude of any of these planets at any desired date. The Ptolemaic theory of longitudes is the same for all these planets; only the numerical parameters differ. (Ptolemy's theory of Mercury, which is not addressed in this book, contains an extra complication.) The parameters of Jupiter and Saturn may be obtained by procedures similar to those used in the case of Mars. A wholly different approach is required for Venus, because this planet has no oppositions. The parameters given in Table 7.4 were calculated using more precise procedures than those described here.[72] The parameters for Mars recorded in the table therefore differ slightly from those obtained above by approximate, graphical techniques. In our further work we shall use the more accurate parameters given in table 7.4.

7.20 EXERCISE: PARAMETERS OF JUPITER

Using the method illustrated for Mars in section 7.19, derive all the necessary parameters for the theory of longitude of Jupiter.

1. From your table of oppositions for Jupiter (sec. 7.5), deduce values for the mean daily motion in longitude and in epicyclic anomaly.
2. Compare your results from problem 1 with values resulting from the Babylonian rule for Jupiter: 6 tropical periods = 65 synodic periods = 71 years.
3. Using your table of oppositions, plot a graph of the average angular distance between oppositions as a function of longitude. Locate the apogee of Jupiter's deferent at the minimum of the graph.
4. Choose the opposition lying nearest to either the apogee or the perigee, and one other opposition lying from 60° to 120° away, and use these to determine the eccentricity.
5. Determine the mean longitude and the mean epicyclic anomaly at one opposition.
6. Use these results, and the rates of mean motion, to determine $\bar{\lambda}$ and $\bar{\mu}$ for our standard epoch, January 0.5, 1900 (J.D. 241 5020.0).
7. From table 7.1, choose one observation of Jupiter about 90 days after the opposition that you used in problem 4, and use this to establish the radius of Jupiter's epicycle.

7.21 GENERAL METHOD FOR PLANET LONGITUDES

In this section we shall see how to find the longitude of Venus, Mars, Jupiter, or Saturn in Ptolemy's theory for any desired date. We shall illustrate the method by giving a sample calculation for Mars.

Example: Calculate the longitude of Mars for May 31, A.D. 1585, Greenwich noon (Gregorian calendar).

First Step: Preparing the Ground

In our work on the theory of Mars, we confined ourselves to observations made over a relatively short interval of time. We were therefore able to neglect precession, which proceeds at the slow rate of 1° in about 72 years. In Ptolemy's theory (*Almagest* IX, 5), the line of apsides does not remain fixed at a constant longitude, but rather remains *fixed with respect to the stars*. Thus, a planet's apogee behaves like a fixed star: it moves forward in longitude at the rate of precession.[73] Ptolemy's estimate of this rate was 1° in 100 years. We adopt the more accurate modern value. Our method for calculating longitudes must take into account the motion of the line of apsides. Fortunately, this is only

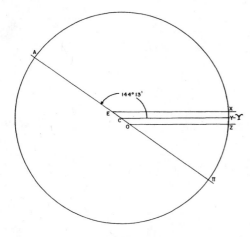

FIGURE 7.41. Longitude of Mars, May 31, A.D. 1585. First step.

a slight complication, requiring only a minor change in the procedure we have used all along.

A. Find the longitude of the apogee on the desired date. Because the apogee moves so slowly, it is sufficiently accurate to work to the nearest year:

Date	1585
Epoch	1900
Difference	−315 years

Now we apply the rule

$$A = A_\circ + \Delta t \times f_p.$$

$A_\circ$ is the longitude of the apogee at epoch 1900. f_p is the rate at which the apogee moves—that is, the rate of precession. Both of these numbers may be found in the table of modern Ptolemaic parameters (table 7.4). Thus,

$$A = 148°37' - 315 \text{ years} \times 0.838'/\text{year}$$
$$= 148°37' - 4°24'$$
$$= 144°13'.$$

The motion was subtractive because 1585 fell before the epoch.

B. Prepare a large piece of paper for the geometrical solution. This is done in exactly the manner to which we already are accustomed. Draw a circle about center C with a radius equal to the radius of the deferent of the Ptolemaic slats (the distance between T and center H of the grommet hole). Then, as in figure 7.41, draw a line through C to represent the zero of longitude. This line cuts the circle at Y.

Place the center of a protractor at C and lay out the line of apsides along direction 144°13', cutting the circle at the apogee A and perigee Π.

Along the line of apsides, mark the location of the equant E and the Earth O. These must be placed so that $CE = CO = $ the eccentricity × the radius of the deferent. For Mars, the eccentricity is 0.103.

Draw lines from E and from O parallel to line CY. These new lines cut the circle at X and Z and serve as the zeros of longitude for angles measured at E or at O.

Second Step: $\bar{\lambda}$ and $\bar{\mu}$

A. Find the Julian day number of the desired date and use this to calculate the time elapsed since epoch.

From tables 4.2–4.4 we have

1500	226	8923
85	3	1046
May 31		151
Total	230	0120

The epoch, for which we know the planet's mean longitude and mean epicyclic anomaly, is 1900 January 0.5 (J.D. 241 5020.0). The time elapsed since epoch is therefore

$$\Delta t = \begin{array}{r} 2,300,120 \\ -2,415,020 \\ \hline -114,900 \text{ days.} \end{array}$$

The minus sign indicates that the desired date fell before the epoch.

B. Calculate the planet's mean longitude at the desired date:

$$\bar{\lambda} = \bar{\lambda}_o + \Delta t \times f_\lambda.$$

In table 7.4 we find the epoch mean longitude $\bar{\lambda}_o$ and the mean daily motion f_λ, with the result

$$\bar{\lambda} = 293.6° + (-114,900 \text{ days}) \times 0.5240712°/\text{day}$$
$$= 293.6° + - 60,215.8°$$
$$= -59,922.2°$$

Now we must add enough whole circles of 360° to make the mean longitude come out as a positive number between 0 and 360. The easy way to do this is the following:

$$-59,922.2°/360° = -166.45 \text{ whole circles.}$$

Thus, the mean longitude is minus a bit more than 166 whole circles. Adding 167 circles will make the mean longitude come out in the desired range:

$\bar{\lambda}$	$-59,922.2°$	
Plus 167 whole circles	$60,120.0°$	$(167 \times 360°)$
$\bar{\lambda}$ in desired range	$197.8°$	

C. Calculate the planet's mean epicyclic anomaly:

$$\bar{\mu} = \bar{\mu}_o + \Delta t \times f_\mu.$$

Taking the necessary numbers from the table of parameters, we obtain

$$\bar{\mu} = 346.1° + (-114,900 \text{ days}) \times 0.4615762°/\text{day}$$
$$= -52,689.0°$$
$$= 231.0°, \text{ with addition of 147 complete circles.}$$

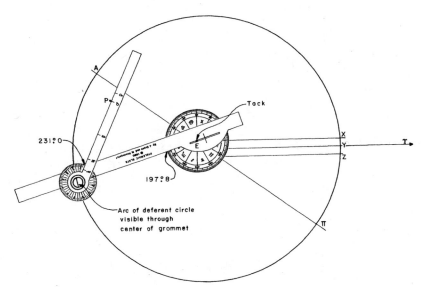

FIGURE 7.42. Longitude of Mars, May 31, A.D. 1585. Second step.

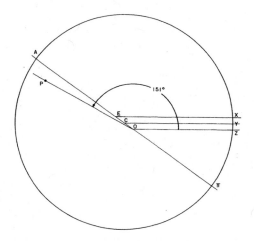

FIGURE 7.43. Longitude of Mars, May 31, A.D. 1585. Third step.

Note that in computing the changes in either $\bar{\lambda}$ or $\bar{\mu}$ over long time intervals, it is important to use the full precision in the figures for f_λ and f_μ provided by table 7.4.

D. Lay the Ptolemaic slats on the diagram: Push a tack through the equant E from underneath the paper. Place the center of the large paper protractor over the tack and turn the protractor so that the 0° mark falls on line EX, as in figure 7.42. Place the slot in the deferent slat over the tack. Turn the deferent slat until it comes to mean longitude of 197.8° on the paper protractor. Pull the slat in or out as required to place the grommet at the epicycle's center directly over the deferent circle. When you look through the grommet you should be able to see a small arc of this circle passing through the middle of the field of view. Now turn the epicycle slat until it comes to the correct mean epicyclic anomaly of 231.0°, as read on the little protractor built into the slats. Then, next to the mark on the epicycle slat labeled with the sign for Mars, place a dot P on the paper. This is the position of the planet at Greenwich noon, May 31, 1585.

Third Step: Finding the Longitude

Remove the Ptolemaic slats and protractor. Draw a line of sight OP from the Earth to the planet, as in figure 7.43. Place the center of a protractor at O and measure angle ZOP, finding about 151°:

Longitude of Mars at Greenwich noon, May 21, 1585 = 151°.

7.22 EXERCISE: CALCULATING THE PLANETS

1. Determine the longitude of Mars on May 30, 1982 (at Greenwich noon), using our modern Ptolemaic theory and the Ptolemaic slats. Step-by-step guidelines follow.

   A. Compute the number of days elapsed since the epoch, noon, January 0, 1900. (Answer: 30,100 days.)
   B. Calculate the longitude of the Martian apogee on May 30, 1982. (Answer: 149°46′.)
   C. Compute the planet's mean longitude and mean epicyclic anomaly on this date. (Answer: $\bar{\lambda} = 228°06′$, $\bar{\mu} = 199°36′$.)
   D. Draw a figure and lay out the Ptolemaic slats to represent this situation, as explained in section 7.21, and determine the longitude of Mars. Compare your result with the actual longitude of the planet given in table 7.1.

2. Use the modern Ptolemaic theory and the Ptolemaic slats to work out the longitude of Jupiter on August 26, A.D. 1597 (Julian calendar). Compare your final answer with Jupiter's actual longitude on this date, as listed in Tuckerman's *Planetary, Lunar and Solar Positions: A.D. 2 to A.D. 1649*: 75°.

3. According to the *American Ephemeris and Nautical Almanac* for 1948, a conjunction of Venus and Saturn occurred on October 8 of that year at 8 P.M. Greenwich time. That is, at the stated moment the longitudes of these two planets were the same. Test our modern Ptolemaic theory by working out the longitudes of the two planets for the date of this conjunction.

7.23 TABLES OF MARS

The goal of a planetary theory is, for any desired date, to anwer the question, What is the position of the planet? If a quick answer is wanted, and if rough accuracy will suffice, then the Ptolemaic slats suffice. If great precision is demanded, then a strict trigonometric calculation is required. The tedium involved in a strict calculation can be reduced with the aid of *planetary tables*. An example of such tables for Mars is provided by tables 7.5–7.7. These tables for computing the longitude of Mars are modeled on Ptolemy's in the *Almagest* and the *Handy Tables*. However, they are based on the modern values for Mars's parameters in table 7.4. Let us begin by describing the contents of the tables in a general way.

The Quantities Contained in the Tables

Table of the Mean Motion of Mars   The table of mean motion (table 7.5) gives the motion in mean longitude ($\bar{\lambda}$) and mean epicyclic anomaly ($\bar{\mu}$) for various time intervals, including minutes, hours, days, and groups of 30 days. The table also gives the motion for 1, 2, or 3 common years (years of 365 days).

The year used in the tables for "Julian Years by Fours" and "Julian Years by Hundreds" is the Julian year of 365.25 days. Since this year is used in the table only in multiples of four, a whole number of leap days are always included. Four Julian years represent a complete four-year leap-day cycle of (365 + 365 + 365 + 366) days. (In our tables of the sun [tables 5.1–5.3], all time intervals were reduced to days. Here an alternative arrangement is presented.)

The values of $\bar{\lambda}_o$ and $\bar{\mu}_o$ for Mars at epoch (Greenwich noon, January 0, 1900) are given at the end of the table of mean motion.

Table for the Longitude of the Martian Apogee   Table 7.6 gives the longitude of the apogee of Mars's deferent circle at hundred-year intervals, and the motion of the apogee for ten-year intervals.

Table of Equations   The motion of Mars appears irregular for two reasons (see fig. 7.32). First, the center K of the epicycle does not move at a uniform angular speed as seen from the Earth O. Second, the motion of the planet P on the epicycle causes P to be seen alternately ahead of and behind K. The irregularities in the motion of the planet thus produced are called the first and the second inequalities, respectively.

Figure 7.44 illustrates the geometry of these inequalities. The Earth is at O. The center K of the epicycle moves around the deferent at a uniform angular speed as seen from E. Thus, $\bar{\lambda}$ (the mean longitude) changes uniformly with time. Now draw a line $O\bar{P}$ parallel to EK. $\bar{P}$ marks the *mean direction of the planet*. That is, the planet would be seen along the direction $O\bar{P}$ if there were no inequalities at all. If there were no second inequality, the radius of the epicycle would be zero and P would coincide with K. If there were no first inequality, E would coincide with O, so EK would fall on top of $O\bar{P}$. Thus, in the absence of any inequalities, the planet would be seen in the direction $O\bar{P}$.

Because of the first inequality, K is not seen along line $O\bar{P}$. Rather, K is displaced from the planet's mean position by angle $\bar{P}OK$, which is called the *equation of center*, denoted q.

Because of the second inequality, P is not seen in the same direction as K. Rather, the planet is shifted away from line OK by angle KOP, which we call the *equation of the epicycle*, denoted θ.

From figure 7.44, the planet's actual longitude is

$$\lambda = \bar{\lambda} + q + \theta,$$

or, in words,

true longitude = mean longitude + equation of center

+ equation of the epicycle.

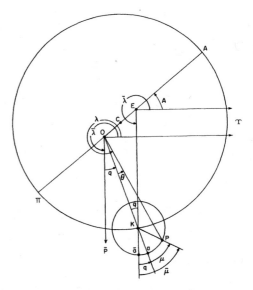

FIGURE 7.44. Illustrating the corrections that must be applied to the mean longitude: $\lambda = \bar{\lambda} + q + \theta$. q is the equation of center. θ is the equation of the epicycle.

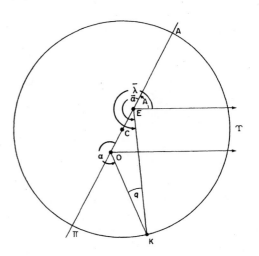

FIGURE 7.45.

TABLE 7.5. Mean Motion of Mars

Julian Years By Hundreds	Longitude	Epicyclic Anomaly	Days By Thirties	Longitude	Epicyclic Anomaly
100	61°42.0′	299°04.1′	30	15°43.3′	13°50.8′
200	123°24.0′	238°08.3′	60	31°26.7′	27°41.7′
300	185°06.0′	177°12.4′	90	47°10.0′	41°32.5′
400	246°48.0′	116°16.5′	120	62°53.3′	55°23.3′
500	308°30.0′	55°20.7′	150	78°36.6′	69°14.2′
600	10°12.0′	354°24.8′	180	94°20.0′	83°05.0′
700	71°54.0	293°28.9′	210	110°03.3′	96°55.9′
800	133°36.0′	232°33.1′	240	125°46.6′	110°46.7′
900	195°18.0′	171°37.2′	270	141°30.0′	124°37.5′
1000	257°00.0′	110°41.4′	300	157°13.3′	138°28.4′
1100	318°42.0′	49°45.5′	330	172°56.6′	152°19.2′
			360	188°39.9′	166°10.0′

Julian Years By Fours	Longitude	Epicyclic Anomaly	Days Singly	Longitude	Epicyclic Anomaly
4	45°40.1′	314°21.8′	1	0°31.4′	0°27.7′
8	91°20.2′	268°43.5′	2	1°02.9′	0°55.4′
12	137°00.2′	223°05.3′	3	1°34.3′	1°23.1′
16	182°40.3′	177°27.1′	4	2°05.8′	1°50.1′
20	228°20.4′	131°48.8′	5	2°37.2′	2°18.4′
24	274°00.5′	86°10.6′	6	3°08.7′	2°46.2′
28	319°40.6′	40°32.4′	7	3°40.1′	3°13.9′
32	5°20.6′	354°54.1′	8	4°11.6′	3°41.6′
36	51°00.7′	309°15.9′	9	4°43.0′	4°09.3′
40	96°40.8′	263°37.7′	10	5°14.4′	4°36.9′
44	142°20.9′	217°59.4′-	11	5°45.9′	5°04.6′
48	188°01.0′	172°21.2′	12	6°17.3′	5°32.3′
52	233°41.0′	126°43.0′	13	6°48.8′	6°00.0′
56	279°21.1′	81°04.7′	14	7°20.0′	6°27.7′
60	325°01.2′	35°26.5′	15	7°51.7′	6°55.4′
64	10°41.3′	349°48.2′	16	8°23.1′	7°23.1′
68	56°21.4′	304°10.0′	17	8°54.6′	7°50.8′
72	102°01.4′	258°31.8′	18	9°26.0′	8°18.5′
76	147°41.5′	212°53.5′	19	9°57.4′	8°46.2′
80	193°21.6′	167°15.3′	20	10°28.9′	9°13.9′
84	239°01.7′	121°37.1′	21	11°00.3′	9°41.6′
88	284°41.8′	75°58.8′	22	11°31.8′	10°09.3′
92	330°21.8′	30°20.6′	23	12°03.2′	10°37.0′
96	16°01.9′	344°42.4′	24	12°34.7′	11°04.7′
			25	13°06.6′	11°32.4′

Common Years			Days Singly	Longitude	Epicyclic Anomaly
1	191°17.2′	168°28.5′	26	13°37.6′	12°00.1′
2	22°34.3′	336°57.0′	27	14°09.0′	12°27.8′
3	213°51.5′	145°25.6′	28	14°40.4′	12°55.4′
			29	15°11.9′	13°23.1′
			30	15°43.3′	13°50.8′

Hours	Longitude	Epicyclic Anomaly	Hours	Longitude	Epicyclic Anomaly
1	0°01.3′	0°01.2′	13	0°17.0′	0°15.0′
2	0°02.6′	0°02.3′	14	0°18.3′	0°16.2′
3	0°03.9′	0°03.5′	15	0°19.7′	0°17.3′
4	0°05.2′	0°04.6′	16	0°21.0′	0°18.4′
5	0°06.6′	0°05.8′	17	0°22.3′	0°19.6′
6	0°07.9′	0°06.9′	18	0°23.6′	0°20.7′
7	0°09.2′	0°08.1′	19	0°24.9′	0°21.9′
8	0°10.5′	0°09.2′	20	0°26.2′	0°23.1′
9	0°11.8′	0°10.3′	21	0°27.5′	0°24.2′
10	0°13.1′	0°11.5′	22	0°28.8′	0°25.4′
11	0°14.4′	0°12.7′	23	0°30.1′	0°26.5′
12	0°15.7′	0°13.8′	24	0°31.4′	0°27.7′

Minutes	Longitude	Epicyclic Anomaly
10	0°00.2′	0°00.2′
20	0°00.4′	0°00.4′
30	0°00.7′	0°00.6′
40	0°00.9′	0°00.8′
50	0°01.1′	0°01.0′
60	0°01.3′	0°01.2′

At epoch 1900 Jan 0.5 GMT
(J.D. 241 5020, Greenwich noon):
Mean longitude = 293°33′.0;
Mean epicyclic anomaly = 346°08′.8

374

TABLE 7.6. Longitude of the Martian Apogee

Year	Longitude	Year	Longitude	Year	Longitude	Ten-Year Periods	Motion
801 B.C.	99°34′ ·	200 A.D.	117°38′	1200 A.D.	135°42′	10	0°11′
701	101°23′	300	119°26′	1300	137°30′	20	0°22′
601	103°11′	400	121°15′	1400	139°18′	30	0°33′
501	105°00′	500	123°03′	1500	141°07′	40	0°43′
401	106°48′	600	124°51′	1600	142°55′	50	0°54′
301	108°36′	700	126°40′	1700	144°43′	60	1°05′
201	110°25′	800	128°28′	1800	146°32′	70	1°16′
101	112°13′	900	130°16′	1900	148°20′	80	1°27′
1 B.C.	114°01′	1000	132°05′	2000	150°08′	90	1°38′
100 A.D.	115°50′	1100	133°53′	2100	151°57′		

TABLE 7.7. Equations for Mars

Common Argument	Equation of Center (Arg $\bar{\alpha}$)	Equation of the Epicycle (Argument μ)			Interpolation Coefficient (Arg $\bar{\alpha}$)
		Diminution at Apogee	Equation at Mean Distance	Augmentation at Perigee	
0° (360)	−(+) 0°00′	0′	+(−) 0°00′	0	Dim 1.000
5° (355)	0°56′	07′	1°59′	08	0.998
10° (350)	1°51′	14′	3°58′	16	0.990
15° (345)	2°46′	21′	5°56′	24	0.978
20° (340)	3°40′	28′	7°54′	32	0.961
25° (335)	4°33′	35′	9°52′	40	0.939
30° (330)	5°24′	43′	11°49′	48	0.911
35° (325)	6°13′	50′	13°45′	57	0.879
40° (320)	7°00′	58′	15°41′	65	0.841
45° (315)	7°44′	65′	17°35′	74	0.799
50° (310)	8°26′	73′	19°28′	83	0.750
55° (305)	9°04′	82′	21°20′	93	0.697
60° (300)	9°39′	90′	23°10′	103	0.638
65° (295)	10°10′	99′	24°58′	113	0.573
70° (290)	10°37′	109′	26°44′	124	0.504
75° (285)	11°00′	118′	28°27′	136	0.428
80° (280)	11°18′	129′	30°07′	148	0.348
85° (275)	11°32′	140′	31°44′	161	0.262
90° (270)	11°41′	151′	33°17′	175	0.171
95° (265)	11°45′	163′	34°44′	189	Dim 0.075
100° (260)	11°43′	176′	36°07′	205	Aug 0.019
105° (255)	11°23′	190′	37°22′	223	0.096
110° (250)	11°36′	205′	38°30′	241	0.176
115° (245)	11°05′	221′	39°27′	262	0.258
120° (240)	10°41′	238′	40°14′	284	0.340
125° (235)	10°11′	256′	40°46′	309	0.423
130° (230)	9°36′	274′	41°01′	336	0.505
135° (225)	8°56′	293′	40°53′	365	0.584
140° (220)	8°11′	312′	40°19′	397	0.660
145° (215)	7°21′	329′	39°09′	430	0.732
150° (210)	6°27′	342′	37°15′	462	0.798
155° (205)	5°29′	346′	34°24′	488	0.856
160° (200)	4°27′	334′	30°21′	499	0.906
165° (195)	3°23′	299′	24°54′	476	0.946
170° (190)	2°17′	231′	17°52′	394	0.976
175° (185)	1°09′	127′	9°23′	231	0.994
180° (180)	−(+) 0°00′	0′	+(−) 0°00′	0	Aug 1.000

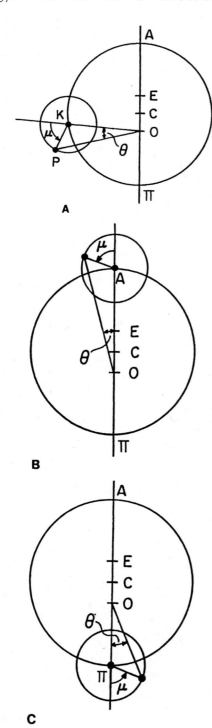

A

B

C

FIGURE 7.46. Dependence of the equation of the epicycle θ on the position of the epicycle. In A–C the true epicyclic anomaly μ is the same. θ is considerably larger when the epicycle is at the perigee of the eccentric (C) than when it is at the apogee (B).

Equation of Center: The equation of center (angle $\bar{P}OK$) is also equal to angle OKE (since EK and $O\bar{P}$ are parallel). If one thinks of the equation of center as OKE, it is easy to see how this equation varies as K moves around the deferent. The equation of center is zero when K lies in either the apogee A, or the perigee Π, of the eccentric deferent circle. It is therefore convenient to define a new angle called the *mean eccentric anomaly* $\bar{\alpha}$ (see fig. 7.45):

$$\bar{\alpha} = \bar{\lambda} - A.$$

$\bar{\alpha}$ is the angular distance of K from the eccentric circle's apogee, as measured at the equant. The angular distance of K from the apogee, as measured at the Earth, is called the *true eccentric anomaly* and is denoted α. The word *mean* will always indicate a quantity measured with respect to the equant, while the word *true* will indicate that the quantity is measured with respect to the Earth. The mean and the true eccentric anomaly differ by the equation of center:

$$\alpha = \bar{\alpha} + q.$$

q may be either positive or negative.

The equation of center is a function of the mean eccentric anomaly. q is zero when $\bar{\alpha} = 0$ or $180°$. q reaches its greatest magnitude when $\bar{\alpha}$ is approximately $90°$ or $270°$.

The Equation of Center in the Tables: In the table of equations (table 7.7), the first and second columns may be used to determine the equation of center. The left column (common argument) then represents various values of the mean eccentric anomaly $\bar{\alpha}$ in $5°$ steps. The second column (equation of center) gives the corresponding value of the equation. For example, if $\bar{\alpha} = 270°$, then $q = +11°41'$, but if $\bar{\alpha} = 90°$, then $q = -11°41'$.

Equation of the Epicycle: As we have seen, the equation of center q depends on a single variable, the mean eccentric anomaly. The equation of the epicycle θ, however, depends on two variables.

First, θ depends on the planet's position on the epicycle. Until now, we have specified the planet's position on the epicycle in terms of the *mean epicyclic anomaly* $\bar{\mu}$ (see fig. 7.32). This is the angular distance of the planet from the mean apogee $\bar{a}$ of the epicycle. However, for the construction of tables, the *true epicyclic anomaly* μ is more useful. μ is the angular distance of the planet from the true apogee a of the epicycle (see fig. 7.44). The equation of the epicycle θ is zero whenever μ is 0 or $180°$. Finally, the true and the mean epicycle anomaly differ by the equation of center:

$$\mu = \bar{\mu} - q.$$

Note the sign: $\alpha = \bar{\alpha} + q$, but $\mu = \bar{\mu} - q$.

Second, the equation of the epicycle also depends on the position of K on the deferent circle (see fig. 7.46). If K is in the apogee A of the deferent, the epicycle, as viewed from Earth, will appear diminished in size, and the equation of the epicycle will be somewhat reduced. But if K is in the perigee Π of the deferent, the epicycle will look larger and the equation of the epicycle will be magnified. The simplest way of specifying the position of K is in terms of the mean eccentric anomaly $\bar{\alpha}$.

The Equation of the Epicycle in the Tables: Since the equation of the epicycle depends on both μ and $\bar{\alpha}$, one could construct a table to double entry, with μ running horizontally, say, and $\bar{\alpha}$ running vertically. For each possible pair of values the equation θ could be given. This would, however, require an enormous table and it was not the scheme adopted by Ptolemy. Rather,

Ptolemy used an ingenious interpolation scheme that permitted the table for the equation of the epicycle to be reduced to manageable size.

Consider the column labeled "Equation at Mean Distance." This column gives the value of the equation of the epicycle for the given values of μ under the assumption that the epicycle is at its mean distance from the Earth. The epicycle is at its mean distance from the Earth if OK is equal to the radius CK of the deferent. In the case of Mars, this situation obtains when the mean eccentric anomaly is approximately 99° (or 261°). For example, suppose that the epicycle is located at its mean distance, as in figure 7.46A. Suppose that the true epicyclic anomaly μ is 70°. Entering the left column (common argument) with 70°, and going across to the column labeled "Equation at Mean Distance," we find that the equation of the epicycle is θ = +26°44′. (Note that when μ < 180°, θ is positive; when μ > 180°, θ is negative.)

Now consider the case when K is at the apogee of the deferent, as in figure 7.46B. Let μ be 70°, as before. Then θ will be smaller than 26°44′. The column labeled "Diminution at Apogee" gives the amount by which θ is smaller than the value of θ found at mean distance. For μ = 70°, the diminution is 109′. That is, when K is at the apogee of the deferent circle and μ = 70°, then the equation of the epicycle is 26°44′ − 109′ = 24°55′.

Similarly, if the epicycle's center lies at the perigee of the deferent (fig. 7.46C), then θ will be larger than when the epicycle's center is at mean distance. The column labeled "Augmentation at Perigee" gives the amount by which the equation of the epicycle is increased over its value at mean distance. For μ = 70°, we find that the augmentation is 124′. That is, when K is at the perigee of the deferent and μ = 70°, then the equation of the epicycle is 26°44′ + 124′ = 28°48′.

We have explained how the table may be used to obtain the equation of the epicycle when the epicycle's center is at mean distance, at apogee, or at perigee. The last column of table 7.7, labeled "Interpolation Coefficient," is used for intermediate cases. For example, at mean distance, with μ = 70°, the equation of the epicycle was 26°44′; with K at perigee, the equation was 124′ greater. For positions of K intermediate between mean distance and perigee, the equation will be augmented by some fraction of the 124′. The fraction is supplied by the column of interpolation coefficients.

Suppose that ᾱ = 145°, intermediate between mean distance and perigee. Again, let μ = 70°. Going into the table with ᾱ, we find that the interpolation coefficient is 0.732. Thus, the equation of the epicycle is

$$\theta = 26°44′ + (0.732 \times 124′)$$

$$= 26°44′ + 91′$$

$$= 28°15′.$$

Ptolemy's interpolation scheme provides an elegant way of giving the equation of the epicycle in a table of compact size. This interpolation scheme is not exact but involves an approximation. The value of the the equation of the epicycle obtained from the tables may therefore differ slightly from the value that would be obtained by strict trigonometric calculation. If the highest precision is needed, there is no recourse but to perform all the computations strictly, as Ptolemy himself remarks.

Precepts for the Use of the Tables of Mars

1. The date should be expressed in terms of the Gregorian calendar and referred to the meridian of Greenwich.

Determine the time elapsed from the epoch (1900 Jan 0, Greenwich mean noon) to the date of interest, and express the interval in terms of completed calendar years, days, and hours. The chief difficulty in time reckoning is that the calendar years do not all contain the same number of days—there can be either 365 or 366. However (with only a few exceptions, to be addressed soon), every interval of four successive calendar years contains precisely one leap day, for a total interval of 365.25 × 4 days.

Express the number of completed calendar years as a multiple of four plus a remainder. That is, write the number n of calendar years in the form

$$n = 4m + r,$$

where the remainder r is 0, 1, 2, or 3. The $4m$ calendar years contain m leap days and are therefore equivalent to $4m$ Julian years. The r years in the remainder are common years of 365 days, plus perhaps at most one leap day. Whether or not these r years contain a leap day is easily determined by inspection. In any event, the time interval is expressed as $4m$ Julian years + r common years + the odd days and hours.

Finally, one must correct for the anomalies of the Gregorian calendar. Three century years out of every four are not leap years in the Gregorian calendar. These are the years

A.D.   900   1300   1700   2100   2500
       1000   1400   1800   2200   2600
       1100   1500   1900   2300   2700,   etc.

One day must be subtracted from the time interval for each of these years that the interval contains. In particular, for all dates of the twentieth century (except those preceding March 1, 1900) it will be necessary to subtract one day, since the year 1900 was not a leap year.

2. Finding the mean motions: Enter with the number of Julian years completed ($4m$), the common years completed (r), and the odd days, and take out the corresponding motions in mean longitude and in mean epicyclic anomaly. Find also the motions for the hours and minutes, if required. The total motion in each quantity is the sum of all.

If the date of interest falls after the epoch, add the mean motion in each quantity to the value of that quantity at epoch, but if the date falls before the epoch, subtract the motion from the epoch value. Subtract or add as many multiples of 360° as needed to make each quantity positive and less than 360°. Round to the minute of arc. The results are the planet's mean longitude $\bar{\lambda}$ and mean epicyclic anomaly $\bar{\mu}$ at the date of interest.

3. Finding the longitude of the apogee: Enter with the century year immediately preceding the required year. For example, for A.D. 1583, use 1500; for 183 B.C., use 201 B.C. Add to this longitude the motion of the apogee during the interval from the century year to the required year. It is sufficient to work to the nearest decade. For example, for 1583 add 80 years' motion; for 183 B.C. add 20 years' motion. The sum is the longitude A of the eccentric's apogee at the required date.

Calculate also the mean eccentric anomaly $\bar{\alpha}$:

$$\bar{\alpha} = \bar{\lambda} - A$$

If $\bar{\alpha}$ should turn out negative, add 360°.

4. Equation of center: Enter with $\bar{\alpha}$ as argument and take out the equation of center q. Note that q is negative if $\bar{\alpha}$ falls between 0 and 180°, and positive if $\bar{\alpha}$ falls between 180° and 360°. The interpolation should be done with care.

5. Find the true epicyclic anomaly μ by subtracting the equation of center from the mean epicyclic anomaly:

$$\mu = \bar{\mu} - q$$

Note that μ will be larger than $\bar{\mu}$ if q is negative, and smaller if q is positive.

6. Now that $\bar{\alpha}$ and μ are known, the equation of the epicycle may be determined. This requires several steps.

A. Enter the column for the equation of the epicycle at mean distance, with the true epicyclic anomaly μ as argument, and take out the equation. Take out also the diminution at apogee and the augmentation at perigee—although only one of these two will be used.

B. Enter the column for the interpolation coefficient, with the mean eccentric anomaly $\bar{\alpha}$ as argument, and take out the coefficient. Note whether it is a coefficient of diminution or of augmentation: it is a coeffiecient of diminution if $\bar{\alpha}$ is either less than 99° or greater than 261°. It is a coefficient of augmentation if $\bar{\alpha}$ is between 99° and 261°.

C. If $\bar{\alpha}$ is less than 99° or greater than 261°, form the equation of the epicycle θ according to the rule

$$|\theta| = \frac{\text{equation at}}{\text{mean distance}} - \frac{\text{diminution}}{\text{at apogee}} \times \frac{\text{interpolation}}{\text{coefficient.}}$$

But if $\bar{\alpha}$ is between 99° and 261°, form θ as follows:

$$|\theta| = \frac{\text{equation at}}{\text{mean distance}} + \frac{\text{augmentation}}{\text{at perigee}} \times \frac{\text{interpolation}}{\text{coefficient}}$$

These rules give the absolute value of θ only. The whole equation of the epicycle thus formed is positive if μ lies between 0° and 180°, and negative if μ lies between 180° and 360°.

7. Finally, the longitude of the planet at the required date is calculated:

$$\lambda = \bar{\lambda} + q + \theta.$$

Example

Problem: Calculate the longitude of Mars on October 9, A.D. 1971, 0^h (midnight, Greenwich mean time).

Solution:

1. Elapsed time from the epoch A.D. 1900 Jan 0, 12^h Greenwich mean time:

From 1900 Jan 0, 12^h *to* 1971 Jan 0, 12^h is 71 calendar years.
From 1971 Jan 0, 12^h *to* 1971 Oct 8, 12^h is 281 days.
From 1971 Oct 8, 12^h *to* 1971 Oct 9, 0^h is 12 hours.

The 71 calendar years are handled as follows: 71 = 68 + 3, that is, 17 four-year cycles plus 3 years left over. The 17 four-year cycles dispose of the time from the beginning of 1900 to the beginning of 1968. The three whole years remaining are 1968, 1969, and 1970, the first of which was a leap year. One extra day must therefore be added to time interval, for the leap day in 1968. Finally, we must subtract one day from the time interval because 1900 was not a leap year in the Gregorian calendar.

The total time elapsed is therefore

	68	Julian years
+	3	common years
+	281	days
+	1	day (for 1968)
−	1	day (for 1900)
+	12	hours

68 Julian years + 3 common years + 281 days + 12 hours.

In this example, the leap day at the end of the time interval (1968) canceled out the missing leap day of 1900. This will not always happen.

2. Mean motions:

		$\bar{\lambda}$		$\bar{\mu}$	
68	Julian years	56°	21.4′	304°	10.0′
3	common years	213	51.5	145	25.6
270	days	141	30.0	124	37.5
11	days	5	45.9	5	04.6
12	hours	0	15.7	0	13.8
	Total	415°	164.5′	578°	91.5′
	Epoch value	293	33.0	346	08.8
		708°	198′	924°	100′
or		351°	18′	205°	40′

3. Longitude of eccentric's apogee and mean eccentric anomaly:

1900	148°20′
70 years	1°16′
A	149°36′
$\bar{\lambda}$	351°18′
$-A$	149°36′
$\bar{\alpha}$	201°42′

4. Equation of center:

$$(\text{argument } \bar{\alpha} = 201°42′) \quad q = + 4°48′.$$

5. True epicyclic anomaly:

$\bar{\mu}$	205°40′
$-q$	$-4°48′$
μ	200°52′

6. Equation of the epicycle:

A. Since $\bar{\alpha} = 201°42′$, the epicycle's center lies between mean distance and perigee. The equation will therefore have to be augmented above its mean value. Both the mean value and the augmentation may now be taken out of the table (argument $\mu = 200°52′$).

Equation at mean distance: 31°03′
Augmentation at perigee: 497′

B. Now take out the interpolation coefficient:

$$(\text{argument } \bar{\alpha} = 201°42′) \quad \text{interpolation coefficient} = 0.889$$

C. Form the absolute value of the equation of the epicycle:

$$|\theta| = 31°03′ + 497′ \times 0.889$$

$$= 38°25′.$$

Since μ lies between 180° and 360°, θ is negative.

7. Longitude of the planet:

$$\begin{array}{ll} \bar{\lambda} & 351° \ 18' \\ +q & + 4° \ 48' \\ +\theta & -38° \ 25' \\ \hline \lambda & 317° \ 41' \end{array}$$

Historical Specimens

Figures. 7.47 and 7.48 are photographs of the planetary tables in a ninth-century parchment *Almagest*. (This manuscript was described in sec. 2.13.) Figure 7.47 is the beginning of the table of mean motion. The top part of the figure has been translated in figure 7.49. This page of the manuscript table is devoted to the mean motion of Saturn in longitude (motion of the epicycle's center around the deferent) and in anomaly (motion of the planet around the epicycle). The two motions are given for time intervals of 18 Egyptian years and multiples thereof. For example, in 18 years, Saturn's mean motion in longitude is $220° \ 01' \ 10'' \ 57''' \ 09^{iv} \ 04^{v} \ 30^{vi}$. The high precision (six sexagesimal places in the fractional degree) is unnecessary. (In the *Handy Tables*, compiled after the *Almagest*, the mean motions are given only in degrees and minutes.) Succeeding pages of the tables give the mean motions of Saturn for periods of from 1 to 18 years, for months from 1 to 12, for days from 1 to 30, and for hours from 1 to 24. At the top of the ancient table we also find the mean longitude, mean epicyclic anomaly and the longitude of

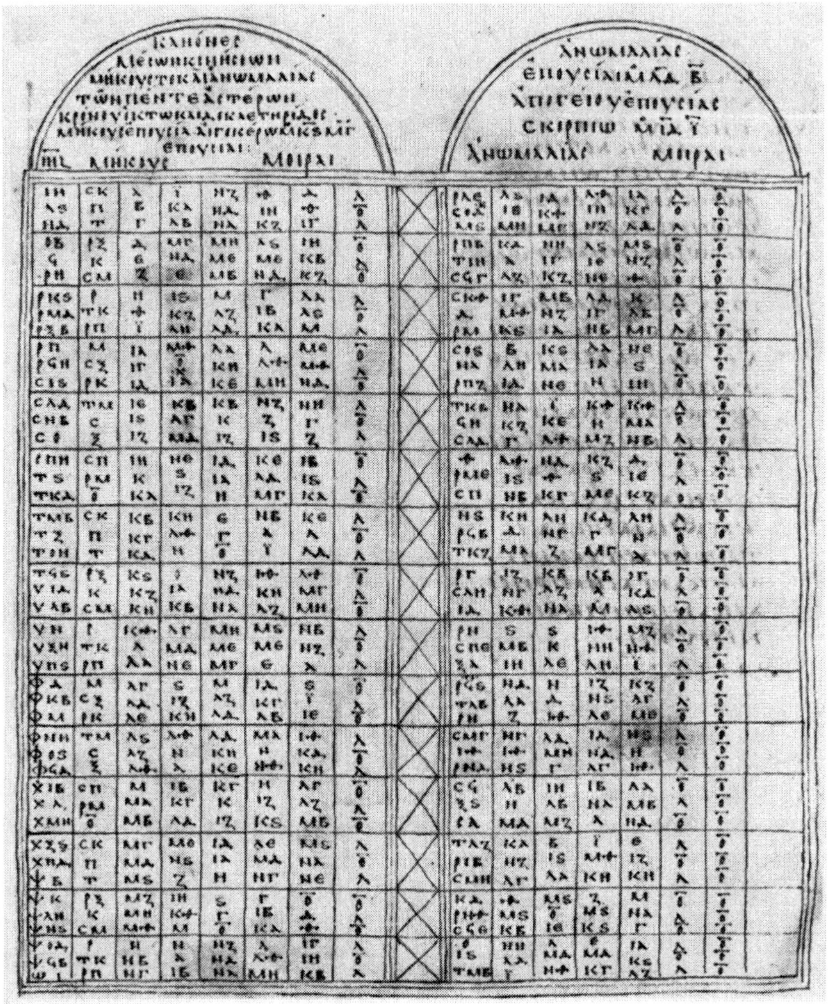

FIGURE 7.47. Beginnning of the table of mean motion for Saturn in a ninth-century parchment *Almagest*. Bibliothèque Nationale, Paris (MS. Grec 2389, fol. 247v).

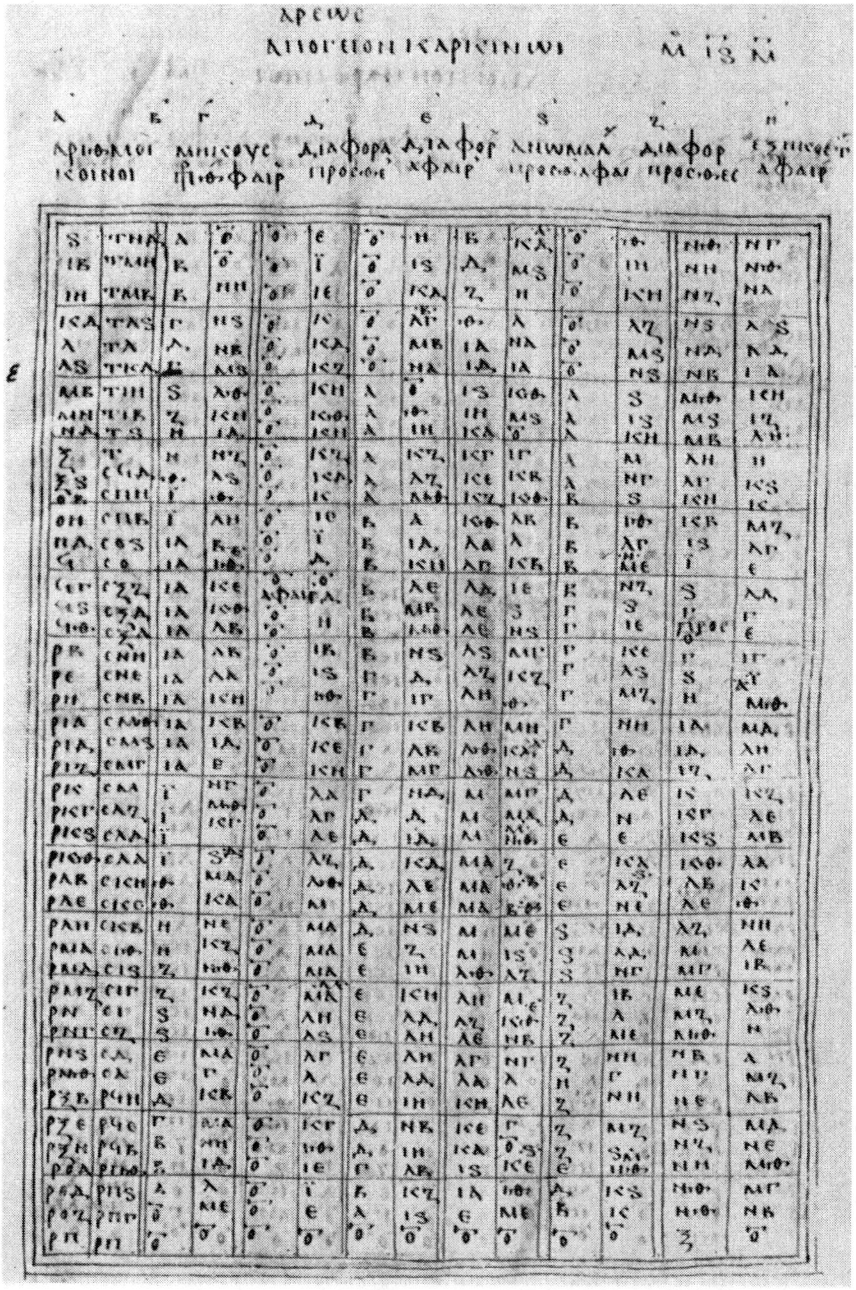

FIGURE 7.48. The table of equations for Mars in a ninth-century parchment *Almagest*. Bibliothèque Nationale, Paris MS. Grec 2389, fol. 310v).

the apogee of Saturn at Ptolemy's adopted epoch, the beginning of the reign of Nabonassar.

Figure 7.48 (partially translated in fig. 7.50) is the table of equations for Mars, taken from the same manuscript. These figures can be compared with our own table of equations for Mars (table 7.7). Columns 1 and 2 of the ancient table correspond to the common argument columns of our table. In our table, the argument runs by 5° intervals from 0 to 180°. Ptolemy uses 6° intervals for the first part of the table (0° to 90°) but 3° intervals for the rest (90° to 180°). The equations change more rapidly near perigee than near apogee and Ptolemy felt that the 3° intervals would therefore give better precision. In the *Handy Tables*, the equations are given for each single degree from 0 to 180°.

Columns 3 and 4 of Ptolemy's table correspond to the single column *equation of center* in table 7.7. Indeed, the equation of center is obtained by adding Ptolemy's columns 3 and 4. Thus, for a mean eccentric anomaly of

18's	Degrees of Longitude						Degrees of Anomaly						
18	220	1	10	57	9	4	30	135	36	14	39	11	30	0
36	80	2	21	54	18	9	0	271	12	29	18	23	0	0
54	300	3	32	51	27	13	30	46	48	43	57	34	30	0
72	160	4	43	48	36	18	0	182	24	58	36	46	0	0
90	20	5	54	45	45	22	30	318	1	13	15	57	30	0
108	240	7	5	42	54	27	0	93	37	27	55	9	0	0

Tables of mean motions in longitude and anomaly of the five stars. Saturn: Eighteen-year periods. Longitude at epoch: Goat 26° 43'. Anomaly at epoch: 34° 2'. Apogee at epoch: Scorpion 14° 10'.

FIGURE 7.49. Translation of the beginning of the Ptolemaic table of mean motion shown in Figure 7.47.

30°, the equation of center is 5°16′ (= 4°52′ + 0°24′). This compares closely with the 5°24′ in our own table. Our equation of center differs slightly from Ptolemy's because we adopted a slightly larger value for the eccentricity of the Martian deferent. Ptolemy's purpose in splitting the equation of center into two columns was partly pedagogical. Column 3 represents the equation of center in the intermediate model of figure 7.27A, in which the equant point and the center of the deferent coincide. Column 4 represents the change in the equation of center that is produced by separating the equant point from the center of the deferent and placing it halfway between the center of the deferent and the Earth (fig. 7.32). In all practical computation in the final model, one has need only of the sum of columns 3 and 4. In the *Handy Tables*, there is but a single column for the equation of center, as in Table 7.7. The separation of the equation of center into two parts in the tables of the *Almagest* reflects the newness of the equant point, introduced by Ptolemy himself.

Column 6 of the ancient table gives the equation of the epicycle at mean distance. Columns 5 and 7 give the diminution at apogee and the augmentation at perigee, exactly as in Table 7.7. Ptolemy's numerical values at 30° and at 60° of epicyclic anomaly are nearly the same as those of Table 7.7. The small discrepancies are due to our choice of slightly different values for the radius of the epicycle and the eccentricity of the deferent.

Mars

Apogee: Crab 16°40'

1. 2. Common Numbers		3. Equation in Longitude		4. Difference in Equation		5. Subtractive Diff.		6. Equation in Anomaly		7. Additive Difference		8. Sixtieths of Difference	
6	354	1	0	0	5	0	8	2	24	0	9	59	53
12	348	2	0	0	10	0	16	4	46	0	18	58	59
18	342	2	58	0	15	0	24	7	8	0	28	57	51
24	336	3	56	0	20	0	33	9	30	0	37	56	36
30	330	4	52	0	24	0	42	11	51	0	46	54	34
36	324	5	46	0	27	0	51	14	11	0	56	52	11
42	318	6	39	0	28	1	0	16	29	1	6	49	28
48	312	7	28	0	29	1	9	18	46	1	16	46	17
54	306	8	14	0	28	1	18	21	0	1	28	42	38
60	300	8	57	0	27	1	27	23	13	1	40	38	8

FIGURE 7.50. Translation of the beginning of the Ptolemaic table of equations shown in Figure 7.48.

Column 8 of Ptolemy's table gives the interpolation coefficient, expressed in sixtieths rather than in decimal fractions. Thus, for mean eccentric anomaly 30°, we find for the interpolation coefficient $54/60 + 34/3600 = 0.9094$, which compares well with the 0.911 in Table 7.7.

After Ptolemy's time, virtually all planetary equation tables were constructed according to his convenient scheme. One sees minor changes in arrangement and minor adjustments of the underlying numerical parameters, but the basic principles do not change. Even Copernicus's tables of equations (A.D. 1543) are of essentially the same form.[74]

7.24 EXERCISE: USING THE TABLES OF MARS

Use tables 7.5–7.7 to calculate the position of Mars on the following three dates:

1. June 4, A.D. 1983 (near a conjunction with the Sun).
2. April 4, A.D. 1984 (near a station).
3. May 19, A.D. 1984 (near the middle of a retrogradation).

Compare your results with the actual positions of Mars, taken from table 7.1.

If you wish to put Ptolemy's theory (and our tables of Mars) to the most demanding test, try calculating some positions of Mars during a retrogradation that occurred when the epicycle was near the perigee of the deferent (e.g., during the retrogradation of 1971). Errors are magnified in this situation, for then Mars is closest to Earth.

7.25 PTOLEMY'S COSMOLOGY

So far, we have delt with Ptolemy as the culminating figure of Greek technical astronomy. However, Ptolemy was equally influential as a cosmological thinker who tried to determine the structure of the whole universe.[75] In his cosmology, Ptolemy attempted to satisfy the demands of planetary astronomy as well as the requirements of sound physics, as he perceived them. This resulted in a unified worldview that dominated cosmological thought throughout the entire medieval period. Although the *Almagest* does provide some insight into Ptolemy's physical assumptions, Ptolemy's cosmological speculations are mostly confined to a separate, short work called *Planetary Hypotheses*.[76]

Overview

Ptolemy's cosmology is based on two fundamental assumptions. First, Ptolemy assumes that the deferent-and-epicycle models of the *Almagest* represent the actual machinery of the universe. However, the planets cannot be carried by infinitely thin, two-dimensional circles. Rather, the deferent circles and epicycles must be envisioned as "equator circles" of solid, three-dimensional spheres. These spheres, invisible to us, are made of the fifth element (the ether), like the planets themselves. Thus, Ptolemy's worldview involves a merging of deferent-and-epicycle astronomy with the old solid-sphere cosmology of Eudoxus and Aristotle.

Ptolemy's second fundamental assumption is that the cosmos contains no empty space. The mechanism (deferent circle and epicycle) that produces a planet's motion fills a spherical shell. The thickness of this shell is determined by the eccentricity of the planet's deferent circle and by the radius of the planet's epicycle. The shells for all the celestial bodies are arranged one within another in the standard order.

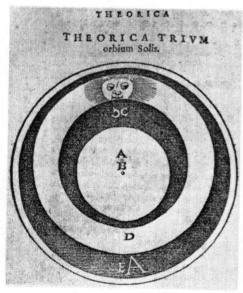

FIGURE 7.51. Ptolemy's three-dimensional system for explaining the motion of the Sun. The system requires three etherial orbs, nested one within another. Two of the bodies, C and E, are black in the diagram. The Sun is embedded in the middle body D, which is white in the diagram. This diagram is from a sixteenth-century textbook, the Paris, 1553, edition of Georg Peurbach's *Theoricae novae planetarum*. Courtesy of Special Collections Division, University of Washington Libraries (Negative UW 13653).

Figures 7.51 and 7.52 illustrate Ptolemy's cosmology as adapted by Georg Peurbach, an important figure in the Renaissance of European astronomy. Figure 7.51 shows Peurbach's system for the Sun. The Sun's system requires three orbs, labeled C, D, and E. The Earth is point B, the center of the cosmos. Point A is the center of the circle that the Sun travels in the course of the year. Orb C (black in the diagram) has its inner surface centered on B and its outer surface centered on A. The Sun is embedded in orb D (white in the diagram). This orb turns around once in the course of the year. This is how the Sun's annual motion around point A is effected. The outer orb E (black) has its inner surface centered on A and its outer surface centered on B. The two black orbs thus act as spacers for the orb carrying the Sun. Also, the inner hollow, bounded by the inner surface of C, serves as the receptacle into which the system for Venus is inserted. The system for Mars would be placed just outside orb E.

Figure 7.52 shows Peurbach's systems for the Sun and Venus. The three orbs for the Sun are all labeled A. They are exactly as in figure 7.51. The Sun is shown as a circle with a dot in it, embedded in the white solar orb. Three orbs for Venus are labeled B. Venus itself is the asterisk located on the epicycle embedded in the middle (white) orb of the Venus system. The epicycle is actually a solid sphere, which rotates inside a recess in the white orb. Points D, C, and H are, respectively, the Earth, the center of Venus's deferent circle, and Venus's equant point. The boundary between Venus's system and the Sun's system is the thin white crack between the outermost B orb and the innermost A orb (both black). Peurbach simplified the picture by omitting some technical details. (We shall take a closer look at Ptolemy's own description of the nested-spheres cosmology near the end of this section.) Nevertheless, Peurbach's illustrations preserve the essential features of the system described by Ptolemy in the *Planetary Hypotheses*.

Ptolemy also worked out numerical values for the thicknesses of all the nested planetary systems. Ptolemy's numerical values were only slightly modified by those who followed. Figure 7.53 illustrates Ptolemy's cosmos to scale, based on the parameters in the *Planetary Hypotheses*. The Earth is the small dot. Concentric spherical shells are assigned to the individual planets. The space between the Earth and the sphere of the Moon is filled by the lighter terrestrial elements, namely, air and fire. Each planetary shell in figure 7.53 is, of course, made up of a number of orbs, as in figures 7.51 and 7.52. Let us see how Ptolemy arrived at the thicknesses of these shells.

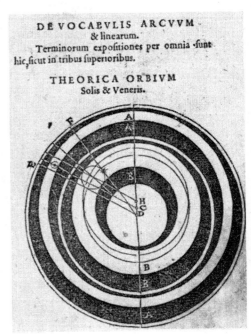

FIGURE 7.52. Peurbach's simplified view of Ptolemy's three-dimensional system for Venus, nested inside the system for the Sun. The three etherial bodies for the Sun are all labeled A. Three etherial bodies for Venus are labeled B. Venus is the asterisk set in the middle (white) body of the Venus system. From Georg Peurbach, *Theoricae novae planetarum* (Paris, 1553). Courtesy of Special Collections Division, University of Washington Libraries (Negative UW 13654).

Distances of the Moon and Sun in the Almagest

A complete cosmology required working out the absolute distances of the all the planets from the Earth. Deferent-and-epicycle astronomy provided no help here. For each planet, all that was astronomically deducible was the ratio of the size of the epicycle to that of the deferent. Thus, it was impossible to decide by observation which planet was closest to the Earth. As we saw in section 7.15, Ptolemy therefore had to fall back on physical or philosophical arguments. In the end, he adopted the standard order of his day: Moon, Mercury, Venus, Sun, Mars, Jupiter, Saturn.

Only for the Moon could the methods of ancient astronomy yield a measurement of the absolute distance. Modifications of Aristarchus's method by Hipparchus and Ptolemy resulted in accurate values for the Moon's distance. Ptolemy, for example, makes the Moon's mean distance from the center of the Earth at new or full Moon equal to 59 Earth radii, which is quite a good value. In contrast, the distance of the Sun resulting from all versions of the ancient method was very poor. The Sun's parallax is so small that it lay well below the level of precision of ancient astronomy. Nevertheless, the Greek astronomers *thought* they knew the Sun's distance. Aristarchus found the Sun

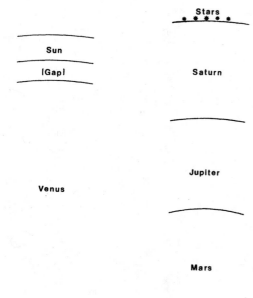

FIGURE 7.53. Thicknesses of the nested planetary spheres in Ptolemy's cosmology, drawn to scale. The scale changes by a factor of fifteen between the two figures. That is, if A were shrunk by a factor of fifteen, it could be inserted into B. This figure is based on the values in table 7.10. The gap between the spheres of Venus and the Sun reflects the numbers in table 7.10, but Ptolemy probably did not believe in the existence of an empty zone.

to be about nineteen times farther from us than the Moon is. Ptolemy adopted a figure not very different.

Successes and Failures of Ptolemy's Lunar Theory   The goal of Ptolemy's lunar theory in the *Almagest* was to permit prediction of the Moon's position in the zodiac. The absolute distance of the Moon from the Earth is irrelevant to the construction of such a theory. Ptolemy's lunar theory was very successful, in that it did accurately represent the Moon's position in the zodiac at all times of the month. In this it was a considerable advance over the lunar theory of Hipparchus.

However, Ptolemy's lunar model (made up of deferent circle, epicycle, and a special "crank" mechanism) greatly exaggerated the monthly variation in the Moon's distance from the center of the Earth. The ratio of the greatest to the least distance deduced from the model is nearly 2:1. In fact, the Moon's distance varies by only about 10% in the course of the month. Ptolemy is curiously silent about this defect of his lunar theory. But, again, this defect did not interfere with accurate prediction of angular positions.

Absolute Distance of the Moon   The distance of the Moon was the fundamental measuring stick by which the scale of the whole universe had to be judged. Moreover, the distance of the Moon did have some practical significance: it was needed for a proper treatment of parallax, which affects the visibility of solar eclipses. For both these reasons, Ptolemy begins the construction of a cosmological distance scale by determining the distance of the Moon.

In *Almagest* V, 11–13, Ptolemy attempts to find the distance of the Moon by parallax methods. He compares a position of the Moon observed from Alexandria with a position computed from his deferent-and-epicycle theory of the Moon's motion. This parallax measurement served to fix the absolute scale of the Moon's system. When the parallax measurement was combined with the deferent-and-epicycle theory, Ptolemy could then deduce the greatest and least distances of the Moon from the center of the Earth in absolute units (Earth radii, say). (See table 7.8.) The numerical values, taken from the *Planetary Hypotheses*, are rounded versions of the numbers in the *Almagest*.

Ptolemy's measurement of the Moon's parallax is problematical. He obtained a value for the parallax that was a good deal too large, making the Moon substantially too close to the Earth at the time of his observation. It is likely that he "pushed" or "fudged" the parallax measurement a bit to make it fit with his theoretical notion of the monthly variation in the Moon's distance. This problem of the Moon's distance is one of the least satisfactory parts of Ptolemy's astronomy. He measured the angular diameter of the Moon himself, so he clearly knew it did not double in the course of the month. Yet, paradoxically, in constructing his cosmology, he took this 2:1 variation in distance seriously. According to Ptolemy, the Moon's greatest distance is 64 1/6 Earth radii. (This figure is rounded in table 7.8.) From this it follows that the Moon's horizontal parallax (when at greatest distance) is $P_M = 53'35''$. (Ptolemy gives $53'34''$ in his table of parallaxes in *Almagest* V, 18.)

Angular Sizes of the Moon, Sun, and Earth's Shadow   To determine the absolute distance of the Sun, Ptolemy used the method of the eclipse diagram (fig. 1.47 and sec. 1.17), due originally to Aristarchus of Samos. As a preliminary

TABLE 7.8.   Absolute Distances of Sun and Moon

	Least Distance	Greatest Distance
Moon	33 Earth radii	64 Earth radii
Sun	1,160 Earth radii	1,260 Earth radii

(*Almagest*, V 14), Ptolemy determined the angular diameter of the Moon when it was at its greatest distance from the Earth and found it to be 31′20″. Moreover, he took the angular diameter of the Sun to be the same. Thus, the angular radius of the Sun (σ in fig. 1.44) is half this, or 15′40″. For the angular radius of the Earth's shadow as seen on the Moon during a lunar eclipse (with the Moon at greatest distance), Ptolemy found 40′40″ (τ in fig. 1.44). But in later calculation, he says the shadow is 2 3/5 times as big as the Moon, which would make τ = 40′44″.

Some of Ptolemy's predecessors had measured the angular diameter of the Moon by sighting it with a dioptra, or by timing with a water clock how long the Moon takes to rise. Ptolemy judged these methods fraught with error and difficulty. He therefore devised a clever method based on the comparison of two lunar eclipses of different degrees of totality. But Ptolemy's method is better in theory than in practice. In any case, Ptolemy's values for σ and τ did not differ much from those of his predecessors.

Absolute Distance of the Sun   Now let us take up the eclipse diagram (fig. 1.44). As proved in section 1.17,

$$\sigma + \tau = P_M + P_S.$$

We already have Ptolemy's results for σ, τ, and P_M. If we substitute numerical values (15′40″, 40′44″, and 53′35″, respectively), we get P_S = 2′49″ for the Sun's horizontal parallax. Then the distance of the Sun may be found from $d = r/\sin P_S$. Ptolemy obtains 1,210 Earth radii for the Sun's distance from the center of the Earth. (More accurate computation from Ptolemy's values for σ, τ, and the Moon's distance would give 1,218 Earth radii. The difference is due largely to rounding in Ptolemy's calculation.) Ptolemy adopts 1,210 Earth radii as the Sun's average or mean distance.

Ptolemy's solar theory involves an eccentric circle (fig. 5.8). If the radius of the circle is taken as 60 (arbitrary units), Ptolemy's value for the eccentricity OC is 2 1/2. The greatest and least distances of the Sun from the center of the Earth in these arbitrary units are 62.5 and 57.5. To pass over from the arbitrary units to absolute units (Earth radii), we multiply by the scale factor 1,210/60. Thus, the greatest distance of the Sun is 1,260 Earth radii, as in table 7.8. The Sun's least distance is 1,160 Earth radii.

Scale of Cosmic Distances in the Planetary Hypotheses

For the planets, all that Ptolemy could determine astronomically was the ratio of least to greatest distance. For a concrete example, let us examine the case of Mars (refer to fig. 7.32). Let us define some symbols:

R, radius of the deferent (= CA),
r, radius of the epicycle (= KP),
e, eccentricity of the deferent (= $CO = CE$).

Mars is closest to Earth when the center of the epicycle is at the perigee of the deferent and the planet is also at the perigee of the epicycle. Thus, the least distance of Mars = $R - r - e$. Similarly, Mars's greatest distance from Earth is $R + r + e$. In the *Almagest*, for each planet, Ptolemy arbitrarily chooses $R = 60$ units. In these terms, his values for the parameters of Mars are $r = 39.5$, $e = 6$. Thus, for Mars, the least and greatest distances are 14.5 and 105.5, in the arbitrary units. The ratio of greatest to least distance is then about 7.3 to 1, which Ptolemy rounds to 7:1 in the *Planetary Hypotheses*.

In a similar way, one may work out the ratio of greatest to least distance for each of the remaining planets. Table 7.9 presents Ptolemy's results in the *Planetary Hypotheses*. To proceed further, Ptolemy had to supplement the

TABLE 7.9. Astronomical Distance Ratios

Planet	Ratio of Least to Greatest Distance
Mercury	34 : 88
Venus	16 : 104
Mars	1 : 7
Jupiter	23 : 37
Saturn	5 : 7

astronomy of the *Almagest* with physical and cosmological premises. He assumes the order of the planets discussed above. Further, he assumes that the mechanisms of neighboring planets are nested one above the other, with no empty space between them.

Mercury is next above the Moon. Thus, Mercury's least distance must be 64 Earth radii, equal to the Moon's greatest distance in table 7.8. Then, using Table 7.9, Mercury's greatest distance = 64 Earth radii × 88/34 = 166 Earth radii, as listed in Table 7.10.

Then, Venus's least distance is also 166. To get the greatest distance, we again use Table 7.9. Venus's greatest distance = 166 × 104/16 = 1,079 Earth radii, as listed in Table 7.10.

And here it was possible for Ptolemy to perform a crucial check on the procedure. If the cosmological premises were correct, the Sun's least distance should also be 1,079 Earth radii. Now, the Sun's least distance, found by combining the method of the eclipse diagram with the eccentric circle theory of the Sun, was 1,160 Earth radii (table 7.8). This seemed too good to be true!

Pure astronomy fixed the maximum distance of the Moon at 64 Earth radii and the minimum distance of the Sun at 1,160. It turned out that the interval between these distances was almost the right size to be filled by the mechanisms of Mercury and Venus, with no empty space left over. Of course, it did not work out quite perfectly. For the maximum distance of Venus turned out to be only 1,079 Earth radii. Thus, there was a gap (between 1,079 and 1,160) that Ptolemy could not account for. But he points out that if the distance of the Moon is increased a little, the distance of the Sun will automatically be decreased a little. This is clear from the relation

$$\sigma + \tau = P_M + P_S$$

The left side of the equation (σ and τ) is fixed by relatively simple observations. If we make P_M (the Moon's horizontal parallax) a little smaller, then P_S (the Sun's parallax) must be made a little bigger to compensate. In this way, it might be possible to fill that small gap between the shells for Venus and the Sun. In any case, Ptolemy does not attempt to modify his numbers.

TABLE 7.10. Cosmological Distance Scale in the *Planetary Hypotheses*

	Least Distance (Earth radii)	Greatest Distance (Earth radii)
Moon	33	64
Mercury	64	166
Venus	166	1,079
Sun	1,160	1,260
Mars	1,260	8,820
Jupiter	8,820	14,187
Saturn	14,187	19,865

TABLE 7.11. Sizes of the Stars in the *Planetary Hypotheses*

	Mean Distance (Earth radii)	Angular Diameter (fraction of Sun)	Linear Diameter (Earth diameters)
Moon	48	1 1/3	7/24
Mercury	115	1/15	1/27
Venus	622 1/2	1/10	3/10
Sun	1,210	1	5 1/2
Mars	5,040	1/20	1 1/7
Jupiter	11,504	1/12	4 43/120
Saturn	17,026	1/18	4 3/10
1st magnitude stars	20,000	1/20	4 11/20

The least and greatest distances of the outer planets are easily filled in. Ptolemy puts Mars's least distance equal to the Sun's greatest distance of 1,260 Earth radii. Then Mars's greatest distance is 1,260 × 7/1 = 8,820 Earth radii. He finds the least and greatest distances of Jupiter and Saturn in the same way. The fixed stars lie just beyond the sphere of Saturn, at 19,865 Earth radii, which Ptolemy later rounds to 20,000 Earth radii. Ptolemy then converts all of these distances into stades, starting from his value of 180,000 stades for the circumference of the Earth.

Ptolemy concludes this portion of his discussion by reiterating his assumption that the nested spheres are contiguous, "for it is not conceivable that there be in nature a vacuum, or any meaningless and useless thing." But, he says, if there is space or emptiness between the spheres, the distances cannot be any smaller than those he has set down.

Sizes of the Stars and Planets

Ptolemy reports some angular sizes of the planets, compared to the disk of the Sun. For example, the angular diameter of Venus is one-tenth that of the Sun. From the angular diameter and the absolute distance, he works out the actual diameter of each planet. Ptolemy's results are displayed in table 7.11. (The mean distances in the table are averages of the least and greatest distances in table 7.10.) Ptolemy even goes on to compute the volumes of the planets, in comparison with the volume of the Earth.

The Ptolemaic System of Nested Spheres

The system of nested planetary spheres, based on deferent-and-epicycle astronomy and supplemented by the cosmological distance scale, is often referred to as the Ptolemaic system. In the *Almagest*, the theories of the planets are elaborated in terms of circles, not solid spheres. The circles are all that is required for practical astronomical calculation. But it is clear that Ptolemy always regarded the spheres as physically necessary.[77] In the second book of the *Planetary Hypotheses*, Ptolemy squarely faced the problem of reconciling deferent-and-epicycle astronomy with the solid-sphere cosmology of Aristotle and Eudoxus.

To account for the daily rotation of the whole cosmos, Ptolemy surrounds the sphere of the stars with a spherical shell of ether, as in figure 7.54. *AB* is the axis of the daily rotation. Line *CD* passes through the poles of the ecliptic. The Earth lies in the middle of the diagram, at the intersection of *AB* and *CD*. The exterior ether shell (1) turns once a day, from east to west, about axis *AB*. The sphere of stars (2) is pierced by axles (*CE* and *ZD*) set into the rotating ether shell. Thus, the westward daily motion of sphere 1 carries the sphere of stars around with it. Meanwhile, the sphere of stars turns slowly to

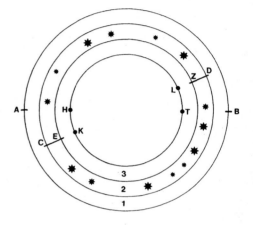

FIGURE 7.54. The intervening ether shells described by Ptolemy in the *Planetary Hypotheses*. Surrounding the sphere of stars 2 is an ether shell 1. Ether shell 3 intervenes between the sphere of stars and the system for Saturn. The intervening ether shells are responsible for communicating the daily rotation to the spherical systems of the individiual planets.

the east about axis *CD*: this is the precession of the fixed stars, which Ptolemy put at 1° in 100 years.

Inside the starry sphere is an ether shell 3, which rotates about *EZ* in such a way that sphere 3 remains stationary with respect to the outermost sphere 1. Thus, points *H* and *T* on sphere 3 remain directly under the corresponding points *A* and *B* of the outermost rotating ether shell. The system for Saturn may then be plugged into sphere 3 at points *K* and *L*. In other words, the daily rotation of the intervening ether shell 3 is responsible for carrying the Saturn system around once a day in exactly the same way as sphere 1 carries around the sphere of stars. The spherical system for each of the other planets is surrounded by a similar ether shell. There are seven such intervening shells (one each around the systems for the Sun, Moon, and five planets), plus the outermost ether sphere (surrounding the sphere of fixed stars), for a total of eight. However, Ptolemy is not certain that the intervening shells are necessary. As we have seen above, he ignored the intervening shells when he worked out his scale of cosmological distances.

Ptolemy then describes the individual systems for the Sun, Moon, and planets in detail. The simplest case is that of the Sun. Ptolemy's description is essentially the same as that given above in connection with figure 7.51.

More complex are the systems for the planets. Figure 7.52, discussed above, omits a number of details. Let us therefore look at Ptolemy's own description of his system for the planets. For an example let us consider figure 7.55, which illustrates Ptolemy's system for Mars, Jupiter, or Saturn. Earth is at *O*. Surrounding the Earth are three bodies 1, 2, and 3, more or less as in figure 7.52. The exterior surface of body 1 is a sphere centered on the Earth *O*. The inner surface of 1 is also spherical, but has its center at point *C*. Inside body 1 is a spherical shell 2, centered on *C*. Inside body 2 is body 3. The outer surface of body 3 is a sphere centered on *C*; the inner surface of 3 is a sphere centered on the Earth *O*.

Body 2 has a hollow that contains a spherical shell 4, which in turn contains a sphere 5. Embedded in sphere 5, near its surface, is the planet *P* itself. The

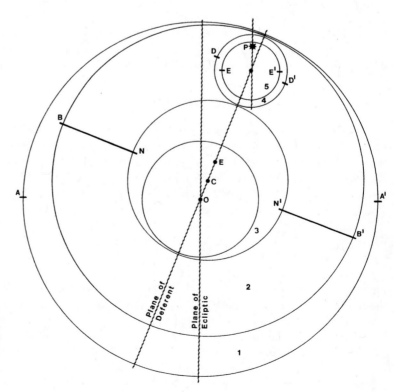

FIGURE 7.55. Solid-sphere mechanism for the superior planets, as described by Ptolemy in the *Planetary Hypotheses*. If we regard this figure as representing the system for Saturn, it can be plugged into figure 7.54, with axis *AA'* of this figure going into points *K* and *L* of figure 7.54.

spherical shell 4 and the sphere 5 are required to incorporate the planet's epicycle into the three-dimensional scheme.

Let us look now at the details of the motion. In Ptolemy's theory of the superior planets, the planes of the deferent circles are slightly inclined the plane of the ecliptic. These inclinations are required to explain the planets' latitudes, that is, their slight departures from the plane of the ecliptic. To show these inclinations, Ptolemy draws his figures with the ecliptic perpendicular to the plane of the diagram. (This is different from the case of fig. 7.52, in which the ecliptic lies in the plane of the figure.)

Thus, in figure 7.55, axis AA' passes through Earth O and the poles of the ecliptic. The ecliptic plane is therefore perpendicular to AA' and to the plane of the page. Body 2 (carrying 4 and 5 with it) rotates slowly about axis BB', completing one rotation during the planet's tropical period. Axis BB' is tilted slightly with respect to AA'. The tilt has been exaggerated in the figure. The epicycle spheres (4 and 5) are therefore carried around a circle centered on C. Of course, the angular motion of epicycle around the deferent is supposed to be uniform with respect to the equant point E, and not C or O. But Ptolemy does not provide a mechanical realization of the equant in his three-dimensional theory.

Now let us take up the epicycle spheres 4 and 5. In Ptolemy's theory of the superior planets, the plane of the deferent is tilted with respect to the ecliptic, but the plane of the epicycle is again parallel to the ecliptic plane. Ptolemy provides spherical shell 4, rotating about axis DD', to "cancel out" the rotation about the tilted axis BB'. Thus, DD' is parallel to BB'. Sphere 4 rotates about DD' at the same rate as 2 rotates about BB', but in the opposite direction. The net result is that 4 is carried in a circular translation (i.e., without rotation) about axis BB'. Sphere 5 (carrying the planet P) rotates about axis EE', which is parallel to AA'. The rotation of 5 is made in the course of the planet's synodic period.

After explaining the solid-sphere models for each of the planets, Ptolemy totals up the contents of the universe. Each of the superior planets requires five bodies, while the Sun requires three. The theory of Venus is similar to that of the superior planets—hence, five ethereal bodies are required. It also happens that Mercury needs seven and the Moon needs four. The fixed stars need a sphere of their own. Also, eight intervening ether shells are required. The total therefore ought to come to 43. But for some reason Ptolemy assigns only one ethereal body to the Sun and thus reaches a total of 41. In an alternative version of the system, Ptolemy replaces most of the three-dimensional orbs by rings or tambourines. In this version, Ptolemy reckons that only 29 bodies are required. However, as remarked above, he is not sure that the eight intervening shells are necessary. One is certainly required, surrounding the whole cosmos. But perhaps the other seven can be eliminated. Thus, Ptolemy conjectures that only 34 bodies are necessary $(41 - 7)$ in the first version of his cosmology. In the second version, the elimination of seven ether shells brings the total down to 22 $(29 - 7)$.

Ptolemy's Place in the History of Cosmological Thought

The Ptolemaic system was a harmonious blend of several lines of ancient thought. The basic physical principles of the system derived from the fourth century B.C.: Aristotle's physics and Eudoxus's cosmology of three-dimensional spheres. However, neither Aristotle nor Eudoxus had provided a planetary theory with quantitative, predictive power. Quantitative planetary astronomy became possible only after the development of deferent-and-epicycle theory by Apollonius of Perga and Hipparchus (third and second centuries B.C., respectively). The final, very successful models for the motions of the Moon and planets were due to Ptolemy himself. Ptolemy's problem as a cosmological thinker was to combine accepted physical principles, inherited from Aristotle and Eudoxus, with the successful planetary theory of the later astronomers.

The incorporation of deferent-and-epicycle astronomy into solid-sphere cosmology was not original with Ptolemy. Theon of Smyrna, who was perhaps a generation older than Ptolemy, had already discussed the way that deferent-and-epicycle theory could be incorporated into a world view based on solid, nested spheres.[78] Indeed, it is likely that the originator of deferent-and-epicycle theory, Apollonius of Perga, discussed it in terms of solid spheres.[79] But it was Ptolemy who showed how to incorporate all the technical details and who worked out a complete scale of cosmic distances on the basis of the models. In any case, the standard cosmology of the whole medieval period derived directly from Ptolemy: the technical astronomy of the *Almagest* supplemented by the solid spheres and the distance scale of the *Planetary Hypotheses*. For 1,400 years, people were lucky enough to understand the whole structure of the universe![80]

Ptolemy did not ascribe the same certainty to every feature of his universe. The deferent-and-epicycle theory of the *Almagest* was something he took quite seriously. The planetary models were all based on observation and trigonometric demonstration. They were therefore relatively certain. And they do work very well, after all. But even in the *Almagest*, as when discussing alternative models for the motion of the Sun, Ptolemy shows ample awareness that astronomical observation cannot answer every question. There remains a certain freedom in the selection of models; the astronomer must therefore fall back on physics or philosophy to guide his choice. Nevertheless, Ptolemy clearly felt that the models of the *Almagest* could not be very far from the truth. From a modern point of view, they still look pretty accurate as descriptions of the apparent motions.

In the *Planetary Hypotheses* Ptolemy shows much more caution. He is not certain about the total number of orbs and spheres. He does not even assert that the principle of no empty space is true. He only uses it to deduce the minimum possible distances of the planets. The whole cosmological system is proposed only as a plausible idea: the universe must be more or less like this, but Ptolemy cannot vouch for all the details.

7.26 ASTRONOMY AND COSMOLOGY IN THE MIDDLE AGES

In large measure, the planetary theory and cosmology of the Middle Ages descend from four works of Greek antiquity. For planetary theory, the essential work was Ptolemy's *Almagest*. For practical computing, Theon of Alexandria's edition of Ptolemy's *Handy Tables* served as prototype. In cosmology, two works had a profound influence. The underlying philosophy of nature was that of Aristotle, especially as embodied in *On the Heavens*, while the technical cosmology of the Middle Ages was based on Ptolemy's *Planetary Hypotheses*.

Astronomy and Cosmology in Islam

In the seventh and eighth centuries A.D., Islam expanded with remarkable speed. By the year 710, a new spiritual empire, under the rule of the Umayyad caliphate at Damascus, stretched from the borders of India, through the Middle East, all across the north coast of Africa to Spain. In 750 the Umayyads were overthrown by the Abbasids, who moved the capital to the newly founded city of Baghdad. The monolithic character of the empire was soon tempered, as local rulers at the fringes began to assert their independence. At first the greatest energies were expended in military conquest and religious conversion. But by the ninth century a renaissance of culture was under way, centered around the Baghdad caliphate of the Abbasids. Patronage for the arts and sciences extended not only to literature, philosophy, and medicine, but also to astronomy. Thus, a crucial pattern was established at an early date. Not every Islamic ruler was equally supportive of astronomy. But, at intervals, a number of rulers made decisive gestures of support. Many of the most significant astronomers of medieval Islam are known to have received royal patronage of some sort.

The first contact with Greek science was a rather complicated affair. Some Greek works, available in Syriac translations, were translated into Arabic. Later, translations were made directly from the Greek. But Greek astronomical ideas, blended with Babylonian procedures and Indian influences, also came in from the East. How that came about is a remarkable story.

During the Persian period (fifth century B.C.), when the Achaemenid dynasty ruled not only Persia and Mesopotamia but also northwest India, some techniques of Babylonian astronomy filtered into India. These included a number of period relations, the use of the tithi as a unit of time, and the use of arithmetic progressions for calculating the length of the day. During the Seleucid period, Greek astronomical ideas, with Babylonian features, also entered Indian astronomy. Thus, Hipparchus's length of the year was transmitted, along with procedures for computing positions of the Sun and Moon from arithmetic progressions. Indian planetary texts based on deferent-and-epicycle theory preserve features of Greek astronomy from before the time of Ptolemy—before, for example, the invention of the equant. Modern scholars thus attempt to use the Indian material to reconstruct the development of Greek astronomy between the time of Hipparchus and that of Ptolemy.

Arabic astronomers came into contact with this remarkable mixture of Indian, Greek, and Babylonian astronomy at about the same time as they began to acquire the classics of the Greek tradition. Although the Indian material had the defect of inconsistency—not surprising, in view of its heterogeneous origin—it also afforded easier, numerical methods for computing planetary phenomena than did the Ptolemaic methods. Ptolemy's astronomy soon won out, but not without a brief period of competition.[81]

Astronomy in Service to Islam   Islamic religion posed a number of practical problems for astronomers. In principle, a new month begins in the Muslim lunar calendar with the first visibility of the crescent Moon in the west just after sunset. It is important that religious festivals be celebrated on the right day by the actual Moon. For example, it is important that the fasting for the month of Ramadan begin on the correct day. Tables of lunar visibility had a long tradition, going back to India and even to Babylonia. But the Islamic astronomers of the early Middle Ages worked out new theoretical methods for predicting the first visibility of the crescent.

A second service that astronomy could render to religion was the calculation of the *qibla*, that is, the direction to Mecca. In prayer, a Muslim is supposed to face toward Mecca. In the later Middle Ages, this was interpreted as facing along the shortest or great-circle direction to Mecca. Determining the direction of the *qibla* from the latitude and longitude of one's city is a nontrivial problem in spherical trigonometry.

Yet a third service that astronomy could render to religion was the calculation of the correct times of prayer during the day. As we have seen (sec. 3.7 and fig. 3.42), auxiliary curves for prayer times are sometimes found on Islamic astrolabes. By the thirteenth century it was not uncommon for a mosque to maintain a trained astronomer on the staff in the role of time keeper. The *madrasas* (schools) associated with some mosques provided for the teaching of astronomy, though usually only on an extracurricular basis.

To some extent, then, religious patronage complemented the support lent to astronomy by political rulers. Of course, religious authorities did not always—or even often—take the advice offered by astronomers. Thus, the actual practices followed by religious leaders in reckoning months, in orienting mosques, and in timing the daily prayers showed considerable variability. Nevertheless, the problems of the visibility of the lunar crescent, the direction of the *qibla*, and the times of prayer are the subjects of a large body of specialized astronomical literature.[82]

However, it would be easy to overstate the importance of religious utility in explaining the development of astronomy in Islam. Although the religious motive was important for many astronomers it often had deeper roots than mere practical utility. Many astronomers saw their studies as a way of understanding God's plan for the world and of glorifying him by exalting his works. For others, the main benefit of astronomical knowledge was not religious at all, but the power it lent to the practicing astrologer. But perhaps the most constant and significant impetus to the study of astronomy was the Hellenistic ideal of science for its own sake.

The Zīj The handbook of practical astronomy known as a *zīj* held a central place in the Arabic astronomical tradition. The ancient prototype of a *zīj* is the *Handy Tables*. Thus, a typical *zīj* includes a complete set of tables: tables for problems associated with the diurnal motion (such as a table of ascensions), as well as tables for the ecliptic motions of the Sun, Moon, and planets. Naturally, the tables must be accompanied by a set of canons.

An influential early *zīj* was that of al-Khwārizmī, who worked at Baghdad in the early ninth century. Al-Khwārizmī was a capable mathematician as well as an astronomer. The English word *algorithm* is a corruption of his name (which indicates that he was a native of Khwārizm in west-central Asia). Al-Khwārizmī's *zīj* was based on the *Handy Tables*, but it incorporated a lot of Indian and Persian material. Despite these inconsistencies, it had a long life and was reworked by Arabic astronomers in Spain, then translated into Latin by Adelard of Bath early in the twelfth century. The demise of Al-Khwārizmī's heterogeneous methods is clearly reflected in the fact that his *zīj* now survives only in this Latin translation.[83]

Al-Khwārizmī's *zīj* was criticized already in the ninth century. Newer works in this genre, notably that of al-Battānī, turned decisively toward pure Ptolemaic methods. We saw in section 6.9 that ninth-century Arabic astronomy was already far enough advanced to improve on Ptolemy's solar eccentricity and to reveal the decrease in the obliquity of the ecliptic. Arabic astronomers also found, by measuring the lengths of the seasons, that the apogee of the Sun's eccentric circle had shifted and that it was not fixed with respect to the equinoctial point as Ptolemy had claimed. Throughout the medieval period the common assumption was that the Sun's apogee was, like the apogees of the planets, fixed with respect to the stars and that it therefore participated in precession (and trepidation, too, if one subscribed to that theory). Al-Battānī's *zīj* was important not only for its greater theoretical self-consistency, but also for the incorporation of new results, such as al-Battānī's own values for the obliquity of the ecliptic and the longitude of the Sun's apogee. The *zīj*, then, was not necessarily merely a slavish imitation of the *Handy Tables*; it could, and sometimes did, incorporate original astronomy.[84]

Hundreds of *zījes* are preserved in libraries today, spanning the period from the ninth to the fifteenth century. The great majority of Arabic *zījes* are based on Ptolemaic methods. The numerical parameters might differ a little. There might or might not be a table of trepidation. The Sun's apogee might be movable, rather fixed with respect to the equinox. But for the most part these texts are in the tradition of the *Almagest* and the *Handy Tables*. Among the later *zījes*, the most influential for the development of European astronomy were the eleventh-century *Toledan Tables*, mentioned in section 6.9.[85]

The Almagest *Tradition* The *Almagest* was translated into Arabic on several occasions—once already by the beginning of the ninth century, at the request of the vizier of the Abbasid caliph Hārūn al-Rashīd. Two new translations were made of the *Almagest* during the reign of al-Rashīd's son and successor, al-Maʾmūn. One of these translations, made in 827/828 by al-Ḥajjāj, is still extant. The other extant Arabic version is the excellent and influential translation made around 892 by Isḥāq ibn Ḥunayn and later revised by Thābit ibn Qurra.[86]

Scholars could immediately see the need for easier works to introduce students to Ptolemy. Thābit ibn Qurra himself wrote several elementary accounts of Ptolemaic astronomy and cosmology. These had titles like *The Almagest Simplified, Introduction to the Almagest,* and *Résumé of the Almagest.*[87] Thābit's were not the first such works, but they helped establish a genre of elementary astronomical textbooks in Arabic. Many such works were written from the ninth to the sixteenth century. They vary greatly in length, quality, and level of detail. These introductions to astronomy can be considered the Arabic counterparts of the Greek manuals by Geminus, Theon of Smyrna, and Proclus. Like Proclus's *Hypotyposis* (as opposed to the earlier works by Geminus and Theon of Smyrna), they are based directly or indirectly on Ptolemy, even when they do not cite him explicitly.

Besides translations of the *Almagest* and elementary textbooks intended to introduce students to Ptolemaic astronomy, there was one other genre of Arabic astronomical writing that constituted part of the *Almagest* tradition. These were the commentaries on the *Almagest*. A typical commentary followed the *Almagest* (or, more often, a restricted portion of it) more closely than did a general introduction to astronomy. The point of a commentary was not only to explain difficult passages, but also to offer alternative demonstrations or new data, or even to question some of Ptolemy's assumptions. Thus, the *Almagest* commentary, like the *zīj*, could be a genre for publishing the results of original astronomical investigations.[88]

Ptolemaic Cosmology in Islam The essential features of Ptolemaic cosmology are the nested three-dimensional spheres and the cosmological distance scale. These were both described by Ptolemy in his *Planetary Hypotheses*. But the texts of the *Almagest* and of the *Planetary Hypotheses* had quite different fates. The *Planetary Hypotheses* has been roughly handled by history. Only the first half of the text has survived in Greek. The rest is preserved only in Arabic translation (and a medieval Hebrew translation made from the Arabic). The diagrams have fared even more poorly. It is unlikely that the *Planetary Hypotheses* ever circulated very widely—in contrast to the *Almagest*, which did circulate widely and the text of which has been very well preserved. Often, medieval cosmologists learned the contents of the *Planetary Hypotheses* only second or third hand. Thus, we find writers who discuss the cosmological distance scale without reference to the system of nested spheres. And we find writers who discuss the nested spheres without mentioning the distance scale. We also find writers who seem not to be aware that the system originated with Ptolemy. For example, Proclus (a Greek writer of the fifth century A.D.), in his *Hypotyposis*, ascribes the principle of nested spheres not to Ptolemy, but to "certain people," even though he cites Ptolemy many times in other respects.[89]

Thābit ibn Qurra's *The Almagest Simplified* is a ninth-century example of an Arabic introduction to Ptolemaic astronomy and cosmology. Thābit includes a discussion of Ptolemy's cosmological distance scale, taken directly from the *Planetary Hypotheses*.[90] Thābit's numbers are the same as those in table 7.10, except that Thābit suppresses the gap between the spheres of Venus and the Sun. He does this by making the Sun's least distance equal to 1,079 terrestrial radii. The Sun's greatest distance remains 1,260. Thus, Thābit increases the

thickness of the Sun's sphere to fill the gap, even though this would result
in a solar eccentricity that is far too large. Thābit certainly knew Ptolemy's
Planetary Hypotheses; indeed, he probably made, or revised, the Arabic translation of the *Planetary Hypotheses*. But his use of data derived from the *Planetary
Hypotheses*, in a work that is purportedly an introduction to the *Almagest*,
must have added to the already considerable confusion over the historical
origins of the Ptolemaic system.

An elder contemporary of Thābit, al-Farghānī (ca. 800–870), wrote an
Elements of Astronomy. This elementary survey of Ptolemaic astronomy and
cosmology achieved great popularity. Al-Farghānī seems to have been unaware
of the *Planetary Hypotheses*, but he knew the general principles of the nested-
sphere cosmology. He calculated the greatest and least distances of all the
planets directly from the parameters in the *Almagest*. Al-Farghānī's distances
were therefore a little different from those in Table 7.10.[91] In particular, al-
Farghānī had no gap between the spheres of Venus and the Sun, nor did he
fill it artificially, as Thābit did.

A later book in this same genre was Ibn al-Haytham's (ca. 965–1040)
treatise *On the Configuration of the World*, which provided a survey of astronomy, including an account of the solid-sphere cosmology (though no discussion of the distance scale).[92] Interestingly, although he often cites the *Almagest*,
Ibn al-Haytham never cites the *Planetary Hypotheses*, which he apparently did
not know until somewhat later in his career. Nevertheless, the solid-sphere
cosmology he describes is essentially that of Ptolemy.

Critics of Ptolemy   Although medieval Arabic astronomy remained fundamentally Ptolemaic in its methods and basic assumptions, Ptolemy did have a
number of critics—and more of them as the Middle Ages progressed. The
most common philosophical complaint was that Ptolemy had violated the
basic physical principles of the universe, especially the principle of uniform
circular motion. Ptolemy's introduction of the equant point in his theory of
the planets was often a source of doubt. Ptolemy had included a similar device
in his theory of the motion of the Moon. A good example of skepticism about
these nonuniform motions is found in al-Ṭūsī's (1201–1274) commentary on
the *Almagest*. After describing Ptolemy's lunar model, al-Ṭūsī says, "As for
the possibility of a simple motion on a circumference of a circle, which is
uniform around a point other than the center, it is a subtle point that should
be verified."[93]

Ibn al-Haytham, in addition to his book of Ptolemaic cosmology, *On the
Configuration of the World*, later wrote a more skeptical work called *Doubts
about Ptolemy*. Here he attacked not only the nonuniformity of motion implicit
in the equant, but also the very idea of explaining physically real motions in
terms of artificial geometrical constructs. Ibn al-Haytham doubted that the
motions of real bodies (the planets) could physically be produced by imaginary
lines and planes.[94] Moses Maimonides (1135–1204), in his *The Guide of the
Perplexed*, denied the reality of epicycles and eccentrics, basing his arguments,
like most critics of Ptolemy, on Aristotle's physical principles.[95]

In the later Middle Ages this doubt about Ptolemy's faithfulness to Aristotle
led to concrete proposals for new planetary models. Some of the most original
proposals came from a group of astronomers associated with the observatory
at Marāgha, in northwestern Persia. In the middle of the thirteenth century,
Hūlāgū, the grandson of Genghis Khan and founder of the Ilkhānī dynasty,
conquered most of Persia and Mesopotamia. Hūlāgū was persuaded to found
and support an observatory by the astronomer Naṣīr al-Dīn al-Ṭūsī. The new
observatory at Marāgha was an ambitious undertaking, complete with a library
and a staff of professional astronomers. Around 1272 the astronomers at
Marāgha completed a new *zīj*, the *Ilkhānī Tables*.

Al-Ṭūsī wrote a *Memoir on Astronomy*, which presented a general account and criticism of Ptolemy's astronomy.[96] In this work al-Ṭūsī also proposed some new devices for use in planetary theory. One of them was a proof that a back-and-forth oscillation in a straight line could be produced by a combination of two circular motions. In particular, let one circle roll inside another. If the rolling circle has exactly half the radius of the fixed circle, then a point on the rolling circle will trace out a straight line (a diameter of the fixed circle). This device is today called the "Ṭūsī couple." Al-Ṭūsī and other astronomers of the Marāgha school applied this and other similar devices to construct planetary theories that, while roughly equivalent to Ptolemy's, were physically and philosophically more acceptable.[97]

Perhaps the most ambitious steps in this direction were taken by Ibn al-Shāṭir of Damascus (ca. 1304–1375), who eliminated the equant and replaced it by a minor epicycle. This made it possible to explain the nonuniform motion of the epicycle around the deferent in terms of purely uniform motions.[98] The planetary models of the Marāgha astronomers and of Ibn al-Shāṭir show great cleverness and originality. In the Islamic East, they were the subject of many commentaries and they served as a stimulus to the invention of alternative planetary models from the late thirteenth to the mid-sixteenth century. But they do not appear to have much influenced the direction of practical computational astronomy. Practical computing of planetary positions continued to be done for the most part with standard tables based on Ptolemaic models. However, the constructions of al-Ṭūsī and Ibn al-Shāṭir turn up later in the astronomy of Copernicus, who used their ideas to purge Ptolemy's astronomy of its violations of Aristotelian physics. We shall examine one of these devices in detail in section 7.30 when we study Copernicus's elimination of the equant. How Copernicus learned of them we do not know.

Latin Astronomy

The Early Middle Ages   In the Latin West of the early Middle Ages, astronomy practically ceased to be cultivated. Greek astronomical works were unavailable. The *Almagest* was unknown. The study of astronomy was based almost entirely on a handful of Latin texts of low intellectual quality.[99] Pliny's *Natural History* served as a complete encyclopedia of scientific knowledge. Pliny's book II contained material on the planets, including a discussion of their eccentric deferent circles, but it dated from a time before Ptolemy. In the Middle Ages Pliny therefore represented an out-of-date astronomy and even this the text discussed in only a general way with little technical detail.

Martianus Capella (early fifth century A.D.) wrote an allegory called *The Marriage of Philology and Mercury*, to which he attached seven books on the seven liberal arts.[100] In the early Middle Ages, Capella's book was widely admired as a survey of all the important branches of learning. Capella's *Marriage of Philology and Mercury* has been aptly characterized by W. H. Stahl: "Half classical, half medieval, his work may be likened to the neck of an hourglass through which the classical liberal arts trickled to the medieval world."[101] Capella's book VIII is an introduction to astronomy. Among other things, Capella treats the astronomy of the sphere, including rising and setting times of the zodiac signs (but without being very clear about the latitude to which these apply) and the length of the solstitial day. In his discussion of the Sun, Moon, and planets, he addresses the nonuniform progress of the Sun through the zodiac. In his treatment of the planets, he is disappointingly vague about details of the geometrical models—with one exception. Martianus Capella puts the inferior planets (Mercury and Venus) on epicycles that circle the Sun. Capella's book can be compared with the Greek textbook of Theon of Smyrna. Its scope is similar, but it is less systematic and it often opts for simple arithmetical calculations in lieu of statisfactory discussions of the geometrical models.

A cosmological work of some influence was the Latin translation of part of Plato's *Timaeus* and a commentary thereon by Chalcidius (fourth century A.D.). Plato's work descended from the primitive stage of Greek cosmology—it is pre-Ptolemaic and even pre-Aristotelian in its view of the cosmos. But Chalcidius appended some comments on epicycle theory.[102]

The treatment of astronomy and cosmology in original Latin compositions of the early Middle Ages, such as those by Isidore of Seville (*Etymologies*, seventh century) and Bede (*On the Nature of Things*, eighth century) was slavishly dependent on the protoypes from late antiquity, but mixed in biblical material when it bore on questions of the arrangement of the universe.[103]

The beginnings of a livelier interest in astronomy and cosmology can be seen in the early ninth-century court of Charlemagne. The number of manuscripts touching on astronomical matters increases. And considerable attention is devoted to the *computus*, the art of calendrical reckoning, especially as concerns the luni-solar ecclesiastical calendar.

The Translation Movement   But the real revival of astronomy in the Latin West began only in the twelfth century. A development of paramount importance was the translation of works of philosophy, mathematics, and astronomy from Arabic into Latin. These translations included Greek works (available in Arabic versions) as well as original Arabic treatises. Although many people made translations, one man played a far larger role than anyone else in making Greek and Arabic science available to the Latin West: Gerard of Cremona (ca. 1114–1187).

After Gerard's death, some of his students wrote a memorial to him, compiled a list of all the books he had translated from Arabic into Latin, and appended these to his translation of Galen's *Tegni*. Here is what Gerard's students had to say about how their master came to his task:

> He was trained from childhood at centers of philosophical study and had come to a knowledge of all that was known to the Latins; but for love of the *Almagest*, which he could not find at all among the Latins, he went to Toledo; there, seeing the abundance of books in Arabic on every subject, and regretting the poverty of the Latins in these things, he learned the Arabic language, in order to be able to translate.[104]

Thus, the desire to possess Ptolemy's *Almagest* was a major stimulus to the revival of learning in Europe. Gerard lived at Toledo for many years. He translated some seventy works into Latin, many of which came to serve as the foundations for whole branches of European learning in the next few centuries.

In the philosophy of nature, Gerard translated Aristotle's *Physics* and *On the Heavens*. In astronomy, Gerard's most important translation was, of course, the *Almagest*. But he also translated other works of Greek astronomy, including Theodosius's *On Geographic Places*, Hypsicles' *On Ascensions*, and Autolycus's *On the Moving Sphere*. Gerard also translated a number of Arabic astronomical treatises. The most influential of these were Thābit's *On the Motion of the Eighth Sphere* (which Gerard titled *De motu accessionis et recessionis*, "On the motion of access and recess") and al-Farghānī's *Elements of Astronomy*. Al-Farghānī's version of Ptolemy's cosmology and of the Ptolemaic distance scale thus circulated in Latin Europe from the very beginning of the European revival of learning—but often without Ptolemy's name attached.

During the first wave, practically all the translation was from the Arabic. Only somewhat later did European scholars make many translations directly from Greek into Latin. Perhaps the most active translator from the Greek was William of Moerbeke (ca. 1215–ca. 1286). William translated many of

Archimedes' mathematical works, as well as Aristotle's *On the Heavens*, *Physics*, *Metaphysics*, and *Meteorology*, and much else besides.[105] Thus, in the period of two or three generations, most of the central works of Greco-Arabic astronomy and philosophy of nature became available to Latin Europe.

The Arts Curriculum in the Medieval Universities   Greek philosophy and science came over the Pyrenees like a storm. The figure who commanded the most attention was certainly Aristotle. Many teachers at the newly established universities took up Aristotle with enthusiasm. But the ancient philosophical writers posed grave risks for Christian readers. Most dangerous was Aristotle's doctrine of the eternity of the world. Like most of the ancient Greeks, Aristotle held that nothing can come from nothing. He therefore proved in several different ways that the cosmos always existed and that it cannot pass out of existence. This contradicted the biblical account of the creation of the world by God. A series of crises developed in which Church authorities tried to clamp down on the unruly masters who wanted to teach Aristotle. Good examples are provided by events at Paris in the thirteenth century. In 1210 Aristotle's works on natural philosophy were condemned. The teachers of the arts faculty at Paris were forbidden to read them either in public or in private, under penalty of excommunication.[106]

Despite of this and other efforts to suppress or expurgate Aristotle, by the fourteenth century Aristotle had decisively won. In one of the most remarkable feats of mental gymnastics known to history, several generations of theologians and university professors made the pagan philosophy of nature compatible with Christian theology. The works of Aristotle now formed the core curriculum of the universities.

The curriculum for the bachelor of arts degree became more or less standardized all across Europe. Traditionally it was divided into courses at two levels. The lower-level sequence was called the *trivium*, because it consisted of three parts. (From this derives our word *trivial* for something very easy.) The courses of the trivium were grammar, logic, and rhetoric. The second tier of courses was called the *quadrivium*, because it consisted of four parts: arithmetic, geometry, music theory, and astronomy. Thus, every European city with a university was required to have a scholar who could teach the rudiments of astronomy. These seven liberal arts represent medieval revivals of the school curriculum of late antiquity. These seven had been treated as canonical by Martianus Capella. But in fact they have far older roots. The four mathematical sciences were already recognized as standard divisions by the ancient Pythagoreans.

Latin Textbooks of Astronomy   Although astronomy was a part of the standard university curriculum, it was taught at a very rudimentary level. One essential text was the *Sphere* of Sacrobosco, which treated the theory of the celestial sphere.[107] The professor of astronomy might typically read portions of the text to his students and make demonstrations on a wooden armillary sphere. After an introduction to the celestial sphere, the students might next be given a nontechnical introduction to the planets, including the theory of deferents and epicycles. For this, a commonly used text was the *Theorica planetarum*, an anonymous thirteenth-century text that is sometimes attributed (with inadequate evidence) to Gerard of Cremona.[108] The physical and philosophical principles of the worldview were based on readings from and commentaries on Aristotle's *On the Heavens*.[109]

A medieval university student who had been put through this curriculum would not actually know how to do anything in astronomy, but at least he would have a general introduction to the traditional cosmos of Aristotle and Ptolemy. Of course, in the process of transmission, Christian trappings were added to the pagan cosmos. We have seen (fig. 6.16) that the nested celestial spheres of Aristotle and Ptolemy were embedded in an empyrean sphere, which was the habitation of God. The ancient Greeks had believed that the planets were living, divine things. Ptolemy, in the *Planetary Hypotheses*, therefore conjectured that the planets moved by their own wills, each planet regulating the rotations of its own multiple orbs. In the Middle Ages, the planets lost their divinity and became subject to a single God. But Aristotle's physics required that the motion of every orb be produced by its own unmoved mover.[110] In many Christian commentaries, a compromise is reached that adopts Aristotle's opinion while subjecting all the orbs to a Christian worldview: each of the multiple orbs in each planet's system is turned by an angel of God.[111]

Practical Astronomy   Although the university curriculum scarcely prepared a student for real work in astronomy, practical astronomy did flourish both inside and outside of the universities. The courts of kings and princes often provided patronage for astronomers, who could apply their art to the calculation and interpretation of horoscopes. Astrology verged on heresy, because it seemed to deny human free will and even to call into doubt the omnipotence of God. It was attacked on these grounds by a number of Christian writers.[112] In the early Christian Middle Ages, astrology was not widely practiced. But with the acquisition of astronomical and astrological texts from Spain, astrology grew rapidly in popularity. The fifteenth and sixteenth centuries represent the peak of its popularity and influence.

Although there were always scholars who were interested in understanding the motions of the planets either for their own sake or for insights into God's creation, astrology was widely perceived as the most important practical application of astronomy. And astrology was undoubtedly the greatest stimulus for the copying and refinement of planetary tables. Al-Khwārizmī's tables were translated into Latin by the twelfth century. They were soon superseded by the *Toledan Tables*, which were compiled in Islamic Spain in the eleventh century and translated into Latin by the twelfth. Thus, the earliest planetary tables to circulate in Latin Europe were translations of various Arabic *zījes*.

Planetary theory was an arcane art. Understanding the use of tables separated the master from the dilettante. The astrological application of planetary tables invested the astronomer with an aura of power and mystery. This aspect of medieval European astronomy is clearly reflected in the poetic works of Chaucer. In "The Franklin's Tale," one of Chaucer's central figures is a magician-astrologer of Orleans. Chaucer describes his apparatus and his learning in these terms:

> *His tables Toletanes forth he broght*
> *Ful wel corrected, ne ther lacked noght,*
> *Neither his collect ne his expans yeres,*
> *Ne his rotes ne his othere geres,*
> *As been his centres and his arguments,*
> *And his proporcionels convenients*
> *For his equacions in every thing;*
> *And, by his eighte spere in his wirking,*
> *He knew ful wel how fer Alnath was shove*
> *Fro the heed of thilke fixe Aries above*
> *That in the ninthe speere considered is;*
> *Ful subtilly he calculed al this.*[113]

The "tables Toletanes" are of course the *Toledan Tables*. The lines that follow contain detailed references to the use of the tables, couched in technical astronomical jargon. The "rotes," or roots (Latin, *radix*, plural *radices*) are the initial or epoch values of the angles. "Centre" is what we have called the eccentric anomaly. Similarly, "argument" is what we have called the epicyclic anomaly. The sixth and seventh lines of this passage refer to the proportional parts (i.e., the interpolation coefficient of table 7.7) used in calculating the equation of the epicycle. The last five lines are, of course, a detailed reference to trepidation theory. Alnath is Chaucer's name for α Arietis, the brightest star of the constellation Aries. Knowing how far this star was "shove" from the head of the fixed sign of Aries is equivalent to knowing the equation of trepidation.

The first planetary tables of major significance to originate in Christian Europe were the *Alfonsine Tables*, compiled in Spain around A.D. 1270 under the patronage of Alfonso X, king of Castile. The original Spanish version of the tables does not survive. However, by the 1320s, the *Alfonsine Tables* had arrived in Paris. There they were reworked into more convenient form. Also, several versions of canons were written by astronomers at Paris. The Parisian version of the *Alfonsine Tables* spread rapidly and soon became the standard set of tables everywhere in Christian Europe.[114]

Peurbach and Regiomontanus The time from the twelfth to the fourteenth century was one of gradually increasing activity in astronomy. However, European astronomy remained thoroughly Ptolemaic in all essentials. Although some creativity was shown in the construction of new planetary tables and in the design of new types of astronomical instruments, the astronomy of fourteenth-century Europe was not terribly original. Of the students who completed the arts curriculum at a university, only a small number understood astronomy well enough to use planetary tables. Of this small number, only a tiny fraction was competent to do any original work—for example, to make useful observations or to redesign a planetary table.

In the fifteenth century, the intellectual level of European astronomy rose significantly. In this development, two scholars played key roles, Georg Peurbach (1423–1461) and his student Johann Müller (1436–1477). As we have seen, Spain was the center from which astronomy was disseminated into twelfth-century Europe. By the fourteenth century, the center of actvity had shifted to Paris and, to a lesser extent, to England. In the fifteenth century, the German-speaking lands of central Europe were the focus of the most original and significant work.

Georg Peurbach was an Austrian who received his master's degree at the University of Vienna in 1453. Peurbach served as court astronomer (i.e., astrologer) first to King Ladislaus V of Hungary and later to the German emperor Frederick III. Peurbach also held a chair at the University of Vienna, where he lectured on the classics.[115]

One of the most important works for the dissemination of Ptolemaic astronomy and solid-sphere cosmology in the early Renaissance was a popular textbook written by Peurbach called *Thoricae novae planetarum*.[116] Peurbach composed his text for a series of lectures that he gave in Vienna in 1454. The astronomy was standard Ptolemaic planetary theory. For the solid-sphere version of Ptolemy's cosmology, Peurbach drew on some Arabic source in Latin translation—probably Ibn al-Haytham's *On the Configuration of the World* or a work derived from it. He called his work "NEW Theories of the Planets," not because it contained any new theories, but because he meant it as a replacement for the rather sloppy and unsatisfactory thirteenth-century *Theorica planetarum*, mentioned above. Manuscript copies of Peurbach's work circulated around the universities, but it was not printed until 1472, after Peurbach's death.

Peurbach's work became enormously popular. It was frequently reprinted and was widely used as an elementary university text. No fewer than fifty-six editions, including translations and commentaries, were published between 1472 and 1653. Figures 7.51 and 7.52 are taken from an edition published in 1553, with commentary by Erasmus Reinhold. Copernicus's great book, *On the Revolutions of the Heavenly Spheres*, introduced the new Sun-centered cosmology in 1543. Thus, Ptolemy's cosmology remained in the standard university texts right up to the time of Copernicus, and even afterward.

Two people were destined to have a great influence on Peurbach's life. The first of these was his student Johann Müller, who enrolled at the university of Vienna in 1450 at the age of thirteen. Müller's home town was Königsberg ("King's Mountain"). In the Humanist style, in his own published works he Latinized his name as Joannes de Regio monte. Thus, he has come to be called Regiomontanus. As a student of Peurbach, Regiomontanus kept a notebook in which he copied out Peurbach's *Theoricae novae planetarum*. But soon he became a collaborator with Peurbach in observing eclipses and calculating ephemerides. Regiomontanus was eventually to far outshine his teacher.[117]

The second major influence on Peurbach's astronomical work was Cardinal Johannes Bessarion. In 1460, Bessarion was sent to Vienna by Pope Pius II on a diplomatic mission to smooth out difficulties between Emperor Frederick III and his brother Albert VI of Styria. Bessarion also sought support for a military campaign to recapture Constantinople, which had fallen to the Turks in 1453.

At Vienna, Bessarion met both Peurbach and Regiomontanus. Bessarion was a Greek and was keenly interested in promoting the study of the classics of Greek literature, philosophy, and science. Bessarion himself had a fine collection of manuscripts. He impressed on Peurbach the need for a better Latin translation of the *Almagest* than Gerard of Cremona's version from the Arabic. Peurbach did not read Greek but, according to Regiomontanus, he knew the *Almagest* almost by heart. Bessarion convinced Peurbach to undertake an abridgment of and commentary on the *Almagest* that, Bessarion hoped, would be useful as an advanced textbook of astronomy. Working from Gerard's twelfth-century Latin version, and making use of a rudimentary commentary on Ptolemy then in circulation, Peurbach immersed himself in the task. Peurbach had just reached the end of book VI of Ptolemy's thirteen books when he died in April, 1461, aged only 38. On his deathbed, he extracted a pledge from Regiomontanus to complete the task.

At the end of 1461, when Bessarion returned to Rome, Regiomontanus went with him. Regiomontanus learned Greek and he carried on with Peurbach's abridgment and commentary. This work, the *Epitome of the Almagest*, was completed probably by 1463, though it was not printed until 1496, some twenty years after Regiomontanus's own premature death.

A whole generation of Europeans learned their technical astronomy from the Peurbach-Regiomontanus *Epitome*. It was far more than a mere condensation of the *Almagest*. Regiomontanus was the first European in the Renaissance of astronomy who could face Ptolemy as an equal. Regiomontanus explicated the more difficult derivations in Ptolemy, found alternative ways to do many computations, and added new observations. Regiomontanus also did not hesitate to criticize Ptolemy. He pointed out that, according to Ptolemy's lunar theory, the angular diameter of the Moon should change by a factor of two in the course of the month, which is far greater than the variation actually observed. Arabic astronomers had earlier made the same criticism, but this appears to be the first mention of it in European astronomy.

In 1471, Regiomontanus settled in Nuremberg. He set up a printing press in own house and started a business of publishing mathematical and astronomical books. Other printing establishments had been reluctant to take on scientific works, which could be expensive to produce and risky to market. Regiomontanus's enterprise thus filled a gap. His was the first printing establishment in history that was dedicated to scientific works. The first book off the press was the *New Theories of the Planets* of his deceased friend and teacher, Georg Peurbach.

The second item to be published was Regiomontanus's own *Ephemerides*. This book, printed in 1474, gave the positions of the Sun, Moon, and planets for every day from 1475 to 1506. Ephemerides, calculated from standard Ptolemaic planetary tables, had circulated in manuscript form, for they were essential to the practice of astrology, but Regiomontanus's was the first such work to be printed. Columbus carried a copy of it on his fourth voyage and used its prediction of a lunar eclipse for February 29, 1504, to frighten some natives of Jamaica into supplying food for his men, then in desperate circumstances.[118]

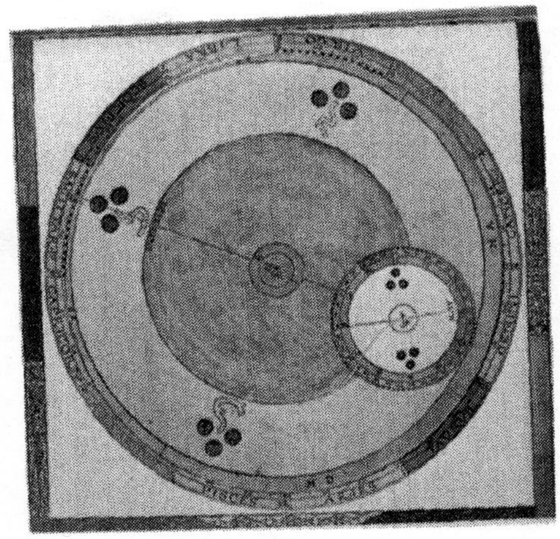

Regiomontanus was also a talented and original mathematician. His book *On All Classes of Triangles* (*De triangulis omnimodis*) was the first stand-alone textbook of trigonometry in the European tradition. Earlier, trigonometry was always treated as a preliminary portion of astronomy. This book, which contained a number of original contributions to trigonometry by Regiomontanus himself, was not printed until 1533.

While Regiomontanus was a capable theoretical astronomer and a good observer, he worked squarely in the Ptolemaic tradition. His most significant work was his masterful *Epitome of the Almagest*, which helped to make European astronomy a living, vital science once again, rather than a revered body of ancient wisdom. At the close of the fifteenth century, European astronomy at last approached the level achieved by the Greeks of the second century. But Renaissance science was endowed with a vitality that far exceeded that of late antiquity. From university professors to court astrologers, from book publishers to instrument makers, there were now hundreds of Europeans engaged with astronomy in a serious way—far more than there had been at any stage of Greek civilization. This new vitality was soon to bring about a revolution in cosmology.

FIGURE 7.56. The equatorium for Saturn in Schöner's *Aequatorium astronomicum* of 1534. By permission of the British Library (Maps c.2.d.10, fol. A6).

7.27 PLANETARY EQUATORIA

If one has frequent occasion for computing planetary positions and if one can tolerate errors of a few degrees, it may be worth the trouble to make a concrete model of paper or wood that can be manipulated to solve problems. Such a concrete model, which functions as a specialized analog computer, is called an *equatorium*. The Ptolemaic slats that we used in section 7.21 provide a modern example of an equatorium. The name of these devices signifies that they supply the *equation*—the difference between the planet's actual and mean positions. The essential feature of an equatorium is that it takes account of the nonuniformity of the planet's motion but nevertheless eliminates the need for trigonometry in making predictions.

Hipparchus's eccentric-circle solar theory is easily realized in concrete form by drawing a solar circle (divided into days of the year) eccentric to a zodiac scale (divided into degrees and signs). As we saw in section 3.7 and figure 3.41, the solar equatorium was a common feature on the backs of European astrolabes in the Middle Ages.

Figure 7.56 shows an equatorium for Saturn, designed by Johann Schöner, published in 1521 and reprinted in unmodified form in 1534. It is easy to identify the main features of the Ptolemaic model. The epicycle with center *A* rides on the deferent circle (DEFERENS). The equant point, the center of the deferent, and the Earth are at *K*, *C*, and *D*, respectively. Strings are attached at *A*, *K*, and *D* as an aid in reading angles. Surrounding everything, and concentric with the Earth *D*, is the zodiac. This equatorium has three movable parts: the epicycle wheel can be turned, as can the deferent wheel, and a circle lying underneath the deferent wheel (which is turned to set the apogee). These movable paper wheels are called *volvelles*.

Tradition of the Equatorium

It is likely that the first planetary equatoria were made by the ancient Greeks, although no specimen or description of a planetary equatorium survives from their time. In his introduction to the *Handy Tables*, Theon of Alexandria explains how to "calculate" positions of the planets according to Ptolemy's theory by drawing scale diagrams.[119] It is but a short step further to make the diagrams reusable, by adding moving parts.

For a genuine equatorium, though only of the Sun, the earliest attestation is that of Proclus (fifth century A.D.). In his *Hypotyposis*, Proclus gives directions for making a solar equatorium.[120] On a wooden board or a bronze plate, one is to draw a zodiac circle and, within it, an eccentric circle, divide both into degrees, and so on. Proclus's astronomical work is a *hypotyposis*, that is, an "outline" or "sketch" of astronomical hypotheses, based on Ptolemy. There is very little original astronomy in Proclus. His idea for a solar equatorium is almost certainly borrowed.

The earliest extant descriptions of planetary equatoria turn up in medieval Spain.[121] In the thirteenth century, the Christian king Alfonso X of Castile acted as patron for a range of scholarly activities with a substantial emphasis on astronomy. One product of his patronage was the collection called *Libros del Saber de Astronomía* (Books of the knowledge of astronomy).[122] This compilation includes translations into Castilian of two eleventh-century Arabic texts on equatoria. The first is a text by Ibn al-Samḥ of Granada (early eleventh century). The second is a treatise by al-Zarqālī (middle of eleventh century), known to medieval Europeans variously as Arzachel, Azarchel, or Azarquiel.[123] Al-Zarqālī is a major figure of medieval astronomy. He is best known for his canons to the *Toledan Tables*.

Equatoria entered medieval Latin astronomy right along with Ptolemaic planetary theory. The first comprehensive introduction to planetary theory written in the Latin West was the *Theorica planetarum* (Theory of the planets) of Campanus of Novara (thirteenth century).[124] The bulk of Campanus's book is devoted to directions for building an equatorium. The instrument resembled an astrolabe, in that the plates for the several planets could be stacked in a single "mother." Disks, presumably of wood, turned in cavities cut into other wooden disks. Campanus's instrument would certainly have worked, but the practical problems posed by its construction have led some to doubt that it was ever built as described. Nevertheless, Campanus's book was the foundation of the equatorium tradition in Latin Europe. Its influence can be seen on nearly all that followed. Campanus himself probably drew on some Arabic treatise, most likely from Spain, but his specific source cannot be identified.

The best manuscripts of Campanus's *Theorica planetarum* contain figures illustrating the parts of the equatorium. Some contain working paper or parchment equatoria with movable volvelles. Makers of equatoria in paper and parchment tended, quite sensibly, to simplify matters by devoting a separate instrument to each planet. Those in the Campanian design are characterized by a stolid fidelity to the geometrical diagram. The Campanian instruments are also characterized by fixed apogees, and thus must be drawn for a particular century.

Although most medieval equatoria were simple paper or parchment constructions, there do exist examples constructed of wood. The most remarkable is that at the monastery of Stams, in the Austrian Tyrol.[125] This wooden equatorium was built in 1428 by Rudolfus Medici, a canon of Augsburg. It is in the form of a table, about 3.40 × 1.13 m. It is divided into three panels, each about 1.13 m square. One panel carries the instruments for Saturn and Jupiter, another those for Venus and the Moon, while the third is devoted to Mercury and Mars. Although the instruments are not exactly of Campanian design, they are, remarkably enough, constructed of wooden disks and rings that fit into recesses in other disks—which was the construction that Campanus recommended. So, perhaps one should not be so sure that Campanus never built the instrument he described.

The oldest text on planetary equatoria in the English language is the anonymous fourteenth-century treatise, *The Equatorie of the Planetis*, which exists in a unique copy in Peterhouse College, Cambridge. Derek J. Price argued—convincingly to many—that it was composed and written by Geoffrey Chaucer. The ascription to Chaucer is based on comparisons of handwriting samples and other evidence.[126] The *Equatorie of the Planetis* gives directions for building and using an equatorium of formidable size—six feet in diameter, so that all the planets may be handled on one large disk. Whether it was ever constructed we do not know. In any case, it is remarkable that the equatorium is present from the very beginning of scientific astronomy in English.

The usual contents of a medieval equatorium treatise consist of directions for building the device and directions for using it. Although there was nearly universal agreement on the details of the underlying planetary theory (Ptolemy's), designers of equatoria had room for individual differences, and hence for creativity, in the physical realization of the theory in wood or paper. Most early equatoria were essentially movable plane diagrams of the Ptolemaic theory. The chief construction problem for the theory of Venus and the superior planets involved the three centers (fig. 7.32): the center of the epicycle must move on a circle whose center is C, but the protractor for measuring mean longitudes must be centered on E, and the protractor for measuring the actual longitude of the planet must be centered at O. The history of equatoria is largely the history of concrete solutions to the difficulties posed by these requirements.

Schöner's Equatoria

The earliest printed equatoria are those of Johann Schöner (1477–1547), a German writer and publisher of astronomical and geographical works. Schöner is also known to historians of cartography for his series of globes, which kept abreast of the latest discoveries in the age of exploration. In his *Aequatorium astronomicum* (actually a series of publications), Schöner provided the first printed planetary equatoria. The user was expected to cut out and assemble the parts provided on Schöner's parts sheets. The assembled equatoria could then be used to predict the positions of the planets according to standard Ptolemaic theory.[127]

Schöner's *Aequatorium astronomicum* of 1521 was printed in large format, the instruments being about 27 cm in diameter. There were nine instruments: an instrument for the motion of the eighth sphere (trepidation), one equatorium each for finding the longitudes of the Sun, the Moon, and the five planets, and a final instrument for reckoning conjunctions and oppositions of the Sun and Moon (useful in eclipse analysis). Figure 7.56 is a photograph of the equatorium for Saturn in a copy of the edition of 1534. The instruments were all hand painted with water colors.

The influence of Campanus is to be seen in minor things. For example, Schöner uses the same letters (D, C, K, A) as Campanus used to label the Earth, the center of the deferent, the equant point, and the center of the epicycle. But Schöner went well beyond Campanus in several respects. Most significantly, Schöner provided for movable apogees. Schöner also found a clever way to eliminate the need for a graduated circle concentric to the equant point. One pulls out a string from the Earth to measure angles on a scale concentric to the Earth. An equal angle measured at the equant point may be set up simply by pulling out the string from the equant point and making it parallel, as judged by the eye, to the string from the Earth. This solution of the old problem is elegant and effective.

On the backs of the equatorium pages are printed tables of mean motions, usually with a given table facing its corresponding planet. (Schöner's tables of mean motion were based on the *Alfonsine Tables*.) Thus, with the brochure open, the tables for Saturn may be seen at the same time as the equatorium for Saturn.

To find the position of Saturn, one proceeds as follows. First, one must find the longitude A of Saturn's apogee. This is done using the equatorium for the eighth sphere and the corresponding tables. Next, one calculates $\bar{\lambda}$, the mean longitude of Saturn, using Schöner's tables. One calculates also $\bar{\lambda}_\odot$, the mean longitude of the Sun from the tables. Then $\bar{\mu}$, the mean epicyclic anomaly of Saturn, is found by subtraction: $\bar{\mu} = \bar{\lambda}_\odot - \bar{\lambda}$ (see sec. 7.12 and fig. 7.19). The calculations require of the user only addition and subtraction. The trigonometry is all performed by the equatorium. One sets the three volvelles at the angles just found, and it is a simple matter to read off the longitude of the planet.

Over the next few years, Schöner published a set of canons for the use of the equatoria, accompanied by worked examples, as well as a more convenient set of tables of mean motions. Because of the successive publication of the various brochures connected with Schöner's *Aequatorium astronomicum*, it is now unusual to find all of them together.

A few years after Schöner's death, his son, Andreas, collected his mathematical and astronomical works and reprinted them, with a few new items, in one large volume, the *Opera mathematica* (1551). This included a substantially reworked edition of the *Aequatorium astronomicum*. Now, for the first time, the equatoria, directions for their assembly, canons, tables, and illustrations showing the use of the equatoria were gathered into one place. New blocks were cut for the parts of the equatoria, in a smaller format. (The equatoria are about 15 cm in diameter, rather than the 27 cm of the first edition.) In section 7.28, the reader will have the opportunity to assemble and use a facsimile of Schöner's equatorium for Mars, based on the edition of 1551.

The universally acknowledged masterpiece of the Renaissance printed equatorium is the *Astronomicum Caesareum*, designed and printed by Petrus Apianus at Ingolstadt in 1540.[128] It has been described as the most sumptuous scientific book ever published. Apianus did not wish to trouble his readers with even performing arithmetic. Thus, the instruments involve multiple additional volvelles. For example, to compute the mean longitude, one looks up in a table the value for the beginning of the century and sets one wheel appropriately. The additions to the mean longitude for the number of whole years elapsed and for the odd months and days are performed by turning one volvelle with respect to another, as in using a circular slide rule. Because of the extra volvelles, Apianus's equatoria appear complicated. But the underlying plantary theory is still strictly Ptolemaic (plus, of course, an Alfonsine trepidation theory).

7.28 EXERCISE: ASSEMBLY AND USE OF SCHÖNER'S *AEQUATORIUM MARTIS*

Using the parts and directions provded here, the reader can assemble a working equatorium for Mars. The parts are reprinted from the *Aequatorium astronomicum* of Johann Schöner, in *Opera mathematica Ioannis Schoneri Carolostadii* (Nuremberg, 1551), courtesy of Cambridge University Library.

Assembly of the Equatorium

1. Photocopy the parts of the equatorium in figures A.6 and A.7. Glue the photocopies to sheets of heavy paper to provide extra strength. To avoid stretching and puckering, use a glue designed for mounting photographs. Some glues of this kind come in stick form.

2. From the photocopy of figure A.7, cut out the four circles numbered 1–4. Label each on its back with its number for future reference.

3. Using scissors, carefully cut out the small (1/2″ diameter) circle in the center of the deferent circle (circle 3). The 1/2″ diameter circle (bearing points K, C, D) thus cut out should be saved. Trim the 1/2″ circle a little so that it will easily fit back into the hole from which it was removed. The 1/2″ circle should turn smoothly, but without too much extra space, inside the hole.

4. Glue the 1/2″ circle cut out in step 3 onto a small scrap of thick paper or manilla file folder. The point of this is to increase its thickness slightly. When the glue has dried, trim around the edges so that the new, bottom layer matches the original circle well. This assembly (consisting of the 1/2″ diameter circle built up in thickness) will be called "the spindle."

5. Using a pin or needle, poke a hole all the way through point D of the spindle. Poke a hole through point D in the center of circle 1. Poke a hole thorough points D and K on circle 4. Poke a hole through point D in the center of the baseplate of the instrument (labeled "Aequatorium Martis").

6. Glue the spindle to circle 1. Points D on the two circles must coincide and the lines through D and C must also match up on the two circles. A good way to guarantee this is to put a needle through hole D in the spindle and then through the corresponding hole D of circle 1 before gluing the circles together. Make sure that the glue goes all the way to the edge of the spindle, so that the spindle is completely bound to circle 1.

7. Using a needle or pin, poke a hole through the center A of the epicycle (circle 2). Also poke a hole through point A on circle 3.

8. Attach circle 2 to circle 3 in the following way. Pass a piece of heavy thread through hole A in circle 2 and then through hole A in circle 3. (Both circles should be face up, with 2 on top of 3.) Tie a knot about 4″ from the end of the thread on the circle-2 side of the assembly. Pull the thread from the back side, so that the knot is snug against circle 2. Trim the thread to about 1/2″ on the back side of the instrument (i.e., the circle-3 side). Then glue the 1/2″ end to the back of circle 3. It may be helpful to glue a small scrap of paper over the 1/2″ end of thread to keep it firmly attached to the back of circle 3.

When you have finished, the epicycle 2 should turn freely about point A while remaining snugly attached to circle 3. There should be about 4″ of thread hanging from point A of the epicycle.

9. Join circle 1 to the baseplate (fig. A.6) in the following way. Place circle 1 face up on the baseplate. Pass a length of heavy thread through hole D of the spindle that has already been glued onto circle 1. Then pass the thread through hole D of the baseplate. Tie a knot about 4″ from the end of the thread on the top side (i.e., on the spindle and circle 1 side). Pull the thread from the back so that the knot is snug against the spindle. Trim the thread to about 1/2″ on the back of the instrument. Glue the 1/2″ length to the back of the baseplate. It may be helpful to glue a small square of paper over the 1/2″ length to keep the thread firmly attached to the back of the baseplate.

When you are finished, circle 1 should turn freely about D but should be held fairly firmly against the baseplate. There should be about 4″ of thread hanging from D.

10. Take a 4″ or 5″ length of heavy thread and pass it through hole K of circle 4, so that about 1/2″ projects through to the back side. Glue the 1/2″ length to the back side of circle 4. There should be about 4″ hanging free on the front side.

11. Place circle 3 on top of circle 1 so that the spindle on 1 fits into the hole in 3. (The thread hanging from point D of the spindle should, of course, be brought through the hole in 3.) Do not glue circle 3 to circle 1. Circle 3 must be free to turn around the spindle.

Attach circle 4 to the spindle on circle 1 in the following way. Take the end of the thread hanging freely from point D of the spindle and bring it through hole D of circle 4 from the back side. Coat the top of the spindle with glue. Coat the top of the spindle completely, but do not get any glue onto circle 3. Place circle 4 onto the spindle, and push it down firmly. Make sure that line DK on circle 4 lies exactly over line DK of the spindle. (This is important. Note that line DK on the spindle will coincide with line CD on circle 1. The ends of this line on circle 1 will be visible and may be used as an aid in aligning circle 4.)

When you are finished, circle 4 will act as a cap to keep circle 3 in place. Circle 3 should turn freely beneath the cap about C as center. Circle 1 should turn freely about D. When circle 1 is turned, it should carry the cap (circle 4) with it.

Using the Equatorium

Before manipulating the equatorium, the user must calculate the values of the three angles necessary for setting the three circles to their positions. In the Middle Ages and the Renaissance, these calculations were facilitated by tables of mean motion. These tables permitted the calculation of the angles by a series of additions: tabular values might be selected for the century, the year, the month, and the day in question and added up. The tables of mean motion thus eliminated the need for multiplication, which was a tedious procedure whenever numbers contained many digits. Today, in the age of the hand calculator, it is quicker to perform a multiplication than a long series of additions. The precepts given here therefore presuppose the use of a calculator.

In only one significant respect have we departed from Renaissance practice: we have suppressed the trepidation of the equinoxes. With trepidation included, the Martian apogee would advance at a variable rate rather than a steady one. We shall use the modern Ptolemaic parameters for Mars, given in table 7.4, except that we adopt a value $f_A = 1.807°$/century for the rate of advance of the line of apsides, which is somewhat faster than the precession rate f_p given in table 7.4. (See n. 73 for a discussion.) Schöner's own parameters were borrowed from the *Alfonsine Tables*.

Worked Example   Find the longitude of Mars at Greenwich noon on May 30, 1982.

Steps A and B give the preliminary calculations that must be performed. Steps 1–7 describe the manipulation of the equatorium.

A. Determine $\Delta t = t - t_0$, the number of days elapsed between the epoch t_0 and the desired date t. If the desired date is after the epoch, Δt will be positive, but if the desired date is before the epoch, Δt will be negative. Divide Δt by 36,525 to determine the number of Julian centuries elapsed since epoch. Denote this ΔT.

From the tables for Julian day number (tables 4.2–4.4), we find

May 30, 1982 Greenwich mean noon J.D.	244 5120.0
Subtract the day number of the epoch	−241 5020.0
Δt	3 0100.0

$$\Delta T = \Delta t / 36{,}525$$

$$= 30{,}100 \text{ days} / 36{,}525$$

$$= 0.82 \text{ century}$$

B. Calculate the longitude A of the apogee, the mean longitude $\bar{\lambda}$, and the mean epicyclic anomaly $\bar{\mu}$:

$$A = A_0 + f_A \Delta T$$

$$\bar{\lambda} = \bar{\lambda}_0 + f_\lambda \Delta t$$

$$\bar{\mu} = \bar{\mu}_0 + f_\mu \Delta t$$

If $\bar{\lambda}$ or $\bar{\mu}$ should be greater than 360°, subtract as many complete cycles of 360° as required to obtain angles that lie between 0 and 360°. If either $\bar{\lambda}$ or $\bar{\mu}$ should be negative (as can happen if the desired date is before the epoch), add as many complete cycles of 360° as required to obtain angles between 0 and 360°. Thus, we have

$$A = 148.33° + 1.807°/\text{century} \times 0.82 \text{ century}$$

$$= 149.8°$$

$$= 4^s\ 29.8° \ (^s \text{ for "signs," each sign being } 30°)$$

$$= 29.8° \text{ within Leo.}$$

$$\bar{\lambda} = 293.55° + 0.524\ 071\ 16°/^d \times 30{,}100^d$$

$$= 16{,}068.1°$$

$$\underline{-15{,}840.0°} \qquad \text{Less 44 complete circles } (44 \times 360°)$$

$$228.1°$$

$$= 7^s\ 18.1°$$

$$= 18.1° \text{ within Scorpius.}$$

$$\bar{\mu} = 346.15° + 0.461\ 576\ 18°/^d \times 30{,}100^d$$

$$= 14{,}239.6°$$

$$\underline{-14{,}040.0°} \qquad \text{Less 39 complete circles } (39 \times 360°)$$

$$199.6°$$

$$= 6^s\ 19.6°.$$

For the manipulation of the instrument, refer to figure 7.57. The circled numbers in figure 7.57 are keyed to the steps described here.

1. Set the apogee of the deferent to longitude A on the zodiac. The apogee of the deferent is the point labeled AUX on circle 1. (*Aux* is the medieval Latin term for the apogee.) In our example, $A = 29.8°$ within Leo.
2. Pull out the string from the Earth D through the point of the zodiac corresponding to the mean longitude $\bar{\lambda}$ (18.1° within Scorpius). (The medieval term for this angle is *medius motus*, the "mean motion.")
3. Pull out the string from the equant K so that it is parallel to the string from D. (Once the string from K is properly placed, it is no longer necessary to hold onto the string from D.)
4. Turn the deferent circle (labeled DEFERENS) until the center A of the epicycle lies directly under the string through K.
5. Turn the epicycle about its own center A until the AUX (or apogee) of the epicycle lies also under the string through K. Thus, these three

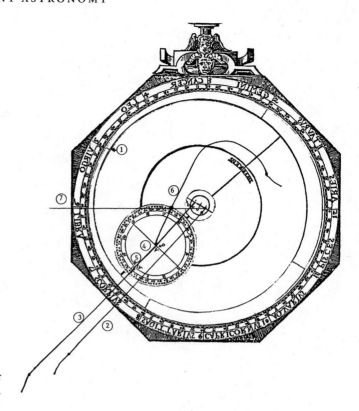

FIGURE 7.57. Manipulation of Schöner's equatorium for
Mars in the version of his *Opera* of 1551.

points will be in a straight line, in the order *K*, *A*, Aux of the epicycle.
(Once the epicycle is properly positioned, it is no longer necessary to
hold onto the string from *K*.)

6. Pull out the string from the epicycle's center and set it at the point on
the rim of the epicycle corresponding to the mean epicyclic anomaly.
(The medieval term for this angle is *argumentum medium*, the "mean
argument.") In our example, $\bar{\mu} = 6^s\ 19.6°$, as indicated in figure 7.57.
The position of Mars is at the outer edge of the dotted circle on the
epicycle (i.e., the one divided into 2° steps) at the place indicated
by $\bar{\mu}$.

7. Now take the string from the Earth *D* and pull it through Mars's
position on the epicycle. The true longitude of the planet in the zodiac
is then read at the place where the string from *D* cuts the zodiac circle.
The longitude of Mars may then be read off as 1° within Libra, or 181°.

Problems

Choose several dates (from table 7.1) for which you know the actual longitude
of Mars. For each of these dates, use Schöner's equatorium to work out the
longitude of Mars according to the Ptolemaic theory. How well did you,
Schöner, and Ptolemy do?

7.29 GEOCENTRIC AND HELIOCENTRIC PLANETARY THEORIES

Modern readers often focus on the Earth-centered nature of the ancient
planetary theory. But, for accurate astronomical prediction, it makes no difference whether the Earth goes around the Sun or the Sun goes around the
Earth. The object taken to be at rest merely reflects the choice of a reference
frame. Sun-centered theories are therefore not intrinsically any more accurate

than Earth-centered theories. The accuracy of a theory depends on the technical details.

Nevertheless, as we have seen, the Sun plays a singular role in Earth-centered planetary theory. In the case of a superior planet, the radius vector of the planet on the epicycle remains parallel to the line from the Earth to the mean Sun. In the case of an inferior planet, the line from the equant point to the epicycle's center remains parallel to the line from the Earth to the mean Sun. Figures 7.34 and 7.35 illustrate these relations in detail. Figure 7.29 shows the general idea but with the picture simplified by suppression of the eccentricities. To put things more simply yet, the Sun controls the motion of a superior planet on its epicycle and the motion of an inferior planet on its deferent. These connections provide the crucial hints that the Sun is actually at the center of the whole system. Let us see just how Sun-and Earth-centered models are related.

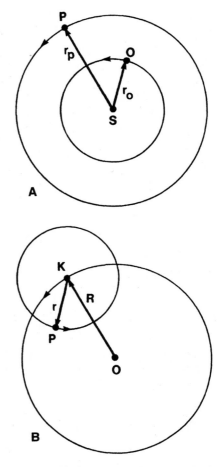

FIGURE 7.58. Superior planet: transformation from the Sun-centered theory (A) to the Earth-centered theory (B).

The Relation of Heliocentric and Geocentric Models

Superior Planet   The discussion is simplified by ignoring the eccentricities. Figure 7.58A shows the Sun-centered theory of a superior planet, such as Mars. The planet P and the Earth O both orbit the Sun S. Thus, vectors **SO** and **SP** both turn about S. The line of sight from the Earth to P is in the direction of vector **OP** = −**SO** + **SP**. But these vectors may be added in either order. Thus, we may also write **OP** = **SP** + −**SO**. The geometry corresponding to the second form of the vector addition is shown in figure 7.58B. Starting from O, we lay out vector **OK**, equal in magnitude and direction to **SP** in figure 7.58A. From K we lay out **KP**, equal in length and opposite in direction to **SO** in figure 7.58A. The two vectors **OK** and **KP** in figure 7.58B turn at the same rates as their counterparts in Fig. 7.58A. Thus, figure 7.58B is *the Ptolemaic theory of a superior planet*. The planet's orbit in the Sun-centered model becomes the deferent circle in the Earth-centered model. And the orbit of the Earth becomes the epicycle.

Inferior Planet   The heliocentric theory of an inferior planet, such as Venus, is illustrated by figure 7.59A. The planet P travels on a smaller orbit about the Sun S than does the Earth O. The vector addition works as before: **OP** = −**SO** + **SP**. Thus, as long as **OK** and **KP** in figure 7.59B are equal to −**SO** and **SP** in figure 7.59A, the Ptolemaic theory will be mathematically equivalent to the Sun-centered theory. The planet's orbit in the Sun-centered model corresponds to the epicycle in the Earth-centered model. And the orbit of the Earth corresponds to the deferent circle. So, the correspondences are reversed in the cases of inferior and superior planets.

Explanatory Advantages of the Sun-Centered System

It is easy to see why a heliocentric theory is not *automatically* more accurate in predicting planet positions than a geocentric theory. Each of the two theories of Mars shown in figure 7.58 involves two circles. Either we work with a deferent circle and an epicycle (in the geocentric model), or we work with the orbit of Mars and the orbit of the Earth (in the heliocentric version). In either case, we really have the same two circles involved. As we saw in section 7.17, a working geocentric theory requires two extra complications: the deferent circle must be slightly eccentric, and we must introduce nonuniform motion according to the law of the equant. An accurate, working heliocentric theory will require similar complications. This is what we meant above when we said that the predictive accuracy of the theory is determined by the technical details.

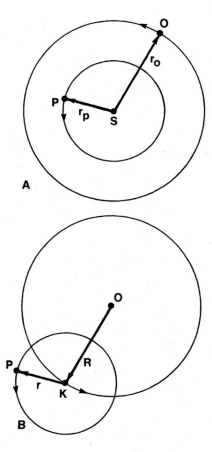

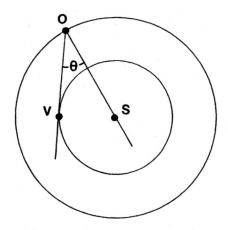

FIGURE 7.59. Inferior planet: transformation from the Sun-centered theory (A) to the Earth-centered theory (B).

FIGURE 7.60. An inferior planet V has a maximum elongation from the Sun S because the orbit radius subtends angle θ as observed from the Earth O.

If a heliocentric theory is not inherently more accurate, what advantages does it offer? There are in fact two major advantages. First, the heliocentric theory explains the weird connections between the motion of the Sun and the motions of the planets. It is easy to see why a superior planet retrogresses when it is in opposition to the Sun (see fig. 7.6). Mars M appears to back up when the Earth E passes by on the inside track. At this moment, Mars and the Sun S are indeed in opposition as viewed from Earth.

Also, we can understand why an inferior planet has limited elongations from the Sun (see fig. 7.60). The heliocentric orbit of Venus V is smaller than that of the Earth O. Thus, the elongation of Venus from the Sun can never be larger than the angle θ under which we see the radius of Venus's orbit.

The reason why the three superior planets all appear to go around on their epicycles in lockstep (fig. 7.29A) is that we, the observers, are actually riding around on a circle once a year. That is, the epicycles of the three superior planets are really one and the same circle: they are all manifestations of the Earth's orbital circle about the Sun.

The reason why the two inferior planets have the same tropical period as the Sun is that their deferent circles are both manifestations of the Earth's heliocentric orbit. Connections that were inexplicable coincidences in the Earth-centered theory (or explicable only in terms of Pythagorean Sun mysticism) find simple geometrical explanations in the Sun-centered theory.

The second major advantage of the Sun-centered theory is that it makes the system of the planets a true system, with a manifest order and coherence. In particular, it allows us to fix the relative sizes of the planets' orbits. In ancient planetary astronomy, the system for each planet is logically independent. That is, we can tell from observations how big Mars's epicycle is compared to its deferent, but we cannot tell how big Mars's deferent is compared to Jupiter's deferent. In his *Planetary Hypotheses*, Ptolemy added the physical assumption that there is no empty space between the system for Mars and the system for Jupiter. It is only this extra assumption (justified by appeal to Aristotlelian physical principles) that allowed Ptolemy to hazard a guess about the relative sizes of the planets' deferent circles.

But if we adopt the transformation to Sun-centered cosmology illustrated by figures 7.58 and 7.59, we see that (instead of arbitrarily insisting that there be no empty space between the geocentric systems of neighboring planets), we should actually make the epicycles of the three outer planets all the same size: they are all simply mainfestations of the Earth's orbital circle. This serves to fix the relative sizes of the heliocentric orbits, with no ambiguity and no arbitrary assumptions. Refer to figure 7.58. Let R and r denote the radius of the deferent and of the epicycle, respectively, in the Ptolemaic system. Let r_p and r_o denote the radius of the planet's orbit and of the Earth's orbit in a Sun-centered theory. Then, to guarantee that two versions of the theory of a superior planet are exactly equivalent, we must require

$$\frac{r_p}{r_o} = \frac{R}{r} \quad \text{(superior planet).}$$

From figure 7.59, the condition for the inferior planets is

$$\frac{r_p}{r_o} = \frac{r}{R} \quad \text{(inferior planet).}$$

Let us choose the radius of the Earth's orbit about the Sun as our unit of measure. That is, we put $r_o = 1$. (Also, note that in table 7.4, the epicycle radii are given for a deferent of radius $R = 1$.) The radii of the planets' orbits about the Sun can then be calculated from the epicycle radii in table 7.4:

Radii of the heliocentric orbits

Planet		r_p
Saturn	1/.105 =	9.54
Jupiter	1/.192 =	5.20
Mars	1/.656 =	1.52
Earth		1
Venus	.723/1 =	0.72
Mercury	.391 =	0.39

(Mercury is included for completeness.) The relative sizes of all the planetary orbits are uniquely determined. This is one of the most striking consequences of the Sun-centered cosmology, which emerged in the first half of the sixteenth century with the work of Nicholas Copernicus. As we shall see in section 7.30, the fixing of a unified scale for the whole system was a feature of heliocentrism to which Copernicus and his followers attached great weight. Figure 7.61 is the diagram of the Sun-centered system from Copernicus's book *On the Revolutions of the Heavenly Spheres* (1543).

Geo-Heliocentric Compromises

It is possible to devise other cosmologies that take some advantage of heliocentrism as an explanatory device but that still keep the Earth at the center of the universe. In this kind of mixed model, some or all of the planets revolve around the Sun, while the Sun revolves around the Earth. The chief philosophical advantage of such geo-heliocentric models is that they retain the centrality of the Earth, in keeping with the physics of Aristotle (and also, later on, with the interpretation of the Bible by the Church fathers).

In Greek antiquity, one such mixed model had already been proposed. As we saw in section 7.15, Theon of Smyrna mentions that is possible that the Sun, Venus, and Mercury all share one deferent. In such a system, then, Venus and Mercury travel around the mean Sun, while the mean Sun circles the Earth. The superior planets are treated just as in the standard (Ptolemaic) cosmology. As Theon points out, in such a system it is especially easy to understand why Mercury and Venus have limited elongations from the Sun. However, this system fails to exploit the advantages of heliocentrism for the superior planets. Historically, it was much easier for people to see the true relation of the planets to the Sun in the case of the inferior planets. In any case, this system never had a very large following, as Ptolemy's arrangement of the planets was accepted by almost everyone in the Middle Ages.

The most important geo-heliocentric system in the Renaissance was that of Tycho Brahe. Brahe could see the great explanatory power of Copernicus's (Sun-centered) cosmology—especially its ability to uniquely determine the relative sizes of the circles. But he was unable to accept the mobility of the Earth. In the 1580s Brahe proposed a system in which all the planets circle the Sun, while the Sun (carrying the planets with it) circles the Earth.[129] Geometrically, this is a trivial transformation of Copernicus's system: Brahe's system is exactly what the world would look like in a Copernican universe, as viewed from the Earth. Thus, Brahe's system possesses all of the explanatory advantages associated with heliocentric cosmology. That is, it explains all the connections between the apparent motions of the planets and the motion of the Sun. It also allows a unique determination of the sizes of the circles.

Figure 7.62 shows a diagram of the so-called Tychonic system of the world from Brahe's *De mundi* of 1588. The Earth is the black dot at the center of the figure. The Sun travels on a circle around Earth, while all the planets circle the Sun. The relative sizes of the circles must be just as in the table above (except that we replace the orbit of the Earth by the orbit of the Sun). Note that since the radius of the orbit of Mars is greater than one but less than two times the radius of the Sun's orbit, the orbit of the Sun and the

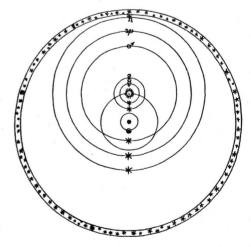

FIGURE 7.61. The diagram of a heliocentric universe, from the first edition of Copernicus's *De revolutionibus* (Nuremberg, 1543).

FIGURE 7.62. The Tychonic system of the world. The planets all circle the Sun while the Sun travels in a circle about a stationary Earth. From Tycho Brahe, *De mundi aetherei recentioribus phaenomenis* (1588).

orbit of Mars must intersect, as shown in the figure. (Mars is not in any danger of hitting the Sun, since the location of the Martian orbit constantly shifts as the Sun—its center—moves around the Earth.)

From the point of view of *planetary theory*—that is, theoretical calculation of planet positions—all of the systems discussed in this section are equally usable. From the point of view of *cosmology*, great advantages attach to the Sun-centered systems. Deciding whether to place the Sun or the Earth at rest then becomes a question of *physics*.

7.30 NICHOLAS COPERNICUS: THE EARTH A PLANET

Nicholas Copernicus (1473–1543) was hardly the person one would have predicted to turn the universe inside out. Copernicus was born at Torun, a city then situated in Prussia but now in the north of Poland. In 1491, Copernicus began to attend the University of Cracow. Although astronomy was a part of the quadrivium, the university curriculum did not prepare a student to do any real astronomy. However, there was at Cracow at competent astronomer, Albert of Brudzewo (1446–1495), who had written a commentary on Peurbach's *New Theories of the Planets*. It is possible that Copernicus received private instruction from Albert of Brudzewo at a more advanced astronomical level, but we have no way of knowing for sure.[130]

In 1496, without having completed his degree, Copernicus left for Italy to study law at the University of Bologna. But already Copernicus's real interests diverged from his plan of study. At Bologna he sought out the astronomer Domenico Maria Novara, from whom he undoubtedly learned more. It was also at Bologna that Copernicus made his first known astronomical observation, on March 9, 1497, of the Moon approaching Aldebaran.

In September, 1500, Copernicus left Bologna for Rome to take part in the jubilee observances proclaimed by Pope Alexander VI. While in Rome, Copernicus gave at least one lecture on astronomy. We do not know the details. In November, still at Rome, Copernicus observed a partial eclipse of the Moon.

The following year, 1501, he was back home in Poland, still without having acquired a university diploma. Some years earlier, his uncle, who was the Bishop of Varmia (also known as Ermland), had obtained for Copernicus the office of canon at the cathedral of Frombork (Frauenburg). A canon is an official who has administrative duties in the cathedral chapter (or staff) but who has not taken holy orders. The cathedral chapter of Frombork had a scholarship fund that provided grants for any canon to complete studies already begun. Copernicus asked for and was granted two years' leave to study medicine at the University of Padua—even though he knew that the medical degree required three years of study.

In 1503, his leave was about to expire and Copernicus had not completed his medical training. Not wanting to return home without a diploma, Copernicus successfully applied to the University of Ferrara to have himself proclaimed a doctor of canon (church) law.

All this was hardly a foreshadowing of greatness to come. Copernicus had lived the life of the vagabond student, wandering from university to university, switching from law to medicine and back to law. When he returned home to Poland, he spent some years in the service of his uncle, the bishop, helping with diplomatic negotiations and church affairs, and acting as personal secretary and private physician.

The Commentariolus

Nevertheless, Copernicus had had the chance to meet astronomers and had acquired a sound understanding of contemporary astronomy. He settled permanently in the little town of Frombork, where his duties as canon left him considerable freedom to pursue his interests in astronomy. His first sketch of a heliocentric planetary theory was worked out by about 1510. In a few sheets, Copernicus described the chief features of his Sun-centered system, and the connections of the planets' apparent motions to the actual motion of the Earth. This document circulated in manuscript among some of Copernicus's friends. But it was not printed until the nineteenth century, under the title *Nicolai Copernici de hypothesibus motuum coelestium a se constitutis commentariolus.* We shall refer to it as the *Commentariolus*—the "Little Commmentary."[131]

This work contained a remarkable statement of seven astronomical and cosmological postulates:

EXTRACT FROM COPERNICUS

Commentariolus

1. There is not one single center for all the celestial orbs or spheres.
2. The center of the Earth is not the center of the world, but only of the heavy bodies and of the lunar orb.
3. All the orbs encompass the Sun which is, so to speak, in the middle of them all, for the center of the world is near the Sun.
4. The ratio of the distance between the Sun and the Earth to the height of the firmament [i.e., the radius of the sphere of stars] is less than the ratio between the Earth's radius and the distance from the Sun to the Earth, in such a manner that the distance from the Sun to the Earth is insensible in relation to the height of the firmament.
5. Every motion that seems to belong to the firmament does not arise from it, but from the Earth. Therefore, the Earth with the elements in its vicinity accomplishes a complete rotation around its fixed poles, while the firmament, or last heaven, remains motionless.
6. The motions that seem to us proper to the Sun do not arise from it, but from the Earth and our [terrestrial] orb, with which we revolve around the Sun like any other planet. In consequence, the Earth is carried along with several motions.
7. The retrograde and direct motions which appear in the case of the planets are not caused by them, but by the Earth. The motion of the Earth alone is sufficient to explain a wealth of apparent irregularities in the heaven.[132]

Postulates 1–3 took the Earth out of the middle of the world and replaced it by the Sun. However, Copernicus's break with the old cosmology and physics was not as complete as it might seem. First of all, for Copernicus, the Sun is only "near" the center of the cosmos. Copernicus's cosmos has its center at the center of the Earth's orbit (which is slightly eccentric to the Sun). Thus, the Earth does continue to hold a somewhat privileged place in the theory. Also, according to postulate 2 the Earth remains the center of the heavy bodies—by which Copernicus means the four Aristotelian elements. By making the Sun the center of the cosmos but retaining the Earth as the center of heaviness, Copernicus bent the rules of Aristotelian physics, but managed to express his discourse in the old terms.

In postulate 4, Copernicus assumes that the radius of the Earth's circle about the Sun is immeasurably small in comparison with the radius of the sphere of stars. Otherwise, Copernicus would be unable to explain why the stars do not suffer any annual parallax due to the motion of the Earth. Long before, Aristarchus had been forced to the same conclusion.

Postulate 5 attributes the apparent daily motion of the firmament to the rotation of the Earth. Here it is clear that Copernicus still regards the stars as fixed to a real celestial sphere.

Postulate 6 attributes the apparent annual motion of the Sun to the motion of the Earth on its circle around the Sun. In his reference to the terrestrial "orb," Copernicus is still making use of solid-sphere cosmology. He does not think of the Earth as moving through empty space on a mathematical circle around the Sun. Rather, the Earth is carried around by an ethereal orb. Here, too, we see Copernicus unable to break entirely away from traditional cosmology and physics. Nevertheless, he asserts quite boldly that "we revolve around the Sun like any other planet."

Copernicus's most significant insight is contained in postulate 7. The complexities of the apparent motions of the other planets are due to the motion of the Earth. Much of the remainder of the *Commentariolus* is devoted to showing just how the motion of Earth affects the apparent motions of the other planets.

While the *Commentariolus* contained the vision of a heliocentric universe, it was far from a finished work of planetary theory. Much remained to be done, including the working out of numerical values for the planetary parameters, and demonstrating how to calculate planet positions from the new theory. This Copernicus set out to do in the work to which he devoted the rest of his life. *De revolutionibus orbium coelestium libri sex* (Six books on the revolutions of the heavenly spheres) was not published until 1543, as Copernicus lay dying.

Rheticus and the Narratio prima

Although Copernicus had not yet published an account of his theory, word of it was spreading among the intellectuals of the Catholic church, as well as in the astronomical community of central Europe. In 1533, the papal secretary Johann Albrecht Widmanstadt explained Copernicus's ideas about the motion of the Earth to Pope Clement VII. Three years later the same Widmanstadt explained the system to Cardinal Nicholas Schönberg, who then sent a letter to Copernicus encouraging him to make his ideas public. (Copernicus printed this letter among the introductory material in *De revolutionibus*.) Copernicus also received moral support from Tiedemann Giese, the Bishop of Chelmno, who was one of his closest friends.

But Copernicus's work, on which he had labored for three decades, might never have been printed were it not for the intervention of a young Lutheran professor of mathematics, Georg Joachim Rheticus (1514–1574). Rheticus, who had taken a leave from Wittenberg University, spent a while traveling about central Europe from one famous scholar to another. In 1538 he was at Nuremberg, where he visited Johann Schöner, the designer of equatoria and printer of mathematical and astronomical books. It was probably Schöner who told Rheticus about Copernicus and his new astronomical system.[133] In 1539, Rheticus paid a visit to Copernicus at Frombork, where he became Copernicus's one and only student and disciple. He wound up staying with Copernicus for over two years. Copernicus allowed him to study the manuscript of *De revolutionibus*, which Rheticus recognized as a work of great significance. Moreover, Rheticus set about writing a shorter account of the new astronomical system.

This work of Rheticus, *Narratio prima* (First account), was the first published description of the heliocentric astronomy.[134] It was printed in 1540 at Gdansk and took the form of an open letter addressed to Johann Schöner. Rheticus spoke of Copernicus as "my teacher" and said that he was worthy of being compared with Ptolemy, for Copernicus had undertaken a reconstruction of the whole of astronomy. Rheticus's book gave a qualitative description of Copernicus's theory of precession and trepidation (now attributed to motions of the Earth), his theory of the annual motion of the Earth, and his theories for the motions of the Moon and planets.

Rheticus could not resist throwing in some astrological prognostications of his own. To accommodate both ancient and modern values for the eccentricity of the Earth's orbit, Copernicus had adopted a theory with a slowly varying eccentricity. Rheticus ventured the guess that there were upheavals in the kingdoms on Earth at critical moments in the cycle. For example, when the eccentricity was at its maximum, the Roman government became a monarchy, but as the eccentricity decreased, Rome declined. When the eccentricity was at its mean value, the Mohammedan faith was established. Rheticus predicted that in another hundred years, when the eccentricity reached its minimum value, the Mohammedan empire would fall with a mighty crash.

While the *Narratio prima* gave a good, readable account of Copernicus's astronomical theories, it was nonmathematical in its treatment of them. Thus, Rheticus's book could not teach an astronomer to apply the new system in practice. For this, there could be no substitute for Copernicus's own book. Rheticus therefore set about seeing that it was printed.

The Publication of De revolutionibus

In 1542 Rheticus took a fair copy of Copernicus's manuscript to Nuremberg and gave it to the printer Johannes Petreius, with whom he was already on friendly terms. As the proof sheets came from the press, Rheticus himself read and corrected them.[135] But Rheticus had to leave Nuremberg before the printing was finished in order to begin teaching at Leipzig University in October, 1542, where he had just been named professor of mathematics. Consequently, the job of seeing Copernicus's book through the press and correcting the proof was turned over to Andreas Osiander (1498–1552), a Lutheran minister at Nuremberg who had some knowledge of astronomy.

This turned out to have unintended consequences. Foreseeing that Copernicus's theory of the motion of the Earth could be objectionable to philosophers and theologians alike, Osiander wrote—without Copernicus's knowledge or approval—and inserted into Copernicus's book an unsigned foreword, "To the reader concerning the hypotheses of this work." In this foreword, Osiander took a staunchly instrumentalist position on the motion of the Earth. Osiander claimed that certainty of knowledge was impossible in astronomy. It was not necessary, therefore, that astronomical hypotheses be true or even probable, as long as they were useful for calculation. Osiander concluded by warning the reader not to take literally the hypothesis of the motion of the Earth, "lest he accept as the truth ideas conceived for another purpose, and depart from this study a greater fool than when he entered it."[136] No doubt Osiander thought he was helping to save Copernicus from unnecessary trouble. But the foreword also reflected views that Osiander genuinely held and that he had expressed on previous occasions. Nevertheless, this unsigned foreword had the effect of negating what Copernicus had intended to be the essential point of his life's work.

According to tradition, Copernicus was presented a copy of the freshly printed book on the day of his death, May 24, 1543. We do not know how he reacted to Osiander's foreword, or even if he saw it. However, Rheticus and Tiedemann Giese were outraged. They tried to institute a legal action with the City Council of Nuremberg to force Petreius to issue a corrected edition with the foreword eliminated. Petreius, however, protested that the foreword was among the rest of the manuscript material given to him for printing. The City Council decided he was not to blame and, consequently, no corrected edition was ever issued.[137]

As a result, many readers of Copernicus's *De revolutionibus* came away with the mistaken idea that Copernicus had not meant the motion of the Earth as a physical hypothesis, but merely as a mathematical device for saving the phenomena. The origin of the anonymous foreword did not become widely known until 1609, when it was announced by Johannes Kepler on the back of the title page of his own *Astronomia nova*.

Copernicus's Intentions

What led Copernicus to his new system? In accounts of the scientific revolution, it is sometimes claimed that Ptolemaic astronomy suffered a crisis in the sixteenth century and that this crisis was manifested in two ways. First, the predictive accuracy of the tables then in circulation (notably the *Alfonsine Tables*), became worse and worse, as slowly accumulating errors made the defects of Ptolemy's theory glaringly obvious. Second, the astronomers responded by adding more and more epicycles in a desperate attempt to rescue Ptolemy until the system became ridiculously complicated. This account makes the astronomical issues very clear: Ptolemaic astronomy fails on the questions of accuracy and simplicity.[138] Unfortunately, this explanation is completely false.

In fact, there was no crisis in astronomy. As pointed out in section 7.29, there is no automatic advantage in predictive accuracy for Sun-centered theories. Thus, the adoption of a Sun-centered cosmology was not the solution to a problem of astronomical accuracy. It is true that a substantial body of accurate observations could suffice to disprove the technical details of Ptolemaic theory—motion on circles, the law of the equant, and so on. Eventually, Tycho Brahe's observations of Mars were used by Kepler with precisely this result. But in Copernicus's time, there was no such body of planetary observations. It is true that from time to time Copernicus's predecessors noted that the circumstances of planetary conjunctions or of eclipses predicted by the *Alfonsine Tables* were not in perfect accord with the events in the sky. But it would be very hard to tell from a few isolated observations whether the errors were due to some fundamental defect of the theory or merely to slightly inaccurate values for the numerical parameters. And it does no good to claim that the errors grew with the centuries until they became intolerably large. Errors that grow with time are due to faulty values for the mean motions (the values of f_λ and f_μ in table 7.4). The way to fix such errors is not to add epicycles, or to shift the center of the system to the Sun, but simply to adopt improved values for the planetary periods. Extra epicycles would produce only small, periodic effects, at the expense of enormously complicating practical calculation. Although a number of late medieval astronomers enjoyed toying with alternative planetary models, practical computation was always based on standard tables. And these, in turn, were invariably based on standard Ptolemaic planetary theory. As Copernicus's reputation grew, people began to construct new planetary tables and to publish ephemerides based on his system. If any further proof were needed that predictive accuracy was not the chief motive behind the new cosmology, this should suffice: the sixteenth-century ephemerides based on Copernicus's theory actually were not much more accurate than the Ptolemaic ephemerides they replaced.[139]

If there was no crisis in astronomy for which heliocentrism was the solution, and if Copernican planetary theory did not immediately lead to greater predictive accuracy, what, then, was the point of Copernicus's work as he saw it himself? The answer is very simple. Copernicus thought he had discovered the true system of the world.

Copernicus certainly believed in his system as physically true. This is clear in the preface to the work, which he wrote himself and dedicated to Pope Paul III. The preface begins with Copernicus's frank acknowledgment that he advocates the motion of the Earth and that some people will therefore repudiate him. He claims that his fear of controversy led him to delay publication of his ideas for a long time. But the constant urging of his friends finally convinced him to go ahead. And here Copernicus mentions Cardinal Nicholas

Schönberg and Bishop Tiedemann Giese. The dedication of the work to the Pope and the publication of the supportive letter from Cardinal Schönberg (mentioned above) were part of a deliberate strategy to forestall criticism based on doctrinal considerations. Moreover, in the preface, Copernicus goes on to say that babblers who are ignorant of astronomy may attempt to attack his work by twisting some piece of Scripture to their purpose. Copernicus rejects the validity of such arguments, saying, "Mathematics is written for mathematicians." This assertion that religious arguments have no bearing on astronomy was not the sign of a man who wished merely to save the phenomena. The diagram of the Sun-centered universe that appeared in book I of *De revolutionibus* (see fig. 7.61) also shows that Copernicus meant the system as physically real.

Moreover, Copernicus's friends were prepared to assist in the fight. Both Giese and Rheticus wrote treatises arguing that the motion of the Earth was compatible with Holy Scripture. Giese's tract is lost. Rheticus's, long believed lost, has recently been rediscovered.[140] This organized campaign by Copernicus and his friends, as well as the angry reaction of Rheticus and Giese to Osiander's interpolated foreword, show that, for them, the actual arrangement of the cosmos was the fundamental issue. Copernicus's friends anticipated trouble with theologians and Aristotelian philosophers. But, rather than pleading not guilty by virtue of instrumentalism, they defended Copernicus's system as physically true.

In the preface to *De revolutionibus*, Copernicus gives us an idea of the aspects of his work he considered most important. He criticizes the old astronomy on the grounds that it could not solve the most important problem, that is, to determine the structure of the universe and the commensurability of its parts. Copernicus compares his predecessors to an artist who takes "from various places hands, feet, a head, and other pieces, very well depicted, it may be, but not for representation of a single person; since these fragments would not belong to one another at all, a monster rather than a man would be put together from them."[141] Here Copernicus is probably referring to the independence of each planet's system in Ptolemaic theory. The systems for Mars, Jupiter, Venus, and so on, are like the separate parts of a monstrous body, with no fixed scale determining their relative proportions. The recognition of the Earth's motion is what makes the cosmos a unified whole, as we have explained above.

Copernicus also lays great stress on some technical aspects of his planetary theory—in particular, on his fidelity to the principle of uniform circular motion. In this matter, Copernicus was a much stricter Aristotelian than Ptolemy. In the preface to *De revolutionibus*, Copernicus admits that the theories of his predecessors (i.e., Ptolemy) based on eccentric circles are satisfactory for computing the apparent motions. But he objects that they contradict the principle of uniformity of motion. Here Copernicus is voicing his dissatisfaction with Ptolemy's equant point (and a similar device used in Ptolemy's lunar theory). He comes back to this subject in several other places. In the beginning of his discussion of the lunar theory, Copernicus is particularly adamant.[142] Uniformity of motion is an axiom of astronomy. Moreover, uniformity defined artificially with respect to some point other than the center of the circle is no sort of uniformity at all.

For Copernicus, the equant was a physical and philosophical absurdity. But he had to replace it with something else that was more or less equivalent. As we shall see below, Copernicus found that a minor epicycle would allow him to account for apparent nonuniformity of motion (associated with the equant) by a combination of motions that really were uniform and circular about the centers of their circles. The importance that Copernicus attaches to this technical departure from Ptolemy shows, once again, that he was seeking a planetary theory that was physically and philosophically more accept-

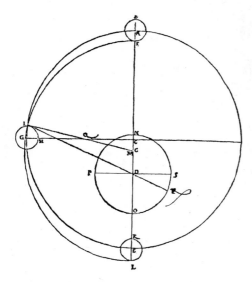

FIGURE 7.63. Copernicus's theory of the superior planets. *NPO* is the orbit of the Earth. *AGB* is the deferent circle of a superior planet, such as Mars. Mars itself moves on a small epicycle which is responsible for producing an anomaly of motion more or less equivalent to that produced by Ptolemy's equant. From *De revolutionibus* V, 4 (Nuremberg, 1543).

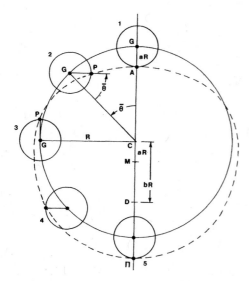

FIGURE 7.64. Copernicus's minor epicycle, a replacement for Ptolemy's equant.

able. While this stress on a coherent system served Copernicus very well in the shift to Sun-centered cosmology, it led him astray in technical matters. For it turns out that the planets really do move nonuniformly and that Ptolemy's equant theory was closer to the mark than Copernicus's "improvement" on it.

Copernican Planetary Theory

A good sense of Copernicus's astronomy can be obtained by examining his theory for the superior planets. Copernicus himself placed a high value on this work, which he believed improved on Ptolemy. Here we must confront not only Copernicus's use of a moving Earth, but also his method of accounting for the planets' nonuniformity of motion.

For the orbit of the Earth, Copernicus chose an eccentric circle: the Earth moves at uniform speed on a circle that is eccentric to the Sun. The model is essentially the same as the solar theory of Ptolemy. For computation of positions it makes no difference whether the Earth or the Sun moves. The essence of the model is uniform circular motion on an off-center circle.

For the superior planets, Copernicus adopted an eccentric circle plus a modified form of the Ptolemaic equant. As we have seen, Copernicus could not abide the equant. But he had, of course, to replace it with something else. He found that a minor epicycle could perform very nearly the same function.

Figure 7.63 is a diagram from the first edition of *De revolutionibus*, illustrating Copernicus's theory of the superior planets. The Earth travels around the annual circle *NPO*, which is centered at *D*. The Sun is therefore located near but slightly displaced from *D*. However, the true Sun does not appear in this figure and plays no part in the theory. For this reason, Copernicus's system has been aptly characterized as merely heliostatic, rather than truly heliocentric. The effective center of the whole system is the center *D* of the Earth's orbit, also called the mean Sun.

In figure 7.63, *C* is the center of the deferent circle *AGB* of a superior planet (let us say Mars). Thus, the center of Mars's deferent circle is eccentric to the mean Sun *D*. So far, this resembles Ptolemy's theory. However, Copernicus does not have an equant point. Rather, he places Mars on a small epicycle, shown in the figure. Further, Mars makes a complete counterclockwise orbit on the epicycle while the epicycle's center travels a complete circle around the deferent. Thus, when the epicycle's center is at *A*, Mars is at *F*. When the epicycle's center is at *G*, Mars is at *I*. When the epicycle's center is at *B*, Mars is at *L*. Finally, the radius *GI* of the epicycle is chosen to be one-third of the eccentricity *DC*.

One thing to note is that Copernicus did not eliminate epicycles from planetary theory. However, the large epicycle of Ptolemy *is* gone. Ptolemy's big epicycle was responsible for retrograde motion. In Copernicus's theory of the superior planets (fig. 7.63), this function is taken over by the circle *NPO* of the Earth's annual motion. The minor epicycle *GI* is Copernicus's substitute for Ptolemy's equant point. Let us study this device in more detail.

Refer to figure 7.64, which elaborates on Copernicus's own diagram. The large solid circle of radius *R* is the deferent of Mars, centered at *C*. The deferent circle is eccentric to *D*, the mean Sun, or center of the Earth's orbit. (For simplicity, the Earth's orbit is not shown in this figure.) The dimensionless eccentricity of Mars's deferent circle is $b = CD/R$.

The center *G* of a small epicycle moves counterclockwise and uniformly around the deferent. The planet *P* moves counterclockwise and uniformly on the epicycle whose radius is aR. (Thus, a is a dimensionless number less than 1.) Further, the two angles marked $\bar\theta$ remain equal to one another while increasing uniformly with time. Consequently, while the epicycle's center moves through 180° from position 1 to position 5, the planet revolves through 180° on the epicycle.

The combination of two uniform circular motions for P in figure 7.64 results in a motion that is neither uniform nor circular. The actual path of the planet is indicated by the dashed line. The effective center of the orbit is not C but M, located below C by a distance aR equal to one radius of the epicycle. As Copernicus himself states, the path is not circular but somewhat oblong—the long axis being perpendicular to the line of apsides ΠCA.[143]

Nevertheless, Copernicus's speed rule is virtually indistinguishable from Ptolemy's: the minor epicycle produces a motion that closely approximates equant motion. Refer to figure 7.65. The radius of the epicycle is aR. Let us identify point E on the line of apsides at a distance aR above the center C of the deferent. As already remarked, in Copernicus's model, the rotation of GP is such that angle CGP is always equal to the mean anomaly ACG: both are equal to $\bar{\theta}$. Since also $CE = GP$, it follows that the quadrilateral $ECGP$ is a trapezoid, with sides EP and CG always parallel. Since line CG turns uniformly, it follows that EP turns uniformly, too. In other words, E is an *effective equant point*. The planet P, observed from E, appears to move at uniform angular speed.

Furthermore, Copernicus usually makes the radius of the minor epicycle exactly one-third the eccentricity of the deferent. That is, $b = 3a$. Now, from figure 7.65, $EM = 2aR$, and $MD = bR - aR$, so we get also $MD = 2aR$. Thus, the center M of the effective orbit is exactly midway between D and the effective equant point E. Copernicus, like Ptolemy, bisects the total eccentricity: $EM = MD$ in figure 7.65, just as $EC = CO$ in figure 7.32. An almost perfect equivalence will be established between Ptolemy's eccentric circle with equant point and Copernicus's eccentric circle with minor epicycle if we identify the radius of Copernicus's epicycle with half the Ptolemaic eccentricity e_P; that is, if $a = 1/2\ e_P$. Thus, $b = 3/2\ e_P$.

The combined effect of Copernicus's oblong orbit and hidden equant is illustrated in figure 7.66. M is the center of the solid circle and E represents a Ptolemaic equant point. Thus, if body P moves on the circle according to the law of the equant, $\bar{\theta}$ increases uniformly with time. The dashed curve represents the effective, oblong Copernican orbit. E, then, is also the effective equant point of the Copernican orbit. Thus, when the body is at P according to Ptolemaic hypotheses, it will be at P' according to Copernican principles. For an observer at the equant, P and P' could not be distinguished. But, because of the noncircularity of the Copernican orbit, an observer at D (the center of the Earth's orbit) would see P and P' in directions that differ by a small angle $\Delta\theta$. The eccentricity is greatly exaggerated in figure 7.66. Even in the case of Mars, for which Ptolemy's eccentricity $e_P = 0.1$, the maximum difference $\Delta\theta$ between the directions of P in the two models is only about 3'. Before the work of Brahe and Kepler, the observational consequences of Copernicus's modification of the Ptolemaic equant were nil.

Moreover, Copernicus's values for the eccentricities of the superior planets were borrowed from Ptolemy, as may be seen in the following table:

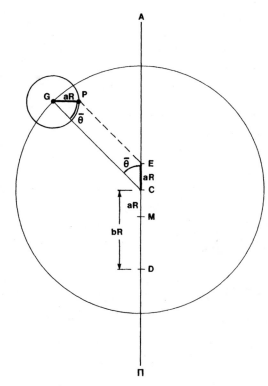

FIGURE 7.65. Copernicus's hidden equant point (E).

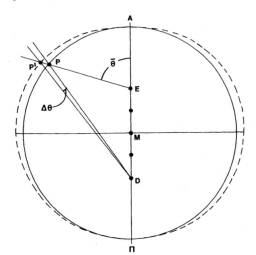

FIGURE 7.66. Comparison of the Copernican model with a Ptolemaic eccentric-with-equant model. The Ptolemaic eccentric circle is drawn in solid line. The oblong Copernican orbit is drawn in dashed line. The Ptolemaic equant point and the hidden, effective equant point of the Copernican model coincide at E. At the same moment (and therefore at the same mean anomaly $\bar{\theta}$) the position of the planet in equant theory is P and the position in Copernican theory is P'. As viewed from the Sun D, there is a small difference $\Delta\theta$ in the directions predicted by the two theories.

Eccentricities of the superior planets

	Ptolemy			Copernicus	
	e_P	$1/2\ e_P$	$3/2\ e_P$	a	b
Mars	0.10000	0.05000	0.15000	0.05000	0.14600
Jupiter	0.04583	0.02292	0.06875	0.02290	0.06870
Saturn	0.05694	0.02847	0.08541	0.02850	0.08540

Column e_P gives Ptolemy's value of the eccentricity for each planet. The columns headed $1/2 e_P$ and $3/2 e_P$ give the appropriate fractions of Ptolemy's eccentricity. As shown above, Copernicus's theory for the superior planets differs insignificantly from Ptolemy's if $a = 1/2e_p$ and $b = 3/2e_p$. The columns headed a and b give the values of these parameters actually adopted by Copernicus in *De revolutionibus*. Only in the case of Mars did Copernicus make a slight change. Believing it necessary to reduce the total eccentricity of Mars slightly below Ptolemy's value, Copernicus decided to leave the radius a of the epicycle unchanged and to effect the reduction wholly in the eccentricity b of the deferent. In the theory of Mars, then, Copernicus departed slightly from a bisection of the eccentricity. But this minute departure from a bisection was without theoretical consequence and, indeed, was not required by the observational material at Copernicus's disposal.

Copernicus's method of reproducing equant motion by means of a minor epicycle is identical to one employed two centuries earlier by Ibn al-Shāṭir of Damascas.[144] We have no way of knowing how Copernicus came by this model. Moreover, from Copernicus's own discussion, it is not clear that he understood how nearly perfectly his model duplicated Ptolemy's. For example, Copernicus never mentions the existence of point E, the effective equant point.[145] The earliest European proof of the existence of an effective equant point in Copernicus's model is contained in a letter of the year 1595 from Michael Mästlin, professor of mathematics at Tübingen, to his former pupil, Kepler.[146]

In sum, Copernicus's theory of the superior planets contains a mixture of radical innovation and conservative astronomical practice. To launch the Earth into orbit was a bold move. The Sun-centered theory does have great explanatory advantages. And it does turn the whole solar system into a unified whole, as Copernicus himself stressed. But in the technical details of his planetary theory, Copernicus remained a part of the Ptolemaic tradition.[147] Nearly every detail of his model has a corresponding element in Ptolemy's model. It was for this reason that Kepler was later to say that Copernicus would have done better if he had interpreted nature, rather than Ptolemy.

The Early Reception of Copernicus's Work

The Astronomers: Mästlin, Reinhold and Brahe   Among astronomers, several kinds of reaction to Copernicus were possible. Some astronomers greeted the new cosmology with enthusiasm. A good example is Michael Mästlin (1550–1631), who taught the heliocentric theory at the University of Tübingen and who helped to develop some of its consequences.[148] It was Mästlin's student, Kepler, who did the most to advance the new astronomy.

Another possible reaction involved agnosticism toward the heliocentric hypothesis, combined with lively interest in the technical details of Copernicus's planetary theory. This was the reaction of Erasmus Reinhold (1511–1553). Reinhold was a Lutheran professor at Wittenberg, who in 1542 published an edition of Peurbach's *New Theories of the Planets*, with new diagrams and a detailed commentary. (Figs. 7.51 and 7.52 show illustrations from a later edition of Reinhold's book.) The Peurbach text was, of course, the standard introduction to Ptolemaic astronomy, a subject that Reinhold taught at Wittenberg. Reinhold was also a competent technical astronomer. After the publication of Copernicus's *De revolutionibus*, Reinhold was inspired to construct a new set of planetary tables, more convenient than Copernicus's own, but based on Copernicus's planetary theory. These *Tabulae prutenicae* (Prussian tables) were printed in 1551 and rapidly became the most respected planetary tables in existence. Reinhold's *Prutenic Tables* stood in the same relationship to *De revolutionibus* as the *Handy Tables* stood to the *Almagest*. The *Prutenic Tables* added to Copernicus's growing fame and helped to win support for his system. From this it would be easy to conclude that Reinhold had become a Copernican. Actually, the choice of the Earth or the Sun as center has no bearing on the construction of planetary tables. Reinhold seems to have been less interested in the cosmological big picture than in the technical details of Copernicus's astronomy—especially in the return to pure uniformity of motion through the elimination of the equant and its replacement by a minor epicycle. Reinhold's marginal notes in his own copy of *De revolutionibus* confirm this impression. The first book (containing the heliocentric hypothesis) is but lightly annotated, in contrast to books III, IV, and V (devoted to solar, lunar, and planetary theory), which are heavily marked up. As Gingerich has noted, the pattern of annotations leaves no doubt about Reinhold's real interests.[149]

Yet another possible reaction combined admiration for Copernicus's technical prowess, recognition of the explanatory power of heliocentrism, but a complete inablity to give up the fixity of the Earth, which was so well supported by Aristotelian physics and by biblical authority. Such was the position of Tycho Brahe. Brahe adopted a geo-heliocentric system (described in sec. 7.29) in which the planets all went around the Sun, while the Sun moved around a stationary Earth.[150]

Brahe even offered observational evidence against heliocentrism. According to Brahe, the angular diameter of a third-magnitude star is about 1'. (Brahe was misled by a characteristic of human vision that makes brighter stars look larger. His figures for the angular diameters were not much different from Ptolemy's, given in table 7.11. In fact, even the nearest stars are effectively pointlike objects.) And, if the Earth moved around the Sun, there should be an annual parallactic shift in the positions of the stars. This Brahe had been unable to detect. He estimated 1' as the upper limit of this undetected annual parallax. Now, the annual parallax is the angular diameter of the Earth's orbit as observed from the closest stars. Thus, if the Earth moved, the orbit of the Earth looked just as large from a nearby star as a third-magnitude star looked from the Earth. Thus, some of the stars were as large as the orbit of the Earth—and some were larger. To Brahe, a world like this, in which some objects were larger than the distances that separated the planets from the Sun, was absurd. So, for Brahe, there were many converging reasons—philosophical, scriptural, and astronomical—for rejecting the heliocentric hypothesis. But while he could not accept the motion of the Earth, Brahe often followed Copernicus in matters of technical detail. Thus, Brahe adopted Copernicus's minor epicycle in place of Ptolemy's equant.

Although the geo-heliocentric compromise may seem a timid half-measure to modern readers, such models played a role in helping European astronomers to abandon the physical celestial spheres of Aristotle, Ptolemy, and Copernicus (who still thought in terms of material orbs). In the Tychonic system of figure 7.62, the circles for Mars and the Sun intersect. For Brahe, this was a proof that the circles did not correspond to physically real spheres. Of course, in the Tychonic system, the circles for Venus and Mercury also cross over that of the Sun, but this did not arouse the same concern. Although they did not represent standard cosmology, circumsolar orbits for Venus and Mercury were familiar ideas with roots in Greek and Roman antiquity. (They had been mentioned by Theon of Smyrna and Martianus Capella, among others.) Besides, in a geo-heliocentric model for the inferior planets, the circles for Venus and Mercury were carried around the Earth, along with the Sun, in the course of a year. Thus, one could always imagine that the orb that carried the Sun was thick enough to hold epicycle orbs for the circles of Venus and Mercury. Mars was a completely different matter. A geo-heliocentric system that included the superior planets thus posed a serious difficulty for material orbs.

Brahe and his contemporaries found other grounds for doubting the reality of the orbs. Brahe's observations of the spectacular comet of 1577 had revealed no parallax. This implied that the comet was farther away than the Moon, in contrast to the view that had prevailed since the time of Aristotle that made comets out to be atmospheric phenomena. Brahe had even sketched a planetary theory to describe the motion of this comet: according to Brahe, it traveled in a (retrograde) circular motion about the Sun, resembling that of Venus and Mercury.[151] At this stage, Brahe had apparently not yet incorporated the outer planets in his geo-heliocentric scheme. In 1578, Michael Mästlin published a similar theory for the motion of the comet, with the comet moving in a circumsolar orbit outside that of Venus. But Mästlin took the extra step of calculating the daily distances of the comet from the Earth. These distances ran from 155 Earth radii to 1,495 Earth radii. If we compare these figures with the entries in Table 7.10, we reach the conclusion that the comet passed through the Ptolemaic spheres of Mercury and Venus. Still, one could always claim that the epicycle orb for the comet was carried around by a solar orb thick enough to contain the epicycles for Mercury, Venus, and the comet. Perhaps it was for this reason that Mästlin did not make as much of his table of distances as he might have. Shortly afterward, however, Christoph Rothmann argued that the motion of comets afforded one of the strongest proofs that the planetary spheres cannot be solid bodies. These ideas were in circulation when Brahe was working out the final version of his geo-heliocentric cosmology. Brahe seized on the comet as confirming what his theory of Mars suggested: the orbs and spheres of the old cosmology were not real things.[152] The planetary spheres had been a standard part of Western cosmology for 1,900 years—since the time of Eudoxus and Aristotle. The dissolution of these spheres was certainly not intended by Copernicus. But this was nevertheless a consequence of Copernicanism.

The Theologians   As we have seen, Copernicus held that "Mathematics are written for mathematicians." However, he and his friends expected criticism from philosophers and theologians. In this expectation they were not to be disappointed. Even before the publication of *De revolutionibus*, leading Protestants had already gone on record as regarding the motion of the Earth a ridiculous conceit. Martin Luther in his *Table Talk* had accused Copernicus (though without mentioning him by name) of turning astronomy upside down out of a mere hunger for notoriety and attention. Luther went on to point out that the scheme was in flat contradiction to Scripture, for Joshua had commanded the Sun and not the Earth to stand still. Philip Melanchthon (1497–1560), one of Luther's principal lieutentants and an influential figure in the reorganization of the German school curriculum, also had nothing good to say about heliocentrism.[153]

However, no one pressed the complaint and the objections of Luther and Melanchthon did not place heliocentric astronomers in jeopardy. As we have seen, it was Rheticus, a Lutheran professor of mathematics, who made the publication of Copernicus's work possible. Nor did this cause any difficulties for Rheticus, who remained on good terms with Melanchthon. And it was Erasmus Reinhold, another Lutheran, who contributed greatly to Copernicus's prestige with the publication of his *Prutenic Tables*.

In the Catholic hierarchy, there was, at first, even less to worry about. Copernicus himself was a canon of a cathedral chapter. His work had been encouraged by Church leaders of high standing. And he almost certainly had secured advance permission to dedicate the work to Pope Paul III.

Thus, Copernicus's book did not produce an immediate upheaval. However, it certainly did offend the sensibilities of conservative religious thinkers, as well as professors of Aristotelian natural philosophy. With the crackdown on freethinking associated with the Catholic Counter-Reformation, any heterodox opinion became more dangerous than it had previously been. But it was not until 1616 that heliocentrism was officially declared erroneous. *De revolutionibus* was placed on the Index of books that were prohibited "until corrected." In principle, *De revolutionibus* could be circulated and read only if erroneous passages (asserting the mobility of the Earth) were removed. Four years later, a list of ten specific corrections was issued. Owen Gingerich has examined nearly all the surviving copies of the 1543 and 1566 editions of *De revolutionibus*, which total more than 500 books. The majority of copies in Italy were censored in conformity with the decree. But the decree had almost no effect elsewhere. Not even in Catholic Spain or Portugal were copies censored.[154] The condemnation of *De revolutionibus* had very little impact on the the acceptance of the heliocentric hypothesis. Even the famous trial of Galileo for continuing to advocate heliocentrism after the condemnation only served to popularize the new cosmology.[155] We should qualify this general proposition with a curious exception. After the condemnation, Jesuit missionaries in China, who introduced the Emperor's astronomers to European astronomy, were forbidden to teach Copernicanism. They therefore continued to teach the Tychonic system long after it had gone out of style in Europe.

The immediate reception of Copernicus's book was thus rather mixed. It did not produce an immediate revolution in astronomy. But, gradually, the new world system won over more and more astronomers. Within a few generations it became the new standard cosmology.

Why 1543?

Why was the new cosmology born when and where it was—in central Europe, in the first half of the sixteenth century? A large literature has grown up around this question. While no unanimity exists among scholars, we can point to several factors that played a part.

To some extent, Copernicus's heliocentric theory can be understood as an internal development in technical planetary theory. Although Copernicus used observations taken by his medieval predecessors, and made some new ones himself, *new observations played no essential role*. The essential facts about the connections between the motions of the planets and the motion of the Sun (illustrated in fig. 7.29) were stated plainly by Ptolemy and were known to all. Copernicus came to his discovery, not by observing the planets more closely, but by understanding Ptolemy more deeply than any of his predecessors. A close study of Copernicus's *De revolutionibus* shows that the single astronomer who exercised the most profound influence on him was Claudius Ptolemy. The single printed work from which he learned the most was the Peurbach-Regiomontanus *Epitome of the Almagest*. Copernicus's work can be understood as a part of the Ptolemaic tradition. Copernicus was one of the last, and one of the most accomplished, of Ptolemaic astronomers.

Why, then, did it take so long? If the Copernican revolution was the culmination of the astronomy of the *Almagest*, why did it take 1,400 years? Here we see most forcibly the inadequacy of a purely internalist explanation: an answer to this question requires several parts. First, we have little trouble understanding why it did not happen in late antiquity. Ptolemy was the last of the original astronomers of the Greeks. The economic and political crises of the third century, the rise of new religions (notably Christianity) that were more focused on the next world than on the sciences of this one, the military pressure of the barbarian tribes, and the division of the Roman empire in the fourth century were all factors that sapped the vitality of ancient science.[156] We should also remember that, in antiquity, scientific work was the activity of a very small number of people. It is noteworthy that there was not a single astronomer comparable to Hipparchus or Ptolemy in the two-and-half-century interval that separated them. Ancient science was a fragile thing, easily disrupted.

It is a little more difficult to understand why there was no "Copernican revolution" in medieval Islam. There, expertise in astronomy reached and remained at a good level much earlier than in Christian Europe. Moreover, from the ninth to the fifteenth century, we recognize many astronomers of ability and creativity. Of course, one must not make the mistake of treating Islamic civilization as if it were a monolithic bloc. Rather, what we see is a succession of cultural flowerings widely separated in time and place. The first flowering of astronomy came in ninth-century Iraq under the patronage of the Abbasid caliphs. By the eleventh century a new center of activity had emerged in Islamic Spain. As we have seen, original and highly speculative work in planetary theory was performed by the astronomers of Marāgha (northwest Persia) in the thirteenth century. Although texts were passed from region to region, this was not always a sure thing. The interruptions of scholarly activity—and the breaks in the chains of teachers and students—that occurred with declines in the economic or political fortunes of royal patrons mean that we should think of Islamic astronomy as involving several intersecting or overlapping traditions, not as one long, continuous development. Finally, there was not in medieval Islam anything corresponding to the European university system with a standardized scientific curriculum. Astronomy was widely cultivated, but astronomy did not assume as prominent a place in Islamic education as it was later to take in the European system of higher education.

The European revival of astronomy began only with the translation movement of the twelfth century. In the fourteenth and fifteenth centuries, European astronomers were still struggling to master the *Almagest*. It is only with Regiomontanus's *Epitome of the Almagest* that we see a work of European astronomy at about the same level as the Greeks had reached in the second century. Thus, one part of the answer is that, in Europe at least, a revolution in astronomy could not have occurred much sooner than the sixteenth century.

Finally we must acknowledge the weighty influence of Aristotle's philosophy of nature. Everywhere in both Islam and Christendom, Aristotle's views, elaborated by medieval commentators, provided the physical basis of cosmology. The element theory and Aristotle's theory of place and of natural motions were so completely integrated with Ptolemaic cosmology that it was difficult to modify the latter without endangering the former. If anything, this posed a more difficult problem in Christendom than in Islam, for the theologians and philosophers of the schools had succeeded so well in Christianizing the cosmos of the Greeks that an attack on Aristotle was, for some, tantamount to an attack on Scripture.

The gradual loosening of these intellectual bonds was an important part of the preparation for the new universe. We have seen that in medieval Islam, Ptolemy was sometimes criticized for his failings—though often his chief failure was perceived to be his lack of fidelity to Aristotle. Alternative planetary models were occasionally proposed. Although these had little immediate effect on practical astronomy, they did subject Ptolemy to critical reevaluation. Some of this criticism made it over the cultural gap between Islamic and European astronomy. Regiomontanus, in the *Epitome of the Almagest*, pointed out (as Islamic astronomers had already done) that according to Ptolemy's lunar theory, the Moon should look twice as large at some times of the month as at others. This vastly exceeds the actual variation in apparent size. Copernicus later echoed this complaint. Moreover, Copernicus also adopted some of the technical details of the planetary theories of the Marāgha school and of Ibn al-Shāṭir (devices for producing irregular motions in terms of uniform circular motions). We have no idea how Copernicus came by these devices. But there are too many of them to be explained in terms of independent discovery.[157] Most likely, Copernicus learned of them from someone in Italy when he was there as a student. Although these technical devices have nothing to do with heliocentrism, it is clear that the late medieval Islamic tradition of critical reassessment of Ptolemy did play some role in the Copernican revolution.

Criticizing Ptolemy was one thing. It was another thing altogether to develop a philosophical alternative to Aristotle. Although Aristotle's works were the core curriculum of the medieval universities, a lively tradition of criticism and disputation meant that in the later Middle Ages Aristotle's views were not simply accepted without close examination.[158] Moreover, in the Renaissance a wave of enthusiasm for Neoplatonism swept through European intellectual circles. Although Neoplatonism offered an alternative to the Aristotelian philosophy of the schools, it was not itself a unified philosophical system. It found a basis in the dialogues of Plato and their interpretations by commentators of late antiquity such as Plotinus, Porphyry, and Proclus. It incorporated a religious element, in which the soul seeks union with the ultimate, universal being through trance or ecstasy. Another important aspect of Neoplatonism was a revival of the Pythagorean Sun mysticism that we encountered in section 7.15.[159] The Sun was identified with excellence of all kinds and, therefore, also with God. This perhaps made it easier to transfer the center of the universe from the Earth to the Sun. No doubt, Neoplatonism did play a role in preparing people to accept heliocentrism. However, it is much harder to assess its influence on Copernicus himself. In *De revolutionibus*, Copernicus's physical arguments for the stability of the Sun and the motion of the Earth mostly take the form of revisions of Aristotelian positions.

One could argue that the astronomical revolution was practically bound to occur. The revival of trade and improvements in agriculture had placed society on a firmer economic basis than at any time in Greek antiquity. Europe could afford to support hundreds of teachers and thousands of students at universities. Moreover, the liberal arts curriculum meant that in every university town someone was responsible for teaching astronomy. The great surge of interest in astrology in the early Renaissance meant that even outside of the universities there were people studying Ptolemy. The number of competent practicing astronomers in sixteenth-century Europe far exceeded the number who had been active at any stage of Greek antiquity. Obviously, it makes no sense to invoke broad economic and social forces to explain Copernicus himself—one man in a tower on the edge of the Baltic, thinking about the planets—but such forces must be taken into account in explaining the great general vitality of sixteenth-century astronomy. Although Copernicus was on the edge of the scientific world, he had studied at good universities and was close enough to a hub of activity in Nuremberg to benefit when the time came to publish.

7.31 KEPLER AND THE NEW ASTRONOMY

While Copernicus had radically transformed the ancient cosmology, he left astronomy largely in the Ptolemaic tradition. The planets still moved in combinations of uniform circular motions. It remained for Johannes Kepler to remove not only circles but also the convention of uniform motion from astronomy. By 1609 Kepler etablished that the paths of the planets are actually ellipses and that the speed of a planet on its path obeys a rule that is somewhat different from the law of the equant. Let us see how Kepler came to make such a radical break with the old astronomy.

The Mystery of the Universe

Kepler came to the study of planetary theory slowly and almost against his will. Kepler went to the University of Tübingen to study philosophy and theology. But there he was lucky enough to encounter a gifted professor of astronomy, Michael Mästlin, a convinced Copernican. Thus, Kepler was among the first generation of university students to learn heliocentrism at school. Mästlin deserves most of the credit for Kepler's interest in astronomy—

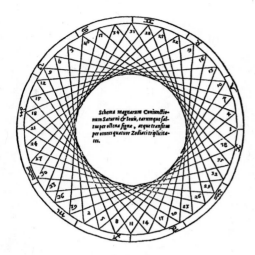

FIGURE 7.67. Kepler's diagram of the great conjunctions of Jupiter and Saturn in *Mysterium cosmographicum*.

and for convincing him to take a teaching job in Graz (Austria) instead of pursuing a church career.[160]

As a teacher, Kepler was a disaster and he attracted few students. He regarded himself as ill-prepared for such work. But it was at Graz that he wrote his first book, *Mysterium cosmographicum* (Cosmographical mystery, or more loosely translated, The mystery of the universe). It came about in the following way. While lecturing to his students about the pattern of conjunctions of Jupiter with Saturn, Kepler noted that the positions in the zodiac of three successive conjunctions formed nearly an equilateral triangle. From one set of three conjunctions to the next, the positions rotated slightly. When he drew them all out, Kepler found that the pattern of rotating triangles generated an inscribed circle (fig. 7.67). This struck Kepler so forcibly that he felt sure it had some deeper significance. Furthermore, the ratio of radius of the circumscribed circle to that of the inscribed circle was nearly the same as the ratio of the radii of the Copernican orbits of Saturn and Jupiter. Excited by his discovery, Kepler tried to include the other planets in the scheme, by inscribing a square in the orbit of Jupiter, and the circle for Mars within the square, and so on. But the numbers did not work out. Moreover, there was no way of knowing which of the infinite number of regular plane polygons (triangle, square, pentagon, etc.) ought to be included in the scheme.[161]

The beauty of Copernican astronomy is that it gives simple geometrical explanations of cosmic facts that were known but inexplicable in terms of Ptolemaic astronomy. However, a large amount of arbitrariness seemed to remain in Copernicus's world system. Why was each planet's orbit the particular size that it was? Why was the eccentricity of each planet's deferent circle just the size that it was? And, moreover, why should there be six planets rather than some other number? Today, we regard all of these facts as accidental results of the initial conditions that prevailed when the solar system was condensing. But Kepler saw them as aspects of God's plan, and he set out to deduce these features of the world from first principles.

It dawned on him that three-dimensional bodies were more suitable than plane figures for his scheme. Moreover, it now became obvious that there could only be six planets *because there were only five regular solids*. A regular solid is one whose faces are all identical, whose edges are all the same length, and whose vertices are all similar. An example is the cube. The regular solids (and the fact that only five of them exist) had been known to the ancient Greeks. Moreover, the regular solids already had a place in the history of cosmological thought.

In his *Timaeus*, Plato identified each of the regular solids with one of the elements.[162] On account of its stability, the cube was identified with earth. Because it is pointed and penetrating, the tetrahedron was identified with fire. The octahedron was identified with air, the icosahedron with water, and the dodecahedron with the cosmos itself. Plato suggested that changes in the physical world were due to the transformations of the elements into one another. This came about through the breakup of the elemental solids into their constituent triangles and the reassembly of these triangles into other solids. Of course, this did not work out in detail, as Aristotle was later to point out.[163] One serious difficulty was that Plato's elemental solids could not really all be transformed into one another. The faces of the tetrahedron are equilateral triangles. The faces of the cube are squares. Each square can be divided in half, yielding two triangles—but these are isosceles, not equilateral.

So, when Kepler tried to associate the regular solids with the intervals between the planets, he was quite in the Pythagorean tradition. The revival of the Pythagorean approach to nature was an aspect of Renaissance Neoplatonism. Kepler went on to become the most outstanding mathematical astronomer of his generation. His greatest gifts were inexhaustible patience, great calculating ability, and a relentless drive to understand. But his motives for astronomical research always involved a quest for higher knowledge. Everywhere, he sought for connections between apparently disparate realms of thought. His wanted to know God's plan for the cosmos. Christian theology played a role in his quest: at one time Kepler saw an analogy between the three things that are always at rest—the Sun, the stars, and the intervening space—and the Trinity of Father, Son, and Holy Spirit. At another stage he revived the Pythagorean doctrine of the harmony of the spheres, in an attempt to associate the speeds of the planets with musical notes. When William Gilbert discovered that the Earth is a giant magnet, Kepler seized on that idea, too, and sought in magnetism an explantion of the motive force that the Sun exerts on the planets.[164] Kepler's approach to nature, combining elements of animism, mathematical mysticism, and physical reasoning, was not at all unusual for the Renaissance. What made Kepler's work different was that he also became a skillful technical astronomer and that he judged his own cosmological speculations very strictly: if he was right, then everything should work out in numerical detail.

We see this aspect of Kepler already present in his first book, *Mysterium cosmographicum*. The basic idea is that the heliocentric spheres for the planets should be inscribed within the regular solids. Spheres and solids form a concentric, nested set, each just touching its inner and outer neighbors, as in figure 7.68. Kepler writes:

> The Earth is the circle which is the measure of all. Construct a dodecahedron round it. The circle surrounding that will be Mars. Round Mars construct

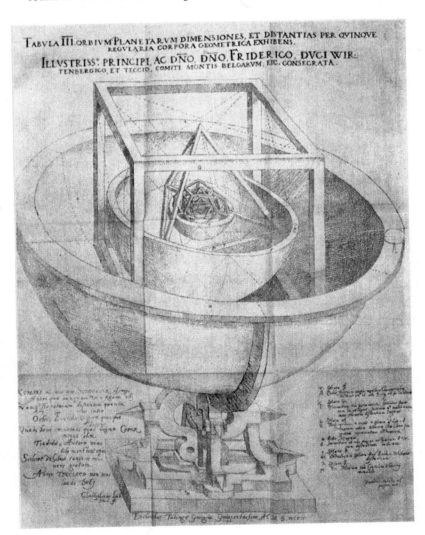

FIGURE 7.68. Kepler's geometrical scheme for deducing the sizes of the planetary spheres in *Mysterium cosmographicum*.

a cube. The circle surrounding that will be Saturn. Now construct an icosahedron inside the Earth. The circle inscribed within that will be Venus. Inside Venus inscribe an octahedron. The circle inscribed within that will be Mercury. There you have the explanation of the number of planets.[165]

The radius of each planetary sphere should correspond to the radius of the planetary orbit in Copernicus's theory. Moreover, each planetary sphere must be assigned a certain thickness, in accordance with the eccentricity of the planetary orbit. In fact, Kepler tried two different versions of the hypothesis. In one, the sphere assigned to the Earth was given a thickness determined by the eccentricity of its annual circle. In the other, the thickness of the Earth's sphere was increased to allow for the orbit of the Moon.

In a way, Kepler's scheme can be regarded as a new version of the old nested-spheres cosmology. Some things were radically different, of course. The places of the Sun and the Earth had been reversed. And, after the work of Tycho Brahe, it was scarcely possible to believe in real, material spheres. Moreover, in the Copernican system, it was impossible to retain the Ptolemaic doctrine of no empty space, for there were vast spaces between the planets. Kepler did not regard his polyhedra as real. These were intellectual constructs, presumably a part of God's plan for determining the spacing between the planets. Nor were the spheres real, material things. Rather, they provided spaces to accommodate the paths of the planets. The material spheres of Aristotle and Ptolemy have now become rather like shadowy, Platonic forms. Nevertheless, they remind us how hard it is to shake off old habits of thought.

When Kepler worked out all the numbers in this geometrical cosmology, he found that his theory agreed very well with Copernicus's figures for the parameters of the planetary spheres—but not perfectly. Kepler suspected that the discrepancies might be due to minor errors in Copernicus's values for the eccentricities and radii of the orbits. Still, the agreement was so good that Kepler felt sure he was onto something important.

Mysterium cosmographicum was seen through the press at Tübingen by Kepler's former teacher, Michael Mästlin. Kepler sent copies of the book to leading scholars and astronomers, including Tycho Brahe in Denmark. As a result, he received an invitation to visit Brahe. Brahe expressed sympathy for Kepler's efforts to understand the arrangment of the universe on an a priori basis, saying that God surely had a harmonious plan for the creation. Brahe suggested that a more decisive test of the nested polyhedra could be made if the true values for the eccentricities, which Brahe himself had sought to determine, were used in place of Copernicus's values. For a while, nothing came of this invitation.

However, Brahe's patron, the king of Denmark, had died and Brahe's relations with the new king and his advisors had been deteriorating. Brahe found a new patron in the German emperor (Rudolph II of the Holy Roman Empire) and moved to Prague with his family, assistants, and instruments. Meanwhile, Kepler's life was becoming more difficult in Graz, where the Catholic authorities had banned Protestants from teaching positions. Kepler at first received a special dispensation, but soon he, too, came under pressure. With Brahe now much closer and with no good reason to remain in Austria, Kepler finally accepted Brahe's invitation—after delicate and protracted negotiations concerning salary and living and working arrangements.

Progress and Failure

What Kepler wanted from Brahe was simply to be given better values for the radii and eccentricities of the planets' circular orbits, so that he could refine his cosmology. He was astonished and disappointed to find that, although Brahe had acquired a huge mass of observational material, he had not yet worked out completed theories for most of the planets. Moreover, Brahe jealously guarded such elements as he had already determined and even seemed to enjoy tormenting Kepler. At dinner, in the course of discussing other matters, Brahe one day mentioned the apogee of one planet, the next day the nodes of another.[166]

Kepler soon found himself immersed in technical planetary theory, attempting to work out a theory of Mars that would agree with Brahe's observations, the most precise and extensive observations of planetary positions in the history of astronomy up to that time. Kepler has a little to say about how this came to pass. According to Kepler, a divine voice leads humans to the study of astronomy. And this voice is expressed, not in words or symbols, but in the world itself, in the conformity of the human intellect and the human senses with the dispositions of the celestial bodies. But there is also a kind of fate. When Kepler arrived to work with Brahe, Brahe's principal assistant was busy observing Mars, on the occasion of its opposition to the Sun, and with preparing a theory of the motion of Mars. Christen Sørensen worked with Brahe from 1589. Son of a farmer from the village of Lomborg in the northwest of Jutland, he Latinized his name and the name of his native village, styling himself Christianus Severinus Longomontanus.[167] As Kepler says, "Had Christian been treating a different planet, I would have started on it as well."[168]

Of the naked-eye planets, only Mercury and Mars have eccentricities large enough to make the departures of their orbits from perfect circles (and of their motions from equant motion) apparent through the analysis of naked-eye observations. But Mercury is so near the Sun that one cannot easily obtain a sufficient number of observations to work with. This means, practically speaking, that Kepler's discoveries could only have been made through a study of the motion of Mars. As Kepler put it,

> I therefore once again think it to have happened by divine arrangement, that I arrived at the same time in which he was intent upon Mars, whose motions provide the only possible access to the hidden secrets of astronomy, without which we would remain forever ignorant of those secrets.[169]

Mars, then, despite its astrologically maleficent character, was friendly to astronomers: let us recall (sec. 7.17) that Ptolemy probably discovered the equant point while grappling with Mars's retrogradations.

Providence, fate, or accident, it was a long series of circumstances that led Kepler to Mars. If Mästlin hadn't been at Tübingen, if Brahe hadn't fallen out with the new king, if Kepler hadn't been forced out of Gratz, if Longomontanus hadn't been working on Mars. . . .

Equants or Epicycles? The planets do not move on paths that all have the same center. And they do not move uniformly. These two facts of nature must be confronted regardless of whether we choose a Sun-or an Earth-centered cosmology. Let us therefore step away from the cosmological question and simply consider the devices available to Renaissance theoreticians for dealing with apparent nonuniformity of motion.

Figure 7.69 shows a generalized equant model. A celestial body B moves around a circle that is centered on C. Let D be eccentric to C. In a Sun-centered theory, D would be the mean Sun and B the planet itself. In Earth-centered astronomy, D would be the Earth and B the center of the planet's epicycle. In Ptolemy's theory $EC = CD$. That is, the two eccentricities e_1 and e_2 are equal. But now we wish to allow for the possibility that these two are not the same.

We call e_1 the eccentricity of the eccentric. Because D is separated from the center of the circle by distance e_1R, point B would *appear* to move around D nonuniformly (in degrees per day), even if the actual motion on the circle (in miles per day) were uniform. This apparent variation in speed is an optical effect, due to the varying distance of B from D.

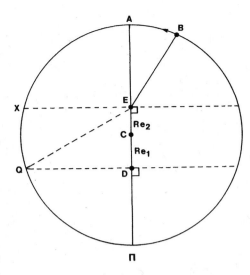

FIGURE 7.69. A generalized eccentric-and-equant model.

We call e_2 the eccentricity of the equant. Because B moves at a uniform *angular* speed about E, but E is not the center of the circle, it follows that B physically speeds up and slows down (in miles per day) while traveling around the circle.

To sum up, e_2 produces a *physical* variation in speed of point B; e_1 produces an additional *apparent* variation in speed of B as observed from D. The total irregularity of apparent motion is the sum of the two effects. We call $e_1 + e_2$ the total eccentricity. The maximum value of the equation of center (the maximum apparent departure of the planet from where it would be if it moved uniformly) is approximately determined by the total eccentricity. Again, in Ptolemy, *the total eccentricity is bisected*, so that $e_1 = e_2$.

As we saw in section 7.30, Copernicus rejected the physical variation in speed due to the equant. He replaced the equant by a minor epicycle of radius aR, moving on a deferent circle displaced from D by distance bR (fig. 7.64). But if the total eccentricity is kept the same (so that $a + b = e_1 + e_2$) and if we put $b = 3a$, then Copernicus's model differs little from Ptolemy's. Thus, Copernicus, too, could be said to have adopted a bisection of the eccentricity.

But now let us allow for the possibility that e_1 and e_2 are not equal (fig. 7.69). How are e_1 and e_2 related to a and b in this more general situation? In figure 7.65, we see that in a generalized Copernican model, the effective center M of the planet's path is located a distance away from D equal to $bR - aR$. And the hidden equant point E of the Copernican model is located a distance away from M equal to $2aR$. Thus, comparing figures 7.65 and 7.69, we see that the two models will be approximately equivalent if

$$e_1 = b - a,$$

$$e_2 = 2a.$$

Turning this around, we can also express a and b in terms of e_1 and e_2:

$$a = \frac{1}{2}e_2$$

$$b = e_1 + \frac{1}{2}e_2 .$$

If these relations hold, the generalized Copernican device (eccentric plus minor epicycle) will be roughly equivalent to the generalized Ptolemaic device (eccentric plus equant). The two models are not *precisely* equivalent, but it would take very accurate observations to tell which rule a real planet was following. Before Kepler, the choice between these devices was based on nonastronomical criteria. For example, Copernicus rejected the equant because it involved a physical variation in speed.

Tycho Brahe's Theory of Mars   When Kepler arrived in Prague as Brahe's guest early in the year 1600, he found Brahe and Longomontanus occupied with Mars. Brahe and Longomontanus had worked out a theory of Mars based on the Copernican model of figure 7.65. The total eccentricity $(a + b)$ had been determined from three oppositions of Mars to the mean Sun. The model was checked against the observations of ten successive oppositions from 1580 to 1600. Brahe and Longomontanus, however, had found it necessary to divide the eccentricity differently than had Copernicus.[170]

Indeed, Brahe and Longomontanus put

$$a = 0.0378, \quad b = 0.1638.$$

So, $b/a = 13/3$, rather than $9/3$ as with Copernicus. Thus, Brahe made the radius a of the epicycle a bit smaller and the eccentricity b of the deferent a bit larger. The total eccentricty $a + b$ came to 0.2016, which differed very little from the values adopted by Copernicus and Ptolemy. The equivalent values of e_1 and e_2 in an equant model are

$$e_1 = b - a$$

$$= 0.1260,$$

$$e_2 = 2a$$

$$= 0.0756,$$

so $e_1/e_2 = 5/3$. To speak in terms of equants, while Ptolemy and Copernicus had assigned half the total eccentricity to each e_1 and e_2, Brahe and Longomontanus gave 5/8 of the total to e_1 and 3/8 to e_2. Brahe and Longomontanus found that their theory agreed with the observed longitudes of the ten mean oppositions to within $2'$.

A bisection of the eccentricity would not have been capable of such excellent agreement with the oppositions—a fact that Brahe and Longomontanus must certainly have discovered before their adoption of the unequal division. The 5:3 division on which they settled is, in fact, the best possible for matching the kind of observation they were using.[171] As their equivalent values for e_1 and e_2 stand exactly in the ratio 5:3, it is clear that this division was the result of an a priori decision, but it was probably arrived at only after several trials.

Although the theory was in perfect accord with the observed longitudes of ten mean oppositions, it failed completely to give the proper latitudes at opposition. Nor did it give satisfactory longitudes in situations away from the oppositions. And there Longomontanus was stuck.

Kepler's Physical Intuitions   Kepler worked with Brahe only for a short time, for Brahe died quite suddenly in 1601. As he lay dying, Brahe extracted a pledge from Kepler not to abandon his cosmological system—the geo-heliocentric compromise. Kepler was appointed to Brahe's position as imperial mathematician under Rudolph II. However, Kepler soon faced severe difficulties. His imperial stipend was rarely paid on time. Even more problematical, he had trouble with Brahe's heirs, who were for a long while unwilling to give up Brahe's notebooks of observations.

Kepler persevered. In his struggle with Mars, Kepler was guided constantly by physical principles, as he perceived them. The book in which he recorded these struggles and in which he announced the ellipticity of the orbit and the law of areas was titled *Astronomia nova* ΑΙΤΙΟΛΟΓΗΤΟΣ, *seu physica coelestis* (A new astronomy, founded on causes, or celestial physics). Again and again, Kepler stresses that he seeks a physical, and not merely an astronomical, solution to the problem of the planets.

This new insistence of physical causes entailed a more thoroughgoing heliocentrism than that of Copernicus or Brahe. These two, following Ptolemy, had referred all motions to the mean Sun rather than to the true Sun. Thus, in Copernicus, the lines of apsides of all the planets pass through the mean Sun, that is, through the center of the Earth's orbit. Kepler insisted that the lines of apsides should all pass through the body of the Sun itself. This went a step further than Copernicus had gone in depriving the Earth of special privileges. It also resulted in small corrections to the directions of the lines of apsides of the planets.

Kepler's Theory of the Earth's Motion   In his treatment of the motion of the Earth, Kepler was able to demonstrate that there is an equant point in the Earth's orbit. Let us recall that Hipparchus and Ptolemy had let the Sun move around the Earth at constant speed on a simple eccentric circle. Copernicus,

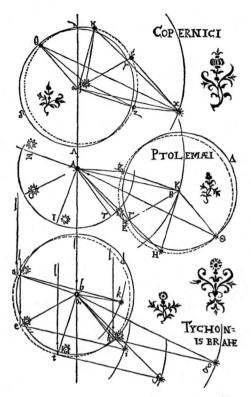

FIGURE 7.70. Kepler's diagram from *Astronomia nova* illustrating the necessity of bisecting the eccentricity of the Earth's orbit and, thus, of treating the Earth in the same way as the other planets.

while reversing the roles of the Sun and Earth, had adopted the same model. Unlike the other planets, Copernicus's Earth did not have a minor epicycle. In proving that the Earth's orbit about the Sun must be treated in exactly the same way as the orbits of the other planets, Kepler took another decisive step forward.

The introduction of an equant point into the Earth's orbit meant putting $e_1 = e_2$ in figure 7.69 for the Earth, just as for all the other planets. Ptolemy and Copernicus had effectively put $e_2 = 0$ but had compensated by making e_1 twice as large. Thus, in Kepler's theory of the motion of the Earth (as well as in the modern theory), the variation in the distance of the Earth from the Sun in the course of the year is only half as great as in Ptolemy or Copernicus. Kepler came to this conclusion, not by direct observation of the Sun, but through a study of the effect of the Earth's orbit on the observed positions of Mars.

Kepler's method of determining the orbit of the Earth involved a clever method of triangulation. Among Brahe's extensive observations, Kepler was able to find several pairs of observations of Mars separated by one Martian orbital period (about 687 days). That is, at the times of several different observations, Mars was known to be at the same point x of its orbit, as in the upper portion of figure 7.70. Successive positions of the Earth at times 687 days apart are labeled θ, η, ε, and ζ. Analysis showed that the Earth's actual circle (shown in dashed line) was only half as eccentric to the Sun as Copernicus's version of the Earth's circle (solid line). But if the eccentricity e_1 were only half as large as formerly believed, it was necessary to introduce an equant point and a physical variation in speed to make up the full known inequality in the lengths of the seasons.

Incidentally, figure 7.70 shows how faithfully Kepler kept his pledge to Brahe not to let the Tychonic system disappear. In the first portions of *Astronomia nova*, Kepler patiently shows how everything goes in three different world systems, Copernican, Ptolemaic, and Tychonic. Thus, no matter which system we adopt, the eccentricity of the annual circle (of the Earth or the Sun) must be cut in half. These parallel exercises proved beneficial to Kepler, for they helped him appreciate more clearly the real advantages of the Copernican system. But he soon gave them up: in the later portions of *Astronomia nova* he was content to work purely in heliocentric terms.

Kepler's Vicarious Hypothesis for the Motion of Mars   In his study of the motion of Mars, Kepler found yet another way to insist on physical plausibility. In his treatment of the inequality in Mars's motion about the Sun, Kepler returned to the principle of equant motion. The outer planets, such as Saturn, traveled more slowly than the inner planets. It was not unreasonable to suppose, then, that the speed of a given planet does physically vary with the planet's distance from the Sun—the planet traveling more slowly when farther from the Sun and more rapidly when nearer. The principle of equant motion directly addressed this variation in speed. In the nearly equivalent epicycle model of Copernicus, the physical variation in speed was hidden behind the guise of uniform circular motion.

Ptolemy had bisected the eccentricity of the superior planets without explicit justification, and Copernicus had followed him. Brahe and Longomontanus had favored a 5:3 division. Kepler was unwilling to assume any a priori division, but sought to determine e_1 and e_2 directly from observation. This problem—to determine e_1, e_2, and the direction of the line of apsides—was more difficult than Ptolemy's version of the problem, in which e_1 and e_2 were assumed to be equal. Kepler's more general problem required the use of four oppositions rather than three. And while Ptolemy had used a clever method of iterations, in which successive corrections to an approximate solution were calculated, Kepler had to proceed more nearly by trial and error. It is in the description of this procedure that Kepler makes his famous remark, that if the reader finds the discussion tedious and difficult, he should pity the author who had to perform the same calculation seventy times before arriving at an answer.[172]

The calculations were based on the oppositions of 1587, 1591, 1593, and 1595. Assuming that Mars moves on a circle eccentric to the true Sun, with a speed that varies in accordance with the law of the equant, Kepler found that the angular position of the planet was reproduced at each of the four oppositions if

$$e_1 = 0.11332, \quad e_2 = 0.07232,$$

while the total eccentricity $e_1 + e_2$ came to 0.18564. This theory of the motion of Mars Kepler called his "vicarious hypothesis" (because it worked well and he was therefore he was able to use it for some purposes even when he knew it was not the final answer).

The values that Kepler had so laboriously found for e_1 and e_2 were not very different from those adopted by Brahe and Longomontanus. Indeed, $e_1/e_2 = 4.7/3$, while Brahe and Longomontanus had put $e_1/e_2 = 5/3$. When Kepler checked the new theory against the 12 oppositions between 1580 and 1604, he found that in only four cases was the discrepancy more than 1', the worst being but 2'12". Discrepancies of this size could be attributed to the finite angular diameter of the planet and to the fact that the planet's parallax and the correction for atmospheric refraction were not known with precision.

And yet this theory that had cost so much labor, and that was so well confirmed by the dozen oppositions, was completely false! Kepler demonstrated the falsity of the vicarous hypothesis most directly by an investigation of Mars's *latitudes* while in opposition. Especially useful were oppositions near the northern and southern limits of the planet's orbit. Refer to figure 7.71. Let OO' represent the plane of the Earth's orbit seen from the edge. Let NN' represent the plane of the orbit of Mars. Both planes pass through the Sun S. The inclination i of the two planes Kepler had already determined. Consider an opposition of Mars when the Earth is at O and Mars is in the northern limit N of its orbit. Thus, S, O, and N lie in a plane that is perpendicular to the plane of the Earth's orbit. The latitude β of the planet is found by observation. Distance SO is given by the already completed theory of the Earth's motion. Thus, in triangle SON, the angles at S and O are known and side SO is known, so side SN may be calculated.

Similarly, consider a second opposition of Mars located 180° farther around in longitude, when the Earth is at O' and Mars is at the southern limit N' of its orbit. A measurement of Mars's latitude β' permits the calculation of SN'. Then it is possible to compare SN with SN'. The difference between SN and SN' leads directly to a value for the distance between the Sun and the midpoint of NN'—hence a value for e_1.

The results depended on the details of the adopted theory of the motion of the Earth, which affected distances SO and SO'. But Kepler concluded that

$$0.08000 \leq e_1 \leq 0.09943,$$

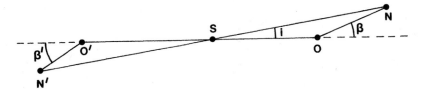

FIGURE 7.71. Illustrating Kepler's method of determining e_1 for Mars from the latitudes measured when the planet was simultaneously in opposition to the Sun and in the northern or southern limit of its orbit.

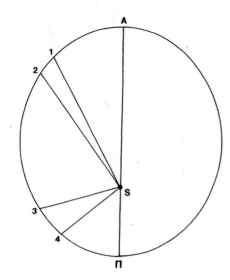

FIGURE 7.72. Kepler's law of areas.

while the vicarious hypothesis (supported by the longitudes of a dozen oppositions) required $e_1 = 0.11332$. *Two methods of determining e_1 led to two different results.* The vicarious hypothesis was false.

Now, the total eccentricity in the vicarious hypothesis was 0.18564. Half this was 0.09282, a value that fell near the middle range of the possible values for e_1 determined by the latitudes. Thus, Kepler wondered whether a bisection of the eccentricity might be justified after all. Kepler conjectured that Ptolemy must have settled on a bisection after experiencing similar difficulties with the latitudes.

With frustration, Kepler turned back to the equant model with bisected eccentricity. The model with a bisected eccentricity, with $e = 0.09282$ (half the total eccentricity in the vicarious model), will reproduce the longitudes of the oppositions located near aphelion or perihelion, as well as those located $\pm 90°$ from these points. But when Kepler examined oppositions in the *octants* ($\pm 45°$ from either aphelion or perihelion), he found that that the bisection model disagreed by 8' or 9' with the observations of Tycho Brahe. As Kepler remarks, Ptolemy did not claim to have observed with precision better than 10'. The uncertainty in Ptolemy's observations therefore exceeded the error in the model.

At this point Kepler remarks that, as "divine benevolence has vouchsafed us Tycho Brahe, a most diligent observer, from whose observations the 8' error in this Ptolemaic computation is shown, it is fitting that we with thankful mind both acknowledge and honor this benfit of God. . . . These eight minutes alone . . . led the way to the reformation of all of astronomy."[173]

Principles of the New Astronomy

Kepler set out to do what no one else in the history of astronomy had done—to determine from scratch the varying motion of Mars around its orbit and the shape of the orbit itself, with no assumptions about what was appropriate to a celestial object. Needless to say, this would have been impossible without Tycho Brahe's observations. But Kepler himself brought essential qualities to the task—especially his relentless drive and his intuition to seek a physical solution to the problem of the planets and not merely a mathematical one.

The Law of Areas   As mentioned above, Kepler's physical intuition opened him up to the reality of a physical variation in speed. But when two different versions of the equant model (with bisected eccentricity and with the 5:3 division) proved incapable of agreement with Brahe's observations, Kepler was forced to abandon the equant. After considerable difficulty and many false steps, Kepler arrived at the true speed rule of planetary motion, the law of areas.

Refer to figure 7.72. The planet moves in an orbit about the Sun S. The planet physically speeds up and slows down, so that it travels most rapidly at perihelion Π and most slowly at aphelion A. Further, *the speed varies in such a way that the radius vector from the Sun to the planet sweeps out equal areas in equal times.* As the planet moves from position 1 to position 2, the radius vector sweeps out the quasi-triangular area S12. Suppose that the planet moves from 1 to 2 in one day. Consider one day's worth of motion at some other place on the path. Thus, let the planet move from position 3 to position 4 in one day. The radius vector sweeps out the quasi-triangular area S34. Then, according to Kepler, the areas S12 and S34 are equal. This law of areas is usually called Kepler's second law, though it was the first to be discovered.

Elliptical Orbits   By itself, the area law could not eliminate all discrepancies between theory and the observations of Tycho Brahe. It remained for Kepler to determine the actual shape of the path. As we have seen, he had begun with the universal assumption that the paths are eccentric circles and only gave up this assumption when it failed after repeated trials. He soon became convinced that the actual path was somewhat oval shaped. It took much longer to determine the precise character of the oval and Kepler arrived at the answer only after making many bad guesses.

Kepler's method of investigation was, again, triangulation. At two times when Mars was at the same point of its orbit, but the Earth was at different places on its own orbit, the apparent directions of Mars as seen from the Earth provided two lines of sight. Where these sight lines crossed was the position of Mars. This procedure had of course to be repeated for multiple positions of Mars.[174]

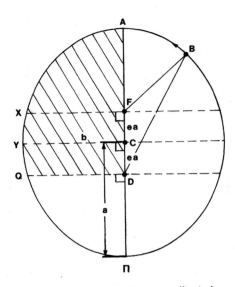

FIGURE 7.73. Mars travels on an elliptical orbit, one focus of which is at the Sun.

Kepler's conclusion was that the orbit of Mars was an ellipse, with one of its two foci located at the Sun. Refer to figure 7.73. The Sun is at D, one of the two foci of the ellipse. The other focus is F. Distance $a = AC = \Pi C$ is called the semimajor axis of the ellipse. Both foci are separated from the center C of the ellipse by distance ae, where e is called the eccentricity of the ellipse. The semiminor axis of the ellipse is distance CY, which we denote b.

The ellipse was a curve known from Greek antiquity. It belongs to a class of curves called conic sections, because they can be produced by slicing a right circular cone. But perhaps the simplest way to define the ellipse is in terms of its focal-point property. In figure 7.73, let B be any point on the ellipse. Then, for all such points, the sum of the two distances FB and DB is constant (and in fact equal to $2a$, as may be seen by considering the situation when B coincides with A or Π).

A simple way to draw an ellipse on paper is to place two tacks at the foci F and D and to tie the ends of a string of length $2a$ to the tacks. (Thus, in fig. 7.73, the string is represented by FB and BD.) If one places the point of a pencil inside the loop of string, as at B, and draws around the two foci while keeping the string taut, an ellipse will be scribed on the paper.

Publication of Astronomia nova   Kepler's discoveries were published in 1609 in his *Astronomia nova* (New astronomy). The introductory portions of the book include (1) Kepler's dedicatory letter to Emperor Rudolf II, in which he constantly describes his astronomical endeavors as a war against Mars, with Tycho Brahe as commander-in-chief; (2) Kepler's ode to Tycho Brahe ("Smooth is the road now, that formerly no one could travel for ages . . . "); (3) Kepler's announcement that it was the theologian Andreas Osiander, and not Copernicus himself, who inserted the notorious preface on hypotheses into opernicus's *De revolutionibus*; and (4) perhaps the most interesting of all, Kepler's own long introduction, which sets out the astronomical, physical, philosophical, and religious reasons for believing in heliocentrism.

Astronomia nova was a difficult book for others to read, a fact that Kepler was well aware of. In the introduction, Kepler laments the dilemma confronted by the author of a mathematical book. Unless the author maintains the rigorous sequence of proposition, construction, demonstration, and conclusion, the book will not be mathematical, but maintaining that sequence makes the book tiresome to read. Moreover, adds Kepler, "there are very few suitably prepared readers these days." He admits that he finds it wearying to reread his own work and that he vacillates between two opposite faults—lack of clarity through too little explanation and lack of clarity through too much verbosity.

Kepler did not do a very good job of highlighting his most important results—the law of areas and the elliptical shape of the orbit. These are buried deep in the book. In large part, this was a result of Kepler's method of composition. Much of *Astronomia nova* had been written before Kepler even arrived at his most important discoveries. Moreover, the later chapters of the book are needlessly burdened with Kepler's speculations about magnetic motive forces. Of course, Kepler would not have been Kepler without these speculative asides. As we have seen, his quest for a physical solution to the problem of the planets motivated all of his work in technical astronomy. While some readers found the magnetic philosophy attractive, it did not add anything to the strength of Kepler's geometrical demonstrations and may, in fact, have weakened them by making them seem to depend on dubious hypotheses.[175]

Kepler's Later Work

The Harmonic Law Kepler never abandoned the search for the secret of the universe that he had begun in his first book. He returned to the polyhedra again in his *Harmonice mundi* (Harmony of the world), published in 1619. But now Kepler tried to tie these geometrical figures to the Pythagorean notion of a celestial harmony. The new twist is that, instead of basing the theory of harmony on arithmetic ratios, he tries to build it upon proportions obtained from the geometry of polyhedra. In book V of the *Harmonice*, Kepler takes up a more standard version of Pythagoreanism and associates musical notes with the motions of the planets. The pitch (high or low) is associated with the planet's speed. The range of notes emitted by a planet is connected with the eccentricity of the orbit—the more eccentric the orbit, the greater the musical range.[176]

In the course of this investigation, Kepler happened upon his third great discovery. This is called Kepler's third law or, sometimes, the harmonic law. Kepler's first two laws describe the shape of an individual planet's orbit and the variation of the planet's speed on the orbit. The third law states a relationship that connects the properties of the orbits of different planets. Already as early as Plato and Aristotle, people had conjectured that there must be some relation between the orbital periods and the radii of the planets' circles. It is Kepler's harmonic law that provides the connection.

The rule is this: the ratio of the cube of the semimajor axis to the square of the (sidereal) orbital period is constant for all bodies orbiting the Sun. Expressed in symbolic language, we may write

$$\frac{a^3}{T^2} = \text{constant}.$$

To illustrate the meaning of this rule, let us use it to calculate the orbital period of Mars from its semimajor axis. The constant (which applies to all of the planets) is most easily evaluated using the Earth. Let a_o denote the astronomical unit, that is, the semimajor axis of the Earth's orbit. Let T_o denote the orbit period of the Earth, that is, the sidereal year. Similarly, let a_M and T_M denote the semimajor axis and the orbital period of Mars. Then, according to Kepler's harmonic law,

$$\frac{a^3_M}{T^2_M} = \frac{a^3_o}{T^2_o}.$$

Since we know that $a_M = 1.52\ a_o$ (see sec. 7.29), it is easy to calculate the sidereal period of Mars:

$$T_M = \sqrt{1.52^3} \text{ years}$$

$$= 1.88 \text{ years}.$$

In the seventeenth century the most important application of the harmonic law was the reverse of our example—the determination of the semimajor axis of a planet's orbit from its orbital period. The orbital periods could be detemined to high precision from very simple observations extended over long time intervals. Derivation of the semimajor axis directly from position measurements was a much more delicate business.

The harmonic law, today regarded as one of the three essential facts of planetary motion, was buried in the *Harmonice* as the eighth of thirteen points "necessary for the contemplation of celestial harmonies." Kepler, always seeking for deeper mysteries, sometimes failed to perceive what was most significant and lasting in his own work. As was the case with the first two laws in *Astronomia nova*, Kepler did not make it easy for readers to extract the essential results.

The Epitome of Copernican Astronomy   Fortunately, Kepler did emphasize his laws of planetary motion in later publications. In his *Epitome of Copernican Astronomy*, published in stages between 1618 and 1621, Kepler provided the first systematic textbook of the new astronomy, setting out in patient detail all of the advantages of Copernicanism, as well as Kepler's own principles of planetary motion. To make it more accessible, Kepler cast the whole into the form of a catechism—all questions and answers. The three laws of planetary motion were given ample attention, and clear directions were provided for doing practical calculations in the new elliptical astronomy. Of course, Kepler could not resist including his polyhedra and "physical" explanations of the anomalies of planetary motion in terms of solar magnetism.

Rudolphine Tables   Kepler had long contemplated the construction of new planetary tables, based on the elliptical orbits and the law of areas. (He had already published some ephemerides.) The tables were not published until 1627, under the title *Tabulae Rudolphinae*, in honor of Kepler's patron, Emperor Rudolph II. This work had as much influence as any of Kepler's others in winning acceptance for the new astrononomy, for it rapidly became clear that the *Rudolphine Tables* were the most accurate planetary tables in existence. Kepler's planetary tables were also the first to make use of logarithms (newly discovered by Napier and considerably advanced by Kepler himself), which provided major savings in labor for the user of the tables.

In the decades following 1630, Keplerian astronomy gradually became accepted. The elliptical orbit and the law of areas were not adopted immediately by everyone. Some astronomers had a hard time letting go of the circles. Also, it is possible to generate an ellipse by means of an epicycle that rotates at the appropriate speed while riding around a deferent circle. Thus, astronomers who wanted to hold onto the old physics could at least make a reasonable argument. Already in 1607, before the publication of *Astronomia nova*, the astronomer David Fabricius wrote to Kepler to urge him not to abandon the principle of uniform circular motion. Fabricius added, "You can excuse the ellipse by another small circle."[177] Unfortunately, an elliptical motion generated in this way does not take place with the correct variation in speed.

Another popular dodge was to accept the elliptical shape of the orbit but to reject the area law. Ellipses were bad enough, but at least they could be constructed from circles. The area law had the double disadvantage of lacking an accepted physical basis and of being inconvenient for calculation. One alternative was to place a Ptolemaic equant point at the empty focus of an elliptical orbit. Such a theory was advocated by the English astronomer Seth Ward and by the French astronomer Emile-François de Pagan. This theory is very close to Kepler's but contains errors of longitude that reach a maximum in the octants. Neither Ward nor Pagan offered any observational evidence in support of their views.[178] By the 1660s Kepler's astronomy had carried the day.

Envoi

Kepler's work completed what Copernicus had started—the construction of a new planetary astronomy. As we have seen, Copernicus was a reluctant revolutionary. He argued that his universe was truer to Aristotelian physics than Ptolemy's was. But in the process Copernicus actually undermined Aristotle. Reversing the places of the Sun and Earth played havoc with the element theory and with the theory of natural motions. After Kepler, the circles and the principle of uniform circular motion were gone. The discovery of a new astronomy therefore resulted in the complete destruction of the old physics.

Much of seventeenth-century natural philosophy can be understood as a search for a new physics and a new philosophy of nature to replace the wreckage of the old. The leaders of this search were Galileo, Descartes, and Newton. Kepler's laws of planetary motion made possible Newton's discovery of the laws of motion and the law of universal gravitation. The area law, which is equivalent to the principle of conservation of angular momentum, provided the crucial clue that the force exerted on a planet by the Sun is directed radially inward toward the body of the Sun itself, and not tangentially around the orbit, as Kepler and his contemporaries had supposed. The elliptical shape of the orbit and, even more directly, the harmonic law provided the clues that the attractive force exerted by the Sun on a planet varies as the inverse square of the distance between them.

For the ancient Greeks, planetary astronomy had been a branch of mathematics. Kepler's constant goal was to provide a physical basis for astronomy. In his *Mathematical Principles of Natural Philosophy* of 1687, Isaac Newton showed how to deduce Kepler's laws of planetary motion from his own laws of motion and the law of universal gravitation: it was Newton who realized Kepler's dream of making planetary astronomy into a branch of physics. And thus humanity stepped from the medieval to the modern world. Newton learned of Kepler's laws, not by reading the impenetrable Kepler, but by stumbling across them in various English and Dutch authors.[179] Moreover, by Newton's time, Kepler's version of "physical explanation" seemed hopelessly medieval and obscurantist. Thus, it is ironic but understandable that Newton does not even mention Kepler by name in the first book of the *Principia*, where Newton develops the laws of planetary astronomy from the new physics.

From the Venus tablets of Ammi-saduqa to the elliptical astronomy of Kepler is a long way. It took the human race three thousand years to figure out the solar system. For us, who live in an era of ceaseless scientific progress, this can be an astonishing and even a daunting thought. When we look back over the history of this endeavor—from the Babylonian Jupiter theory of system A and the spheres of Eudoxus to the singularly successful theory of Ptolemy—it is easier to understand why scientific change is not guaranteed to be a steady and continuous thing. The world as we perceive it through our science is not simply given to us by observation. Rather, it is a complex amalgam of observation, invention, mathematical convenience, mystic insight, and philosophical prejudice. The truly astonishing thing is that such an amalgam—which has as much to do with us as with the external universe—should be so successful.

The cosmos of the Greeks was a beautiful accomplishment, developed over five hundred years from the time of Eudoxus and Aristotle to that of Ptolemy. Their cosmos was a blend of solid fact, detailed calculation, and unrestrained flights of imagination. A large part of Ptolemy's worldview was close to the mark, but a large part turned out only to have been a fantastic dream. This universe was radically transformed in the century and a half that began with Copernicus and ended with Newton. The development of a new astronomy and a new physics surely brought us closer to understanding the external world, but just as surely, it involved the same complex mixture of mathematics and mystery, observation and inner vision. Nowhere is this clearer than in the tortuous path of Kepler the dreamer. We cannot say with any assurance what part of our own world picture will survive the next few centuries and what part will melt away. Much of our own science surely will be vindicated, but some part of it will turn out only to have been a dream.

Mathematical Postscript: Kepler and Ptolemy

As we have seen, the choice of putting the Sun or the Earth in the center of the system has no immediate effect on the accuracy of a theory. Thus, Copernicus was not guaranteed to reach better predictive power than Ptolemy had. And even though the orbits are really ellipses, they depart from perfect circles by only a very small amount. This is why astronomers through 1,500 years were able to calculate reasonably accurate planet positions using circles rather than ellipses.

Moreover, Ptolemy's speed law (uniform angular motion about an equant point) is a very good approximation to Kepler's area law. This is the claim we investigate in this mathematical postscript. We will compare more closely the generalized equant model of figure 7.69 with the Keplerian model of figure 7.73, to see under just what circumstances the Ptolemaic model will perform satisfactorily.

The true theory of the motion of a planet around the Sun is shown in Fig. 7.73. The planet B moves on a Keplerian ellipse. D is the Sun, C is the center of the ellipse, and F is the empty focus; a is the semimajor axis, b is the semiminor axis, and e is the eccentricity. The planet moves on the ellipse in accordance with the law of areas: the radius vector from the Sun to the planet sweeps out equal areas in equal times.

Figure 7.69 presents a heliocentric eccentric-and-equant theory, such as Kepler's vicarious hypothesis. A circle of radius R is described about center C, which is eccentric to the Sun D. The planet B moves around the circle in accordance with the law of the equant: angle BEA changes uniformly with time. The eccentricity of the orbital circle is $e_1 = CD/R$. The eccentricity of the equant is $e_2 = CE/R$.

Let us calculate, in each model, the time required for the planet to travel from apogee A to quadrature Q. While moving from A to Q, the planet travels through 90° as seen from the Sun D. But the time required will be more than one-fourth of the orbital period.

Kepler Motion   In Kepler motion, the speed of the planet is regulated by the law of areas:

$$\text{time of travel} = \frac{T}{\pi ab} \times \text{area swept out by radius vector.}$$

The constant of proportionality, $T/\pi ab$, is just the orbital period T divided by the area of the ellipse. The time T_{AQ} required for the planet to travel from apogee to quadrature is therefore proportional to the crosshatched area in figure 7.73. This crosshatched area is one-quarter of the area of the ellipse plus the area of the zone $DCYQ$. The quarter-ellipse ($CAXY$) has area $\pi ab/4$. If the eccentricity of the ellipse is small, we may treat $DCYQ$ as a rectangle of area $CD \times CY = eab$. The area of the whole crosshatched region is then approximately $\pi ab/4 + eab$. Thus, the time of travel from A to Q is

$$T_{AQ} = \frac{T}{\pi ab} \times \left(\frac{\pi ab}{4} + eab \right)$$

$$= \frac{T}{4} + \frac{e}{\pi} T.$$

That is, the time from apogee to quadrature is greater than one-fourth of the orbital period by eT/π.

Eccentric Circle with Equant Point   In equant motion, the time of travel is proportional to the angle subtended by the path at the equant. The constant of proportionality is just the constant angular speed $2\pi/T$. (We shall work with the angles in radian measure.) Thus, in figure 7.69, the time of travel from A to Q is

$$T_{AQ} = \frac{T}{2\pi} \times (\text{angle } AEQ)$$

$$= \frac{T}{2\pi} \times (AEX + XEQ)$$

$$= \frac{T}{2\pi} \times \left(\frac{\pi}{2} + XEQ\right).$$

Now, $XEQ = EQD$. And

$$EQD = \tan^{-1} (ED/DQ)$$

$$\simeq ED/DQ$$

$$\simeq (Re_1 + Re_2)/R$$

$$= e_1 + e_2,$$

provided that the eccentricities are small. Thus, the time from apogee to quadrature becomes, approximately,

$$T_{AQ} = \frac{T}{2\pi} \times \left(\frac{\pi}{2} + e_1 + e_2\right)$$

$$= \frac{T}{4} + \frac{e_1 + e_2}{2\pi}T.$$

Comparing the two expressions for T_{AQ} (for Kepler motion and for equant motion on an eccentric circle), we see that the Ptolemaic model will agree with the actual facts of planetary motion if

$$e_1 + e_2 = 2e.$$

That is, the total eccentricity $e_1 + e_2$ must be twice the eccentricity of the Keplerian ellipse.

It follows that distance DE in figure 7.69 must be equal to distance DF in figure 7.73. That is, *the empty focus of the elliptical orbit corresponds to the equant point of the eccentric-circle model.* Furthermore, this correspondence is quite good. An imaginary observer standing at the empty focus of Mars's elliptical orbit would see Mars moving around the zodiac very nearly at uniform angular speed. An observer standing at the Sun would not see Mars move at uniform angular speed. For an observer at the Sun, the maximum departure of Mars from its uniformly moving mean position is about $11°$—a very considerable departure from uniform angular motion. (This is roughly the maximum size of the equation of center in the Ptolemaic model—see table 7.7). For an observer at Mars's empty focus, the planet's greatest departure from its uniformly moving position is about $0.1°$. Thus, at the level of precision that was achievable in antiquity, the empty focus of a Keplerian ellipse is indistinguishable from an equant point.[180]

To put things a little differently, the planets really do move nonuniformly and Ptolemy's rule of equant motion was a very close approximation to the actual facts of planetary motion. Ptolemy's introduction of the equant point into astronomy represented the deepest insight into the nature of planetary motion before the time of Kepler. In the Middle Ages and the Renaissance, Ptolemy's equant was often perceived as an unfortunate rupture of the rules of Aristotelian physics. This dissatisfaction with the physical aspects of Ptolemaic planetary theory did help to bring about the revolution in astronomy of the sixteenth century. But this is somewhat ironic since most of the supposed remedies (such as the minor epicycle that stood in for the equant in the astronomies of Ibn al-Shāṭir and Copernicus) represented a step backward from Ptolemy.

Shedding Light on Kepler's Dilemma   If the total eccentricity $e_1 + e_2$ is equal to $2e$, the eccentric-and-equant model will guarantee that the planet reaches four points at its orbit at precisely the right times—apogee, perigee, and both quadrants. But, of course, the equant is not *precisely* equivalent to the law of areas. Thus, Kepler found that when very precise observations were used, there was no way to split up the total eccentricity between e_1 and e_2 that would give perfect agreement between theory and observation.

It turns out that if one adopts the 5:3 division (as in the vicarious hypothesis) so that $e_1 = 5/4\ e$ and $e_2 = 3/4\ e$, the planet will also reach the four octants of its orbit at the right times.[181] This is why Kepler found that the vicarious hypothesis could match the heliocentric longitudes of ten oppositions nearly perfectly. But then any prediction that depends on the *actual distance* of Mars from the Sun will be ruined.

This is because the distance of the center of the orbit from the Sun should be ea. So, to get the distance effects right, one must put $e_1 = e$ and adopt Ptolemy's bisection of the eccentricity. This fact Kepler discovered by using the observed latitudes of Mars. But then the eccentric-and-equant model will fail to give the right heliocentric longitudes in the octants (the famous 8′ discrepancy).

Should one put $e_1 = 5/4\ e$ and $e_2 = 3/4\ e$, or should one put $e_1 = e$ and $e_2 = e$? Neither choice is perfect, which is why Kepler had to abandon the eccentric circle and equant model. So, again, we see that Kepler was the first to push the fundamental accuracy of a planetary theory beyond what was possible in Ptolemaic astronomy.

Incidentally, we can also see now why Hipparchus's eccentric-circle theory of the Sun works so well—and also why Kepler had to modify it by inserting an equant point. Hipparchus's model may be considered a special case of the eccentric-plus-equant model, with $e_2 = 0$. Since the total eccentricity must still be equal to $2e$, we require $e_1 = 2e$. That is, if the eccentricity of the eccentric circle is twice the eccentricity of the Keplerian ellipse, Hipparchus's model will do a good job of matching the angular progress of the Earth (or the Sun) around the orbit. However, the variation in distance will then be too large by a factor of two. The apparent variation in the size of the Sun's disk in the course of the year is only half as large as predicted by Hipparchus's theory—but this is an observation that was not possible until modern times. Kepler was led to bisect the eccentricity to account for the impact of the distance effect on the observed positions of other planets.

To engage in philosophical discussion with everyone is unseemly, but with Eratosthenes, Hipparchus, Posidonius, Polybius and others of such kind, it is a beautiful thing.

Strabo, *Geography* I, 2, 1.

FIGURE A.1.

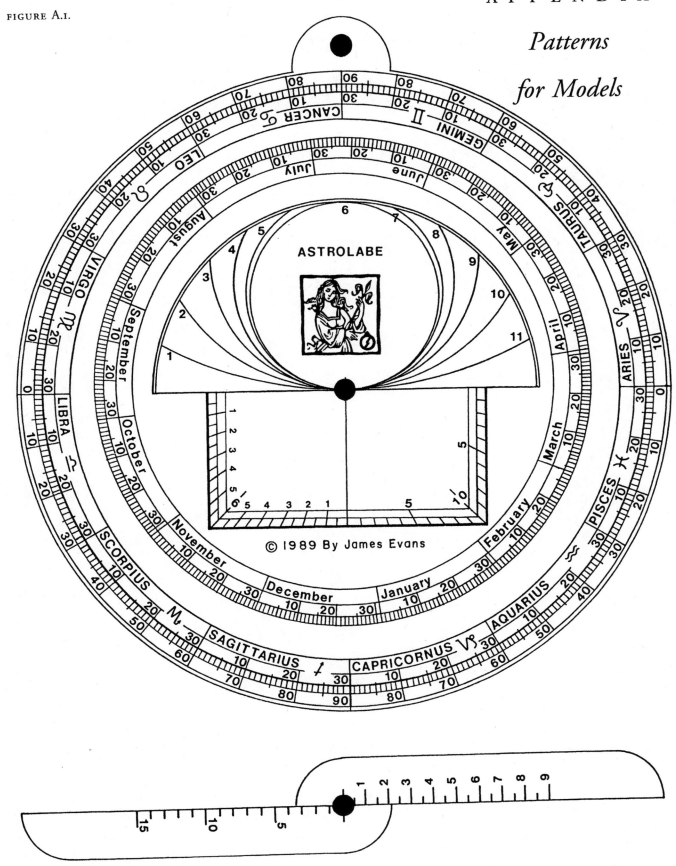

ASTROLABE

© 1989 By James Evans

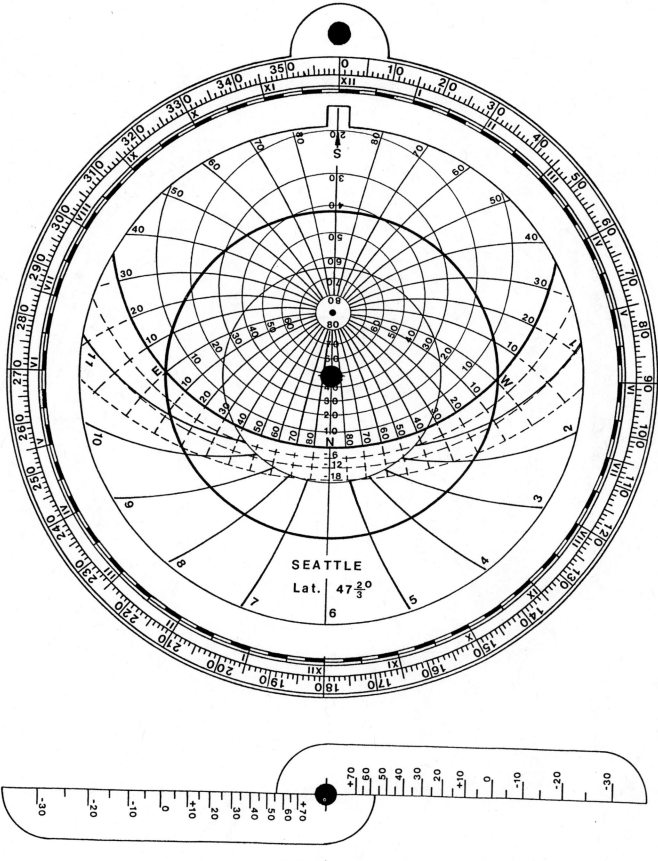

FIGURE A.2.

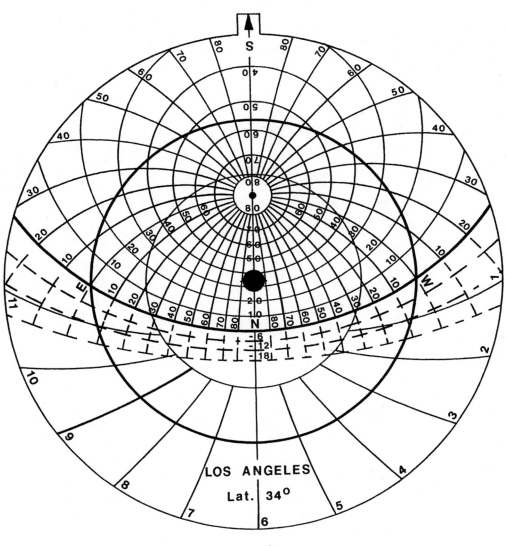

FIGURE A.3.

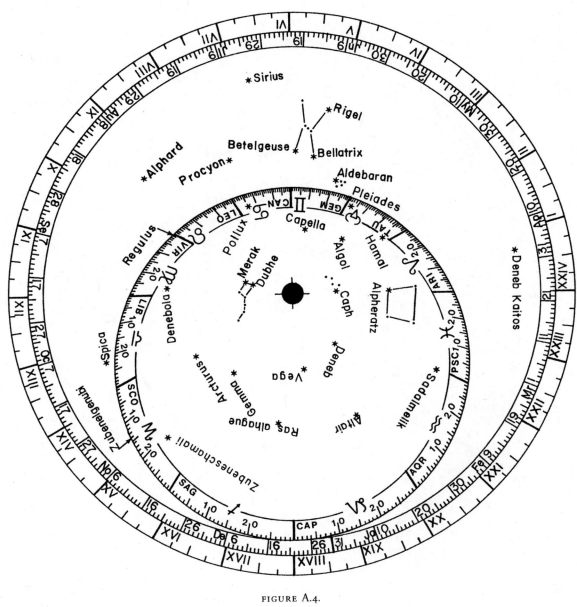

FIGURE A.4.

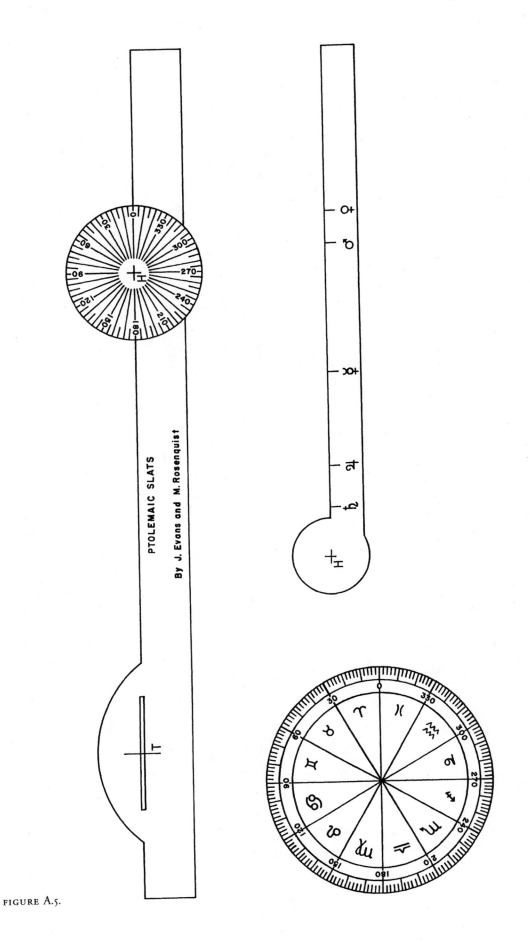

PTOLEMAIC SLATS

By J. Evans and M. Rosenquist

FIGURE A.5.

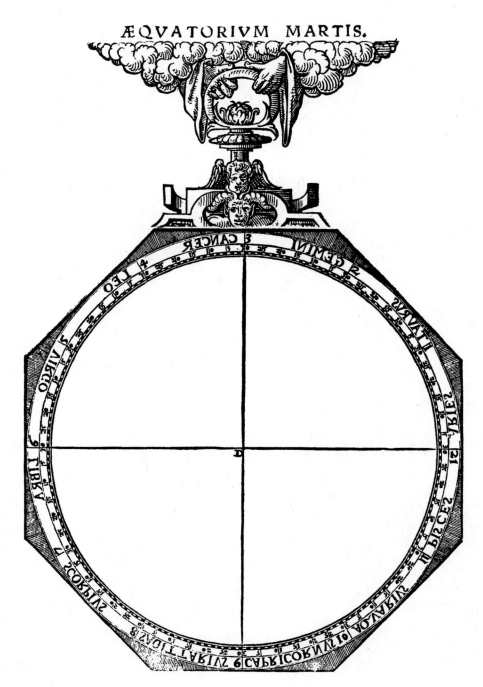

FIGURE A.6.

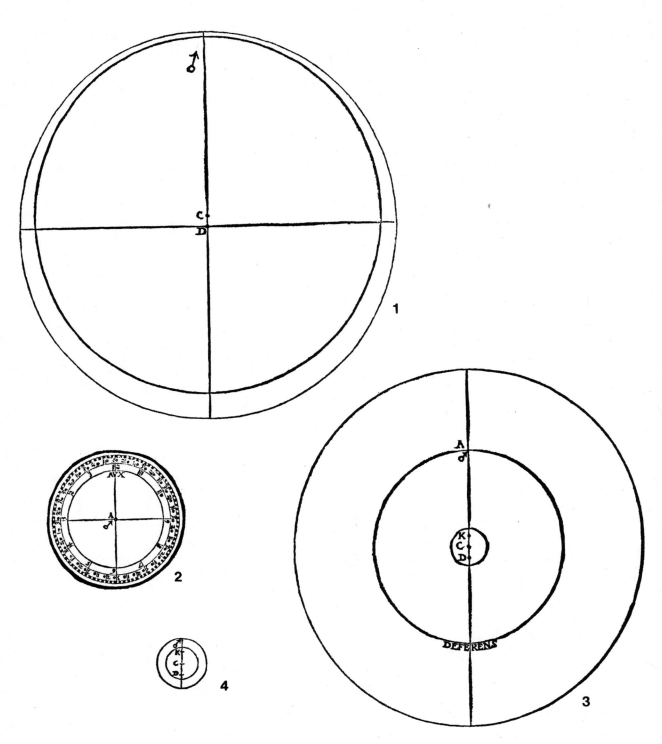

FIGURE A.7.

Notes

CHAPTER I

1. Homer, *Iliad* XVIII, 478, 483–489, 606–607 (trans. Lattimore). For a more detailed discussion of the astronomy in Homer and Hesiod, see Dicks (1970).

2. *Iliad* XXII, 29. *Odyssey* V, 272.

3. *Odyssey* V, 277.

4. *Odyssey* III, 1–2; XV, 329; XVII, 565. *Iliad* XVII, 425.

5. *Iliad* V, 5–6 (trans. Lattimore).

6. *Iliad* XXII, 29–31 (trans. Lattimore).

7. *Iliad* XXII, 317–318; XXIII, 226. *Odyssey* XIII, 93–94.

8. Hesiod, *Works and Days*, 383–384.

9. *Works and Days*, 479–482.

10. *Works and Days*, 564–569.

11. *Works and Days*, 571–572.

12. *Works and Days*, 582–588.

13. *Works and Days*, 609–610.

14. *Works and Days*, 615–618.

15. MUL.APIN, tablet I, column i, lines 1–5. Translated by Hunger and Pingree (1989), pp. 18–19.

These stars probably correspond to the following:

Plow	α and β Trianguli and γ Andromedae
Wolf	α Trianguli
Old Man	roughly our Perseus
Great Twins	α and β Geminorum and nearby stars

16. MUL.APIN I, ii 36–40. Hunger and Pingree (1989), pp. 40–41.

17. MUL.APIN I, iii 13–16. Hunger and Pingree (1989), pp. 47–48.

18. Aratus, *Phenomena*, line 569.

19. MUL.APIN I, iii 34–38. Hunger and Pingree (1989), pp. 53–54.

20. For a French translation of this papyrus, see Tannery (1893), pp. 283–294.

21. MUL.APIN II, A 10–11. Hunger and Pingree (1989), p. 90.

22. For an excellent account of the role of astronomy in Babylonian magic and divination, see Reiner (1995). MUL.APIN II, iii 30, 33; II, B 7; II, iv 1. Hunger and Pingree (1989), pp. 112, 118, 119.

23. These 36-star lists are discussed in detail in van der Waerden (1974), pp. 64–67.

24. A good introduction is Thorkild Jacobsen, "*Enuma Elish*—'The Babylonian Genesis,'" reprinted in Munitz (1957), pp. 8–20. See also Wayne Horowitz, "Mesopotamian Accounts of Creation," in Hetherington (1993a), pp. 387–397. For a complete translation of *Enuma Elish*, see Dalley (1989). For an introduc-

tion to the Babylonian world view, see Francesca Rochberg-Halton, "Mesopotamian Cosmology," in Hetherington (1993), pp. 398–407.

25. *Enuma Elish* V, 3–4. Translated by Dalley (1989).

26. Herodotus, *Histories* II, 109.

27. This example is borrowed from van der Waerden (1974), p. 46. A remarkably useful introduction to cuneiform writing is Labat (1976).

28. This example is from Jensen (1969), p. 94. For a readable account of the development of cuneiform writing, see Hooker (1990).

29. Oates (1986) is an excellent brief introduction to Babylonian civilization. A standard source for chronology is Parker and Dubberstein (1956).

30. Langdon and Fotheringham (1928), p. 15. This is the standard edition of the Venus tablets, but the conclusions regarding Babylonian chronology are no longer accepted.

31. For a discussion of Babylonian eclipse records, see van der Waerden (1974), pp. 99–100.

32. For a good introduction to Babylonian mathematics, see van der Waerden (1975).

33. Strabo (*Geography* XVII, 1, 8) gives a disappointingly brief description of the Alexandria Museum. A good account of the Alexandrian renaissance and the museum is given in Sarton (1970), Vol. 2, pp. 3–34. On the Alexandria Library, see Canfora (1990) and Parsons (1967).

34. For a good account of the decipherment of the Babylonian sources, see Neugebauer (1969).

35. Herodotus, *Histories* II, 109.

36. Vitruvius, *On Architecture* I, 6, 12–13 (trans. Morgan).

37. Geminus, *Introduction to the Phenomena* XII, 1–4.

38. *Introduction to the Phenomena* I, 4.

39. For Hero's dioptra, see A. G. Drachman, "Hero's Dioptra and Levelling Instrument," in Singer (1954–1970), Vol. 3, pp. 609–612. See also Drachman (1950).

40. According to some, Plato asserted the rotation of the Earth on its axis in *Timaeus* 40B. However, the passage is notoriously obscure, and the bulk of the argument rests on the translation of a single problematical word. Aristotle mentions this passage (*On the Heavens*, 293b30) and seems to attribute to Plato the idea that the Earth is not immobile. The dispute over the interpretation of this passage from Plato thus began with the ancients

The Earth thus began with the ancients

themselves. It has continued to our day and has generated a considerable literature. See Dreyer (1906), pp. 71–79, and Dicks (1970), pp. 132–137. Current consensus is against attributing to Plato any idea of the Earth's rotation. Certainly, it would be odd if Plato had broached so interesting an idea in a single sentence, only to drop it without any further discussion.

41. Aëtius, *Opinions of the Philosophers* III, 13, 3. Translated by Heath (1932), p. 94.

42. Simplicius, *Commentary on Aristotle's On the Heavens*, quoted by Heath (1932), p. 93.

43. This account of Aristotle's physical doctrines is drawn from *On the Heavens*. For the fifth element (the ether), see *On the Heavens* I, 2–3. For forced and natural motions, see III, 2. For the elements, their number and their transformations, see III, 3–6. For weight and lightness of the four sublunar elements, IV, 4–5. The particular arguments against the mobility of the Earth are paraphrased from II, 14.

44. Ptolemy, *Almagest* I, 7 (trans. Toomer).

45. For the Babylonian constellations, the standard reference is Gössmann (1950). A good brief account is given in van der Waerden (1974).

46. The attribution to Eratosthenes is doubted by some scholars. See Neugebauer (1975), pp. 577–578. For most constellations, the number of stars in the *Catasterisms* is smaller than in a constellation list attributed to Hipparchus (published in Boll, 1901); thus, the *Catasterisms* probably predates Hipparchus. On the other hand, as the text mentions Coma Berenices, it cannot be earlier than 247 B.C. Thus, there seems no good reason to doubt that it really is by Eratosthenes. The work was already attributed to Eratosthenes in antiquity, as is clear in the Latin *Astronomy* of Hyginus.

47. Callimachus, *Aetia* 110.

48. Catullus, poem 66.

49. Geminus, *Introduction to the Phenomena* III.

50. Allen (1899) is still useful as a guide to classical and Renaissance sources on the constellations (but not to be trusted regarding Arabic sources). For the Arabic names, Kunitzsch and Smart (1986) is convenient.

51. Callimachus, *Epigram* XX (trans. Mair and Mair), p. 150.

52. Aratus, *Phenomena*, 903–908 (trans. Mair), p. 277.

53. For a modern Arabic edition of al-Ṣūfī's book, see Nizāmuʾd-Din (1954). For a French translation, see Schjellerup (1874). For more on al-Ṣūfī, see the article by Paul Kunitzsch in *DSB* as well as "The Astrono-

mer Abuʾl-Ḥusayn al-Ṣūfi and his Book on the Constellations" in Kunitzsch (1989).

54. Heath (1932), p. 20.

55. Vitruvius, *On Architecture* IX, 2, 1–2.

56. *On Architecture* IX, 6, 2.

57. Plato, *Cratylus* 409A.

58. Kirk and Raven (1960), pp. 391–392; Heath (1932), pp. 26–28. Hippolytus, a Christian theologian who lived in Rome in the third century A.D., wrote a *Refutation of All Heresies* in nine books. He attacked Christian heresies by claiming that they were revivals of pagan philosophy. In his first book he gave biographical accounts of the early philosophers.

59. Diogenes Laërtius, *Lives and Opinions of Eminent Philosophers* II, 6–16. Diogenes Laërtius wrote this history of Greek philosophy in Greek, probably in the third century A.D. Diogenes' treatment is anecdotal rather than analytical. His book is an album of quotations from earlier writers. His history remains, however, a valuable source for the study of early Greek philosophy because he had access to many primary sources and early secondary compilations that are now lost.

60. The fragments have been collected by Kirk and Raven (1966) and by Heath (1932). Useful discussions have also been provided by Dreyer (1906), pp. 9–52, and by Dicks (1970), pp. 39–91.

61. "Further, we are told that he [Pythagoras] was the first to call the heaven the cosmos and the Earth spherical, though Theophrastus says it was Parmenides, and Zeno that it was Hesiod." Diogenes Laërtius, *Lives and Opinions of Eminent Philosophers* VIII, 48.

62. Aristotle, *On the Heavens* II, 14, 297b23–298a21, (trans. Guthrie, slightly modified).

63. See Diller (1949), and Dicks (1960), pp. 42–46.

64. Theon of Smyrna, *Mathematical Knowledge Useful for Reading Plato* III, 2.

65. Strabo, *Geography* I, 1, 12.

66. Ptolemy, *Geography* I, 4.

67. Ptolemy's discussion of this famous eclipse is in *Geography* I, 4. Further on in the text, in the list of cities, Ptolemy gives the longitude of Carthage as 34°50′ (*Geography* IV, 3, 7) and the longitude of Arbela as 80° (*Geography* VI, 1, 5). (In the *Geography*, Ptolemy uses the meridian through the Islands of the Blessed [the Canary Islands] as the zero of longitude; these islands were the most westerly known land.) The difference in longitude between the two cities thus comes to 45°10′, in good agreement with the longitudinal difference deduced from the eclipse observations. In the table of cities

found in the *Handy Tables* (composed by Ptolemy and revised by Theon of Alexandria in the fourth century A.D.), the longitudes of Carthage and Arbela are the same as in Ptolemy's *Geography*. The 45° longitudinal distance between these cities thus became standard in the geographical tradition.

68. Ptolemy, *Almagest* I, 4 (trans. Toomer).

69. Cleomedes, *On the Elementary Theory of the Heavenly Bodies* I, 8, 7.

70. Theon of Smyrna, *Mathematical Knowledge Useful for Reading Plato* III, 3.

71. The earliest mention of the argument from sailing ships appears to be that of Strabo (*Geography* I, 1, 20).

72. Theon of Smyrna, *Mathematical Knowledge Useful for Reading Plato* III, 3.

73. Strabo, *Geography* I, 1, 8 (trans. Jones).

74. For Aëtius's account of Anaximander's cosmology, see Kirk and Raven (1960), pp. 135–137, and Heath (1932), pp. 6–7.

75. Ptolemy, *Almagest* III, 1.

76. MUL.APIN II, i 1–8. Translated by Hunger and Pingree (1989), pp. 70–71.

77. MUL.APIN I, iv 31–39.

78. MUL.APIN II, A 1–4. Translated by Hunger and Pingree (1989), pp. 88–89.

79. For the little we know of Oenopides' geometry, see Heath (1921), Vol. I, pp. 174–176.

80. Theon of Smyrna, *Mathematical Knowledge Useful for Reading Plato* III, 40.

81. Euclid, *Elements* IV, 16.

82. Proclus, *Commentary on the First Book of Euclid's Elements*, Proposition viii (trans. Morrow), p. 210.

83. Vitruvius, *On Architecture* IX, 7, 1 (trans. Morgan).

84. Pliny, *Natural History* II, 182 (trans. Rackham).

85. Strabo, *Geography* I, 4, 4 (trans. Jones).

86. *Geography* II, 2, 3; II, 5, 43.

87. Strabo discusses the various ways of defining the zones in *Geography* II, 2, 3 to II, 3, 2.

88. Aristotle, *On the Heavens* 298a16.

89. Heath (1897), p. 222.

90. Dreyer (1906), pp. 173–174, makes a case for the attribution to Dicaearchus.

91. Strabo, *Geography* I, 2, 2.

92. Cleomedes, *On the Elementary Theory of the Heavenly Bodies* I, 10, 3–4. For an English translation of this passage, see Heath (1932), pp. 109–112, or Kish (1978), pp. 74–75.

93. The ancient geographical writers all accept this as a fact. Strabo, *Geography* II, 5, 7; Ptolemy, *Almagest* II, 6.

94. Strabo, *Geography* XVII, 1, 48.

95. Strabo (*Geography* II, 5, 34) says that Eratosthenes reckoned the circumference of the Earth at 252,000 stades and that Hipparchus accepted the same figure.

96. *Geography* II, 5, 7.

97. *Geography* II, 2, 2.

98. Thorndike (1949), p. 122.

99. Dreyer (1906), pp. 249–250. See also Sayili (1960).

100. Morison (1974), p. 30.

101. Cleomedes, *On the Elementary Theory of the Heavenly Bodies*, I, 10, 2. Translated from Todd's edition, in which this passage is given the numeration I, 7. Throughout this book, passages of Cleomedes are cited according to Ziegler's numeration, which is also followed in Goulet's French translation.

102. See Heath (1913), pp. 299–316, for a discussion of all the ancient testimony on Aristarchus.

103. Translation of Heath (1897), p. 222, slightly modified.

104. The source for Cleanthes' opinion is Plutarch (*On the Face in the Orb of the Moon*, 922F–923A; see Plutarch, *Moralia*, Vol. XII). Diogenes Laërtius provides a list of Cleanthes' writings, among which is a book called *Against Aristarchus*. It was no doubt in this work that Cleanthes criticized Aristarchus for impiety.

105. Translated by Heath (1913), pp. 353–355.

106. Heath (1897), p. 223. In the same passage, Archimedes describes his own measurements of the angular diameter of the Sun, using a dioptra of the type later described by Ptolemy in *Almagest* V, 1.4. Archimedes finds that the diameter of the Sun is between 1/164 and 1/200 of a right angle, i.e., between 0.549° and 0.45°. See also Dijksterhuis (1987), pp. 364–366.

107. Van Helden (1985), pp. 10–13. This book also provides a good account of later efforts to find the distances of celestial bodies.

CHAPTER 2

1. Neugebauer (1975), p. 577.

2. Ptolemy, *Almagest* I, 3.

3. *Almagest* I, 2.

4. Geminus introduces the celestial sphere in his *Introduction to the Phenomena* (I, 23). Theon of Smyrna (*Mathematical Knowledge Useful for Reading Plato* III, 1) presents the astronomical hypotheses in clear fashion: the cosmos and the Earth are both spheres, the Earth is a mere point located at the center of the cosmos, etc.

5. Hipparchus, *Commentary on the Phenomena of Aratus and Eudoxus* I, 1.3–8. For

an English translation of these brief passages, see Heath (1932), pp. 116–121.

6. Aratus, *Phenomena* 462–558.

7. The best study of early celestial globes is Savage-Smith (1985).

8. It has been suggested that Hipparchus used a globe as an "analog computer" in compiling the "phenomena" section of his *Commentary on the Phenomena of Aratus and Eudoxus*. See Nadal and Brunet (1984) and (1989); also Grasshoff (1990), pp. 190–191. Similarly, Ptolemy is believed to have used a globe in "computing" much of the data concerning heliacal risings and settings of the stars in his *Phaseis*. In this work, Ptolemy gave the dates of the four heliacal risings and settings, for each of 30 stars, as observed in each of 5 different latitudes. This amounts to $4 \times 30 \times 5 = 600$ calculations. To have done this trigonometrically would have been a great labor, which would hardly have been worth the little extra precision resulting from an exact treatment. Ptolemy's use of a globe is inferred from the scatter in some of his results. See Neugebauer (1975), pp. 928–931.

9. Gaius Sulpicius Gallus, of an influential Roman family, fought in Macedon under Aemilius Paulus, was consul in 166 B.C., and played an important role in the affairs of Greece and Asia. He had a reputation for scientific, especially astronomical, knowledge. He is famous for relieving the fears of Paulus's army on the occasion of a solar eclipse, in 168 B.C., by explaining its cause before an assembly of the troops. On the following day, the army defeated the forces of King Perseus at Pydna in Macedonia (Pliny, *Natural History* II, ix). His discussion of celestial globes is quoted by Cicero in *De re publica* I, 14.

10. Plato, *Timaeus*, 36 B–D (trans. Lee).

11. Plutarch, *Marcellus* XVII, 3–5 (see Plutarch, *Lives*, Vol. V).

12. Pappus of Alexandria, at the beginning of book VIII of his *Mathematical Collection*; trans. Hultsch, Vol. 3, p. 1026; trans. Ver Eecke, Vol. 2, pp. 813–814.

13. Ovid, *Fasti* VI, 277–280.

14. Cicero, *De re publica* I, xiv. Even Cicero does not say he has seen Archimedes' globes. Rather, he has them mentioned by characters in a dialogue, and the dialogue is set in the year 129 B.C., more than twenty years before Cicero's birth.

15. *De re publica* I, xiv (trans. Keyes). Archimedes' orrery is mentioned by Cicero also in *Tusculan Disputations* I, xxv.

16. Cicero, *On the Nature of the Gods* II, xxxv.

17. Theon of Smyrna, *Mathematical Knowledge Useful for Reading Plato* III, 16.

18. Price (1974).

19. Field and Wright (1984).

20. Proclus, *Commentary on the First Book of Euclid's Elements* (trans. Morrow), pp. 31–35. This passage has also been discussed by Heath (1921), Vol. 1, pp. 10–18, and Tannery (1887), pp. 38–42.

21. A famous example is the *mesolabes*, or *mesolabium*, an instrument invented by Eudoxus and Archytas to solve the following problem: given the two extreme lines, find the two mean lines of a continued proportion. That is, given the numbers a and d, find b and c such that $a/b = b/c = c/d$. See Plutarch's life of Marcellus.

22. Ptolemy, *Planetary Hypotheses* I, 1, *Opera*, Vol. II, pp. 70–71. Nevertheless, when Ptolemy goes on in the same work to describe his cosmology, he remarks that "he will use the simplest method, so that it will be easy to construct instruments," by which he clearly means working models of the universe (*Opera*, Vol. II, pp. 72–73). Thus, Ptolemy did not object to *sphairopoiïa* itself, but only to the shortcomings of its practitioners.

23. Photographs of Renaissance globes and armillary spheres are found in many books on the history of scientific instruments. Examples are Guye and Michel (1971), Michel (1967), and Wynter and Turner (1975). Stevenson (1921) is the most extensive general history of globes. Tycho Brahe's sixteenth-century illustrations of his own armillary spheres are reproduced in Raeder, Stromgren, and Stromgren (1946).

24. The most recent edition (1979) of Autolycus's works is that of Aujac (Greek text with facing French translation). Bruin and Vondjidis provide an English translation (1971) of Autolycus's works, based on the Greek text (1950) of Mogenet. But this translation must be used with care; for a highly critical review, see Neugebauer (1973).

25. Struik (1948), p. 59.

26. These propositions are quoted from Euclid's *Phenomena* (trans. Berggren and Thomas).

27. Copernicus, *On the Revolutions of the Heavenly Spheres* I, 6.

28. A list of the contents of this manuscript is given in Aujac's 1979 translation of Autolycus, pp. 29–30. For a summary of the contents of the works of Theodosius, of Euclid's *Optics*, and of Menelaus's *Spherics*, see Delambre (1817), Vol. 1, pp. 234–243, 58–60, and 243–246, respectively.

29. See, e.g., Tannery (1893), p. 35.

30. Neugebauer (1975), pp. 768–769.

31. For Geminus's date, see Neugebauer (1975), pp. 579–581. Formerly, a date of

about 70 B.C. was ascribed to the *Isagoge*, but this was due to a confusion between the Egyptian and the Alexandrian calendar.

32. Translated into English by James Evans from Aujac's 1975 Greek text.

33. Aratus, *Phenomena*, 559–562 (trans. Mair and Mair).

34. *Phenomena*, 569–572.

35. Hipparchus, *Commentary on the Phenomena of Aratus and Eudoxus* II, 1, 4–6. From Hipparchus's remarks, it is clear that some of his contemporaries (including Attalus, another commentator on Aratus) were still making this assumption.

36. Ptolemy, *Geography* (trans. Stevenson), p. 34.

37. Neugebauer (1975), p. 788.

38. For a modern edition of the *Alfonsine Tables*, see Poulle (1984).

39. Tannery (1893), p. 138.

40. Strabo, *Geography* II, 5, 34.

41. Ptolemy, *Tetrabiblos* III, 2. (trans. Robbins), pp. 228–231. A good brief introduction to the history of Greek astrology is provided by Tester (1987).

42. For details on the trigonometry of the *Almagest*, see Pedersen (1974), pp. 65–78, 94–121.

43. For more detail on Hypsicles' procedure, see Neugebauer (1975), pp. 715–718. See also Heath (1921), Vol. 2, pp. 213–218.

44. The Babylonian material is discussed fully by Neugebauer (1975), pp. 366–371.

45. Cleomedes, *On the Elementary Theory of the Heavenly Bodies* I, 6.1.

46. Manilius, *Astronomica* III, 275–294.

47. *Astronomica* III, 443–482.

48. See Jones (1991) for a more extensive treatment of the adaptation of Babylonian methods by Greek astronomers.

49. For the armillary sphere as an instrument of observation in the Renaissance, see Dreyer (1890), Thoren (1990), and Raeder, Stromgren, and Stromgren (1946).

CHAPTER 3

1. The authoritative survey of Greek and Roman sundials is Gibbs (1976).

2. This dial is no. 4008G in Gibbs (1976).

3. No. 4001G in Gibbs (1976).

4. Stuart and Revett ([1762] 1968). For the Tower of the Winds, see Vol. 1, pp. 12–25. A modern photograph of the tower and some of the engravings of Stuart and Revett are reproduced in Watson (1956).

5. Price (1967).

6. *Peri Analemmatos*. The Greek text (which is only partly preserved) is printed in Ptolemy, *Opera*, Vol. 2, pp. 187–223. For

a discussion, see Neugebauer (1975), pp. 839–856.

7. Adapted from the English translation (1914) of Morgan.

8. Bilfinger (1886). Bilfinger's construction of a horizontal dial is repeated with full explanation in Soubiran (1969). Drecker (1925) has applied the analemma to a variety of plane and cylindrical dials.

9. The hour lines theoretically ought not to be straight, even though the ancients always drew them so. Straight lines are a good approximation, especially for the hours near noon.

10. These measurements are from Gibbs (1976), p. 332.

11. Gibbs (1976), p. 324.

12. No. 16 in the collection of the Time Museum (Rockford, Ill.). See Turner (1985), pp. 124–127.

13. No. 186 in the collection of the National Museum of American History. See Gibbs (1984), pp. 139–140.

14. To be sure, there are some variations in the fronts. Early Islamic astrolabes, e.g., usually have no rule and no scale of equinoctial hours on the limb of the mater. Time was reckoned in seasonal hours, and the seasonal hour curves on the plate sufficed for this. The rule, which is very helpful in working with equinoctial hours, has no function in reckoning seasonal hours.

15. No. 87 in the collection of the National Museum of American History. See Gibbs (1984), pp. 132–134.

16. For the reader who wishes to sleuth out the functions of the various scales on the back of this instrument, it should be mentioned that the sines are in base 60. Also, the numerals are in Arabic alphanumeric notation. I.e., the numbers are represented by letters of the Arabic alphabet:

1	2	3	4	5	6	7	8	9	10	11	12
١	٢	٣	٤	٥	٦	٧	٨	٩	١٠	١١	١٢

This is similar in spirit to the Greek system of alphanumerics discussed in Sec. 4.5.

The numerals that we use today have a different ancestry. They represent medieval European variations of Arabic modifications of number symbols that originated in India. Our numerals are sometimes called "Arabic," but "Indian" or "Hindu/Arabic" would be more appropriate.

17. Mathematical proofs of these properties of stereographic projection can be found in Neugebauer (1975). Although the ancient and medieval writers implicitly use conformality (as when laying out the azimuth circles of the latitude plate), it appears that they never directly asserted that stereographic projection is angle preserving. In

contrast, the preservation of circles was directly stated and constantly used by the ancient writers.

18. Listed in Turner (1985), p. 14.

19. On the early history of stereographic projection and of the astrolabe, see Neugebauer (1975), pp. 857–879.

20. Vitruvius, *Ten Books on Architecture* IX, 8.8–15.

21. For references to the literature on these two disks see Neugebauer (1975), p. 870, n. 5, 6. For a drawing of the disk from Grand (which carries month names but no constellation figures), see King (1978), p. 12.

22. Ptolemy, *Opera*, Vol. II.

23. Printed with a French translation and commentary by Segonds (1981).

24. There also exists a table of contents of Theon's treatise on the astrolabe preserved in Ya'qūbī's *History of the World*, written around A.D. 880. Sebokht's treatise corresponds closely to the contents of Theon's lost work as summarized by Ya'qūbī. Philoponus shows more independence but clearly draws on the same source. For a detailed comparison of these three works, see Neugebauer (1949), reprinted in Neugebauer (1983b). For the adaptation of these sources by Arabic astronomers, see "Observations on the Arabic Reception of the Astrolabe" in Kunitzsch (1989).

25. See Turner (1985), p. 15, for a photograph of this astrolabe, now at the Royal Scottish Museum in Edinburgh.

26. Māshā'allāh (or Messahalla, as spelled by Latin writers of the Middle Ages and Renaissance) was an eighth-century Jewish astrologer. He does not seem actually to have written on the astrolabe. See "On the Authenticity of the Treatise on the Composition and Use of the Astrolabe Ascribed to Messahalla" in Kunitzsch (1989).

27. The thirteenth-century Latin treatise on the astrolabe by Jordanus of Nemore has been edited and provided with an English translation by Thomson (1978). A Renaissance English treatise on the astrolabe has also been reprinted recently. See Blagrave, *Mathematical Jewel*.

28. The most ambitious guide to the astronomy and astrology in Chaucer's works is North (1988). For Chaucer's treatise on the astrolabe, see Skeat (1872). For a translation into modern English, see Gunther (1929).

29. Directions for making other parts of an astrolabe can be found in Knox (1976) and Saunders (1984).

CHAPTER 4

1. Originally, the Roman month did coincide with a lunation; indeed, *month* and

Moon derive from the same word root. The old Roman calendar, eliminated by the Julian reform, was of the luni-solar type.

2. Herodotus (*Histories* VIII, 51), speaking of the Persians, says, "After the crossing of the Hellespont, from which they began to march—having consumed one month in the crossing into Europe—in three more months they came into Attica while Kalliades was archon among the Athenians."

3. Diodorus Siculus, *Histories* XVIII, ii, 1.

4. This account is based mostly on Bickerman (1980), pp. 43–51; Samuel (1972), pp. 153–170; and the article "The Calendar" in the *Explanatory Supplement to the American Ephemeris* (1974), pp. 407–442.

5. In the historical period, the Roman calendar always had twelve months. But, at an even earlier time (before the fifth century B.C.), there were only ten months, and the year began with March. This explains the names of the months September through December.

6. The details of the intercalation scheme in the early Roman calendar are a matter of dispute. See Samuel (1972), pp. 160–164.

7. Note that the months (March, May, July, and October) that always had 31 days are precisely those that kept their Nones on the 7th and their Ides on the 15th. The other months (of 29 days before the reform) kept their Nones on the 5th and their Ides on the 13th even after they were increased in length by the reform.

8. For an account of the astrological method of deducing the order of the days of the week from the order of the planets, see Bickerman (1980), p. 61, or Sarton (1970), Vol. 2, p. 332.

9. On Friday morning the Jews would not enter Pilate's hall, lest they be defiled and so prevented from eating the paschal meal (John 18:28). Friday was the "day of preparation" for the Passover (John 19:14, Luke 23:54). The Sabbath, beginning at sundown on Friday, therefore coincided with Passover, for which reason it was called a "great Sabbath" (John 19:31).

10. That Christ is said to have risen "on the third day" (Sunday) following the crucifixion (Friday) is another example of inclusive counting.

11. Archer (1941), p. 39.

12. For an account of the Easter controversy in Britain in the seventh century, see Bede's *Ecclesiastical History* (e.g., II 2, II 19, III 4, III 25, IV 5, V 15, V 19, V 21).

13. Tables for calculating the date of Easter under either the Julian or the Gregorian calendar are provided in the *Explana-tory Supplement to the American Ephemeris* (1974), pp. 420–429.

14. For an interesting account of medieval discussion of the inadequacy of the calendar, see "The Western Calendar: *Intolerabilis, Horribilis et Derisibilis*; Four Centuries of Discontent," in North (1989).

15. For portraits of Lilius, Clavius, and Gregory, as well as notes on the opposition to the reform by such notable scholars and astronomers as Scaliger, Viète, and Mästlin, see Moyer (1982). For a technical discussion of the reform, see Delambre (1821), Vol. 1, pp. 1–84.

16. Archer (1941), p. 39.

17. Hughes (1926), p. 16.

18. Drake (1978), p. 436.

19. Westfall (1983), p. 40.

20. On Scaliger and his construction of the Julian period, see Reese and Everett (1981).

21. Copernicus, *On the Revolutions of the Heavenly Spheres* III, 13.

22. For example, in *On the Revolutions of the Heavenly Spheres* III, 13, Copernicus writes that he observed the "autumn equinox at Frauenburg in the year of Our Lord 1515 on the 18th day before the Kalends of October, but according to the Egyptian calendar it was the 1840th year after the death of Alexander on the 6th day of the month of Phaophi, half an hour after sunrise."

23. This cycle of 1461 Egyptian years is sometimes called the *Sothic cycle*, after *Sothis*, the Egyptian name for the star Sirius. The morning rising of Sirius, in July, marked its first reappearance from the rays of the Sun. This event coincided fairly closely with the beginning f the annual flood of the Nile and so marked the beginning of the agricultural year. The morning rising of Sirius thus fell later and later in the calendar, working its way through all twelve months in one Sothic cycle. See Bickerman (1980), pp. 41–42.

24. The *Handy Tables* were edited and provided with a French translation by Halma (1822–1825).

25. Bickerman (1980), pp. 109–110.

26. In the Greek manuscripts of the astronomical canon, the names of the kings are written in the genitive (possessive) case. Thus, for example, Xerxes is written ξερ-ξου, signifying that the 21 years opposite his name are the 21 years *of Xerxes*. In the list shown in fig. 4.2, as in many others of its period, the symbol ȣ is used for the dipthong ου.

27. The letters are assigned numerical values in accordance with their order in the alphabet. However, two letters that dropped out of the alphabet by classical times retained their places and their conventional uses as numerals. These are ϛ (6), called *digamma*, phonetically equivalent to our v or w, and ϙ (90), the *koppa*, equivalent to our q. The Phoenician script from which the Greek letters derived represented a Semitic language and thus had need of more sibilants than did Greek. One of these surplus sibilants was used as the symbol for 900. (Not all scholars agree on this orgin for the symbol.) Its Byzantine form is shown in the list of correspondences. In Byzantine times, it was called *sampi*.

28. This list is adapted from Bickerman (1980), p. 20. The standard source on the calendars of the Greek cities is Samuel (1972), pp. 57–138.

29. Bickerman (1980), p. 36; Samuel (1972), p. 58.

30. Samuel (1972), p. 58.

31. Pritchett and Neugebauer (1947) is the the most thorough recent treatment of the Athenian calendar.

32. For the ancient lists of reigns, see Samuel (1972) and Bickerman (1980).

33. Diodorus Siculus, *Histories* XII, 77, 1 (trans. Oldfather et al.).

34. Neugebauer (1975), p. 617.

35. Fotheringham (1924) and van der Waerden (1960).

36. Ptolemy, *Almagest* VII, 3 (trans. Toomer), p. 334, slightly modified.

37. Somewhat different practices are followed in transliterating the month names. This list follows Sachs and Hunger (1988), pp. 13–14, with some additions from Neugebauer (1983a), p. 38.

38. Parker and Dubberstein (1956), pp. 1–2.

39. The evidence for all known intercalations is collected in Parker and Dubberstein (1956). See van der Waerden (1974), pp. 103–104, for a discussion.

40. Parker and Dubberstein (1956).

41. These circular visibility diagrams were introduced by Schmidt (1949).

42. Ptolemy, *Phaseis* 5; *Opera*, Vol. II p. 8.

43. Theon of Smyrna, *Mathematical Knowledge Useful for Reading Plato* III, 14.

44. Ginzel (1906–1914), Vol. 2, pp. 520–522. Ginzel's tables are reprinted in Bickerton (1980), pp. 112–114.

45. Trans. by J. Evans from Aujac's (1975) text of Geminus.

46. Diels and Rehm (1904).

47. An excellent survey of Greek and Latin parapegmata is provided by Albert Rehm, "Parapegma," in *RE*.

48. Trans. adapted from Grenfell and Hunt (1906), p. 152.

49. Neugebauer (1975), p. 706.

50. Ptolemy, *Phaseis*; *Opera*, Vol. II.

CHAPTER 5

1. Ptolemy, *Almagest* III, 1; Diodorus Siculus, *Histories* XII, 36, 1–2.

2. For the opposite conclusion—that probably Hipparchus used a quadrant and Ptolemy an equatorial ring—see Britton (1992), pp. 12–17.

3. A warp in the ring was proposed as a cause of the double equinox by Rome (1937), pp. 233–234.

4. Multiple equinoxes produced by refraction would be common rather than exceptional. See Britton (1992), p. 28. The effect of refraction has also been analyzed by Bruin and Bruin (1976).

5. At least one ancient writer was aware of atmospheric refraction. Cleomedes mentions an eclipse of the Moon in which the Sun and the Moon were both slightly above the horizon and offers as an explanation the refraction of the visual ray by the air (Cleomedes, *On the Elementary Theory of the Heavenly Bodies* II, 6, 7–10).

6. Eudemus of Rhodes, a pupil of Aristotle who wrote histories of mathematics and astronomy, said that Thales (sixth century B.C.) was the first to find that "eclipses of the Sun and its returns to the solstices do not always take place after equal times." Apparently, this indicates a suspicion that the year might be of variable length. Eudemus's works are lost, and we owe this sentence to Theon of Smyrna, *Mathematical Knowledge Useful for Reading Plato* III, 40.

7. Theon of Smyrna (*Mathematical Knowledge Useful for Reading Plato* III, 12) gives the following values for the wanderings of the planets in latitude: Sun, 1° (i.e., ±1/2°); Moon, 12°; Mercury, 8°; Mars and Jupiter, 5°; Saturn, 3°. The longevity of this mistaken idea no doubt stemmed from its distinguished pedigree: the solar theory (homocentric spheres) of Eudoxus and Callippus incorporated a motion in latitude for the Sun, analogous to, but smaller than, the Moon's motion in latitude.

8. See, e.g., Delambre (1817), Vol. I, pp. xxv–xxvi; (1819), pp. lxvii–lxix, 36.

9. R. R. Newton (1977) has argued that Ptolemy fabricated data. For a thorough study of Ptolemy's equinoxes and solstices, see Britton (1992).

10. Callippus's season lengths, according to the papyrus known as "The Art of Eudoxus," are, beginning with summer, 92, 89, 90, and 94 days; see Tannery (1893), p. 294. The same papyrus gives season lengths for Euctemon (ca. 430 B.C.) that appear to show a recognition of the inequality of the seasons: 90, 90, 92, 93. But it is wrong to interpret these figures this way. Rather, Euctemon's seasons follow from a uniform distribution of 360 days over the four seasons, with the five extra days arbitrarily divided as evenly as possible between winter and spring—a matter of arithmetical convenience, not of observation. Callippus's are the first Greek data that incontestably show a recognition of the solar anomaly. See Neugebauer (1975), pp. 627–629.

11. Surviving proofs of the equivalence may be found in Ptolemy (*Almagest* III, 3) and in Theon of Smyrna (*Mathematical Knowledge Useful for Reading Plato* III, 26). For a reconstruction of Apollonius's lost proof of the equivalence, see Neugebauer (1959).

12. Strabo, *Geography* I, 1, 12; II, 1, 41. The complete title was probably *Three Books against the "Geography" of Eratosthenes*. See Dicks (1960).

13. Ptolemy, *Almagest* IX, 2.

14. On the Babylonian solar theory, see Neugebauer (1975), pp. 371–379. For a less technical introduction to Babylonian astronomy, see Neugebauer (1969), pp. 97–144.

15. The 8° Babylonian convention is mentioned by Geminus: "The summer solstitial point, according to the practice of the Greeks, is in the first part of the Crab; but according to the practice of the Chaldeans, in the eighth degree. The case goes similarly for the remaining points" (*Introduction to the Phenomena* I, 9). The 8° convention persists in many Roman writers, e.g., Vitruvius (*On Architecture* IX, 3) and Pliny (*Natural History* XVIII, 221). In the same way, Columella (*De re rustica* XI, 2.31) says that the spring equinox follows the Sun's entry into the sign of the Ram by seven or eight days. See Neugebauer (1975), pp. 593–600, for a discussion of all the ancient conventions for the signs.

16. Neugebauer (1969), p. 115.

17. Theon of Smyrna, *Mathematical Knowledge Useful for Reading Plato* III, 30.

18. G. J. Toomer, "Hipparchus and Babylonian Astronomy," in Leichty, Ellis, and Gerardi (1988), pp. 353–362.

19. See van der Waerden (1974), pp. 295–298.

20. Jones (1991).

21. Theon of Smyrna, *Mathematical Knowledge Useful for Reading Plato* III, 26.

22. *Mathematical Knowledge Useful for Reading Plato* III, 34.

23. Ptolemy, *Almagest* III, 4.

24. For an English translation, see Duhem (1969).

25. Dreyer (1906), pp. 196, 201.

26. Sambursky (1962), p. 146.

27. Dijksterhuis (1961), p. 67.

28. Koestler (1963), p. 74.

29. G. E. R. Lloyd has shown that Duhem's arguments are often based on misreadings or mistranslations of the Greek sources; Lloyd (1978), reprinted in Lloyd (1991), pp. 254–280. For a criticism of the traditional view, supported by Duhem, that Plato was responsible for introducing the convention of uniform motion into Greek astronomy, see Knorr (1989).

30. Trans. by J. Evans from Diels's edition of Simplicius's *Commentary on Aristotle's Physics*, pp. 291–292.

31. In 140 B.C. the longitude of the solar apogee was 66°18′. For the eccentricity Hipparchus should have found 0.0352.

32. In modern astronomical writing, what we call the *equatorial mean Sun* is called simply the *mean Sun*. Following ancient practice, we use the latter for the fictitious body that travels on the ecliptic and whose longitude is always equal to the Sun's mean longitude (see sec. 5.7).

33. Ptolemy, *Almagest* IV, 6 (trans. Toomer), p. 191, slightly modifed.

34. Theon of Alexandria, in his *Big Commentary* on the *Handy Tables* (see Tihon, 1985, 1991) mentions a certain Serapion in connection with the equation of time. Some historians identify this Serapion with a man of the same name mentioned by Cicero. If this identification is correct, then the Greek astronomers worked out numerical tables for the equation of time by about 50 B.C. However, Anne Tihon (1985, pp. 288–290) has shown convincingly that Theon's Serapion was a commentator on Ptolemy. Thus, Ptolemy's mathematical treatment of the equation of time is not only the oldest that we have, but also the oldest of which we know.

35. For the *Handy Tables*, see Halma (1822–1825). Although Ptolemy's tables have not survived, his introduction, including directions for the use of the tables, has come down to us. For the Greek text, see Ptolemy, *Opera*, Vol. II, pp. 156–185. Comparison of Ptolemy's directions with Theon's version of the tables shows that the tables we have could not differ in any important way from those that came from Ptolemy's hand. Theon also wrote two commentaries on the *Handy Tables*, including directions for their use, no doubt intended to replace Ptolemy's rather concise instructions. For the *Little Commentary*, see Tihon (1978). For the *Big Commentary*, see Tihon (1985) and (1991).

CHAPTER 6

1. Isaac Newton, *Philosophiae Naturalis Principia Mathematica*, 1687. The precession is deduced in book I, in corollaries xx–xxii to proposition LXVI on the three-body problem. See Cajori's translation [Newton (1934)], Vol. I, pp. 186–189.

2. See Smart (1977) or Woolard and Clemence (1966).

3. Aristotle, *On the Heavens* I, 2–3. This separation of the substance of the heavens from the four base elements (earth, water, air, and fire) was already made by Plato (*Timaeus* 22–23).

4. Aristotle, *On the Heavens* 270b13–17 (trans. Guthrie).

5. Pliny, *Natural History* II, 95.

6. Ptolemy's alignments are quoted from Toomer (1984). In calculating the accuracy of the alignments, I have used the star places for 130 B.C. (and A.D. 100) given by Peters and Knobel (1915).

7. Pedersen (1974), p. 237.

8. The calculations of modern alignments are based on the positions in Eichelberger (1925).

9. The drawing is a reconstruction by P. Rome and A. Rome, based on Ptolemy's description of the armillary sphere in *Almagest* V, I, and the remarks by Pappus in his *Commentary* on this section of the *Almagest*. See Rome (1927).

10. Włodarczyk (1987) provides an interesting account of the measurement of star places using a replica of Ptolemy's armillary sphere.

11. On Timocharis, see Goldstein and Bowen (1989).

12. For arguments that Hipparchus did indeed adopt a larger value for the precession rate, see Tannery (1893), pp. 195, 266–268; Neugebauer (1975), pp. 297–298.

13. Ptolemy, *Almagest* III, 1.

14. Delambre (1817); see the discussion in Vol. 1, beginning on p. xxv.

15. R. R. Newton (1977).

16. For a good introduction, which touches on many issues other than the precession rate, see the lively exchange between Owen Gingerich (one of Ptolemy's ablest defenders) and Robert R. Newton (one of Ptolemy's most severe critics) in the *Quarterly Journal of the Royal Astronomical Society* **20** (1979) 383–394, **21** (1980) 253–266, **21** (1980) 388–399, and **22** (1981) 40–44.

17. Dreyer (1906), p. 203.

18. Peters and Knobel (1915).

19. Boll (1901).

20. Dreyer (1917) and (1918).

21. Vogt (1925).

22. There are smaller numbers of frac-

tions in 1/4 and 3/4, i.e., 15′ and 45′. We need not consider these here.

23. Rawlins (1982).

24. Delambre (1817), Vol. 2, p. 284.

25. For an edition of Ulugh Beg's star catalog, see Knobel (1917). The canons accompanying the tables were given a commentary and translated into French in Sédillot (1853). As far as I know, the tables themselves have never been published.

26. Sayili (1960), pp. 260–289.

27. This account of the history of Brahe's star catalog is based on Dreyer (1890), pp. 227, 265–266.

28. For the 777-star catalog, see Brahe, *Tychonis Brahe Dani Opera omnia*, Vol. II, pp. 258–280; for the 1,000-star catalog, Vol. III, pp. 344–373.

29. Evans (1987), pp. 258–271.

30. The + is our translation of Ptolemy's "greater." Thus, γ Arae has a magnitude greater than 4; i.e., the star is a bit brighter than fourth magnitude.

31. Grasshoff (1990).

32. Evans (1992).

33. Shevchenko (1990).

34. This value is attributed to the astronomers of al-Ma'mūn by Thābit ibn Qurra; see Neugebauer (1962a), p. 293. Although this value is a little low, it is much more accurate for its time than Ptolemy's value of the obliquity. According to Ibn Yūnus, the first value of the obliquity to be obtained after the Greeks was 23°31′, observed between A.H. 160 and 170 (A.D. 782–792); see Delambre (1819), p. 100. Many of the medieval solar observations used for determining the obliquity of the ecliptic were of fairly high precision—certainly much better than the Greek observations; see Said and Stephenson (1995).

35. For the principal work of Ibn Yūnus, *The Large Hakemite Zīj*, see Caussin (1804). The longitudes of Regulus are discussed in Delambre (1819), p. 87.

36. Hartner, "Al-Battānī," in *DSB*.

37. Al-Battānī, quoted by Thābit ibn Qurra in Neugebauer (1962a), p. 294.

38. Doubts that Thābit was the author of this treatise are expressed by Morelon (1987), p. xix. For a detailed discussion of the evidence, see Ragep (1993), pp. 400–408. Ragep's own view is that the treatise was written by Thābit's grandson, Ibrāhīm ibn Sinān. See also Ragep (1996).

39. The most recent edition of *De motu octave spere* is Carmody (1960). However, much of the commentary and analysis accompanying this edition is unreliable. For an English translation of *De motu* with explanatory notes, see Neugebauer (1962a).

Still useful is the analysis of *De motu* by Delambre (1819), pp. 73–75, 264–281.

40. On Thābit ibn Qurra, see the article by B. A. Rosenfeld and A. T. Grigorian in *DSB*. For a survey of Thābit's astronomy, see Morelon (1994). Thābit's extant astronomical works in Arabic are collected with French translations and commentaries in Morelon (1987).

41. Carmody (1960), pp. 45–46.

42. Halma (1822–1825), Vol. 1, p. 53.

43. Dreyer (1906), p. 204.

44. The Arabic calendar is purely lunar. That is, the months follow the phases of the Moon but the years have nothing to do with the Sun. Each calendar year consists of twelve months. Full and hollow months alternate, but eleven times in thirty years, a month that would have been hollow is made full. Thus, the average length of the Arabic year over the thirty-year cycle is 354 11/30 days $\simeq$ 354.367 days.

45. An exact calculation results in

$$\tan \Upsilon C = \frac{\sin r}{\sin \varepsilon_0} \sin \beta \sqrt{1 + \tan^2 r \cos^2 (\beta - \varepsilon_0)}$$

There is no evidence that Thābit made such a calculation.

46. The obliquity of the ecliptic ε is given by

$$\sin^2 \varepsilon = \frac{\sin^2 \varepsilon_0}{1 + \tan^2 r \cos^2(\beta - \varepsilon_0)} + \sin^2 r \sin^2 \beta.$$

47. On the *Toledan Tables*, see Toomer (1968). Although al-Zarqālī accepted Thābit's tables for calculating trepidation, he also made studies of alternative models. For texts and commentaries, see Millás-Vallicrosa (1950). See also "Trepidation in al-Andalus in the 11th Century" in Samsó (1994).

48. Although the *Alfonsine Tables* give radices (initial values) for a number of different epochs, including the Incarnation, the fundamental epoch of these tables was the first year of the reign of Alfonso X, i.e., January 1, A.D. 1252. For a modern edition and study of the *Alfonsine Tables*, see Poulle (1984).

49. For details, see Noel Swerdlow, "On Copernicus's Theory of Precession," in Westman (1975), pp. 49–98.

50. For an English translation of Peurbach's *Theoricae novae planetarum*, see Aiton (1987).

51. The best biography of Tycho Brahe is Thoren (1990). Still very useful for its discussion of Brahe's technical work is Dreyer (1890).

52. For a discussion of Brahe's instruments, see Thoren (1990), pp. 144–191. The illustrations in Thoren are drawn from Brahe's own published description of his instruments, *Tychonis Brahe Astronomiae instauratae mechanica* (Wandsbeck, 1598). (The title means "Mechanics of the reform of astronomy.") For an English translation of this work see Raeder, Stromgren, and Stromgren (1946).

53. Tycho Brahe, translated by Raeder, Stromgren, and Stromgren (1946), p. 113. Brahe did give a detailed account of his investigation, which involved "correcting" Copernicus's observations, in *Tychonis Brahe Dani Astronomiae instauratae progymnasmata* (1602–1603). This book was written and printed in stages by Brahe in the years following 1582. Parts of it were circulated among Brahe's friends and correspondents, but it was not published during his lifetime. It was Kepler who gathered it into final form and saw that it was published after Brahe's death in 1601. The *Progymnasmata* is available in Dreyer's edition of Brahe's *Opera*. For Brahe's discussion of precession, see *Opera*, Vol. II, pp. 253–257.

54. On the so-called Tychonic system (in which the planets all go around the Sun while the Sun goes around the Earth), see Christine Schofield, "The Tychonic and Semi-Tychonic World Systems," in Taton and Wilson (1989), pp. 33–44.

55. Brahe's discussion of the changes in the latitudes of the stars is found in the first part of his *Progymnasmata, Opera*, Vol. II, pp. 234–247.

56. Moesgaard (1989).

57. Halley (1717–1719).

CHAPTER 7

1. Table 7.1 is adapted from Stahlman and Gingerich, (1963); used by permission. Stahlman and Gingerich produced their ephemeris by computer calculation, using the tables of P. V. Neugebauer, which are reprinted in their book. While Neugebauer's tables appear to be good to the nearest degree, Stahlmann and Gingerich include errors in the positions of Mars about the oppositions that rise occasionally to 2° or 3°. In table 7.1 the longitudes of Mars around the times of retrograde motion have been corrected with data from the *American Ephemeris*. Several positions of Jupiter have also been corrected.

2. Diogenes Laërtius, *Lives and Opinions of Eminent Philosophers*, VIII, 14; IX, 23. Aëtius, *Opinions of the Philosophers*, II, 15.4; See Heath (1932), pp. 11, 20.

3. Plato discusses the appropriateness of the divine associations in *Epinomis* 986A–988E.

4. Hunger and Pingree (1989), pp. 85–86.

5. Hunger and Pingree (1989), pp. 117–118, translation slightly modifed.

6. Sachs (1952).

7. Sachs and Hunger (1988) include texts and translations of all known diaries.

8. Sachs and Hunger (1988), Vol. 1, p. 43.

9. Recommended is the SC1 Constellation Chart from Sky Publishing Corporation, 49 Bay State Rd., Cambridge, Mass. 02238.

10. The moment of the planet's opposition to the Sun could occur a day or two before or after the planet reaches the midpoint of its retrograde arc. In addition, we should point out that Ptolemy's planetary theory was based on oppositions to the mean, rather than the true, Sun. (The true Sun is not a good keeper of time—it appears to travel around the ecliptic at a varying speed.) The differences between the mean and the true oppositions, and the differences between either and the center of the retrograde arc, are small enough that we may ignore them, since our planetary data are good only to the nearest degree.

11. Aristotle, *Metaphysics* 1073b17–1074a15; Simplicius, *Commentary on Aristotle's De caelo* (ed. Heiberg), pp. 488, 493–506. English translations of portions of these passages are given in Heath (1932), pp. 65–70. A French translation of a longer extract from Simplicius is given in Verdet (1993), pp. 155–171.

12. Schiaparelli (1875). For a German translation see Schiaparelli (1877).

13. Dreyer (1906), pp. 87–122; Heath (1913), pp. 190–224. Dicks (1970), pp. 175–208, follows Schiaparelli, but expresses skepticism about details of the reconstruction, especially where Schiaparelli has had to "correct" the ancient accounts. Neugebauer (1975), pp. 677–685, provides a full discussion based on Schiaparelli's reconstruction, but is more cautious than Schiaparelli in introducing unsupported assumptions.

14. Translation adapted from Heath (1932), pp. 65–67.

15. Theon of Smyrna, *Mathematical Knowledge Useful for Reading Plato* III, 12.

16. An example of such a problem is the construction by Hippias (late fifth century B.C.) of the curve called the *quadratrix*; See Heath (1921), Vol. 1, pp. 226–230.

17. Heath (1921), Vol. 1, pp. 246–249. The authority for Eudoxus's relation to Archytas is Diogenes Laërtius, *Lives and Opinions of Eminent Philosophers* VIII, 86.

18. Sachs (1948).

19. For a list of the Babylonian normal stars, see Sachs (1974), p. 46.

20. For an argument, due to P. Huber, supporting this explanation of the use of two different great cycles, see van der Waerden (1974), p. 109.

21. For an English translation of this text (British Museum, Shemtob 135), see van der Waerden (1974), pp. 107–108.

22. Although the 18-year eclipse cycle is attested in the Babylonian material, the name "saros" is of modern (probably seventeenth-century) origin; see Neugebauer (1975), p. 497, n. 2. Since the saros contains 6,585 1/3 days, we can form a period three times longer, which will contain a whole number of days. Such a period was called by the Greeks the *exeligmos*; see Geminus, *Introduction to the Phenomena* XVIII, and Ptolemy, *Almagest* IV, 2. After one exeligmos, not only do most circumstances of the lunar eclipse recur, but the eclipse even takes place at about the same time of day.

23. The material is collected with translations and commentary in Neugebauer (1983a). For detailed discussions, see Neugebauer (1975) and van der Waerden (1974). A good introduction to Babylonian planetary theory is Aaboe (1958). For an ambitious attempt to reconstruct the development of Babylonian planetary theory, see Swerdlow (1998b).

24. Pliny, *Natural History* VI, 121–123.

25. See Kugler (1907–1924). The story of the decipherment is told in engaging detail in Neugebauer (1969).

26. Herodotus, *Histories* I, 181–183.

27. Pliny, *Natural History* VI, 121–122.

28. For the Uruk colophons, see Neugebauer (1983a), Vol. I, pp. 11–26. See also Francesca Rochberg, "The Cultural Locus of Astronomy in Late Babylonia," in Galter (1993), pp. 31–45.

29. ACT 18. The tablets of Babylonian mathematical astronomy will be cited according to their ACT numbers, i.e., the numbers assigned in Neugebauer (1983a), *Astronomical Cuneiform Texts*.

30. ACT 123a.

31. ACT 122.

32. Strabo, *Geography* XVI, 1, 6. Pliny mentions Cidenas in *Natural History* II, 39. Vettius Valens, the Greek astrological writer of the second century A.D., mentions Kidenas and Sudines; see Vettius Valens, *Anthology* (ed. Kroll), p. 354, 4–7.

33. ACT 600. Neugebauer (1983a), Vol. 3, p. 176, with commentary in Vol. 2, p. 339. Some numerical values have been restored by Neugebauer in damaged portions of the tablet. However, these are quite se-

cure, since they are computed by a well-understood scheme.

34. ACT 813 and ACT 814.

35. ACT 810. Neugebauer (1983a), Vol. 2, p. 377.

36. ACT 605. Neugebauer (1983a), Vol. 3, p. 182; see also Vol. 2, p. 342. Some of the longitudes of the second stations have been restored by Neugebauer to fill in damaged portions of the text.

37. The modern values for the longitudes of the second stations are from Tuckerman (1962).

38. ACT 612. Neugebauer (1983a), Vol. 2, pp. 344–345; Vol. 3, p. 189.

39. ACT 610 and ACT 613a. See the discussion in Neugebauer (1983a), Vol. 2, p. 345.

40. ACT 620.

41. The theorem is stated by Ptolemy and attributed to Apollonius in *Almagest* XII, 1. For a discussion and proof of the theorem, see Pedersen (1974), pp. 331–338, or Neugebauer (1975), pp. 267–270.

42. For the importance of genre in Greek mathematics, see the discussion of Geminus's classification of the mathematical arts in sec. 2.2.

43. Pliny, *Natural History* II, 63–64.

44. There is a vestige of intermediate model A in Ptolemy's treatment of the superior planets. In the final model (separate equant point), the calculation of the eccentricity from the observed times and positions of three oppositions is extremely difficult. Therefore, Ptolemy begins by supposing that the equant point coincides with the center of the deferent. That is, he temporarily assumes intermediate model A. After obtaining an approximate value of the eccentricity from this calculation, he calculates the correction to the eccentricity produced by passing over to the final model in which the equant and the center of the deferent are separate (*Almagest* X, 7; XI, 1; XI, 5). In the first stage of the calculation we see, no doubt, a traditional method of determining the eccentricity in the intermediate model that Ptolemy had received from his predecessors.

45. Sachs (1952).

46. Geminus, *Introduction to the Phenomena* II, 5; Vitruvius, *On Architecture* IX, 6.2; Strabo, *Geography* XVI, 1.6; Sextus Empiricus, *Against the Mathematicians* V, 2.

47. Neugebauer and van Hoesen (1959).

48. A good brief introduction to the Western astrological tradition is Tester (1987). For the ancient period see Barton (1994). Also useful is G. P. Goold's introduction to his translation of the *Astronomica* of Manilius. But the best introduction is Ptolemy's own *Tetrabiblos*.

49. Vettius Valens, *Anthology* I, 20 (ed. Kroll), pp. 33–36. For a discussion, see Neugebauer (1975), pp. 793–801.

50. A useful introduction to this material is Alexander Jones, "A Classification of Astronomical Tables on Papyrus," in Swerdlow (1998a). Recent detailed studies are Jones (1997a), (1997b), and (1997c).

51. O. Neugebauer, "A Babylonian Lunar Ephemeris from Roman Egypt," in Leichty, Ellis, and Geradi (1988).

52. Jones (1990). For a survey of the evidence for Greek use of Babylonian planetary theory, see Jones, "Evidence for Babylonian Arithmetical Schemes in Greek Astronomy," in Galter (1993), pp. 77–94.

53. Van der Waerden (1974), pp. 295–298.

54. Pliny, *Natural History* VII, 123.

55. For a good general discussion of the problem of transmission of Babylonian astronomy to the Greeks, see Jones (1991).

56. For Theon's dependence on Adrastus, see Theon of Smyrna, *Mathematical Knowledge Useful for Reading Plato* III, 17, 23, 26. For Theon's criticism of Chaldaean astronomy, see III, 30.

57. Aristotle, *On the Heavens* 292a4.

58. *On the Heavens* 291a29–291b10.

59. Plato, *Republic* X, 616b–617d.

60. Theon of Smyrna, *Mathematical Knowledge Useful for Reading Plato* III, 15.

61. *Mathematical Knowledge Useful for Reading Plato* III, 33.

62. Pliny, *Natural History* II, 12–13 (Trans. Rackham). In the opening line, where Rackham has translated *medius* by "in the midst," I have substituted the more definite "in the middle." Pliny places three planets above the Sun and three below.

63. Ptolemy, *Almagest* IX, 1.

64. Ptolemy, *Planetary Hypotheses* I, 2.2.

65. *Planetary Hypotheses* I, 2.3.

66. Aristotle, *On the Heavens* 291b24–293a14.

67. Theon of Smyrna, *Mathematical Knowledge Useful for Reading Plato* III, 33.

68. For a study of the evidence and a detailed refutation of the attribution of this system to Herclides, see Eastwood (1992).

69. The following discussion is based on Evans (1984).

70. For alternative views of how Ptolemy arrived at the idea of the equant, see Pedersen (1974), pp. 277–279, 306–307; Neugebauer (1975), p. 155. For a criticism, see Evans (1984), pp. 1087–1088.

71. Ptolemy derives the parameters of Mars from observations in *Almagest* X, 7–10. For a discussion of his procedure, see Neugebauer (1975), pp. 172–182. Pedersen (1974), pp. 269–290, gives a detailed ac-count of Ptolemy's methods using Saturn as example.

72. We explain here how the parameters for the superior planets in table 7.4 were established. The mean motions are based on modern values. The radius of the epicycle was taken to be the ratio of the major axis of the Earth's elliptical orbit to that of the superior planet. When the radius of the epicycle is derived from observational data in Ptolemy's manner, the result varies somewhat, depending on the observations used. This reflects the fact that the epicycle model is not quite in perfect accord with the actual motions. The epicycle radii in table 7.4 perform well, but slightly different radii might perform equally well. The eccentricity and the longitude of the apogee of the deferent were determined in proper Ptolemaic fashion by requiring the model to reproduce three oppositions of the planet to the mean Sun. The oppositions used were:

Planet	Date		Longitude
Mars	1971 Aug	9.40	317.24
	1973 Oct	23.78	32.04
	1978 Jan	22.45	121.44
Jupiter	1969 Mar	23.63	180.93
	1972 Jun	25.18	273.41
	1975 Oct	11.93	19.88
Saturn	1961 Jul	19.01	296.57
	1967 Oct	1.11	9.16
	1976 Jan	20.97	119.47

73. In fact, the apogees of the planets all move at slightly different rates, and somewhat more rapidly than precession. Thus, the apogees cannot be considered as fixed with respect to the stars. One of the first to realize this was Copernicus, who compared the positions of the apogees determined from recent observations with the postions of the apogees determined by Ptolemy. In this way he found that Saturn's apogee had a motion in longitude of about 1° in 100 years over and above precession. Similarly, Copernicus found for Jupiter's apogee a motion of about 1° in 300 years with respect to the stars, and for Mars's apogee, about 1° in 130 years (*De revolutionibus* V, 7, 12, 16). Copernicus's rates for the motions of the planetary apogees with respect to the stars are all too rapid, but they are of the right order of magnitude.

In using our modern Ptolemaic theory (and the Ptolemaic slats) to calculate planetary postions many centuries before or after A.D. 1900, we will get slightly better results if we use good values for the rates of motion of the apogees. In sec. 7.21, we follow Ptolemy in assuming that all the apogees advance at the rate of precession. To do better, we can use the following rates:

Planet	Motion of Apogee
Venus	1.425′ per year
Mars	1.084
Jupiter	0.915
Saturn	1.237

The epoch 1900 positions of the apogees in table 7.4 are left unchanged.

74. For Ptolemy's method of constructing his tables of planetary equations, see Neugebauer (1975), pp. 183–186, or Pedersen (1974), pp. 291–294.

75. This section is based on J. Evans, "Ptolemy's Cosmology," in Hetherington (1993a), pp. 528–544.

76. The *Planetary Hypotheses* is conventionally divided into two books. Only the first part of the first book survives in Greek. However, the entire work, apart from some numerical tables that have not survived, is preserved in a medieval Arabic translation. The Teubner edition of Ptolemy's *Opera* (Vol. II) contains the Greek text and a German translation of book I, part 1, together with a German translation of the Arabic version of book II. A part of considerable cosmological interest (book I, part 2, containing the distance scale) was inadvertently omitted from the Teubner edition. For the entire Arabic text and an English translation of book I, part 2, see Goldstein (1967).

77. On circles and spheres in Western cosmology, see Aiton (1981). On Ptolemy's attitude toward the physical elements of his theory, see Murschel (1995). For a discussion of the ethical and philosophical roots of Ptolemy's cosmology, see Taub (1993). Taub stresses (and perhaps overstresses) the Platonist elements of Ptolemy's thought.

78. Theon of Smyrna, *Mathematical Knowledge Useful for Reading Plato* III, 32–33.

79. The case that Apollonius discussed deferent-and-epicycle theory in terms of solid spheres is made in Evans (1996).

80. For a study of Ptolemy's cosmological distance scale and its modification by medieval writers, see Swerdlow (1968). A good brief account is given by Van Helden (1985), which traces the history of the effort to find the distances of the planets down to the early eighteenth century.

81. For a survey of Indian astronomy, see the article by Pingree in *DSB*. See also Pingree (1976). A good introduction to Indian astronomy is available in North (1995), pp. 162–174. On the early Arabic translations from Syriac, Greek, and Sanskrit, see Pingree (1973).

82. For an introduction to this literature, see "On the Astronomical Tables of the Islamic Middle Ages" in King (1986),

as well as King (1993). For the variety of ways in which the *qibla* was defined in the early Middle Ages, see King (1995).

83. For an edition of al-Khwārizmī's *zīj*, see Suter (1914). For an English translation, see Neugbauer (1962b).

84. The Arabic text of al-Battānī's *zīj* has been published with Latin translation by Nallino (1899–1907). An account of al-Battānī's work can be found in Willy Hartner's article "Al-Battānī" in *DSB*. Although, as mentioned above, the solar apogee was usually held to be fixed with respect to the stars, some later astronomers asserted that it had an additional proper motion. See Toomer (1969).

85. For an overview of the *zīj* tradition, see Kennedy (1956). For a study of the *Toledan Tables*, see Toomer (1968).

86. On the Arabic translations of the *Almagest*, see Saliba (1994), pp. 143–144, and Toomer (1984), p. 2.

87. For a discussion of Thābit's astronomical works, as well Arabic texts and French translations of those extant, see Morelon (1987).

88. On the role of the *Almagest* commentaries, and on that of al-Ṭūsī in particular, see Saliba (1994), pp. 143–160.

89. Proclus, *Hypotyposis* VII, 19–23 (ed. Manitius), pp. 220–224.

90. Morelon (1987), pp. 13–15.

91. Van Helden (1985), p. 30.

92. For an English translation of and commentary on Ibn al-Haytham's *On the Configuration of the World* (*Maqālah fī hay'at al-ᶜālam*), see Langermann (1990).

93. Saliba (1994), p. 151.

94. Langermann (1990), pp. 8–10.

95. For an English translation of this passage (as well as quotations from medieval defenders of Ptolemy), see Grant (1974), pp. 516–529.

96. *Al-Tadhkira fī ᶜilm al-hay'a*. This work has been published with English translation and commentary by Ragep (1993).

97. A recent discussion of the Ṭūsī couple is Di Bono (1995), which includes full references to the growing literature on this subject. Although, for simplicity of expression, we have described the motion as a "rolling" of the smaller circle within the larger one, al-Ṭūsī does not speak of it in these terms. For him, the smaller circle is carried within the larger, with its rate of rotation matched to that of the larger. In medieval cosmology there is no rolling (although the two representations are mathematically equivalent). It is perhaps significant that Aristotle had explicitly said that "the stars do not roll" (*On the Heavens* 290a25).

98. On Ibn al-Shāṭir see Kennedy and Roberts (1959) and Kennedy and Ghanem (1976). See also David King's article "Ibn al-Shāṭir" in *DSB*.

99. For a survey of European astronomy in the early Middle Ages, see Stephen C. McCluskey, "Astronomies in the Latin West from the Fifth to the Ninth Centuries," in Butzer and Lohrmann (1993), pp. 139–160.

100. For a translation and study of Martianus Capella, see Stahl, Johnson, and Burge (1971–1977).

101. Stahl, Johnson, and Burge (1971–1977), Vol. 1, p. ix.

102. An excellent introduction to Latin astronomy in the early Middle Ages may be found in the publications of Bruce Eastwood, who has made a close study of the manuscripts circulating in Carolingian Europe. See Eastwood, "The Astronomies of Pliny, Martianus Capella and Isidore of Seville in the Carolingian World," in Butzer and Lohrmann (1993), pp. 161–180, as well as Eastwood (1987), (1994), and (1995).

103. Extracts from Isidore of Seville on astronomical and cosmological matters are available in Grant (1974), pp. 11–16, 25–27. For a useful study and a wider range of extracts, see Brehaut (1912). For a discussion of Bede's *On the Nature of Things*, see "Bede's Scientific Achievement" in Stevens (1995).

104. Grant (1974), p. 35.

105. For a complete list, see Grant (1974), pp. 39–41.

106. For translations of documents relating to the thirteenth-century condemnations of Arisotle, see Grant (1974), pp. 42–50.

107. For an edition and translation of Sacrobosco's *Sphere*, as well as the texts of some of his medieval commentators, see Thorndike (1949). Part of Thorndike's translation of Sacrobosco is reprinted in Grant (1974), pp. 442–451.

108. For a translation of the *Theorica planetarum*, see Grant (1974), pp. 451–465.

109. A thorough account of the medieval cosmology curriculum is given in Grant (1994).

110. Aristotle, *Metaphysics* 1074a14–18.

111. Grant (1994), pp. 526–545.

112. For Nicole Oresme's fourteenth-century attack on astrology, see Grant (1974), pp. 488–494.

113. Chaucer, *Canterbury Tales*, "Franklin's Tale," 545–556.

114. For a modern edition of the *Alfonsine Tables* with commentary and worked examples, see Poulle (1984). An extract from the *Alfonsine Tables*, sufficient for calculat-

ing eclipses, is available in Grant (1974), pp. 465–487. On Alfonso's role, see "Alfonso X as a Patron of Astronomy" in Gingerich (1993), pp. 115–128.

115. See the article "Peurbach" by C. Doris Hellman and Noel M. Swerdlow in *DSB*.

116. For an English translation of Peurbach's *Thoricae novae planetarum*, see Aiton (1987).

117. A good introduction to Regiomontanus's work is the article "Regiomontanus, Johannes," by Edward Rosen in *DSB*.

118. Morison (1942), pp. 653–654.

119. Halma (1822–1825), Vol. 1, p. 10.

120. Proclus, *Hypotyposis* III, 66–72 (ed. Manitius), pp. 72–77.

121. For a comprehensive survey of European equatoria in the Middle Ages and the Renaissance, see Poulle (1980). For a good brief account, see North (1976), Vol. 2, pp. 249–286. For discussions of particular instruments, see North (1976), Price (1955), Kennedy (1952) and (1960), and Benjamin and Toomer (1971).

122. Rico-Sinobas (1863–1867). The two tracts on equatoria are in Vol. 3. For a discussion of the *Libros del Saber*, see "Alfonso X as a Patron of Astronomy" in Gingerich (1993). For the dependence of this material on Arabic sources, see "Alfonso X and Arabic Astronomy" in Samsó (1994).

123. The Arabic text of a second treatise on the equatorium by al-Zarqālī has survived. For a Spanish translation, see Millás-Vallicrosa (1950), pp. 460–479. The Arabic text is concerned mostly with the use of the instrument, while the Castilian text is concerned mostly with its construction. There are some minor differences in the forms of the instruments described.

124. For an edition, translation, and commentary, see Benjamin and Toomer (1971). This should not be confused with the anonymous *Theorica planetarum* that was widely used as a university text (sec. 7.26).

125. Poulle (1980), Vol. 1, pp. 279–294, with photographs in Vol. 2, plates 24 and 25.

126. Price (1955). For a recent reassessment of the evidence, see Rand Schmidt (1993).

127. First publication of the equatorium: Schöner (1521/1534); canons with worked examples: Schöner (1522); new tables of initial values and mean motions accompanied, for the first time, by diagrams illustrating the worked examples: Schöner (1524); revised edition of equatorium with complete tables, canons, and illustrated worked examples: Schöner (1551/1561).

128. For a description of Apianus's *Astronomicum Caesareum* with many photographs of the instruments, see Ionides (1936). See also Poulle (1980) and Gingerich (1971).

129. See Christine Schofield, "The Tychonic and Semi-Tychonic World Systems," in Taton and Wilson (1989), pp. 234–247.

130. Two biographies of Copernicus are Rosen (1971) and Swerdlow and Neugebauer (1984).

131. An English translation of the *Commentariolus* is available in Rosen (1971) and in *Nicholas Copernicus: Complete Works. Minor Works*. For a translation with a detailed technical commentary, see Swerdlow (1973).

132. Quoted from M.-P. Lerner and J.-P. Verdet, "Copernicus," in Hetherington (1993b), pp. 152–153.

133. On Rheticus's relations with Schöner and Petreius, see Swerdlow (1992).

134. An edition of *Narratio prima*, accompanied by a French translation and a full commentary, is available in Huggonard-Roche and Verdet (1982). For an English translation, see Rosen (1971).

135. For details about the printing of the book, see "*De revolutionibus*: An Example of Renaissance Scientific Printing" in Gingerich (1993), pp. 252–268.

136. [Osiander], in Copernicus, *De revolutionibus* (trans. Rosen), p. xx.

137. Rosen (1971), pp. 403–406.

138. This untenable explanation of the Copernicus's innovation is adopted in the otherwise very learned and readable book by Kuhn (1957).

139. For a comparison of several sixteenth-century ephemerides, see "'Crisis' versus Aesthetic in the Copernican Revolution" in Gingerich (1993), pp. 193–204.

140. Hooykaas (1984).

141. Copernicus, *De revolutionibus* (trans. Rosen), Preface.

142. *De revolutionibus* IV, 2.

143. *De revolutionibus* V, 4. The flattened character of the orbit is easily seen by comparison of the planet's distance from the orbit center M at $0°$ and at $90°$ of mean anomaly, i.e., at positions 1 and 3 in fig. 7.64. In position 1 we have $MP = CG - GP + CM = R - aR + aR = R$. But in position 3 we have $MP = [CG^2 + (GP + CM)^2]^{1/2} = R(1 + 4a^2)^{1/2}$, which is somewhat greater than R. Note, however, that the long axis of the Copernican orbit coincides with the *minor* axis of the Keplerian ellipse. In this respect, Copernicus's model represents a step backward from Ptolemy's.

144. For Ibn al-Shāṭir's planetary theory, see Kennedy and Roberts (1959). This article is reprinted with other studies in Kennedy and Ghanem (1976).

145. For a guess at how Copernicus could have worked with the model without fully realizing its equivalence to Ptolemy's, see Evans (1988), n. 10.

146. Grafton (1973), pp. 526–528. Kepler wrote back to his teacher, "Now at last I understand this business" about the equant.

147. For more detail on Copernicus's planetary theory, see Swerdlow and Neugebauer (1984). For a comparison of Copernicus's model with those of Ptolemy, Brahe, and Kepler, see Evans (1988).

148. On Mästlin's teaching of Copernicus's theory, see Methuen (1996). A good brief discussion of the reception of Copernicus's cosmology up to the year 1600 is given by Lerner and Verdet, "Copernican Cosmology," in Hetherington (1993a), pp. 71–90. For a book-length treatment centered on the role of Chistoph Clavius, see Lattis (1994).

149. "Erasmus Reinhold and the Dissemination of the Copernican Theory," in Gingerich (1993). See also "Early Copernican Ephemerides" in the same work, pp. 205–220.

150. On the Tychonic system, see Thoren (1990), pp. 236–264. Brahe was not the only Renaissance astronomer to advocate a geo-heliocentric system. In fact, Brahe got into a nasty dispute over the credit for this system. See Jardine (1984), Rosen (1986), and Gingerich and Westman (1988).

151. Thoren (1990), pp. 123–127.

152. Thoren (1990), pp. 257–258.

153. For a selection from sixteenth- and seventeenth-century writers reacting to Copernicus, see Crowe (1990), pp. 174–188. For an extended discussion, see Kuhn (1957).

154. "The Censorship of Copernicus's *De revolutionibus*" in Gingerich (1993), pp. 269–285.

155. A standard account of the Galileo affair is Santillana (1955). A good brief account is available in Sharratt (1994).

156. On the decline of science in late antiquity, see Lloyd (1973), pp. 154–178.

157. See Swerdlow (1973), pp. 467–469, and Saliba (1994), pp. 291–305. For an interesting argument that some of these models were circulating in central Europe shortly before Copernicus's time, see Dobrzycki and Kremer (1996).

158. Grant (1994).

159. Kuhn (1957), pp. 126–133, makes a case for the role of Neoplatonism in the Copernican revolution.

160. For Kepler's life, see Caspar (1993). For an overview of Kepler's work, see Gin-

gerich, "Johannes Kepler," in Taton and Wilson (1989), pp. 54–78. Also useful is Koyré (1973).

161. Kepler, Preface to *Mysterium cosmographicum* (trans. Duncan), pp. 62–67.

162. Plato, *Timaeus* 53D–57C.

163. Aristotle, *On the Heavens* 306a1–307b25.

164. For a study of the role of physical speculation in Kepler's astronomy, see Stephenson (1987). The enthusiasm for magnetic motive forces in seventeenth-century astronomy is discussed by Stephen Pumfrey, "Magnetical Philosophy and Astronomy, 1600–1650," in Taton and Wilson (1989), pp. 45–53.

165. Kepler, Preface to *Mysterium cosmographicum* (trans. Duncan), p. 69.

166. Caspar (1993), p. 102.

167. Longomontanus (1572–1647) is best known today for his *Astronomica Danica*, of which there were three printings between 1622 and 1640. Although Longomontanus worked closely with Kepler at Prague, he did not fully accept Kepler's discoveries. He followed Kepler in using the oppositions of the planets to the true, rather than the mean, Sun. But he did not accept the elliptical orbit. For biographical details, see Dreyer (1890), p. 126ff. For a survey of *Astronomica Danica*, see Delambre (1821), Vol. 1, pp. 262–287.

168. Kepler, *Astronomica nova* VII (trans. Donahue), p. 185.

169. *Astronomica nova* VII (trans. Donahue), p. 185.

170. Kepler discusses Longomontanus's theory of Mars in *Astronomia nova* I, iv (p. 15 in the edition of 1609, but mistakenly printed 14). The numerical parameters of the model are again mentioned at II, vii (p. 53), where Longomontanus is explicitly given credit. This working theory of Brahe and Longomontanus is discussed by Small (1804), p. 154.

171. Evans (1988), p. 1019.

172. Kepler, *Astronomia nova* II, xvi (p. 95). Kepler's request for pity is often quoted sympathetically. But Delambre, who estimated that these calculations would run to a mere 200 manuscript pages, repeated them and found them "not so terrible." "One has more calculations to make today to determine the elliptical orbit of a comet" (Delambre, 1821, Vol. 1, p. 417). Delambre adds that "Kepler was sustained by the desire to be right against Tycho, Copernicus, Ptolemy, and all the astronomers of the universe; he tasted this satisfaction, and I do not believe he was so worthy of pity when he made all these calculations."

173. *Astronomia nova* II, xix (trans. Donahue).

174. Kepler's procedure is explained in Koyré (1973) and in mathematical detail in Small (1804). A good brief account is Gingerich, "Johannes Kepler," in Taton and Wilson (1989). The best detailed account is Wilson (1968).

175. This was the reaction of Ismaël Boulliau (1605–1694). See Curtis Wilson, "Predictive Astronomy in the Century after Kepler," in Taton and Wilson (1989), p. 172.

176. For a good account, see Koyré (1973).

177. Curtis Wilson, "Predictive Astronomy in the Century after Kepler," in Taton and Wilson (1989), p. 172.

178. Wilson, "Predictive Astronomy in the Century after Kepler," in Taton and Wilson (1989), p. 178.

179. Curtis Wilson, "The Newtonian Achievement in Astronomy," in Taton and Wilson (1989), p. 238.

180. Evans (1988), p. 1022.

181. For a simple proof, see Evans (1988), pp. 1017–1020.

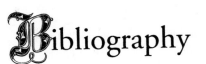

Bibliography

ENCYCLOPEDIC WORKS

DSB. Dictionary of Scientific Biography (1970–). Ed. by C. C. Gillispie. New York: Scribner.

RE. Paulys Real-Encyclopädie der classischen Altertumswissenschaft (1894–). Neue Bearbeitung von G. Wissowa, W. Kroll [*et al.*]. Munich: Druckenmüller.

ANCIENT, MEDIEVAL, AND RENAISSANCE WRITERS

Apianus, Petrus (Caesar's Astronomy). *Astronomicum Caesareum.* Ingolstadt: 1540. Facsimile reprint, Leipzig: Edition Leipzig, 1967.

———. *Cosmographicus liber Petri Apiani.* Landshut: 1524.

Aratus (Phenomena). *Callimachus, Lycophron, Aratus.* Ed. and trans. by A. W. Mair and G. R. Mair. Loeb Classical Library. London: Heinemann; Cambridge, Mass.: Harvard University Press, 1955.

Archimedes (Works). See Heath (1897) and Dijksterhuis (1987).

Aristarchus of Samos (On the Sizes and Distances of the Sun and Moon). See Heath (1913).

Aristotle. *On the Heavens.* Greek text ed. and English trans. by W. K. C. Guthrie. Loeb Classical Library. London: Heinemann; Cambridge, Mass.: Harvard University Press, 1939.

———. *The Metaphysics.* Greek text ed. and English trans. by Hugh Tredennick. 2 vols. Loeb Classical Library. London: Heinemann; Cambridge, Mass.: Harvard University Press, 1933–1935.

Autolycus (Works). *Autolyci De sphaera quae movetur. De ortibus et occasibus.* Greek text ed. and Latin trans. by F. Hultsch. Leipzig: Teubner, 1885.

——— (Works). *Autolycus de Pitane. Histoire du texte suivie de l'édition critique.* Ed. by Joseph Mogenet. Universit de Louvain, Recueil de travaux d'histoire et de philologie, 3e série, fasc. 37, 1950.

——— (Works). *The Books of Autolykos: On a Moving Sphere and On Risings and Settings.* Trans. by Frans Bruin and Alexander Vondjidis. Beirut: American Uninversity of Beirut, 1971.

——— (Works). *Autolycos de Pitane: Le sphère en mouvement. Levers et couchers héliaques.* Greek text ed. and French trans. by Germaine Aujac. Paris: Les Belles Lettres, 1979.

Al-Battānī (Zīj). See Nallino (1899–1907).

Bede (Ecclesiastical History). *A History of the English Church and People.* Trans. by Leo Sherley-Price, revised by R. E. Latham. Harmondsworth: Penguin Books, 1968.

Blagrave, John. *The Mathematical Jewel.* London: 1585. Reprint, Amsterdam: Da Capo Press, 1971.

Brahe, Tycho (Opera). *Tychonis Brahe Dani Opera omnia.* Ed. by J. L. E. Dreyer. 15 vols. Copenhagen: Libraria Gyldendaliana, 1913–1929.

Callimachus of Cyrene (Hymns and Epigrams). See Aratus.

Campanus of Novara (Theorica planetarum). See Benjamin and Toomer (1971).

Chaucer (Equatorie of the Planetis). See Price (1955).

———. (Treatise on the Astrolabe). See Gunther (1929) and Skeat (1872).

Cicero (De re publica). *Marcus Tullius Cicero. De re publica. De legibus.* Trans. by Clinton Walker Keyes. Loeb Classical Library. London: Heinemann; New York: Putnam, 1928.

——— (On the Nature of the Gods). *De natura deorum; Academia.* Ed. and trans. by H. Rackham. Loeb Classical Library. London: Heinemann; New York: Putnam (1933).

———. *Tusculan Disputations.* Ed. and trans. by J. E. King. Loeb Classical Library. London: Heinemann; New York: Putnam, 1928.

Cleomedes (On the Elementary Theory of the Heavenly Bodies). *Cléomède. Théorie élémentaire.* French trans. and commentary by Richard Goulet. Paris: Librairie Philosophique J. Vrin, 1980.

———. *Cleomedis Caelestia* (METEΩPA). Ed. by Robert Todd. Leipzig: Teubner, 1990.

———. *Cleomedis De motu circulari corporum caelestium libriduo.* Ed. by Hermann Ziegler. Leipzig: Teubner, 1891.

Columella (De re rustica). *Lucius Junius Moderatus Columella. On Agriculture and Trees.* Latin text ed. and English trans. by E. S. Forster and E. D. Heffner. 3 vols. Loeb Classical Library. London: Heinemann; Cambridge, Mass.: Harvard University Press, 1968.

Copernicus. (Commentariolus). See Rosen (1971).

——— (De revolutionibus). *On the Revolutions of the Heavenly Spheres*. Trans. by A. M. Duncan. New York: Barnes & Noble, 1976.

——— (Works). *Nicholas Copernicus. Complete Works*. Trans. by Edward Rosen. 2 vols.: *On the Revolutions and Minor Works*. Baltimore: Johns Hopkins University Press, 1992. (Originally published 1978–1985)

Diodorus Siculus (Histories). *Diodorus Siculus*. Trans. by C. H. Oldfather et al. 12 vols. Loeb Classical Library. London: Heinemann; New York: Putnam, 1933.

Diogenes Laërtius (Lives and Opinions of Eminent Philosophers). *Lives of Eminent Philosophers*. Trans. by R. D. Hicks. 2 vols. Loeb Classical Library. Cambridge, Mass.: Harvard University Press; London: Heinemann, 1925; reprint 1972.

Eratosthenes (Catasterisms). See Halma (1821).

Euclid (Elements). See Heath (1926).

——— (Opera). *Euclidis opera omnia*. Ed. by J. L. Heiberg and H. Menge. 8 vols. Leipzig: Teubner, 1883–1916.

——— (Phenomena). *Euclid's Phaenomena: A Translation and Study of a Hellenistic Treatise in Spherical Astronomy*. Trans. by J. L. Berggren and R. S. D. Thomas. New York and London: Garland, 1996.

Geminus (Introduction to the Phenomena). See Manitius (1898) and Aujac (1975).

Herodotus. *The Histories*. Trans. by Aubrey de Sélincourt, revised by A. R. Burn. Harmondsworth (England): Penguin 1972.

Hesiod (Works and Days). *Hesiod, The Homeric Hymns and Homerica*. Greek text ed. and English trans. by H. G. Evelyn-White. Loeb Classical Library. London: Heinemann; Cambridge, Mass.: Harvard University Press, 1936.

Hipparchus (Commentary on the Phenomena of Aratus and Eudoxus). *Hipparchi in Arati et Eudoxi Phaenomena commentariorum libri tres*. Greek text ed. and German trans. by Carolus Manitius. Leipzig: Teubner, 1894.

Hippolytus (Refutation of all Heresies). *Hippolytus. Philosophumena or the Refutation of all Heresies*. Trans. by F. Legge. 2 vols. London: 1921.

Homer (Iliad). *The Iliad of Homer*. Trans. by Richmond Lattimore. Chi-

cago: University of Chicago Press, 1961.

———. *The Odyssey of Homer*. Trans. by Richmond Lattimore. New York: Harper & Row, 1967.

Hyginus (Astronomy). *Hygin. L'astronomie*. Latin text ed. and French trans. by André Le Boeuffle. Collections des Universités de France. Paris: Les Belles Lettres, 1983.

Ibn al-Haytham (On the Configuration of the World). See Langermann (1990).

Ibn Yūnus (Hakemite Zīj). See Caussin (1804).

Kepler, Johannes (Astronomia Nova). *New Astronomy*. Trans. by William H. Donahue. Cambridge: Cambridge University Press, 1992.

———. *Epitome of Copernican Astronomy*, books IV and V. Trans. by Charles Glenn Wallis, in *Great Books of the Western World*, Vol. 16. Chicago: Encyclopaedia Britannica, 1952.

——— (Harmonice Mundi). *The Harmony of the World*. Trans. and commentary by E. J. Aiton, A. M. Duncan and J. V. Field. *Memoirs of the American Philosophical Society*, **209** (1997).

———. *Mysterium Cosmographicum: The Secret of the Universe*. Trans. by A. M. Duncan with commentary by E. J. Aiton. New York: Abaris Books, 1981.

——— (Rudolphine Tables). *Jean Kepler. Tables Rudolphines*. French trans. by Jean Peyroux. Paris: Blanchard, 1986.

——— (Works). *Johannes Kepler Gesammelte Werke*. Ed. by W. von Dyck, M. Caspar and F. Hammer. Munich: C. H. Beck, 1937–.

Al-Khwārizmī (Zīj). See Neugebauer (1962b) and Suter (1914).

Manilius (Astronomica). *Manilius. Astronomica*. Ed. with English trans. by G. P. Goold. Loeb Classical Library. London: Heinemann; Cambridge, Mass.: Harvard University Press, 1977.

Martianus Capella (Marriage of Philology and Mercury). See Stahl, Johnson, and Burge (1971–1977).

Ovid (Fasti). *Ovid's Fasti*. Trans. by James G. Frazer, Loeb Classical Library. London: Heinemann; New York: Putnam, 1931.

Pappus (Commentary on the Almagest). See Rome (1931–1943).

——— (Mathematical Collection). *Pappi Alexandrini Collectionis quae supersunt*. Greek text ed. by F. Hultsch. 3 vols. Berlin: Weidmann, 1876–1878; Amsterdam: Hakkert, 1965.

——— (Mathematical Collection). *Pap-

pus d'Alexandrie. La Collection Mathématique*. French trans. by Paul Ver Eecke. Paris: Albert Blanchard, 1933.

Peurbach, Georg (New Theories of the Planets). *Theoricae novae planetarum*. See Aiton (1987).

Philoponus, John (Treatise on the Astrolabe). See Segonds (1981).

Plato. *The Collected Dialogues*. Ed. by Edith Hamilton and Huntington Cairns. Princeton: Princeton University Press, 1963.

——— (Timaeus). *Timaeus and Critias*. Trans. by Desmond Lee. Harmondsworth: Penguin Books, 1971.

Pliny (Natural History). *Pliny. Natural History*. Latin text ed. and English trans. by H. Rackham. 10 vols. Loeb Classical Library. London: Heinemann; Cambridge, Mass.: Harvard University Press, 1949.

Plutarch (Lives). *Plutarch's Lives*. Greek text ed. and English trans. by Bernadotte Perrin. 11 vols. Loeb Classical Library. London: Heinemann; Cambridge, Mass.: Harvard University Press, 1958–1962.

——— (Moralia). *Plutarch. Moralia*. Greek text ed. and English trans. by Frank Cole Babbitt. 15 vols. Loeb Classical Library. London: Heinemann; Cambridge, Mass.: Harvard University Press, 1958–.

Proclus (Commentary on Euclid). *Proclus. A Commentary on the First Book of Euclid's Elements*. Trans. by Glenn R. Morrow. Princeton, N.J.: Princeton University Press, 1970.

——— (Hypotyposis). *Procli Diadochi Hypotyposis astronomicarum positionum*. Ed. by C. Manitius. Leipzig: Teubner, 1909.

Ptolemy (Almagest). See Toomer (1984).

——— (Geography). *The Geography of Claudius Ptolemy*. Trans. by E. L. Stevenson. New York: Fischer, 1932.

——— (Geography: reproductions of the maps from Codex Lat. V F.32, National Library, Naples). *Claudii Ptolemaei Cosmographia Tabulae*. Introduction by Lelio Pagani. Leicester: Magna Books, 1990.

——— (Opera). *Claudii Ptolemaei Opera quae exstant omnia*. Greek text ed. by J. L. Heiberg et al. Leipzig: Teubner. Vol. I (2 parts): *Syntaxis mathematica* [the *Almagest*], ed. by J. L. Heiberg, 1898 and 1903. Vol. II: *Opera astronomica minora*, ed. by J. L. Heiberg, 1907. Vol. III, Part 1: *Apotelesmatika*, ed. by F. Boll and A. Boer, 1954. Vol. III, Part 2: *Peri kriteriou kai hegemonikou*, ed. by F. Lammert; *Karpos*, ed. by A. Boer; 2nd ed., 1961.

——— (Phaseis). See *Opera*, Vol. II, pp. 1–67.

——— (Planetary Hypotheses). *Opera*, Vol. II. See also Goldstein (1967) and Morelon (1993).

——— (Tetrabiblos). *Tetrabiblos*. Greek text ed. and English trans. by F. E. Robbins. Loeb Classical Library. London: Heinemann; Cambridge, Mass.: Harvard University Press, 1940.

Reinhold, Erasmus (Prutenic Tables). *Prutenicae tabulae coelestium motuum*. Tübingen: 1551.

Rheticus, Georg Joachim (Narratio prima). See Rosen (1971) and Huggonard-Roche and Verdet (1982).

Sacrobosco (Sphere). See Thorndike (1949).

Schöner, Johann. *Aequatorium astronomicum*. Bamberg: 1521; reprint, Nuremberg: 1534.

———. *Equatorii astronomici omnium ferme uranicarum theorematum explanatorii canones*. Nuremberg: 1522.

———. *Opera mathematica Ioannis Schoneri Carolstadii*. Nuremberg: 1551, reprint, 1561.

———. *Tabule radicum . . . cum demonstrationibus exemplaribus pro motibus planetarum ex equatorio aucupandis*. Timiripa [i.e., Kirchehrenbach], 1524.

Sextus Empiricus (Against the Mathematicians). *Sextus Empiricus*. Greek text ed. and English trans. by R. G. Bury. 4 vols. Loeb Classical Library. London: Heinemann; Cambridge Mass.: Harvard University Press, 1936.

Simplicius (Commentary on Aristotle's Physics). *Simplicii in Aristotelis physicorum libros quattuor priores commentaria*. Commentaria in Aristotelem Graeca, Vol. 9. Ed. by H. Diels. Berlin: Reimer, 1882.

——— (Commentary on Aristotle's De Caelo). *Simplicii in Aristotelis De caelo commentaria*. Commentaria in Aristotelem Graeca, Vol. 7. Ed. by J. L. Heiberg. Berlin: Reimer, 1894.

Strabo (Geography). *The Geography of Strabo*. Greek text ed. and English trans. by H. L. Jones. 8 vols. Loeb Classical Library. London: Heinemann; New York: Putnam, 1917.

Al-Ṣūfī (Book on the Constellations of the Fixed Stars). See Niẓāmuʾd-Din (1954) and Schjellerup (1874).

Thābit ibn Qurra. See Carmody (1960), Morelon (1987), and Neugebauer (1962).

Theodosius of Bithynia (sometimes called

Theodosius of Tripoli) (Spherics). *Les Sphériques de Théodose de Tripoli*. Trans. by Paul Ver Eecke. Paris: Blanchard (1959).

Theon of Alexandria (Commentary on the Almagest). See Rome (1931–1943).

——— (Handy Tables). See Halma (1822–1825) and Tihon (1978), (1985), and (1991).

Theon of Smyrna (Mathematical Knowledge Useful for Reading Plato). *Théon de Smyrne, Philosophe platonicien. Exposition des connaissances mathématiques utiles pour la lecture de Platon*. Trans. by J. Dupuis. Paris: Hachette, 1892; reprint, Bruxelles: Culture et Civilisation, 1966.

Al-Ṭūsī (Memoir on Astronomy). See Ragep (1993).

Vettius Valens (Anthology). *Vettii Valentis Anthologiarum libri*. Greek text ed. by Guilelmus Kroll. Berlin: Weidmann, 1908.

Vitruvius (Ten Books on Architecture). *On Architecture*. Latin text ed. and English trans. by Frank Granger. 2 vols. Loeb Classical Library. London: Heinemann; Cambridge, Mass.: Harvard University Press, 1931–1934.

———. *Vitruvius. The Ten Books on Architecture*. Trans. by Morris Hicky Morgan. Cambridge, Mass.: Harvard University Press, 1914; reprint, New York: Dover, 1960.

———. *Vitruve. De l'Architecture. Livre IX*. Latin text ed. and French trans. by Jean Soubiran. Collection des Universités de France. Paris: Les Belles Lettres, 1969.

MODERN WRITERS

Aaboe, Asger (1958). "On Babylonian Planetary Theories." *Centaurus* 5, 209–277.

Aiton, E. J. (1981). "Celestial Spheres and Circles." *History of Science* 19, 75–114.

——— (1987). "Peurbach's *Theoricae novae planetarum*. A Translation with Commentary." *Osiris* (2nd Ser.), 3, 5–43.

Allen, Richard Hinkley (1899). *Star Names and Their Meanings*. New York: G. E. Stechert. Reprinted as *Star Names: Their Lore and Meaning*. New York: Dover, 1963.

Archer, Peter (1941). *The Christian Church and the Gregorian Reform*. New York: Fordham University Press.

Aujac, Germaine (1975). *Géminos. Introduction aux phénomènes*. Paris: Les Belles Lettres.

Barton, Tamsyn (1994). *Ancient Astrology*. London: Routledge.

Benjamin, Francis S., Jr., and G. J. Toomer (1971). *Campanus of Novara and Medieval Planetary Theory*. Madison: University of Wisconsin Press.

Bickerman, E. J. (1980). *Chronology of the Ancient World*. 2nd ed. Ithaca, N.Y.: Cornell University Press.

Bilfinger, Gustave (1886). *Die Zeitmesser der Antiken Völker*. Stuttgart.

Boll, Franz (1901). "Die Sternkataloge des Hipparch und des Ptolemaios." *Bibliotheca Mathematica* 3(ii), 185–195.

Brehaut, Ernest (1912). *An Encyclopedist of the Dark Ages: Isidore of Seville*. New York: Columbia University. Reprint, New York: Burt Franklin Reprints, 1964.

Britton, John P. (1992). *Models and Precision: The Quality of Ptolemy's Observations and Parameters*. New York: Garland.

Bruin, Frans, and Margaret Bruin (1976). "The Equatorial Ring, Equinoxes, and Atmospheric Refraction." *Centaurus* **20**, 89–111.

Budge, E. A. W. (1904). *The Gods of the Egyptians*. 2 vols. Chicago: Open Court; London: Methuen. Reprint, New York: Dover, 1969.

Butzer, P.L., and D. Lohrmann, eds. (1993). *Science in Western and Eastern Civilization in Carolingian Times*. Basel: Birkhäuser.

Canfora, Luciano (1990). *The Vanished Library*. Trans. by Martin Ryle. Berkeley: University of California Press.

Carmody, Francis J. (1960). *The Astronomical Works of Thabit b. Qurra*. Los Angeles: University of California Press.

Caspar, Max. (1993). *Kepler*. Trans. by C. Doris Hellman. New York: Dover.

Caussin, C. (1804). *Le livre de la grande table Hakemite par Ebn Iounis*. Paris: An XII.

Crowe, Michael J. (1990). *Theories of the World from Antiquity to the Copernican Revolution*. New York: Dover.

Dalley, Stephanie (1989). *Myths from Mesopotamia. Creation, The Flood, Gilgamesh, and Others*. Oxford: Oxford University Press.

Delambre, J. B. J. (1817). *Histoire de l'astronomie ancienne*. 2 vols. Paris. Reprint, New York: Johnson Reprint, 1965.

——— (1819). *Histoire de l'astronomie du moyen âge*. Paris. Reprint, New York: Johnson Reprint, 1965.

——— (1821). *Histoire de l'astronomie moderne*. 2 vols. Paris. Reprint, New York: Johnson Reprint, 1965.

Di Bono, Mario (1995). "Copernicus, Amico, Fracastoro and Ṭūsī's Device: Observations on the Use and Transmission of a Model." *Journal for the History of Astronomy* **26**, 133–154.

Dicks, D. R. (1960). *The Geographical Fragments of Hipparchus*. London: Athlone Press.

———— (1970). *Early Greek Astronomy to Aristotle*. Ithaca, N.Y.: Cornell University Press.

Diels, Hermann (1924). *Antike Technik*. 3rd ed. Leipzig: Teubner.

Diels, H., and A. Rehm (1904). "Parapegmenfragmente aus Milet." *Sitzungsberichte der königlich Preussischen Akademie der Wissenschaften*, Jahrgang 1904, 92–111.

Dijksterhuis, E. J. (1961). *The Mechanization of the World Picture*. Trans. by C. Dikshoorn. London: Oxford University Press.

———— (1987). *Archimedes*. Trans. by C. Dikshoorn. Revised ed. Princeton, N.J.: Princeton University Press.

Diller, Aubrey (1949). "Ancient Measurements of the Earth." *Isis* **40**, 6–9.

Dobrzycki, Jerzy, and Richard L. Kremer (1996). "Peurbach and Marāgha Astronomy? The Ephemerides of Johannes Angelus and Their Implications." *Journal for the History of Astronomy* **27**, 187–237.

Drachman, A.G. (1950). "A Detail of Heron's Dioptra." *Centaurus* **13**, 241–247.

Drake, Stillman (1978). *Galileo at Work: His Scientific Biography*. Chicago: University of Chicago Press.

Drecker, Joseph (1925). *Die Theorie der Sonnenuhren*. Berlin: W. de Gruyter.

Dreyer, J. L. E. (1890). *Tycho Brahe. A Picture of Scientific Life and Work in the Sixteenth Century*. Edinburgh: Black. Reprint, New York: Dover, 1963.

———— (1906). *History of the Planetary Systems from Thales to Kepler*. Cambridge: Cambridge University Press. Reprinted as *A History of Astronomy from Thales to Kepler*. New York: Dover, 1953.

———— (1917). "On the Origin of Ptolemy's Catalogue of Stars." *Monthly Notices of the Royal Astronomical Society* **77**, 528–539.

———— (1918). "On the Origin of Ptolemy's Catalogue of Stars. Second Paper." *Monthly Notices of the Royal Astronomical Society* **78**, 343–349.

Duhem, Pierre (1969). *To Save the Phenomena: An Essay on the Idea of Physical Theory from Plato to Galileo*. Trans.

by Edmund Doland and Chaninah Maschler. Chicago: University of Chicago Press.

Eastwood, Bruce (1987). "Plinian Astronomical Diagrams in the Early Middle Ages." In Edward Grant and John E. Murdoch, eds., *Mathematics and Its Applications to Science and Natural Philosophy in the Middle Ages*. Cambridge: Cambridge University Press.

———— (1992). "Heraclides and Heliocentrism: Texts, Diagrams, and Interpretations." *Journal for the History of Astronomy* **23**, 233–260.

———— (1994). "The Astronomy of Macrobius in Carolingian Europe: Dungal's Letter of 811 to Charles the Great." *Early Modern Europe* **3**, 117–134.

———— (1995). "Celestial Reason: The Development of Latin Planetary Astronomy to the Twelfth Century." In Susan J. Ridyard and Robert G. Benson, eds., *Man and Nature in the Middle Ages*. Sewanee, Tenn.: University of the South Press.

Eichelberger, W. S. (1925). "Positions and Proper Motions of 1504 Standard Stars for the Equinox of 1925.0." *Astronomical Papers Prepared for the Use of the American Ephemeris and Nautical Almanac*, **10**, Part 1.

Evans, James (1984). "On the Function and the Probable Origin of Ptolemy's Equant." *American Journal of Physics* **52**, 1080–1089.

———— (1987). "On the Origin of the Ptolemaic Star Catalogue." *Journal for the History of Astronomy* **18**, 155–172, 233–278.

———— (1988). "The Division of the Martian Eccentricity from Hipparchos to Kepler: A History of the Approximations to Kepler Motion." *American Journal of Physics* **56**, 1009–1024.

———— (1992). "The Ptolemaic Star Catalogue." *Journal for the History of Astronomy* **23**, 64–68.

———— (1996). "A Cosmogenesis: The Origins of Ptolemy's Universe." In F. De Gandt and C. Vilain, eds., *Histoire et Actualité de la Cosmologie*, Vol. 2. Paris: Observatoire de Paris.

Explanatory Supplement to The Astronomical Ephemeris *and* The American Ephemeris and Nautical Almanac. London: Her Majesty's Stationery Office, 1974.

Field, J. V., and M. T. Wright (1984). "Gears from the Byzantines: A Portable Sundial with Calendrical Gearing." *Annals of Science* **42**, 87–138.

Fotheringham, J. K. (1924). "The Meto-

nic and Callippic Cycles." *Monthly Notices of the Royal Astronomical Society* **84**, 383–392.

Galter, Hannes D., ed. (1993). *Die Rolle der Astronomie in den Kulturen Mesopotamiens*. Grazer Morgenländische Studien, 3. Graz.

Gibbs, Sharon (1976). *Greek and Roman Sundials*. New Haven and London: Yale University Press.

———— (1984). *Planispheric Astrolabes from the National Museum of American History*. Smithsonian Studies in History and Technology, No. 45. Washington, D.C.: Smithsonian Institution Press.

Gingerich, Owen (1971). "Apianus's Astronomicum Caesareum and Its Leipzig Facsimile." *Journal for the History of Astronomy* **2**, 168–177.

———— (1993). *The Eye of Heaven: Ptolemy, Copernicus, Kepler*. New York: American Institute of Physics.

Gingerich, Owen, and Robert S. Westman (1988). "The Wittich Connection: Conflict and Priority in Late Sixteenth-Century Cosmology." *Transactions of the American Philosophical Society*, Vol. 78, Part 7.

Ginzel, Friedrich Karl (1906–1914). *Handbuch der mathematischen und technischen Chronologie*. 3 vols. Leipzig: Hinrichs.

Goldstein, Bernard R. (1967) "The Arabic Version of Ptolemy's Planetary Hypotheses." *Transactions of the American Philosophical Society* (n.s.) **57**, Part 4.

Goldstein, Bernard R., and Alan C. Bowen (1989). "On Early Hellenistic Astronomy: Timocharis and the First Callippic Calendar." *Centaurus* **32**, 272–293.

Gössmann, Felix (1950). *Planetarium Babylonicum oder Die Sumerisch-Babylonischen Stern-Namen*. Ed. by Anton Deimel. *Sumerisches Lexikon*, Teil IV, Band 2. Rome: Verlag des Päpstlichen Bibelinstituts.

Grafton, Anthony (1973). "Michael Maestlin's Account of Copernican Planetary Theory." *Proceedings of the American Philosophical Society*, **117**(6), 523–550.

Grant, Edward (1974). *A Source Book in Medieval Science*. Chicago: University of Chicago Press.

———— (1994). *Planets, Stars, and Orbs. The Medieval Cosmos, 1200–1687*. Cambridge: Cambridge University Press.

Grasshoff, Gerd (1990). *The History of Ptolemy's Star Catalogue*. New York: Springer.

Grenfell, Bernard P., and Arthur S. Hunt (1906). *The Hibeh Papyri. Part 1.* London: Egypt Exploration Fund.

Gunther, R. T. (1929). *Early Science in Oxford, Vol. V: Chaucer and Messahalla on the Astrolabe* [Middle English text with modern English translation]. Oxford: Oxford University Press.

Guye, Samuel, and Henri Michel (1971). *Time and Space: Measuring Instruments from the 15th to the 19th Century.* Trans. by Diana Dolon. New York: Praeger.

Halley, Edmund (1717–1719). "Considerations on the Change of the Latitudes of Some of the Principal Fixt Stars." *Philosophical Transactions of the Royal Society of London* **30**, 736–739.

Halma, [Nicolas] (1821). *Les Phénomènes, d'Aratus de Soles, et de Germanicus César, avec les Scholies de Théon, les Catastérismes d'Eratosthène, et la Sphère de Leontius.* Paris.

———— (1822–1825). *Commentaire de Théon d'Alexandrie sur les tables manuelles astronomiques de Ptolémée.* 3 vols. Paris.

Heath, T. L. (1897). *The Works of Archimedes.* Cambridge: Cambridge University Press.

———— (1913). *Aristarchus of Samos.* Oxford: Clarendon Press.

———— (1921). *A History of Greek Mathematics.* 2 vols. Oxford: Clarendon Press.

———— (1926). *The Thirteen Books of Euclid's Elements.* 2nd ed. 3 vols. Cambridge: Cambridge University Press.

———— (1932). *Greek Astronomy.* London: Dent.

Hetherington, Norriss S., ed. (1993a). *Encyclopedia of Cosmology.* New York: Garland.

————, ed. (1993b). *Cosmology: Historical, Literary, Philosophical, Religious and Scientific Perspectives.* New York: Garland.

Hinke, William J. (1907). *A New Boundary Stone of Nebuchadrezzar I from Nippur.* Vol. 4 of H. V. Hilprecht, ed., *The Babylonian Expedition of The University of Pennsylvania,* Series D. Philadepelphia: University of Pennsylvania.

Hooker, J. T. (1990). *Reading the Past. Ancient Writing from Cuneiform to the Alphabet.* Berkeley: University of California Press.

Hooykaas, R. (1984). *G. J. Rheticus' Treatise on Holy Scripture and the Motion of the Earth.* Amsterdam: North-Holland.

Houlden, Michael A., and F. Richard Stephenson (1986). *A Supplement to the Tuckerman Tables.* Memoirs of the American Philosophical Society **170**.

Huggonard-Roche, Henri, and Jean-Pierre Verdet (1982). *Georgii Joachimi Rhetici Narratio prima.* Studia Copernicana XX. Wrocław: Maison d'Edition de l'Académie Polonaise des Sciences.

Hughes, Ruppert (1926). *George Washington: The Human Being and the Hero.* New York: William Morrow.

Hunger, Hermann, and David Pingree (1989). *MUL.APIN. An Astronomical Compendium in Cuneiform.* Archiv für Orientforschung, Beiheft 24. Horn, Austria: Ferdinand Berger.

Ionides, S. A. (1936). "Caesar's Astronomy (Astronomicum Caesarium) by Peter Apian, Ingolstadt 1540." *Osiris* **1**, 356–389.

Jardine, N. (1984). *The Birth of History and Philosophy of Science: Kepler's* A Defense of Tycho against Ursus *with Essays on Its Provenance and Significance.* Cambridge: Cambridge University Press.

Jensen, Hans (1969). *Sign, Symbol and Script.* 3rd ed. Trans. by George Unwin. New York: Putnam.

Jones, Alexander (1990). "Babylonian and Greek Astronomy in a Papyrus Concerning Mars." *Centaurus* **33**, 97–114.

———— (1991). "The Adaptation of Babylonian Methods in Greek Numerical Astronomy." *Isis* **82**, 441–453.

———— (1997a). "Studies in the Astronomy of the Roman Period I. The Standard Lunar Scheme." *Centaurus* **39**, 1–36.

———— (1997b). "Studies in the Astronomy of the Roman Period II. Tables for Solar Longitude." *Centaurus* **39**, 211–229.

———— (1997c). "Studies in the Astronomy of the Roman Period III. Planetary Epoch Tables." *Centaurus* **40**, 1–41.

Kennedy, E. S. (1952). "A Fifteenth-Century Planetary Computer: Al-Kāshī's 'Ṭabaq al-Manāṭeq.'" *Isis* **43**, 42–50.

———— (1956). "A Survey of Islamic Astronomical Tables." *Transactions of the American Philosophical Society (n.s.)* **46**, Part 2.

———— (1960). *The Planetary Equatorium of Jamshīd Ghiyāth al-Dīn al-Kāshī (d. 1429).* Princeton, N.J.: Princeton University Press.

Kennedy, E. S., and Imad Ghanem (1976). *The Life and Work of Ibn al-Shāṭir.* Aleppo: Institute for the History of Arabic Science.

Kennedy, E. S., and Victor Roberts (1959). "The Planetary Theory of Ibn al-Shāṭir." *Isis* **50**, 227–235.

King, David A. (1986). *Islamic Mathematical Astronomy.* London: Variorum Reprints.

———— (1987). *Islamic Astronomical Instruments.* London: Variorum Reprints.

———— (1993). *Astronomy in the Service of Islam.* Aldershot, Hampshire: Variorum.

———— (1995). "The Orientation of Medieval Islamic Religious Architecture and Cities." *Journal for the History of Astronomy* **26**, 253–274.

King, Henry C. (1978). *Geared to the Stars. The Evolution of Planetariums, Orreries and Astronomical Clocks.* Toronto: University of Toronto Press.

Kirk, G. S., and J. E. Raven (1960). *The Presocratic Philosophers.* Cambridge: Cambridge University Press.

Kish, George (1978). *A Source Book in Geography.* Cambridge, Mass.: Harvard University Press.

Knobel, Edward Ball (1917). *Ulugh Beg's Catalogue of Stars.* Carnegie No. 250. Washington, D.C.: Carnegie Institution of Washington.

Knorr, Wilbur R. (1989). "Plato and Eudoxus on the Planetary Motions." *Journal for the History of Astronomy* **20**, 313–329.

Knox, Richard (1976). *Experiments in Astronomy for Amateurs.* New York: St. Martin's Press.

Koestler, Arthur (1963). *The Sleepwalkers.* New York: Universal Library.

Koyré, Alexandre (1973). *The Astronomical Revolution.* Trans. by R. E. W. Maddison. Ithaca, N.Y.: Cornell University Press.

Kugler, Franz Xaver (1907–1924). *Sternkunde und Sterndienst in Babel.* 3 vols. in 6 parts. Münster: Aschendorff.

Kuhn, Thomas (1957). *The Copernican Revolution.* Cambridge, Mass.: Harvard University Press.

Kunitzsch, Paul (1989). *The Arabs and the Stars.* Northampton: Variorum Reprints.

Kunitzsch, Paul, and Tim Smart (1986). *Short Guide to Modern Star Names and Their Derivations.* Wiesbaden: Otto Harrassowitz.

Labat, René (1976). *Manuel d'épigraphie akkadienne.* 5th ed. Paris: Librairie Orientaliste Paul Geunthner.

Langdon, S., and J. K. Fotheringham (1928). *The Venus Tablets of Ammizaduga.* Oxford: Oxford University Press; London: Humphrey Milford.

Langermann, Y. T. (1990). *Ibn al-Haytham's* On the Configuration of the World. New York: Garland.

Lattis, James E. (1994). *Between Copernicus and Galileo: Christopher Clavius and the Collapse of Ptolemaic Cosmology.* Chicago: University of Chicago Press.

Leichty, E., M. J. Ellis, and P. Gerardi, eds. (1988). *A Scientific Humanist: Studies in Memory of Abraham Sachs.* Philadelphia: Occasional Publications of the Samuel Noah Kramer Fund.

Lloyd, G. E. R. (1973). *Greek Science after Aristotle.* New York: Norton.

—— (1978). "Saving the Appearances." *Classical Quarterly* **28**, 202–222.

—— (1991). *Methods and Problems in Greek Science.* Cambridge: Cambridge University Press.

Manitius, C. (1898). *Gemini Elementa astronomiae.* Leipzig: Teubner.

Methuen, Charlotte (1996). "Maestlin's Teaching of Copernicus: The Evidence of His University Textbook and Disputations." *Isis* **87**, 230–247.

Michel, Henri (1967). *Scientific Instruments in Art and History.* Trans by R. E. W. Maddison and F. R. Maddison. London: Barrie and Rockliff.

Millás-Vallicrosa, José M. (1950). *Estudios sobre Azarquiel.* Madrid: Escuelas de Estudios Arabes.

Moesgaard, Kristian Peder (1989). "Tycho Brahe's Discovery of Changes in Star Latitudes." *Centaurus* **32**, 310–323.

Morelon, Régis (1987). *Thâbit ibn Qurra: Oeuvres d'Astronomie.* Paris: Les Belles Lettres.

—— (1993). "La version arabe du *Livre des Hypothèses* de Ptoémée." Institut Dominicain d'Etudes Orientales du Caire, *Mélanges* **21** (*Mideo*).

—— (1994). "Tābit b. Qurra and Arab Astronomy in the 9th Century." *Arabic Sciences and Philosophy* **4**, III–139.

Morison, Samuel Eliot (1942). *Admiral of the Ocean Sea: A Life of Christopher Columbus.* Boston: Little, Brown.

—— (1974). *The European Discovery of America: The Southern Voyages,* A.D. *1492–1616.* New York: Oxford University Press.

Moyer, Gordon (1982). "The Gregorian Calendar." *Scientific American,* **246**(5), 144–152.

Munitz, Milton K. (1957). *Theories of the Universe: From Babylonian Myth to Modern Science.* New York: Free Press.

Murschel, Andrea (1995). "The Structure and Function of Ptolemy's Physical Hypotheses of Planetary Motion." *Journal for the History of Astronomy* **26**, 33–61.

Nadal, R., and J.-P. Brunet (1984). "Le 'Commentaire' d'Hipparque. I. La sphère mobile." *Archive for History of Exact Sciences* **29**, 201–236.

—— (1989). "Le 'Commentaire' d'Hipparque. II. Position de 78 étoiles." *Archive for History of Exact Sciences* **40**, 305–354.

Nallino, Carlo (1899–1907). *Al-Battānī sive Albetenii Opus astronomicum.* 3 vols. Milan: Hoeplium.

Neugebauer, Otto (1949). "The Early History of the Astrolabe." *Isis* **40**, 240–256.

—— (1959). "The Equivalence of Eccentric and Epicyclic Motion According to Apollonius." *Scripta Mathematica* **24**, 5–21.

—— (1962a). "Thabit Ben Qurra 'On the Solar Year' and 'On the Motion of the Eighth Sphere.'" *Proceedings of the American Philosophical Society,* **106**(3), 264–298.

—— (1962b). *The Astronomical Tables of al-Khwārizmī.* Historiskfilosofiske Skrifter udgivet af Det Kongelige Danske Videnskabernes Selskab, Bind **4**, nr. 2. Copenhagen: Ejnar Munksgaard.

—— (1969). *The Exact Sciences in Antiquity.* New York: Dover.

—— (1973). "Notes on Autolycus." *Centaurus* **18**, 66–69.

—— (1975). *A History of Ancient Mathematical Astronomy.* 3 vols. Berlin: Springer.

—— (1983a). *Astronomical Cuneiform Texts.* 2nd ed. 3 vols. New York: Springer.

—— (1983b). *Astronomy and History. Selected Essays.* New York: Springer.

Neugebauer, O., and H. B. van Hoesen (1959). *Greek Horoscopes.* Memoirs of the American Philosophical Society **48**.

Newton, Isaac (1934). *Sir Isaac Newton's Mathematical Principles of Natural Philosophy and His System of the World. Translated into English by Andrew Motte in 1729. The translations revised . . . by Florian Cajori.* Berkeley: University of California Press.

Newton, Robert R. (1977). *The Crime of Claudius Ptolemy.* Baltimore: Johns Hopkins University Press.

Nizāmuʾd-Din, M. (ed.) (1954). Abuʾl-Husayn ʿAbduʾr-Rahmān As-Sūfi, *Suwaruʾl-Kawākib, or Uranometry.* Hyderabad-Deccan: Osmania Oriental Publications Bureau.

North, J. D. (1976). *Richard of Wallingford.* 3 vols. Oxford: Oxford University Press.

—— (1988). *Chaucer's Universe.* Oxford: Clarendon Press.

—— (1989). *The Universal Frame: Historical Essays in Astronomy, Natural Philosophy and Scientific Method.* London: Hambledon Press.

—— (1995). *The Norton History of Astronomy and Cosmology.* New York: Norton.

Oates, Joan (1986). *Babylon.* London: Thames and Hudson.

Parker, Richard A. and Waldo H. Dubberstein (1956). *Babylonian Chronology 626 B.C.–A.D. 75.* Providence, R.I.: Brown University Press.

Parsons, Edward A. (1967). *The Alexandrian Library.* New York: American Elsevier.

Pedersen, Olaf (1974). *A Survey of the Almagest.* Acta Historica Scientiarum Naturalium et Medicinalium, No. 30. Odense: Odense University Press.

—— (1993). *Early Physics and Astronomy.* 2nd ed. Cambridge: Cambridge University Press.

Peters, C. H. F., and E. B. Knobel (1915). *Ptolemy's Catalogue of Stars: A Revision of the Almagest.* Carnegie No. 86. Washington, D.C.: The Carnegie Institution of Washington.

Pingree, David (1973). "The Greek Influence on Early Islamic Astronomy." *Journal of the American Oriental Society* **93**, 32–43.

—— (1976). "The Recovery of Early Greek Astronomy from India." *Journal for the History of Astronomy* **7**, 109–123.

Poulle, Emmanuel (1980). *Les instruments de la théorie des planètes ʒelon Ptolémée. Equatoires et horlogerie planétaire du XIIIe au XVIe siècle.* 2 vols. Geneva: Librarie Droz.

—— (1984). *Les Tables Alphonsines avec les canons de Jean de Saxe.* Paris: Editions du Centre National de la Recherche Scientifique.

Price, Derek J. de Solla (1955). *The Equatorie of the Planetis.* Cambridge: Cambridge University Press.

—— (1967). "Piecing Together an Ancient Puzzle. The Tower of the Winds." *National Geographic* April, 586–596.

—— (1974). "Gears from the Greeks. The Antikythera Mechanism—A Calendar Computer from ca. 80 B.C." *Transactions of the American Philosophical Society,* (n.s.) **64**, Part 7.

Pritchett, W. Kendrick, and Otto Neugebauer (1947). *The Calendars of Athens.* Cambridge, Mass.: Harvard University Press.

Raeder, H., E. Stromgren, and B. Stromgren. (1946). *Tycho Brahe's Description of his Instruments and Scientific Work as*

Given in Astronomiae instauratae mechanica. Copenhagen: Munksgaard.

Ragep, F. J. (1993). *Naṣīr al-Dīn al-Ṭūsī's Memoir on Astronomy (al-Tadhkira fī ʿilm al-hayʾa)*. 2 vols. New York: Springer.

——— (1996). "Al-Battānī, Cosmology, and the Early History of Trepidation in Islam." In Josep Casulleras and Julio Samsó, eds., *From Baghdad to Barcelona. Essays on the History of the Exact Sciences in Honour of Prof. Juan Vernet.* Barcelona: Universidad de Barcelona, Facultad de Filología.

Rand Schmidt, Kari Anne (1993). *The Authorship of* The Equatorie of the Planetis. Chaucer Studies, 19. Rochester, N.Y.: Brewer.

Rawlins, Dennis (1982). "An Investigation of the Ancient Star Catalog." *Publications of the Astronomical Society of the Pacific* **94**, 359–373.

Reese, Ronald L., and Stephen M. Everett (1981). "The Origin of the Julian Period: An Application of Congruences and the Chinese Remainder Theorem." *American Journal of Physics* **49**, 658–661.

Reiner, Erica (1995). "Astral Magic in Babylonia." *Transactions of the American Philosophical Society*, Vol. 85, Part 4.

Rico-Sinobas, M. (1863–1867). *Libros del saber de astronomía del rey Alfonso X de Castilla.* 5 vols. Madrid: E. Aguado.

Rome, A. (1927). "L'astrolabe et le météoroscope d'après le *Commentaire* de Pappus sur le 5e livre de l'*Almageste*." *Annales de la Société scientifique de Bruxelles.* (sér. A), **47**, 77–102.

——— (1931–1943). *Commentaires de Pappus et de Théon d'Alexandrie sur l'Almageste. Tome 1:* Pappus d'Alexandrie, *Commentaire sur les livres 5 et 6 de l'Almageste,* Studi e Testi **54**. Rome: Biblioteca Apostolica Vaticana, 1931. *Tome 2:* Théon d'Alexandrie, *Commentaire sur les livres 1 et 2 de l'Almageste,* Studi e Testi **72** (1936). *Tome 3:* Théon d'Alexandrie, *Commentaire sur les livres 3 et 4 de l'Almageste,* Studi e Testi **106** (1943).

——— (1937–1938). "Les observations d'équinoxes et de solstices dans le chapitre 1 du livre 3 du Commentaire sur l'Almageste par Théon d'Alexandrie." *Annales de la Société Scientifique de Bruxelles* **57** (1937), 213–236 and **58** (1938), 6–26.

Rosen, Edward (1971). *Three Copernican Treatises.* 3rd ed. New York: Octagon Books.

——— (1986). *Three Imperial Mathematicians: Kepler Trapped between Tycho Brahe and Ursus.* New York: Abaris Books.

Sachs, A. (1948). "A Classification of the Babylonian Astronomical Tablets of the Seleucid Period." *Journal of Cuneiform Studies* **2**, 271–290.

——— (1952). "Babylonian Horoscopes." *Journal of Cuneiform Studies* **6**, 49–75.

——— (1974). "Babylonian Observational Astronomy." *Philosophical Transactions of the Royal Society of London* (Ser. A) **276**, 43–50.

Sachs, A. J., and H. Hunger (1988). *Astronomical Diaries and Related Texts from Babylonia.* 2 vols in 4 parts. Philosophisch-Historische Klasse. Denkschriften, Band 195. Wien: Verlag der Österreichische Akademie der Wissenschaften.

Said, S. S., and F. R. Stephenson (1995). "Precision of Medieval Islamic Measurements of Solar Altitudes and Equinox Times." *Journal for the History of Astronomy* **26**, 117–132.

Saliba, George (1994). *A History of Arabic Astronomy. Planetary Theories during the Golden Age of Islam.* New York: New York University Press.

Sambursky, S. (1962). *The Physical World of Late Antiquity.* Princeton, N.J.: Princeton University Press.

Samsó, Julio (1994). *Islamic Astronomy and Medieval Spain.* Hampshire: Variorum.

Samuel, Alan E. (1972). *Greek and Roman Chronology: Calendars and Years in Classical Antiquity.* Handbuch der Altertumswissenschaft, Series 1, Part 7. Munich: Beck.

Santillana, Giorgio de (1955). *The Crime of Galileo.* Chicago: University of Chicago Press.

Sarton, George (1970). *A History of Science.* 2 vols. New York: Norton. (Originally published 1952–1959)

Saunders, Harold N. (1984). *All the Astrolabes.* London: Senecio.

Savage-Smith, Emilie (1985). *Islamicate Celestial Globes: Their History, Construction, and Use.* Smithsonian Studies in History and Technology **46**.

Sayili, Aydin (1960). *The Observatory in Islam.* Ankara: Turkish Historical Society.

Schiaparelli, Giovanni V. (1875). *Le sfere omocentriche di Eudosso, di Callippo e di Aristotele.* Pubblicazioni del R. Osservatorio di Brera in Milano, **9**.

——— (1877). "Die homocentrischen Sphären des Eudoxus, des Kallippus und des Aristoteles." (German translation by W. Horn.) *Abhandlungen zur Geschichte der Mathematik,* Leipzig: Erstes Heft.

Schjellerup, H. C. F. C. (1874). *Description des étoiles fixes . . . par Abd-al-Rahman al-Sûfi.* St. Petersburg: Académie Impériale des Sciences.

Schmidt, Olaf (1949). "Some Critical Remarks about Autolycus' *On Risings and Settings.*" Den 11. *Skandinaviske Matematikerkongress i Trondheim* 22–25 August 1949, pp. 27–32.

Schott, Albert (1934). "Das Werden der babylonisch-assyrischen Positions-Astronomie und einige seiner Bedingungen." *Zeitschrift der Deutschen Morgenländischen Gesellschaft* **88**, 302–337.

Sédillot, L. P. (1853). *Prolégomènes des tables astronomiques d'Oloug-Beg.* Paris: Firmin Didot Frères.

Segonds, A. P. (1981). *Jean Philopon. Traité de l'Astrolabe.* Astrolabica 2. Paris: Alain Brieux.

Sharratt, Michael (1994). *Galileo: Decisive Innovator.* Oxford: Blackwell.

Shevchenko, M. (1990). "An Analysis of Errors in the Star Catalogues of Ptolemy and Ulugh Beg." *Journal for the History of Astronomy* **21**, 187–201.

Singer, Charles (1954–1970). *A History of Technology.* 7 vols. New York: Oxford University Press.

Skeat, Walter W. (1872). *A Treatise on the Astrolabe, Addressed to His Son Lowys by Geoffrey Chaucer.* A.D. 1391. Early English Text Society. Extra Series, 16. London: Trübner.

Small, Robert (1804). *An Account of the Astronomical Discoveries of Kepler.* London: T. Gillet. Reprint, Madison: University of Wisconsin Press, 1963.

Smart, William Marshall (1977). *Textbook on Spherical Astronomy.* 6th ed., revised by R. M. Green. Cambridge: Cambridge University Press.

Soubiran, Jean (1969). *Vitruve. De l'Architecture. Livre IX.* Paris: Les Belles Lettres.

Stahl, William Harris, Richard Johnson, and E. L. Burge (1971–1977). *Martianus Capella and the Seven Liberal Arts. Vol. 1, The Quadrivium of Martianus Capella: Latin Traditions in the Mathematical Sciences, 50 B.C.–A.D. 1250. Vol. 2, The Marriage of Philology and Mercury.* New York: Columbia University Press.

Stahlman, W. D., and O. Gingerich (1963). *Solar and Planetary Longitudes for Years −2500 to +2000 by 10-Day In-*

tervals. Madison: University of Wisconsin Press.

Stephenson, Bruce (1987). *Kepler's Physical Astronomy*. New York: Springer.

Stevens, Wesley M. (1995). *Cycles of Time and Scientific Learning in Medieval Europe*. Hampshire: Variorum.

Stevenson, Edward Luther (1921). *Terrestrial and Celestial Globes*. 2 vols. New Haven, Conn.: Yale University Press.

Struik, Dirk J. (1948). *A Concise History of Mathematics*. 3rd ed. New York: Dover.

Stuart, James, and Nicholas Revett (1968). *The Antiquities of Athens*. New York: Benjamin Bloom. (Facsimile reprint of London, 1762 ed.)

Suter, H. (1914). *Die Astronomischen Tafeln des Muḥammed ibn Mūsā al-Khwārizmī*. Copenhagen: Host.

Swerdlow, Noel M. (1968). *Ptolemy's Theory of the Distances and Sizes of the Planets: A Study in the Scientific Foundations of Medieval Cosmology*. Unpublished Ph.D. dissertation, Yale University.

Swerdlow, Noel, ed. (1998a). *Ancient Astronomy and Celestial Divination*.

——— (1973). "The Derivation and First Draft of Copernicus's Planetary Theory: A Translation of the *Commentariolus* with Commentary." *Proceedings of the American Philosophical Society*, **117**(6), 423–512.

——— (1992). "Annals of Scientific Publishing: Johannes Petreius's Letter to Rheticus." *Isis* **83**, 270–274.

——— (1998b). *The Babylonian Theory of the Planets*. Princeton: Princeton University Press.

Swerdlow, N., and O. Neugebauer (1984). *Mathematical Astronomy in Copernicus's De Revolutionibus*. New York: Springer.

Tannery, Paul (1887). *La géométrie grecque*. Paris: Gauthier-Villars.

——— (1893). *Recherehes sur l'histoire de l'astronomie ancienne*. Paris: Gauthier-Villars.

Taton, René, and Curtis Wilson, eds. (1989). *The General History of Astronomy. Vol. 2, Planetary Astronomy from the Renaissance to the Rise of Astrophysics. Part A: Tycho Brahe to Newton.* Cambridge: Cambridge University Press.

Taub, Liba C. (1993). *Ptolemy's Universe*. Chicago: Open Court.

Tester, Jim (1987). *A History of Western Astrology*. New York: Ballantine Books.

Thomson, Ron B. (1978). *Jordanus de Nemore and the Mathematics of Astrolabes: De Plana Spera*. Toronto: Pontifical Institute of Mediaeval Studies.

Thoren, Victor E. (1990). *The Lord of Uraniborg. A Biography of Tycho Brahe*. Cambridge: Cambridge University Press.

Thorndike, Lynn (1949). *The Sphere of Sacrobosco and Its Commentators*. Chicago: University of Chicago Press.

Thureau-Dangin, F. (1922). *Tablettes d'Uruk*. Musée du Louvre, Département des Antiquités Orientales, Textes Cunéiformes, Tome 6. Paris: Paul Geuthner.

Tihon, Anne (1978). *Le "Petit Commentaire" de Théon d'Alexandrie aux Tables Faciles de Ptolémée*. Studi e Testi **282**. Città del Vaticano: Bibliotheca Apostolica Vaticana.

——— (with Joseph Mogenet) (1985). *Le "Grand Commentaire" de Théon d'Alexandrie aux Tables Faciles de Ptolémée*, Livre I. Studi e Testi **315**. Città del Vaticano: Bibliotheca Apostolica Vaticana.

——— (1991). *Le "Grand Commentaire" de Théon d'Alexandrie aux Tables Faciles de Ptolémée*, Livres II et III. Studi e Testi **340**. Città del Vaticano: Bibliotheca Apostolica Vaticana.

Toomer, G. J. (1968). "A Survey of the Toledan Tables." *Osiris* **15**, 1–174.

——— (1969). "The Solar Theory of az-Zarqāl: A History of Errors." *Centaurus* **14**, 306–336.

——— (1984). *Ptolemy's Almagest*. London: Duckworth.

Tuckerman, Bryant (1962). *Planetary, Lunar and Solar Positions: 601 B.C. to A.D. 1*. Memoirs of the American Philosophical Society **56**.

——— (1964). *Planetary, Lunar and Solar Positions: A.D. 2 to A.D. 1649*. Memoirs of the American Philosophical Society **59**. [See also Houldon and Stephenson (1986).]

Turner, A. J. (1985). *The Time Museum: Catalogue of the Collection . Vol. 1., Part 1, Astrolabes and Astrolabe Related Instruments*. Rockford, Ill.: Time Museum.

Van der Waerden, B. L. (1960). "Greek Astronomical Calendars and Their Relation to the Athenian Civil Calendar." *Journal of Hellenic Studies* **80**, 168–180.

——— (1974). *Science Awakening II. The Birth of Astronomy*. New York: Oxford University Press.

——— (1975). *Science Awakening I*. 4th ed. Dordrecht: Noordhoff.

Van Helden, Albert (1985). *Measuring the Universe. Cosmic Dimensions from Aristarchus to Halley*. Chicago: University of Chicago Press.

Verdet, Jean-Pierre (1993). *Astronomie & Astrophysique*. Textes Essentiels. Paris: Larousse.

Vogt, H. (1925). "Versuch einer Wiederherstellung von Hipparchs Fixsternverzeichnis." *Astronomische Nachrichten* **224**, cols. 17–54.

Watson, E. C. (1956). "Reproductions of Prints, Drawings, and Paintings of Interest in the History of Physics, 76. An Early Meteorological Observatory and Town Clock." *American Journal of Physics* **24** 455–458.

Westfall, Richard S. (1983). *Never at Rest: A Biography of Isaac Newton*. Cambridge: Cambridge University Press.

Wilson, Curtis (1968). "Kepler's Derivation of the Elliptical Path." *Isis* **59**, 5–25.

Włodarczyk, J. (1987). "Examining the Armillary Sphere." *Journal for the History of Astronomy* **18**, 173–195.

Woolard, Edgar William, and Gerald Maurice Clemence (1966). *Spherical Astronomy*. New York: Academic Press.

Wynter, Harriet, and Anthony Turner (1975). *Scientific Instruments*. New York: Scribner.

Abbasid caliphate, 25, 392–393, 395, 426
Abélard, 157
accession and recession. *See* trepidation of the equinoxes
Achaemenid dynasty, 393
Achilles Tatius, 263
aeon-tables, 345
Adrastus, 346
Aëtius, 35–36, 46, 56, 67
Agrippa, 261
Akkadian language, 12
Alexander, 17, 21, 51, 164, 188, 312, 346
era of, *see* era
Alexandria, 21, 37, 49, 61, 64, 65, 66, 67, 73, 115, 133, 165, 182, 186, 203, 207, 238–239, 272–273
Alfonsine Tables, 104, 107–108, 279, 280, 350, 401, 406, 418
Alfonso X of Castile, 107, 401, 404
alidade, 146
almucantar, 144
altitude, 28–29, 144
measurement of, 149–150
of the pole, 33
Ammi-saduqa, 14–15
analemma, 133–137
anaphoric clock, 132, 155–156
Anaxagoras, v, 46, 56, 68, 94
Anaximander, 47, 56
Anaximenes, 47
Ancona, 61
Andronikos of Kyrrhos, 131
anomaly (nonuniformity of motion), 211
anomaly (angular distance from apogee), 226–227, 376
Antoninus (emperor of Rome), 177, 178
Anu or An, 12, 319
Anu, way of, 8–10, 57
Apianus, Petrus, 33, 80, 280, 406
apogee
of an eccentric, 356
of an epicycle, 337
motion of 215
of Sun 211
Apollonius of Perga, 22, 90, 212, 216, 337–341, 392
Apollonius's theorem, 340–341
apsides, line of, 211, 356
Arabic astronomy, 25–26, 65
and the *Almagest*,
and the astrolabe 154–157
and criticism of Ptolemy, 396–397
early medieval, 392–393
motion of solar apogee in, 215
and precession and trepidation, 274–279
star names, 43–44
the *zij*, 394–395

Aratus of Soli, 6, 18, 21, 40–42, 66, 75–76, 79, 92, 98, 115, 200, 298
Arbela, 51
Archimedes, 22, 51, 63, 67, 69, 81–82, 132, 399, 455n106
Archytas, 132
arctic circle, 62, 92, 94
Aristarchus of Samos, 22, 213
his invention of a sundial, 135
on motion of the Earth, 36, 38, 67–68
his observation of a summer solstice, 209
on sizes and distances of Sun and Moon 67–73, 89, 90
Aristotle, 20, 52, 58, 75, 217, 218, 219, 248, 305
correlates distances of planets with their periods, 347–348
criticizes Plato's identification of elements with regular polyhedra, 428
on elements and natural motions 36–37
on Eudoxus's planetary theory, 306–307
on the Milky Way 94–95
modifies Eudoxus's planetary theory, 310–311
his physical doctrines 19, 247, 280
on shape of Earth 47–49
on size of Earth 63
suppresion of his works, 399
translations of his works, 398–399
on weather and celestial bodies 200
Aristophanes, 183
Aristyllos, 21, 103, 260
arithmetic progressions, 11
for day lengths, 121–125, 202–203
armillary sphere, 78–80,
as instrument of observation 125–127, 251, 255–256
Arsacid dynasty, 17
Art of Eudoxus (papyrus), 6, 458n10
ascensions, tables of, 109–125
arithmetic methods for, 121–125
trigonometric methods for, 118–119
Ashurbanipal, 9, 16
Assyria, 16
astrolabe, 141–161, 445–448
history of, 153–158
making a latitude plate for, 158–161
use of, 147–153
"astrolabe" (Babylonian star list)
circular, 8–11
rectangular, 15
astrology, 16, 343–344, 400
Astronomical Canon, 176–177
astronomical reckoning of years, 163
Athens, 35, 46, 61, 63, 131, 133, 184, 185, 205, 207, 305
atomists, 19
Augustus, 165, 180

Index

Autolycus of Pitane, 21, 22, 24, 89, 91, 213
  attempted to explain variation in distance of planets, 312
  On the Moving Sphere, 87–88
  On Rising and Settings 190–197
azimuth, 144–145

Babylon, 8, 12, 17, 238–239, 317–318
Babylonian astronomy
  and astronomical diaries 16, 298–299, 302, 315
  day length in, 10–11, 124–125
  early 5–11, 14–16
  in India, 393
  initial points of zodiac signs, 213–214
  planet names, 297
  zodiac, 56–57, 103–104
Babylonian planetary theory, 17, 22–23, 297–298
  character of, 320–321
  goal-year texts, 312–317
  mean solar speed, 326
  social setting of, 319–320
  three versions of for Jupiter, 321–334
Baghdad, 25, 275
al-Battāni, 275, 279, 282, 394
Bayer, Johann, 44
Bede, 398
Bēl, 319
Berenice, 41
Berosus, 45–46, 135, 346
Bessarion, Johannes, 402
Bilfinger, Gustave, 135
Bithynia, 79, 215
Black Sea, 62
Boll, Franz, 266
Borysthenes (Dnieper) River, 62
boundary stones, 39–40
Brahe, Tycho
  discovers changes in latitudes of the stars, 283–286
  his life, 281
  on precession, 282
  his refutation of trepidation, 281–283
  and stability of the Earth, 423–424
  his star catalog, 269, 271–272
  his system of the world, 413–414
  his theory of Mars, 432–433
Britain, 62, 167
Byzantium (Istanbul), 62

Caesar, 163, 165, 204, 306
calendar
  Alexandrian, 179, 203
  Athenian, 104, 182–183
  Babylonian, 6, 187–188
  of Boeotia, 182
  Christian ecclesiastical, 188
  conversion between Egyptian and Julian, 178–179
  of Delos, 182–183
  Egyptian, 175–182

Gregorian, 166–171
  Islamic, 459n44
  Jewish, 163, 167, 188
  Julian, 163–166
  luni-solar, 7, 163, 167
  Roman, 164–165, 204
  of Thessaly, 182
Callimachus of Cyrene, 41, 43
Callippus, 199–201, 203, 204, 210, 212, 226
  modifies Eudoxus's planetary theory, 306–307, 310–311
  his season lengths, 458n10
Callisthenes, 312
Campanus of Novarra, 404, 405
caput Arietis, 277
Carthage, 51
Cassini, J., 287
Catullus, 41
celestial sphere, 29
  aspects of, 32–33
  daily rotation, 31–34
  description of by Geminus, 91–93
  historical origins of, 75–76
Censorinus, 262
Chalcidius, 262, 349, 398
Chaldaean dynasty, 16
Chaldaeans as astronomers and astrologers, 16, 318–319
Chaucer, Geoffrey
  Canterbury Tales, 400–401
  Equatorie of the Planetis, 405
  Treatise on the Astrolabe 152, 157
Chios, 57
Christianity, 25–26, 424–425
Cicero, 66, 82
circumpolar stars, 3, 32
Clavius, Christopher, 168
Cleanthes of Assos, 68, 455n104
Cleomedes, 24, 49, 262
  and atmospheric refraction, 458n5
  on day lengths, 124
  on shape of Earth, 49–52
  on size of Earth, 64–65
Cleopatra, 17, 177
clime, 92, 94, 96–97, 111
colophon, 319
Columbus, Christopher, 53, 66, 175, 403
Columella, 18
comet of 1577, 423–424
compass declination, 28
conjunction, 300
Conon, 41
Constantinople, 167, 177
constellations. *See also* stars and constellations
  Andromeda, 41, 76
  Aquarius (water-pourer), 40, 267
  Aquilla, 200
  Argo, 66, 79
  Aries, 6, 39, 41–42, 245, 249
  Auriga (charioteer), 76, 79, 199, 249
  Boötes, 3, 39, 266

Cancer, 34, 42, 76, 79, 98
  Canis Major, 33, 79
  Capricornus, 34, 39, 40
  Cassiopeia, 41
  Carina, 33, 48
  Centuarus, 41
  Cepheus, 33
  Cetus, 41
  Claws (= Libra), 249
  Coma Berenices, 41
  Corona Borealis (the crown), 98, 202, 204
  Corvus, 79
  Crater, 79
  Cygnus (the bird), 76, 199
  Draco, 266
  Equuleus, 41
  Gemini, 39, 42, 76, 79
  Hercules, 79
  Hyades, 3, 5, 39, 44, 200, 248, 249
  Hydra, 79
  Leo, 42, 79, 289
  Libra, 12, 249
  Microscopium, 42
  Ophiuchus, 76
  Orion, 3, 5, 33, 39, 44, 200, 201, 249
  Pegasus, 76
  Perseus, 76, 203
  Piscis Austrinus, 987
  Sagittarius (archer), 40, 267
  Scorpius, 200
  Taurus, 39, 42, 44, 57, 79, 245, 248, 249
  Telescopium, 42
  Thyrsus-lance, 41
  Triangulum, 249
  Ursa Major or Bear (including Big Dipper or Wain), 32–33, 39, 79, 92, 94, 249, 266
  Ursa Minor, 33, 264, 266, 267
  Virgo, 76, 199, 201
coordinates
  celestial, 100–105, 144
  conversion between equatorial and ecliptic, 105, 115
  geographical, 99–100, 102–103
  orthogonal, 99, 273
Copernicus, Nicholas, 26, 116, 175, 282
  and Aristotle's physics, 415, 427
  Commentariolus, 414–416
  on decrease in obliquity of ecliptic, 285
  his dependence on Ptolemy, 422, 425
  life of, 414–418
  planetary theory of, 420–422
  publication of *De revolutionibus*, 417–418
  rejects Ptolemy's equant, 419
  his star catalog, 265
  and trepidation, 280
  and unified scale for solar system, 413, 419
  uses midseason points to find solar eccentricity, 223

Cordoba, 157
Cos, 46, 346
cosmology, 347
  and astronomy, 351, 414
  in the Middle Ages, 395–396, 399–400
Council of Nicaea, 167
Council of Trent, 168
culmination, 114
cuneiform writing, 12–14
cycles, luni-solar, 58, 182–190
  eight-year, 184–185, 315
  nineteen-year (Metonic), 16, 20, 167, 171, 185–186, 188–190, 207
  seventy-six-year (Callippic), 186–187
Cyprus, 48
Cyrene, 63
Cyrus (of Persia), 16

Damascus, 392
Darius III of Persia, 51
day, length of, 11, 112–113, 119–120, 148, 203
days of week, 166, 174
daylight savings time, 28
declination, 60–61, 100–102, 115
  of Sun, 105–108
deferent circle, 212, 337
deferent sphere, 311
Delos, 54, 130, 131, 140, 141, 184
Democritus, 19, 40, 94, 204
Dendera, round zodiac of, 39–40
Delambre, J.-B.-J., 262
Descartes, René, 440
diaries, astronomical, 16, 298–299, 302, 315
Dicaearchus of Messina, 51–52, 63
Dijksterhuis, E. J., 217
Diodorus of Sicily, 184
Diogenes Laërtius, v, 454n59
Dionysius Exiguus, 166
dioptra, 34–35, 84, 387, 455n106
dock-pathed star, 194
Dositheus, 199, 204
doubly-visible star, 196
Dreyer, J. L. E., 217, 266, 276
Duhem, Pierre, 217

Ea, way of, 8–10, 57
Earth
  centrality of, 76–77
  shape of, 47–53
  size of, 63–67, 77
Easter, 167–168
eccentric anomaly, 376
eccentric circle, 211, 355
eccentricity
  bisection of, 421–422, 432
  in Kepler's vicarious theory, 434–436
  of a planet's deferent, 356
  of the Sun, 220–226, 433–434

eclipses, lunar
  and Aristarchus's eclipse diagram, 68–69
  and atmospheric refraction, 458n5
  Babylonian records of, 23, 176–178, 238–239
  cause of, 45–46
  and ecliptic circle, 55
  and Greek geography, 50–51, 454n67
  Hipparchus's use of in discovery of precession, 259
  observed by Ptolemy, 182
  their use in measurement of star longitudes, 251
  used by Aristotle to prove sphericity of Earth, 47–48
eclipses, solar
  annular, 311–312
  cause of, 45–46
  use of by Aristarchus, 69
ecliptic, 54–56, 76
  obliquity of, 54, 59–60, 274, 279
  rotation of plane of, 283
Ecphantus the Pythagorean, 35–36
eighth sphere, 275, 280
Egypt, 17, 20, 21, 48, 61, 65, 155, 180, 201–203
Ekur-zākir, 319
elements, 37
elongation, 295, 300
empyrean sphere, 280, 400
Enlil, way of, 8–10, 57
Enuma Anu Enlil, 8, 15
Enuma Elish, 10, 14
epagomenal days, 175
epact, 325–326
ephemeris, 318
  for Jupiter, Babylonian, 321–323, 332–334
  of Regiomontanus, 403
epicycle, 212, 337
epicyclic anomaly, 337, 356, 376
eponymous year, 183
Epping, J., 318
equant, 355–356
  discovery of by Ptolemy, 357–358
  its equivalence to Copernicus's minor epicycle, 420–422, 431–432
  its equivalence to empty focus of Kepler ellipse, 442–443
  and Mästlin, 422
  rejected by Copernicus, 419–420
equation of center
  in Ptolemy's planetary theory, 373, 376
  in solar theory, 226–227
equation of the epicycle, 373, 376–377
equation of time, 235–236, 458n34
  cause of, 237–238
  computation of, 242–243
  Ptolemy's treatment of, 240–242
equator, celestial, 31
equatorial mean Sun, 237–238

equatorial ring, 206–207
equatorium
  planetary, 403–410, 450–451
  solar, 215–216
equinoxes, 4, 53, 56
  location of variously defined, 104
  methods of observation of, 205–207
era
  of Alexander (or Philippos), 177–178
  of Antoninus, 178
  Christian, 163
  Diocletian, 166
  of Hadrian, 178
  of Nabonassar, 176–178
  Seleucid, 188
Eratosthenes, 50, 52, 66, 132, 443
  Catasterisms, 41, 64, 454n46
  and obliquity of ecliptic, 59
  on size of Earth, 20, 51, 63–65
Esangila, 8, 16, 187
etesian winds, 199–200
Euclid, 21, 22, 24, 34, 57, 59, 77, 83, 89–91, 156
  Phenomena, 88–89
Euctemon
  his parapegma, 40, 199–201, 204
  his season lengths, 226, 458n10
  his summer solstice, 20, 56, 205, 209, 259
Eudemus, 58, 306, 458n6
Eudoxus of Cnidus, 22, 48, 90, 212, 218, 455n21
  on celestial sphere (Phenomena), 21, 40, 75–76
  his parapegma, 199, 201, 204
  placed equinox at midpoint of sign, 104, 203
  his theory of planets (On Speeds), 81, 305–312, 341
Euphrates River, 12
Eusebius, 346
evening star, 4, 295
exeligmos, 460n22

Fabricius, David, 439
al-Farghānī, 396, 398
Farnese Atlas, 78–79
finger, 249, 299
Fortunate Islands, 102
Frauenberg (Frombork), 175, 414

Galileo, 170, 425, 440
gears, 83
Geminus, 24, 103, 104, 262, 350
  on Babylonian lunar theory, 345
  on celestial circles, 91–93
  his classification of the sciences, 83–84
  on day lengths, 124
  on dioptra, 34–35
  on globes, 79, 80

Geminus (*continued*)
  on methods of astronomy and physics,
    217–219
  on nineteen-year cycle, 186–187
  his parapegma, 199–201
  on variation in the *nychthemeron*,
    239
Gemma Frisius, 33, 80
genre in Greek mathematics, 83–84, 341
geo-heliocentric planetary systems, 349,
    413–414
Gerard of Cremona, 26, 398, 399,
    402
Giese, Tiedemann, 419
Gilbert, William, 429
Gingerich, Owen, 423, 425
Ginzel, F. K., 198
globe
  on coins, 215
  Hipparchus's, 250
  history of, 78–82
  manipulation of, 85–87
  Ptolemy's, 79, 95, 246
  used as calculator, 79, 203, 455n8
gnomon, 27–31, 61–62, 77, 84, 129, 133,
    139–141, 205
goal-year texts, 312–317
golden number, 190
Graz, 428, 430
great cycles of planets, 313
Gregory XIII (Pope), 168
Grenfell, Bernard P., 201–202

Hadrian, 49, 177, 178
Halley, Edmund, 285–287
Hammurapi, 12, 14
Handy Tables, 104, 215
  and Astronomical Canon, 176, 181
  contrasted with *Almagest*, 381–383
  and equation of time, 240–242
  as prototype of *zij*, 394
Hartmann, Georg, 157
Hārūn al-Rashīd, 395
heliacal risings and settings. *See* phases of
    fixed stars
Hellespont, 73
Héloïse, 157
Heraclides of Pontos, 35–36, 38, 219,
    349
Hermann the Dalmatian, 156
Hero of Alexandria, 35
Herodotus, 11, 27, 319
Hesiod, 4–5, 7, 17–18, 39, 42, 47, 56, 297,
    454n61
Hibeh Papyri, 202–203
Hicetas of Syracuse, 36
Hipparchus, 20, 22, 24, 35, 52, 62, 90,
    116, 155, 282, 443
  advocated astronomicial methods in ge-
    ography, 50
  Against Eratosthenes, 213
  and Callippic cycle, 187

*On the Change of the Tropic and Equi-
    noctial Points*, 208, 249
Commentary on Aratus and Eudoxus, 41,
    75, 98, 103, 213, 250, 267, 273
  constellations of, 41–42
  his declinations of stars, 260–261, 283
  his discovery of precession, 246, 248,
    259
  his equinoxes and solstices, 205–209
  on lengths of seasons, 210
  his longitude for Spica, 251
  and new star, 247–248
  preferred epicycle to eccentric in solar
    theory, 216–217
  on size and distance of Moon, 72–73
  his solar theory, 210–216
  and star alignments, 248–250
  on the tropical year, 208–209
Hippolytus, 46, 454n58
hippopede, 310
Homer, 3–4, 39, 64
horizon, 87
horoskopos, 116
hour
  conversion of, 113
  equinoctial, 96
  seasonal, 95, 130–131, 136
House of Wisdom, 25
Hunt, Arthur S., 201–202
Hven, 271, 272, 281
Hypatia, 156
Hypsicles, 90, 115, 344, 398
  Anaphorikos, 121–125

Ibn al-Haytham, 396, 401
Ibn al-Ṣaffār, 157
Ibn al-Samḥ, 404
Ibn al-Shāṭir, 397, 422, 426
Ibn Yūnus, 274–275, 279
ideogram, 12, 187
ides, 164
Ilkhānī Tables, 396
inclusive counting, 164
Index of prohibited books, 424
India, 48, 393
indiction, 172
inequality (of a planet) with respect to
    the Sun, 340
instrumentalism, 216–219, 417
interpolation coefficient (in Babylonian
    planetary theory), 324
Iraq, 155
Isḥāq ibn Ḥunayn, 275, 276, 395
Ishtar, 297
Isidore of Seville, 398
Islam and astronomy, 25–26, 393–394.
    See also Arabic astronomy

Jesus, 166, 167
Jones, Alexander, 214, 345
Josephus, 346
Julian day number, 171–175

kalends, 164
Kassite dynasty, 15, 39–40
Kepler, Johannes, 205
  Astronomia Nova, 437–438
  and Brahe, 430–433, 437
  Epitome of Copernican Astronomy, 439
  Harmonice Mundi, 438–439
  and laws of planetary motion, 436–437,
    438
  his life, 427–428, 430–431
  and magnetic motive forces, 429, 437,
    439
  Mysterium cosmographicum, 428–430
  and physical causes, 433, 436
  Rudolphine Tables, 439
  his theory of the Earth's motion,
    433–434
  his vicarious hypothesis, 434–436
al-Khwārizmī, 394
Kidinnu, 320
Knobel, E. B., 266
Koestler, Arthur, 217
Kugler, F. X., 318

La Caille, Nicolas Louis de, 42, 287
latitude
  celestial, 101–104
  terrestrial, 33–34, 51, 60, 99, 102–103
latitude plate (of an astrolabe), 143
Lesbos, 30
Leucippus, 19
library of Alexandria, 21, 37, 64
Libros del Saber de Astronomía, 404
Little Astronomy, 89–91
local apparent time, 235
local mean time, 235
longitude, celestial, 101–104
  methods of measuring, 250–259
longitude, terrestrial, 51, 99, 102–103,
    243
Longomontanus (Christen Sørensen),
    431–433, 464n167, 464n170
Lucretius, 19
Luther, Martin, 424

Macrobius, 165, 262, 349
madrasa, 393
magnitude (of a star), 264, 272–273
al-Maʾmūn, 25, 65, 274, 278–279, 395
Manilius, 124, 262, 343
Maragha, 396–397, 426
Marcellus, 82
Mardokempad, 238
Marduk, 8, 10, 12, 14, 297
  in omens, 298
Marinus, 102
Martianus Capella, 262, 349, 397, 423
Massilia (Marseille), 62, 63
Mästlin, Michael, 422, 424, 427, 430
mater (of an astrolabe), 146
mathematics, branches of, 49, 83–84
Mayer, Tobias, 287

mean epicyclic anomaly, 373
mean longitude of a planet, 337, 373
mean Sun, 226
    its relation to the planets, 337–338,
        358–359
Melanchthon, Philip, 424
Menelaus, 90, 261
meridian, 29
    standard, 236–237
    time correction for change of, 236–239
Messahalla, 157, 456n26
Meton, 20, 40, 56, 185, 205, 207, 209, 259
Miletus, 201
Milky Way, 92–95
mina, 11
Mithradates II of Parthia, 17
month
    anomalistic, 316
    draconitic, 316
    full or hollow, 182, 186, 189
    synodic, 163, 315
    tropical, 315
Moon
    angular size of, 67, 311–312,
    eclipses of, 45–46, 48
    motion of, 57–58
    node of, 307, 316
    phases of, 5, 44–46
    size and distance of, 68–74, 385–386
    tables for visibility of, 393
morning star, 295
Moses Maimonides, 396
MUL.APIN, 5–8, 57, 188, 190, 297–298
museum of Alexandria, 21, 64, 67
music and astronomy, 429, 438
Mytilene, 30

Nabonassar, 15–16,
    era of, see era
Naburimannu, 320
Neoplatonism, 427
Nergal, 297
Neugebauer, Otto, 124, 186, 318, 321
Newton, Isaac, 170, 246, 247, 286, 440
Newton, Robert R., 262, 267–269, 273
Nicaea, 167, 215
night, length of, 97, 112–113, 119–120
night-pathed star, 195
Nile River, 65, 207
Nineveh, 16
node of Moon's orbit, 311, 316
nones, 164
noon, local, 27–28
normal stars, 299, 314
north, 28
North Star. See Stars and constellations
Novara, Domenico Maria, 414
numerals
    Arabic, 456n16
    cuneiform, 13
    Greek, 181, 264
    medieval Latin, 108

Nuremberg, 157, 416–417, 427
Nut, 75
nychthemeron, 239

obliquity of ecliptic, 54, 59–60
    decrease in, 274, 279, 283
obliquity, table of, 105–109
occultation, 186, 261
Ocean, 3–4, 52
Oenopides of Chios, 57–58
oikumene, 92–94
Olympiads, 183–184
omens, 8, 14
oppositions of planets to Sun, 295, 302
    table of for Mars, 302–303
orreries, 82
Osiander, Andreas, 417, 419, 437
Osiris, 202
Ovid, 18, 82

Pagan, Emile-Franois de, 439
Pakistan, 155
Palatine Anthology, v
Palmyra, 65
Pappus, 25, 73, 82, 90, 206
papyri 24, 345
    Hibeh, 201–203
    Oxyrhyncus, 345–346
parallax
    annual, 67–68
    diurnal (horizontal), 69
    of the Moon, 251–253
parapegma, 24, 39–40, 58, 98, 190
    of Geminus, 199–201
    in Hibeh papyri, 201–203
    in MUL.APIN, 6
    of Ptolemy, 203–204
    on stone, 201–202
Paris, 108, 399
Parmenides, 45, 47, 75, 296, 454n61
Parthians, 17
Passover, 167
Paul III (Pope), 168, 418, 424
Pericles, 46
perigee
    of an epicycle, 337
    of Sun, 211
period relations, 304–305
    in Babylonian Jupiter theory, 327–
        329
    and deferent-epicycle theory, 337–338
Persia, 16, 50, 155, 393, 426
Peters, C. H. F., 266
Petreius, Johannes, 417
Peurbach, Georg, 280, 384, 401–402, 422,
    425
phases
    of fixed stars, 4–5, 190–199
    of inferior planets, 300–301
    of planets in Babylonian diaries,
        298–299
    of superior planet, 302

Phenomena
    of Aratus, 18, 21, 40, 75–76
    of Euclid, 34, 88–89
    of Eudoxus, 21, 75
Philoponus, 156, 181
phonogram, 12
physics and astronomy, 218–219, 351,
    414
pillars of Heracles, 48
planetary theory
    of Apollonius, 337–341
    arithmetic versions of among the
        Greeks, 344–347
    Babylonian, 22–23, 321–335
        accuracy of, 330–332
        for Jupiter, System A, 321–328
        for Jupiter, System A', 328–329
        for Jupiter, System B, 332–334
    of Brahe, 413–414, 423–424
    of Copernicus, 413–422
    of Eudoxus, 305–312
    geo-heliocentric, 413–414, 423–424
    heliocentric vs. geocentric, 410–413
    Hipparchus on, 342–343
    intermediate model, 341–342, 461n44
    of Kepler, 433–439, 441–443
        its relation to Ptolemy's, 441–443
    of Ptolemy, 355–359
        method for planet longitudes in,
            369–372
        tables for, 372–384
planets
    apogees of, 461n73
    on Babylonian astrolabes, 9
    their connections with the Sun, 337–
        338, 358–359
    distances of, 387–389, 412–413
    great cycles of, 313
    how to observe, 301
    inequalities of motion of, 340
    inferior, 295, 299–301
    Jupiter, 8, 181, 289
        Babylonian theories of, 321–335
        omens from, 298
        latitudes of, 76, 458n7
    Mars, 289–296
        parameters of, 362–369
        periods of, 304–305
        retrograde arcs of, 340
        synodic cycle of in MUL.APIN,
            297
        tables of, 372–384
    Mercury, 295
        direction of on epicycle, 338–39
        motion of, 295, 299–301
        radius of epicycle of, 339
    in MUL.APIN, 57
    names of, 296
    order of, 347–351
    parameters of, 362–369, 461n72
    Saturn, 181–182
    sizes of, 389

planets (*continued*)
  superior, 295
  symbols for, 350–351
  tropical and synodic periods of, 295,
    305, 309
  Venus, 4, 14–15, 103, 179
    eight-year great cycle of, 313–315

Plato, 19, 31, 35, 46, 49, 83, 84, 132, 305
  on colors of planets, 301
  did not assert rotation of Earth,
    453n40
  and divine planet names, 297
  on order of planets, 348
  and regular polyhedra, 428
  Timaeus, 81, 398
Pliny, 24, 26, 262, 397
  on apogees of planets, 343
  on Babylonian astronomy and astrol-
    ogy, 318, 319, 346
  and equinoctial shadows, 61, 63
  on Hipparchus's new star, 247–248
  on power of Sun, 348
Plotinus, 427
Plutarch, 81
Polaris. *See* stars and constellations
pole, celestial, 31
Polybius, 62, 443
Porphyry, 427
Posidonius, 19, 52, 218–219, 443
  his orrery, 82
  and size of Earth, 65–66
  on zones, 62
prayer lines (on an astrolabe), 154
Prague, 281
precession, 96, 103, 245–247
  Brahe's treatment of, 282–283
  discovery of by Hipparchus, 248, 259
  effect of on dates of star phases,
    197–198
  in late antiquity, 262
  Ptolemy's treatment of, 259–262
  See also trepidation
Price, Derek J., 405
procedure text, 319, 328
Proclus, 59, 83, 262, 395, 404, 427
prograde motion, 289
proper motion, 249–250
Prutenic Tables, 422
Ptolemaic slats, 352, 449
Ptolemaic system, 389–391
Ptolemaios I Soter, 17, 21
Ptolemaios II Philadelphos, 21, 67
Ptolemaios III Euergetes, 41, 64, 179, 207
Ptolemaios IV Philopater, 64
Ptolemy (Claudius Ptolemaeus), v, 20, 22,
    25, 35, 67, 90, 108, 177–179, 282
  Almagest, 23–24, 26
  On the Analemma, 133
  his armillary sphere, 255–256
  on Callippic calendar, 187
  his celestial globe, 79, 95, 246

his cosmology, 384–392
his declinations of stars, 260–261, 283
and dioptra, 35, 218
discovers the equant, 357–358
on diurnal rotation, 37–38
on equation of time, 240–242
on fundamental propositions of astron-
    omy, 75–77
Geography, 50, 102–103
Handy Tables, 176, 240–242, 381–383
and initial point of zodiac, 104
on length of year, 208–20
his life 23, 37
and measurements of star coordinates,
    251–256
and meridian of Alexandria 238–239
on obliquity of ecliptic, 59–60, 107
his observations of equinoxes and sol-
    stices, 205–207
on order of planets, 348–349
Phaseis and parapegma, 180, 194, 196,
    197, 203–204
on planet observations, 302
Planetary Hypotheses, 218, 349, 350,
    384–391, 412
his planetary tables, 381–384
his planetary theory, 355–359
Planisphere, 156
on precession, 248, 259–262
preferred eccentric to epicycle in solar
    theory, 217
as a realist, 218
on shape of Earth, 49–52
on size and distance of Sun and Moon,
    73
on size of Earth's shadow, 71
solar tables 231
solar theory, 210–216
and *sphairopoiïa*, 84
star alignments, 248–250
star catalog, 41–43, 103, 264–274
table of ascensions, 109–111, 116–118
Tetrabiblos, 343–344
translations of his works, 25, 26, 395–
    396, 398–399
on zones, 62
Pythagoras, 47, 75, 132, 296, 454n61
Pythagoreans, 94, 412
Pythas of Massilia, 62

quadrant, 59, 205–206
quadrivium, 399
qibla, 393–394

Rawlins, Dennis, 268, 271
realism, 216–219, 418–420
refraction, 207, 458n5
Regiomontanus (Johann Müller), 401–
    403, 425, 426
regular polyhedra (or solids), 428–430
Reinhold, Erasmus, 401, 422–423, 424
rete, 142, 153

retrograde motion, 289
Revett, Nicholas, 131, 132
Rheticus, Georg Joachim, 416–417, 419,
    424
Rhodes, 48, 61, 65, 66, 82, 93, 116, 133,
    272–273
right ascension, 101–102, 115
right sphere, 32, 111
Rome, 49, 54, 61, 82, 129, 130, 140, 167
Rothmann, Cristoph, 281, 286, 424
Rudolfus Medici, 404
Rudolph II (of Holy Roman Empire),
    281, 430, 437, 439
rule (of an astrolabe), 145

Sacrobosco, 399
Salzburg, 156
Samarkand, 269
Sambursky, S., 217
saros, 316, 460n22
saving the phenomena, 217–218
Scaliger, Joseph, 171–172
Schiaparelli, G. V., 306
Schiller, 152
Schönberg, Nicholas, 416, 418–419
Schöner, Johann, 405–410, 416
seasons
  cause of, 53
  lengths of, 210, 221, 224–225, 458n10
secular accelerations of Moon and Earth,
    267
Seleucia, 17
Seleucid dynasty, 17, 188
separation mark, 323
Severus Alexander, 215
Severus Sebokht, 156
sexagesimal numbers
  computation with, 98–99
  in division of time and angle, 11, 100
  notation for, 11, 13–14
shadow box (of an astrolabe), 147
shadow length as measure of latitude,
    61–62
shadow plot, 27, 53–54, 63
Shevchenko, M., 274
Simplicius, 36, 218
  on Eudoxus's planetary theory, 306,
    309–312
Sin-leqē-unninnī, 319
Sippar, 318
solar theory
  and anomaly of motion, 210–211
  Babylonian, 213–214
  early history of, 212–215
  eccentric-circle version of, 211
  epicycle model version of, 212
  equation of center in, 226
  equatorium for, 215–216
  equivalence of eccentric and epicycle
    versions of, 212
  of Hipparchus and Ptolemy, 210–216
  motion of apogee in, 215, 394

parameters of, 211, 220–223
and relations among lengths of seasons, 224–226
tables for, 226–235
solstices, 4, 53, 56, 220–221
methods of observation of, 205–209
Sørensen, Christen, 431–433
Sosigenes (1st century BC), 165
Sosigenes the Peripatetic (2nd century AD), 306
Sothic cycle, 457n23
Spain, 50, 156, 279, 426
Speusippus, 35
sphairopoiïa, 78–84
spherics, 129
stade, 48, 65
Stahl, W. H., 397
Stars. *See also* stars and constellations
  γ And (Andromeda's left foot), 249
  β and γ Ara, 273
  g, h, i and k Cen, 272
  ε and η CMa, 272
  α Cru, 273
  α Lib, 249, 250, 261
  π¹ Ori, 248
  τ Pup, 273
  γ, δ and ζ Sag, 272
  β Sco, 261
  ε Tau, 248
  β Tri, 249
  γ Vir, 181, 257
  δ, ζ, ι and κ Vir, 257
  η Vir, 103, 257
  θ Vir, 257–258
  Alcor (g Uma), 33
  Aldebaran (Palilicium, α Tau), 248, 249, 260, 270, 271, 284–287, 414
  Altair, (α Aqu)42, 260, 284
  Antares (α Sco), 42, 261, 270, 284
  Arcturus (α Boo), 4, 5, 33, 35, 42, 43, 152, 190, 196, 197, 199, 200, 201, 202, 249, 250, 261, 287
  the Asses (γ and δ Can), 42, 43, 181
  Bellatrix (Orion's western shoulder, γ Ori), 204, 249, 260, 283, 284
  Betelgeuse (Orion's eastern shoulder, α Ori), 42, 44, 191, 193–195, 198, 199, 260, 284, 286–287
  Canopus (α Car), 42, 48, 65, 66, 273
  Capella (goat, α Aur), 42, 199, 200, 204, 249
  Castor (head of western twin, α Gem), 204, 284, 285
  Denebola (β Leo), 43, 153, 192
  El Nath (β Tau), 249
  Errai (γ Cep), 245
  Fomalhaut (α PsA), 272
  Hamal (ram's jaw, α Ari), 249, 401
  the Kids (ζ and η Aur), 42, 43, 200
  Kochab (β Umi), 33, 35
  Manger (M44), 43
  Miaplacidus (β Car), 33, 35

Mintaka (β Ori), 33, 35
Mizar (ζ UMa), 33, 249, 250
Pleiades, 3–7, 9, 39, 42, 43, 44, 57, 179, 186, 190, 198, 200, 201, 202, 260, 261
Polaris (North Star, Pole Star, α UMi), 32, 33, 34, 39, 245, 264
Pollux (β Gem), 260, 284, 285
Procyon (α CMi), 42, 246
Regulus (α Leo), 42, 192, 254–255, 261, 270, 274, 282, 284, 285
Rigel (star common to River and Orion's foot, β Ori), 44, 153, 204
Rigil Kentaurus (α Cen), 180, 273
Sirius (Dog Star, α CMa), 3–5, 33, 35, 42, 43, 157, 195–196, 198, 200, 201, 203, 204, 207, 260, 272, 287
Spica (wheat-ear, α Vir), 42, 199, 201, 249, 257–259, 261, 270, 282, 283, 284
Vega (α Lyr), 42, 43, 198
Vendemiatrix (ε Vir), 42, 43, 199, 200
stars and constellations. *See also* constellations; stars
  alignments of, 248–250
  Babylonian names for, 5–7, 56–58
  history of nomenclature for, 39–44
  in Homer and Hesiod, 3–5
station (of a planet), 289
stereographic projection, 154–155, 158–161
Stoic philosophy, 19
Strabo, 50, 65, 116, 267, 320, 443
  criticizes Eratosthenes, 64
  on possibility of circumnavigating the Earth, 52–53
  on zones, 62–63
Strassmaier, J. N., 318
Strato of Lampsacus, 67
Stuart, James, 131, 132
style (for beginning of year), 170
Sudines, 320
al-Ṣūfī, 44, 79
Sulpicius Gallus, 80, 82, 455n9
Sun
  as heart and mind of cosmos, 348
  motion of, 53–55
  progress of through zodiac, 96
  size and distance of, 68–74, 385–387
  See also solar theory
sundials, 29, 54, 67, 77, 84, 126, 129–141
  conical, 129, 135
  how to make, 135–140
  plane, 129–133, 135
  spherical, 129, 135
Syene, 20, 64
Symmachus (Pope), 167
synodic arc, 323, 325
synodic cycle
  of inferior planet, 300–301
  of superior planet, 302
Syracuse, 36, 82

Syria, 155
Syriac language, 393
Syrus, 37, 344
Sumerian language, 12

Tannery, Paul, 116
Tarentum, 61, 133
Thābit ibn Qurra, 275–279, 283, 395, 398
Thales, 47, 80
Theodosius of Bithynia, 24, 89–90, 398
theology and astronomy, 399, 424–425, 429
Theon (friend of Ptolemy), 49, 103, 179, 271
Theon of Alexandria, 206, 215, 262
  and Astronomical Canon, 176
  Commentary on *Almagest*, 25
  on graphical computation, 404
  and *Handy Tables*, 240
  treatise on the astrolabe, 156
  on trepidation, 275–276
Theon of Smyrna, 24, 58, 77, 197, 216, 218, 262, 346
  on heliocentric theory for Venus and Mercury, 349, 413, 423
  on inadequacy of Babylonian planetary theory, 214, 219
  on latitudes of planets and Sun, 458n7
  made a model of the cosmos, 83
  Mathematical Knowledge Useful for Reading Plato, 49
  on order of planets, 348
  on shape of Earth, 49–52
  on solid-sphere planetary theory, 392
Theophrastus, 67
Theorica planetarum
  anonymous, 399
  of Companus of Novarra, 404
Thureau-Dangin, François, 321
Tiglath-Pilesar III, 16
Tigris River, 12, 16, 17
time-telling
  with an armillary sphere, 126
  with an astrolabe, 150–152
  by risings of zodiac constellations, 95–99
  in terms of apparent or mean time, 235–243
time zones, 236–237
Timocharis, 21, 103, 186, 187, 248, 251, 259–261, 282, 283
tithi, 322, 393
Toledan Tables, 279, 394, 400–401, 404
Toledo, 26, 398
Toomer, G. J., 214
Tower of the Winds, 131–132
translation movement, 398–399
trivium, 399
trepidation of the equinoxes, 274–280
tropic circles, 53, 62
al-Ṭūsī, 396–397
twilight, 145

Ulugh Beg, 269–270, 281
Umayyad caliphate, 392
universities and astronomy in Middle
    Ages, 399–400
unrolling sphere, 311
Uraniborg, 281
Uruk, 317–321, 332

van der Waerden, 214, 346
Venezia, 61
Venus tablets of Ammi-saduqa, 14–15
Vettius Valens, 343, 345
Virgil, 18
Vitruvius, 24, 45–46, 67, 346
    on anaphoric clock, 155–156
    his directions for finding meridian,
        29–31
    and shadow lengths, 61, 63
    on sundials, 132–135

Vogt, Heinrich, 267

Ward, Seth, 439
Washington, George, 169
watch (division of the night), 11
water clock, 10–11. *See also* anaphoric
    clock
week, 165–166, 174
Wiegand, 201
Wilhelm of Hesse, 281
William of Moerbeke, 398
winds, 30, 130–132

Xenocrates, 35
Xenophanes, 47
Xerxes, 16, 164

year
    Gregorian, 171

    Julian, 166
    sidereal, 247
    tropical, 166, 175, 205, 207–209,
        246–247

al-Zarqālī, 279, 404, 463n123
zenith, 29
zenith distance, 28–29
zigzag function, 214, 333
zīj, 394
zodiac, 23, 39, 56–58, 76, 93
zodiac signs, 39, 76
    Babylonian names for, 323
    conventions for their beginnings, 104,
        203, 213–214, 276, 458n15
    symbols for, 104–105
zodiacal inequality, 340
zone time, 236–237
zones, terrestrial, 62–63